OXFORD SERIES ON SYNCHROTRON RADIATION

Series Editors

J. CHIKAWA J. R. HELLIWELL
S. W. LOVESEY

OXFORD SERIES ON SYNCHROTRON RADIATION

1. S. W. Lovesey, S. P. Collins: *X-ray scattering and absorption by magnetic materials*
2. J. R. Helliwell, P. M. Rentzepis: *Time-resolved diffraction*
3. P. J. Duke: *Synchroton radiation: production and properties*
4. J. A. Clarke: *The science and technology of undulators and wigglers*
5. M. J. Cooper, P. Mijnarends, N. Shiotani, N. Sakai, A. Bansil: *X-ray Compton scattering*
6. D. M. Paganin: *Coherent X-ray optics*
7. W. Schülke: *Electron dynamics by inelastic X-ray scattering*

Electron Dynamics by Inelastic X-ray Scattering

Winfried Schülke

Institute of Physics
University of Dortmund

OXFORD
UNIVERSITY PRESS

Great Clarendon Street, Oxford OX2 6DP

Oxford University Press is a department of the University of Oxford.
It furthers the University's objective of excellence in research, scholarship,
and education by publishing worldwide in

Oxford New York

Auckland Cape Town Dar es Salaam Hong Kong Karachi
Kuala Lumpur Madrid Melbourne Mexico City Nairobi
New Delhi Shanghai Taipei Toronto

With offices in

Argentina Austria Brazil Chile Czech Republic France Greece
Guatemala Hungary Italy Japan Poland Portugal Singapore
South Korea Switzerland Thailand Turkey Ukraine Vietnam

Oxford is a registered trade mark of Oxford University Press
in the UK and in certain other countries

Published in the United States
by Oxford University Press Inc., New York

© Oxford University Press 2007

The moral rights of the author have been asserted
Database right Oxford University Press (maker)

First published 2007

All rights reserved. No part of this publication may be reproduced,
stored in a retrieval system, or transmitted, in any form or by any means,
without the prior permission in writing of Oxford University Press,
or as expressly permitted by law, or under terms agreed with the appropriate
reprographics rights organization. Enquiries concerning reproduction
outside the scope of the above should be sent to the Rights Department,
Oxford University Press, at the address above

You must not circulate this book in any other binding or cover
and you must impose the same condition on any acquirer

British Library Cataloguing in Publication Data

Data available

Library of Congress Cataloging in Publication Data

Data available

Typeset by Newgen Imaging Systems (P) Ltd., Chennai, India
Printed in Great Britain
on acid-free paper by
Biddles Ltd., King's Lynn, Norfolk

ISBN 978-0-19-851017-8

1 3 5 7 9 10 8 6 4 2

Preface

It was as early as 1988 when S. W. Lovesey, one of the editors of the *Oxford Series on Synchrotron Radiation*, persuaded me to write a monograph on the application of inelastic X-ray scattering to the investigation of electron dynamics for the above series. At that time, synchrotron radiation had been used only in a few cases for research in this field. Synchrotron radiation sources of the third generation which, in the following years, have proven to be the most important motive power for further development of inelastic X-ray scattering, were just under construction. Moreover, some branches, such as resonant inelastic X-ray scattering or X-ray Raman scattering, were still completely underdeveloped so that one could not foresee the particular importance they would acquire for solving fundamental problems of excitations in condensed matter physics. Thus it might be considered as favorable circumstances that various obligations in research, teaching and administration did not leave me enough time to attack this book project seriously. The field of inelastic X-ray scattering could grow in importance but also quantitatively without being inopportunely fixed by a snapshot. However, in the same proportion also my bad conscience grew because I had not fulfilled the obligations to Oxford University Press. Not until my retirement from the University in 2001 did I get time enough to sit down and write this monograph, nevertheless again with a forced break because of surgery. One has to only look to the references in this book to imagine how fast the field of inelastic X-ray scattering has developed even within these four years. Therefore, it was clear from the very beginning that I had to restrict the subject-matter of this book in two ways: (i) I have excluded the application of inelastic X-ray scattering to ionic dynamics; the book deals only with electron dynamics investigated by the various branches of inelastic X-ray scattering. (ii) Even within this field I have concentrated on the scattering of hard X-rays, since I believe that it is the bulk information on electronic excitations, free of disturbing surface effects, which makes inelastic hard X-ray scattering superior to soft X-ray investigations, and will also open up in the future, due to the rapid development of synchrotron radiation sources on the way to the fourth generation, a wealth of applications, especially in material sciences.

The aim of this book is to provide the growing community of researchers in the field of inelastic X-ray scattering with accounts of the various experimental methods, synchrotron-based instrumentation and data analysis, widespread examples of successful investigations, and additionally a theoretical framework for the interpretation of the measured results. It might be considered as a peculiarity of this presentation that both experimental and theoretical issues are treated to the same extent. Thus I hope to provide a firm basis from which researchers can hopefully move forward.

In order to also give newcomers the access to the applications of inelastic X-ray scattering, the book starts with an introductory survey, which facilitates a review using simple arguments without going too deeply into the theoretical background. Nevertheless, this chapter also presents a list of representative examples of applications in the various fields with corresponding references. The following four chapters are each devoted to one main branch of inelastic X-ray scattering: (i) Nonresonant inelastic X-ray scattering: regime of characteristic valence electron excitations. (ii) Nonresonant inelastic X-ray scattering: regime of core-electron excitation (X-ray Raman scattering). (iii) The Compton scattering regime. (iv) Resonant inelastic X-ray scattering (RIXS) spectroscopy. Finally I provide the interested reader with a theoretical foundation where the relations used in the foregoing chapters to interpret experiments are derived by starting from basic principles, so that a unifying concept binds together the various branches of inelastic X-ray scattering.

Winfried Schülke
University of Dortmund, Institute of Physics   January 2007

Contents

	List of important symbols	ix
1	**Introductory survey**	**1**
	1.1 Introduction and historical background	1
	1.2 Orientation	4
	1.3 Survey of experiments	17
	1.4 References	58
2	**Characteristic valence electron excitations**	**71**
	2.1 Introduction	71
	2.2 Instrumentation with synchrotron radiation	77
	2.3 Measurement of the dynamic structure factor $S(\mathbf{q},\omega)$	99
	2.4 Study of specific solid state properties	145
	2.5 Coherent inelastic X-ray scattering	155
	2.6 Inverting $S(\mathbf{q},\omega)$ measurements into time and space	162
	2.7 Static structure factor $S(\mathbf{q})$	164
	2.8 References	179
3	**Core-electron excitation (X-ray Raman scattering (XRS))**	**186**
	3.1 Basic relations	186
	3.2 Core-hole–electron interaction in X-ray Raman scattering	190
	3.3 Special instrumentation with synchrotron radiation. Data processing	194
	3.4 Early XRS experiments	197
	3.5 Momentum transfer dependence in XRS experiments	203
	3.6 Core-hole–electron interaction in XRS experiments	211
	3.7 XRS studies on molecular gases, liquids and solids	215
	3.8 High-pressure XRS experiments	228
	3.9 Coincidence experiments	231
	3.10 References	234
4	**The Compton scattering regime**	**237**
	4.1 The impulse approximation	237
	4.2 The reciprocal form factor	240
	4.3 Electron exchange and correlation in Compton scattering	243
	4.4 Compton scattering versus positron annihilation	246
	4.5 Special instrumentation for the Compton scattering regime	250
	4.6 Compton scattering of atoms and molecules	267
	4.7 Compton scattering and chemical bonds	271

4.8	Compton scattering of metals	293
4.9	Compton scattering of special systems	322
4.10	Compton scattering beyond the impulse approximation	334
4.11	Coherent Compton scattering	343
4.12	γ-eγ experiments	346
4.13	Magnetic Compton scattering	352
4.14	References	365

5 Resonant inelastic X-ray scattering (RIXS) — 377

5.1	Basics of RIXS	378
5.2	Instrumentation for RIXS	382
5.3	Investigation of core excitations using RIXS	384
5.4	RIXS applied to mixed valent systems	401
5.5	Symmetry selective RIXS	405
5.6	Coherence of absorption and re-emission process in RIXS	417
5.7	Excitations in the intermediate state	442
5.8	Magnetic circular and linear dichroism in RIXS	467
5.9	References	479

6 Theoretical foundation — 486

6.1	Hamiltonian for the photon–electron interaction	487
6.2	The generalized Kramers–Heisenberg formula	489
6.3	Correlation functions	492
6.4	Response functions and dielectric functions	496
6.5	Lindhard polarization function	501
6.6	Density response and macroscopic dielectric function	508
6.7	Nonresonant spin- and orbital-magnetic scattering	510
6.8	The Compton scattering regime	516
6.9	Relativistic treatment of Compton scattering	542
6.10	Resonant inelastic scattering	553
6.11	Coherent inelastic scattering	573
6.12	References	579

Index — 583

List of important Symbols

A	photon vector potential
A_0, A_h	amplitudes of the photon Bloch wave
$A_l(\mathbf{q})$	generalized matrix element
$A(\mathbf{k}, E)$	spectral density function
$a_\nu(\mathbf{k}, \mathbf{g})$	plane wave expansion coefficient of the Bloch state $|\mathbf{k}, \nu\rangle$
b	asymmetry factor
$b_l(r)$	expansion coefficients of $B(\mathbf{r})$ into lattice harmonics
C	vector function of the orbital magnetic cross-section
$C_m^{(l)}$	spherical tensor operator
$C_{L'M'LM}^{z\zeta}$	Clebsch–Gordan coefficients
c	velocity of light
c, c^+ (a, a^+) (d, d^+)	annihilation and creation operators
$B(\mathbf{r})$	reciprocal form factor
D	vector function of the spin magnetic cross-section
$\mathbf{D}(\mathbf{q}, \omega)$	Fourier component of the displacement-field vector
$\mathbf{d}_{\eta\mu}$	dipole matrix element
$dg_{nl}/d\varepsilon$	oscillator denstiy
E	photon electric field vector
$\mathbf{E}(\mathbf{q}, \omega)$	Fourier component of the electric field vector
E_i, E_n, E_f	energy of the initial, intermediate, and final state of the target
E_K	K-shell binding energy
$E_\mathbf{k}$	energy of an electron with wavevector $\mathbf{k}$
$E_\nu(\mathbf{k}), E_{\mathbf{k}\nu}, E(\mathbf{k}, \nu)$	energy of a Bloch electron in the state $|\mathbf{k}\nu\rangle$
E_F	Fermi energy
E_1, E_2	energy of the electron before and after scattering
$E_{xc}[\rho]$	exchange-correlation energy functional
e	unit of charge for an electron
e_c	ground-state correlation energy
$\mathbf{e}_1, \mathbf{e}_2$	unit vector in the direction of the polarization of the incident and scattered photon
$|F\rangle$	final state of the photon–electron system
$F(\mathbf{q})$	form factor
$F_j(\mathbf{k})$	scattering amplitude of an atom of the jth shell
$F_l(\theta, \phi)$	lattice harmonics
$F_{\nu n}(\omega_1)$	**RIXS** amplitude of the optical excitation $n \to \nu$

$f(\mathbf{k},\nu)$	occupation number of a Bloch electron $|\mathbf{k},\nu\rangle$
$|f\rangle$	final state of the target
$f_{\mathrm{xc}}[\rho](\mathbf{r}t,\mathbf{r}'t')$	dynamical exchange-correlation kernel
$f_{\mathrm{xc}}(\mathbf{q},\mathbf{q}',\omega)$	Fourier transform of the dynamical exchange-correlation kernel
$f_{nl}(r)$	radial part of the position space wavefunction
G	one-particle Green's function
$G_{\mathrm{c}}, G_{\mathrm{m}}$	charge and magnetic contribution to the scattering amplitude
$\tilde{G}(\mathbf{q},\omega)$	local-field correction factor
$\mathbf{G}_{\mathrm{c}}(\mathbf{r},t)$	space-time correlation function
$\mathbf{G}$	scattering amplitude operator
$g(\mathbf{r}',\mathbf{r}-\mathbf{r}'),\ (g(r))$	pair correlation function for inhomogeneous (homogeneous) systems
$\mathbf{g}$	reciprocal lattice vector
g_i	probability of the state $|i\rangle$
$g_{nl,n'l'}$	oscillator strength of the transition $(nl)^{-1} \to (n'l')^{-1}$
H	Hamiltonian
$\hbar$	Planck's constant$/2\pi$
$|I\rangle$	initial state of the photon–electron system
i_{c}	critical angle
$|\mathrm{i}\rangle$	initial state of the target
$J(p_{\mathrm{q}}),\ J(p_z)$	Compton profile
$J_m(p_z)$	magnetic Compton profile
$j_l(kr)$	spherical Bessel function
$\mathbf{K}_1,\ \mathbf{K}_2$	wavevector of the incident and scattered photon
$\mathbf{k}$	electron (Bloch) wavevector
$\mathbf{k}_{\mathrm{e}},\mathbf{k}_{\mathrm{h}}$	Bloch wavevector of an electron and a hole
k_{B}	Boltzmann constant
k_{F}	Fermi wavevector
l_{c}	mean interparticle distance
M	transition matrix element
m	rest mass of the electron
$|m\rangle$	intermediate state of the target
$\mathbf{m}$	unit vector in the direction of the preferred magnetic axis
$|N\rangle$	intermediate state of the photon–electron system
$|n\rangle$	intermediate state of the target
$n(\mathbf{p})$	electron momentum density
$n(\mathbf{r})$	ground-state number density
$n_0^{\mathrm{h}}(p)[\rho(\mathbf{r})]$	momentum density of an interacting electron gas
$n_0^{\mathrm{f}}(p)[\rho(\mathbf{r})]$	momentum density of a noninteracting electron gas
$n_{\nu\nu'}(\mathbf{k})$	elements of the occupation number density matrix
$n_2(\mathbf{r},\mathbf{r}',t)$	time dependent two-particle density correlation function
$P_{1,2,3}$	Stokes parameters

List of important Symbols

$P(\mathbf{r})$	Patterson function
$P(\mathbf{p})$	reciprocal Patterson function
$\mathbf{p}$	electron momentum vector (operator)
p_F	Fermi momentum
$\mathbf{p}_1, \mathbf{p}_2$	momentum of the electron before and after scattering
p_i	probability of the initial state $|i\rangle$
$\mathbf{q} = \mathbf{K}_1 - \mathbf{K}_2$	scattering vector
$\mathbf{q}_r$	reduced scattering vector
q_c	plasmon cutoff vector
$\mathbf{R}$	lattice translation vector
R_i	integrated reflectivity
R_c	core-hole decay rate
R_{ph}	electron–phonon scattering rate
$\mathbf{r}$	position vector
r_j	distance of an atom of the jth shell from the emitting atom
r_s	electron density parameter
r_0	classical radius of the electron
S	overlap integral
$S(\mathbf{q})$	static structure factor
$S_m(\mathbf{q})$	magnetic static structure factor
$S(\mathbf{q}, \omega)$	(diagonal) dynamic structure factor
$S_m(\mathbf{q}, \omega)$	magnetic dynamic structure factor
$S(\mathbf{q}, \mathbf{q}+\mathbf{g}, \omega)$	off-diagonal dynamic structure factor
s	scattering power
$\mathbf{s}$	spin vector (operator)
$\mathbf{s}(\mathbf{r}, t)$	spin density
$\mathbf{s}(\mathbf{q}, t)$	Fourier transform of the spin density
T	transmission
T	temperature
$T_{\mathbf{gg}'}$	element of the matrix $\mathbf{T}$
t	time
t_{exB}	extinction length of the Bragg case
U	potential
$u_{nl}(p)$	radial part of the momentum space wavefunction
$u_{\nu\mathbf{k}}(r)$	lattice periodic part of the Bloch wavefunction
V	normalization volume
V	Coulomb potential operator
$V_\mathbf{g}$	$\mathbf{g}$th Fourier component of the periodic lattice potential
$v(\mathbf{q})$	Fourier component of the Coulomb potential
$v_{xc}[\rho]$	exchange-correlation potential
$w^{(ab)r}$	ground state double tensor operators (multipolar moments)
$Y_{lm}(\theta, \phi)$	spherical harmonics
$Y_{l,m}(\theta, \phi)$	real representation of the spherical harmonics
$\mathbf{Y}_{LM}$	vector spherical harmonics

List of important Symbols

Z_F	renormalization constant
α	angle between lattice planes and crystal surface
α_j	phase shift at the jth shell
α_l	threshold exponent
$\alpha_\nu(\mathbf{k}+\mathbf{g})$	plane wave expansion coefficient of a Bloch wavefunction
$\Gamma_{n(f)}$	lifetime energy broadening of the intermediate (final) state
Γ_K	energy width of the **K**-shell hole
$\Gamma(\boldsymbol{p},\boldsymbol{p}')$	vertex function
$\Gamma_1(\mathbf{r}\|\mathbf{r}')$	one-particle density matrix in position space
$\Gamma_1(\mathbf{p}\|\mathbf{p}')$	one-particle density matrix in momentum space
$\Delta\theta_{\text{in,out}}$	Darwin width for incident, outgoing radiation
γ_{nl}	screening constant
ε	efficiency
$\varepsilon_{\mathbf{k}}, \varepsilon(\mathbf{k})$	kinetic energy of a free particle with momentum $\mathbf{k}$
$\varepsilon(\mathbf{q},\omega)$	dielectric function
$\varepsilon_1, \varepsilon_r$	real part of the dielectric function
$\varepsilon_1, \varepsilon_i$	imaginary part of the dielectric function
$\varepsilon_0^h[\rho(\mathbf{r})], \varepsilon_0^f[\rho(\mathbf{r})]$	ground-state energy per electron of interacting and noninteracting electrons
$\varepsilon_{\mathbf{gg}'}(\mathbf{q}_r,\omega)$	element of the dielectric matrix $\mathbf{1}+\mathbf{T}$
θ	scattering angle
θ	step function
θ_A	analyzer Bragg angle
θ_M	monochromator Bragg angle
θ_B	Bragg angle
κ_1, κ_2	incident and scattered photon four-vector
λ	wavelength
μ	linear absorption coefficient
μ	chemical potential
μ	spin magnetic moment
μ	density matrix
ν	band index
ν	Poisson number
Ξ_{sc}	two-body interaction operator
π_1, π_2	four-vector of a free electron before and after the scattering process
ρ	averaged electron density
$\rho(\mathbf{r})$	position space electron density
$\rho(\mathbf{p})$	momentum space electron density
$\rho_p(\mathbf{p})$	electron–positron pair momentum density
$\rho(\mathbf{g}), \rho_{\mathbf{g}}$	Fourier component of the electron density
$\rho(\mathbf{r},t)$	(operator of the) time-dependent electron density in position space

List of important Symbols

$\rho(\mathbf{q},t)$	electron density fluctuation operator
Σ	self-energy operator
σ	cross-section
σ	spin variable
$\boldsymbol{\sigma}$	spin vector operator
σ_K	K-shell contribution to the photoelectric absorption cross-section
σ_j	Debye–Waller factor of atoms in the jth shell
$\sigma_{x,z}$	synchrotron radiation source dimension in x,z direction
$\sigma'_{x,z}$	synchrotron radiation source divergence in x,z direction
σ'_R	natural divergence of the synchrotron radiation
$\Phi(\mathbf{r})$	solutions of the Kohn–Sham equation
$\phi_{\text{ext}}(\mathbf{q},\omega)$	Fourier component of an external potential
$\phi_{\text{ind}}(\mathbf{q},\omega)$	Fourier component of the induced potential
$\phi_{\mathbf{k},\nu}(\mathbf{r})$	one-electron Bloch wavefunction
$X_{\mathbf{gg'}}$	element of the polarization matrix $\mathbf{X}$
$X^0_{\mathbf{gg'}}$	element of the proper polarization matrix $\mathbf{X}^0$
$X^L_{\mathbf{gg'}}$	element of the Lindhard-type polarization matrix $\mathbf{x}^L$
$\chi(\mathbf{r},\mathbf{r'},t-t')$	density–density response function
$\chi(\mathbf{q},\omega)$	polarization function
χ_h	$\mathbf{g}_h^{\text{th}}$ Fourier component of the crystal dielectric susceptibility
$\chi_0(\mathbf{q},\omega)$	proper polarization function
$\chi_L(\mathbf{q},\omega)$	Lindhard polarization function
$\chi(\mathbf{q}_r+\mathbf{g'},\mathbf{q}_r+\mathbf{g},\omega)$	element of the polarization matrix $\mathbf{X_{gg'}}$
$\chi(\mathbf{p})$	momentum space wavefunction
$\psi(\mathbf{r})$	position space wavefunction
$\psi_+(\mathbf{r})$	positron wavefunction
$\phi_c(\mathbf{r}), \phi_n(\mathbf{r})$	atomic wavefunction
Ω	solid angle
Ω_{nl}	threshold frequency
ω_1, ω_2	energy of the incident and scattered photon$/\hbar$
ω	transferred energy$/\hbar$, $\omega_1 - \omega_2$
ω_e	energy of the ejected electron
ω_p	plasmon frequency
ω_F	Fermi frequency

1

Introductory survey

1.1 Introduction and historical background

Inelastic X-ray scattering experiments, including both energy and directional analysis of scattered photons, comprises a very powerful tool for investigating the correlated motion of electrons in a many-particle system by looking at the excitation of this system left behind in the inelastic scattering process.

The purpose of this chapter is to give an introductory survey of the dynamics and excitations of electronic systems in condensed matter that have been recently investigated by inelastic X-ray scattering. Hence, I cover the nature of the relevant techniques, namely nonresonant inelastic scattering in the low-momentum transfer regime and nonresonant inelastic scattering in the regime of high momentum and energy transfer (Compton limit), as well as resonant inelastic scattering, and representative examples of materials that have been investigated. Later chapters expand on all aspects of the field, ranging from experimental methods (including preparation of special initial photon states and utilization of the spin-dependent part of the cross-section), properties of the synchrotron photon sources, and instrumentation through details of the theoretical framework helpful for interpretation of the measured data at an atomic level. For the moment, though, it is my goal to provide a short historical background and an overview of the fields, neglecting finer points.

The following historical remarks will focus on those events where steps were done towards the essentials for the further development of the field.

Only a few years after the discovery and the correct interpretation of inelastic X-ray scattering (Compton 1923, Debye 1923), inelastic scattering experiments in the Compton limit on beryllium by DuMond (1929) verified the prediction of the then new Fermi statistics and were at variance with the classical Maxwell–Boltzmann distribution law. Thus, the very first application of inelastic X-ray scattering in condensed matter physics led to the first direct evidence of a fundamental physical principle. Also, further experimental results of DuMond and coworkers (e.g. DuMond and Kirkpatrick 1931a,b, 1937, DuMond and Hoyt 1931) were of great physical significance and set up milestones in verifying experimentally the quantum theory of many-electron systems.

One must not forget the tremendous experimental difficulties that these experimentalists had to overcome since they had to rely on weak conventional X-ray tubes and photographic registration of the spectra. Only DuMond's exceptional experimental skill (see, e.g. his multicrystal spectrometer, DuMond and Kirkpatrick 1930) could handle the low photon flux in the scattered beam. Therefore, one can understand that, during the following 25 years, experimentalists were deterred from this difficult technique. Only the advent of rotating anode generators and of highly efficient scintillation counters led to a revival of inelastic X-ray scattering experiments in the Compton limit in the mid-1960s, prompted by the pioneering work of Cooper *et al.* (1965). Compton scattering experiments utilizing unmonochromatized characteristic X-rays and crystal dispersive analysis of the scattered radiation were further developed by Phillips and Weiss (1968) and Schülke *et al.* (1972). Another direction of development, introduced by Eisenberger and Reed (1972), is inelastic scattering of γ-rays or monochromatized characteristic X-rays with primary energies in most cases above 50 keV, and using solid state detectors for energy analysis of the scattered radiation. This technique was widely used in the following 20 years and made Compton scattering is a valuable tool for investigating the electron momentum density, a ground-state property of atoms, molecules and solids, even if the momentum space resolution was rather poor; see the review of these developments by Cooper (1985).

It was predicted by Platzman and Tzoar (1970) that, by utilizing an interference term between inelastic X-ray charge (Thomson) scattering and X-ray magnetic scattering, one can obtain information about the momentum distribution of unpaired-spin electrons, where the incident radiation must be circularly polarized. The first experiments of this kind, designated magnetic Compton scattering, using the circularly polarized γ-emission from oriented nuclei, were reported by Sakai and Ono (1976).

Whereas the developments of inelastic X-ray scattering studies described so far were devoted to the Compton limit thus giving information about ground-state properties (electron momentum distribution), it was first pointed out by Nozieres and Pines (1959) that the spectral analysis of inelastically scattered X-rays, when performed with rather small energy and momentum transfer can give valuable information about the frequency and wavevector dependent so-called dynamic structure factor of electrons, a quantity which is intimately related to the two-particle density correlation function thus describing the correlated motion of electrons in a many-particle system. Decisive for developing this X-ray technique, which will be designated inelastic X-ray scattering (IXS) spectroscopy in what follows, was its advantage over **e**lectron **e**nergy **l**oss **s**pectroscopy (EELS) which is due to a much smaller contribution of multiple scattering of the former, especially for larger momentum transfer. The first experiments on simple low-Z metals (Schülke *et al.* 1969, Priftis 1970, Eisenberger *et al.* 1973), still performed with conventional X-ray sources, were mainly devoted to measuring the dispersion of the plasmon loss or to demonstrate the role of Coulomb correlation in the dynamic structure factor (Schülke *et al.* 1969). But there also came up studies of the dynamic structure factor in the region of intermediate momentum transfer initiated by Platzman and Eisenberger (1974a) and Schülke and Lautner (1974). The detection of a two-peak or multipeak fine structure of the dynamic structure factor and its interpretation by Platzman and Eisenberger (1974b) as an indication of an incipient Wigner electron lattice has prompted much theoretical work on the dynamic structure of electrons in nearly homogeneous systems.

A special very promising branch of nonresonant IXS is connected with excitation of inner shell electrons rather than valence electrons, as in the above mentioned cases. As first stressed by Mizuno and Ohmura (1967), this technique, often designated as X-ray Raman scattering, renders a spectral distribution of scattered radiation similar to a soft X-ray absorption spectrum but with the great advantage that it can be obtained using rather hard, well penetrating X-rays. The first pioneering X-ray Raman scattering experiments were performed by Suzuki and collaborators (Suzuki 1967, Suzuki et al. 1970, Suzuki and Nagasawa 1975).

It was pointed out by Schülke (1981, 1982) that the information both about electron momentum density in the Compton limit and about the dynamic electron structure with IXS can be extended, so that also access to nondiagonal components of the momentum space density matrix and of the response function, respectively, can be obtained, if the initial photon state of the inelastic scattering experiment is prepared to be a standing-wave field (so-called coherent inelastic X-ray scattering). First experiments of this kind in the Compton limit were reported by Golovchenko et al. (1981) and Schülke et al. (1981).

Resonant inelastic X-ray scattering (RIXS) spectroscopy now earns, thanks to the tunability of modern synchrotron radiation sources, increasing attention, and was introduced by Sparks (1974). We owe the first comprehensive theoretical treatment of RIXS, at that time under the designation "resonant X-ray Raman scattering", to Tulkii and Åberg (1980). RIXS offers selectivity of electronic excitation under investigation with respect to spin (first reported by Hämäläinen et al. 1992), crystal momentum (initiated by Ma et al. 1992), and symmetry (utilized for the first time for the excitation and the re-emission process of RIXS by Cowan et al. 1986) of the electronic states involved and is additionally element specific.

A big step forward for the whole field of inelastic X-ray scattering was made possible by the availability of strong sources of synchrotron X-rays in the mid-1970s. Since then, depending on the tunable primary X-ray energy, these have been studies of resonant inelastic X-ray scattering, performed by Eisenberger et al. (1976a,b), who have utilized synchrotron radiation for the first time in the field. Only then were the first Compton scattering experiments performed with synchrotron radiation by Holt et al. (1978) using energy analysis by a solid state detector, and by Loupias and Petiau (1980) using crystal dispersive analysis of the scattered radiation. The application of synchrotron radiation, especially its circularly polarized component in the off-plane emission, was essential for the further development of magnetic Compton scattering as pioneered by Cooper et al. (1986). Synchrotron radiation based IXS experiments for investigating the dynamic structure factor of valence electrons with 1 eV resolution were initiated by Schülke and Nagasawa (1984), and Schülke et al. (1984).

Synchrotron radiation was introduced for the first time into nonresonant Raman scattering studies by Nagasawa *et al.* (1989). Coherent inelastic X-ray scattering has benefited by synchrotron radiation since the first experiments by Schülke and Kaprolat (1991a).

All these experiments have demonstrated the large potential of synchrotron radiation in this field, where in most cases the pioneering experiments were, in the first instance, performed at bending magnet beamlines. Nevertheless, it is clear that for further progress only the much higher spectral flux of insertion devices together with the low emittance of dedicated rings of the third generation will bring about the full utilization of all techniques of inelastic scattering and will thus open a new field of condensed matter research. Of course, the advent of X-ray lasers in the 1 Å wavelength region, possibly within the next decade, will certainly create new applications of inelastic X-ray scattering, which are more than only the extension of today's techniques to higher resolution and accuracy.

1.2 Orientation

This book covers four main topics of inelastic scattering, namely (i) nonresonant inelastic X-ray scattering with low and intermediate momentum and energy transfer with the goal to get information about the dynamic structure factor of electrons; (ii) nonresonant inelastic X-ray scattering connected with core-electron excitations (nonresonant X-ray Raman scattering); (iii) nonresonant inelastic X-ray scattering in the Compton limit, which delivers the momentum distribution of scattering electrons; and (iv) resonant inelastic X-ray scattering with its potential to look very specifically for electronic excitations, which are intrinsically influenced by electron dynamics. These four topics have two common aspects. They are based on the same kinematics of the scattering experiment and the observed quantity is the double differential scattering cross-section to be deduced from a Hamiltonian, which describes the interaction of the X-ray electromagnetic field with the electron system.

The basic kinematics of a typical inelastic scattering experiment are sketched in Fig. 1.1. A photon of energy $\hbar\omega_1$, wavevector $\mathbf{K}_1$ and polarization (unit)

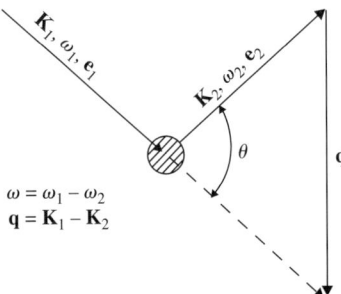

Fig. 1.1. Principle of an inelastic scattering experiment.

vector $\mathbf{e}_1$ impinging upon a target, which is characterized by a state vector $|i\rangle$ and energy E_i, is scattered by an angle θ into a photon of energy $\hbar\omega_2$, wavevector $\mathbf{K}_2$ and polarization vector $\mathbf{e}_2$, leaving the target in a state $|f\rangle$ with energy E_f. The energy $\hbar\omega \equiv \hbar(\omega_1 - \omega_2)$ and the momentum $\hbar\mathbf{q} \equiv \hbar(\mathbf{K}_1 - \mathbf{K}_2)$ are transferred to the target, where energy conservation requires

$$\hbar\omega = E_f - E_i. \tag{1.1}$$

The modulus of the transferred momentum is connected with the scattering angle θ by

$$q = (\omega_1^2 + \omega_2^2 - 2\omega_1\omega_2 \cos\theta)^{1/2}/c. \tag{1.2}$$

As long as $\omega \ll \omega_1$, equation (1.2) reduces to

$$q \approx 2K_1 \sin(\theta/2). \tag{1.3}$$

A typical inelastic X-ray scattering experiment consists of producing a well-collimated beam of monochromatic photons, of selecting a certain solid angle element $d\Omega_2$ of the scattered beam (thus fixing $\mathbf{q}$ according to (1.3)), and of energy-analyzing this part of the scattered radiation to a resolution $d\hbar\omega_2$. In other words, the double differential scattering cross-section (DDSCS) $d^2\sigma/(d\Omega_2 d\hbar\omega_2)$ is measured as a function of $\mathbf{q}$ and ω, where the DDSCS is defined as follows:

$$\frac{d^2\sigma}{d\Omega_2 d\hbar\omega_2} \equiv \frac{\text{current of photons scattered into the solid angle element } [\Omega_2, d\Omega_2] \text{ and into the range of energy } [\hbar\omega_2, d\hbar\omega_2]}{\text{current density of the incident photons} \times d\Omega_2 \times d\hbar\omega_2}. \tag{1.4}$$

The different branches of inelastic scattering, as mentioned above, differ from another according to the Hamiltonian describing the interaction of the electromagnetic field represented by its vector potential $\mathbf{A}$ with the nonrelativistic electron system (Blume 1985). Deriving this interaction Hamiltonian H_{int} from the kinetic energy operator $[(\mathbf{p} - e\mathbf{A}/c)^2/2m]$, it contains an $\mathbf{A}^2$ and a $\mathbf{p}\cdot\mathbf{A}$ term. Additionally, there is a Zeeman interaction between an electron spin $\mathbf{s}$ and the photon magnetic field rot $\mathbf{A}$, which is linear in $\mathbf{A}$, and a second magnetic term $\mathbf{s}\cdot(\mathbf{A}\times\mathbf{E})$, which, in the classical limit, is deduced from the potential energy of the spin magnetic moment in the relativistic magnetic field of the form $(\mathbf{p} - e\mathbf{A}/c)\times\mathbf{E}$, whose contribution $\mathbf{s}\cdot(\mathbf{p}\times\mathbf{A})$ is the spin–orbit interaction. Because the electric field $\mathbf{E}$ is proportional to the time derivative of $\mathbf{A}$, this term is quadratic in $\mathbf{A}$. Since always two photon fields are involved in the inelastic scattering process only the terms quadratic in $\mathbf{A}$ deliver contributions to the scattering amplitude in a first-order perturbation treatment; this is the Thomson charge scattering amplitude resulting from the $\mathbf{A}^2$ term and a spin originated magnetic scattering amplitude arising from the second magnetic term quadratic in $\mathbf{A}$. Terms linear in $\mathbf{A}$ will contribute to the scattering amplitude only in a second-order perturbation treatment, giving rise to a resonant term, well known as the Kramers–Heisenberg dispersion formula, the basic relation to interpret resonant inelastic scattering. But this Kramers–Heisenberg formula must be generalized for scattering of magnetic materials in order to include the Zeeman interaction term linear in **A**. The final treatment of the total scattering amplitude including first and second order terms is somewhat complicated through the fact that the magnetic part of the second-order term contributes to the scattering amplitude also off-resonance. Therefore, the magnetic scattering amplitude contains, also off-resonance, a spin and an orbital contribution.

It is the Thomson charge scattering amplitude which constitutes scattering from time-dependent density fluctuations of the electron system and forms the basis of that branch of inelastic X-ray scattering which brings about direct experimental access to the time-dependent two-particle density correlation, the main ingredient of the dynamic structure factor $S(\mathbf{q},\omega)$ (see Schülke 1991), via

$$\frac{d^2\sigma}{d\Omega_2 d\hbar\omega_2} = r_0^2 (\mathbf{e}_1 \cdot \mathbf{e}_2)^2 \left(\frac{\omega_2}{\omega_1}\right) S(\mathbf{q},\omega), \quad r_0 = \frac{e^2}{mc^2} \tag{1.5}$$

$$S(\mathbf{q},\omega) = \frac{1}{2\pi\hbar} \int_{-\infty}^{\infty} dt \exp(-i\omega t) \int d^3 r'$$

$$\times \int d^3 r \exp(-i\mathbf{q}\cdot\mathbf{r}) n_2(\mathbf{r},\mathbf{r}',t), \tag{1.6}$$

where $n_2(\mathbf{r},\mathbf{r}',t)$ is the time-dependent two-particle density correlation function, defined as

$$n_2(\mathbf{r},\mathbf{r}',t) \equiv \langle \rho(\mathbf{r}',t=0)\rho(\mathbf{r}'+\mathbf{r},t)\rangle, \tag{1.7}$$

$$\rho(\mathbf{r},t) \equiv \sum_j \delta(\mathbf{r}-\mathbf{r}_j(t)). \tag{1.8}$$

$\langle\ \rangle$ denotes the ground-state expectation value of the density operator product and the summation j is over all scattering particles of the system, so that, in the classical limit, $n_2(\mathbf{r},\mathbf{r}',t)$ gives the probability of finding an electron at time t at $\mathbf{r}'+\mathbf{r}$, if there was detected an electron (the same or another one) at time $t=0$ at $\mathbf{r}'$.

These last three equations associate inelastic X-ray scattering with scattering from time-dependent electron density fluctuations. It is the well-known fluctuation–dissipation theorem, which connects scattering from fluctuations with excitations of the system (connected with dissipation of energy) in the course of inelastic scattering as far as the description of the scattering process is concerned. It will be derived later on in greater detail that, in the sense of this theorem, the DDSCS can also be related to excitations of the system from the (many-particle) initial ground state $|i\rangle$ with energy E_i into all excited final states $|f\rangle$ with energy E_f, that are allowed according to energy conservation, by writing

$$\frac{d^2\sigma}{d\Omega_2 d\hbar\omega_2} = r_0^2 (\mathbf{e}_1 \cdot \mathbf{e}_2)^2 \left(\frac{\omega_2}{\omega_1}\right) \sum_{i,f} p_i \left| \langle f| \sum_j \exp(i\mathbf{q} \cdot \mathbf{r}_j) |i\rangle \right|^2$$

$$\times \delta(E_f - E_i - \hbar\omega), \qquad (1.9)$$

where p_i is the probability of the initial state $|i\rangle$.

In this sense investigation of electron dynamics always means also the study of electronic excitations and vice versa. Equation (1.9) reveals access to a certain symmetry selectivity of nonresonant inelastic X-ray scattering spectroscopy, since the matrix element is subject to symmetry-related selection rules.

It is equations (1.6)–(1.8) which facilitate the understanding of the physical information that can be extracted from measurements of $S(\mathbf{q},\omega)$ in the different branches of inelastic X-ray scattering experiments. Especially, the Compton limit becomes intuitively clear without referring to derivations, which will be given later. Assume that the reciprocal momentum transfer q^{-1} is large compared with characteristic distances of the scattering electron system, l_c, say the interparticle distance. Then the phase factor $\exp(i\mathbf{q} \cdot \mathbf{r})$ accentuates the correlated motion of different particles of the system, since it is the interference between amplitudes scattered from different particles which mainly defines $S(\mathbf{q},\omega)$, so that it is the collective behavior of the system one is investigating. Of course, since one is observing the phase correlation at different times, the result of a scattering experiment also depends on how the transferred energy compares with characteristic frequencies ω_c of the system. Therefore, one can learn, for example, something about the collective motion of the scattering system if the transferred ω one is looking at is near the frequency of that collective motion (for instance the plasma frequency). On the contrary, let the reciprocal momentum transfer q^{-1} be very small compared with the interparticle distances. Then the interference between amplitudes scattered at different particles has a negligible influence on $S(\mathbf{q},\omega)$, so that it is the single-particle properties of the system one is restricted to. In more detail, one is observing the correlation between single-particle positions at very small time distances. This means that one gets information about the momenta of single particles, more precisely, about the electron momentum distribution, provided the time-scale of the experiment is short enough, or the energy transfer $\hbar\omega$ is large enough, to prevent the remaining system becoming rearranged. These are exactly the prerequisites of the so-called impulse approximation (Eisenberger and Platzman 1970), which later will define the Compton limit, where the DDSCS is proportional to the projection of the electron momentum density $\rho(\mathbf{p})$ on the direction of the momentum transfer $\mathbf{q}$, the so-called Compton profile $J(p_z)$:

$$\frac{d^2\sigma}{d\Omega_2 d\hbar\omega_2} = r_0^2(\mathbf{e}_1 \cdot \mathbf{e}_2)^2 \left(\frac{\omega_2}{\omega_1}\right) \int d^3p\, \rho(\mathbf{p})\, \delta\left(\frac{\hbar \mathbf{p} \cdot \mathbf{q}}{m} - \frac{\hbar^2 q^2}{2m} - \hbar\omega\right)$$

$$= r_0^2(\mathbf{e}_1 \cdot \mathbf{e}_2)^2 \left(\frac{\omega_2}{\omega_1}\right)\left(\frac{m}{\hbar q}\right) \int\int \rho(\mathbf{p}) dp_x dp_y;\; p_z = \frac{\hbar q}{2} + \frac{\omega m}{q}$$

$$= r_0^2 \left(\frac{1}{2}\right)(1 + \cos^2\theta)\left(\frac{\omega_2}{\omega_1}\right)\left(\frac{m}{\hbar q}\right) J(p_z), \qquad (1.10)$$

where the p_z-axis is chosen to coincide with the $\mathbf{q}$-direction. Additionally, in the last row of (1.10), we have assumed that the polarization of the scattered beam is equally distributed parallel and perpendicular to the scattering plane. We will show later that the above described possibility to decompose the double differential scattering cross-section into one part that only depends on properties of the scattering system $[S(\mathbf{q},\omega)]$ and one part that is only determined by properties of the probe is in principle lost in a strict relativistic calculation (Jauch and Rohrlich 1955). But Ribberfors (1975) has formulated experimental conditions where such a decomposition can be accomplished to a good approximation.

Thus summarizing the above statements, one can, by looking at the magnitude of q and ω, distinguish crudely between four types of nonresonant inelastic charge scattering experiments, a classification that is not only helpful as a guide to the significance of physical information but also defines the experimental requirements:

(1) Scattering by valence electron density fluctuations, or, in the sense of the fluctuation–dissipation theorem, by valence electron excitations, is investigated with $qr_c \approx 1$ and $\omega \approx \omega_p$, where r_c is the interparticle distance and ω_p the plasma frequency.

(2) Scattering by inner-shell excitations (X-ray Raman scattering regime) is investigated with $qa \leq 1$, $\hbar\omega \approx E_B$, where a is the inner-shell orbital radius and E_B the binding energy of that orbital. It will turn out that this kind of spectroscopy will deliver access to the fine structure of excitation edges of low-Z materials, comparable with absorption edges, but using hard X-rays with a high penetration depth rather than soft X-rays as is necessary in the case of absorption spectroscopy.

(3) Scattering by collective ion excitations (phonons) is followed up with $qd \approx 1$, $\omega \approx \omega_{\text{ph}}$, where d is the interionic spacing and ω_{ph} the phonon frequency. (This branch of inelastic X-ray scattering will not be further considered in this book, since it is connected with electron dynamics only

through the fact that X-rays are coupling primarily to electrons, which are assumed to follow instantaneously the lattice vibrations (adiabatic approximation), so that the DDSCS will offer information about phonon dispersion, provided the experiment gives access to energy transfers of the order of the phonon energy. The reader is referred to the literature, for instance to Burkel (1991).

(4) The Compton scattering regime (Compton limit), where the electron momentum density is the goal of the experiment. This regime is defined by $qr_c \gg 1$, $\hbar\omega \gg E_0$, where E_0 is a characteristic energy of the system (binding energy for inner-shell sates, Fermi energy for valence electrons); for a review see Cooper (1985).

Figure 1.2 illustrates the different categories of experimental nonresonant inelastic X-ray scattering spectra, for small and large momentum transfer,

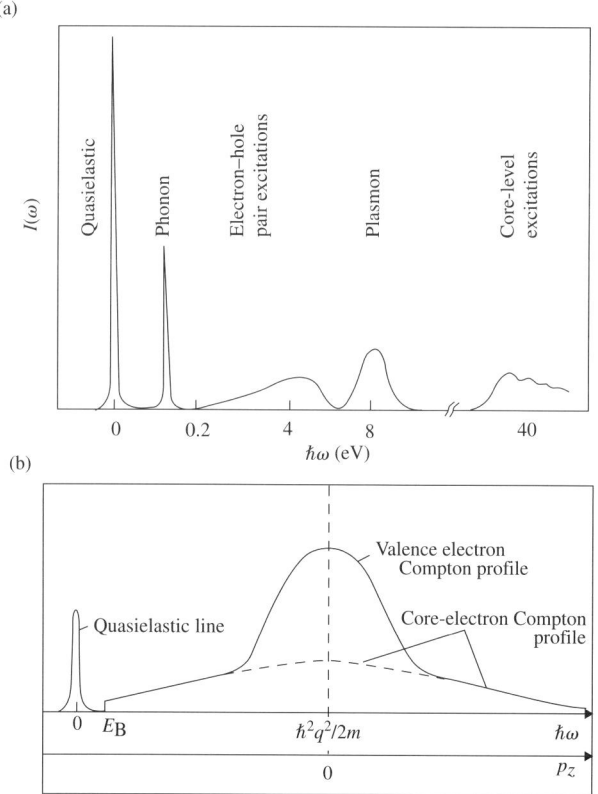

Fig. 1.2. (a) Schematic of an excitation spectrum for $qr_c \leq 1$. (b) Schematic of an excitation spectrum in the Compton limit: $qr_c \gg 1$; E_B is the core-level binding energy.

respectively. As already mentioned, the total nonresonant DDSCS contains, in addition to the Thomson charge scattering part which has already been discussed, a magnetic orbital and spin contribution with a rather complicated polarization dependence (Blume 1985). It should be presented in the "excitation picture", analogous to (1.10), together with the Thomson charge scattering part:

$$\frac{d^2\sigma}{d\Omega_2 d\hbar\omega_2} = r_0^2 \left(\frac{\omega_2}{\omega_1}\right) \sum_{i,f} p_i \Bigg| \langle f| \sum_j \exp(i\mathbf{q}\cdot\mathbf{r}_j)|i\rangle (\mathbf{e}_1 \cdot \mathbf{e}_2)$$

$$- i\left(\frac{\hbar\omega_1}{mc^2}\right) \langle f| \sum_j \exp(i\mathbf{q}\cdot\mathbf{r}_j) \left(i\left(\frac{\mathbf{q}\times\mathbf{p}_j}{\hbar K_1^2}\right) \cdot \mathbf{C} + \mathbf{s}_j \cdot \mathbf{D}\right) |i\rangle \Bigg|^2$$

$$\times \delta(E_f - E_i - \hbar\omega) \tag{1.11}$$

where the components of the spin vector $\mathbf{s}_j$ on the jth electron are the well-known Pauli matrices, and (in the quasi elastic limit)

$$\mathbf{C} = \mathbf{e}_2 \times \mathbf{e}_1 \tag{1.12}$$

$$\mathbf{D} = (\mathbf{e}_2 \times \mathbf{e}_1) + (\hat{\mathbf{K}}_2 \times \mathbf{e}_2)(\hat{\mathbf{K}}_2 \cdot \mathbf{e}_1) - (\hat{\mathbf{K}}_1 \times \mathbf{e}_1)(\hat{\mathbf{K}}_1 \cdot \mathbf{e}_2)$$

$$- (\hat{\mathbf{K}}_2 \times \mathbf{e}_2) \times (\hat{\mathbf{K}}_1 \times \mathbf{e}_1). \tag{1.13}$$

$\hat{\mathbf{K}}$ means a unit vector in the $\mathbf{K}$-direction. Obviously, the electronic state vectors have to include spin coordinates.

The first part of the second (magnetic) term of (1.11) is the orbital, the second part the spin contribution to the magnetic scattering amplitude, and the different polarization factors of these two contributions offer, at least in principle, the possibility to distinguish experimentally between the two.

One can easily realize that the second term of (1.11) gives rise to a pure magnetic contribution to the DDSCS which is by $(\hbar\omega_1/mc^2)^2$ (i.e. for 10 keV X-rays by three to four orders of magnitude) smaller than the pure charge scattering term so that it will be extremely difficult to obtain, in this very direct manner, information about spin or orbital correlations. However, (1.11) can give rise to an interference term between charge and magnetic scattering amplitudes, which should be much easier to identify experimentally, provided the phase shift of the magnetic amplitude by 90° (factor i) relative to the charge scattering amplitude can be compensated. This can be accomplished either by using circularly polarized incident radiation so that for radiation propagating in the z-direction $\mathbf{e}_1$ can be written as

$$\mathbf{e}_1 = \tfrac{1}{\sqrt{2}}(\hat{\mathbf{u}}_x \pm i\hat{\mathbf{u}}_y) \tag{1.14}$$

or by analyzing the degree of circular polarization of the scattered radiation.

By making use of the former possibility in the course of investigating ferromagnetic materials in the Compton limit, as first proposed by Platzman and Tzoar (1970), one can change the sign of the interference term either by switching the handedness of the circular polarization or by rotating the direction of magnetization. Thus the interference term can be separated from the charge scattering contribution by subtracting the measured DDSC with a different sign for the interference term ending up with:

$$\left(\frac{d^2\sigma}{d\Omega_2 d\hbar\omega_2}\right)_{\text{interf}} = -2\left(\frac{m}{q}\right) r_0^2 \left(\frac{\hbar\omega_1}{mc^2}\right) P_2(1-\cos\theta)$$

$$\times (\hat{\mathbf{K}}_1 \cos\theta + \hat{\mathbf{K}}_2) \cdot \mathbf{m} \, J_{\text{m}}(p_z), \qquad (1.15)$$

where $\mathbf{m}$ is a unit vector in the direction of the preferred magnetic axis and P_2 is the second Stokes parameter, which describes the degree of circular polarization of the incident beam (Lipps and Tolhoek 1954). $J_{\text{m}}(p_z)$ denotes the so-called magnetic Compton profile (Cooper et al. 1986), which is the projection on the scattering vector of the momentum density of electrons with unpaired spins:

$$J_{\text{m}}(p_z) \equiv \int\int [\rho_\uparrow(\mathbf{p}) - \rho_\downarrow(\mathbf{p})] \, dp_x \, dp_y, \qquad (1.16)$$

whereas the charge Compton profile, defined in (1.10), of course can be written in the following way:

$$J(p_z) \equiv \int\int [\rho_\uparrow(\mathbf{p}) + \rho_\downarrow(\mathbf{p})] \, dp_x \, dp_y. \qquad (1.17)$$

The two equations (1.15) and (1.16) are the result of a longer derivation, which shows that the orbital contribution in (1.11) is negligibly small compared with the spin contribution, whenever the prerequisites of the impulse approximation are valid (Lovesey 1993). Relativistic effects are only included up to first order in $(\hbar\omega_1/mc^2)$ by taking into account the charge–magnetic interference term. Moreover, the quantity $(1 + K_2/K_1)/2$ has been set equal to one.

The information on the correlated motion of electrons and their momentum distribution, respectively, can be considerably expanded, if the initial photon state of a nonresonant inelastic X-ray scattering experiment is prepared to be the coherent superposition of two plane waves, whose wavevectors differ by a reciprocal lattice vector $\mathbf{g}$. In that case the DDSCS exhibits, due to the coherence of the two participating plane waves, an additional interference term, which contains the so-called off-diagonal structure factor

$$\mathbf{S}(\mathbf{q}, \mathbf{q}+\mathbf{g}, \omega) = \frac{1}{2\pi\hbar} \int dt \exp(-i\omega t) \int_{-\infty}^{\infty} d^3r' \exp[i(\mathbf{q}+\mathbf{g})\cdot\mathbf{r}']$$

$$\times \int d^3r \exp(-i\mathbf{q}\cdot\mathbf{r}) \, n_2(\mathbf{r}, \mathbf{r}', t), \qquad (1.18)$$

so that, in inhomogeneous systems, also the dependency of the two-particle density correlation function on the reference coordinate $\mathbf{r}'$ is probed (Schülke 1982).

In the Compton limit the interference term contains, under certain experimental symmetry requirements, information about projections of nondiagonal elements of the one-particle density matrix in momentum space, $\Gamma_1(\mathbf{p}|\mathbf{p}+\mathbf{g})$, on the scattering vector (Schülke and Mourikis 1986).

Until now it was tacitly assumed that the recoil electron of a Compton scattering process was not a subject of investigation neither with respect to its energy nor in regard to its direction. If one arranges an experiment in such a way that the photons scattered in a distinct direction are energy-analyzed and the recoil electrons are selected in coincidence with the scattered photon with respect of their direction, one is measuring the triple differential scattering cross-section $d^3\sigma/d\omega_2 d\Omega_\gamma d\Omega_e$, which reads, within the limits of the relativistic impulse approximation (Bell et al. 1991),

$$\frac{d^3\sigma}{d\hbar\omega_2 d\Omega_\gamma d\Omega_e} = \left(\frac{\omega_1}{\omega_2}\right) p' \left(\frac{d\sigma}{d\Omega_\gamma}\right)_{KN} \rho(\mathbf{p}). \qquad (1.19)$$

Momentum conservation demands $\mathbf{p} = \mathbf{p}' - \mathbf{q}$, if $\mathbf{p}'$ is the momentum of the recoil electron. The Klein–Nishina scattering cross-section for linear polarized photons, $(d\sigma/d\Omega_\gamma)_{KN}$, is given by

$$\left(\frac{d\sigma}{d\Omega_\gamma}\right)_{KN} = \left(\frac{e^4}{4m^2c^4}\right) \left(\frac{K_2}{K_1}\right)^2 \left[\left(\frac{K_2}{K_1}\right) + \left(\frac{K_1}{K_2}\right) + 4(\mathbf{e}_1 \cdot \mathbf{e}_2)^2 - 2\right]. \qquad (1.20)$$

Insertion of the Klein–Nishina cross-section neglects a weak $\mathbf{p}$-dependence of the factor which connects the triple differential scattering cross-section and the electron momentum density in (1.19). Thus, the complete 3D electron momentum density $n(\mathbf{p})$ becomes accessible.

Let us now discuss that part of the DDSCS which is the result of a second-order perturbation treatment of the interaction Hamiltonian linear in the vector potential $\mathbf{A}$. The diagrams of Fig. 1.3 have to be evaluated, where only the first

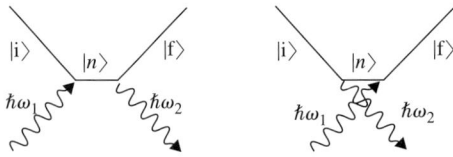

Fig. 1.3. Schematic of two-photon processes to be treated in second-order perturbation theory.

one will give rise to a real resonating term, so that only this should be considered in our first overview. In this first diagram the incident photon $(\mathbf{K}_1, \omega_1, \mathbf{e}_1)$ excites the electron system from its ground state $|i\rangle$ into an intermediate state $|n\rangle$, which decays after a short time into the final state $|f\rangle$ emitting the energy difference between $|n\rangle$ and $|f\rangle$ as the photon $(\mathbf{K}_2, \omega_2, \mathbf{e}_2)$. It should be stressed that, due to the short lifetime of the intermediate state $|n\rangle$, the excitation process $|i\rangle \to |n\rangle$ is not limited by energy conservation. Only the total resonant scattering process has to satisfy energy conservation. This process is once more sketched in single-particle approximation in Fig. 1.4, which reflects the real experimental situation one is confronted with, when utilizing this resonant term for investigating excitations in condensed matter. One sees that the intermediate state $|n\rangle$ is connected with a hole in a deeper lying inner-shell level and an electron in a formerly unoccupied level. The final state is described by a hole in a higher lying level (inner-shell level or a valence electron state) and an electron in a formerly unoccupied level. Thus the final state is the same as that which would occur when the excitation was driven by the A^2 term as given in equation (1.9). However, the inclusion of an inner-shell level and the intimate connection between the excitation $|i\rangle \to |n\rangle$ and the deexcitation (re-emission) process $|n\rangle \to |f\rangle$ produces a much greater selectivity of the resonant inelastic scattering process, when compared with the nonresonant one. This fact might be considered as a disadvantage, as long as one is interested in utilizing the direct connection between excitation and fluctuation expressed in the fluctuation–dissipation theorem (see the connection between (1.5) and (1.9)), in order to investigate, in a very direct manner, electron correlations via the dynamic structure factor. On the other hand, the greater selectivity offers very valuable tools to study specific properties of condensed matter as will be discussed later. The formal treatment in second-order perturbation theory of the interaction Hamiltonian of the first diagram of Fig. 1.3

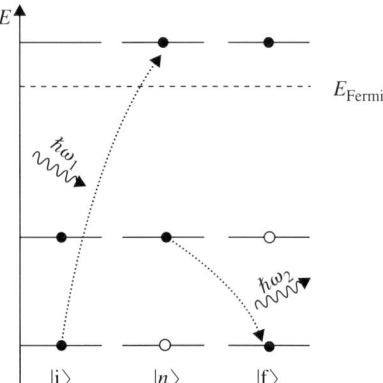

Fig. 1.4. Energy-level scheme of a resonant inelastic scattering process.

leads to the extended Kramers–Heisenberg formula (Blume 1985)

$$\frac{\mathrm{d}^2\sigma}{\mathrm{d}\Omega_2 \mathrm{d}\hbar\omega_2} = r_0^2 \left(\frac{\omega_2}{\omega_1}\right)^2 \sum_{i,f} p_i \left| \left(\frac{\hbar^2}{m}\right) \right.$$

$$\times \sum_n \frac{\langle f| \sum_j [\mathbf{e}_2^* \cdot \mathbf{p}_j/\hbar - i(\mathbf{K}_2 \times \mathbf{e}_2) \cdot \mathbf{s}_j] \exp(-i\mathbf{K}_2 \cdot \mathbf{r}_j)|n\rangle}{E_n - E_i - \hbar\omega_1 - i\Gamma_n/2}$$

$$\times \langle n|\mathbf{e}_1 \cdot \sum_j [\mathbf{p}_j/\hbar + i(\mathbf{K}_1 \times \mathbf{e}_1) \cdot \mathbf{s}_j]|i\rangle \exp(i\mathbf{K}_1 \cdot \mathbf{r}_j) \Big|^2 \delta(E_f - E_i - \hbar\omega)$$

(1.21)

where $|i\rangle$, $|f\rangle$ and $|n\rangle$ are the initial, final and an intermediate state of the scattering electron system, respectively, with their corresponding energies E_i, E_f and E_n, as sketched in Fig. 1.4. $(\Gamma_n/\hbar)^{-1}$ is the lifetime of the intermediate state. If the incident photon energy approaches the resonance condition

$$\hbar\omega_1 = E_n - E_i \qquad (1.22)$$

the resonant part of the DDSCS as given in (1.21) can be greater by several orders of magnitude than the nonresonant part as given in (1.11), so that the latter can be neglected.

We will first discuss the more general consequences of (1.21), namely those of the resonance denominator, which constitutes resonant inelastic scattering (RIXS) spectroscopy.

Let the intermediate state $|n\rangle$ be a core hole with an excited electron in a quasicontinuum of unoccupied states, and let the state $|f\rangle$ be a hole in occupied valence or corestates. Then, as will be shown in detail later, the interplay of the δ-function and the denominator of (1.21) shifts the energy of the scattered photon to higher energies with increasing energy of the incident photon, due to the finite lifetime $(\Gamma_n/\hbar)^{-1}$, where the shift depends linearly on the incident photon energy as long as $\hbar\omega_1 < E_B$ (E_B = binding energy of the core electron involved in $|n\rangle$) and levels out for $\hbar\omega_1 > E_B$. This shift is called, in what follows, the Raman shift (Eisenberger et al. 1976a,b). If, on the other hand, $|n\rangle$ is discrete, the linear Raman shift persists over the whole range of the incident photon energy. Of course, the spectral contribution of this discrete state $|n\rangle$ to the DDCS is modulated by the denominator of (1.21).

This property of RIXS can be utilized in three different ways: (i) to measure Γ_n, the core-hole lifetime broadening; (ii) to emphasize the contributions of transitions $|i\rangle \Rightarrow |n\rangle$ to the photon absorption cross-section, which are too weak to be detectable in conventional X-ray absorption spectroscopy (XAS); and (iii) to eliminate, at least within the limits of a one-electron theory, the broadening of X-ray absorption spectra due to the lifetime of the excited core state.

As in the case of nonresonant scattering (equation 1.11), the spin-dependent contribution to the scattering amplitude, is smaller by roughly two orders of magnitude smaller than the $\mathbf{e} \cdot \mathbf{p}$ term. Thus, for getting a first orientation, it might be enough to comprise only the consequences of the $\mathbf{e} \cdot \mathbf{p}$ terms in the following discussion.

(1) The $\mathbf{e} \cdot \mathbf{p}\exp(i\mathbf{K} \cdot \mathbf{r})$ terms in the matrix elements, both for the excitation into the intermediate state and for the deexcitation from the intermediate into the final state, make resonant inelastic scattering highly symmetry selective when one adjusts the polarization vectors and/or the wavevectors of the incident and/or of the scattered radiation relative to distinct directions of the contributing orbitals. This way certain symmetries of the orbitals involved can be emphasized and other symmetries can be suppressed. Even if the scattering sample does not possess orientational order like a gas target, a predominant excitation into a specific intermediate state can select molecules with a certain orientation relative to both the polarization and the wavevector of the incident beam. This way the re-emission takes place from a set of aligned molecules, so that a polarization analysis of the scattered beam brings about a symmetry assignment of the re-emitted radiation (Cowan *et al.* 1986). It is the intimate coupling between excitation and re-emission in a RIXS process which opens up this unique access to the symmetry of electronic states in disordered materials. Moreover, since the excitation starts from inner-shell levels, this spectroscopy probes the local symmetry of the orbitals into which the inner-shell electron is excited, local with respect to the position of the contributing inner-shell level. The same argument applies to the deexcitation process.

If one considers magnetic materials, the $\mathbf{e} \cdot \mathbf{p}\exp(i\mathbf{K} \cdot \mathbf{r})$ terms especially should offer selectivity with respect to transitions connected with distinct changes of the magnetic quantum number, ΔM, both in excitation and re-emission. It is the relative orientation of $\mathbf{e}_1$, $\mathbf{e}_2$ and of $\mathbf{e}_1^* \times \mathbf{e}_2$, respectively, to the quantization axis on the one hand and the dependence of the transition matrix elements on ΔM, which make the DDSCS selective to magnetic polarization of unoccupied and occupied bands involved either in the excitation or the re-emission process. It should be stressed that this possibility to investigate magnetic polarization does not make use of the magnetic part of the resonant DDSCS but is only connected with the $\mathbf{e} \cdot \mathbf{p}$ term. This selectivity has been widely used in elastic resonant scattering under the name exchange scattering (Hannon *et al.* 1988; see for a review Lovesey and Collins 1996). A thorough theoretical analysis of the RIXS process by Carra *et al.* (1995) and van Veenendaal (1996) will open up this type of selectivity also for resonant inelastic scattering studies.

(2) Additional selectivity arises from the fact that the excitation and the re-emission process as described by (1.16) has to be considered as an undivided inelastic scattering process, so that the coherence of excitation and re-emission processes requires momentum conservation and also spin conservation for the whole scattering process. Therefore, if the excitation of an electron occurs into a formerly unoccupied Bloch state with Bloch wavevector $\mathbf{k}_e$ and the deexcitation produces a hole in the valence band with a Bloch wavevector $\mathbf{k}_h$, momentum

conservation requires that

$$\mathbf{k}_h - \mathbf{k}_e = \mathbf{q} + \mathbf{g}, \qquad (1.23)$$

where $\mathbf{q}$ is the transferred momentum and $\mathbf{g}$ a reciprocal lattice vector. This way resonant inelastic scattering spectroscopy is Bloch-$\mathbf{k}$ vector selective as found the first time by Ma *et al.* (1992).

(3) Spin conservation offers an interesting possibility to investigate the spin polarization of energy bands by resonant inelastic X-ray scattering provided the excitation occurs into spin polarized formerly unoccupied states and, additionally, also the final hole state is spin polarized (spin split in energy) (Hämäläinen *et al.* 1992). Both spin polarizations must be the result of an exchange interaction with a spin aligned occupied state, the so-called internal spin reference, for instance the spin aligned 4f state in a rare earth element or a spin aligned 3d state in a transition metal element. Let the re-emission process produce a hole in a shallow core level, where an electron with a certain orientation of the spin relative to the spin orientation of the internal reference is missing. Then this re-emission is only possible if the excitation process has transferred an electron to a formerly unoccupied state with the same spin orientation relative to the internal spin reference. This way one can probe the spin polarization of unoccupied bands by fixing the energy of the scattered photons successively to one of the two components of the spin polarized re-emission line. Since the exchange interaction with the internal spin reference is strictly local this spectroscopy does not need a magnetically ordered phase to investigate the spin polarization of unoccupied bands.

(4) The intimate coupling of excitation and re-emission mediated by the intermediate state of the resonant inelastic scattering process enables the utilization of shakeup processes in the intermediate state to study excitation processes otherwise difficult to investigate (Platzman and Isaacs 1998). As sketched in Fig. 1.5,

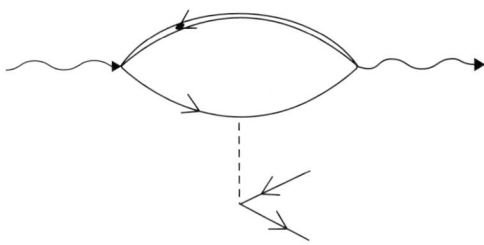

Fig. 1.5. Shakeup process in the intermediate state. The double line is the propagator of the core hole, the single line is the particle propagator, and the dashed line means Coulomb interaction with the surrounding medium producing a particle–hole pair excitation. Reprinted with permission from Platzman and Isaacs (1998); copyright (1998) by the American Physical Society.)

the incident photon creates an electron–hole pair with a hole in an inner-shell level, which is very localized. This pair is bound as an exciton by the Coulomb interaction, takes up the momentum of the incident photon and scatters off the valence electron system producing an excitation. When the exciton recombines, the emitted photon reflects the energy and the momentum imparted to the system. This means that the dispersion of the shakeup excitation becomes accessible.

1.3 Survey of experiments

In this section we provide a brief survey of inelastic X-ray scattering experiments, which of course is too short to give a full review. The aim is to expand on the orientation of our field by looking at a range of physical properties. We feature all the topics that appear in the book, namely inelastic X-ray scattering (IXS) spectroscopy connected with valence electron excitation, briefly "valence electron IXS", inner-shell IXS, IXS in the Compton limit, briefly "Compton scattering", and resonant IXS, briefly "RIXS".

1.3.1 *Valence electron IXS*

The aim of most valence electron IXS experiments was to measure the dynamic structure factor $S(\mathbf{q},\omega)$ in the whole relevant range of energy transfer ω and for a larger set of momentum transfers q, in order to cover a wide field of correlation distances one is looking at. The first experiments of this kind were still performed using conventional X-ray sources (CXS). Among those, the measurements of Platzman and Eisenberger (1974a), Eisenberger and Platzman (1976), and of Eisenberger *et al.* (1975) gain special attention, since the interpretation of a double peak structure (Platzman and Eisenberger 1974b) found in the electron–hole pair continuum of the $S(\mathbf{q},\omega)$ spectra to be an indication of an incipient Wigner electron lattice has initiated a fruitful discussion among theoreticians on the dynamic structure of electrons in nearly homogeneous systems. Synchrotron radiation was first used by Schülke *et al.* (1984) at a measurement of $S(\mathbf{q},\omega)$ on single crystals of Li, where it was shown that the measured intensity distribution in the first instance arbitrarily normalized can be brought to an absolute scale with respect to the dynamic structure factor by utilizing the so-called f-sum rule. Moreover, this study and further investigations on Li single crystals (Schülke *et al.* 1986) have shown that at least part of the double-peak structure found in previous studies of the electron–hole pair continuum can be traced back to excitation gaps originated from the removal of the degeneracy at Brillouin zone boundaries, so-called zone boundary collective states (ZBCS) (Foo and Hopfield 1968). IXS studies aiming to cover the whole respective range of energy transfer and a large domain of transferred momenta yielded valuable insight into the role of electron correlation in simple metals and s-p-semiconductors, where the interpretation has gone beyond the random phase approximation taking into

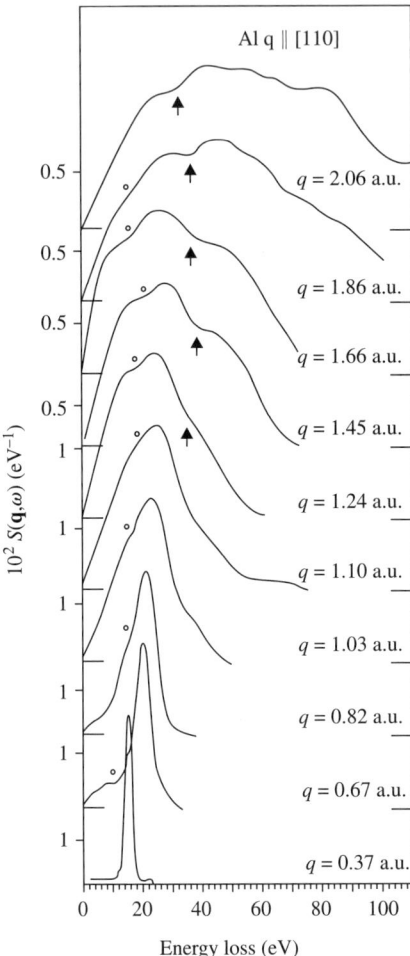

Fig. 1.6. Synchrotron radiation based measurements of the dynamical structure factor $S(\mathbf{q},\omega)$ of Al metal for $\mathbf{q} \parallel [110]$ and various values of q as indicated (a.u. = atomic units: $\hbar = e = m = 1$). Points and arrows mark lattice-induced fine structure. (Reprinted with permission from Schülke et al. (1993); copyright (1993) by the American Physical Society.)

account self-energy, local-field effects and the interaction of collective excitations with electron–hole pair excitations, the so-called plasmon-Fano resonances (Sturm et al. 1992). As an example of these studies, Fig. 1.6 presents $S(\mathbf{q},\omega)$ measurements on single-crystal Al (Schülke et al. 1993). The sharp structures for small q have to be attributed to plasmon excitation, whereas the broader features for larger q are due to electron–hole pair excitations.

Another section of valence electron IXS experiments was more devoted to investigating interband transitions of special interest, where selection rules imposed upon the matrix elements of (1.9) were utilized. These studies made use of the Kramers–Kronig relation in order to extract the dielectric function, especially its imaginary part, from the IXS data (e.g. Schülke et al. 1988b). In this connection a study (Caliebe et al. 2000) should be stressed, where the crystal momentum corresponding to an indirect band gap was read directly from the spectra taken on diamond.

Finally, also studies utilizing the coherent superposition of two plane waves as the primary photon state of an inelastic scattering experiment fall among valence electron IXS. As a prominent result of those investigations the existence of a plasmon band structure in Si was demonstrated (Schülke and Kaprolat 1991) and the size of the plasmon band gap at the (111) zone boundary was measured (Sturm and Schülke 1992).

So far IXS has investigated electron dynamics in the frequency and momentum representation. However, the fundamental relations (1.6) between the dynamic structure factor $S(\mathbf{q},\omega)$ and the two-particle density correlation function $n_2(\mathbf{r'},\mathbf{r},t)$ should enable us, at least in principle, to convert experimental results from the momentum–frequency into the space–time domain. This possibility is the more interesting, since it would mean experimental access to correlated movements on the attosecond time-scale (1 as $= 10^{-18}$ s) and probing correlation distances of one Å and less. Abbamonte et al. (2004) were the first to utilize the momentum flexibility of inelastic X-ray scattering to invert its loss function with the aim of imaging electron density disturbances in water.

Table 1.1 contains a selection of materials which have been successfully investigated by valence electron IXS.

1.3.2 *Inner-shell IXS*

Inelastic X-ray scattering spectra, which reflect excitations of electrons from an inner-shell level into unoccupied states above the Fermi level, contain the same information about the density of unoccupied states as X-ray absorption spectra, provided they satisfy the condition of the dipole approximation, namely $qa < 1$, where a is the inner-shell orbital radius (Nagasawa et al. 1989). As an example Fig. 1.7 shows the comparison of Li inner-shell IXS spectra, taken with different sizes of q, with a soft X-ray absorption spectrum (Haensel et al. 1970). Under the conditions of the dipole approximation, the vector $\mathbf{q}$ of momentum transfer takes over the same role as the electric field vector $\mathbf{E}$ of the photon annihilated in the X-ray absorption spectroscopy. This means that the experimentalist can play with the direction of $\mathbf{q}$ in order to investigate different symmetry projections of the unoccupied density of states (DOS) (see Schülke et al. 1988b) in the same manner as an X-ray absorption spectrum yields different symmetry projections depending on the direction of $\mathbf{E}$. However, inner-shell IXS has the great advantage over soft X-ray absorption spectroscopy that one can probe inner-shell

Table 1.1. Representative examples of materials investigated by valence electron IXS using conventional X-ray sources (CXS) and synchrotron radiation sources (SRS).

Material	Ref.	X-ray source	Chief object
Li	(1) (2) (3)	CXS	$S(\mathbf{q},\omega)$
Li	(4) (5)	SRS	$S(\mathbf{q},\omega)$, ZBCS, self-energy
Li	(6)	SRS	empirical local-field effects
Li	(7)	SRS	plasmon-Fano resonances
Be	(8) (9) (10)	CXS	$S(\mathbf{q},\omega)$, plasmon dispersion
Be	(11) (12)	SRS	$S(\mathbf{q},\omega)$, ZBCS, self-energy
C (diamond)	(13)	SRS	indirect band gap
C (graphite)	(14)	CXS	plasmon dispersion anisotropy
C (graphite)	(15)	SRS	$\varepsilon(\mathbf{q},\omega)$, interband transitions
LiC_6	(16)	SRS	$\varepsilon(\mathbf{q},\omega)$, interlayer band shift
KC_8	(17)	SRS	$\varepsilon(\mathbf{q},\omega)$, interlayer states
Al	(18)	CXS	$S(\mathbf{q},\omega)$, double peak structure
Al	(19) (20)	SRS	$S(\mathbf{q},\omega)$ fine structure
Al	(21)	SRS	empirical local field effects
Si	(22) (23)	SRS	$S(\mathbf{q},\omega)$, plasmon-Fano resonances
Si	(24) (25)	SRS	$S(\mathbf{q},\mathbf{q}+\mathbf{g},\omega)$, plasmon bands
LiF	(13)	SRS	$S(\mathbf{q},\omega)$, electron–hole interaction
V_2O_3	(26)	SRS	charge transfer exciton, Hubbard gap
C_{60}	(27)	SRS	$S(\mathbf{q},\omega)$, σ- and π-plasmons
Ti, TiC	(28)	SRS	$S(\mathbf{q},\omega)$; ZBCS
SiC	(29)	SRS	$S(\mathbf{q},\omega)$; full dielectric matrix
BN	(30)	SRS	$S(\mathbf{q},\omega)$, electron–hole interaction
MgB_2	(31)	SRS	$S(\mathbf{q},\omega)$; plasmon dispersion
Li, Na, Al	(32) (33)	SRS	$S(\mathbf{q},\omega)$; liquid metals
Li-NH_3	(34)	SRS	$S(\mathbf{q},\omega)$; r_s dependence
H_2O	(35)	SRS	inverting $S(\mathbf{q},\omega)$ into real space and time
Na, Al	(36)	SRS	$S(\mathbf{q},\omega)$; double plasmon excitation
TiO_2	(37)	SRS	$S(\mathbf{q},\omega)$; role of crystal local fields
Sc, Cr	(38)	SRS	$S(\mathbf{q},\omega)$; d-d transitions

(1) Schülke and Lautner (1974); (2) Eisenberger et al. (1975); (3) Priftis et al. (1978); (4) Schülke et al. (1984); (5) Schülke et al. (1986); (6) Schülke et al. (1996a); (7) Höppner et al. (1998); (8) Eisenberger et al. (1973); (9) Platzman and Eisenberger (1974a); (10) Vradis and Priftis (1985); (11) Schülke et al. (1987); (12) Schülke et al. (1989); (13) Caliebe et al. (2000); (14) Eisenberger and Platzman (1976); (15) Schülke et al. (1988b); (16) Schülke et al. (1988a); (17) Schülke et al. (1991b); (18) Platzman and Eisenberger (1974b); (19) Platzman et al. (1992); (20) Schülke et al. (1993); (21) Larson et al. (1996); (22) K. Sturm et al. (1992); (23) Schülke et al. (1995); (24) Schülke and Kaprolat (1991a); (25) Sturm and Schülke (1992); (26) Isaacs et al. (1996); (27) Isaacs et al. (1992); (28) Macrander et al. (1996); (29) Montano et al. (2002); (30) Galambosi et al. (2001); (31) Galambosi et al. (2005); (32) Hill et al. (1996); (33) Sternemann et al. (1998); (34) Burns et al. (2002); (35) Abbamonte et al. (2004); (36) Sternemann et al. (2005); (37) Gurtubay et al. (2004); (38) Gurtubay et al. (2005).

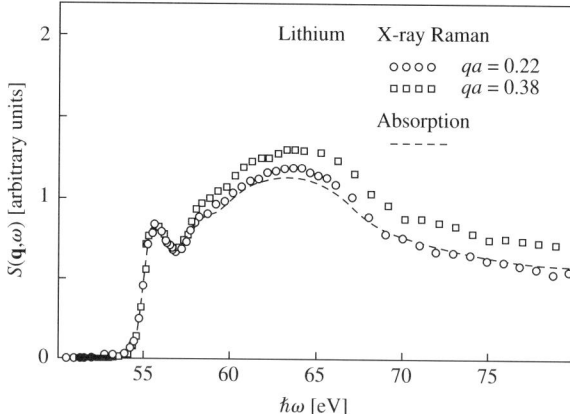

Fig. 1.7. Circles and squares: synchrotron radiation based experimental inner-shell inelastic X-ray scattering spectrum of Li metal for two values of momentum transfer q Dashed line: reduced soft X-ray absorption curve (Haensel et al. 1970). (Reprinted with permission from Nagasawa et al. (1989); copyright (1989) by the Physical Society of Japan.)

thresholds with energies between 20 and 1000 eV (K-shell edges of low-Z elements or L/M shells of higher Z-elements) using hard X-rays with energies between 10 and 20 keV with a corresponding large penetration depth, whereas soft X-rays are restricted to very surface-near portions of the sample and need a vacuum environment for the sample. Thus the low-energy thresholds including their extended X-ray **a**bsorption **f**ine **s**tructure (EXAFS) become accessible in samples containing other highly absorbing constituents in arbitrary environments, e.g. reacting gases. One may wonder why these unique properties of inner-shell IXS have been seldom used. Table 1.2 presents a selection of materials investigated with inner-shell IXS.

A further advantage of inner-shell IXS over soft X-rax absorption spectroscopy is the free choice of the size of q, so that the transition from the dipole approximation to higher multipole excitation can be studied (see Nagasawa et al. 1997, Krisch et al. 1997).

In cases, where excitation spectra of very deep-lying inner-shell levels are superimposed by contributions of more shallow ones, coincidence techniques were utilized and first introduced by Fukamachi and Hosoya (1972) and further worked out by Namikawa and Hosoya (1984). The scattered radiation from a γ-ray source, energy-analyzed by means of a solid state detector, is measured in coincidence with the fluorescence radiation produced by the refilling of the deep-lying inner-shell hole. This way scattering events from one deep-lying inner-shell excitation are separated from excitations of all the other levels. But it has been pointed out by Manninen (1986) that these early experiments might be falsified by false coincidences. If one intends to utilize monochromatized

Table 1.2. Representative selection of materials investigated with inner-shell IXS.

Element	Material	Ref.		Edge	Chief object
^{4}He	solid, hcp	(1)		K	Frenkel-type exciton
Li	Li metal	(2)	(3)	K	comparison with soft XAS
Li	Li metal	(4)	(5)	K	dipole to multipole transition
Li	LiC_6	(6)		K	van Hove singularities, interlayer states
Be	Be metal	(3)		K	**q** orientation dependence
C	graphite	(3)	(7)	K	symmetry projections of the DOS
C	graphite	(8)		K	EXAFS
C	fused-ring aromatics	(9)		K	near-edge fine structure
C	C_{60}	(10)		K	sp_2 carbon network
O	liquid H_2O, ice	(11) (13)	(12) (14)	K	EXAFS, hydrogen bond network
B	BN	(15)		K	near-edge fine structure
B, N	BN	(16)		K	near-edge fine structure, high-pressure
B	B_4C	(17)		K	core exciton, **q**-dependence
F	LiF	(18)		K	near-edge fine structure
Be	BeO	(19)		K	core exciton
Fe,Cu	metal	(20)		K	γ-source, coincidence technique
Cu,Zr	metal	(21)		K	γ-source, coincidence technique
Cu	metal	(22)		K	W K-radiation, coincidence techn.
Cu	metal	(23)		K	synchr. rad., coincidence technique
Ag	metal	(24)		K	synchr. rad., coincidence technique

(1) Schell et al. (1995); (2) Schülke et al. (1986); (3) Nagasawa et al. (1989); (4) Nagasawa et al. (1997); (5) Krisch et al. (1997); (6) Schülke et al. (1991c); (7) Schülke et al. (1988b); (8) Tohji and Udagawa (1987); (9) Gordon et al. (2003); (10) Rueff et al. (2002); (11) Bowron et al. (2000); (12) Bergmann et al. (2002); (13) Wernet et al. (2004); (14) Cai et al. (2005); (15) Watanabe et al.; (16) Meng et al. (2004); (17) Feng et al. (2004); (18) Hämäläinen et al. (2002); (19) Soininen et al. (2001); (20) Namikawa and Hosoya (1984); (21) Manninen et al. (1990); (22) Laukkanen et al. (1996); (23) Marchetti et al. (1987); (24) Laukkanen et al. (1998).

synchrotron X-radiation, which is superior to γ-ray sources with respect to the total flux, the bunch character of synchrotron radiation emission must be taken into account, when the rate of false coincidences should be estimated (Laukkanen et al. 1998). Contrary to the above listed IXS experiments dealing with inner-shell level binding energies of less than 1 keV, these studies utilizing coincidence technique were mainly devoted to finding deviations from the impulse approximation for $qa \leq 1$. A few representative experiments of that type are summarized at the end of Table 1.2.

1.3.3 IXS in the Compton limit

1.3.3.1 Compton charge scattering
As shown in Section 1.2, equation (1.10), Compton scattering, i.e. IXS in the Compton limit, is aiming to get information about the ground-state electron density $\rho(\mathbf{p})$ in momentum space, more specifically information about the projection $\int \mathrm{d}^3 p \, \rho(\mathbf{p}) \delta(\mathbf{p} \cdot \mathbf{q}/m - q^2/2m - \hbar\omega)$ of $\rho(\mathbf{p})$ on the direction of the momentum transfer $\mathbf{q}$. As already mentioned in Section 1.1, this type of inelastic X-ray scattering experiments were the very first successful ones and led to conclusive confirmation of the Fermion character of solid state electrons by looking at the width of $\rho(\mathbf{p})$, as reflected by the Compton profiles measured on Be metal (DuMond 1929). The first Compton scattering experiments after the revival of this technique by Cooper et al. (1965) were performed by using conventional X-ray sources (sealed tubes or rotating anodes). Compton scattering as a tool for investigating the electron density in momentum space got a rapid increase after the introduction of γ-ray sources by utilizing the then new high-resolution solid state detectors. A huge number of condensed matter systems were investigated, some selected examples of which are summarized in Table 1.3, which do not claim completeness, since this volume is mainly devoted to the application of synchrotron radiation. Five aspects of these Compton scattering techniques were in the forefront:

(i) Free atoms and molecules: In the early days of the revival of Compton scattering experiments the investigation of free atomic and free molecular systems were a point of main effort, since those measurements offer the possibility to test the adequacy of Hartree–Fock calculations and the role of correlation effects. Only a few examples of these very ambitious theoretical efforts are added to the selected experiments listed in Table 1.3.

(ii) Chemical bonding and related phenomena: Contrary to the measurement of the electron density in real space as performed by means of X-ray diffraction, where the contribution of the valence electrons is rather difficult to pursue, since it is visible only in a few diffraction peaks belonging to the smallest reciprocal lattice vectors, the contribution of the valence electrons to the Compton profile is concentrated around its peak, thus dominating its shape and its $\mathbf{q}$-orientation dependence. Therefore, features of chemical bonding, such as the amount of covalency and ionicity, bond directionality, models for the implementation of hydrogen into metals, and charge transfer upon alloying, are strongly reflected

Table 1.3. Some selected examples of Compton studies performed by means of conventional X-ray or γ-ray sources, and corresponding calculations related to the investigation of (i) free atoms and molecules, (ii) chemical bonding, (iii) Fermi surface and momentum space density, (iv) correlation effects, and (v) coherent Compton scattering. CXS: conventional X-ray source; γ: γ-source; CD: crystal dispersive energy analysis of the scattered radiation; SSD: energy analysis by means of a solid state detector.

Material	Ref.	Source	Energy analysis	Chief object of investigation
He, H_2	(1)	CXS	CD	momentum density, comparison with Hartree–Fock (HF) calcul.
N_2, O_2, Ne	(2) (3)	CXS	CD	momentum densities, comparison with HF calcul. includ. correlation
Ne, Ar, Kr	(4)	theory		calculation, correlation on local density approx. (LDA) level included
N_2	(5)	theory		configuration interaction calculation
CH_4, C_2H_4, C_2H_6	(6)	CXS	CD	transferability of bond profiles
Pd-H, V-H	(7)	γ	SSD	models for hydrogen in metals
C(diamond), Si	(8)	γ	SSD	molecular orbit model, $B(\mathbf{r})$
FeAl, CoAl, NiAl	(9)	γ	SSD	charge transfer upon alloying
Li_3N	(10)(11)	γ	SSD	2D reconstruction of $B(\mathbf{r})$, orthogonality of ionic orbitals

Material	Ref.	Source	Energy analysis	Chief object of investigation
LiOH	(12)	γ	SSD	3D reconstruction of B(**r**)
GaP	(13)	γ	SSD	degree of ionicity
MgO	(14)	γ	SSD	degree of ionicity
LiH	(15)	γ	SSD	polarization and covalency
α-oxalic acid	(16)	γ	SSD	hydrogen bonds
Li	(17) (18)	CXS	CD	momentum density, reconstructed Fermi surface
Be	(19)	CXS	CD	momentum density, Fermi surface
Be	(20)	γ	SSD	momentum density, Fermi surface
Si, Ge, C(diamond)	(21)(22)	γ	SSD	momentum density, 3D reconstruction
V	(23) (24)	γ	SSD	momentum density, Fermi surface
Cu	(25) (26)	γ	SSD	LDA, directional correlation correction
Cr	(27) (28)	γ	SSD	directional correlation correction
Si	(29),(30),(31)	CXS	SSD	coherent Compton scattering nondiagonal elements of $\gamma(\mathbf{p}|\mathbf{p}+\mathbf{g})$

(1) Eisenberger (1970); (2) Eisenberger (1972); (3) Eisenberger et al. (1972); (4) Tong and Lam (1978); (5) Thakkar et al. (1986); (6) Eisenberger and Marra (1971); (7) Lässer and Lengeler (1978); (8) Pattison et al. (1981); (9) Manninen et al. (1981); (10) Pattison and Schneider (1980); (11) Pattison et al. (1984); (12) Heuser-Hoffmann and Weyrich (1985); (13) Rao et al. (1985); (14) Causa et al. (1986); (15) Asthalter and Weyrich (1993); (16) Weisser and Weyrich (1993); (17) Eisenberger et al. (1972); (18) Schülke (1977); (19) Currat et al. (1971); (20) Hansen et al. (1979); (21) Reed and Eisenberger (1972); (22) Mueller (1977); (23) Rollason et al. (1983); (24) Wakoh et al. (1976); (25) Bauer and Schneider (1984); (26) Bagayoko et al. 1980; (27) Cardwell et al. (1989); (28) Wakoh and Matsumoto (1989); (29) Golovchenko et al. (1981); (30) Schülke et al. (1981); (31) Schülke and Mourikis (1986).

by Compton measurements preferably on single crystals. Eisenberger and Marra (1971) were the first to analyze the Compton profiles of various hydrocarbons in terms of invariant bond profile contributions, whose transferability was claimed by these authors. Epstein (1970) and Epstein and Lipscomb (1970) have questioned the universality of bond transfer. Nevertheless, Epstein and Tanner (1977) have introduced principles which support the interpretation of chemical bonding from a momentum viewpoint. On the other hand, the introduction of the concept of Fourier transformed Compton profiles and the discussion of their properties in terms of the so-called reciprocal form factor $B(\mathbf{r})$ (Pattison et al. 1977, Schülke 1977a,b, Weyrich et al. 1979) was rather helpful. The Fourier transformed Compton profile is directly related to $B(z)$, where the z-direction coincides with the direction of $\mathbf{q}$. $B(\mathbf{r})$ is given by the autocorrelation of Wannier functions, describing the electron states of a solid. An impressive example of evaluating six directional Compton profiles (Pattison et al. 1984) measured in the basal plane of Li_3N in terms of the 2D reciprocal form factor $B(\mathbf{r})$ is shown in Fig. 1.8. The positive (solid lines) and negative (broken lines) contours reflect the orthogonality requirements of the ionic orbitals, centered at the sites as depicted in the plan below. This demonstrates how, in spite of the rather limited momentum space resolution, γ-ray Compton scattering is sufficient to offer evidence about chemical bonding, especially between nearest neighbors.

(iii) Fermi surface and other lattice induced features of the momentum density: It will be shown later in greater detail that, within the independent-particle model, the momentum density of electrons in solids is determined by two ingredients, namely by the shape of the occupied $\mathbf{k}$-space in the extended zone scheme (in the case of metals, the $\mathbf{k}$-space enclosed by the Fermi surface in the extended zone scheme, often denoted the "primary" Fermi surface), and by the strength of the plane-wave components of the Bloch waves being classed with the reciprocal lattice vectors. The modulus squared of the amplitude $\mathbf{a}_\nu(\mathbf{k},\mathbf{g})$ of such a plane-wave component, where $\mathbf{k}$ is the reduced Bloch wavevector, $\mathbf{g}$ a reciprocal lattice vector, and ν the band index, determines the probability to find the electron with momentum $\hbar(\mathbf{k}+\mathbf{g})$ in a Bloch state $|\mathbf{k},\nu\rangle$. We call momenta with $\mathbf{g} \neq \mathbf{0}$ the higher-momentum components. In this context the evaluation of Compton profiles in terms of the reciprocal form factor $B(\mathbf{r})$ helps to separate these two ingredients, at least for metals in the independent particle model, since $B(\mathbf{R})$, the values of the reciprocal form factor at lattice translation vectors $\mathbf{R}$, can be considered as the Fourier components of the occupation number densities in the first Brillouin zone summed up over all occupied bands (Schülke 1977a). Therefore, in order to make full use of these properties, one has to measure a larger set of directional Compton profiles for different $\mathbf{q}$-directions and has to utilize one of the reconstruction procedures proposed by several authors (Mueller 1977, Mijnarends 1977, Hansen 1980). Many simple and $3d$ metals as well as s-p-semiconductors have been investigated by Compton scattering along these lines, where in this case the rather low momentum space resolution is a real handicap and prevents the detection of finer details of the Fermi surface or of

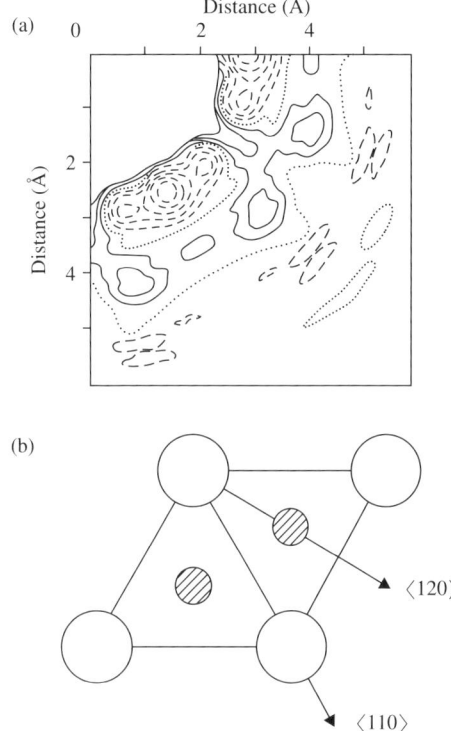

Fig. 1.8. (a) Contour diagram of the two-dimensional reciprocal form factor $B(\mathbf{r})$ as reconstructed from six Compton profiles measured in the basal plane of Li_3N, which is shown in (b) in plan form: dashed circles, Li; open circles, N. Negative contours in (a) (dashed lines) are peaked around the N-N direction, while peaks in the positive contours (solid lines) are along the N-Li directions, reflecting the orthogonality requirements of the ionic orbitals on different sites. (Reprinted with permission from Pattison *et al.* (1984); copyright (1984) by the International Union of Crystallography.)

the distribution of higher momentum components in momentum space. Some selected examples are again summarized in Table 1.3.

(iv) Correlation effects: It was first pointed out by Lam and Platzman (1974) that the momentum density $\rho(\mathbf{p})$ of an electron system as deduced from Fourier transformed solutions ψ_i of the Kohn–Sham equation (Kohn and Sham 1965) according to

$$\rho(\mathbf{p}) = \sum_{i \text{ occp.}} |\langle \mathbf{p} | \psi_i \rangle|^2 \tag{1.24}$$

needs a correction, which takes into account correlation. Within the limits of the local density approximation (LDA) of the density functional theory this correction $\Delta\rho(\mathbf{p})$ (so-called Lam–Platzman correction) is given by

$$\Delta\rho(p) = \int \rho(\mathbf{r})(n_0^h(p)[\rho(\mathbf{r})] - n_0^f(p)[\rho(\mathbf{r})])\mathrm{d}\mathbf{r}. \qquad (1.25)$$

Here $n_0^h(p)[\rho(\mathbf{r})]$ and $n_0^f(p)[\rho(\mathbf{r})]$ are the momentum density of an interacting and a noninteracting homogeneous electron system of local density $\rho(\mathbf{r})$, respectively. $n_0^h(p)[\rho(\mathbf{r})]$ can be calculated within different stages of approximation, e.g. within the random phase approximation (RPA) by Daniel and Vosko (1960). Independent of the approximation chosen, $n_0^h(p)$ is characterized by a reduction of momentum density within the Fermi sphere and additional momentum density outside the Fermi sphere, where the break of the momentum density at the Fermi momentum survives but is reduced. By comparing experimental γ-Compton profiles from transition metals with corresponding calculations performed within the limits of the LDA, several authors have shown that the Lam–Platzman correction cannot fully account for the correlation effects. An example is shown in Fig. 1.9 (Bauer and Schneider 1985; this figure will be more thoroughly discussed in Section 4.8.2 together with Fig. 4.43), where the difference between the LDA calculated Compton profile of Cu, which includes the Lam–Platzman correction (Bagayoko et al. 1980), and the experiment is confronted with the Fermi surface of Cu in the repeated zone scheme. One can easily attribute the peaks of the oscillations to regions in momentum space that are strongly occupied due to the higher momentum components of the momentum density. This finding leads to the conclusion that correlation effects transport momentum density not only out of the region enclosed by the so-called primary Fermi surface around $\mathbf{g}=\mathbf{0}$ as in the homogeneous case, but also out of the regions comprised by the secondary Fermi surfaces of the higher momentum components around reciprocal lattice vectors $\mathbf{g}\neq\mathbf{0}$. This way the correlation correction becomes orientation dependent, as taken into account by Wakoh and Matsumoto (1989) in a semiempirical scheme.

(v) Coherent Compton scattering: By preparing the initial photon state of a Compton scattering process to be the coherent superposition of two plane waves (standing waves within a crystal in the Bragg position), the resulting Compton profile contains an interference term, which is, under certain experimental conditions, the projection of non-diagonal elements $\Gamma(\mathbf{p}\,|\,\mathbf{p}+\mathbf{g})$ of the one-particle density matrix in momentum space on the scattering vector $\mathbf{q}$, where $\mathbf{g}$ is the reciprocal lattice vector of the corresponding Bragg reflection (Schülke 1981). Until now only a few coherent Compton scattering studies on nearly perfect Si crystals have been performed, listed at the end of Table 1.3.

The advent of strong sources of high-energy synchrotron X-radiation has led to a tremendous improvement of momentum space resolution by a factor of three and more compared to the γ-Compton experiments by utilizing

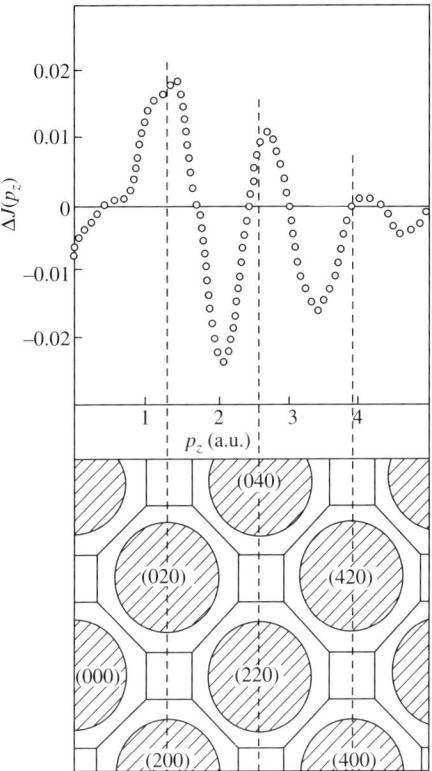

Fig. 1.9. Upper part: Difference between the LDA calculated Cu [110] Compton profile (Bagayoko et al. 1980) and the corresponding γ-measurement Bauer and Schneider (1985) (theory has been convoluted with the experimental resolution). Lower part: Plan of the Brillouin zone and the Fermi surface of Cu in the repeated zone scheme. (Reprinted with permission from Bauer and Schneider (1985); copyright (1985) by the American Physical Society.)

crystal dispersive analysis. This way hitherto hidden Fermi surface signatures, orientation-dependent correlation effects and even the minute influence of final states and of temperature became accessible. Figure 1.10 (Blaas et al. 1995) demonstrates how the improvement of resolution reveals finer details of Fermi surface features in the directional differences of Compton profiles of FeAl. Especially in the field of intercalation compounds of graphite and of C_{60} nonrigid band behavior was studied with high accuracy. First measurements of Compton profiles at high pressure were performed. Also the investigation of chemical bonding was extended to new systems. Representative examples of those investigations together with theoretical studies important for the interpretation of experiments are listed in Table 1.4.

Table 1.4 Representative selection of materials investigated by IXS in the Compton limit with high momentum space resolution utilizing synchrotron radiation sources (SRS). CD: crystal dispersive energy analysis of the scattered radiation; SSD: energy analysis by means of a solid state detector.

Material	Ref.	Source	Energy analysis	Chief object
Li	(1)	SRS	CD	Fermi surface anisotropy
Li	(2) (3)	SRS	CD	reconstr. of 3D e-momentum density
Li	(4) (5) (6)	SRS	CD	final state and temperature effects
Li	(1) (2) (7)	SRS	CD	e-correlation
Li	(8) (9) (10) (11)		theory	e-correlation
Li	(12)	SRS	SSD	high pressure
Be	(13) (14) (15)	SRS	CD	e-momentum density, Fermi surface
Be	(16)	SRS	CD	e-correlation, final state effects
Be	(17)	SRS	CD	HMC, temperature effect
Na	(18)	SRS	SSD	high pressure
Al	(19) (20) (5)	SRS	CD	e-momentum density e-correlation, temperature effect
Si	(19) (21) (22)	SRS	CD	e-momentum density self-interaction correction
V	(23)	SRS	CD	Fermi surface
Cu	(24)	SRS	CD	Fermi surface, e-correlation
Cr	(25), (26)	SRS	CD	2D projected Fermi surface
Y	(27)	SRS	CD	3D reconstr., comparison with ACAR
H_2O (ice)	(28)	SRS	CD	covalency of hydrogen bond
LiH	(29)	SRS	CD	chemical bonding
$Li_{100-x}Mg_x$	(30)	SRS	CD	reconstruction 3D e-momentum density
FeAl	(31)	SRS	CD	Fermi surface
a-Si:H	(32)	SRS	CD	transition amorph-crystalline
$CoSi_2$	(33)	SRS	CD	comparison with self consistent linear muffin tin orbitals
$Sn_{0.8}In_{0.2}$	(34)	SRS	CD	Fermi momentum in [100] direction, comparison with ACAR

Material	Ref.	Source	Energy analysis	Chief object
$Cu_{72.5}Pd_{27.5}$	(35)	SRS	CD	Fermi surface nesting
$Ti_{48.5}Ni_{51.5}$	(36)	SRS	CD	Fermi surface nesting
$Cu_{84.2}Al_{15.8}$	(37)	SRS	CD	comparison with pure Cu
$Al_{97}Li_3$	(38)	SRS	CD	comparison with pure Al
$CO(NH_2)_2$ urea	(39)	SRS	CD	hydrogen bond
LiC_6	(40) (41)	SRS	CD	nonrigid band behavior
KC_8	(42)	SRS	CD	band structure features
LiC_{12}	(43)	SRS	CD	polarization of the graphite valence band orbitals
C_{60}	(44)	SRS	CD	delocalization of ground-state charge density
K_3C_{60}, K_4C_{60} $CsRb_2C_{60}$, Rb_4C_{60}	(45)	SRS	CD	modification of the C_{60} density
$KHCO_3$	(46)	SRS	CD	hydrogen bonds, LCAO calc.
$YBa_2Cu_3O_{7.6}$ $Bi_2Sr_2CaCu_2O_8$	(47)	SRL	SSD, CD	transition into supercond. state
$PrBa_2Cu_3O_7 \cdot \delta$	(48)	SRS	CD	hole depletion and localization
$Al_{72}Ni_{12}Co_{16}$ decagonal	(49)	SRS	CD	charge transfer
$Cd_{84}Yb_{16}$ icosahedral	(50)	SRS	CD	Hume–Rothery mechanism

(1) Sakurai et al. (1995b) (2) Schülke et al. (1996b) (3) Tanaka et al. (2001) (4) Sternemann et al. (2000) (5) Sternemann et al. (2001) (6) Chen et al. (1999) (7) Schülke et al. (2001) (8) Kubo (1997) (9) Baruah et al. (1999) (10) Filippi and Ceperley (1999) (11) Dugdale and Jarlborg (1998) (12) Oomi and Ito (1993) (13) Loupias et al. (1980) (14) Hämäläinen et al. (1996) (15) Itou et al. (1998) (16) Huotari et al. (2000) (17) Huotari et al. (2002) (18) Hämäläinen et al. (2000) (19) Shiotani et al. (1989) (20) Suortti et al. (2000) (21) Schmitz et al. (1993) (22) Kubo et al. (1997) (23) Shiotani et al. (1992) (24) Sakurai et al. (1999) (25) Dugdale et al. (2000) (26) Tanaka et al. (2000) (27) Kontrym Sznaid (2002) (28) Isaacs et al. (1999) (29) Loupias and Chomilier (1986) (30) Stutz et al. (1999) (31) Blaas et al. (1995) (32) Bellin et al. (1997) (33) Bellin et al. (1995) (34) Manuel et al. (1996) (35) Matsumoto et al. (2001) (36) Shiotani et al. (2004) (37) Kwiatkowska et al. (2004) (38) Suortti et al. (2001) (39) Shukla et al. (2001) (40) Loupias et al. (1985) (41) Chou et al. (1986) (42) Loupias et al. (1988) (43) Rabii et al. (1989) (44) Moscovici et al. (1995) (45) Marangolo et al. (1999) (46) Ahuja et al. (1994) (47) Manninen et al. (1999) (48) Shukla et al. (1999) (49) Okada et al. (2002) (50) Okada et al. (2003).

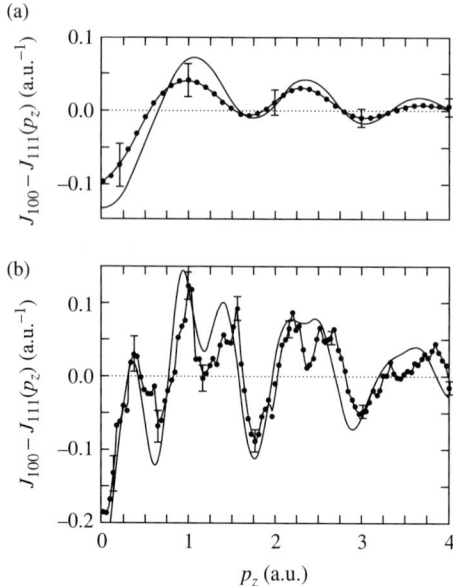

Fig. 1.10. Experimental and theoretical difference Compton profile [100]–[111] of FeAl: (a) Bullets: low-resolution (Mo Kα_1 X-rays, energy dispersive analysis) experiment; solid line: theory (full-potential linearized augmented plane wave (FLAPW) method) convoluted with the low experimental resolution. (b) Bullets: high-resolution (synchrotron radiation crystal dispersive analysis) experiment; solid line: theory like in (a) but convoluted with the high experimental resolution. (Reprinted with permission from Blaas et al. (1995); copyright (1995) by the American Physical Society.)

1.3.3.2 *(γ, eγ) experiments* As shown in Section 1.2 the total 3D electron momentum density $\rho(\mathbf{p})$ becomes experimentally accessible, if one is measuring the triple differential scattering cross-section; this means scattering of monochromatized photon beams in a certain direction, energy-analyzing the scattered photons in coincidence with the detection of the recoil electrons whose direction has to be fixed by the experimental conditions, which leads to the experimental setup sketched in Fig. 1.11 (Metz et al. 1999b).

Looking more precisely at the kinematics of the underlying processes, scans through the electron momentum density can be performed in two different ways.

(i) One measures the coincidence countrate as a function of the Doppler broadening $\Delta\hbar\omega_2 \equiv \hbar(\omega_2 - \omega_{20})$ ($\hbar\omega_{20}$ is the scattered photon energy for $\mathbf{p}=\mathbf{0}$) for a fixed direction of the recoil electrons. For kinematical reasons one is looking at the distribution of electron momentum components

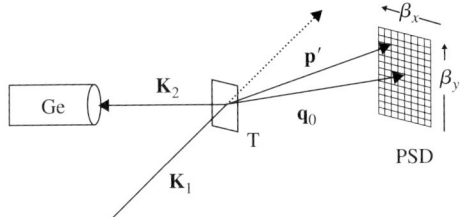

Fig. 1.11. Experimental setup for $(\gamma, e\gamma)$ spectroscopy: Ge, Ge diode for the energy analysis of the scattered photon; T, target; PSD, two-dimensional position-sensitive detector for fixing the direction of the recoil electrons with momentum $\mathbf{p}'$. $\mathbf{q}_0$ is the momentum transfer: $\mathbf{q}_0 = \mathbf{K}_1 - \mathbf{K}_{20}$, where $\mathbf{K}_{20}$ is the wavevector of the photon scattered from electrons at rest. Ge and PSD are working in coincidence. (Reprinted with permission from Metz *et al.* (1999b); copyright (1999) by the American Physical Society.)

parallel to the momentum transfer vector $\mathbf{q}$. Of course, in this respect a measurement of the recoil electron energy distribution ΔE_2 is equivalent to the measurement of the Doppler broadening, since owing to energy conservation $\Delta \hbar \omega_2 = -\Delta E_2$.

(ii) One measures the coincidence countrate for a fixed value of the Doppler broadening $\Delta \hbar \omega_2$ by changing the direction of the registered recoil electrons. In that case the distribution of electron momentum components perpendicular to $\mathbf{q}$ is studied.

The most critical point of these experiments is the mismatch between the recoil electron escape depth in solids which is only a few nanometers and the penetration depth of several mm of hard X-rays necessary to keep the requirements of the impulse approximation. Therefore, in order to prevent too many false coincidences, one has to restrict the sample thickness to between 10 and 100 nanometers. This compromise, of course, leads to rather low coincidence countrates, and demands high flux sources of high-energy synchrotron X-radiation and prohibits a crystal dispersive energy analysis of the scattered photons, so that solid state detectors with much lower energy resolution have to be used. The angular analysis of the recoil electrons can be facilitated by installing one- or two-dimensional detector arrays. Nevertheless, the statistical accuracy of those experiments is still rather poor and the time for the experiments tremendous, so that their much higher information content (3D momentum density) compared with that of conventional Compton scattering studies has been, until now, not yet completely exploited, especially if one has in mind the possibilities of 3D reconstruction in conventional Compton scattering experiments.

A further problem of $(\gamma, e\gamma)$ measurements is the strong incoherent elastic scattering of the electrons, which smears the electron directions and disturbs the evaluation of the electron momentum density. Going to thinner samples with thicknesses in the order of the average free pathlength of the recoil electrons (a few nanometers) can diminish this multiple scattering effect. Rollason and Woolf (1995) have developed Monte Carlo simulation schemes to correct experimental $(\gamma, e\gamma)$ data with respect to electron multiple scattering.

The first $(\gamma, e\gamma)$ measurements were done in the early days of Compton scattering by Bothe and Geiger (1925), but their goal was not the evaluation of the electron momentum density but to prove the validity of energy and momentum conservation in the individual photon–electron encounter.

The pioneering synchrotron radiation based $(\gamma, e\gamma)$ experiment aiming to show the influence of the intrinsic electron momentum density of bound electrons is due to Rollason et al. (1989 a,b) on Al foils of 1.4 μm thickness measuring the triple scattering cross-section in the above-described mode (i) using a solid state detector for measuring the Doppler broadening and fixing the recoil electron direction by means of a single solid state detector for the electrons counted in coincidence with the photon detector. In the following years some other solids were investigated as summarized in Table 1.5.

Improvements of the experimental technique were achieved by replacing the single electron detector by a linear (Kurp et al. 1997a) and still more effectively by a two-dimensional (Metz et al. 1999a) array of electron detectors. As an example the result of a $(\gamma, e\gamma)$ measurement on Al (Metz et al. 1999b) is shown in Fig. 1.12.

A further step forward in the $(\gamma, e\gamma)$ technique was made by Itou et al. (1999), who measured, on graphite, not only the Doppler broadening of the scattered photon energy but, in coincidence, also the energy distribution of the recoil electrons by means of a time-of-flight analysis. By means of the plot shown in Fig. 1.13, which correlates the energy of the scattered photons and the energy of the recoil electrons measured in coincidence, the influence of the binding energy of the 1s electrons of carbon becomes evident. This way a first step on the way to measuring the electron momentum density state selectively has been done, so that $(\gamma, e\gamma)$ experiments can yield similar results as the $(e, 2e)$ techniques, where in an inelastic electron scattering experiment the scattered electron and the recoil electron are analyzed in coincidence both with respect to their direction and their energy (for a recent review see Vos and McCarthy 1995). A comparison of $(\gamma, e\gamma)$ with $(e, 2e)$ techniques was performed by Kurp et al. (1997b).

1.3.3.3 *Magnetic Compton scattering* It has been shown in Section 1.2 that the spin contribution to the double differential scattering cross-section (see equation 1.11) can be utilized to obtain what we have called the magnetic Compton profile $J_m(p_z)$, which is the projection on the scattering vector of the momentum density of electrons with unpaired spins. In order to achieve this goal one needs, first of all, an incident beam with a sufficiently high degree of circular polarization P_2 (see equation 1.15), and moreover, an incident photon energy as high as possible. In their first experiments on magnetic Compton scattering, Sakai and Ono (1976) used a Co^{57} radioactive γ-source. By implanting

Table 1.5 Representative examples of $(\gamma, e\gamma)$ measurements of the 3D electron momentum density (3D-EMD) of solids using synchrotron radiation. SSD: solid state detector; TF: time-of-flight.

Material	Ref.	Photon energy analysis	Electron detection	Electron energy analysis	Comments
Al	(1) (2)	SSD	single SSD	no	3D-EMD
Cu	(3)	SSD	single SSD	no	3D-EMD
C(graphite)	(4) (5)	SSD	single SSD	no	e-multiple scattering Monte Carlo estim.
C(graphite)	(6)	SSD	linear detector array	no	3D-EMD
C(graphite)	(7)	SSD	linear detectors array	no	comparison with (e, 2e) spectroscopy
C(graphite) C_{60}	(8)	SSD	2-D detector array	no	3D-EMD; comparison between graphite and C_{60}
Al	(9)	SSD	2-D detector array	no	correlation effects
C(graphite)	(10)	SSD	single detector	yes TF	indication of state selectivity
$Cu_{0.5}Al_{0.5}$	(11)	SSD	2-D detector array	no	comparison between CuNi alloy and CuNi sandwiches

(1) Rollason et al. (1989) (2) Bell et al. (1990) (3) Bell et al. (1991) (4) Tschentscher et al. (1993) (5) Kurp et al. (1996) (6) Kurp et al. (1997a) (7) Kurp et al. (1997b) (8) Metz et al. (1999) (9) Metz et al. (1999) (10) Itou et al. (1999) (11) Metz et al. (1999).

the radioactive nuclei into a ferromagnetic host lattice of Fe at very low temperatures (40 mK), a high degree of circular polarization was attained for the 122 keV γ-rays, when the source sample was placed in a large (10 kG) external field parallel to the direction of the γ-emission. The γ-rays were Compton scattered

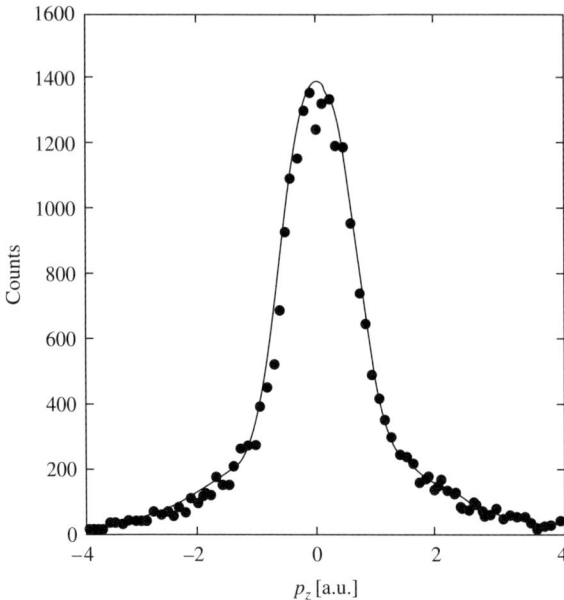

Fig. 1.12. Experimental electron momentum density $\rho(0,0,p_z)$ (dots) of Al, as obtained by $(\gamma, e\gamma)$ spectroscopy, compared with theory (solid line) including the instrumental resolution. (Reprinted with permission from Metz *et al.* (1999b); copyright (1999) by the American Physical Society.)

into an energy-dispersive solid state detector. The magnetic Compton profile was isolated by periodically reversing the magnetizing field direction and subtracting the corresponding data sets. The resulting first magnetic Compton profile shown in Fig. 1.14 suffers from very low statistics although measured for 139 h in total. Nevertheless, it exhibits a prominent dip around $p_z = 0$, which, according to calculations of Wakoh and Kubo (1977) (also shown in Fig. 1.14), must be attributed to the negative spin polarization of the s,p-like itinerant electrons in opposition to the positive spin polarization of the more localized d electrons.

The advent of synchrotron radiation sources with photon energies in the 100 keV range changed the situation with magnetic Compton scattering dramatically. The pioneering work of Cooper *et al.* (1986) utilized the fact that the synchrotron radiation emission from a bending magnet is linearly polarized in the plane of the circulating electron, but elliptically polarized, with a high degree of circular polarization, if the source is viewed under an angle with respect to the orbital plane. The handedness of the elliptic polarization changes on passing through the orbital plane. Since the degree of circular polarization increases but the emitted intensity rapidly decreases with increasing viewing angle, the optimum viewing angle is always a compromise between intensity

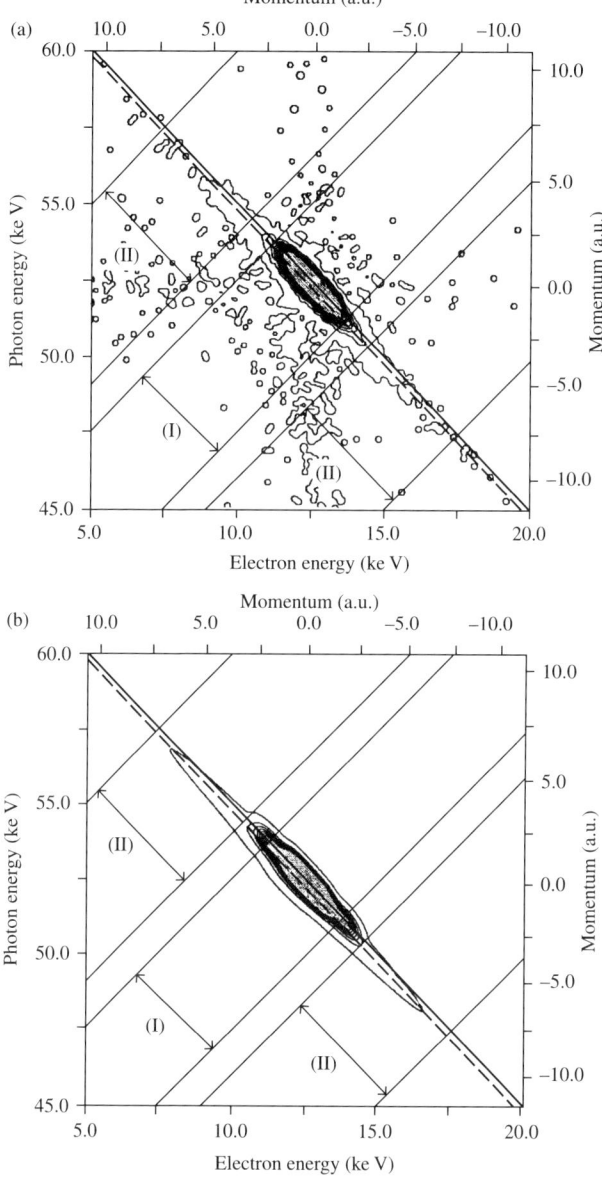

Fig. 1.13. A contour map showing the coincident events distribution from a graphite foil plotted between the energy of the Compton scattered photons $\hbar\omega_2$ and the recoil electron energy T. The solid line represents $\hbar\omega_2 + T = \hbar\omega_1 = 65.0$ keV; the dashed line is $\hbar\omega_2 + T = \hbar\omega_1 - E_b = 64.7$ keV, thus demonstrating the visible influence of the C1s electron binding energy E_b. (Reprinted with permission from Itou et al. (1999); copyright (1999) by the Physical Society of Japan.)

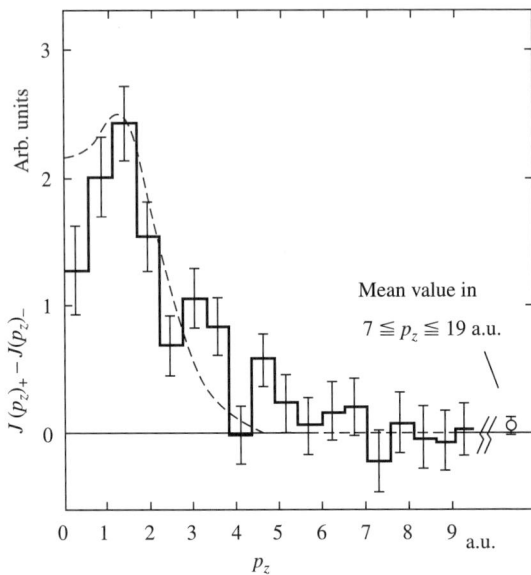

Fig. 1.14. Magnetic Compton profile measured using the γ-technique by Sakai and Ono (1976) together with a calculation of Wakoh and Kubo (1977). (Reprinted with permission from Sakai and Ono (1976); copyright (1976) by the American Physical Society.)

and polarization and is of the order of a few tenths of a mrad for hard X-rays. The relative low intensity of the elliptically polarized incident beam together with the good statistics necessary to extract the magnetic Compton profile did not allow a crystal-dispersive energy analysis of the scattered beam. Likewise in the γ-experiments an energy-dispersive solid state detector was applied. Since, according to (1.15), the charge–spin interference term of the double differential scattering cross-section increases with increasing energy of the incident photons, it was rather helpful to use the synchrotron radiation emission of a high-field wavelength shifter rather than that of a bending magnet (Laundy 1990).

The application of a so-called quarter-wave phase plate put into the path of a linearly polarized synchrotron beam as emitted in the orbital plane offers a second possible method to obtain a high degree of circular polarization. Quarter-wave phase plates for the X-ray region make use of the fact that in the Laue case of diffraction the anomalous weakly absorbed wave field consists of two orthogonally polarized components with slightly different effective indices of refraction. By orientating the diffraction plane in such a way relative to the polarization vector of the incident beam that both components are excited with equal amplitudes, a sufficiently long beam path within the crystal provides a circularly polarized beam at the exit surface (Hart 1978, Golovchenko et al. 1986). Mills (1987) used this technique for the first time to measure magnetic Compton profiles of iron, cobalt, nickel and gadolinium, where the iron and the nickel data are in good agreement with corresponding calculations of Wakoh and Kubo (1977) and Rennert *et al.* (1983), respectively.

A marked increase of the intensity of the elliptically polarized incident beam could be achieved by employing an elliptical multiple wiggler (Yamamoto *et al.* 1989). This insertion device combines the periodic array of vertical magnetic fields with an array of weaker horizontal fields. The electron trajectory is no longer a two-dimensional sine wave as in conventional wigglers but transformed into a three-dimensional flattened helix. The weak vertical field shifts the emission of the left-hand turns above the orbital plane and the emission of the right-hand turns below. This way both turns emit elliptically polarized radiation with the same handedness parallel to the orbital plane. Additionally segmented solid state detectors were utilized to prevent saturation of the countrate, thus making the energy analysis of the scattered radiation much more efficient, so that magnetic Compton profiles could be measured with good statistics in a much shorter data collection time. A large number of polycrystalline and single-crystalline systems were investigated with the elliptical multipole wiggler; a representative selection of these is summarized in Table 1.6.

A set of directional magnetic Compton profiles of iron is shown in Fig. 1.15. This set was used to produce the first 3D spin density map in momentum space (Tanaka *et al.* 1993).

This summary of the instrumentation of magnetic Compton scattering must not end without mentioning the only experiment where crystal-dispersive energy analysis of the scattered radiation has been installed, with the goal to bring the momentum space resolution of the magnetic Compton profiles of Fe-5.8 at% Si down to 0.12 a.u. (Sakurai *et al.* 1994).

Three further aspects of magnetic Compton scattering have to be stressed:

(i) Equations (1.16) and (1.17) seem to indicate that the experimental majority-spin band Compton profile $J(\uparrow, p_z) \equiv \int \int \rho_\uparrow(\mathbf{p}) \, dp_x \, dp_y$ and the minority spin one $J(\downarrow, p_z)$ can straightforwardly be obtained by normalizing the charge Compton profiles to the total electron number and the magnetic Compton profile to the number of unpaired electrons, whenever they are known from an independent experiment. But both the majority-spin band and the minority-spin band Compton profile are dominated by the same nonmagnetic core-electron contribution. Therefore, it is more useful to apply this separation of majority- and minority-spin band contributions to directional differences rather than total Compton profiles as first proposed by Collins *et al.* (1989). The spherical core contributions cancel upon the subtraction of two directional Compton profiles.

(ii) As already stated in Section 1.2, magnetic Compton scattering is only sensitive to the spin magnetization, so that the integral of the magnetic

Table 1.6 Representative examples of magnetic Compton profile measurements. CP = Compton profile; MS = momentum space; BM = bending magnet; WS = wavelength shifter; IV = inclined view; PP = quarter-wave phase plate; EMW = elliptical multipole wiggler; SSD = solid state detector; CD = crystal dispersive analysis of the Compton scattered radiation.

Material	Ref.	Source/production of circular polar.	Energy analysis	Chief object of investigation
Fe Mn-Ferrit	(1) (2)	Co^{57} in Fe γ-source/40 mK; 10 kG	SSD	magn. CP
Fe	(3)	BM/IV	SSD	magn. CP
Fe	(4)	BM/IV	SSD	directional magn. CP
Fe, Co, Ni, Gd	(5)	BM/PP	SSD	magn. CP
Gd	(6)	BM/IV	SSD	magn. CP
Ni	(7)	WS/IV	SSD	directional magn. CP
Fe	(8)	WS/IV	SSD	majority- and minority-spin directional differences, magn. CP
Gd	(9)	EMW	segm. SSD	magn. CP
$HoFe_3$	(10)	EMW	segm. SSD	temperature dependence of the Ho spin moment
$HoFe_3$	(11)	EMW	segm. SSD	separation of spin moments of Ho and Fe
Fe	(12)	EMW	segm. SSD	3D reconstruction of spin density in MS
FeSi	(13)	EMW	CD	directional magn. CP with 0.12 a.u. resolution
UTe	(14)	EMW	segm. SSD	combin. with measurem. of magnetiz./orbital magn mom.
$Mn_{1.1}Sb$	(15)	EMW	segm. SSD	directional magn. CP

Material	Ref.	Source/production of circular polar.	Energy analysis	Chief object of investigation
$DyFe_2$	(16)	EMW	segm. SSD	temperature dependence of the Dy, Er, Fe spin moments
$ErFe_2$				
$CeFe_2$	(17)	EMW	segm. SSD	separation of Ce and Fe spin moments
USe	(18)	EMW	segm. SSD	conduction like orbital moments
UFe_2	(19)	EMW	segm. SSD	anomalous high spin moment at U site
$CeRh_3B_2$	(20)	WS	SSD	anomalous high Ce5d spin and orbital moment
Cu_2MnAl	(21)	WS	SSD	separation of spin moments at different sites
Ni	(22)	WS	SSD	directional magn. CP comp. with gen. grad. appr.
Gd	(23)	WS	SSD	directional magn. CP
Gd-Y alloy	(24)	WS	SSD	induced spin moment
$SmMn_2Ge_2$	(25)	WS	SSD	temperature dependence of 3d, 4f spin moments
Fe_3Si	(26)	WS	SSD	magn. CP
Fe_3Al				
$La_{2-2x}Sr_{1+2x}Mn_2O_7$	(27)	EMW	SSD	population of e_g orbitals

(1) Sakai and Ono (1976) (2) Sakai and Sekizawa (1987) (3) Cooper et al. (1986) (4) Cooper et al. (1988) (5) Mills (1987) (6) Brahmia et al. (1989) (7) Timms et al. (1990) (8) Collins et al. (1989) (9) Sakai et al. (1991) (10) Cooper et al. (1993) (11) Zukowski et al. (1993) (12) Tanaka et al. (1993) (13) Sakurai et al. (1994) (14) Sakurai et al. (1995a) (15) Nakamura et al. (1995) (16) Lawson et al. (1995) (17) Cooper et al. (1996) (18) Hashimoto et al. (1997) (19) Lawson et al. (1997) (20) Yaouanc et al. (1998) (21) Zukowski et al. (1997) (22) Dixon et al. (1998) (23) Duffy et al. (1998) (24) Duffy et al. (2000) (25) McCarthy et al. (2000) (26) Zukowski et al. (2000) (27) Koizumi et al. (2001).

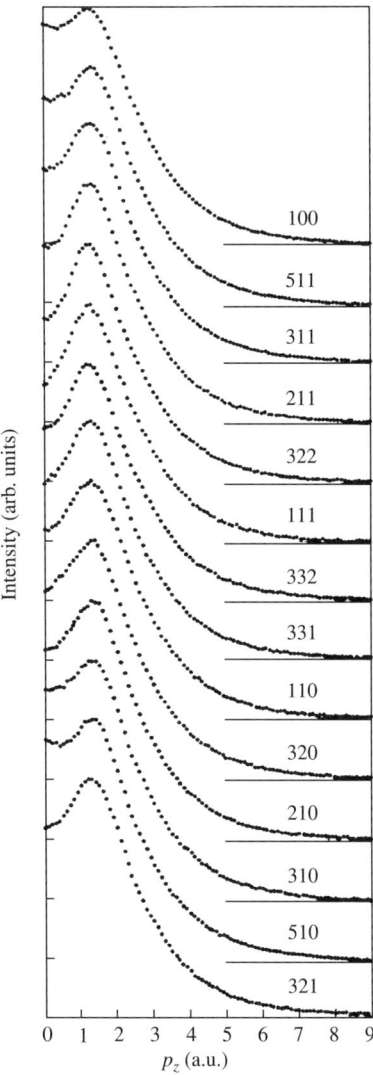

Fig. 1.15. Experimental synchrotron radiation based directional magnetic Compton profiles of Fe + 3 wt% Si. (Reprinted with permission from Tanaka *et al.* (1993); copyright (1993) by the American Physical Society.)

Compton profile with respect to p_z has a well-known relation to the number of unpaired electrons and is thus proportional to the spin-induced magnetic moment. If, therefore, the total magnetic moment is known from measurements of the magnetic induction, the orbital contribution to the total magnetic moment can be deduced.

(iii) Magnetic Compton scattering offers, by means of a line-shape analysis, the possibility to separate the contribution to the spin-induced magnetic moment of each component, since they have a different characteristic momentum distribution, such as s,p-like, 3d or 4f electrons. In doing so, the more localized 3d and 4f contributions are assumed to follow the free atomic momentum distribution, whereas the s,p contribution can be modeled by the momentum distribution of a free electron system. The result of such an analysis on ferromagnetic HoFe$_2$ (Zukowski *et al.* 1993) will later be shown in Fig. 4.69, where spin moments of $1.70\mu_\text{B}$, $-2.19\mu_\text{B}$, and $0.80\,\mu_\text{B}$ for iron (3d), holmium (4f) and diffuse electrons, respectively, were determined.

In Table 1.6 representative examples of magnetic Compton measurements on different ferromagnetic and ferrimagnetic systems are summarized.

1.3.4 *Resonant inelastic X-ray scattering (RIXS) spectroscopy*

1.3.4.1 *Core-level resonant Raman spectra* It was first shown by Hämäläinen *et al.* (1989) that the cross-section for a resonantly excited KL transition (initial hole in the K-shell), which is derived by integrating the DDCS of (1.21) with respect to ω_2, and which can be measured by tuning the incident photon energy across the K-edge, is given by

$$\left(\frac{d\sigma}{d\Omega_2}\right)_\text{KL} = \left[\frac{E_\text{K} + \omega_\text{e}}{4\pi^2 \hbar \omega_1}\right] \sigma_K(E_\text{K} + \omega_\text{e}) \tan^{-1}\left(\frac{\Gamma_\text{K}}{2\Delta E}\right), \qquad (1.26)$$

where E_K is the K-shell binding energy, ω_e is the average energy of the ejected electron, σ_K is the K-shell contribution to the photoelectric absorption, ΔE is the incident energy relative to the K-edge ($\Delta E \equiv E_\text{K} - \hbar\omega_1$), and Γ_K the K-shell hole width, due to its finite lifetime. The latter quantity can be experimentally determined by utilizing (1.26): one has to measure the scattered intensity into a scattering angle of nearly 90° by means of a solid state detector, which integrates over the energy distribution of the Kα and Kβ emission, tuning the incident photon energy with an energy width of ~ 1 eV across the K-edge. The incident photons should be highly linearly polarized (synchrotron radiation) in order to suppress, together with the 90° scattering angle, the nonresonant scattering. This way Γ_K of Cu, Zn and $\Gamma_{\text{L}_{\text{III}}}$ of Ho (Hämäläinen *et al.* 1989) and $\Gamma_{\text{L}_{\text{III}}}$ of Yb and Ta (Hämäläinen *et al.* 1990) were determined with an accuracy of 0.1 eV.

It was first claimed by Hämäläinen *et al.* (1991) that by using a selected wavelength of the resonantly scattered (re-emitted) radiation as a signal, which indicates the strength of the absorption process when scanning the incident photon energy across an edge, the edge spectra with their information about the density of unoccupied states should be free from being convoluted by the lifetime broadening of the excited corelevel. The total energy resolution of this type of edge spectroscopy would then depend only on the energy resolution of the monochromator for the incident beam, the energy resolution of the analyzer and

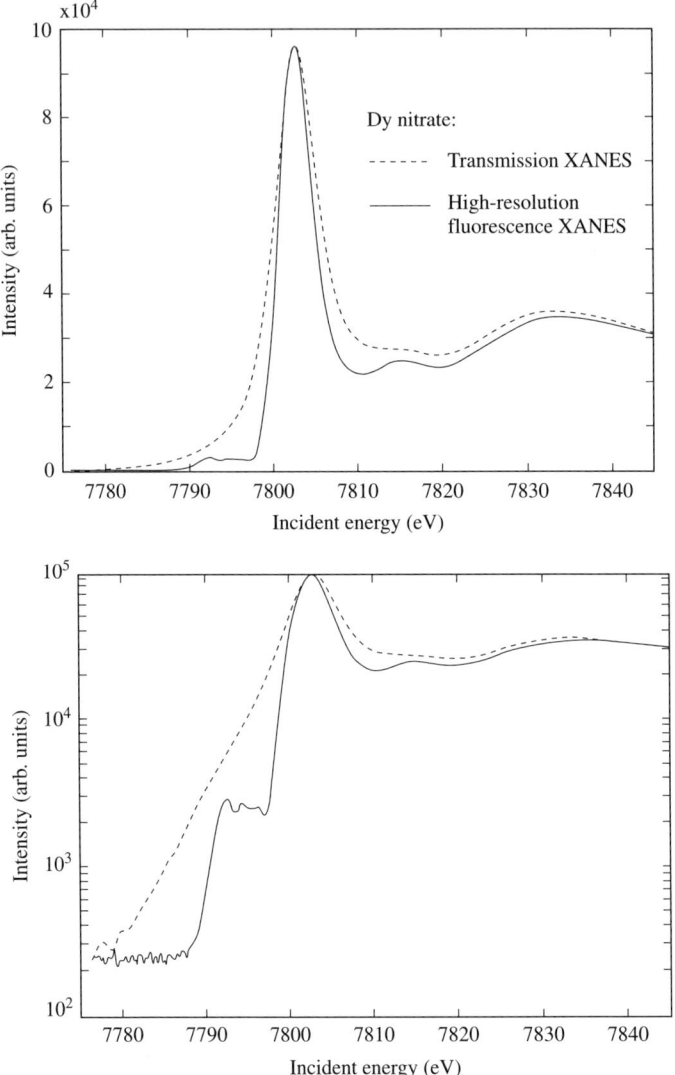

Fig. 1.16. Upper part: solid line: Dy L_{III}-edge spectrum of Dy(NO$_3$)$_3$ using the high-resolution fluorescence technique; dashed line: the same edge but with conventional transmission technique. Lower part: the same spectra as in the upper part but in a logarithmic scale in order to emphasize that part of the fluorescence spectrum attributed to quadrupolar 2p-4f transitions not recognizable in the conventional spectrum. Note that both measurements were accomplished using the same sample and the same incident photon resolution. (Reprinted with permission from Hämäläinen *et al.* (1991); copyright (1991) by the American Physical Society.)

the lifetime broadening of the core hole left behind in the re-emission process. An impressive example is shown in Fig. 1.16. As will be shown in detail later, this remarkable fact is again, like the Raman shift, a consequence of the interplay of the δ-function and the denominator in (1.21). Of course, it has been stressed by Carra *et al.* (1995) that this simple picture holds only within the limits of a one-electron theory. Many-particle interactions can change the involved energy levels, so that, e.g. the fluorescence line can be shifted.

If one performs, in each case for one incident photon energy below the excited core-level binding energy, an energy analysis of the complete spectrum of the resonantly scattered radiation originating from the re-emission from a core level, then these spectra often reveal a multipeak structure. Each peak can be attributed to the excitation into a certain set of unoccupied levels of the conduction band, and due to their individual Raman shift the energy distances of these peaks are equal to the differences of the corresponding excitation energies. The intensity of each peak goes through resonance if the incident energy just reaches the excitation energy of the corresponding set of unoccupied levels. Thus the trace of rather weak excitations, as for instance quadrupolar excitations, not visible as distinct structures in conventional absorption spectroscopy, can be found. This was first demonstrated by Krisch *et al.* (1995), who were able to find evidence for a quadrupolar excitation channel at the L_{III} edge of gadolinium, not detectable in conventional X-ray absorption spectroscopy. A more extensive study of this quadrupolar channel at the L_{III} edge of a series of rare earth compounds has been performed by Bartolome *et al.* (1997). Such a series of multipeaks in the spectra of inelastically scattered photons, namely the $K\alpha_1$ spectrum of Cu in CuO for incident photon energies between 8977 and 8984 eV, all below the K-shell binding energy of Cu, $E_K = 8985$ eV, is exemplarily presented in Fig. 1.17 (Döring *et al.* 2004). The possibility to separate clearly the quadrupolar excitation channel from the dipolar one by looking at the peak structure of the energy-resolved fluorescence was used by Caliebe *et al.* (1996) to distinguish between the quadrupolar and the dipolar contribution to the X-ray magnetic circular dichroism.

1.3.4.2 *Symmetry selectivity of RIXS* As pointed out in Section 1.2, RIXS spectroscopy can be made highly symmetry selective by utilizing the properties of the matrix elements (1.21) containing the $\mathbf{e} \cdot \mathbf{p} \exp(i\mathbf{K} \cdot \mathbf{r})$ terms. First of all, both excitation and re-emission in the RIXS process are dominated by the dipole selection rule. This has been well known since the earliest days of X-ray spectroscopy, both in conventional X-ray absorption and in conventional X-ray emission spectroscopy. Defining experimentally the polarization of the incident beam and defining the direction or measuring the polarization of the emitted beam relative to distinct axes of the sample, respectively, helped to distinguish between the symmetry of electronic states involved. This way also higher multipole contributions were analyzed. Representative examples in the field of X-ray absorption spectroscopy have been published by Dräger *et al.* (1988),

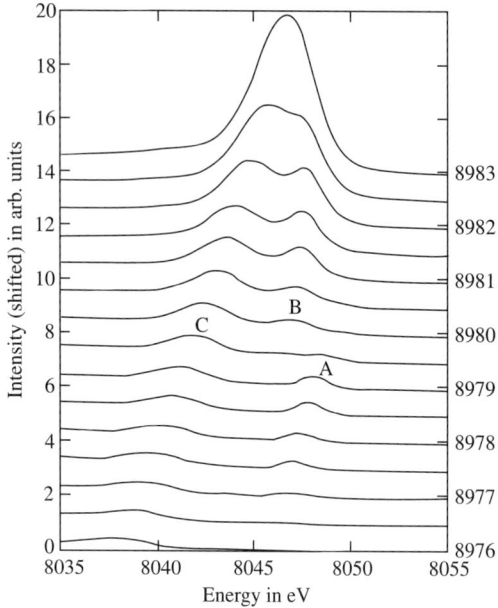

Fig. 1.17. Cu Kα_1 fluorescence spectra of Cu for increasing excitation energies as indicated (in eV), all below the Cu K-shell binding energy (8985 eV). The various peaks can be attributed to different quadrupolar and dipolar excitation energies. (Reprinted with permission from Döring et al. (2004); copyright (2004) by the American Physical Society.)

Bocharov et al. (1998, 2001), and Sipr et al. (1999). As selected examples of symmetry selectivity in X-ray emission spectroscopy the work of Dräger and Brümmer (1984), of Simunek and Wiech (1987), and of Eisberg et al. (1988) should be mentioned. RIXS can make use of this type of symmetry selectivity either in the excitation or in the re-emission part of the whole process. The former case became important when intermediate states of certain symmetries were selected in Nd_2CuO_4, from which shakeup processes can occur (Hill et al. 1998, Hämäläinen et al. 2000). The latter case was utilized by Carlisle et al. (1995) in order to distinguish between the σ/π character of the states contributing to the Bloch-**k** selective RIXS emission of graphite, and by Nilsson et al. (1997) who were the first to detect new orbitals in the RIXS emission spectra of N atoms in N_2 molecules on a Ni(100) surface due to the interaction with the Ni 3d band, and of C atoms and O atoms for CO on Ni(100).

The full potential of RIXS in this respect was only brought about by combining the symmetry selectivity of the excitation part with the symmetry selectivity of the re-emission part of a RIXS process, thus making use of the intimate interconnection of both processes. The most prominent work in this direction goes back to Cowan et al. (1986) and Lindle et al. (1988) performed on molecular

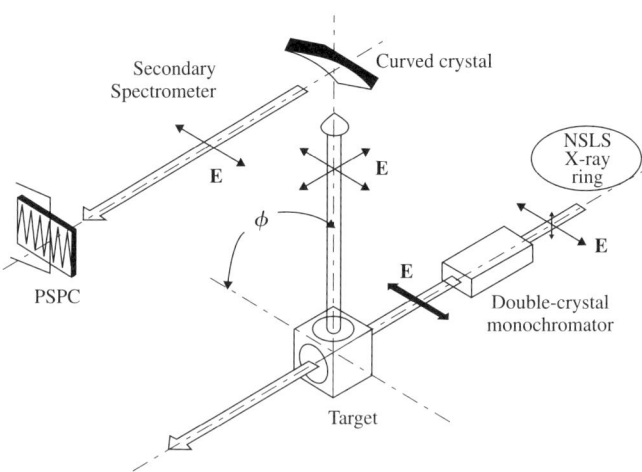

Fig. 1.18. Schematic of the instrumentation used for the measurement of the angular distribution of polarized X-ray emission from gas targets. PSPC = position − sensitive photon counter. (Originally published by Southworth *et al.* (1991); copyright (1991) by the American Physical Society.)

gases. Although the molecules are randomly oriented, an aligned ensemble of excited molecules can be created by means of resonant excitation with polarized X-rays. This resonant excitation depends strongly upon the instantaneous molecular orientation. This way a novel opportunity for studying molecular symmetries and the structure of matter in disordered phases was presented. A sketch of the experimental apparatus used in those experiments is shown in Fig. 1.18 together with the result of a measurement on CF_3Cl in Fig. 1.19. The alignment of molecules achieved in this way can be utilized in two different manners in order to get information about the symmetry of electronic states involved in the re-emission: either by investigating the polarization of the re-emitted radiation, e.g. by means of Bragg reflection under Bragg angles of 45° (Lindle *et al.* 1988), or by inspecting the angular distribution of the re-emitted intensity (Southworth *et al.* 1991). Further studies of molecular gases utilizing the excitation induced alignment were performed on CO (Skytt *et al.* 1997), where the angular distribution of the emission peaks due to 1π and 5σ electrons filling the core hole, respectively, depends on whether the core electron is promoted to a σ or π orbital upon excitation. The parity selection rule that governs the full resonant inelastic scattering process was tested with investigations on O_2 (Glans *et al.* 1996) and N_2 (Agren and Gel'mukhanov 1996). Parity conservation means that, if a core electron is promoted to an orbital of *gerade* (*ungerade*) symmetry only electrons from *gerade* (*ungerade*) orbitals are allowed to fill the core hole. But it has been shown by Skytt *et al.* (1996) that breaking of the inversion symmetry in polyatomic molecules, as e.g. in CO_2 due to vibrating along the asymmetrical stretch mode, can lead to marked violations of the parity selection rule.

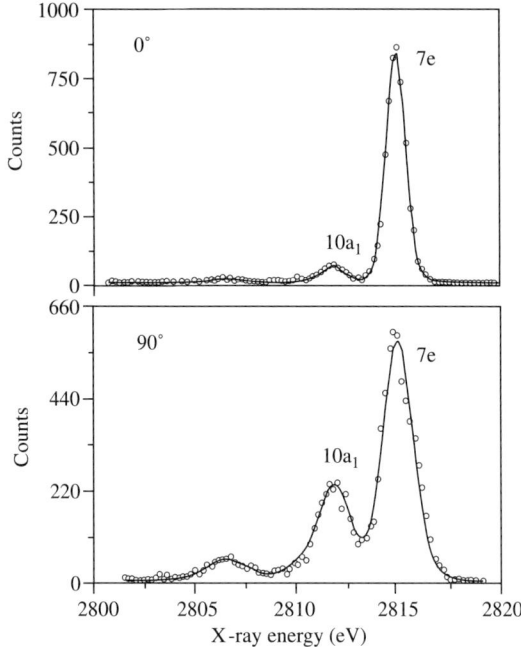

Fig. 1.19. Cl Kβ emission from CF$_3$Cl recorded on resonance at emission angles (see Fig. 1.18) $\theta = 0°$ (top) and $90°$ (bottom). In both cases the spectrometer was positioned to detect X-rays polarized parallel to the plane which contains the incident polarization vector and which is normal to the propagation direction of the incident X-rays. The solid lines are the result of a fitting procedure. (Originally published by Southworth *et al.* (1991); copyright (1991) by the American Physical Society.)

1.3.4.3 *Bloch-***k** *selective RIXS* As first shown by Ma *et al.* (1992) on diamond and by Miyano *et al.* (1993) on Si, in both cases in the soft X-ray regime, and by Enkisch *et al.* (1999) on a NiAl alloy in the hard X-ray regime, the shape of the resonantly excited valence fluorescence spectra from single crystals strongly depends on both the energy of the incident photons and, in the case of harder X-rays, on the direction as well as the amount of the transferred momentum **q**. It was shown in a thorough treatment by Ma (1994) that this behavior of the resonantly excited fluorescence can be traced back to momentum conservation within the resonant inelastic scattering process by making full use of the coupling of absorption and re-emission via the intermediate state as documented in (1.21). By treating the RIXS process as a transition from the ground state into the intermediate state, consisting of a hole in the core state $|c\rangle$ with energy E_c, and an excited electron in the Bloch state $|\mathbf{k}_e\rangle$ with energy $E(\mathbf{k}_e)$, followed by the decay of the intermediate state by transition of a valence electron from the Bloch state $|\mathbf{k}_h\rangle$ and energy $\mathbf{E}(\mathbf{k}_h)$ to the core hole, one ends up with

$$\frac{d^2\sigma}{d\Omega_2 d\hbar\omega_2} = \sum_{\mathbf{k}_e, \mathbf{k}_h} |M_{\mathbf{k}_c,\mathbf{k}_h} M_{\mathbf{k}_c,\mathbf{k}_e}|^2 \, \delta(E(\mathbf{k}_e) - E_c - \hbar\omega_1)$$

$$\times \delta_{\mathbf{g},(\mathbf{K}_1-\mathbf{K}_2+\mathbf{k}_h-\mathbf{k}_e)} \delta(E(\mathbf{k}_h) - E_c - \hbar\omega_2), \quad (1.27)$$

where the core state was represented in a tight binding ansatz ($\mathbf{R}$ = lattice translation vector, ϕ_c atomic wave function),

$$|c\rangle = \sum_{\mathbf{R}} \exp(i\mathbf{k}_c \cdot \mathbf{R}) \phi_c(\mathbf{r} - \mathbf{R}), \quad (1.28)$$

$M_{\mathbf{k}_c,\mathbf{k}_h}$ denotes the matrix element

$$M_{\mathbf{k}_c,\mathbf{k}_h} = \langle \phi_c | \mathbf{e}_2^* \cdot \mathbf{p} \, \exp(-i\mathbf{K}_2 \cdot \mathbf{r}) | \mathbf{k}_h \rangle, \quad (1.29)$$

and $M_{\mathbf{k}_c,\mathbf{k}_e}$ the matrix element

$$M_{\mathbf{k}_c,\mathbf{k}_e} = \langle \mathbf{k}_e | \mathbf{e}_1 \cdot \mathbf{p} \, \exp(i\mathbf{K}_1 \cdot \mathbf{r}) | \phi_c \rangle. \quad (1.30)$$

Equation (1.27) reflects both the energy conservation connected with the absorption and the re-emission process and the crystal momentum conservation, which governs the whole resonant inelastic scattering process, so that the momentum transfer $\mathbf{q} = \mathbf{K}_1 - \mathbf{K}_2$ must be equal to the vector difference $\mathbf{k}_e - \mathbf{k}_h$ between the Bloch vectors of the excited electron and that of the hole left behind, modulo a reciprocal lattice vector $\mathbf{g}$, as already stated with (1.23). This property makes RIXS Bloch-$\mathbf{k}$ vector selective, as demonstrated schematically in Fig. 1.20, which refers to the soft X-ray case, where $\mathbf{q}$ is practically zero. The incident photon energy (here denoted $h\nu_{\text{in}}$) defines those Bloch states of the unoccupied band structure, which can be reached by the excited core electron. Then distinct Bloch states with the same $\mathbf{k}$-vectors as those involved in the excitation process are contributing to the re-emission from the occupied valence band. It must be stressed that Fig. 1.20 applies in a straightforward manner only to the situation where $h\nu_{\text{in}}$ is above the absorption threshold. If $h\nu_{\text{in}}$ is below the threshold virtual excitations due to the finite lifetime of the core hole are spread over a volume in Bloch-$\mathbf{k}$ space just above the Fermi momentum. A series of RIXS spectra of Si measured for excitation energies above the Si$2p_{3/2}$ threshold (Eisebitt *et al.* 1998) is shown in Fig. 1.21 compared with simulations based on band structure calculations, given in the insert. On top of this series a spectrum of the Si valence fluorescence is shown as excited with photon energies far above threshold. One can easily see how peaks are evolving and dispersing with the incident photon energy thus indicating scanning of the band structure. Table 1.7 summarizes representative examples of Bloch-$\mathbf{k}$ selective RIXS studies, many of them aiming to yield band structure information (for a review see also Eisebitt and Eberhardt 2000).

In contrast to **a**ngle **r**esolved **p**hotoemission **s**pectroscopy (ARPES), which exhibits a similar Bloch-$\mathbf{k}$ vector selectivity, RIXS is both element specific, thus

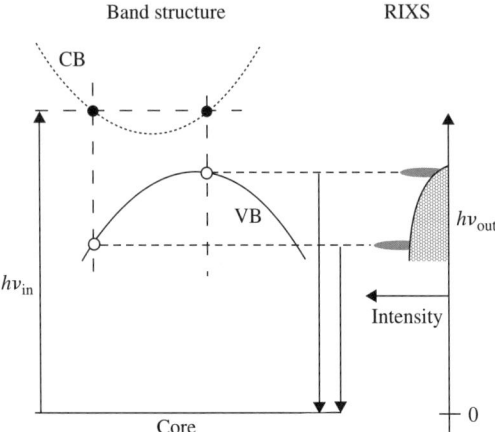

Fig. 1.20. Schematic illustrating the Bloch-**k** selectivity of RIXS: CB: conduction band, VB: valence band (for further comments see text). The broad background in the RIXS spectrum is caused by Bloch-**k** unselective processes. (Reprinted with permission from Eisebitt and Eberhardt (2000); copyright (2000) by Elsevier B.V.)

offering a higher degree of selectivity for compounds, and symmetry selective, since the matrix elements (1.29) and (1.30) prefer certain symmetries of the conduction and of the valence band states due to the strong preference of dipole transitions. Moreover RIXS does not suffer from charging and is insensitive to electric and magnetic fields at the sample.

But it is a shortcoming of RIXS, when compared with ARPES, that relaxation processes in the intermediate state can partly abolish the Bloch-**k** vector selectivity, where the relaxation can either be due to Coulomb interaction between the core hole and the excited electron as discussed by van Veenendaal and Carra (1997), or, as shown by Eisebitt and Eberhardt (2000), by electron–phonon or electron–electron scattering. The electron–phonon contribution to relaxation is mainly determined by the ratio of the core-hole decay rate R_c and the electron–phonon scattering rate R_{ph}, resulting in a **k**-selective fraction f:

$$f = \frac{R_c}{R_c + R_{\mathrm{ph}}}. \tag{1.31}$$

If one compares soft with hard X-ray RIXS, the latter has a much higher f according to (1.31) since $R_c \gg R_{\mathrm{ph}}$, but the large R_c of the core hole excited by hard X-rays causes a rather big broadening of the spectra. On the other hand, hard X-ray RIXS probes the real bulk rather than the surface near part of the sample, and offers the momentum transfer **q** as an additional degree of freedom in the hands of the experimentalist.

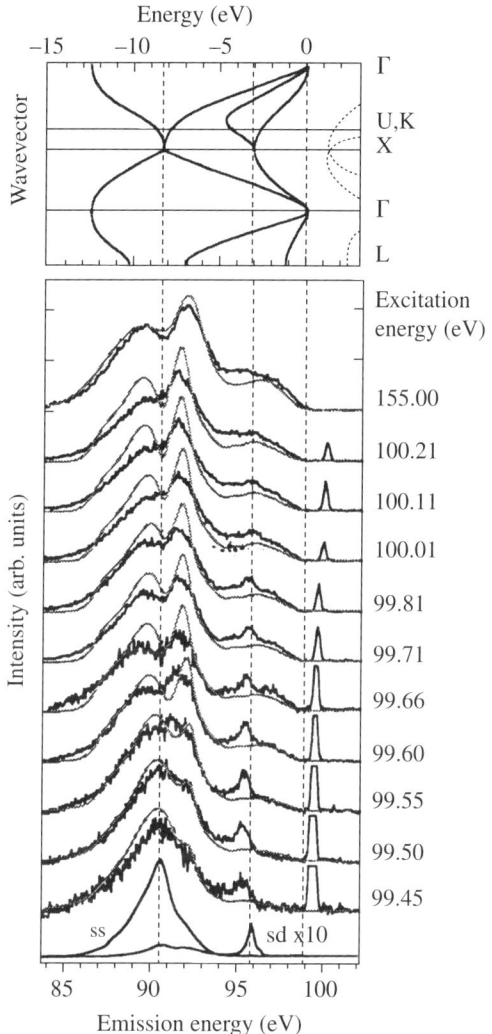

Fig. 1.21. Si $2p_{3/2}$ RIXS data (solid lines) compared to calculated spectra based on band structure computations (top panel and dotted lines in the bottom panel). The contribution of d-states in the scattering is shown in the bottom lines as a result of a separate simulation. (Reprinted with permission from Eisebitt *et al.* (1998a); copyright (1998) by Elsevier B.V.)

1.3.4.4 *Spin-selective RIXS, internal spin reference* It was first shown by Hämäläinen *et al.* (1992) that RIXS can be used to investigate the density of unoccupied states spin selectively. This effect is demonstrated schematically in Fig. 1.22 for the case of the resonantly excited Mn $K\beta$ (1s3p) emission in MnO.

Table 1.7 Representative examples of Bloch-**k** selective RIXS studies.

Material	Ref.	Core level	Comment
Si	(1)	Si 2p	no interpretation on the basis of **k**-selectivity
C (diamond)	(2)	C 1s	first interpretation on the basis of **k**-selectivity
c-Si; a-Si	(3)	Si 2p	excitation above threshold
C (diamond)	(4)		theory in connection to (2)
C (graphite)	(5)	C 1s	experiment and theory, angle variation
h-BN,		B 1s	
cub BN	(6) (7) (8)	N 1s	experiment and theory, excitation below threshold
c-Si	(9)	Si 2p	experiment 18 K
BaTiO$_3$, TiO$_2$			
TiO, Ti	(10)	Ti 2p$_{3/2}$ 2p$_{1/2}$	experiment, excitation below and above threshold
C (graphite)	(11)		theory in connection to (5)
cub. SiC	(12)	C 1s Si 2p	experiment, band mapping
Cu	(13)	Cu 1s	experiment, excitation above threshold
Ge	(13)	Ge 1s	experiment, excitation below threshold
c-Si	(14)	Si 2p	experiment and theory
single-wall carbon nanotubes	(15)	C 1s	experiment
3C-,4H-,6H-SiC	(16)	C 1s, Si 2p	experiment, theory, XPS
AlN, GaN, InP	(17)	N 1s, P 1s	experiment, theory
NiAl	(18)	Ni 1s	experiment, theory
Nd$_2$O$_3$	(19)	Nd 3d	excitation through the 3d-4f resonance
TiO$_2$	(20)	Ti 2p	experiment, theory
KNbO$_3$	(21)	Nb 3d$_{3/2,5/2}$	experiment, theory
Dy$_2$O$_3$	(22)	Dy 4d	excitation through the 4d-4f resonance

(1) Rubensson et al. (1990) (2) Ma et al. (1990) (3) Miyano et al. (1993) (4) Johnson and Ma (1994) (5) Carlisle et al. (1995) (6) Jia et al. (1996) (7) Agui et al. (1996) (8) Agui et al. (1997) (9) Shin et al. (1996) (10) Jimenez-Mier et al. (1996) (11) van Veenendaal and Carra (1997) (12) Lüning et al. (1997) (13) Kaprolat and Schülke (1997) (14) Eisebitt et al. (1998a) (15) Eisebitt et al. (1998b) (16) Lüning et al. (1999) (17) Eisebitt et al. (1999) (18) Enkisch et al. (1999) (19) Moewes et al. (1999b) (20) Finkelstein et al. (1999) (21) Moewes et al. (1999c) (22) Moewes et al. (1999a).

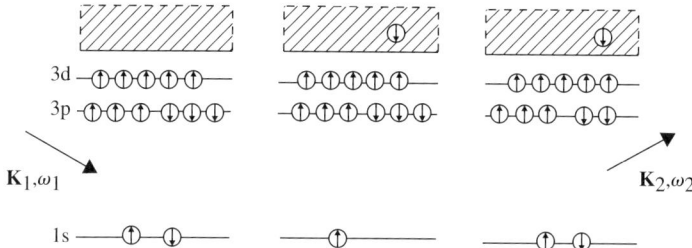

Fig. 1.22. Simplified atomic picture of the spin selectivity making use of the MnO $K\beta$ fluorescence excited just above threshold, where the photoelectron has minority spin. The initial, intermediate, and the final states are shown from left to right, respectively. (Reprinted with permission from Hämäläinen et al. (1992); copyright (1992) by the American Physical Society.)

The Mn $K\beta$ emission consists of a main line and a satellite, by 17 eV lower in energy than the main line. A similar structure is also found in the 3p photo-emission with the same final state as the $K\beta$ fluorescence, and which has been interpreted as being due to atomic multiplets of the $3p^5 3d^5$ final state configuration (Hermsmeier et al. 1988). As a consequence of the exchange interaction between the 3p hole and 3d electrons the main line is dominated by 3p holes where electrons with spins antiparallel to the 3d spins (minority spin, spin down) are missing, whereas the satellite corresponds to 3p holes with missing majority spin (spin up) electrons. This spin assignment is based on the analogy of the Mn $K\beta$ fluorescence with the Mn 3p photoemission (Sinkowic and Fadley 1985). Let the incident photon, as shown in Fig. 1.22, excite an electron from the Mn 1s core level into the MnO conduction band, where the spin of this electron might be antiparallel (spin down) to the aligned 3d spins of Mn, which act as an internal (atomic) spin reference. Since we are claiming spin conservation for the whole resonant scattering process, the core hole can be filled up only by an electron with spin down, so that the emission is into the main line. If, on the other hand, an electron from the 1s core level is excited into a spin-up state of the conduction band, the core hole must be filled up by an electron with spin up, so that the emission is into the satellite. Therefore, by setting the analyzer energy on the main line and by scanning the incident energy over the Mn K edge one probes the spin down (minority) DOS at the Mn atom, whereas, with the analyzer energy on the satellite, the spin up (majority) DOS is investigated. Of course, this simplified description neglects the energy dependence of the contributing dipole matrix elements (1.29) and (1.30). Figure 1.23 shows the result of such a measurement on MnO at room temperature, which means in the paramagnetic phase. It should be stressed that, contrary to measurements utilizing magnetic circular dichroism (MCD), first introduced by Schütz et al. (1987), where ferromagnetically ordered samples are necessary, this spin-selective probing of the unoccupied local DOS does not need a spin-ordered state, since the spin orientation of the state under

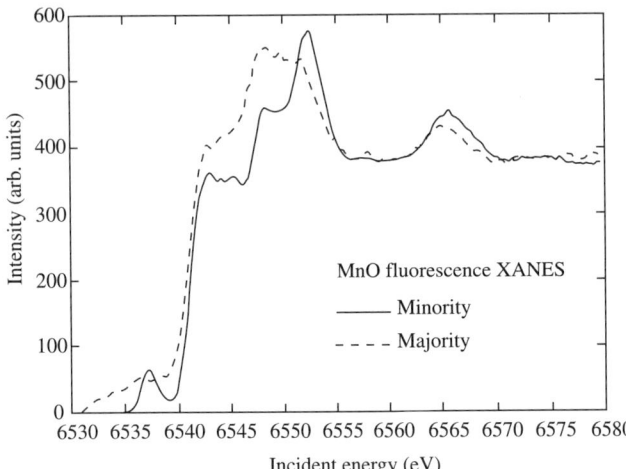

Fig. 1.23. Spin-selective Mn K near-edge fine structure of MnO in the paramagnetic phase: minority spin, solid line, majority spin, dashed line. Note that the quadrupolar pre-edge structure as caused by 1s-3d transitions is only visible in the minority spin contribution, due to the spin saturation of the occupied 3d-states. (Reprinted with permission from Hämäläinen et al. (1992); copyright (1992) by the American Physical Society.)

investigation is related to an internal **atomic** spin reference, here the aligned 3d spins of the Mn ion.

Further applications of this technique to MnP by de Groot et al. (1995), and a corresponding theoretical analysis of spin-selective RIXS spectra of MnO and MnF$_2$ by Soldatov et al. (1994) should be mentioned, especially since in the former study the comparison between the spin-selective RIXS spectra with the magnetic circular dichroism spectra, measured both at the Mn K edge, allows the determination of the energy dependence of the Fano factor (Fano 1969).

It was demonstrated by Peng et al. (1994) that also the Fe (1s3p) fluorescence emission and by Schülke (2001) that the Eu (2p4d) fluorescence emission can likewise be used to get information about the spin polarization of unoccupied states.

1.3.4.5 *Shakeup processes in the intermediate state of RIXS* As already discussed in Section 1.2 shakeup processes in the intermediate state, as sketched by the diagram of Fig. 1.5, opens up an additional window for investigating electronic excitations. To give an example let us point out how Abbamonte et al. (1999) have treated such a shakeup process interpreting RIXS spectra of La$_2$CuO$_4$ taken with the incident energy tuned near the Cu K absorption edge, as shown in Fig. 1.24. Applied to this system, the diagram of Fig. 1.5 stands

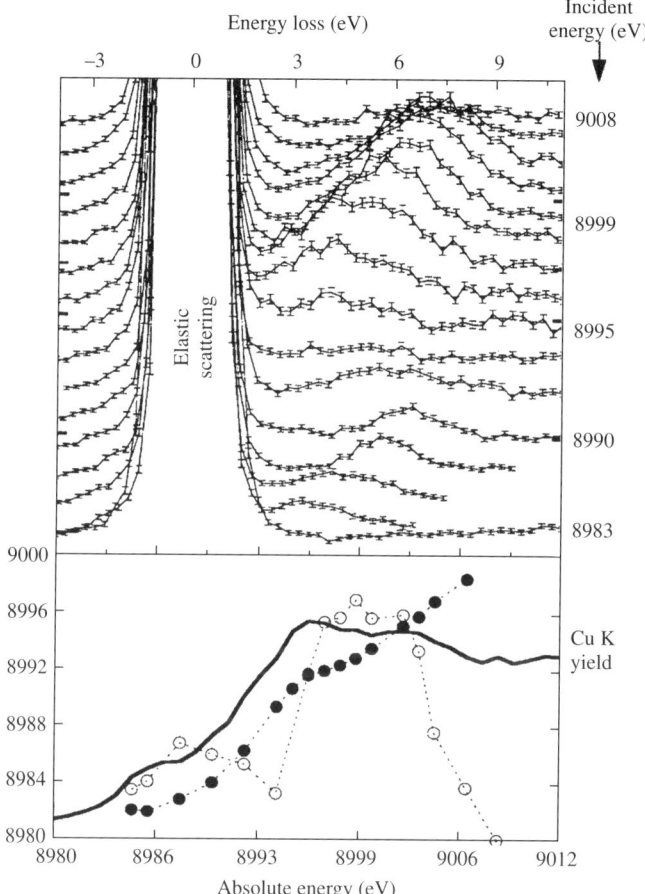

Fig. 1.24. Upper panel: RIXS spectra of La_2CuO_4 with $\mathbf{e}_1 \parallel \mathbf{a}$ and $\mathbf{q} = 1.27\,\text{Å}^{-1}$ parallel to $\mathbf{c}$ for different values of the excitation energy plotted against the energy loss. Lower panel: Open and filled circles are the peak height of the structures related to shakeup processes and their position, respectively, plotted against the incident energy. The line is the total fluorescence yield which peaks at the Cu 1s-4p transition. Note that the filled circles do not follow a strict linear relationship as expected for a simple Raman shift. (Reprinted with permission from Abbamonte *et al.* (1999); copyright (1999) by the American Physical Society.)

for the following processes. The incident photon creates a virtual 1s4p pair on a copper site (1s means a 1s core hole). This pair is bound as an exciton by the Coulomb interaction, and by taking up the momentum of the incident photon it can be scattered by the valence electron system leaving behind an excitation, e.g. a particle–hole excitation. After the recombination of the exciton, the emitted photon carries away the information about the energy and the momentum imparted to the system. Therefore, the scattering amplitude was calculated in third-order perturbation theory (second order in $\mathbf{p} \cdot \mathbf{A}$ and first order in the Coulomb interaction), which yields a double resonance denominator. This double resonance denominator gives rise to deviations from the simple linear relationship between incident energy and scattered photon energy in a RIXS process for incident energies smaller than the threshold, as depicted in Section 1.2 under the designation Raman shift.

If one accepts this interpretation of structures in the RIXS spectra, resonant inelastic X-ray scattering opens a unique possibility to investigate element specifically and resonantly enhanced electronic excitations and especially their dispersion, since, as stated above, the full momentum $\mathbf{q} = \mathbf{K}_1 - \mathbf{K}_2$ is imparted to the shakeup excitation, if other relaxation processes such as phonon emission are neglected. Thus Tsutsui et al. (1999) have proposed to utilize this momentum dependence of shakeup satellites of RIXS spectra to probe the charge gap in a Mott insulator (insulating cuprates) through the creation of a hole in the occupied Zhang–Rice band (ZRB) (Zhang and Rice 1988), thereby promoting an electron across the gap to the unoccupied upper Hubbard band (UHB) with a finite momentum transferred to the system tunable in size and direction. These experiments have been performed by Hasan et al. (2000) and have shown, in good agreement with the predictions of Zhang and Rice (1988), a partial doublet nature of the excitations and a strongly anisotropic dispersion. The results of these experiments could help to fill the gap of knowledge about the momentum-resolved electronic structure of the UHB, which is of importance for understanding the physics of n-type superconductors, since the doped electrons occupy the UHB.

RIXS experiments, somewhat earlier performed than those mentioned above and which were also devoted to investigate spectral features due to excitations connected with charge transfer in NiO (pioneering experiment of Kao et al. 1996) and Nd_2CuO_4 (Hill et al. 1998 and Hämäläinen et al. 2000), have been interpreted in another way even if the physics behind seems not to be so different to the former. The authors didn't see the need to treat the scattering process in third-order perturbation theory, but remain in second order, thus having only a single resonance denominator and a simple linear Raman shift. Their interpretation of charge transfer features in the RIXS spectra is based on the Anderson impurity model (Tanaka et al. 1991, Tanaka and Kotani 1993), where, for the case of the cuprates, the Cu $3d^9$ configuration hybridizes with $3d^{10}\underline{L}$, where $\underline{L}$ represents an O 2p ligand hole, so that the ground state is the bonding state with about 60% of the $3d^9$ configuration, whereas the antibonding state, by Δ apart in energy from the ground state, is mainly $3d^{10}\underline{L}$ with a continuous band between them. If in the intermediate state a Cu 1s electron is excited to the Cu $4p\pi/\sigma$ conduction band, the core-hole potential reverses the balance between the $3d^9$ and the $3d^{10}\underline{L}$ configurations, so that the bonding state with predominately

$\underline{1s}3d^{10}\underline{L}4p$ is lower in energy than the antibonding state $\underline{1s}3d^94p$. If either of these intermediate states decays into the antibonding excited state an energy loss of Δ would result. This energy loss is found in the experiment. Since the energy shift of the RIXS spectra with the incident energy is simple Raman-like in the above mentioned cases the interpretation of the spectra within the limits of second-order perturbation theory seems to be conclusive.

It was stressed by Döring et al. (2004) that RIXS experiments, when performed for getting information about excitations connected with shakeup processes, will exhibit polarization induced features, which follow the predictions of Hannon et al. (1988).

Finally it must be mentioned that also excitations connected with changes of orbital ordering can be investigated by means of RIXS, more precisely, by looking at shakeup processes in the intermediate state. Since the highest occupied and the lowest unoccupied electronic states of orbitally ordered insulators have different orbital character, the particle–hole excitation across the Mott gap changes

Table 1.8 Representative examples of shakeup processes in the intermediate state of RIXS. BACT = bonding–antibonding charge transfer excitation; QLDMI = excitations in quasilow-dimensional Mott insulators; OE = orbital excitations.

Materials	Ref.	Type of excitation	Core level
NiO	(1)	BACT	Ni 1s
Nd_2CuO_4	(2) (3)	BACT	Cu 1s
La_2CuO_4	(4)	BACT	Cu 1s
CuO	(5)	BACT	Cu 1s
copper oxygen plaquettes in various compounds	(6)	BACT	Cu 1s
NiO	(7)	BACT, high pressure	Ni 1s
$Ca_2CuO_2Cl_2$	(8)	QLDMI	Cu 1s
$Sr_2CuO_2Cl_2$	(4)	QLDMI	Cu 1s
La_2CuO_4	(9)	QLDMI	Cu 1s
$SrCuO_2$, Sr_2CuO_3	(10) (11)	QLDMI	Cu 1s
Li_2CuO_2	(12)	QLDMI	Cu 1s
$LaMnO_3$	(13)	OE	Mn 1s
$La_{1-x}Sr_xMnO_3$ $LaMnO_3$	(14)	OE	Mn 1s
$LaMnO_3$	(15)	OE	Mn 1s

(1) Kao et al. (1996) (2) Hill et al. (1998) (3) Hämäläinen et al. (2000) (4) Abbamonte et al. (1999) (5) Döring et al. (2004) (6) Kim et al. (2004a) (7) Shukla et al. (2003) (8) Hasan et al. (2000) (9) Kim et al. (2002) (10) Hasan et al. (2002) (11) Kim et al. (2004b) (12) Hasan et al. (2003) (13) Inami et al. (2003) (14) Ishii et al. (2004) (15) Grenier et al. (2004).

the symmetry of the electronic system. The first experimental investigation of this individual orbital excitation was performed by Inami et al. (2003) on LaMnO$_3$ single crystals. A more extended study on orbital excitation by RIXS was presented by Ishii et al. (2004) on hole-doped manganites La$_{1-x}$Sr$_x$MnO$_3$ ($x = 0.2, 0.4$) and compared with the results on undoped LaMnO$_3$.

Table 1.8 presents representative examples for shakeup processes in the intermediate state of RIXS, where three applications are stressed, namely bonding–antibonding charge transfer excitation (BACT), excitations in quasilow-dimensional Mott insulators (QLDMI), and finally, orbital excitations (OE).

1.4 References

Abbamonte, P., C.A. Burns, E.D. Isaacs, P.M. Platzman, L.L. Miller, S.W. Cheong, and M.V. Klein (1999). *Phys. Rev. Lett.* **83** 860

Abbamonte, P., K.D. Finkelstein, M.D. Collins, and S.M. Gruner (2004). *Phys. Rev. Lett.* **92** 237401

Agren, H. and F. Kh. Gel'mukhanov (1996). *Phys. Rev. A* **54** 379

Agui, A., S. Shin, M. Fujisawa, Y. Tezuka, T. Ishii, Y. Maramatsu, O Mishima, and K. Era (1996). *J. El. Spec. Rel. Phen.* **79** 191

Agui, A., S. Shin, M. Fujisawa, Y. Tezuka, T. Ishii, Y. Maramatsu, O Mishima, and K. Era (1997). *Phys. Rev. B* **55** 2073

Ahuja, B.L., C. Bellin, J. Moscovici, E. Zukowski, G. Loupias, and M.J. Cooper (1994). *J. Phys. Condens. Matter* **6** 8701

Asthalter, T. and W. Weyrich (1993). *Z. Nat.forsch. A* **48A** 303

Bagayoko, D., D.G. Laurent, S.P. Singhal, and J. Callaway (1980). *Phys. Lett.* **76** 187

Bartolome, F., J.M. Tonnerre, L. Seve, D. Raoux, J. Chaboy, L.M. Garcia, M.H. Krisch, and C.-C. Kao (1997). *Phys. Rev. Lett.* **79** 3775

Baruah, T., R.R. Zope, and A. Kshirsagar (1999). *Phys. Rev. B* **60** 10770

Bauer, G.E.W. and J.R. Schneider (1985). *Phys. Rev. B* **31** 681

Bell, F., A.J. Rollason, J.R. Schneider, and W. Drube (1990). *Phys. Rev. B* **41** 4887

Bell, F., T. Tschentscher, J.R. Schneider, and A.J. Rollason (1991). *J. Phys. Condens. Matter* **3** 5587

Bellin, C., G. Loupias, A.A. Manuel, T. Jarlborg, Y. Sakurai, Y. Tanaka, and N. Shiotani (1995). *Solid State Commun.* **96** 563

Bellin, Ch., P. Roca, I. Cabarrocas, K. Zellama, M.L. Theye, and G. Loupias (1997). *Solid State Commun.* **104** 193

Bergmann, U., Ph. Wernet, P. Glatzel, M. Cavalleri, L.G.M. Petterson, A. Nilsson, and S.P. Cramer (2002). *Phys. Rev. B* **66** 092107

Blaas, C., J. Redinger, S. Manninen, V. Honkimäki, K. Hämäläinen, and P. Suortti (1995). *Phys. Rev. Lett.* **75** 1984

Blume, M. (1985). *J. Appl. Phys.* **57** 3615

Bocharov, S., G. Dräger, D. Heumann, A. Simunek, and O. Sipr (1998). *Phys. Rev. B* **58** 7668

Bocharov, S., Th. Kirchner, G. Dräger, O. Sipr, and A. Simunek (2001). *Phys. Rev. B* **63** 045104

Bothe, W. and H. Geiger (1925). *Z. Phys.* **32** 639

Bowron, D.T., M.H. Krisch, A.C. Barnes, J.L. Finney, A. Kaprolat, and M. Lorenzen (2000). *Phys. Rev. B* **62** R9223

Brahmia, A., M.J. Cooper, D.N. Timms, S.P. Collins. A.P. Kane, and D. Laundy (1989) *J. Phys. Condens. Matter* **1** 3879

Burkel, E. (1991) *Inelastic Scattering of X-rays with Very High Energy Resolution* (Springer, Berlin) pp. 85–896

Burns, C.A., P. Giura, A. Said, A. Shukla, G. Vanko, M. Tuel-Benckendorf, E. D. Isaacs, and P.M. Platzman (2002). *Phys. Rev. Lett.* **89** 236404

Cai, Y.Q., H.-K. Mao, P.C. Chow, J.S. Tse, Y. Ma, S. Patchkovskii, J.F. Shu, V. Stuzhkin, R.J. Hemley, H. Ishii, C.C. Chen, I. Jarrige, C.T. Chen, S.R. Shieh, E.P. Huang, and C.-C. Kao (2005). *Phys. Rev. Lett.* **94** 025502

Caliebe, W.A., C.-C. Kao, L.E. Berman, J.B. Hastings, H.M. Krisch, F. Sette, and K. Hämäläinen (1996). *J. Appl. Phys.* **79** 6509

Caliebe, W.A., J.A. Soininen, E.L. Shirley, C.-C. Kao, and K. Hämäläinen (2000). *Phys. Rev. Lett.* **84** 3907

Cardwell, D.A., M.J. Cooper, and S. Wakoh (1989). *J. Phys. Condens. Matter* **1** 541

Carlisle, J.A., E.L. Shirley, E.A. Hudson, L.E. Terminello, T.A. Callcot, J.J. Jia, D.L. Ederer, R.C.C. Perera, and F.J. Himpsel (1995). *Phys. Rev. Lett.* **74** 1234

Carra, P., M. Fabricius, and B.T. Thole (1995). *Phys. Rev. Lett.* **74** 3700

Causa, M., R. Dovesi, C. Pisani, and C. Roetti (1986). *Phys. Rev. B* **34** 29

Chen, K.J., V. Caspar, Ch. Bellin, and G. Loupias (1999). *Solid State Commun.* **110** 357

Chou, M.Y., M.L. Cohen, and S.G. Louie (1986). *Phys. Rev. B* **33** 6619; (1986). *Phys. Rev. Lett.* **49** 1452

Collins, S.P., M.J. Cooper, D. Timms, A. Brahmia, D. Laundy, and P.P. Kane (1989). *J. Phys. Condens. Matter* **1** 9009

Compton, A.H. (1923). *Phys. Rev.* **21** 207, 483

Cooper, M.J. (1985). *Rep. Prog. Phys.* **48** 415

Cooper, M.J., J.A. Leake, and R.J. Weiss (1965). *Philo. Mag.* **12** 797

Cooper, M.J., D. Laundy, D.A. Cardwell, D.N. Timms, R.S. Holt, and G. Clark (1986). *Phys. Rev. B* **34** 5984

Cooper, M.J., S.P. Collins, D.N. Timms, A. Brahmia, P.P. Kane, R.S, Holt, and L. Laundy (1988). *Nature* **333** 151

Cooper, M.J., E. Zukowski, D.N. Timms, R. Armstrong, F. Itoh, Y. Tanaka, M. Ito, H. Kawata, and R. Bateson (1993). *Phys. Rev. Lett.* **71** 1095

Cooper, M.J., P.K. Lawson, M.A. Dixon, E. Zukowski, D.N. Timms, F. Itoh, H. Sakurai, H. Kawata, Y. Tanaka, and M. Ito (1996). *Phys. Rev. B* **54** 4068

Cowan, P.L., S. Brennan, R.D. Deslattes, A. Henins, T. Jach, and E.G. Kessler (1986). *Nucl. Instrum. & Methods A* **246** 154
Currat, R., P.D. DeCicco, and R. Kaplow (1971). *Phys. Rev. B* **3** 243
Daniel, E. and S.H. Vosko (1960). *Phys. Rev.* **120** 2041
Debye, P. (1923). *Phys. Z.* **24** 165
Dixon, M.A.G., J.A. Duffy, S. Gardelis, J.E. McCarthy, M.J. Cooper, S.B. Dugdale, T. Jarlborg, and D.N. Timms (1998). *J. Phys. Condens. Matter* **10** 2759
Döring, G., C. Sternemann, A. Kaprolat, A. Mattila, K. Hämäläinen, and W. Schülke (2004). *Phys. Rev. B* **70** 085115
Dräger, G. and O. Brümmer (1984). *Phys. Stat. Sol. (b)* **124** 11
Dräger, G., R. Frahm, G. Materlik, and O. Brümmer (1988). *Phys. Stat. Sol. (b)* **146** 287
Duffy, J.A., S.B. Dugdale, J.E. McCarthy, M.A. Alam, M.J. Cooper, S.B. Palmer, and T. Jarlborg (2000). *Phys. Rev. B* **61** 14331
Duffy, J.A., J.E. McCarthy, S.B. Dugdale, V. Honkimäki, M.J. Cooper, M.A. Alam, T. Jarlborg, and S.B. Palmer (1998). *J. Phys. Condens. Matter* **10** 10391
Dugdale, S.B. and J. Jarlborg (1998). *Solid State Commun.* **105** 283
Dugdale, S.B., H.M. Fretwell, K.J. Chen, Y. Tanaky, A. Shukla, T. Buslaps, Ch. Bellin, G. Loupias, M.A. Alam, A.A. Manuel, P. Suortti, and N. Shiotani (2000). *J. Phys. Chem. Solids* **61** 361
DuMond, J.W.M. (1929). *Phys. Rev.* **33** 643
DuMond, J.W.M. and A. Hoyt (1931). *Phys. Rev.* **37** 1443
DuMond, J.W.M. and H.A. Kirkpatrick (1930). *Rev. Sci. Instrum.* **1** 88
DuMond, J.W.M. and H.A. Kirkpatrick (1931a). *Phys. Rev.* **37** 136
DuMond, J.W.M. and H.A. Kirkpatrick (1931b). *Phys. Rev.* **38** 1094
DuMond, J.W.M. and H.A. Kirkpatrick (1937). *Phys. Rev.* **52** 419
Eisberg, R., G. Wiech, and R. Schlögl (1988). *Solid State Commun.* **65** 705
Eisebitt, S. and W. Eberhardt (2000) *J. El. Spec. Rel. Phen.* **110–111** 335
Eisebitt, S., J. Lüning, J.-E. Rubensson, A. Settels, P.H. Dederichs, W. Eberhardt, S.N. Patitsas, and T. Tiedje (1998a). *J. El. Spec. Rel. Phen.* **93** 245
Eisebitt, S., A. Karl, W. Eberhardt, J.E. Fischer, C. Sathe, A. Agui, and J. Nordgren (1998b). *Appl. Phys. A* **67** (1998b) 89
Eisebitt, S., J. Lüning, J.-E. Rubensson, A. Karl, and W. Eberhardt (1999). *phys. stat. sol. (b)* **215** 803
Eisenberger, P. (1970). *Phys. Rev. A* **2** 1078
Eisenberger, P. (1972). *Phys. Rev. A* **5** 628
Eisenberger, P. and W.C. Marra (1971). *Phys. Rev. Lett.* **27** 1413
Eisenberger, P. and P.M. Platzman (1970). *Phys. Rev. A* **2** 415
Eisenberger, P. and P.M. Platzman (1976). *Phys. Rev. B* **13** 934
Eisenberger, P., and A.A. Reed (1972). *Phys. Rev. A* **5** 415
Eisenberger, P., W.H. Henneker, and P.E. Cade (1972a). *J. Chem. Phys.* **56** 1207

Eisenberger, P., L. Lam, P.M. Platzman, and P. Schmidt (1972b). *Phys. Rev. B* **6** 3671

Eisenberger, P., P.M. Platzman, and K.C. Pandy (1973). *Phys. Rev. Lett.* **31** 311

Eisenberger, P., P.M. Platzman, and P. Schmidt (1975). *Phys. Rev. Lett.* **34** 18

Eisenberger, P., P.M. Platzman, and H. Winick (1976a). *Phys. Rev. Lett* **36** 623

Eisenberger, P., P.M. Platzman, and H. Winick (1976b). *Phys. Rev. B* **13** 2377

Enkisch, H., A. Kaprolat, W. Schülke, M.H. Krisch, and M. Lorenzen (1999). *Phys. Rev. B* **60** 8624

Epstein, I.R. (1970). *J. Chem. Phys.* **53** 4425

Epstein, I.R. and W.N. Lipscomb (1970). *J. Chem. Phys.* **53** 4418

Epstein, I.R. and A.C. Tanner (1977). *Compton Scattering*, ed. B.C. Williams (McGraw-Hill, New York) p. 209

Fano, U. (1969). *Phys. Rev.* **178** 131

Feng, Y., G.T. Seidler, J.O. Cross, A. T. Macrander, and J.J. Rehr (2004). *Phys. Rev. B* **69** 125402

Filippi, C. and D.M. Ceperley (1999). *Phys. Rev. B* **59** 7907

Finkelstein, L.D., E.Z. Kurmaev, M.A. Korotin, A. Moewes, B. Schneider, S.M. Butorin, J.-H. Guo, J. Nordgren, D. Hartmann, M. Neumann, and D.L. Ederer (1999). *Phys. Rev. B* **60** 2212

Foo, E. Ni, and J.J. Hopfield (1968). *Phys. Rev.* **173** 635

Fukamachi, T. and S. Hosoya (1972). *Phys. Lett.* **38A** 341

Galambosi, S., J.A. Soininen, K. Hämäläinen, E.L. Shirley, and C.-C. Kao (2001). *Phys. Rev. B* **64** (2001) 024102

Galambosi, S., J.A. Soininen, A. Mattila, S. Huotari, S. Manninen, Gy. Vanko, N.D. Zhigadlo, J. Karpinski, and K. Hämäläinen (2005). *Phys. Rev. B* **71** 060504(R)

Glans, P., K. Gunnelin, P. Skytt, J.-H. Guo, N. Wassdahl, J. Nordgren, H. Agren, F.Kh. Gel'mukhanov, T. Warwick, and Eli Rotenberg (1996). *Phys. Rev. Lett.* **76** 2448

Golovchenko, J., D.R. Kaplan, B.M. Kincaid, R.A. Levesque, A.E. Meixner, M.P. Robbins, and J. Felsteiner (1981). *Phys. Rev. Lett.* **46** 1454

Golovchenko, J.A., B.M. Kincaid, R.A. Levesque, A.E. Meixner, and D.R. Kaplan (1986). *Phys. Rev. Lett.* **57** 203

Grenier, S., J.P. Hill, V. Kiryukhin, W. Ku, Y.-Y. Kim, K.J. Thomas, S.-W. Cheong, Y. Tokura, Y. Tomioka, D. Casa, and T. Gog (2005). *Phys. Rev. Lett.* **94** 047203

de Groot, F.M.F., S. Pizzini, A. Fontaine, K. Hämäläinen, C.-C. Kao, and J.B. Hastings (1995). *Phys. Rev. B* **51** 1045

Gurtubay, I.G., W. Ku, J.M. Pitarke, A.G. Eguiluz, B.C. Larson, J. Tischler, and P. Zschak (2004). *Phys. Rev. B* **70** 201201(R)

Gurtubay, I.G., J.M. Pitarke, W. Ku, A.G. Eguiluz, B.C Larson, J. Tischler, P. Zschak, and K.D. Finkelstein (2005). *Phys. Rev. B* **72** 125117

Haensel, R., G. Keitel, B. Sonntag, C. Kunz, and P. Schreiber (1970). *Phys. Stat. Sol. (a)* **2** 85

Hämäläinen, K., S. Manninen, P. Suortti, S.P. Collins, M.J. Cooper, and D. Laundy (1989). *J. Phys. Condens. Matter* **1** 1

Hämäläinen, K., S. Manninen, S.P. Collins, and M.J. Cooper (1990). *J. Phys. Condens. Matt.* **2** 5619

Hämäläinen, K., D.P. Siddons, J.B. Hastings, and L.E. Berman (1991). *Phys. Rev. Lett.* **67** 2850

Hämäläinen, K., C.-C. Kao, J.B. Hastings, D.P. Siddons, L.E. Berman, V. Stojanoff, and S.P. Cramer (1992). *Phys. Rev. B* **46** 14274

Hämäläinen, K., S. Manninen, C.-C. Kao, W. Caliebe, J.B. Hastings, A. Bansil, S. Kaprzyk, and P.M. Platzman (1996). *Phys. Rev. B* **54** 5453

Hämäläinen, K., S. Huotari, J. Laukkanen, A. Soininen, S. Manninen, C.-C. Kao, T. Buslaps, and M Mezour (2000a). *Phys. Rev. B* **62** R735

Hämäläinen, K., J.P. Hill, S. Huotari, C.-C. Kao, L.E. Berman, A. Kotani, T. Ide, J.L. Peng, and R.L. Greene (2000b). *Phys. Rev. B* **61** 1836

Hämäläinen, K., S. Galambosi, J.A. Soininen, E.L. Shirley, J.-P. Rueff, and A. Shukla (2002). *Phys. Rev. B* **65** 155111

Hannon, J.P., G.T. Trammel, M. Blume, and D. Gibbs (1988). *Phys. Rev. Lett.* **61** 1245

Hansen, N.K. (1980). *Reconstruction of the EMD from a Set of Compton Profiles* (Hahn Meitner Institut, Berlin Rep. HMI B) p. 342

Hansen, N.K., P. Pattison, and J.R. Schneider (1979). *Z. Phys. B* **35** 215

Hart, M. (1978). *Phil. Mag. B* **38** 41

Hasan, M.Z., E.D. Isaacs, Z.-X. Shen, L.L. Miller, K. Tsutsui, T. Tohyama, and S. Maekawa (2000). *Science* **288** 1811

Hasan, M.Z., P.A. Montano, E.D. Isaacs, Z.-X. Shen, H. Eisaki, S.K. Sinha, Z. Islam, N. Motoyama, and S. Ushida (2002). *Phys. Rev. Lett.* **88** 177403

Hasan, M.Z., Y.-D. Chuang, Y. Li, P.A. Montano, Z. Hussain, G. Dhalenne, A. Revocolevschi, H. Eisaki, N. Motoyama, and S. Uchida (2003). *Int. J. Mod. Phys. B* **17** 3519

Hashimoto, H., H. Sakurai, H. Oike, F. Itoh, A. Ochiai, H. Aokic, and T. Suzukic (1997). *Sci. Rep. Res. Inst. Tohoku Univ. A Phys. Chem. Metall.* **45** 93

Hermesmeier, B., C.S. Fadley, M.O. Krause, J. Jimenez-Mier, P. Gerard, and S.T. Manson (1988). *Phys. Rev. Lett.* **61** 2592

Hill, J.P., C.-C. Kao, W.A. Caliebe, D. Gibbs, and J.B. Hastings (1996). *Phys. Rev. Lett.* **77** 3665

Hill, J.P., C.-C. Kao, W.A. Caliebe, M. Matsubara, A. Kotani, J.L. Peng, and R.L. Greene (1998). *Phys. Rev. Lett.* **80** 4967

Hiraoka, N., H. Ishii, I. Jarrige, and Y.Q. Cai (2005). *Phys. Rev. B* **72** 075103

Höppner, K., A. Kaprolat, and W. Schülke (1998). *Eur. Phys. J. B* **5** 53

Holt, R.S., M.J. Cooper, and K.R. Lea (1978). *J. Phys. E* **11** 68

Huotari, S., K. Hämäläinen, S. Manninen, S. Kaprzyk, A. Bansil, W. Caliebe, T. Buslaps, V. Honkimäki, and P. Suortti (2000). *Phys. Rev. B* **62** 7956

Huotari, S., K. Hämäläinen, S. Manninen, C. Sternemann, A. Kaprolat, W. Schülke, and T. Buslaps (2002). *Phys. Rev. B* **66** 085104

Inami, T., T. Fukuda, J. Mizuki, S. Ishihara, H. Kondo, H. Nakao, T. Matsumara, K. Hirota, Y. Murakami, S. Maekawa, and Y. Endoh (2003). *Phys. Rev. B* **67** 045108

Isaacs, E.D., P.M. Platzman, P. Zschack, K. Hämäläinen, and A.R. Kortan (1992). *Phys. Rev. B* **46** 12910

Isaacs, E.D., P.M. Platzman, P. Metcalf, and J.M. Honig (1996). *Phys. Rev. Lett.* **76** 4211

Isaacs, E.D., A. Shukla, P.M. Platzman, D.R. Hamann, B. Barbiellini, and C.A. Tulk (1999). *Phys. Rev Lett.* **82** 600

Ishii, K., T. Inami, K. Ohwada, K. Kuzushita, J. Mizuki, Y. Murakami, S. Ishihara, Y. Endoh, S. Maekawa, K. Hirota, and Y. Moritomo (2004). *Phys. Rev. B* **70** 224437

Itou, M., Y. Sakurai, T. Ohata, A. Bansil, S. Kaprzyk, Y. Tanaka, K. Kawata, and N. Shiotani (1998). *J. Phys. Chem. Solids* **59** 99

Itou, M., K. Kishimoto, H. Kawata, M. Ozaki, H. Sakurai, and F. Itoh (1999). *J. Phys. Soc. Jpn.* **68** 515

Jauch, J.M. and F. Rohrlich (1976). *The Theory of Photons and Electrons* (Springer Verlag, Berlin)

Jia, J.J., T.A. Callcott, E.L. Shirley, J.A. Carlisle, L. Terminello, A. Asfaw, D.L. Ederer, F.J. Himpsel, and R.C.C. Pererar (1996). *Phys. Rev. Lett.* **76** 4054

Jimenez-Mier, J., D.L. Ederer, U. Diebold, A. Moewes, T.A. Callcott, L. Zhou, J.J. Jia, J. Carlisle, E. Hudson, L.E. Terminello, F.J. Himpsel, and R.C.C. Perera (1996). Proceedings of the Workshop *Raman Emission by X-Rays*, ed. D.L. Ederer and J.H. McGuire (World Scientific Publishing, Singapore)

Johnson, P.D. and Y. Ma (1994). *Phys. Rev. B* **49** 5024

Kao, C.-C., W.A.L. Caliebe, J.B. Hastings, and J.-M. Gillet (1996). *Phys. Rev. B* **54** 16361

Kaprolat, A. and W. Schülke (1997). *Appl. Phys. A* **65** 169

Kim, Y.-J., J.P. Hill, C.A. Burns, S. Wakimoto, R.J. Birgeneau, D. Casa, T. Gog, and C.T. Venkataraman (2002). *Phys. Rev. Lett.* **89** 177003

Kim, Y.-J., J.P. Hill, G.D. Gu, F.C. Chou, S. Wakimoto, R.J. Birgeneau, S. Komya, Y. Ando, N. Motoyama, K.M. Kojima, S. Uchida, D. Casa, and T. Gog (2004a). *Phys. Rev. B* **70** 205128

Kim, Y.-J., J.P. Hill, H. Benthien, H.H.L. Essler, E. Jeckelman, H.S. Choi, T.W. Noh, N. Motoyama, K.M. Kojima, S. Uchida, D. Casa, and T. Gog (2004b). *Phys. Rev. Lett.* **92** 137402

Kohn, W. and L.J. Sham (1965). *Phys. Rev.* **140** A1133

Koizumi, A., S. Miyaki, Y. Kakutami, H. Koizumi, N. Hiraoka, K. Makoshi, and N. Sakai (2001). *Phys. Rev. Lett.* **86** 5589

Kontrym-Sznajd, G., M. Samsel-Szekala, A. Pietraszko, H. Sormann, S. Manninen, S. Huotari, K. Hämäläinen, R.N. West, and W. Schülke (2002). *Phys. Rev. B* **66** 155110

Krisch, M., C.-C. Kao, F. Sette, W.A. Caliebe, K. Hämäläinen, and J.B. Hastings (1995). *Phys. Rev. Lett.* **74** 152

Krisch, M.H., F. Sette, C. Masciovecchio, and R. Verbeni (1997). *Phys. Rev. Lett.* **78** 2843

Kubo, Y. (1997). *J. Phys. Soc. Jpn.* **66** 2236

Kubo, Y., Y. Sakurai, Y. Tanaka, T. Nakamura, H. Kawata, and N. Shiotani (1997). *J. Phys. Soc. Jpn.* **66** 2777

Kurp, F.F., Th. Tschentscher, H. Schulte-Schrepping, J.R. Schneider, and F. Bell (1996). *Europhys. Lett.* **35** 61

Kurp, F.F., A.E. Werner, J.R. Schneider, Th. Tschentscher, P. Suortti, and F. Bell (1997a). *Nucl. Instrum. & Methods B* **122** 269

Kurp, F.F., M. Vos, Th. Tschentscher, A.S. Kheifats, J.R. Schneider, E. Weigold, and F. Bell (1997b). *Phys. Rev. B* **55** 5447

Kwiatkowska, J., F. Maniawski, I. Matsumoto, H. Kawata, N. Shiotani, L. Litynska, S. Kaprzyk, and A. Bansil (2004). *Phys. Rev. B* **70** 075106

Lam, L. and P.M. Platzman (1974). *Phys. Rev. B* **9** 5122

Larson, B.C., J.Z. Tischler, E.D. Isaacs, P. Zschack, A. Fleszar, and A.G. Eguiluz (1996). *Phys. Rev. Lett.* **77** 1346

Lässer, R. and B. Lengeler (1978). *Phys. Rev. B* **18** 637

Laukkanen, J., K. Hämäläinen, and S. Manninen (1996). *J. Phys. Condens. Matter* **8** 2153

Laukkanen, J., K. Hämäläinen, S. Manninen, and V. Honkimäki (1998). *Nucl. Instrum. & Methods A* **416** 475

Laundy, D. (1990). *Nucl. Instrum. & Methods A* **290** 248

Lawson, P.K., J.E. McCarthy, M.J. Cooper, E. Zukowski, D.N. Timms, F. Itoh, H. Sakurai, Y. Tanaka, H. Kawata, M. Ito (1995). *J. Phys., Condens. Matter* **7** 389

Lawson, P.K., M.J. Cooper, M.A.G. Dixon, D.N. Timms, E. Zukowski, F. Itoh, and H. Sakurai (1997). *Phys. Rev. B* **56** 3239

Lindle, D.W., P.L. Cowan, R.E. LaVilla, T. Jach, R.D. Deslattes, B.A. Karlin, J.A. Sheehy, T.J. Gil, and P.W. Langhoff (1988). *Phys. Rev. Lett.* **60** 1010

Lipps, F.W. and H.A. Tolhoek (1954). *Physica* **20** 395

Loupias, G. and J. Chomilier (1986). *Z. Phys. D: Atoms, Mol. and Clusters* **2** 297

Loupias, G. and J. Petiau (1980). *J. Phys. (Paris)* **41** 265

Loupias, G., J. Chomilier, and D. Guerard (1985). *Synth. Met.* **12** 257

Loupias, G., J. Chomilier, J. Tarbes, and D. Guerard (1988). *Synth. Met.* **23** 179

Lovesey, S.W. (1993). *Rep. Progr. Phys.* **56** 257

Lovesey, S.W. and S.P. Collins (1996). *X-ray Scattering and Absorption by Magnetic Materials* (Clarendon Press, Oxford)

Lüning, J., J.-E. Rubensson, C. Ellmers, S. Eisebitt, and W. Eberhardt (1997). *Phys. Rev. B* **56** 13147

Lüning, J., S. Eisebitt, J.-E. Rubensson, and W. Eberhardt (1999). *Phys. Rev. B* **59** 10573

Ma, Y. (1994). *Phys. Rev. B* **49** 5799
Ma, Y., N. Wassdahl, P. Skytt, J. Guo, J. Nordgren, P.D. Johnson, J.E. Rubensson, T. Boske, W. Eberhardt, and S. Kevan (1992). *Phys. Rev. Lett.* **69** 2598
Macrander, A.T., P.A. Montano, D.L. Price, V.I. Kushnir, R.C. Blasdell, C.C. Kao, and B.R. Cooper (1996). *Phys. Rev. B* **54** 305
Manninen, S. (1986). *Phys. Rev. Lett.* **57** 1500
Manninen, S., B.K. Sharma, T. Paakkari, S. Rundquist, and M.W. Richardson (1981). *Phys. Stat. Solidi b* **107** 749
Manninen, S., K. Hämäläinen, and J. Graeffe (1990). *Phys. Rev. B* **41** 1224
Manninen, S., K. Hämäläinen, M.A.G. Dixon, M.J. Cooper, D.A. Cardwell, and T. Buslaps (1999). *Physica C* **314** 19
Manuel, A.A., D. Vasumathi, A. Shukla, P. Suortti, A. Yu. Rumiatsev, and A.S. Ivanow (1996). *Helv. Phys. Acta* **69** 33
Marangolo, M., Ch. Bellin, G. Loupias, S. Rabii, S.C. Erwin, and Th. Buslaps (1999). *Phys. Rev. B* **60** 17084
Marchetti, V. and C. Franck (1987). *Phys. Rev. Lett.* **59** 1557
Matsumoto, I., H. Kawata, and N. Shiotani (2001). *Phys. Rev. B* **64** 195132
McCarthy, J.E., J.A. Duffy, C. Detlefs, M.J. Cooper, and P. Canfield (2000). *Phys. Rev. B* **62** R6073
Meng, Y., H.K. Mao, P.J. Eng, T.P. Trainor, M. Newville, M.Y. Hu, C.C. Kao, J.F. Shu, D. Häusermann, and R.J. Hemley (2004). *Nature Mat.* **3** 111
Metz, C., Th. Tschentscher, P. Suortti, A.S. Kheifets, D.R. Lun, T. Sattler, J.R. Schneider, and F. Bell (1999a). *J. Phys. Condens. Matter* **11** 3933
Metz, C., Th. Tschentscher, P. Suortti, A.S. Kheifets, D.R. Lun, T. Sattler, J.R. Schneider, and F. Bell (1999b). *Phys. Rev. B* **59** 10512
Metz, C., Th. Tschentscher, T. Sattler, K. Höppner, J.R. Schneider, K. Wittmaak, D. Frischke, and F. Bell (1999c). *Phys. Rev. B* **60** 14049
Mijnarends, P.E. (1977). *Compton Scattering*, ed. B.G. Williams (McGraw-Hill, New York) chapter 10
Mills, D.M. (1987). *Phys. Rev. B* **36** 6178
Miyano, K.E., D.L. Ederer, T.A. Callcott, W.L. O'Brien, J.J. Jia, L. Zhou, Q.Y. Dong, Y. Ma, J.C. Woicik, and D.R. Mueller (1993). *Phys. Rev. B* **48** 1918
Mizuno, Y. and Y. Ohmura (1967). *J. Phys. Soc. Japan* **22** 445
Moewes, A., M.M. Grush, T.A. Callcott, and D.L. Ederer (1999a). *Phys. Rev. B* **60** 15728
Moewes, A., D.L. Ederer, M.M. Grush, and T.A. Callcott (1999b). *Phys. Rev. B* **59** 5452
Moewes, A., A.V. Postnikov, B. Schneider, E.Z. Kurmaev, M. Matteucci, V.M. Cherkashenko, D. Hartmann, H. Hesse, and M. Neumann (1999c). *Phys. Rev. B* **60** 4422
Montano, P.A., D.L. Price, A.T. Macrander, and B.R. Cooper (2002). *Phys. Rev. B* **66** 165218

Moscovici, J., G. Loupias, S. Rabii, S. Erwin, and A. Rassat (1995). *Europhys. Lett.* **31** 87

Mueller, F.M. (1977). *Phys. Rev. B* **15** 3039

Nagasawa, H., S. Mourikis, and W. Schülke (1989). *J. Phys. Soc. Japan* **58** 710

Nagasawa, H., S. Mourikis, and W. Schülke (1997). *J. Phys. Soc. Japan* **66** 3139

Nakamura, J., T. Takeda, K. Asai, N. Yamada, Y. Tanaka, N. Sakai, M. Ito, A. Koizumu, and H. Kawata (1995). *J. Phys. Soc. Jpn.* **64** 1385

Namikawa, K. and S. Hosoya (1984). *Phys. Rev. Lett.* **53** 1606

Nilsson, A., N. Wassdahl, M. Weinelt, O. Karis, T. Wiell, P. Bennich, J. Hasselström, A. Fölisch, J. Stöhr, and M. Samant (1997). *Appl. Phys. A* **65** 147

Nozieres, P. and D. Pines (1959). *Phys. Rev.* **113** 1254

Okada, T., Y. Watanabe, Y. Yokoyama, N. Hiraoka, M. Itou, Y. Sakurai, and S. Nanao (2002). *J. Phys. Condens. Matter* **14** L43

Okada, T., Y. Watanabe, S. Nanao, R. Tamura, S. Takeuchi, Y. Yokoyama, N. Hiraoka, M. Itou, and Y. Sakurai (2003). *Phys. Rev. B* **68** 132204

Oomi, G. and F. Itoh (1993). *Jpn. J. Appl. Phys. Suppl.* **32** Sup. 32–1, 352

Pattison, P. and J. R. Schneider (1980). *Acta Cryst. A* **36** 390

Pattison, P., W. Weyrich, and B.G. Williams (1977). *Solid State Commun.* **21** 967

Pattison, P., N.K. Hansen, and J.R. Schneider (1984). *Acta Cryst. B* **40** 38

Peng, G., X. Wang, C.R. Randall, J.A. Moore, and S.P. Cramer (1994). *Appl. Phys. Lett.* **65** 2527

Phillips, W.C. and R.J. Weiss (1968). *Phys. Rev.* **171** 790

Platzman, P.M. and P. Eisenberger (1974a). *Phys. Rev. Lett.* **33** 152

Platzman, P.M. and P. Eisenberger (1974b). *Solid Stat Commun.* **14** 1

Platzman, P.M. and E.D. Isaacs (1998). *Phys. Rev. B* **57** 107

Platzman P.M. and N. Tzoar (1970). *Phys. Rev. B* **2** 3556

Priftis, G.D. (1970). *Phys. Rev. B* **2** 54

Priftis, G.D., J. Boviatsis, and A. Vradis (1978). *Phys. Lett. A* **68** 482

Rabii, S., J. Chomilier, and G. Loupias (1989). *Phys. Rev. B* **40** 10105

Rao, M.N., D.P. Mohapatra, B.K. Panda, and H.C. Padhi (1985). *Solid State Commun.* **55** 241

Reed, W.A. and P. Eisenberger (1972). *Phys. Rev. B* **6** 4596

Rennert, P., G. Carl, and W. Herget (1983). *Phys. Stat. Sol. B* **120** 273

Ribberfors, R. (1975). *Phys. Rev. B* **12** 3136

Rollason, A.J. and M.B. Woolf (1995). *J. Phys. Condens. Matter* **7** 7939

Rollason, A.J., R.S. Holt, and M.J. Cooper (1983). *J. Phys. F: Metal Phys.* **13** 1807

Rollason, A.J., F. Bell, and J.R. Schneider (1989a). *Nucl. Instrum & Methods A* **281** 147

Rollason, A.J., F. Bell, J.R. Schneider, and W. Drube (1989b). *Solid State Commun.* **72** 297

Rubensson, J.-E., D. Mueller, R. Shuker, D.L. Ederer, C.H. Zhang, J. Jia, and T.A. Callcott (1990). *Phys. Rev. Lett.* **64** 1047

Rueff, J.P., Y. Joly, F. Bartolome, M. Krisch, J.L. Hodeau, L. Marques, M. Mezouar, A. Kaprolat, M. Lorenzen, and F. Sette (2002). *J. Phys., Condens. Matter* **14** 11635

Sakai, N. and K. Ono (1976). *Phys. Rev. Lett.* **37** 357

Sakai, N. and H. Sekizawa (1987). *Phys. Rev. B* **36** 2164

Sakai, N., Y. Tanaka, F. Itoh, H. Sakurai, H. Kawata, T. Iwazumi (1991). *J. Phys. Soc. Jpn.* **60** 1201

Sakurai, Y., Y. Tanaka, T. Ohata, Y. Watanabe, S. Nanao, Y. Ushigami, T. Iwazumi, H. Kawata, and N. Shiotani (1994). *J. Phys. Condens. Matter* **6** 9469

Sakurai, Y., H. Hashimoto, A. Ochiai, T. Suzuki, M. Ito, and F. Itoh (1995). *J. Phys. Condens. Matter* **7** (1995a) L599

Sakurai, Y., T. Tanaka, A. Bansil, S. Kaprzyk, A.T. Stewart, Y. Nagashima, T. Hyodo, S. Nanao, H. Kawata, N. Shiotani (1995b). *Phys. Rev. Lett.* **74** 2252

Sakurai, Y., S. Kaprzyk, A. Bansil, Y. Tanaka, G. Stutz, H. Kawata, and N. Shiotani (1999). *J. Phys. Chem. Solids* **60** 905

Schell, N., R.O. Simmons, A. Kaprolat, W. Schülke, and E. Burkel (1995). *Phys. Rev. Lett.* **74** 2535

Schmitz, J.R., H. Schulte-Schrepping, A. Berthold, S. Mourikis, and W. Schülke (1993). *Z. Nat.forsch. A* **48A** 279

Schülke, W. (1977a). *Phys. Stat. Sol. (b)* **80** K67

Schülke, W. (1977b). *Phys. Stat. Sol. (b)* **82** 229

Schülke, W. (1981). *Phys. Lett. A* **83** 451

Schülke, W. (1982). *Solid State Commun.* **43** 863

Schülke, W. (1991). 'Inelastic scattering by electronic excitations', in *Handbook on Synchrotron Radiation*, ed. G. Brown and D.E. Moncton, vol. 3, pp. 565–637 (Elsevier, Amsterdam)

Schülke, W. (2001). *J. Phys. Condens. Matter* **13** 7557

Schülke, W. and A. Kaprolat (1991). *Phys. Rev. Lett.* **67** 879

Schülke, W. and W. Lautner (1974). *Phys. Stat. Sol. (b)* **66** 211

Schülke, W. and S. Mourikis (1986). *Acta Cryst. W* **42** 86

Schülke, W. and H. Nagasawa (1984). *Nucl. Instrum. & Methods* **222** 203

Schülke, W., U. Berg, and O. Brümmer (1969). *Phys. Stat. Sol.* **35** 227

Schülke, W., O. Brümmer and U. Berg (1972). *Reinststoffe in Wissenschaft und Technik*, ed. Balarin (Akademie-Verlag, Berlin) p. 83

Schülke W., U. Bonse, and S. Mourikis (1981). *Phys. Rev. Lett.* **47** 1209

Schülke, W., H. Nagasawa, and S. Mourikis (1984). *Phys. Rev. Lett* **52** 2065

Schülke, W., H. Nagasawa, S. Mourikis, and P. Lanzki (1986). *Phys. Rev. B* **33** 6744

Schülke, W., U. Bonse, H. Nagasawa, S. Mourikis, and A. Kaprolat (1987). *Phys. Rev. Lett.* **59** 1361

Schülke, W., A. Berthold, A. Kaprolat, and H.-J. Güntherodt (1988a). *Phys. Rev. Lett.* **60** 2217

Schülke, W., U. Bonse, H. Nagasawa, A. Kaprolat, and A. Berthold (1988b). *Phys. Rev. B* **38** 2112

Schülke. W., H. Nagasawa, S. Mourikis, and A. Kaprolat (1989). *Phys. Rev. B* **40** 12215

Schülke, W., A. Berthold, H. Schulte-Schrepping, and K.-J. Gabriel (1991a). *Solid State Commun.* **79** 661

Schülke, W., K.-J. Gabriel, A. Berthold, and H. Schulte-Schrepping (1991b). *Solid State Commun.* **79** 657

Schülke, W., H. Schulte-Schrepping, and J.R. Schmitz (1993). *Phys. Rev. B* **47** 12426

Schülke, W., J.R. Schmitz, H. Schulte-Schrepping, and A. Kaprolat (1995). *Phys. Rev. B* **52** 11721

Schülke, W., K. Höppner, and A. Kaprolat (1996a). *Phys. Rev. B* **54** 17464

Schülke, W., G. Stutz, F. Wohlert, and A. Kaprolat (1996b). *Phys. Rev. B* **54** 14381

Schülke, W., C. Sternemann, A. Kaprolat, and G. Döring (2001). *Z. Phys. Chem.* **215** 1353

Schütz, G., W. Wagner, W. Wilhelm, P. Kienle, R. Zeller, R. Frahm, and G. Materlik (1987). *Phys. Rev. Lett.* **58** 737

Shin, S., A. Agui, M. Watanabe, M. Fujisawa, Y. Tezuka, and T. Ishii (1996). *Phys. Rev. B* **53** 15660

Shiotani, N., N. Sakai, M. Ito, O. Mao, F. Itoh, H. Kawata, Y. Amemiya, and M. Ando (1989). *J. Phys., Condens. Matter* **1** Suppl. A p. 27

Shiotani, N., Y. Tanaka, M. Ito, N. Sakai, Y. Sakurai, H. Sakurai, F. Itoh, T. Iwazumi, H. Kawata, and M. Ando (1992). *Mat. Sci. Forum* **105–110** 833

Shiotani, N., I. Matsumoto, H. Kawata, J. Katsuyama, M. Mizuno, H. Araki, and Y Shirai (2004). *J. Phys. Soc. Japan* **73** 1627

Shukla, A., B. Barbiellini, A. Erb, A.A. Manuel, T. Buslaps, V. Honkimäki, and P. Suortti (1999). *Phys. Rev. B* **59** 12127

Shukla, A., E.D. Isaacs, D.R. Hamann, and P.M. Platzman (2001). *Phys. Rev. B* **64** 052101

Shukla, A., J.-P. Rueff, J. Badro, G. Vanko, A. Mattila, F.M.F. de Groot, and F. Sette (2003). *Phys. Rev. B* **67** 081101

Simunek, A. and G. Wiech (1987). *Solid State Commun.* **64** 1375

Sinkovic, B. and C.S. Fadley (1985). *Phys. Rev. B* **31** 4665

Sipr, O., A. Simunek, S. Bocharov, Th. Kirchner, and G. Dräger (1999). *Phys. Rev. B* **60** 15115

Skytt, P., P. Glans, J.-H. Guo, K. Gunnelin, J. Nordgren, F.Kh. Gel'mukhanov, A. Cesar, and H. Agren (1996). *Phys. Rev. Lett.* **77** 5035

Skytt, P., P. Glans, K. Gunnelin, J.-H. Guo, and J. Nordgren (1997). *Phys. Rev. A* **55** 146K.

Soldatov, A.V., T.S. Ivanchenko, A.P. Kovtun, S. Dealla Longa, and A. Bianconi (1995). *Phys. Rev. B* **52** 11757

Southworth, S.H., D.W. Lindle, R. Meyer, and P.L. Cowan (1991). *Phys. Rev. Lett.* **67** 1098

Sparks Jr., C.J. (1974). *Phys. Rev. Lett.* **33** 262

Sternemann, C., A. Kaprolat, and W. Schülke (1998). *Phys. Rev. B* **57** 622

Sternemann, C., K. Hämäläinen, A. Kaprolat, A. Soininen, G. Döring G, C.-C. Kao, S. Manninen, and W. Schülke (2000). *Phys. Rev. B* **62** R7686

Sternemann, C., T. Buslaps, A. Shukla, P. Suortti, G. Döring, and W. Schülke (2001). *Phys. Rev. B* **63** 094301

Sternemann, C., S. Huotari, G. Vanko, M. Volmer, G. Monaco, and W. Schülke (2005). *Phys. Rev. Lett.* **95** 157401

Sturm, K. and W. Schülke (1992). *Phys. Rev. B* **46** 7193

Sturm, K., W. Schülke, and J.R. Schmitz (1992). *Phys. Rev. Lett.* **68** 228

Stutz, G., F. Wohlert, A. Kaprolat, W. Schülke, Y. Sakurai, V. Tanaka, M. Ito, H. Kawata, N. Shiotani, S. Kaprzyk, and A. Bansil (1999). *Phys. Rev. B* **60** 7099

Suortti P., T. Buslaps, V. Honkimäki, C. Metz, A. Shukla, Th. Tschnetscher, J. Kwiatkowska, F. Maniawski, A. Bansil, S. Kaprzyk, A.S. Kheifets, D.R. Lun, T. Sattler, J.R. Schneider, and F. Bell (2000). *J. Phys. Chem. Solids* **61** 397

Suortti, P., T. Buslaps, V. Honkimäki, A. Shukla, J. Kwiatkowska, F. Maniawski, S. Kaprzyk, and A. Bansil (2001). *J. Phys. Chem. Solids* **62** 2223

Suzuki, T. (1967). *J. Phys. Soc. Jpn.* **22** 1139

Suzuki, T. and H. Nagasawa (1975). *J. Phys. Soc. Jpn.* **39** 1579

Suzuki, T., T. Kishimoto, J. Kaji, and T. Suzuki (1970). *J. Phys. Soc. Jpn.* **29** 730

Tanaka, S. and A. Kotani (1993). *J. Phys. Soc. Jpn.* **62** 464

Tanaka, S., K. Okada, and A. Kotani (1991). *J. Phys. Soc. Jpn.* **60** 3893

Tanaka, Y., N. Sakai, Y. Kubo, and H. Kawata (1993). *Phys. Rev. Lett.* **70** 1573

Tanaka, Y, K.J. Chen, C. Bellin, G. Loupias, H.M. Fretwell, A. Rodrigues-Gonzales, M.A. Alam, S.B. Dugdale, A.A. Manuel, A. Shukla, T. Buslaps, P. Suortti, N. Shiotani (2000). *J. Phys. Chem. Solids* **61** 365

Tanaka, Y., Y Sakurai, A.T. Stewart, N. Shiotani, P.E. Mijnarends, S. Kaprzyk, and A. Bansil (2001). *Phys. Rev. B* **63** 045120

Thakkar, A.J., J.W. Liu, and W.J. Stevens (1986). *Phys. Rev. A* **34** 4695

Timms, D.N., A. Brahmia, M.J. Cooper, S.P. Collins, S. Hamonda, D. Laundy, C. Kilbourne, M.-C. Saint Lager (1990). *J. Phys. Condens. Matter* **2** 3427

Tohji, K. and Y. Udagawa (1988). *Phys. Rev. B* **36** 9410

Tong, B.Y. and L. Lam (1978). *Phys. Rev. A* **18** 552

Tschentscher, T., J.R. Schneider, and F. Bell (1993). *Phys. Rev. B* **48** 16965

Tsutsui, K., T. Tohyama, and S. Maekawa (1999). *Phys. Rev. Lett.* **83** 3705

Tulkki, J. and T. Åberg (1980). *J. Phys. B: Atom. Molec. Phys.* **13** 3341

van Veenendaal, M. and P. Carra (1997). *Phys. Rev. Lett.* **78** 2839

van Veenendaal, M., P. Carra, and B.T. Thole (1996). *Phys. Rev. B* **54** 16010

Vos, M. and I.E. McCarthy (1995). *Rev. Mod. Phys.* **67** 713

Vradis, A. and G.D. Priftis (1985). *Phys. Rev. B* **32** 3556
Wakoh, S. and Y. Kubo (1977). *J. Magnetism and Magnet. Mater.* **5** 202
Wakoh, S. and M. Matsumoto (1989). *J. Phys. Condens. Matter* **2** 797
Wakoh, S., Y. Kubo, and J. Yamashita (1976). *J. Phys. Soc. Jpn.* **40** 1043
Watanabe, N., H. Hayashi, Y. Udegawa, K. Takeshita, and H. Kawata (1996). *Appl. Phys. Lett.* **69** 1370
Weisser, M. and W. Weyrich (1993). *Z. Nat.forsch A* **48A** 315
Wernet, Ph., D. Nordlund, U. Bergmann, M. Cavalleri, M. Odelius, H. Ogasawara, L.A. Näslund, T.K. Hirsch. L. Ojamäe, P. Glatzel, L.G.M. Pettersson, and A. Nilsson (2004). *Science* **304** 995
Weyrich, W., P. Pattison, and B.G. Williams (1979). *Chem. Phys.* **41** 271
Yamamoto, S., H. Kawata, H. Kitamura, M. Ando, N. Sakai, and N. Shiotani (1989). *Phys. Rev. Lett.* **62** 2672
Yaouanc, A., P. Dalmas de Roetier, J.P. Sanchez, T. Tschentscher, and P. Lejay (1998). *Phys. Rev. B* **57** R681
Zhang, F.C. and T.M. Rice (1988). *Phys. Rev. B* **37** 3759
Zukowski, E., S.P. Collins, M.J. Cooper, D.N. Timms, F. Itoh, H. Sakurai, H. Kawata, Y. Tanaka, and A. Malinowski (1993). *J. Condens. Matter* **5** 4077
Zukowski, E., A. Andrejczuk, L. Dobrzynski, M.J. Cooper, M.A.G. Dixon, S. Gardelin, P.K. Lawson, T. Buslaps, S. Kaprzyk, K.-U. Neumann, and K.R.A. Ziebeck (1997). *J. Phys. Condens. Matter* **9** 10993
Zukowski, E., A. Andrejczuk, L. Dobrzynski, S. Kaprzyk, M.J. Cooper, J.A. Duffy, and D.N. Timms (2000). *J. Phys. Condens. Matter* **12** 7229

2

Nonresonant inelastic X-ray scattering; regime of characteristic valence electron excitations

2.1 Introduction

Nonresonant inelastic X-ray scattering is a photon-in photon-out process and based on that part of the photon–electron interaction which is expressed in the $\mathbf{A}^2$ term of the corresponding interaction Hamiltonian. In this chapter we shall deal with scattering processes, which are connected with a momentum transfer q characterized by $2\pi/q \approx l_c$, where l_c is the mean electron–electron distance, and with an energy transfer, or energy loss, characteristic for excitations of the scattering valence electron system. We designate this special type of scattering process the "regime of characteristic energy losses". It is the very fundamental relation between excitations of an electron system on the one hand and the time-dependent density–density correlation in such a system, as expressed in the so-called dissipation–fluctuation theorem, which provides unique access to fundamental properties of correlated many-particle systems. In this introduction we shall describe the basic properties and excitations of many-electron systems that can be deduced from inelastic X-ray scattering experiments within the characteristic energy-loss regime. We shall do this both in the "excitation picture" and in the "correlation picture", and shall have recourse to relations derived in Chapter 6 of this book. Then the requirements are defined which the experiments have to meet. Important details of synchrotron radiation based inelastic scattering experiments are presented in the following section. Further sections are devoted to key results of experiments in the characteristic energy-loss regime.

The quantity we are measuring in an inelastic scattering experiment is the double differential scattering cross-section DDSCS as defined in equation (1.4), which tells us how many photons within an energy range $d\hbar\omega_2$, related to a given incident photon flux, are scattered into a solid angle element $d\Omega_2$ around the photon wavevector $\mathbf{K}_2$, which has an angular distance θ from the incident wavevector $\mathbf{K}_1$. The DDSCS for nonresonant inelastic X-ray scattering is deduced from the $\mathbf{A}^2$ term of the Hamiltonian describing the interaction of the photon field with the electron system (see equation 6.5)

$$H_i = (e^2/2mc^2) \sum_j \mathbf{A}^2(\mathbf{r}_j), \qquad (2.1)$$

in which the summation is over the electrons of the system, and the vector potential operator writes in terms of c^+ and c, the creation and annihilation operators, respectively

$$\mathbf{A}(\mathbf{r}) = \sum_{\mathbf{K}\lambda} \left(\frac{2\pi\hbar c^2}{V\omega_\mathbf{K}}\right)^{1/2} [\mathbf{e}(\mathbf{K}\lambda)c(\mathbf{K}\lambda)\exp(i\mathbf{K}\cdot\mathbf{r} - i\omega_\mathbf{K}t)$$
$$+ \mathbf{e}^*(\mathbf{K}\lambda)c^+(\mathbf{K}\lambda)\exp(-i\mathbf{K}\cdot\mathbf{r} + i\omega_\mathbf{K}t)], \quad (2.2)$$

where λ counts the two orthogonal polarization states of the photon field corresponding to the wavevector $\mathbf{K}$, the frequency $\omega_\mathbf{K}$ and polarization unit vector $\mathbf{e}(\mathbf{K}\lambda)$. Using Fermi's golden rule to calculate, in first-order perturbation theory, the probability for a transition of the electron system from its ground state $|i\rangle$ with energy E_i into a final state $|f\rangle$ with energy E_f under the influence of the interaction Hamiltonian (2.1), one ends up with the following expression for the DDSCS in the excitation picture

$$\frac{d^2\sigma}{d\Omega_2 d\hbar\omega_2} = r_0^2 \left(\frac{\omega_2}{\omega_1}\right) |\mathbf{e}_1 \cdot \mathbf{e}_2^*|^2 \sum_{if}\sum_{jj'} g_i \langle i| \exp(-i\mathbf{q}\cdot\mathbf{r}_j)|f\rangle$$
$$\times \langle f| \exp(i\mathbf{q}\cdot\mathbf{r}_{j'})|i\rangle \delta(E_i - E_f + \hbar\omega), \quad (2.3)$$

where $\mathbf{q} = \mathbf{K}_1 - \mathbf{K}_2$. We have averaged over all initial states by introducing the probability of initial state $|i\rangle$, given in the usual manner by

$$g_i = Z^{-1} \exp(-E_i/k_B T), \quad (2.4)$$

Z being the partition function and k_B the Boltzmann factor. The second summation is over all final states allowed by energy conservation. r_0 is the so-called classical electron radius $r_0 \equiv e^2/mc^2$. In this way the DDSCS is determined, on the one hand, by the Thomson differential scattering cross-section

$$\left(\frac{d\sigma}{d\Omega_2}\right)_{\text{Th}} \equiv r_0^2 \left(\frac{\omega_2}{\omega_1}\right) |\mathbf{e}_1 \cdot \mathbf{e}_2^*|^2, \quad (2.5)$$

that describes the strength of the photon–electron coupling, and, on the other hand, by the strength of the excitation the scattering system is undergoing, expressed in terms of the so-called dynamic structure factor

$$S(\mathbf{q},\omega) \equiv \sum_{if}\sum_{jj'} g_i \langle i| \exp(-i\mathbf{q}\cdot\mathbf{r}_j)|f\rangle\langle f| \exp(i\mathbf{q}\cdot\mathbf{r}_{j'})|i\rangle \delta(E_i - E_f + \hbar\omega). \quad (2.6)$$

We will designate this representation of the dynamical structure factor to depict the excitation picture, since it is directly related to the allowed excitations of the scattering electron system.

We now introduce the electron density operator

$$\rho(\mathbf{r}) \equiv \sum_j \delta(\mathbf{r} - \mathbf{r}_j) \quad (2.7)$$

and its Fourier transform, the electron density fluctuation operators

$$\rho(\mathbf{q}) = \int d^3 r \sum_j \exp(-i\mathbf{q}\cdot\mathbf{r})\,\delta(\mathbf{r}-\mathbf{r}_j) = \sum_j \exp(-i\mathbf{q}\cdot\mathbf{r}_j). \qquad (2.8)$$

Then by transforming the operators $\rho(\mathbf{q})$ from their Schrödinger representation into the time-dependent operators in the Heisenberg representation $\rho(\mathbf{q},t)$ (van Hove 1954), and using additionally the integral representation of the energy-conserving delta function in (2.6), we can write the dynamical structure factor in terms of the so-called correlation picture:

$$S(\mathbf{q},\omega) = \frac{1}{2\pi\hbar}\int_{-\infty}^{\infty} dt\, \exp(-i\omega t)\langle \rho(\mathbf{q},0)\,\rho^+(\mathbf{q},t)\rangle,$$

$$= \frac{1}{2\pi\hbar}\int_{-\infty}^{\infty} dt\,\exp(-i\omega t)\Big\langle \sum_{j,j'} \exp[-i\mathbf{q}\cdot\mathbf{r}_j(0)]\exp[i\mathbf{q}\cdot\mathbf{r}_{j'}(t)]\Big\rangle, \qquad (2.9)$$

where the time-dependent correlation functions of the electron density fluctuation operators in (2.9) are defined as follows:

$$\langle \rho(\mathbf{q},0)\,\rho^+(\mathbf{q},t)\rangle \equiv \sum_i g_i \langle i|\rho(\mathbf{q},0)\,\rho^+(\mathbf{q},t)|i\rangle. \qquad (2.10)$$

In this way scattering is interpreted as being caused by density fluctuations of the scattering electron system, namely by the correlation of the electron density in one space/time point with the density in another point of space and time. Another way to present the correlation picture is to introduce the so-called van Hove space–time correlation function of an electron system

$$G_c(\mathbf{r},t) = \sum_{jj'} \int d^3 r' \langle \delta(\mathbf{r}'-\mathbf{r}_j(0))\,\delta(\mathbf{r}'+\mathbf{r}-\mathbf{r}_{j'}(t))\rangle,$$

$$= \int d^3 r'\, n_2(\mathbf{r}',\mathbf{r},t) \qquad (2.11)$$

where $n_2(\mathbf{r}',\mathbf{r},t)$, the two-particle density correlation function, determines, in the classical limit, the probability to find, at the time t, a particle at the vector distance $\mathbf{r}$ from the reference point $\mathbf{r}'$, if at the time $t=0$ a particle (the same or another one) is at the reference position $\mathbf{r}'$. The correlation function is then the result of averaging over all reference points. Inserting $G_c(\mathbf{r},t)$ into (2.9) we obtain

$$S(\mathbf{q},\omega) = \frac{1}{2\pi\hbar}\int_{-\infty}^{\infty} dt\,\exp(-i\omega t)\int d^3 r\,\exp(i\mathbf{q}\cdot\mathbf{r})\,G_c(\mathbf{r},t). \qquad (2.12)$$

The two possibilities to represent the dynamic structure factor, namely in the excitation and in the correlation picture, is nothing else than a specific form of the well-known fluctuation–dissipation theorem, which connects the dissipation of energy in a many-particle system, due to excitations, with fluctuations of characteristic properties of this system in its ground state (for example the particle density).

By limiting, in what follows, the momentum transfer $\mathbf{q}$ of the scattering experiment to values smaller than $2\pi/l_c$, where l_c is the mean interparticle distance, one investigates, according to (2.9), long-range correlations of the electron system, so that especially collective correlation phenomena are to be considered. Moreover, we will restrict our experimental observations to an energy transfer $\hbar\omega$, so that $\omega \approx \omega_c$, where ω_c is a characteristic frequency of the electron system, especially either the plasmon frequency ω_p, or the Fermi frequency ω_F. Then (2.9) tells us that interference between scattered waves from an electron at position $\mathbf{r}_j$ and at time 0, and waves from an electron position at $\mathbf{r}_{j'}$ and at time t dominates the observed scattering process. We will denote experiments performed within this range of $\mathbf{q}$ and ω to belong to the "regime of characteristic energy losses".

The excitation picture of inelastic scattering can still be extended, so that the dynamic structure factor may be connected with the polarization function $\chi(\mathbf{q},\omega)$ on the one hand and with the reciprocal macroscopic response function $1/\varepsilon(\mathbf{q},\omega)$ on the other. In this way, the results of inelastic X-ray scattering experiments can be related to optical properties.

When, for a homogeneous electron system, the change $\delta\langle\rho(\mathbf{q})\rangle$ of the Fourier component of the electron charge density is assumed to be linearly related to the corresponding Fourier component in space and time of an external potential, $\phi_{\text{ext}}(\mathbf{q},\omega)$, the polarization function $\chi(\mathbf{q},\omega)$ is defined as

$$\delta\langle\rho(\mathbf{q})\rangle(t) = (-e)\lim_{\varepsilon\to 0_+}\chi(\mathbf{q},\omega)\phi_{\text{ext}}(\mathbf{q},\omega)\exp[-\mathrm{i}(\omega+\mathrm{i}\varepsilon)t]. \qquad (2.13)$$

Then the dynamic structure factor can be deduced in terms of the imaginary part of the polarization function with help of a first-order time-dependent perturbation treatment:

$$S(\mathbf{q},\omega) = -\frac{1}{\pi}\left(\frac{1}{1-\exp(-\hbar\omega/k_B T)}\right)\operatorname{Im}\chi(\mathbf{q},\omega), \qquad (2.14)$$

or by inserting into (2.14) the following relation between the dielectric function $\varepsilon(\mathbf{q},\omega)$ and the polarization function

$$\frac{1}{\varepsilon(\mathbf{q},\omega)} = 1 + v(\mathbf{q})\,\chi(\mathbf{q},\omega), \qquad (2.15)$$

where

$$v(\mathbf{q}) = \frac{4\pi e^2}{q^2} \qquad (2.16)$$

is the Fourier transform of the Coulomb potential. We end up with

$$S(\mathbf{q},\omega) = \left(\frac{1}{1-\exp(-\hbar\omega/k_B T)}\right)\left(\frac{q^2}{4\pi^2 e^2}\right)\operatorname{Im}\left[\frac{-1}{\varepsilon(\mathbf{q},\omega)}\right]. \qquad (2.17)$$

Appropriate modeling of the polarization function is a challenging task for many-particle theoreticians. One fruitful approach is the self-consistent field approximation (SCF), where the so-called proper polarization function $\chi_0(\mathbf{q},\omega)$ is introduced. This function determines the density fluctuation $\delta\langle\rho(\mathbf{q},\omega)\rangle$ of the charge density's Fourier coefficient as brought about both by the external potential $\phi_{\text{ext}}(\mathbf{q},\omega)$ together with the potential $\phi_{\text{ind}}(\mathbf{q},\omega)$, self-consistently induced in the many-particle system, precisely by $\delta\langle\rho(\mathbf{q},\omega)\rangle$ via the Poisson equation, so that we can write, in analogy to (2.13)

$$\delta\langle\rho(\mathbf{q},\omega)\rangle = (-e)\chi_0(\mathbf{q},\omega)[\phi_{\text{ext}}(\mathbf{q},\omega) + \phi_{\text{ind}}(\mathbf{q},\omega)]. \tag{2.18}$$

Running the self-consistency cycle leads then to the following SCF polarization function in terms of the proper polarization function

$$\chi^{\text{SCF}}(\mathbf{q},\omega) = \frac{\chi_0(\mathbf{q},\omega)}{1 - v(\mathbf{q})\chi_0(\mathbf{q},\omega)}, \tag{2.19}$$

and hence

$$\left[\frac{1}{\varepsilon(\mathbf{q},\omega)}\right]^{\text{SCF}} = 1 + \left(\frac{v(\mathbf{q})\chi_0(\mathbf{q},\omega)}{1 - v(\mathbf{q})\chi_0(\mathbf{q},\omega)}\right). \tag{2.20}$$

It is the different approaches to the proper polarization function that characterize the approximation, which will be compared with experimental results in what follows. Taking the Lindhard polarization function as given explicitly in (6.74) to act as a proper polarization function we speak, for historical reasons, of the random phase approximation (RPA), and use RPA to characterize this approximation.

Measurements of the DDSCS (see equation 2.3) in relative units can be brought to an absolute scale by utilizing (2.17) together with the following sum rule

$$\int_0^\infty d\omega\, \omega\, \text{Im}\left[\frac{-1}{\varepsilon(\mathbf{q},\omega)}\right] = \frac{\pi}{2}\omega_p^2, \tag{2.21}$$

where $\omega_p = (4\pi\rho e^2/m)^{1/2}$ is the so-called plasmon frequency (ρ = spatially averaged electron density). Having then $S(\mathbf{q},\omega)$ and consequently also $\text{Im}[1/\varepsilon(\mathbf{q},\omega)]$ on an absolute scale, one can furthermore utilize the so-called Kramers–Kronig dispersion relation (Brauer 1972)

$$\text{Re}\left[\frac{1}{\varepsilon(\mathbf{q},\omega)}\right] - 1 = \frac{1}{\pi}\text{P}\int \frac{d\omega'\text{Im}[1/\varepsilon(\mathbf{q},\omega)]}{\omega' - \omega}, \tag{2.22}$$

in order to obtain the full complex $1/\varepsilon(\mathbf{q},\omega)$ and accordingly the full complex dielectric function $\varepsilon(\mathbf{q},\omega)$, which allows contact with optical data (P means the principal value of the integral). Obviously, the utilization of the sum rule as well as the application of the Kramers–Kronig relation presupposes measurements on a sufficiently wide frequency scale, so that the contributions of the electron system under investigation (for example the system of valence electrons) to the dynamic structure factor are fading out within the range of attainable frequencies. Furthermore the relevant frequency ranges covered by the dynamic structure factor of different systems (for example the valence electrons and the core electrons) must be separable, a demand which cannot be satisfied in all cases.

It must be stressed that both the definition of the polarization function (2.13) and the relation between the polarization function and the (macroscopic) inverse dielectric function (2.15) are valid only for a homogeneous electron system. Consequently, this holds true also for the expressions (2.14) and (2.17) of the dynamic structure factor. Considering an electron system under the influence of a periodic potential (periodic solids), whose electron density can be expanded into the following Fourier series ($\mathbf{g}$ is a reciprocal lattice vector)

$$\rho(\mathbf{r}) = \sum_{\mathbf{g}} \rho(\mathbf{g}) \exp(i\mathbf{r} \cdot \mathbf{g}), \tag{2.23}$$

then a Fourier component of the external potential $\phi_{\text{ext}}(\mathbf{q}_r + \mathbf{g}, \omega)$ can also induce density fluctuations belonging to another reciprocal lattice vector $\mathbf{g}'$ ($\mathbf{q}_r$ is restricted to lie in the first Brillouin zone), so that one has to replace (2.13), within the limits of linear response theory, by

$$\delta \langle \rho(\mathbf{q}_r + \mathbf{g}') \rangle(t) = (-e) \lim_{\varepsilon \to 0_+} \sum_{\mathbf{g}} \chi(\mathbf{q}_r + \mathbf{g}', \mathbf{q}_r + \mathbf{g}, \omega)$$
$$\times \phi_{\text{ext}}(\mathbf{q}_r + \mathbf{g}, \omega) \exp[-(i\omega + \varepsilon)t]. \tag{2.24}$$

This means that the polarization function transforms into a $(\mathbf{g}, \mathbf{g}')$ polarization matrix $X_{\mathbf{gg}'} \equiv \chi(\mathbf{q}_r + \mathbf{g}', \mathbf{q}_r + \mathbf{g}, \omega)$. In analogy to (2.18) we can also define the proper polarization matrix $X^0_{\mathbf{gg}'} \equiv \chi_0(\mathbf{q}_r + \mathbf{g}', \mathbf{q}_r + \mathbf{g}, \omega)$ by

$$\delta \langle \rho(\mathbf{q}_r + \mathbf{g}', \omega) \rangle = (-e) \sum_{\mathbf{g}} \chi_0(\mathbf{q}_r + \mathbf{g}', \mathbf{q}_r + \mathbf{g}, \omega)$$
$$\times (\phi_{\text{ext}}(\mathbf{q}_r + \mathbf{g}, \omega) + \phi_{\text{ind}}(\mathbf{q}_r + \mathbf{g}, \omega)). \tag{2.25}$$

Additionally, we define the so-called dielectric matrix $1 + \mathbf{T}$ with its elements

$$\varepsilon_{\mathbf{gg}'} \equiv \delta_{\mathbf{gg}'} - v(\mathbf{q}_r + \mathbf{g}) X^0_{\mathbf{gg}'}. \tag{2.26}$$

Making use of the fact that according to Wiser (1963) the $\mathbf{q}$th Fourier component of any macroscopic quantity is the $\mathbf{q}, \mathbf{g} = \mathbf{0}$ component of the corresponding microscopic quantity, one obtains a relation between the macrosopic reciprocal dielectric function and the proper polarization matrix $X^0_{\mathbf{gg}'}$

$$\mathbf{q} = \mathbf{q}_r + \mathbf{g}_0 \tag{2.27}$$

$$\left[\frac{1}{\varepsilon(\mathbf{q}, \omega)}\right]^{\text{SCF}}_{\text{M}} = [1 + \mathbf{V}\mathbf{X}^0(1 + \mathbf{T})^{-1}]_{\mathbf{g}_0\mathbf{g}_0}, \tag{2.28}$$

where $\mathbf{V}$ is a diagonal matrix built by the Fourier components of the Coulomb potential.

2.2 Instrumentation with synchrotron radiation

We have seen in Section 2.1 how information about the dynamical structure and related quantities characterizing the correlation and dissipation properties of electron systems can be obtained by measuring the double differential scattering cross-section (DDSCS). According to the definition of the DDSCS as given in Section (1.2), the task imposed for this measurement is well outlined: One has to form a collimated monchromatic X-ray beam, whose flux and polarization state should be known and which hits the scattering sample in a certain direction. Out of the scattered radiation a small solid angle has to be selected, so that both a well-defined scattering angle is defined and the direction of the scattered beam is fixed with respect to the scattering sample. The scattered beam must be energy-analyzed to a distinct resolution and its intensity is measured. If the flux of the incoming beam is difficult to estimate, one needs other means (for instance the utilization of sum rules), to obtain absolute values of the DDSCS. Optionally also the polarization state of the scattered beam can be analyzed.

Therefore, the components of instrumentation for an inelastic scattering experiment in the range of characteristic energy losses utilizing synchrotron radiation must meet the following requirements: (i) The momentum transfer **q** should be fixed with respect to the orientation of the scattering sample; (ii) the energy transfer $\hbar\omega$ has to be measurable to an accuracy which is set by the physical information about the scattering system one is seeking, typically between $50\,\text{meV}$ and $1\,\text{eV}$; (iii) the spectral distribution of the scattered radiation should be measured with a minimum statistical accuracy and with the greatest possible signal-to-noise ratio in the shortest possible time; (iv) the polarization state of the incident photon beam should be known and controlled; (v) the physical environment of the sample (temperature, pressure, atmosphere, fields) must be made variable for special purposes.

Thus, the principal experimental setup of an inelastic scattering experiment with synchrotron radiation looks as illustrated in Fig. 2.1. A source of synchrotron radiation is followed by a monochromator, which provides a well-collimated monochromatic incident beam at the scattering sample and can act simultaneously as a polarizer. Upstream of the monochromator a pre-monochromator can be installed, in order to catch the heat-load of the high-flux synchrotron beam. Optionally a special device for producing a distinct polarization state (e.g. circularly polarized radiation) can follow the monochromator. In this case a polarimeter should make this polarization state observable. The intensity of the incident photon beam should be measured by means of a monitor. The sample, adjustable with respect to both the incident and the scattered beam, is housed in a scattering chamber which makes possible the selection of a special physical environment. An energy analyzer connected to a detectors accepts a limited solid angle of radiation scattered by an angle θ, and enables its energy analysis to be carried out, where the spectral intensity must be related to the monitor signal. Additionally, it will be advantageous to use focusing elements in the incident photon beam, often connected with the monochromator.

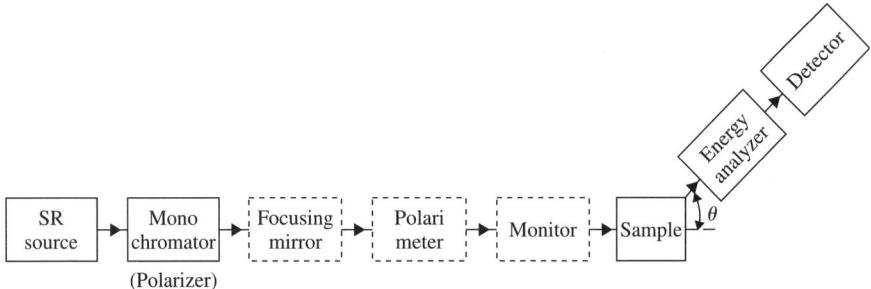

Fig. 2.1. Principle experimental setup of an inelastic X-ray scattering experiment. Boxes framed with a dashed line are optional.

In the following, we discuss the above-mentioned components separately by defining their critical parameters, by giving expressions for these parameters in specific cases and by estimating their actual values in some typical examples. Note that most of the relations derived in Sections 2.2.1–2.2.7 to characterize components of instrumentation are only approximate, but offer the possibility of an "order of magnitude" estimate which can be performed simply when a rough evaluation of an experimental setup is desired.

2.2.1 *Synchrotron radiation sources for inelastic X-ray scattering*

The crucial parameter for estimating a synchrotron radiation source is the spectral flux Φ, defined to be the number of photons per unit time interval emitted into a relative bandwidth $d\lambda/\lambda$, into an angle $d\theta_h$ in the plane of the electron orbit (horizontal plane) and integrated in the vertical plane, measured in units "number of photons per 0.1% bandwidth, per second, per mrad (horizontal)". Another parameter is the spectral brightness I, defined to be the number of photons per unit time interval emitted into a relative bandwidth $d\lambda/\lambda$ and into a solid-angle element $d\Omega$, and measured in units "number of photons per 0.1% bandwidth, per second, per mrad (horizontal), per mrad (vertical)". Finally, the spectral brilliance B may be defined to be the number of photons per unit time interval emitted from an element dS of the source area, emitted into a relative bandwidth $d\lambda/\lambda$ and into a solid-angle element $d\Omega$, and measured in units "number of photons per 0.1% bandwidth, per second, per mm^2, per mrad^2". The above nomenclature follows Buras and Tazari (1984), and is now commonly accepted in this field.

Of course, synchrotron radiation sources are in every case superior to conventional X-ray sources. But nevertheless, one must seek sources with the highest possible flux at the sample or, when utilizing focusing elements in the incident beam, with the highest possible brightness or brilliance.

Multipole wigglers for experiments with higher primary photon energy and undulators for the range of lower primary energy are indispensable for further experiments. If one has the choice between wiggler and undulator, the latter is preferred since one has fever problems with the heat load on the first optical element of the beamline. Additionally, in most cases one does not need focusing elements in the incident beam because of its high collimation.

Another critical parameter of a bending magnet source and a wiggler is its critical energy E_c. As a rule of thumb, one can get useful flux to at least $4E_c$ (1/10 of the peak flux). For an undulator the energy of the first and third harmonic is most important, which depends both on the electron (positron) energy in the ring and on the undulator's magnetic gap.

2.2.2 Monochromators for inelastic X-ray scattering

The appropriate choice of a monochromator structure is determined by two factors: (i) the desired overall energy resolution of the experiment $(\Delta E/E)_{\text{tot}}$ and (ii) by the analyzer's lay-out. In principle, four types of monochromators are used for inelastic X-ray scattering experiments. We will discuss them in what follows with respect to their energy resolution $(\Delta E/E)_M$, which should be roughly $(1/\sqrt{2})(\Delta E/E)_{\text{tot}}$, and their transmission T_M, defined as the fraction of photons impinging on the monochromators within the accepted energy range $(\Delta E)_M$, which is Bragg reflected.

2.2.2.1 Double-crystal monochromator, monochromator cooling
The first type, very common in use, is the double-crystal monochromator in the so-called non-dispersive or parallel setting, where the Bragg-scattering lattice planes of both crystals are identical, parallel to each other, and can be rotated around an axis parallel to the electron orbital plane. As far as these crystals can be estimated to be nearly perfect with a diffraction plane perpendicular to the electron orbit (predominant σ-polarization; the diffraction plane is defined as the plane containing the incident and the Bragg-reflected beam), the monochromator energy resolution is the given by (Burkel, Peisl and Dorner 1987)

$$\left(\frac{\Delta E}{E}\right)_M = \cot\theta_M \, \Delta\theta_M + \frac{|\chi_h|}{(\sqrt{|b|})\sin^2\theta_M}, \tag{2.29}$$

where θ_M is the mean Bragg angle, and $\Delta\theta_M$ is the range of Bragg angles which can be placed at disposal both by the angular divergence of the incident photon beam and by the geometrical shape of the Bragg-reflecting crystals. χ_h is the $\mathbf{g}_h$th Fourier component of the crystal dielectric susceptibility, where $\mathbf{g}_h$ is the reciprocal lattice vector corresponding to the Bragg reflection. The quantity b allows for an asymmetric reflection as sketched in Fig. 2.2. The asymmetry factor b is defined as

$$b \equiv \frac{-\sin(\theta_B - \alpha)}{\sin(\theta_B + \alpha)}. \tag{2.30}$$

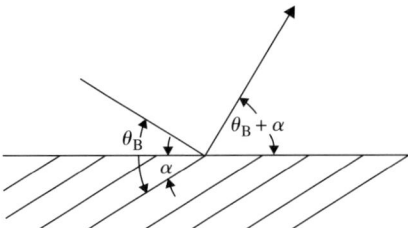

Fig. 2.2. Definition of the asymmetry angle α in the Bragg case of reflection. θ_B = Bragg angle.

The transmission T_M of this double-crystal setting is roughly given by

$$T_M \approx \frac{R_{iM}}{[2.35(\sigma_z'^2 + \sigma_R'^2)^{1/2}]}, \quad (2.31)$$

where R_{iM} is the integrated reflectivity of the monochromator in angular units, given for a thick crystal plate in the nonabsorbing limit by (Zachariasen 1945)

$$R_{iM} = \frac{\pi}{2}\Delta\theta_{in,D}; \quad \Delta\theta_{in,D} = \frac{2|\chi_h|}{(\sqrt{|b|})\sin 2\theta_B}. \quad (2.32)$$

$\Delta\theta_{in,D}$ is the so-called Darwin width, the angular range of the incident beam (index in), where the reflectivity of the nonabsorbing crystal has the value 1 (interference total reflection). σ_z' in (2.31) measures, in the same angular units as used for R_{iM}, the source divergence (one standard deviation), due to the divergence of the electron beam in the orbit at the position of the radiation source. σ_R' is the natural divergence of the synchrotron radiation in the z-direction. (In the following, σ denotes the source dimension and σ' the source divergence with index x in the horizontal (electron orbit) and index z in the vertical plane.) Obviously, if R_{iM} is equal or larger than the denominator in (2.31), then $T_M \approx 1$. The vertical size of the monochromatically illuminated spot on the sample depends both on the vertical diameter σ_z of the electron orbit and on $2.35L(\sigma_z'^2 + \sigma_R'^2)^{1/2}$, where L is the source–sample distance. How the vertical energy distribution in this spot can be utilized to force down the total energy resolution $(\Delta E/E)_{tot}$ far beyond $(\Delta E/E)_M$ by means of the so-called dispersion compensation (Schülke 1986) will be treated in Section 2.2.4. Obviously the maximum throughput of the double-crystal arrangement is reached with the parallel setting of both crystals. Nevertheless, as first shown by Bonse et al. (1976), a slight detuning from the parallel position leads to the highly desirable suppression of higher harmonics in the incident white synchrotron beam, due to the much smaller Darwin width of the higher harmonics (λ^2 dependence of χ_h) compared with the zero component. Certainly this occurs at the expense of the transmitted intensity. It should be mentioned that a double-crystal setting can also be realized by channel-cutting of a single crystal plate (Bonse and Hart 1966). The detuning for suppression of higher harmonics must then be performed by means of a weak link between the interconnected crystal plates.

In a special double-crystal arrangement, the second crystal is a crystal plate cylindrically bent around an axis, which is lying in the scattering plane so that the X-ray beam, diverging in the x-direction, can be sagittally focused to the sample position. Of course cylindrical bending of a crystal plate is always connected with the so-called anticlastic bending around an axis perpendicular to the scattering plane. This anticlastic bending can be minimized by means of stiffening ribs on the backside of a cylindrically bent rectangular plate (Sparks et al. 1982) or by cutting slots into a triangle shaped crystal in order to provide a polygonal approximation to the cylindrical surface, which can be produced by clamping the crystal at the base and by applying a force at the opposite apex as shown by Batterman and Berman (1983).

If the first crystal of a double-crystal setting is the first optical element of a beamline, it has to accept a substantial heat load, a total power of several kW in the case of wiggler sources, and power densities approaching $1\,\text{kW/mm}^2$, when the full cone of a 5 m undulator hits the first optical element. This high heat load must be carried off by appropriate cooling, where the thermal conductivity k of the crystal material (in most cases silicon or diamond), the size and form of the interface between crystal and the coolant, the thickness of the crystal, and the heat transfer properties of the coolant are the critical quantities. Moreover, since it is the goal of all cooling devices to minimize deformations of the first crystal, the coefficient of thermal expansion α of the crystal material and its temperature dependence play a major role in designing cooled optical components. Thus the quotient k/α is a decisive figure of merit. Due to the fact that the thermal expansion coefficient of silicon exhibits a temperature dependence with a zero crossing at 125 K, and also the thermal conductivity k of silicon increases with decreasing temperature, this figure of merit is for silicon more than 50 times larger for 100 K compared with 300 K, and reaches the 300 K value of diamond (the thermophysical data were taken from Blanke (1989)). Thus cryogenic cooling of silicon monochromator crystals with liquid nitrogen as coolant, first proposed by Bilderback (1986), and first realized by Joksch et al. (1991), is today a well established and mostly used cooling method in connection with high-power- density insertion devices, even though the heat transfer properties of liquid nitrogen are worse than those of water. Of course, the figure of merit k/α of diamond at 100 K is nearly 100 times larger than that of silicon at the same temperature. Therefore, whenever the high collimation of undulator radiation allows the application of the rather small sized diamond crystals, these crystals are the preferential choice.

Even if modern cooling methods are applied, a certain thermally induced deformation of the first crystal may remain. Three different deformations can be distinguished: (i) a bending component, which leads to a uniform bending of the lattice planes; (ii) a bump component which is characterized by thermally induced variation of the lattice parameters in the vicinity of the region illuminated by the incident beam; and (iii) a gradient component with a thermally induced gradient of the lattice parameters perpendicular to the crystal surface.

All three components can be expressed in terms of $\Delta\theta_B$, the deviations from the Bragg angle. They turn out to be proportional to α/k and to the X-ray beam power, and should not exceed the Darwin width $\Delta\theta_D$. Much effort has been spent to compensate the thermally induced bending by appropriate mechanical devices (adaptive optics), especially in those cases where no cryogenic cooling of silicon-based monochromators was available. The design of those benders and of appropriately shaped crystals was supported by finite element analysis. A most recent example is presented by Zaeper et al. (2002), where one can also find references to previous works. The gradient component can be compensated by means of two cooling channel systems with different temperatures and flow of the coolant, as demonstrated by Smither (1989).

These considerations about cooling and corrections of thermally induced deformations are of special importance, since silicon double crystal settings are very often used as premonochromators, especially in connection both with the four-bounce monochromator, described in what follows, and with the single plane crystal monochromator for ultrahigh energy resolution.

2.2.2.2 *Four-bounce monochromator* A second monochromator lay-out, first proposed by Ishikawa et al. (1992), and used for higher demands upon energy resolution, is the so-called four-bounce arrangement of four plane crystals which can be considered to be composed of a double-crystal nondispersive setting into which another double-crystal nondispersive arrangement is nested as shown in Fig. 2.3. The outer double-crystal setting is characterized by an asymmetrical cut of the crystal surface with respect to the reflecting lattice planes, so that one has realized a grazing incident beam ($|b| \ll 1$), which can be arranged so that with $\Delta\theta_{in,D}$ according to (2.32) a transmission of nearly 1 is guaranteed, provided the nested double-crystal setting can accept the divergence of the outgoing beam leaving the first crystal of the outer setting. This beam divergence is determined by $\Delta\theta_{out,D}$ which is, according to the Liouville theorem (asymmetric cut), given by

$$\Delta\theta_{out,D} = \frac{2|\chi_h|(\sqrt{|b|})}{\sin 2\theta_B}, \tag{2.33}$$

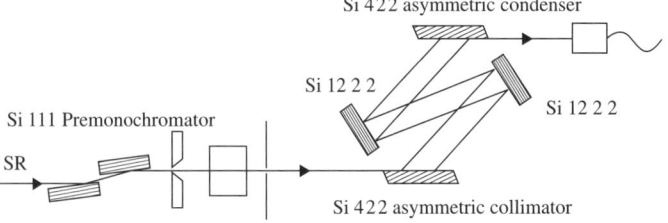

Fig. 2.3. Example of a four-bounce monochromator: A Si (12 2 2) double crystal setting is nested into an asymmetrically cut Si (4 2 2) double crystal arrangement. A Si (1 1 1) premonochromator catches the heat load.

and thus much smaller than $\Delta\theta_{\text{in,D}}$. Therefore, the nested double-crystal setting should satisfy two demands: (1) $\Delta\theta_{\text{out,D}}^1 \approx \Delta\theta_{\text{in,D}}^2$, where the superscripts 1 and 2 refer to the first and second crystal of the four-bounce arrangement; (2) the energy resolution of the symmetrically cut second double-crystal setting, which can be adjusted following (2.29), must be smaller than that of the outer double-crystal setting, so that it determines the desired energy resolution of the whole four-bounce monochromator. By a proper choice of the asymmetry cut of the outer two-crystal setting together with the Bragg angles of the two double-crystal settings the two demands can be satisfied simultaneously, where a Bragg angle near 90° of the second double-crystal setting makes $\Delta E/E \approx 10^{-6}$–$10^{-7}$ feasible. In this way, given a certain vertical divergence of the incident synchrotron X-ray beam, both a transmission of nearly 1 and a high energy resolving power are decoupled unlike the case of a simple double-crystal setting. As far as the size of the illuminated spot on the sample is concerned, the same is valid as in the simple two-crystal case. What both of these monochromator settings distinguish from the following two is the fact that the monochromatized beam proceeds in the same direction as the incident one, so that the sample-analyzing system need not be moved in the horizontal direction when the incident photon energy is changed. By making the distance of the two crystals of the respective double-crystal settings variable with the desired energy by means of an appropriate mechanical drive, also a vertical shift of the illuminated spot on the sample can be avoided (so-called fixed-exit monochromator).

2.2.2.3 Cylindrically bent crystal monochromator

A third type of monochromator equipment used for inelastic X-ray scattering consists of a single cylindrically bent crystal plate with a bending axis perpendicular to the orbital plane. As shown in Fig. 2.4, radiation diverging from the source S on the Rowland circle with radius $R/2$ is focused in the meridional plane to F on the Rowland circle by means of a crystal cylindrically bent to a radius R. For the general case of an asymmetrically cut crystal plate (see Fig. 2.2) one gets the following relation between R, θ_B, α, p and q, where p is the source–crystal distance and q the crystal–focus distance:

$$\frac{2}{R} = \left(\frac{1}{p}\right)\sin(\theta_B + \alpha) + \left(\frac{1}{q}\right)\sin(\theta_B - \alpha). \tag{2.34}$$

Thus the so-called Gunier condition

$$\frac{\sin(\theta_B + \alpha)}{\sin(\theta_B - \alpha)} = \frac{p}{q} \tag{2.35}$$

is valid, and the size of the source S in the meridional plane is magnified/demagnified by a factor q/p into the size of the focus F. Note that exact focusing would occur only if the crystal surface fits the Rowland circle in the meridional plane (so-called Johansson geometry), and if Bragg reflection takes place only at the crystal surface. Therefore, the Johann geometry shown in Fig. 2.4 together

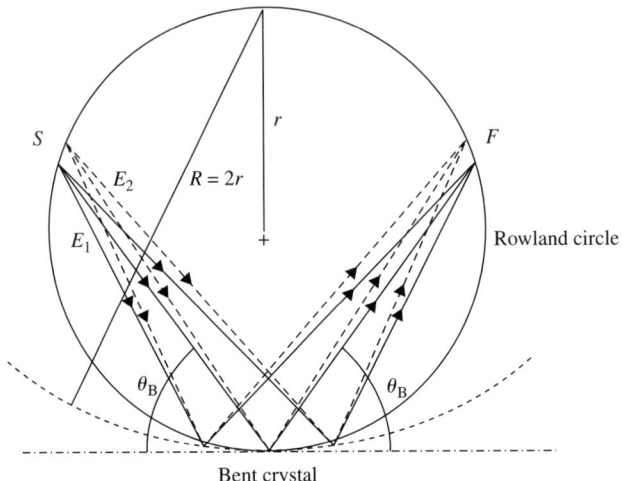

Fig. 2.4. Crystal monochromator (analyzer) cylindrically bent to a radius $R = 2r$ in Rowland geometry (radius of the Rowland circle is r). θ_B is the Bragg angle for photon energy $E_1 > E_2$ (symmetrical Bragg case).

with the finite penetration depth of the incident radiation, and the finite source length perpendicular to the scattering plane, lead to optical aberrations which destroy the point-by-point imaging of the source, and this has been thoroughly discussed by Wohlert (2000). According to (2.29), the energy resolution of the bent crystal monochromator is, on the one hand, determined by the angle σ_x/p under which one point on the bent crystal surface sees the source dimension σ_x in the x-direction, and, on the other hand, by $\Delta\theta_B$, the range of Bragg angles the incident beam finds, when penetrating the bent crystal because the lattice planes change their orientation and their spacing with increasing penetration depth. $\Delta\theta_B$ is given by (Wohlert 2000)

$$\Delta\theta_B = \left(\frac{t}{R}\right)[-\cot(\theta_B + \alpha) + (\nu\cos^2\alpha - \sin^2\alpha)$$
$$\times \tan\theta_B - (1+\nu)\sin\alpha\cos\alpha], \tag{2.36}$$

where ν is the Poisson number, and t the effective thickness of the crystal, determined by photoabsorption and/or extinction along the beam path in the crystal, so that

$$t = \left(\frac{1}{\tau}\right)\left\{\left[\frac{1}{\sin(\theta_B - \alpha)}\right] + \left[\frac{1}{\sin(\theta_B + \alpha)}\right]\right\}. \tag{2.37}$$

τ is the general beam-weakening coefficient assuming, extremely simplifying, a simple exponential attenuating law for the combined action of photoabsorption and/or extinction. Note that, for certain combinations of θ_B, α and ν, the influence of lattice plane rotation and lattice distortion can compensate to $\Delta\theta_B = 0$.

When we additionally assume that a bent crystal can, according to Berthold et al. (1992), be modeled by independent Bragg-reflecting layers, whose thickness t is defined so that the incident beam, penetrating the crystal, finds two successive layers rotated, due to bending, by the Darwin width $\Delta\theta_D$ (2.32), and if t is larger than t_{exB}, the so-called extinction length of the Bragg case given by (v. Laue 1960)

$$t_{exB} = \frac{\lambda \sin\theta_B}{2\pi|\chi_h|}, \qquad (2.38)$$

then the second term on the right-hand side of (2.29) applies also to the case of a curved crystal, so that we end up with

$$\left(\frac{\Delta E}{E}\right)_M = \cot\theta_M \left[\left(\frac{\sigma_x}{p}\right)^2 + (\Delta\theta_B)^2\right]^{1/2} + \frac{|\chi_h|}{(\sqrt{|b|})\sin^2\theta_M}, \qquad (2.39)$$

for the energy resolution of a bent-crystal monochromator. But it must be noted that a bent-crystal monochromator contributes to the total energy resolution of an inelastic X-ray scattering experiment not only by $(\Delta E/E)_M$ of (2.39). On the contrary it is the optical aberrations, due to the Johann geometry of the curved crystal, and due to its finite height in the z-direction, which widens the monochromatically illuminated focal spot, produced by the Bragg-reflected beam at the sample position on the Rowland circle. Thus it depends on the method of energy analysis following the inelastic scattering process, how the finite width in the x-direction of the monochromatically illuminated range on the sample will influence the total energy resolution of the experiment. We will come back to this point in Section 2.2.5.1.

Following Berthold et al. (1992), we can estimate the transmission T_M of a curved-crystal monochromator on the basis of the same layer model described above, ending up with

$$T_M \approx \frac{R_{iM} p}{2.35 \sigma_x}, \qquad (2.40)$$

provided that $\sigma_x/p < (\sigma_x'^2 + \sigma_R'^2)^{1/2}$. The integral reflectivity is, within the limits of the layer model, given by

$$R_{iM} = \left[\frac{\pi \cot(\theta_B + \alpha)}{2\mu R}\right] A^{-1}[1 - \exp(-D\mu A)];$$

$$A = \left[\frac{1}{\sin(\theta_B - \alpha)}\right] + \left[\frac{1}{\sin(\theta_B + \alpha)}\right], \qquad (2.41)$$

if t_{exB} is much smaller than t, and D is the crystal plate thickness. If, on the other hand, t is smaller than the extinction length t_{exB} of the Bragg case, then each layer can be considered to reflect only the fraction $(1 - \exp(-t/t_{exB}))$ of the incident flux, so that R_{iM} reads

$$R_{iM} = \left[\frac{\pi \cot(\theta_B + \alpha)}{2\mu R}\right] A^{-1}[1 - \exp(-D\mu A)](1 - \exp(-t/t_{exB})). \quad (2.42)$$

It is interesting to note that it is the brightness of the source which is critical for the photon flux of the monochromatized beam, if a flat crystal monochromator is used. But is is the brilliance which is decisive for the curved crystal monochromator.

2.2.2.4 Plane crystal monochromator

The simplest monochromator, but nevertheless brought into action as an ultrahigh energy resolution device ($\Delta E/E \approx 10^{-7}$), is a plane crystal plate run at a Bragg angle very near to 90°. Its energy resolution is, according to (2.29), approximately given by

$$\left(\frac{\Delta E}{E}\right)_M \approx (\Delta\theta_M)^2 + |\chi_h|, \quad (2.43)$$

which indicates that one has to make both terms on the right-hand side of (2.43) of order 10^{-7}, the first one by approaching $\theta_M \approx \pi/2 - \Delta\theta_M$, and by reducing $\Delta\theta_M$ using the high collimation of an undulator and/or sufficiently small dimensions of the crystal positioned far away from the source, the second one by choosing an appropriate short wavelength, since $|\chi_h|$ goes like λ^2. These high-resolving plane monochromators must be used in connection with a double-crystal premonochromator in order to catch the heat load. Otherwise the thermal expansion would produce very severe shifts of the monochromator passband. Moreover, these devices are in most cases used in the so-called inverse geometry (see Section 2.2.5), where the energy of the monochromator is detuned with respect to the analyzer by utilizing the linear expansion with temperature, so that a temperature stability of better than 10^{-2} degrees is mandatory.

The transmission of this type of monochromator can be approximated by (2.31), where the integrated reflectivity R_{iM} is given by (symmetrically reflecting crystal in (2.32) and $\sin(2\theta_B) \approx \Delta\theta_M$)

$$R_{iM} = \left(\frac{\pi}{2}\right) \Delta\theta_{in,D}; \quad \Delta\theta_{in,D} = \frac{2|\chi_h|}{\Delta\theta_M}. \quad (2.44)$$

2.2.3 Mirrors in inelastic X-ray scattering beamlines

Grazing-incidence X-ray mirrors in inelastic beamlines have to meet one or, simultaneously, more of the following four demands:

(i) Catching heat load, when followed by a heat-sensitive optical element.
(ii) Function as a low-energy passband in order to reduce the higher harmonics. This is achieved by utilizing the fact that the critical angle for total reflection goes like $1/E$. These two tasks can already be performed

by plane mirrors, whose lengths have to be large enough to accept most of the vertical divergence of the incident beam.

(iii) Reduction of the vertical divergence $2.35(\sigma_z'^2 + \sigma_R'^2)^{1/2}$ of the incident synchrotron radiation beam (see equation 2.31) impinging on the first plane crystal element of a monochromator in order to increase the angular acceptance of the latter. Approximately parabolic bending of mirrors gives rise to these defocusing properties.

(iv) Finally, it is the possibility to focus beams both vertically and sagittally by means of toroidally shaped mirrors, which constitutes the broadest application in the beamline technique. These mirrors are positioned between the monochromator and the scattering sample in order to obtain a high irradiance (defined as the number of photons per unit time interval and per unit irradiated area on the scattering sample.) Vertical focusing is attained in most cases by bending, where the bending radius can reach, due to the rather small angles of the incident beam with the mirror surface (external total reflection), several hundred meters. The demands upon the surface smoothness are rather high and thoroughly discussed by Batterman and Berman (1991). Mirrors with a roughness of only a few Å are now commonly in use. A general description of mirror properties, materials and different design is given by Matsushita and Hashizume (1983). Vertical foussing is achieved by using elliptical mirrors, as shown in Fig. 2.5, where the radius of curvature of the ellipse is given by

$$R_2 = \frac{2F_1 F_2}{(F_1 + F_2)\sin i}, \qquad (2.45)$$

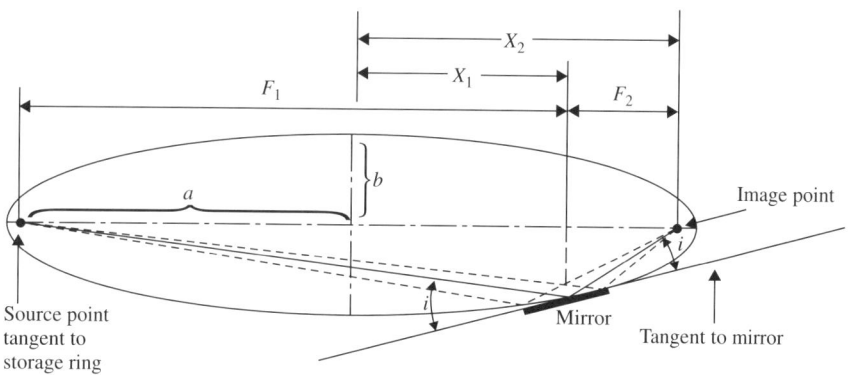

Fig. 2.5. Elliptical mirror. Radiation from the source point at a distance F_1 to the mirror is focussed vertically to the image point at a distance F_2 to the mirror. (Originally published by Russel (1991); copyright (1991) by Elsevier Science Publisher B.V.)

where F_1 and F_2 are the distance between the source point and mirror and between the image point and the mirror, respectively. i is the angle between the grazingly incident beam and the mirror surface and must be smaller than the critical angle i_c of the mirror surface, which, according to Batterman and Berman (1991), can be approximated as

$$i_c E(\text{keV}) = 33 \text{ keV mrad, for low-Z surfaces} \tag{2.46}$$

such as from silicon carbide, silicon, aluminum and float glass, and

$$i_c E(\text{keV}) = 77 \text{ keV mrad, for high-Z surfaces} \tag{2.47}$$

such as gold and platinum coatings. Thus the critical angle is of the order of a few mrad for X-rays in the energy range of 5–30 keV, so that the radius of curvature for vertical focusing R_2, as already said, is up to several hundred meters. On the contrary, the radius of curvature for sagittal focussing

$$R_1 = R_2 \sin^2 i$$

amounts to only a few cm. Therefore, a surface cylindrically polished to a radius R_1 with a cylinder axis parallel to the incident beam, and bent mechanically to a radius R_2 around an axis perpendicular to the beam, is a fair approximation to the ideal toroidal shape. Nevertheless, for photon energies larger than 20 keV, R_1 becomes too small in order to accept a sufficiently large horizontal divergence, so that a sagittally focussing crystal is, for many applications, far superior.

2.2.4 Sample, geometrical considerations

For a quantitative estimation we will consider in Sections 2.2.4 and 2.2.5 a scattering geometry, where the scattering plane, spanned by $\mathbf{K}_1$ and $\mathbf{K}_2$, is always horizontal (i.e. the electron orbit plane). Of course, because of the predominant polarization of the synchrotron radiation in that plane, the useful range of scattering is thus near forward or near backward scattering. In the former case, a symmetrical transmission scattering geometry as sketched in Fig. 2.6 is the most suitable one. If one then defines the scattering power s of the sample as the fraction of incident photons scattered into unit solid angle, s is given in the nonrelativistic case for the transmission scattering geometry as follows,

$$s = \left(\frac{d\sigma}{d\Omega}\right)_{\text{Th}} \rho d \exp\left[-\frac{\mu d}{\cos(\theta/2)}\right], \tag{2.48}$$

where θ is the scattering angle, ρ the electron density of the sample, d the sample thickness and μ the linear absorption coefficient of the sample (neglecting the softening of the scattered radiation). Thus the optimum sample thickness, with respect to the scattering power, is

$$d_{\text{opt}} = \frac{1}{\mu}. \tag{2.49}$$

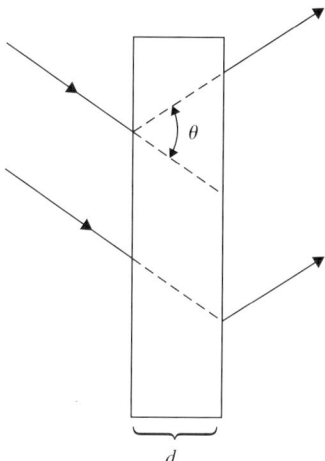

Fig. 2.6. Symmetrical transmission scattering geometry. θ is the scattering angle.

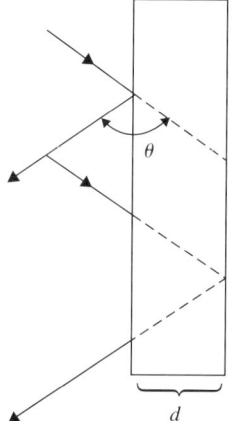

Fig. 2.7. Symmetrical back-scattering geometry. θ is the scattering angle.

For the symmetrical back-scattering geometry, as sketched in Fig. 2.7, the scattering power is

$$s = \left(\frac{d\sigma}{d\Omega_2}\right)_{\mathrm{Th}} \rho \, \sin(\theta/2) \left\{1 - \exp\left[\frac{-\mu d}{\sin(\theta/2)}\right]\right\} \bigg/ 2\mu. \qquad (2.50)$$

For scattering geometries with the scattering plane in a certain angle to the orbit plane, the reader can easily derive corresponding expressions for the scattering power.

2.2.5 Energy analysis of the scattered spectrum

Generally speaking, the energy analyzer in an inelastic X-ray scattering experiment has to accept radiation from an illuminated area on the sample with dimensions a in the z-direction and b in the x-direction, scattered into a solid angle $\Delta\Omega_0 = \Delta\psi_{0h}\,\Delta\psi_{0v}$, and has to sort it according to its energy with an energy resolution $(\Delta E/E)_A$ which again should be roughly $(1/\sqrt{2})(\Delta E/E)_{\text{tot}}$. In this connection $\Delta\psi_{0h}$ and $\Delta\psi_{0v}$ are the allowable horizontal and vertical divergences, respectively, where $\Delta\psi_{0h}$ is limited, in the scattering geometry used in Section 2.2.4, by the required momentum resolution $\Delta q/q$, whereas $\Delta\psi_{0v}$ is mainly determined by the extent to which the **q**-direction has to be fixed.

The efficiency of the energy analysis depends strongly on whether it is performed channel by channel, where each channel has an energy width $(\Delta E)_d = (\Delta E)_A$, or in many channels simultaneously, where $(\Delta E)_d$ may be the total energy width which is processed at the same time. Therefore, it is reasonable to define the transmission of the analyzer, T_A, to be that fraction of the scattered photons which, after being accepted geometrically by the most limiting aperture of the analyzer system, is transmitted by the analyzer into the energy window $(\Delta E)_d$, and accepted by the detector. In this way T_A measures the efficiency of the analyzer.

In what follows we will consider the energy resolution and transmission of five different set-ups used in the characteristic energy loss regime to analyze the inelastically scattered X-ray photons.

2.2.5.1 Focusing crystal analyzer in Rowland geometry

This analyzer consists of a cylindrically bent Johann-type crystal, which acts in full Rowland geometry, as already shown in Fig. 2.4. This means that the scattering sample, illuminated at the focal spot of a curved crystal monochromator (see Section 2.2.2.3), sits on the Rowland circle which is connected with the curved analyzer crystal. The scattered radiation is Bragg-reflected by the analyzer crystal and focused on the detector slit which is also positioned on the Rowland circle. This Rowland geometry must be kept when scanning this spectrometer through the energy E of the scattered radiation. Therefore, the distance between monochromator focus and analyzer crystal has to vary according to $R\sin\theta_A(E)$, where R is the radius of curvature of the analyzer crystal. The analyzer crystal must be rotated around an axis perpendicular to the diffracting plane by $\Delta\theta_A$ in order to change the Bragg angle $\theta_A(E)$. The detector slit, mounted on an arm which rotates around the crystal axis by $2\Delta\theta_A$ must be shifted, so that (with a symmetrically cut analyzer crystal) the crystal–detector slit distance $R\sin\theta_A(E)$ is kept. These movements can either occur mechanically coupled or by means of computer controlled independent elements. Additionally the scattering angle should be made continuously variable by rotating the analyzing Rowland spectrometer around a vertical axis through the focus of the monochromator.

The energy resolution of the Rowland analyzer is mainly determined by the extension ι of the monochromator focus in the plane of the Rowland circle, perpendicular to the scattered beam. Neglecting all aberrations in the imaging process of the curved monochromator, $\iota = 2.35 \, \sigma_x |b| \cos\theta$ (see Wohlert 2000 for a detailed discussion of aberration effects). Using the same layer model as in Section 2.2.2.3 and assuming $t > t_{\text{exB}}$, $(\Delta E/E)_A$ is well approximated (for a symmetrically cut crystal) by

$$\left(\frac{\Delta E}{E}\right)_A = \left(\frac{\iota}{R}\right) \cos\theta_A + \frac{|\chi_h|}{\sin^2\theta_A}, \qquad (2.51)$$

where we have neglected, for sake of simplicity, the change $\Delta\theta_A$ of the analyzer Bragg angle due to finite penetration of the incident radiation into the crystal, as given in (2.37).

The transmission T_A of the Rowland analyzer is, within the limits of the layer model for $t > t_{\text{exB}}$, and with ΔE_A from (2.51),

$$T_A = \frac{R_{iA} \Delta E_A R \sin\theta_A h_s}{2 h_c \iota \delta E} \qquad (2.52)$$

with

$$R_{iA} = \left(\frac{\pi \cos\theta_A}{2\mu R}\right) \left[1 - \exp\left(\frac{-2D\mu}{\sin\theta_A}\right)\right] \qquad (2.53)$$

and

$$R_{iA} = \left[\frac{\pi \cos\theta_A}{2\mu R}\right] \left[1 - \exp\left(\frac{-2D\mu}{\sin\theta_A}\right)\right] \left(1 - \exp\left(\frac{-t}{t_{\text{exB}}}\right)\right) \qquad (2.54)$$

for $t < t_{\text{exB}}$, where δE is the total width of the scattered spectrum, E its centre of gravity, h_c the height in the z-direction of the curved crystal (limiting aperture in the z-direction according to the definition of T_A), and h_s the corresponding height of the detector slit. Moreover, we have assumed that the extension in the z-direction of the focus on the scattering sample is smaller than h_s, which determines $\Delta\psi_{0v}$, and that it is the length of the curved crystal or the width of a corresponding diaphragm which defines $\Delta\psi_{0h}$.

It must be mentioned that the cylindrically bent analyzer crystal is often replaced by a spherically bent crystal analyzer which, for θ_A near $90°$, refocuses also in the vertical direction so that we have to replace the factor $h_s/2h_c$ in (2.52) by 1.

2.2.5.2 *Focusing dispersion compensating analyzer* In order to understand the idea of this device (Schülke 1986), one has to realize that a double-crystal monochromator produces an energy distribution at the illuminated area of the sample with an energy (wavelength) variation in the z-direction, whenever

$$\frac{\sigma_z}{L_s} < \sigma'_{zr}; \quad \sigma'_{zr} = (\sigma'^2_z + \sigma'^2_R)^{1/2}, \qquad (2.55)$$

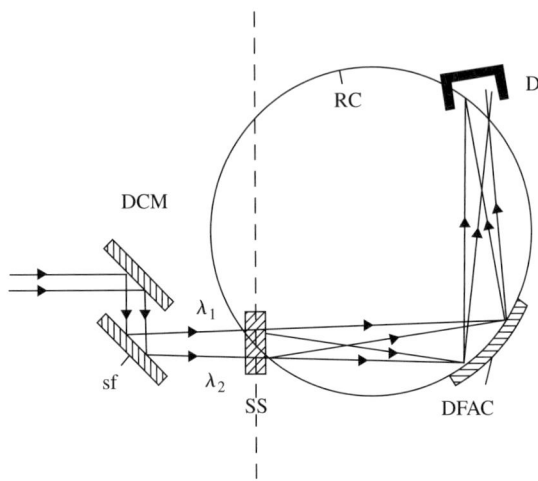

Fig. 2.8. Dispersion compensating focusing analyzer. DCM = double crystal monochromator (second crystal optionally sagittaly focussing (sf)), SS = scattering sample, DFAC = double focusing analyzer crystal, D = detector, RC = Rowland circle. The wavelengths λ_1 and λ_2 are simultaneously Bragg reflected by the DFAC. Note that the drawing plane on the left-hand side of the dashed line is the plane perpendicular to the electron orbital plane which contains the incident photon beam, whereas the drawing plane on the right-hand side of the dashed line is the plane perpendicular to the electron orbital plane which contains the scattered photon beam under analysis.

where L_s is the source–sample distance. This wavelength distribution in the z-direction can be utilized, if the analyzer, as shown in Fig. 2.8, consists of a cylindrically or spherically bent crystal with a Rowland circle in the vertical plane, where both θ_A and R, the bending radius in the vertical plane, are chosen in such a manner that the spatial spectral distribution at the sample in the z-direction just fits the dispersion of the bent analyzer crystal, so that the whole wavelength range of this distribution is Bragg-reflected simultaneously by the bent analyzer crystal (dispersion compensation). In this case the total energy resolution of the scattering experiment $(\Delta E/E)_{\text{tot}}$ is **not** given by

$$\left(\frac{\Delta E}{E}\right)_{\text{tot}} = \left[\left(\frac{\Delta E}{E}\right)_M^2 + \left(\frac{\Delta E}{E}\right)_A^2\right]^{1/2}, \qquad (2.56)$$

where, according to (2.29), $(\Delta E/E)_M$ reads for symmetrically cut crystals

$$\left(\frac{\Delta E}{E}\right)_M = \cot\theta_M(2.35\sigma'_{zr}) + \frac{|\chi_{hM}|}{\sin^2\theta_M}. \qquad (2.57)$$

On the contrary $(\Delta E/E)_M$ in (2.56) has to be replaced by

$$(\Delta E/E)_{M,\text{eff}} = \cot\theta_M(2.35\sigma_z/L_s) + \frac{|\chi_{hM}|}{\sin^2\theta_M}. \tag{2.58}$$

Therefore, if (2.55) holds, one can cut a much larger passband $(\Delta E)_M$ out of the white synchrotron radiation than would be possible, if one had to fit a required total energy resolution according to (2.56). The gain of this method compared to a focusing analyzer without dispersion compensation is given by

$$G = \frac{(\Delta E)_M}{(\Delta E)_{M,\text{eff}}} \tag{2.59}$$

with the full monochromator passband $(\Delta E/E)_M$ according to (2.57).

The condition for dispersion compensation reads as

$$\frac{\cot\theta_M}{\cot\theta_A} = \frac{L_s}{R\sin\theta_A}. \tag{2.60}$$

Note that this condition can be satisfied exactly only for one value of θ_M, if θ_A is held fixed in the course of the spectral analysis of the inelastically scattered radiation (so-called inverse geometry). This means that the spectral resolution is not constant across the spectrum. The energy resolution of the analyzer can be approximated, according to (2.51), by

$$\left(\frac{\Delta E}{E}\right)_A = \left(\frac{\sigma_z}{R}\right)\cos\theta_A + \frac{|\chi_h|}{\sin^2\theta_A}. \tag{2.61}$$

The transmission T_A of a spherically bent dispersion compensating analyzer with θ_A near $\pi/2$ calculates in analogy to (2.52) as

$$T_A = \frac{R_{iA}(\Delta E)_M R\sin\theta_A}{\sigma_z\delta E} \tag{2.62}$$

with R_{iA} according to (2.53) or (2.54).

As far as the total energy resolution $(\Delta E/E)_{\text{tot}}$ and the total transmission (see Section 2.2.6) are concerned, this dispersion compensating system, comprising a double crystal monochromator and a focusing analyzer, is equivalent to combining a curved crystal monochromator with a curved-crystal analyzer, both with the scattering plane perpendicular to the electron orbit. But the dispersion compensating system has two shortcomings: θ_A is not only fixed by the required energy resolution (see equation 2.61), but also by the compensation condition, (2.60), and this condition is dependent on the actual wavelength under analysis. The first restriction sets limits to the transmission of the analyzer which decreases with increasing $\theta_A \to \pi/2$. On the other hand, the dispersion-compensating system offers all the advantages of a double-crystal monochromator as already enumerated in Section 2.2.2.1.

2.2.5.3 *Focusing analyzer and position sensitive detector* It is also possible to utilize the properties of a single- or double-focussing analyzer crystal in connection with a double-crystal monochromator, of course without the advantages of the dispersion compensating method but also without the restrictions this method imposes. In this case the diffracting plane of the analyzer is the electron orbit plane, the sample is not necessarily positioned on the Rowland circle of the analyzer, as depicted in Fig. 2.9, but can be set in an distance smaller (or larger) than $R\sin\theta_A$ from the analyzer crystal, so that the analyzer can expose a section $\delta_s E = E_H - E_L$ of the whole scattering spectrum on the Rowland circle, accepted by a position-sensitive detector. Of course, now the passband of the double-crystal monochromator has to be roughly $(1/\sqrt{2})(\Delta E/E)$, unlike in the dispersion compensating case. Scanning the energy E of the inelastically scattered spectrum, or more precisely, the centre of gravity of the section $\delta_s E$, the following movements have to be coordinated: The analyzer crystal must be rotated around an axis perpendicular to the diffracting plane by $\Delta\theta_A$ in order to change the Bragg angle $\theta_A(E)$. The position-sensitive detector, mounted on an arm which rotates around the crystal axis by $2\Delta\theta_A$, must be shifted, so that (with a symmetrically cut analyzer crystal) the crystal–detector distance $R\sin\theta_A(E)$ is kept.

The individual sections of the spectrum must be composed, where the variation of the sample illumination perpendicular to the incident beam of the

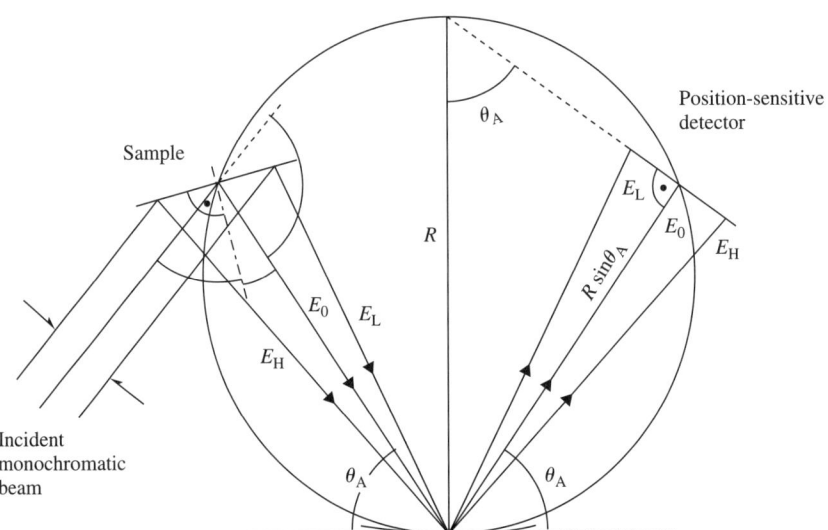

Fig. 2.9. Focussing analyzer in connection with a position-sensitive detector. A photon energy range $\delta_s E = E_H - E_L$ emitted (scattered) by the sample is simultaneously Bragg reflected onto the position-sensitive detector. The sample is not necessarily positioned on the Rowland circle.

analyzer has to be taken into account. The energy resolution of this analyzer system is now mainly controlled by the linear resolution Δl of the position-sensitive detector and, of course, by the intrinsic reflection width of the analyzer crystal:

$$\left(\frac{\Delta E}{E}\right)_A = \left(\frac{\Delta \iota}{R}\right)\cos\theta_A + \frac{|\chi_h|}{\sin^2\theta_A}, \qquad (2.63)$$

where we have assumed that the linear position-sensitive detector is aligned perpendicular to the central Bragg-reflected beam in order to minimize parallax errors due to the finite penetration depth of the position-sensitive detector. On the other hand, the deviation from the Rowland circle causes deteriorations of the resolution for parts of the spectrum which are not taken into account in the estimation of (2.63). The same is true with the aberrations connected with the finite extension in the z-direction of sample, crystal and position sensitive detector, with the Johann-type geometry of the curved crystal, and with the incident beam penetration depth into the crystal, as given in (2.37); see Wohlert (2000) for an exhaustive discussion of all these effects.

The transmission T_A of this analyzer concept can, within the limits of the layer model, be approximated by

$$T_A = \frac{R_{iA} E \cot\theta_A h_d \varepsilon}{2 h_c \delta E}, \qquad (2.64)$$

where E is the center of gravity of the elastically scattered spectrum of total width δE, h_d and h_c are the dimension of the detector and the analyzing crystal perpendicular to the diffraction plane, respectively, and ε is the energy-dependent efficiency of the position sensitive detector. R_{iA} has the same meaning as in Section 2.2.5.1.

Alternatively a position-sensitive detector can also be used in conjunction with a cylindrically bent crystal analyzer in the so-called von Hamos geometry (von Hamos 1938), as sketched in Fig. 2.10. In this geometry the cylindrical axis is in the diffraction plane of the analyzer so that the direction of this axis is also chosen as the dispersion direction where the position-sensitive detectors is

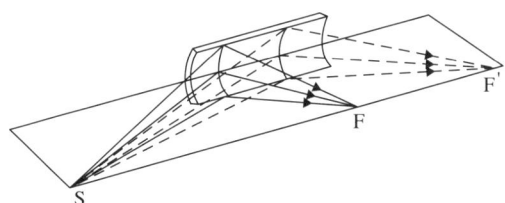

Fig. 2.10. A cylindrically bent crystal analyzer in so-called Hamos geometry. Radiation from a point source S on the sample is dispersed along a line FF′ parallel to the cylindrical axis.

placed. This way a one-to-one focusing of the illuminated area of the sample on the detector is achieved. The energy resolution of such an analyzer setup depends mainly on ι, the linear dimension of the illuminated area on the sample projected on in the dispersion direction, on the linear spatial resolution Δe of the detector in the same direction, and on the intrinsic reflection width (Darwin width):

$$\frac{\Delta E}{E} = \frac{\cot\theta_A [\iota^2 + (\Delta e)^2]^{1/2}}{F_1 + F_2} + |\chi_h| \sin^2\theta_A, \qquad (2.65)$$

where F_1 is the distance between a point S on the sample and the vertex point on the analyzer crystal, and F_2 the distance between vertex and focal point F and F', respectively. We have assumed that the axis of the position-sensitive detector is perpendicular to the central X-ray Bragg-reflected by the analyzer. Note that the mosaic spread of the analyzer crystal does not contribute to the energy resolution. In (2.65) we have neglected other factors, which, for the case of larger Bragg angles are of minor importance for the resolution, for example beam penetration into the crystal, and parallax error of the detector. See Ice and Sparks (1990) for a more detailed discussion.

Utilizing this analyzer setup we can arrange the geometry such that the whole spectrum with width δE can be Bragg-reflected simultaneously so that T_A is, according to its definition, given by

$$T_A = \frac{R_{iA}\varepsilon F_1}{\iota}, \qquad (2.66)$$

where ε has the same meaning as in (2.64), and R_{iA} is the integrated reflectivity of a plane crystal in the Bragg case, given by (2.32) for a nearly perfect crystal or by corresponding relations discussed in detail by Zachariasen (1967) for a mosaic crystal, which guarantee a much higher value of R_{iA} than obtainable with perfect crystals, and which are therefore preferably used (for instance highly oriented pyrolytic graphite (HOPG)). Of course, it is assumed in (2.66) that either the Darwin width or the mean angular mosaic spread is smaller than ι/F_1.

2.2.5.4 *Focusing ultrahigh-energy-resolution analyzer* This analyzer is operated in connection either with an undulator which provides a very well collimated beam for the plane-crystal monochromator of Section 2.2.2.4, the Bragg angle of which is very near to $\pi/2$, as shown in Fig. 2.11, or with a four-bounce monochromator, both capable of providing an energy resolution $(\Delta E/E)_M$ of the order 10^{-7}. The analyzer which fits this monochromator energy resolution would consist, in principle, of a spherically bent (double focussing) crystal, whose Bragg angle is again very near to $\pi/2$. Then, according to (2.42), the energy resolution should be

$$\left(\frac{\Delta E}{E}\right)_A \approx \cot\theta_A \, \Delta\theta_A + |\chi_{hA}|. \qquad (2.67)$$

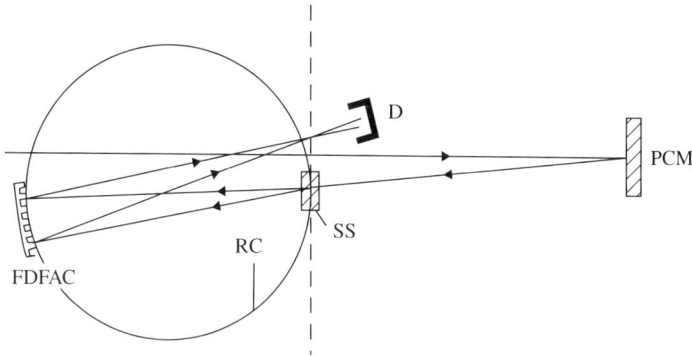

Fig. 2.11. Ultrahigh-resolution spherically bent analyzer in nearly back-scattering geometry, here in connection with a plane-crystal monochromator in nearly back-scattering geometry (optionally also a high-resolution four-bounce monochromator can be used). FDFAC = faceted double focusing analyzer crystal; SS = scattering sample; PCM = plane crystal monochromator; D = detector; RC = Rowland circle. Note that the plane of the Rowland circle must be considered as being rotated by the scattering angle around an axis perpendicular to the electron-orbit plane (dashed vertical line) with respect to the plane containing the incident and the Bragg-reflected beam of the PCM.

Again both terms on the right-hand side can be made of order 10^{-7}. The first one by bringing θ_A as close to $\pi/2$ as the beam-paths allow, and by diminishing the size of the illuminated area on the sample, if possible by means of a focusing mirror between monochromator and sample. The second term can be made of sufficiently low order by choosing an appropriate short wavelength, since χ_h goes like λ^2. But the spherical bending of the analyzer, necessary to accept an ample solid angle of the scattered radiation, introduces lattice distortions which can be expressed as $\Delta d/d$, the relative change of the lattice-plane spacing along the extinction depth t_{exB}; see (2.38). These distortions produce an additional contribution to the energy resolution of the analyzer, namely

$$\left(\frac{\Delta E}{E}\right)_{\mathrm{A\ dist}} = \frac{\Delta d}{d}, \qquad (2.68)$$

which can easily surmount the other two contributions in (2.67), due to the rather large extinction depth for short wavelengths, even if R is chosen to be a few meters. Fujii *et al.* (1982) and Dorner and Peisl (1983) have proposed a way to overcome these difficulties, by faceting the surface of the spherically bent crystal where the depth of the cuts was only slightly less than the crystal thickness. In this way each facet acts upon bending as a nearly unstrained crystal. Recently, Masciovecchio *et al.* (1996) have developed a procedure to mount ~ 21000 independent small crystals obtained from the same silicon wafer on a spherical substrate, where their perfect crystal properties are preserved. This was attained by gluing the faceted surface on the spherical substrate and then etching away, under computer control, the layer, which had connected the facets after the slicing procedure. The transmission of the analyzer T_A amounts approximately to

$$T_A = \frac{R_{iA} R(\Delta E)_A}{a\, \delta E}, \qquad (2.69)$$

where R_{iA} is the integrated reflecting power of (2.43), $(\Delta E)_A$ the transmitted passband according to (2.67), R the bending radius of the spherical analyzer, a the dimension, parallel to the diffraction plane of the analyzer, of the illuminated area on the sample, and δE the total energy width of the scattered spectrum.

This combination of ultrahigh-resolution monochromator and analyzer is runned in inverse geometry. This means the monochromator is scanned through the required energy range and the energy transmitted by the analyzer is held fixed. Tuning of the monochromator can be achieved, in the case of a plane crystal device, either by changing the temperature difference between monochromator and analyzer (both acting on the same Bragg-reflection, so that the thermal expansion is used to detune their lattice spacing) or by subjecting the monochromator to an additional hydrostatic pressure. In the case of a four-bounce monochromator, inverse geometry is realized by detuning the monochromator mechanically with respect to the analyzer.

2.2.5.5 *Double crystal analyzer in dispersive setting* In spite of the fact that a dispersive double crystal setting, as shown in Fig. 2.12, has not yet been used in inelastic X-ray scattering experiment with synchrotron radiation but only in conjunction with conventional sources, we will, for sake of completeness, briefly discuss its transmission and energy resolution. The transmission amounts

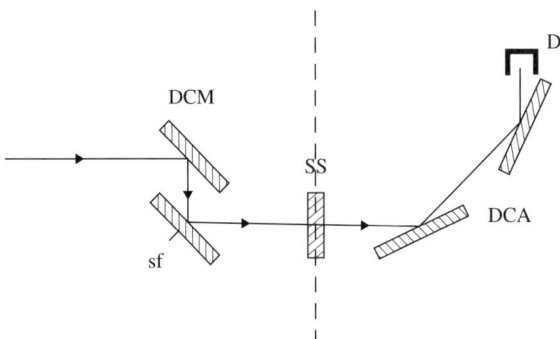

Fig. 2.12. Double crystal analyzer (DCA) in dispersive setting, in connection with a double crystal monochromator (DCM) in nondispersive setting. SS = scattering sample; D = detector. See the note to Fig. 2.8 about the meaning of the vertical dashed line.

approximately to

$$T_A = \frac{R_{iA}^2 E \cot\theta_A}{\Delta\psi_{0v}\,\delta E}, \qquad (2.70)$$

so that the transmission of this device is rather poor since it goes like R_{iA}^2. On the other hand, the energy resolution, which is roughly given by

$$\frac{\Delta E}{E} = R_{iA} \cot\theta_A, \qquad (2.71)$$

can be made very high.

2.2.6 Total transmission

After having provided expressions for the spectral transmission T_M and T_A of different combinations of monochromators and analyzers, one can now give a general formula for calculating the final average countrate N in an inelastic X-ray scattering experiment, if we know the spectral flux Φ of the synchrotron radiation source, as defined in 2.2.1, namely

$$N = \Phi \Delta\phi_h \left(\frac{\Delta E}{E}\right)_{M,\text{eff}} G\,T_M\,s\,T_A\,\Delta\Omega_0. \qquad (2.72)$$

Here $\Delta\phi_h$ is the horizontal divergence of the source, accepted by the monochromator, and G the gain factor, defined in (2.59). The countrate N is of crucial importance for planning an experiment with certain requirements on the statistical accuracy. Of course, equally important is also the total energy resolution

$$\left(\frac{\Delta E}{E}\right)_{\text{tot}} = \left[\left(\frac{\Delta E}{E}\right)_{M,\text{eff}}^2 + \left(\frac{\Delta E}{E}\right)_A^2\right]^{1/2}, \qquad (2.73)$$

since the specific demands are strongly determined by the matter under investigation.

Note that $(\Delta E/E)_{M,\text{eff}}$ is different from $(\Delta E/E)_M$ only in the case of dispersion compensation (see Section 2.2.5.2).

2.3 Measurement of the dynamic structure factor $S(\mathbf{q},\omega)$

The most important quantity to be measured in the characteristic energy-loss regime of inelastic X-scattering is the dynamic structure factor $S(\mathbf{q},\omega)$, first of all on electron systems which are simple enough to study the electron–electron correlation in its fundamental form, that is on nearly-free electron metals and s-p-semiconductors. In these systems the important many-body effects, one can hope, are not yet totally superimposed by band structure phenomena. Just for that reason also the band structure effects have to be studied very carefully. Therefore, we will start with presenting experimental evidence from inelastic X-ray scattering for the collective dynamic behavior of electrons in simple metals, and for the transition into the regime of individual electron excitations, or more precisely into the regime of particle–hole excitations. In the next step, the study of band structure effects shall be placed in the forefront. The influence of band structure or short-range order on the plasmon dispersion is one aspect. Moreover, there are, on the one hand, the consequences of band structure induced excitation gaps, in exceptional cases the so-called zone-boundary collective states, that can be explored. On the other hand, there is the interaction between collective and particle–hole excitation which manifests itself in the so-called plasmon-Fano resonances.

Whereas the collective excitations can still be understood in the framework of the random phase approximation (RPA), it is the special features of the dynamical structure factor in the particle–hole excitation regime which make it necessary to proceed further to more sophisticated approximations, as far as they can be evaluated and selected according to agreement with experiment.

2.3.1 *Plasmons and transition to the particle–hole continuum*

It is common belief that Na and Al are metals whose valence electrons exhibit the most free-electron like behavior. Therefore, it was obvious to study plasmon excitation and the transition to particle–hole excitation with increasing momentum transfer q as predicted by jellium-model calculations by means of inelastic X-ray scattering on these materials.

Figure 2.13 shows the dynamic structure factor $S(\mathbf{q},\omega)$ of Al for $\mathbf{q} \parallel [100]$ for different values of q as obtained with double-crystal monochromatized synchrotron radiation (7.99 keV) from the DORIS storage ring using energy analysis of scattered radiation in inverse geometry by means of a spherically bent analyzer crystal. By utilizing dispersion compensation (see Section 2.2.5.2) a total energy resolution of 1.6 eV could be achieved (Schülke 1993). The measurements were brought to an absolute scale by means of the sum rule (2.21). (Notice that for all plots of experimental results, the absolute values of measured $S(\mathbf{q},\omega)$ thus obtained were further divided by the electron density ρ, so that the **experimental** $S(\mathbf{q},\omega)$ are related to **one** electron rather than to **all** electrons within the quantization volume (here set to be 1), as assumed with the definition (2.6).) The transition from a very sharp structure for $q < 0.8$ a.u. (a.u. = atomic units: $\hbar = e = m = 1$) to a much broader one can be understood as a transition from plasmon excitation to excitation into the particle–hole continuum by looking at Fig. 2.14: For three different values of q, the jellium calculated real, ε_r, and imaginary, ε_i, parts of $\varepsilon^{\mathrm{RPA}}(\mathbf{q},\omega)$ are shown, together with the corresponding $-\mathrm{Im}[1/\varepsilon(\mathbf{q},\omega)]^{\mathrm{RPA}}$ which is proportional to $S^{\mathrm{RPA}}(\mathbf{q},\omega)$, computed using (2.20) along with the Lindhard polarization function (see equation 6.75),

$$\chi_{\mathrm{L}}(\mathbf{q},\omega) = (2/\hbar) \sum_{\mathbf{k}} \frac{\mathrm{f}(E_{\mathbf{k}}) - \mathrm{f}(E_{\mathbf{k}+\mathbf{q}})}{\omega - \hbar(\mathbf{k}\cdot\mathbf{q}/m) - \hbar q^2/2m + i\varepsilon}, \qquad (2.74)$$

the limit as $\varepsilon \rightarrow 0_+$ being understood. $\mathbf{k}$ is the wavevector of electrons with energy $E_{\mathbf{k}}$, and $f(E)$ the Fermi function. This Lindhard polarization function

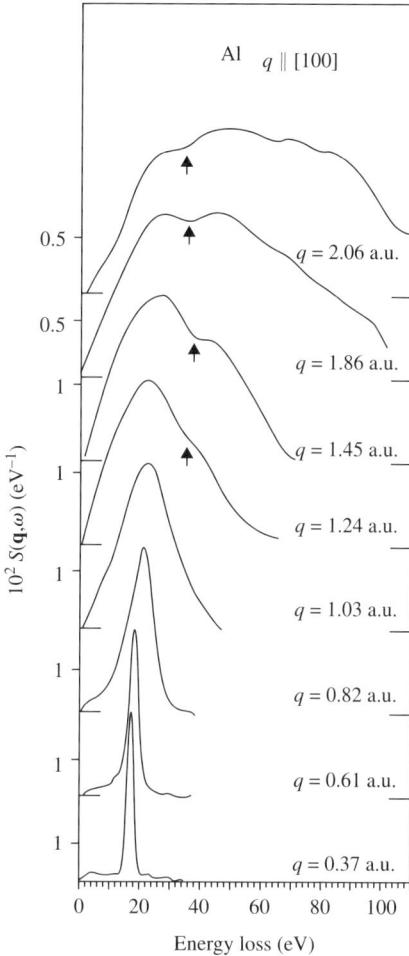

Fig. 2.13. Measured dynamic structure factors $S(\mathbf{q},\omega)$ of aluminum metal (after a cubic spline interpolation of the rough data) for momentum transfer $\mathbf{q} \parallel [100]$ and q in atomic units (a.u.) as indicated. The arrows mark the position of a local zero of the third derivative, indicating the "dip" position of a **q**-orientation independent one-peak-one-shoulder or double-peak fine structure. (Reprinted with permission from Schülke *et al.* (1993); copyright (1993) by the American Physical Society.)

plays, within the RPA scheme, the role of the proper polarization function, and can be calculated utilizing integration in closed form as shown in (6.79).

A sharp structure of $S(\mathbf{q},\omega)$ originates at that frequency ω_p (plasma frequency), where $\mathrm{Re}[\varepsilon^{\mathrm{RPA}}(\mathbf{q},\omega)]$ has a zero passage and where simultaneously also $\mathrm{Im}[\varepsilon^{\mathrm{RPA}}(\mathbf{q},\omega)]$ is nearly equal to zero (plasmon poles of $\varepsilon^{\mathrm{RPA}}(\mathbf{q},\omega)$), both

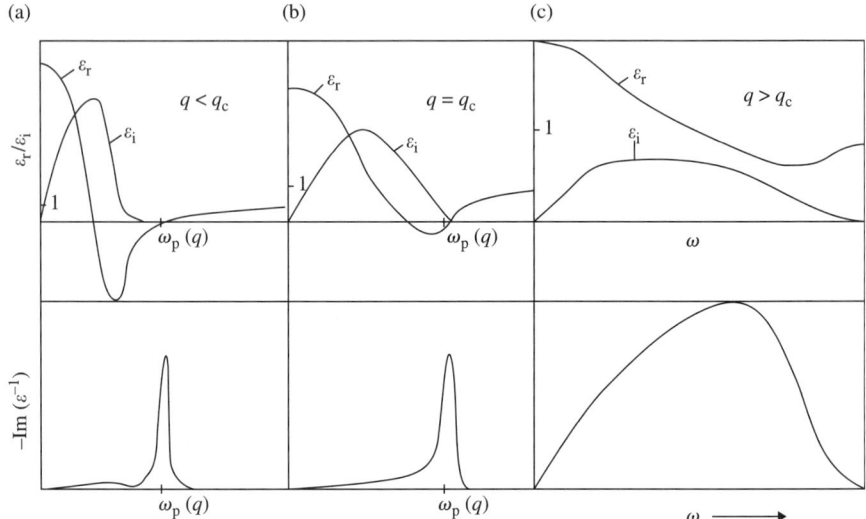

Fig. 2.14. Upper row: real part ε_r and imaginary part ε_i of the free electron RPA dielectric function $\varepsilon(\mathbf{q},\omega)$ (including weak damping the particle–hole excitations) for (a) $q < q_c$ (q_c = plasmon cut-off wave vector); (b) $q = q_c$, and (c) $q > q_c$. Lower row: RPA free-electron response function $-\mathrm{Im}\,\varepsilon^{-1}(\mathbf{q},\omega)$ (including weak damping of the plasmon excitation) (a) $q < q_c$; (b) $q = q_c$, and (c) $q > q_c$.

necessary conditions for the plasma oscillation of the electron system to be a good (nearly undamped) elementary excitation (Mahan 1981). This behavior exists up to the so-called plasmon cut-off momentum transfer vector q_c ($q_c \approx 0.67$ a.u. for Al), where the zero passage of $\mathrm{Re}[\varepsilon^{\mathrm{RPA}}(\mathbf{q},\omega)]$ just coincides with the high-ω edge of $\mathrm{Im}[\varepsilon^{\mathrm{RPA}}(\mathbf{q},\omega)]$. Within the region $q < q_c$, the q-dependence of the $S(\mathbf{q},\omega)$ peak position can be utilized to measure the plasmon dispersion $\omega_p(\mathbf{q})$, as done with inelastic X-ray scattering spectroscopy (IXSS) for Li (Alexandropoulos 1971, still using conventional X-ray sources and a dispersive double-crystal setting; see Section 2.2.5.5; Schülke et al. 1986, and Hill et al. 1996), and for Be (Eisenberger et al. 1973, still using conventional X-ray sources and a dispersive double-crystal setting; see Section 2.2.5.5). Of course, those investigations are the domain of electron energy loss spectroscopy (EELS) (Raether 1980), since the differential cross-section for electron scattering goes like q^{-2}.

Figure 2.15 shows the q-dependence of the $S(\mathbf{q},\omega)$ peak position of Al ($\mathbf{q} \parallel [110]$) for the whole q-range measured, confronted with EELS measurements and different computations. One sees good agreement of the X-ray measurement with the EELS results which are restricted to only $q \leq q_c$. Furthermore it is evident that the RPA calculation cannot account for the $S(\mathbf{q},\omega)$ peak positions, so that one has to look for better approximations of the proper polarization

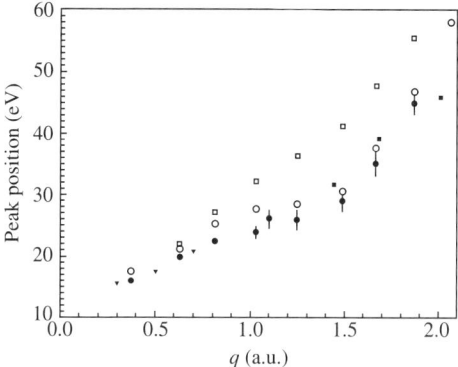

Fig. 2.15. Peak position of the Al $S(\mathbf{q},\omega)$ spectra as a function of momentum transfer q. Full circles: IXSS experimental result for $\mathbf{q} \parallel [110]$. Filled triangles EELS measurements of Spr̈osser-Prou et al. (1989). Open squares: jellium RPA. Open circles: jellium calculations local-field and lifetime corrected. Filled squares: calculated for $r_s = 2$ by Green et al. (1987b) (center of gravity extrapolation). (Reprinted with permission from Schülke et al. (1993); copyright (1993) by the American Physical Society.)

function. We will come to this point later, when considering the lattice effects and treating correlation beyond the RPA scheme.

For $q > q_c$ the plasmons can decay into particle–hole excitations. Thus they become strongly damped. Now particle–hole excitations, which cover a broad ω-continuum, dominate the dynamic structure factor. The same behavior has been found for Li (Alexandropoulos 1971, Schülke et al. 1986, Hill et al. 1996), for Be (Eisenberger et al. 1973, Schülke et al. 1989, and for Na (Hill et al. 1996).

2.3.2 Band structure effects in $S(\mathbf{q},\omega)$

The discussion of band-structure effects in $S(\mathbf{q},\omega)$ have a rather long history and was held rather controversially. It was mainly the double-peak structure of $S(\mathbf{q},\omega)$ for $q > q_c$, which had already been found in early measurement (still performed with conventional X-ray sources) by Platzman and Eisenberger (1974) on Be, Al and graphite, and which has been interpreted by these authors as an indication of an incipient Wigner electron lattice, thus prompted a huge amount of theoretical work. Later measurements with synchrotron radiation using a focusing analyzer of the von Hamos type in conjunction with a position-sensitive detector (see Section 2.2.4.3), performed by Platzman et al. (1992) on Al, which have also revealed the double-peak structure, found their explanation by many-particle arguments invoking multiple excitations in the Fermi liquid. But we will show in what follows that experimental results from inelastic X-ray scattering measurements on many simple metals and s-p-semiconductors along with model- and first-principle calculations have delivered strong evidence that it is band-structure effects, namely excitation gaps, that produce this fine structure. Before doing this we shall have a look on how lattice effects will influence the plasmon dispersion. Finally another band structure effect is treated, which is connected with the Fano-like coupling between the discrete higher plasmon bands with the particle–hole continuum of excitations. These plasmon-Fano resonance need the formalism of dielectric matrices as mentioned in Section 2.1, and discussed in greater destail in Section 6.6.

2.3.2.1 *Plasmon dispersion and plasmon lifetime*   One can easily show, within the limits of simple oscillator model (Raether 1980), that interband transitions in solids, modeled by the excitation of electrons bound with an eigenfrequencey ω_n to the ions, will influence the plasma frequency ω_p of free electrons in a very definite way. If $\omega_n < \omega_p$, the plasmon energy $\hbar\tilde{\omega}_p$ is reduced with respect to the free electron value; if $\omega_n > \omega_p$, the plasmon energy is enhanced. Of course, such a single oscillator model cannot account for the collective behavior of electrons embedded in a periodic ion potential with a large variety of interband transitions with very different oscillator strengths. Therefore, in order to study the influence of the crystal lattice on plasmons, their energy, dispersion and lifetime, we have to compare the results of corresponding inelastic X-ray scattering results with jellium calculations, on the one hand, and with first-principle band structure calculations, on the other. The earliest investigation of plasmon dispersion with X-rays was performed (still with a conventional X-ray source) by Eisenberger *et al.* (1973) on Be single crystals. The results are shown in Fig. 2.16, here plotted as a function of θ^2, and exhibit a remarkable dependence of both the plasmon dispersion and the plasmon linewidth (plasmon lifetime) on the direction of the momentum transfer **q**. This dependence on the **q**-direction is a strong indication of band-structure effects on collective oscillations. Of course, there is only a slight increase (less than 1%) of the measured plasmon energy for $q = 0$ from the RPA value, as found in EELS measurements of Aiyama and Yada (1974), possibly due to interband transition energies well below and above the free-electron plasmon energy. Also the plasmon dispersion, at least that for **q**$\|c$-axis, is not far from the RPA prediction for $q < q_c$. (In Fig. 2.16, the point of the plasmon cutoff momentum transfer on the θ^2 axis is labeled q_c^2.)

The situation is quite different in the case of Al. EELS measurements (recent ones by Sprösser-Prou *et al.* 1989) have revealed a value for $\hbar\omega_p(\mathrm{q}=0) = 15.0\,\mathrm{eV}$, which has to be confronted with 15.8 eV as calculated for the RPA jellium. The experimental dispersion of the plasmon energy for $q < q_c$, is much less steep than the jellium RPA one as shown in Fig. 2.17 taken over from Eguiluz *et al.* (1995). Note that the IXS-data of Schülke *et al.* (1993) fit quite well the EELS data of Sprösser–Prou *et al.* (1989) as demonstrated in Fig. 2.15. The strong deviation of the experiment from the jellium RPA must be traced back to be a band-structure effect as pointed out by calculations of Eguiluz *et al.* (1995) also depicted in Fig. 2.17: A first-principle band structure calculation of the plasmon energy, which remains within the limits of the

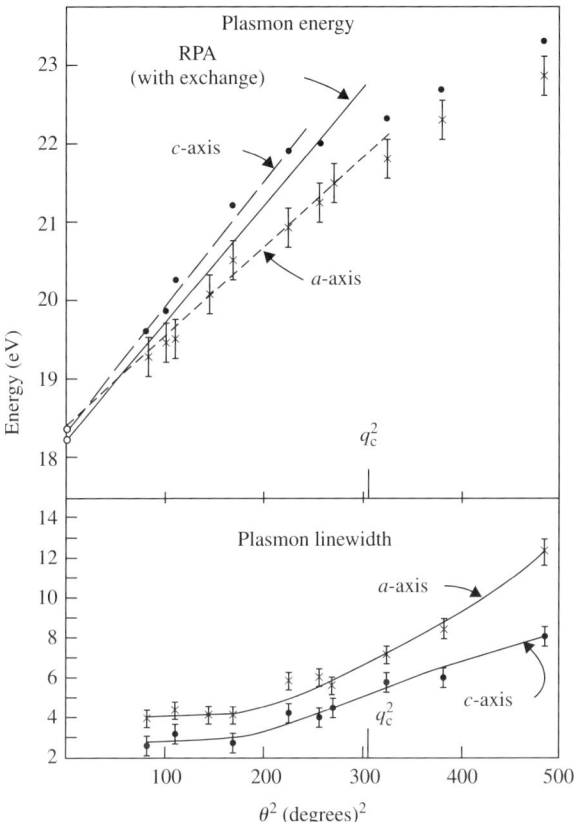

Fig. 2.16. Upper part: Plasmon energy as measured on hcp beryllium metal for $\mathbf{q}\|a$-axis and $\mathbf{q}\|c$-axis, respectively, as a function of θ^2 (θ = scattering angle). Solid line is the RPA result including exchange (local-field correction). q_c is the plasmon cut-off wavevector. Lower part: Plasmon linewidth measured on hcp beryllium metal for $\mathbf{q}\|a$-axis and $\mathbf{q}\|c$-axis as a function of θ^2. (Originally published by Eisenberger et al. (1973); copyright (1973) by the American Physical Society.)

self-consistent (RPA-like) scheme, whose final result is summarized in (6.112) improves the dispersion curve substantially for all wavevectors. In particular, the effect of the band structure on the plasmon frequency for larger q is rather significant (about 4 eV). The remaining discrepancy between calculation and experiment must be attributed to many-particle effects as discussed in Section 2.3.3. It should be mentioned that EELS measurement (Manzke 1978) have also revealed a weak dependence of the plasmon dispersion on the $\mathbf{q}$-direction for single crystals of Al, which could be reproduced by pseudopotential band structure calculations of Bross (1978). Also the finite width of the

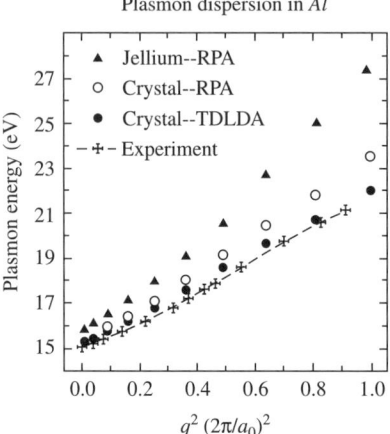

Fig. 2.17. Plasmon dispersion (as a function of q^2 in units of $2\pi/a_0$ (a_0 = lattice parameter) in aluminum metal. Calculated by Eguiluz et al. (1995): closed triangles: jellium RPA; open circles: crystal RPA; filled circles crystal TDLDA (**t**ime-**d**ependent **l**ocal **d**ensity **a**pproximation). EELS measurements by Sprösser–Prou et al. (1989). (Reprinted with permission from Eguiluz et al. (1995); copyright (1995) by the American Physical Society.)

Al-plasmon resonance and its q-dependences found both in EELS and in IXSS experiments, can be traced back at least partly to lattice effects, more precisely to interband transitions, as was shown by band structure calculations of Bross (1978).

IXSS experiments on Li and Na metal upon melting by Hill et al. (1996) and of Sternemann et al. (1998) on Al have especially shed light on the band-structure effects in the plasma resonance. These authors found, as demonstrated in Fig. 2.18 for Li, that both the dispersion of the plasma resonance and also its width is not changed outside experimental error upon melting. Only the plasmon energy E_p itself exhibits a jump at the melting temperature, which can simply be explained by expansion upon melting and the consequent reduction in electron density. This result was all the more surprising, since, as will be shown later, the above authors found the lattice-induced fine structure in the $S(\mathbf{q},\omega)$ spectra for $q > q_c$ to vanish upon melting. This apparent disagreement was solved by the above authors referring to the rather short length-scale probed by the plasmons (1–10 Å), so that the short-range ionic correlations of the liquid are sufficient to mimic the packing arrangement of a solid. This way an effective band structure is built. Those conclusions have also been drawn on the basis of NMR results on liquid and solid by Krishnan and Nordine (1993). Also interband transition have been observed in optical measurements on liquid Al by Zeppenfeld (1969). On the contrary, the lattice-induced fine structure of $S(\mathbf{q},\omega)$ for $q > q_c$ is, as we

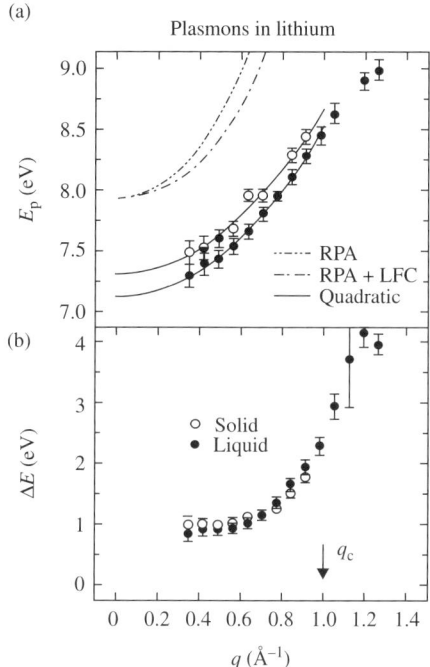

Fig. 2.18. (a) Plasmon energy E_p measured by means of IXSS on Li as a function of q; open circles solid, filled circles liquid. Dotted-dotted-dashed line: jellium RPA; dashed-dotted line: jellium RPA including local-field correction. The solid lines indicate fits of functions quadratic in q to the experimental data. (b) Plasmon linewidth ΔE as a function of q; open circles: solid, filled circles: liquid. (Reprinted with permission from Hill *et al.* (1996); copyright (1996) by the American Physical Society.)

will show later, the result of electron Bragg reflection where a reciprocal lattice with well-defined sharp lattice points is crucial, so that the loss of long-range order upon melting should destroy this fine structure. Nevertheless, it should be stressed, and will be discussed later, that also self-energy effects and particle–hole interaction can contribute remarkably to the plasmon lifetime.

A rather unique nearly free electron system for studying plasmon dispersion and plasmon linewidth is elemental alkali metals, especially Li, dissolved in ammonia. It is well established (see, e.g. Thompson 1976) that at Li concentrations between 4 and 21 mole% metal (MPM) the system is metallic, so that one can easily alter the density parameter $r_s = (4\pi n/3)^{-1/3}/a_0^*$ (n is the free-electron concentration, and $a_0^* = h^2\varepsilon/m^*e^2$ the effective Bohr radius with the static dielectric constant ε) between 7.35 and 11.3 by varying the amount of dissolved metal from 20 to 4 MPM. These r_s values are much larger than those of regular metals, thus offering the opportunity to study the effect of lower electron density on the electron correlation producing deviations from RPA. The first measurements of the plasmon dispersion and plasmon linewidth on Li-NH$_3$ were performed by Burns et al. (1999) on 20 MPM Li-NH$_3$ at the NSLS X21 beamline using a diced spherically bent Ge (733) analyzer in near back-scattering geometry (see Section 2.2.5.4). The total energy resolution was 0.26 eV. The authors found a well-developed plasmon at nearly 2 eV, the dispersion of which was found a little bit below the RPA jellium value. The extrapolated plasmon energy at $q = 0$ was determined to be 1.97 eV. The deviation from the jellium RPA value, which is 2.2 eV, could be attributed to the polarizability of the Li ion cores and the ammonia molecules. In a further publication Burns et al. (2002) have investigated (at beamline ID16 of the ESRF) how the variation of r_s (by changing MPM) influences the deviation of the plasmon dispersion from the jellium RPA behavior. Moreover, they have also measured (again at X21 of NSLS) the plasmon dispersion of the solid phase Li(NH$_3$)$_4$. Figure 2.19 presents the result of this study. The plasmon dispersion becomes weaker with increasing r_s in qualitative agreement with different theoretical models which take into account electron correlation. But this reduction is less pronounced than in the case of regular alkali metals, so that one can conclude that, in the latter case, it is also lattice effects which influence plasmon dispersion. Surprisingly the plasmon dispersion of solid phase is much smaller than that of the corresponding 20 MPM liquid sample despite the larger density of the former. This is in contradiction to the results found at the solid–liquid transition of Li, Na and Al presented above (Hill et al. 1996, Sternemann et al. 1998), where no change in the plasmon dispersion could be observed. The solid Li(NH$_3$)$_4$ data lies much closer to the plasmon dispersion for elemental metals. The plasmon bandwidth also presented in Fig. 2.19 is increasing when going to lower concentrations. This can be explained as a consequence of a less effective screening of the ion potential when the metal–insulator transition (around 4 MPM) is approached, in agreement with the fact that the plasmon width of alkali metals Li, Na and K turns out to be proportional to the square of the Fourier transform of the pseudopotential in a nearly free electron model (Sturm and Oliveira 1981). The increase of the plasmon linewidth when going from the liquid 20 MPM sample to the solid can be attributed to a greater degree of coupling to lattice excitations in the solid.

2.3.2.2 *Excitation gaps, zone-boundary collective states* It is shown in Chapter 6 that the relation (6.75) for the Lindhard polarization function $\chi_L(\mathbf{q}, \omega)$, once more repeated in (2.74), can be modified such that also band structure effects are admitted. The summation over $\mathbf{k}$ in (2.74) has to be replaced by a summation over the reduced Bloch wavevectors $\mathbf{k}$ and over the bands with index ν. Whereas in (2.74) the momentum transfer $\mathbf{q}$ allows only transitions from a free electron state with momentum $\mathbf{k}$ into a state with momentum $\mathbf{k} + \mathbf{q}$, now the probability for a transition from a Bloch state $|\mathbf{k}, \nu\rangle$ into another state $|\mathbf{k}', \nu'\rangle$ is regulated by the square of the matrix element $\langle \mathbf{k}'\nu'| \exp(i\mathbf{q} \cdot \mathbf{r})|\mathbf{k}\nu\rangle$, so that also interband transitions come into play. Therefore, one has to sum additionally

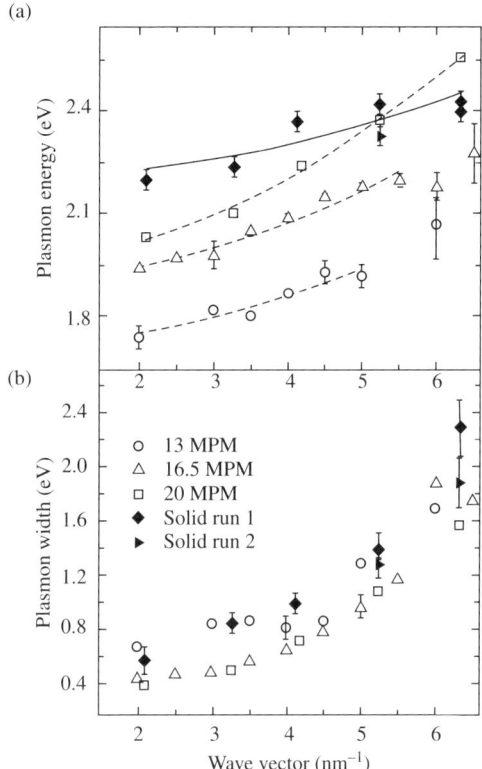

Fig. 2.19. (a) Open symbols: dispersion of the plasmon energy of liquid lithium ammonia of different electron densities corresponding to the lithium concentration in mole % metal (MPM) as indicated. The dashed lines are fits to a dispersion law quadratic in q. Filled circles: dispersion of the plasmon energy of solid Li(NH$_3$)$_4$. The solid line is a fit to a dispersion law quadratic in q. The fitting is in each case for $q < q_\mathrm{c}$. (b) Dispersion of the plasmon linewidth. Symbols are the same as in the upper part. (Reprinted with permission from Burns *et al.* (2002); copyright (2002) by the American Physical Society.)

over $\mathbf{k}'$ and ν', thus ending up with the well-known expression of Ehrenreich and Cohen (1959) for the proper polarization function,

$$\chi_{\mathrm{EC}}(\mathbf{q},\omega) = \lim_{\varepsilon \to 0_+} \sum_{\substack{\mathbf{k},\nu \\ \mathbf{k}',\nu'}} \frac{|\langle \mathbf{k}'\nu'|\exp(\mathrm{i}\mathbf{q}\cdot\mathbf{r})|\mathbf{k}\nu\rangle|^2 [f(\mathbf{k}',\nu') - f(\mathbf{k},\nu)]}{\hbar\omega + E(\mathbf{k},\nu) - E(\mathbf{k}',\nu') + \mathrm{i}\varepsilon}, \tag{2.75}$$

which is equivalent to the Lindhard expression for $\chi_\mathrm{L}(\mathbf{q},\omega)$ and can replace it in the expression (2.19) for the SCF (RPA) polarization function. $E(\mathbf{k},\nu)$ is the energy and $f(\mathbf{k},\nu)$ the occupation number of the one-electron Bloch state $|\mathbf{k},\nu\rangle$. The imaginary part of $\chi_{\mathrm{EC}}(\mathbf{q},\omega)$, which is according to (2.14) and (6.92) essential for the dynamic structure factor $S^{\text{SCF}}(\mathbf{q},\omega)$ in the self-consistent approach

$$\text{Im}\chi_{\text{EC}}(\mathbf{q},\omega) = \pi \sum_{\substack{\mathbf{k},\nu \\ \mathbf{k}',\nu'}} |\langle \mathbf{k}'\nu'|\exp(i\mathbf{q}\cdot\mathbf{r})|\mathbf{k}\nu\rangle|^2$$

$$\times [f(\mathbf{k},\nu) - f(\mathbf{k}',\nu')]\delta[\hbar\omega + E(\mathbf{k},\nu) - E(\mathbf{k}',\nu')] \qquad (2.76)$$

is then directly related to the combined density of occupied and unoccupied states via the energy conserving δ-function: $\delta[\hbar\omega + E(\mathbf{k},\nu) - E(\mathbf{k}',\nu')]$.

The most prominent structure of $S(\mathbf{q},\omega)$, at least for simple metals and s-p-bounded semiconductors, which arises by ion–electron interaction in crystalline solids, can be traced back to excitation gaps for final states on Bragg planes (Ashcroft and Mermin 1976) in the extended zone scheme, perpendicular to $\mathbf{q}$. Within the limits of the so-called empty lattice model (neglect of the periodic lattice potential), final states on the Bragg planes belonging to different bands ν and ν' are degenerate and contribute to the same energy transfer ω. When the Fourier component of the periodic potential belonging to that Bragg plane is switched on, this degeneracy may be removed. This way, an energy gap is opened, which leads to a corresponding excitation gap in the combined density of states, as shown in Fig. 2.20. According to (2.76), this excitation gap must also be present in $\text{Im}\chi_{\text{EC}}(\mathbf{q},\omega)$ and consequently (see equation 6.95) in $\varepsilon_2 \equiv \text{Im}\varepsilon^{\text{SCF}}(\mathbf{q},\omega)$, although this gap can be filled to a certain extent by other transitions. As shown in Fig. 2.21, as the result of a local empirical pseudopotential calculation on Li (Schülke et al. 1986), a gap in ε_2 can produce, via the Kramers–Kronig relation, an additional zero of $\varepsilon_1 \equiv \text{Re}\varepsilon^{\text{SCF}}(\mathbf{q},\omega)$ or at least a strong minimum, which should give rise to a strong peak in $-\text{Im}[1/\varepsilon^{\text{SCF}}(\mathbf{q},\omega)]$. Such a peak, if due to an additional zero of ε_i, can be interpreted as a new collective mode, the so-called zone-boundary collective state (ZBCS), first introduced by Foo and Hopfield (1968) and thoroughly discussed by Sturm and Oliveira (1984) for Al. Even if there is no strong peak in $-\text{Im}[1/\varepsilon^{\text{SCF}}(\mathbf{q},\omega)]$ due to a zero or nearly zero of ε_1, the excitation gap discloses itself by its dip in ε_2.

Excitation gap-induced fine structure of $S(\mathbf{q},\omega)$ for $q > q_c$ was first found in IXS spectra of Li single crystals measured by Schülke et al. (1984, 1986) by means of a focusing dispersion compensating analyzer (see Section 2.2.5.2) in connection with a double crystal monochromator at the DORIS storage ring at DESY/HASYLAB, Hamburg (Schülke and Nagasawa 1984, Schülke 1986). These spectra, a part of them is shown in Fig. 2.22, were discussed on the basis of an empirical local pseudopotential band structure calculation of $S(\mathbf{q},\omega)$. The peak structure of $S(\mathbf{q},\omega)$ around 10 eV can directly be traced back to an excitation gap opened by the potential coefficient V_{200} for final states on the $\mathbf{g}_{200}$ Bragg plane. A ZBCS-like peak structure due to the $\mathbf{g}_{110}$ Bragg plane of Li and its dispersion was also investigated by Schülke et al. (1986). This investigation was repeated with much higher resolution (40 meV) by Burkel (1991) utilizing a focusing ultra high-resolution analyzer (see Section 2.2.5.4) at DESY/HASYLAB. The dispersion of these two ZBCS-like peaks in the $\mathbf{q} \parallel [100]$ and $\mathbf{q} \parallel [110]$ spectra of

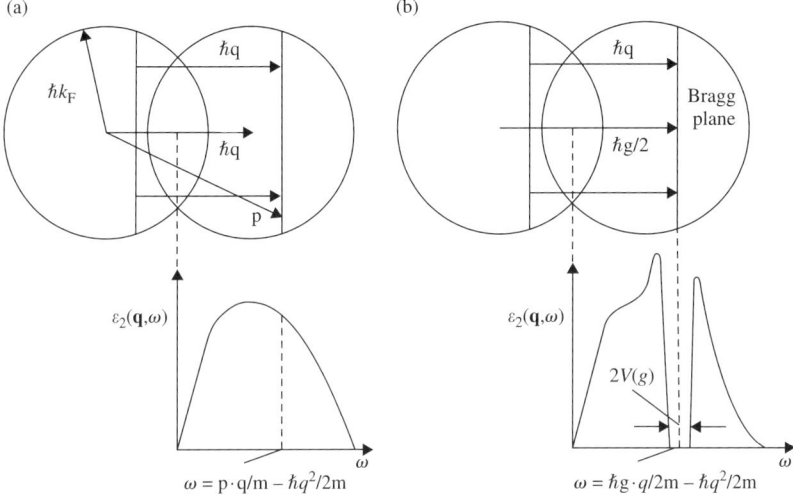

Fig. 2.20. (a) Upper part: transition between one-electron states in momentum space of a homogeneous electron system as induced by an inelastic scattering process transferring momentum $\mathbf{q}$ and energy $\hbar\omega$. Lower part: corresponding $\varepsilon_2(\mathbf{q},\omega)$. (b) Upper part: transition to a Bragg plane in the extended zone scheme of an inhomogeneous electron system. Lower part: corresponding $\varepsilon_2(\mathbf{q},\omega)$ within the limits of a two-band model, exhibiting an excitation gap of width $2V(\mathbf{g})$.

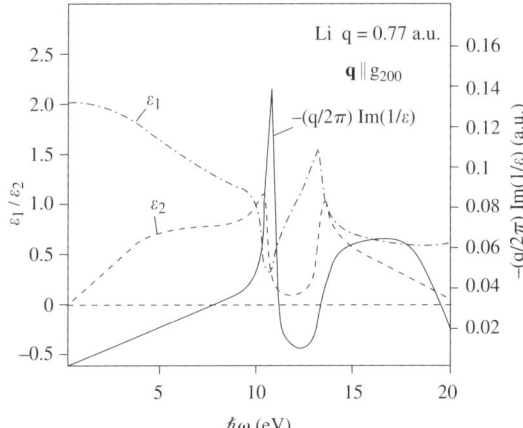

Fig. 2.21. Real part ε_1 (dashed-dotted line) and imaginary part ε_2 (dashed line) of the dielectric function $\varepsilon(\mathbf{q},\omega)$ together with the response function $-\mathrm{Im}\,\varepsilon^{-1}$ (solid line) calculated for the two-band model with $\mathbf{g}\parallel\mathbf{q}$ for Li: $q=0.77$ a.u.; $\mathbf{q}\parallel\mathbf{g}_{200}$; $V_{200}=0.055$ a.u. (Reprinted with permission from Schülke *et al.* (1986); copyright (1986) by the Physical Society of America.)

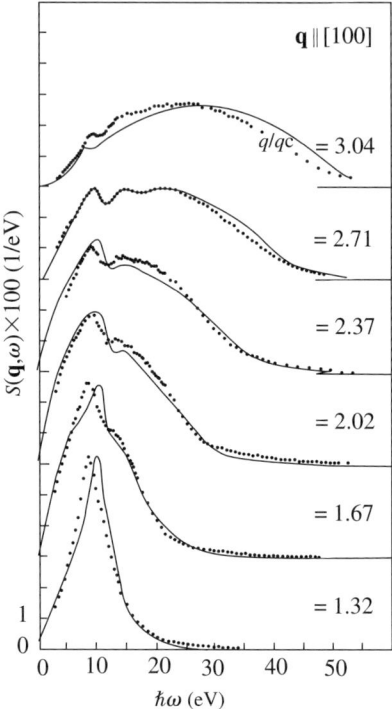

Fig. 2.22. Experimental dynamic structure factor $S(\mathbf{q},\omega)$ of lithium metal for different values of q with $\mathbf{q} \parallel [100]$; $q_c = 0.46$ a.u. Solid lines: model calculated dynamic structure factors (convoluted with the experimental resolution) taking into account lattice effects on the basis of a local pseudopotential scheme, lifetime effects within the limits of the quasiparticle lifetime model and static local-field corrections. (Reprinted with permission from Schülke (1986); copyright (1986) by the American Physical Society.)

Li has been calculated and thoroughly discussed by Sturm and Oliveira (1989) on the basis of a two-band model using parameters (band gap, effective mass, Fermi energy, and Fermi-surface anisotropy) extracted from two different first-principle band-structure calculations (Bross and Bohn 1975, Ching and Callaway 1974). They found a good agreement with experiment over the whole q-range, where these structures were visible, as shown in Fig. 2.23. This agreement must be considered as confirming the principal conception of the ZBCS's. Even a weak dip in the Li spectra found by Schülke et al. (1986) around 14 eV energy loss in the [110] and [111] spectra, and at the same position in polycrystalline spectra by Hill et al. (1996), can probably be ascribed to a lattice induced excitation gap, since this fine structure vanished upon melting (Hill 1996).

IXS spectra very rich in fine structure of $S(\mathbf{q},\omega)$ was found for Be single crystals recorded again by means of a focusing dispersion compensating analyzing

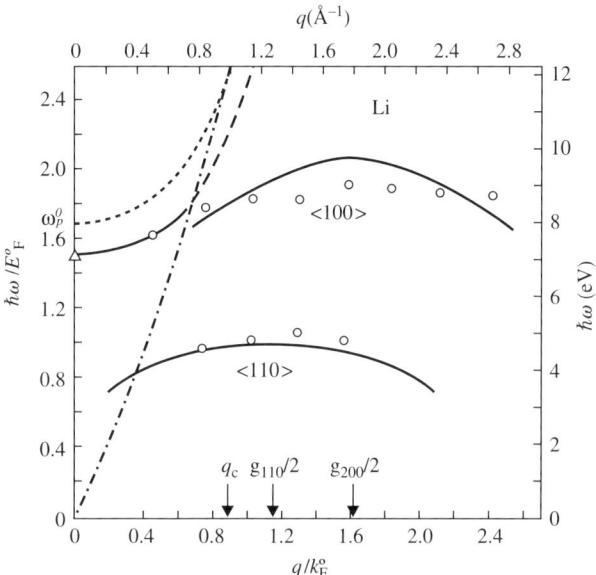

Fig. 2.23. Dispersion of the collective states in Li, both the plasmons and the zone-boundary collective states. Solid lines are the results of the two-band model calculation using parameters from the Bross and Bohn (1975) band-structure calculation for $\mathbf{q} \parallel [100]$ and $[110]$ directions. Dashed line: large-q plasmon that overlaps with the particle–hole excitation spectrum, which is enclosed by the dotted-dashed line. Short-dashed line is the jellium plasmon dispersion ending at q_c. Open circles: X-ray data of Schülke et al. (1986); triangle: EELS data of Raether (1980). (Reprinted with permission from Sturm and Oliveira (1989); copyright (1989) by the American Physical Society.)

system at DESY/HASYLAB (Schülke et al. 1987, Schülke et al. 1989), shown in the survey of Fig. 2.24. Most of the peak-dip or dip structures could be interpreted as being induced by band-structure related excitation gaps using a simple model, where only those Fourier components of the empirical pseudopotential were taken into account in (2.76), whose reciprocal lattice vector are both parallel or antiparallel to the $\mathbf{q}$-orientation of the corresponding experimental spectrum and do not have values larger than $2(q + k_F)$ (k_F = Fermi wavenumber). Whenever the corresponding Bragg plane was surrounded by unoccupied $\mathbf{k}$-space of the extended-zone scheme, the experimental dip position (zero of the third derivative of the experimental $S(\mathbf{q}, \omega)$) was found, within 0.5 eV, at the positions predicted by the model calculation. This is shown in Fig. 2.25 for three selected $\mathbf{q}$-orientations. An exceptional position among the experimental Be results take the $\mathbf{q} \parallel [110]$ spectra, especially that with $q = 0.77$, since it exhibits a very pronounced double-peak structure due to the 110 Bragg plane,

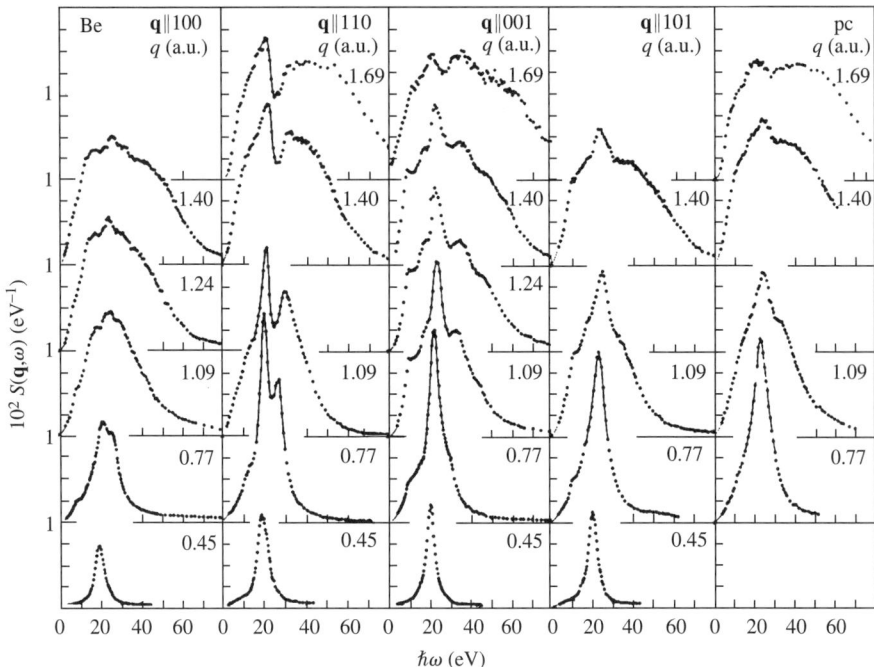

Fig. 2.24. Measured dynamic structure factors of beryllium metal for different directions and values of the momentum transfer **q**, as indicated; pc = polycrystalline. (Reprinted with permission from Schülke *et al.* (1989); copyright (1989) by the American Physical Society.)

and is depicted in Fig. 2.26. A Kramers–Kronig analysis of the experimental data (see for details Section 6.4) reveals that the first peak in Fig. 2.26 can be attributed to a zero-passage of $\mathrm{Re}\varepsilon_{\mathrm{M}}(\mathbf{q},\omega)$ for a momentum transfer $q > q_{\mathrm{c}}$, and must therefore be considered as a pure ZBCS. Until now the interpretation of the $S(\mathbf{q},\omega)$ fine structure of Be to be due to Bragg plane induced excitation gaps was based on simplifying models. But this interpretation found full confirmation by *ab initio* band structure calculations of the dynamic response of Be performed by Maddocks *et al.* (1994a) using a norm-conserving self-consistent pseudopotential of Kerker type (Kerker 1980). The confrontation of the experimental Be data with the calculated ones in Fig. 2.27 (note that the actual size of q in the calculation is the closest possible match with the experimental values) is rather convincing, and shows agreement in many details of the fine structures (peak and dip positions). Only the distribution of weight into these features differs from experiment, possibly due to the neglect of local-field corrections and lifetime effects (see the discussions in Section 2.3.3). Also very early band structure calculations of the Be dielectric response function for $\mathbf{q} \parallel [001]$ within the RPA by Taut

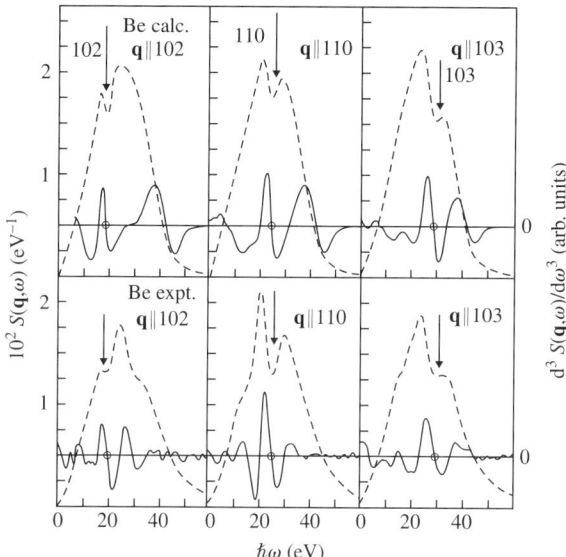

Fig. 2.25. Upper part: dashed lines: local pseudopotential calculated dynamic structure factors of Be for three different directions of $\mathbf{q}$, $q = 1.09$ a.u.. Solid lines: third derivatives of calculated $S(\mathbf{q},\omega)$. Arrows point to excitation-gap-induced dips (zeros of the third derivative), indexing of the arrows according to the Bragg planes involved. Lower part: dashed lines: Smoothed experimental $S(\mathbf{q},\omega)$; solid lines: third derivative of the experimental $S(\mathbf{q},\omega)$. (Reprinted with permission from Schülke et al. (1989); copyright (1989) by the American Physical Society.)

and Hanke (1976) have revealed peak positions in the corresponding $S(\mathbf{q},\omega)$, all in rather good agreement with the experiment, as demonstrated by Schülke et al. (1989).

The situation with Al is somewhat more puzzling. Early IXSS experiments on Al by Platzman and Eisenberger (1974), still performed with conventional X-ray sources (dispersive double-crystal setting, see Section 2.2.5.5) and their repetition with synchrotron radiation using a von Hamos focusing crystal analyzer (see Section 2.2.5.3) by Platzman et al. (1992) revealed a fine structure of $S(\mathbf{q},\omega)$ for $q = 1.7k_\mathrm{F}$ (peak-shoulder with a dip in between), which seems to be $\mathbf{q}$-orientation independent and was therefore attributed to self-energy effects within the jellium model as calculated "on the shell" by Mukhopadhyay et al. (1975) in spite of the fact that Ng and Dabrowski (1986) have put forward a fundamental objection against the application of the "on-shell" approximation. Schülke et al. (1995) have recorded $S(\mathbf{q},\omega)$ spectra by means of a focusing dispersion compensating analyzer. They have interpreted the same fine structure found in their IXS spectra of Al around $\hbar\omega \approx 35$ eV, shown in Fig. 2.13 for $\mathbf{q}\|[100]$, and

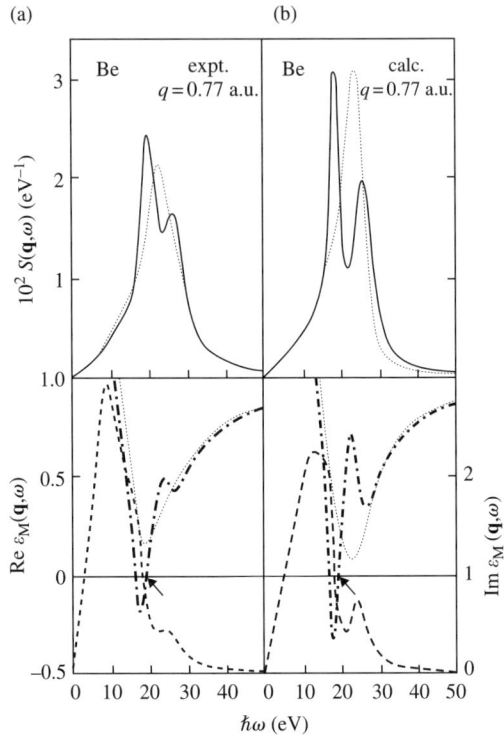

Fig. 2.26. (a) Upper part: thick solid line: experimental $S(\mathbf{q},\omega)$ of Be metal for $\mathbf{q} \parallel [110]$, $q = 0.77$ a.u.; dotted line: experimental $S(\mathbf{q},\omega)$ of polycrystalline Be, $q = 0.77$ a.u. Lower part: thick dashed-dotted line: experimental $\mathrm{Re}\varepsilon_M(\mathbf{q},\omega)$ of single crystal (Kramers–Kronig analysis), note the zero-crossing near 20 eV, indicated by an arrow. Thick dashed line: experimental $\mathrm{Im}\varepsilon_M(\mathbf{q},\omega)$; dotted line: experimental $\mathrm{Re}\varepsilon_M(\mathbf{q},\omega)$ of polycrystalline Be. (Index M means "macroscopic"). (b) Upper part: thick solid line model calculated $S(\mathbf{q},\omega)$ of Be metal for $\mathbf{q}\|110$, $q = 0.77$ a.u.; dotted line: Model calculated $S(\mathbf{q},\omega)$, weighted average over directional spectra of Be in order to simulate polycrystalline Be. Lower part: model calculations; line styles the same as in (a). (Reprinted with permission from Schülke *et al.* (1987); copyright (1987) by the American Physical Society.)

which indeed turned out to be **q**-orientation independent, as being induced by band-structure effects, where the dip of this fine structure could not be attributed to only one Bragg plane but interpreted as a consequence of a downshift of d-like unoccupied bands as predicted by Ojala (1983). Band structure $S(\mathbf{q},\omega)$ calculations of Al performed by Maddocks *et al.* (1994b) and Fleszar *et al.* (1995), both using the time-dependent local density approximation (TDLDA), have clearly shown this fine structure and have given convincing arguments for their

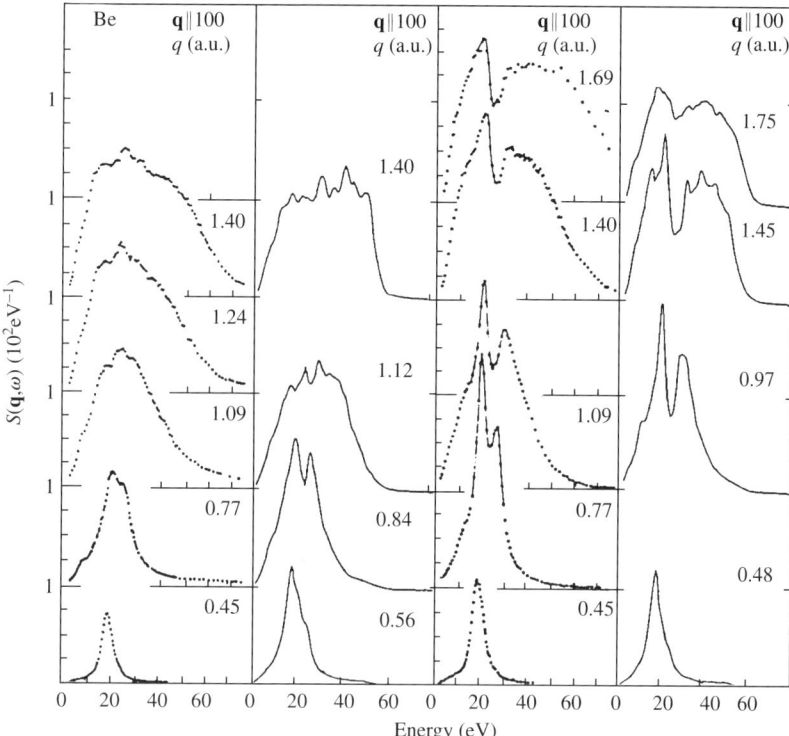

Fig. 2.27. Dots: experimental $S(\mathbf{q},\omega)$ (Schülke et al. 1989) for two different **q**-orientations and several values of q as indicated; full lines: calculated $S(\mathbf{q},\omega)$ (Maddocks et al. 1994a) for two different **q**-orientations; note that the values of q don't match completely the experimental ones. (Originally published by Maddocks et al. (1994); copyright (1994) by the American Physical Society.)

band-structure origin. The Fleszar calculations are presented in Fig. 2.28 together with the Platzman measurement. These calculations contain also those which include local-field corrections (see Section 2.3.3.3) of different degree of approximation. The agreement between experiment and those calculations which include local-field corrections of Brosens et al. (1980) is striking. Nevertheless, the comparison of these calculations with later $S(\mathbf{q},\omega)$ measurements on Al (Tischler et al. 2003) at APS (ID33 UNI-CUT) and at NSLS both performed by means of a spherically bent analyzer have revealed a better overall agreement (at least for $q = 1.7k_F$) with the TDLDA-local-field corrected (see equations (6.86a) and (6.86b)) calculations. This might be traced back to a systematic error in calibrating the inhomogeneous mosaic spread in the pyrolytic graphite energy analyzer used in the Platzman measurements. A further hint at the band structure origin of this fine structure has been delivered by Sternemann et al. (1998) by investigating

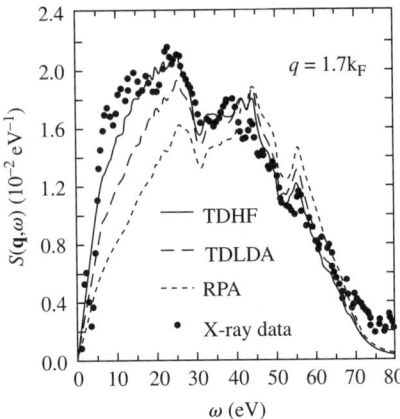

Fig. 2.28. Comparison of the calculated dynamic structure factor $S(\mathbf{q},\omega)$ of Al (Eguiluz et al. 1995) and the X-ray data of Platzman et al. (1992). The theoretical curves correspond to three different choices of the local-field factor $\tilde{G}(\mathbf{q})$: Short dashed: RPA ($\tilde{G}(\mathbf{q}) = 0$); dashed dotted: time-dependent local-density approximation (TDLDA); solid line: $\tilde{G}(\mathbf{q})$ according to Brosens et al. (1980) (TDHF). (Reprinted with permission from Fleszar et al. (1995); copyright (1995) by the American Physical Society.)

its temperature dependence. The stepwise vanishing of this fine structure with increasing temperature over the melting point, as shown in Fig. 2.29, is explained as a consequence of a Debye–Waller factor acting on the pseudopotential coefficients responsible for the appearance of this structure. Moreover, Maddocks et al. (1994) have found another strongly orientation-dependent dip assigned to a (220) Bragg plane for $\mathbf{q} \parallel [110]$, which has also been notified in the IXS spectra of Schülke et al. (1995) for the same $\mathbf{q}$-direction. This must be considered as a strong argument in favor of band-structure induced fine structure of $S(\mathbf{q},\omega)$ in a nearly free electron metal as Al. Finally the IXSS measurements on Si (Schülke et al. 1995) should be mentioned, where an excitation gap fine structure was found for $\mathbf{q}\|[110]$ and could be interpreted as being due to the (220) Bragg plane.

Macrander et al. (1996) have found in their inelastic X-ray scattering study on Ti and TiC performed at the X21 beamline of the NSLS using a spherically bent crystal analyzer in inverse geometry (Macrander et al. 1995) peaks in their TiC spectra, which were traced back, according to ι-projected LDA calculations, to an excitation gap in Im ε, accompanied by a zero-crossing of Re ε. Therefore, the peak structure around 12 eV can unequivocally be attributed to an excitation gap induced collective state.

A classical case for a zone boundary collective state has been predicted by Zhukov et al. (2001), and Ku et al. (2002) for MgB_2. By performing

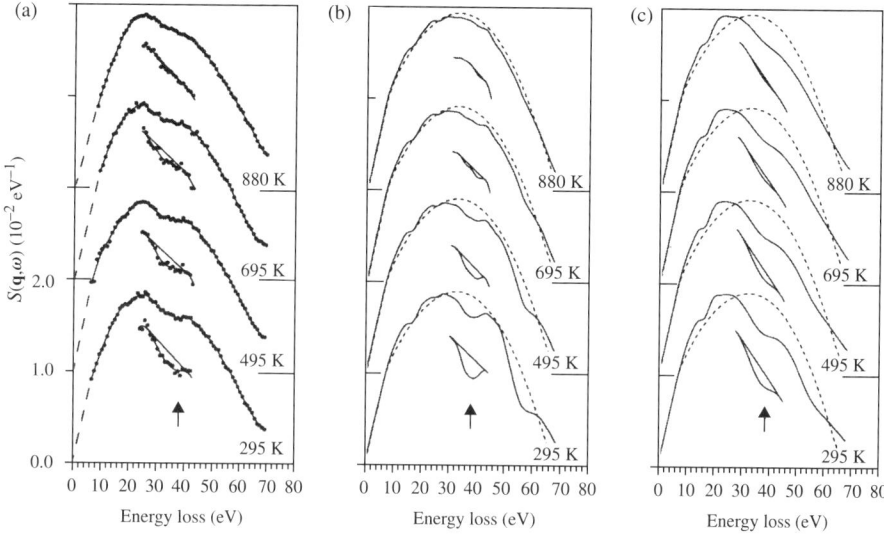

Fig. 2.29. (a) Experimental dynamic structure factor $S(\mathbf{q},\omega)$ of polycrystalline Al with $q=1.5$ a.u. at different temperatures as indicated. Inset: dip region magnified by two. (b) Solid line: empirical pseudopotential (Ojala 1983) calculations (weighted averages of five different **q**-orientations), local-field corrected; temperature effect included via Debye–Waller factor on the pseudopotential coefficients. Dashed line: jellium calculated, local-field corrected $S(\mathbf{q},\omega)$ of Al. (c) Solid line: as in (b), but additionally on-shell self-energy corrected. Dashed line: as in (b). (Reprinted with permission from Sternemann et al. (1998); copyright (1998) by the American Physical Society.)

a band-structure calculation on the basis of the time-dependent local density approximation, Ku et al. (2002) found a large gap of 5 eV width between parallel bands in the ΓM, MK and ΓK directions, so that an excitation gap is opened for **q** along the c-axis. This is demonstrated in Fig. 2.30, where the excitation gap (visible in $\mathrm{Im}\varepsilon(\mathbf{q},\omega)$) between 0 and 5 eV is accompanied by a zero-crossing of $\mathrm{Re}\varepsilon(\mathbf{q},\omega)$ thus producing a sharp peak in $-\mathrm{Im}\varepsilon^{-1}(\mathbf{q},\omega)$, a zone boundary collective state of the first water, which is interpreted by the authors to be the result of charge fluctuations between boron and magnesium sheets. Zhukov et al. (2001) came to nearly the same result. This collective state was detected experimentally by Galambosi et al. (2005) at the beamline ID16 of the ESRF with a total energy resolution of 1 eV and a momentum-transfer resolution of 0.21 a.u. The spectra for q between 0.18 a.u. and 0.77 a.u. with $\mathbf{q} \parallel [001]$ are shown in Fig. 2.31 and exhibit the typical dispersion behavior of a zone-boundary collective state with a forward dispersion to $\mathbf{q}=\mathbf{g}_{001}/2$ followed by a backward dispersion. These findings were verified by Cai et al. (2006) with an IXS study at the BL12XU Spring8 beamline.

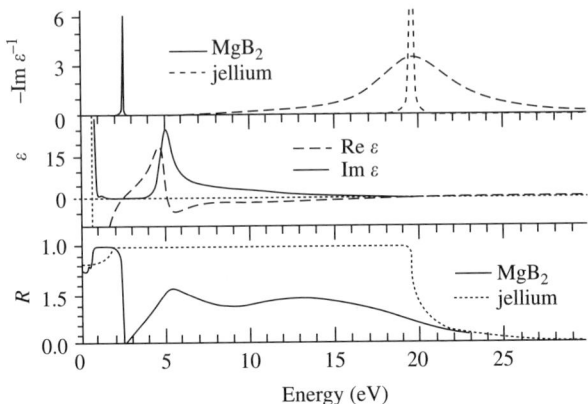

Fig. 2.30. Solid line: MgB$_2$; dashed line: jellium counterpart; **q** $\parallel$ c-axis; $q =$ 0.12 Å^{-1}. Upper panel: Calculated loss function. Middle panel: Real (long dashed) and imaginary part (solid line) of the dielectric function. Lower panel: Reflectivity. (Reprinted with permission from Ku *et al.* (2002); copyright (2002) by the American Physical Society.)

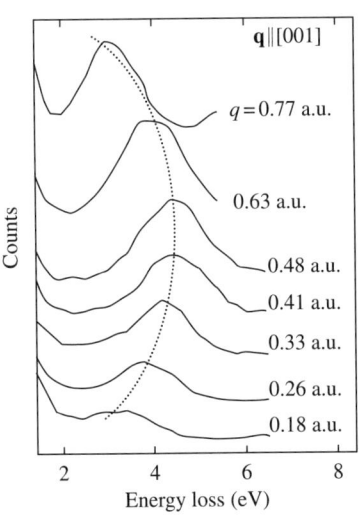

Fig. 2.31. Experimental energy-loss spectra of MgB$_2$ with **q** perpendicular to the hexagonal boron planes. Dotted line shows the position of the various peak maxima (guide to the eye). (Reprinted with permission from Galambosi *et al.* (2005); copyright (2005) by the American Physical Society.)

2.3.2.3 Plasmon bands and the particle–hole continuum, plasmon-Fano(anti) resonances

We have considered in Section 2.1, how an electron system under the influence of a periodic potential may react on an external perturbing potential with a Fourier component $\phi_{\text{ext}}(\mathbf{q}_r + \mathbf{g}, \omega)$, and found that this potential component can induce density fluctuations belonging to another reciprocal lattice vector $\mathbf{g}'$. As shown by Saslow and Reiter (1973), this property of an inhomogeneous electron system also has consequences for the plasmons, the collective excitations of the electron system. The transfer of a momentum $\mathbf{q}_r + \mathbf{g}$ to an inhomogeneous electron system will not only create a plasmon belonging to the wavevector $\mathbf{q}_r + \mathbf{g}$ but also those with wave vector $\mathbf{q}_r + \mathbf{g}'$; $\mathbf{g}' \neq \mathbf{g}$, so that this family $\mathbf{g}' \neq \mathbf{g}$ is establishing the so-called higher plasmon bands, which can be either represented in the reduced zone scheme as $\omega_{p\nu}(\mathbf{q}_r)$, or in the repeated zone scheme as $\omega_p(\mathbf{q}_r + \mathbf{g})$, in the same way as is well-known for one-electron states under the influence of a periodic lattice potential. This plasmon band structure should, in principle, be reflected, when measuring the dynamic structure factor with $q_r < q_c$. According to

$$S(\mathbf{q}, \omega)_{\text{RPA}} = \left[\frac{q_r^2}{4\pi^2 e^2}\right] \text{Im}\left[\frac{-1}{\varepsilon(\mathbf{q}, \omega)}\right]_M^{\text{RPA}} \tag{2.77}$$

and by using (2.28) together with the explicit expressions (6.98) and (6.103a) for the elements $X_{\mathbf{gg}'}^0$ and $T_{\mathbf{gg}'}$ of the respective matrices, we obtain, within the limits of a two-plasmon band model (Oliveira and Sturm 1980; this means by taking into account only one reciprocal lattice vector $\mathbf{g} \neq 0$),

$$S(\mathbf{q}_r, \omega)_{\text{RPA}} = \left[\frac{q_r^2}{4\pi^2 e^2}\right] \text{Im}\left[\frac{-1}{\varepsilon_{00} - \varepsilon_{0\mathbf{g}}\varepsilon_{\mathbf{g}0}/\varepsilon_{\mathbf{gg}}}\right], \tag{2.78}$$

where the elements $\varepsilon_{\mathbf{gg}'}(\mathbf{q}_r, \omega)$ of the so-called dielectric matrix are connected with the elements $\mathbf{X}_{\mathbf{gg}'}^L$ of the Lindhard-type polarization matrix as

$$\varepsilon_{\mathbf{gg}'}(\mathbf{q}_r, \omega) = \delta_{\mathbf{gg}'} - v(\mathbf{q}_r + \mathbf{g})X_{\mathbf{gg}'}^0 \tag{2.79}$$

(the arguments $\mathbf{q}_\mathbf{r}$ and $\boldsymbol{\omega}$ are suppressed in the elements of the dielectric matrix). Equation (2.78) marks a proper description of the plasmon resonance within the two-plasmon band model, insofar as $\varepsilon_{00} - \varepsilon_{0\mathbf{g}}\varepsilon_{\mathbf{g}0}/\varepsilon_{\mathbf{gg}}$ exhibits zero passages at two different frequencies. Additionally remember that, in the one-electron band structure, there exists, at least in a simple two-band model, a band gap at the Brillouin zone boundary equal to $2|V_\mathbf{g}|$, where $V_\mathbf{g}$ is the $\mathbf{g}$th Fourier component of the periodic lattice potential. An analogous situation is true with the plasmon band structure, as depicted in Fig. 2.32: the corresponding gap at the Brillouin zone boundary is, within the limits of a two-plasmon band model equal to $2|\rho_\mathbf{g}|$, where $\rho_\mathbf{g}$ is the $\mathbf{g}$th Fourier component of the electron density ρ.

The existence of Umklapp processes, which lead to plasmon bands, can be proved in an inelastic X-scattering experiment (Sturm et al. 1992b), where one is utilizing the interaction of the **discrete** Umklapp plasmons, as shown in

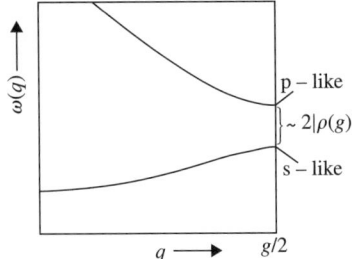

Fig. 2.32. Plasmon bands, two-plasmon band model.

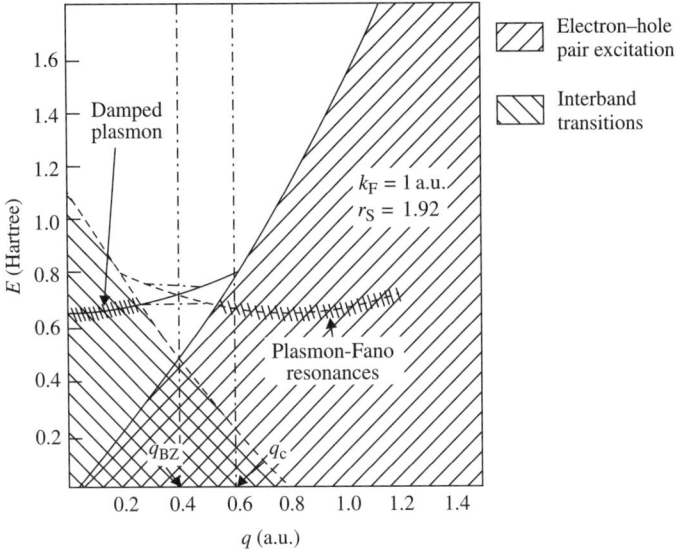

Fig. 2.33. Illustration of the effect of Umklapp processes on both the particle–hole excitation continuum and the plasmon excitation branch in the repeated zone scheme of a two-plasmon band model. Solid lines: regular plasmons of the first band (damped within the range of interband transitions); dashed lines: second plasmon band. Plasmon-Fano resonances, where overlap with the particle–hole continuum takes place; dashed-dotted lines indicate the plasmon band gap at the Brillouin zone (BZ) boundary.

Fig. 2.33 for the repeated zone scheme of a simple two-plasmon band model, with the continuum of particle–hole excitations in the q-range with $q > q_{BZ}$ (q_{BZ} = momentum belonging to the BZ boundary). For general physical reasons such an interaction between discrete excitations and a continuum of excitations should give rise to so-called Fano resonances (Fano 1961). This can be demonstrated within the limits the above discussed two-plasmon band model, which, according to (2.77) together with the explicit expressions (6.98). (6.103a) and (6.112) yields the following result for $S(\mathbf{q},\omega)$ (Sturm et al. 1992b):

$$S(\mathbf{q},\omega) = \left(\frac{q_r^2}{4\pi^2 e^2}\right) \operatorname{Im}\left(\left(\frac{-1}{\varepsilon_{\mathbf{gg}}}\right) + \left(\frac{\varepsilon_{\mathbf{g0}}\varepsilon_{\mathbf{0g}}}{\varepsilon_{\mathbf{gg}}^2}\right)\left[\frac{-1}{\varepsilon(\mathbf{q_r},\omega)}\right]_{\mathrm{M}}^{\mathrm{RPA}}\right) \quad (2.80)$$

(the arguments of $\varepsilon_{\mathbf{gg}}$ and $\varepsilon_{\mathbf{g0}}$ again are suppressed).

This expression consists of two parts in the ( )-brackets. The first one represents the particle–hole excitation spectrum in the range of the Umklapp plasmon. The second one stands for the plasmon coupled to the particle–hole excitation by the factor $\varepsilon_{\mathbf{g0}}\varepsilon_{\mathbf{0g}}/\varepsilon_{\mathbf{gg}}^2$, which is complex with a negative real and imaginary part. Since $\operatorname{Im}\left[-1/\varepsilon(\mathbf{q_r},\omega)\right]_{\mathrm{M}}^{\mathrm{RPA}}$ is always positive and $\operatorname{Re}\left[-1/\varepsilon(\mathbf{q_r},\omega)\right]_{\mathrm{M}}^{\mathrm{RPA}}$ becomes negative for $\omega > \omega_p(\mathbf{q})$ (ω_p = plasmon frequency), this coupling term exhibits a valley–peak structure, typical for a Fano resonance. This valley–peak structure superimposed on the well-known particle–hole contribution was indeed found in $S(\mathbf{q},\omega)$ measurements on Si with $\mathbf{q} \| [111]$ and $q > q_c$ around 18 eV, recorded by means of a focusing dispersion compensating device at DESY/HASYLAB (Sturm et al. 1992b, Schülke et al. 1995), as shown exemplarily in Fig. 2.34. In this figure also calculations using (2.80) and a local empirical pseudopotential are visualized. The agreement with experiment was considerably improved when going beyond the RPA utilizing (6.113) for the reciprocal macroscopic dielectric function (see also the next section 2.3.3). Figure 2.35 from Schülke et al. (1995) demonstrates in convincing form the typical shape of a Fano resonance. Likewise $S(\mathbf{q},\omega)$ measurements on Li with $\mathbf{q} \| [110]$ and $q > q_c$ have exhibited this valley–peak structure (Höppner et al. 1998), which again can be traced back to plasmon-Fano resonances. Here the measurements were compared with

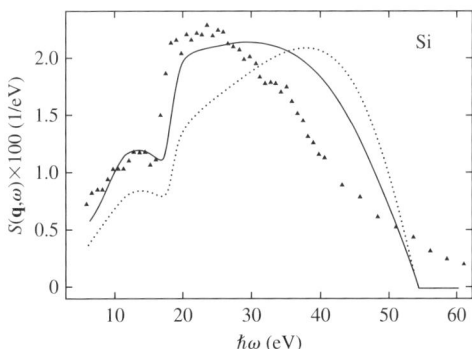

Fig. 2.34. Plasmon Fano (anti)resonance. Triangles: experimental dynamic structure factor of Si for $\mathbf{q} \| [111]$; $q = 1.25$ a.u.; points: calculated dynamic structure factor $S(\mathbf{q},\omega)$, two-plasmon band model; solid line: calculated $S(\mathbf{q},\omega)$ including local-field correction. (Reprinted with permission from Sturm et al. (1992b); copyright (1992) by the American Physical Society.)

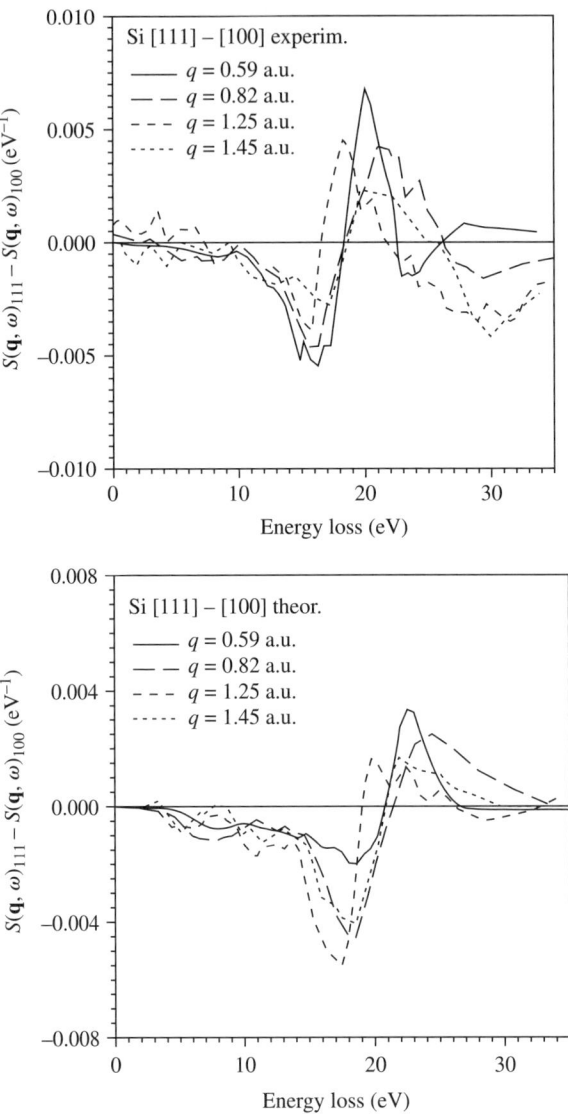

Fig. 2.35. Valley–peak structure typical for plasmon-Fano (anti)resonances in Si. Upper part: experimental differences of Si $S(\mathbf{q},\omega)$ curves between [111] and [100] **q**-orientations for different values of q as indicated in the inset. Lower part: pseudopotential calculated Si $S(\mathbf{q},\omega)$ differences between [111] and [100] **q**-orientations corresponding to the experimental differences. (Reprinted with permission from Schülke et al. (1995); copyright (1995) by the American Physical Society.)

calculations of the full dielectric matrix (Bross and Ehrnsperger 1995). In a study on MgB$_2$, Cai et al. (2006) have found, by going beyond the two-plasmon band model, that the real part of the factor which couples the Umklapp zone boundary collective state to the particle–hole excitation continuum turned out to be positive. Thus this coupling produces a strong peak structure periodic in the reciprocal space.

We have seen that the occurrence of plasmon-Fano (anti)resonances is due to the matrix character of the dielectric function, in other words, they have to be considered as a consequence of the coupling to the short-wavelength density fluctuations described by the nondiagonal elements of the dielectric matrix. Montano et al. (2002) have demonstrated in an inelastic X-ray scattering study on hexagonal and cubic modifications of SiC, that their experimental $S(\mathbf{q}, \omega)$ were only be brought into sufficient agreement with their calculations on the basis of linear combination of muffin-tin orbitals, when the full dielectric matrix (with 29 reciprocal lattice vectors) was taken into account. It is interesting that the difference between their calculation with and without off-diagonal terms exhibits a strong negative contribution (antiresonance) around the plasmon energy position so that an interpretation in terms of a plasmon-Fano antiresonance seems to be appropriate.

2.3.3 Interpretation of $S(\mathbf{q}, \omega)$ beyond RPA

The plasmon features as well as the fine structure of $S(\mathbf{q}, \omega)$ due to band structure effects discussed so far could well be explained within the limits of the RPA at least as far as their energy position is concerned. But it turned out very early that the overall shape of the experimental IXS spectra, especially for $q > q_c$, are far from being described adequately by the RPA. Therefore, more sophisticated corrections to the RPA, as far as the dynamical structure of electron systems is concerned, are necessary, were performed, and there exists a huge amount of literature devoted to this problem. Looking at the time between 1975 and 1990 most of this effort has been triggered by the apparent universal double-peak structure of $S(\mathbf{q}, \omega)$, found in early experiments, and interpreted as being due to a likewise universal property of the jellium dynamic response. Without going through the very exciting history of trying with alternating success to reproduce this double-peak structure along with the overall shape of the experimental $S(\mathbf{q}, \omega)$'s, we will refer to the theoretical results of an Australian group, who started with two fundamental papers (Green et al. 1985a; Green et al. 1985b). These papers and their further detailed studies (Green et al. 1985c; Green et al. 1987a; Green et al. 1987b) can be considered as standing at the very end of this history, and giving the most general answer to the problem under debate, at least for the jellium case. We can, in what follows, only sketch the basis of their studies: The authors start with an expression for the exact ground-state energy of an interacting electron system as a functional of the "fully dressed" one-particle Green's function G (for a definition of the one-particle Green's function G^0, see (6.77), the "fully dressed" Green's function is defined in (2.83)). From the complete set of closed Feynman diagrams, representing the exchange correlation part of this energy functional a subset was retained, whose symmetry ensures that the conservation rules are exactly satisfied. Then the single-particle self-energy $\Sigma[G]$ is the variational derivative of the ground-state energy functional with respect to $(-iG)$. Finally, by taking the second variation with respect to G of the ground-state energy functional, one can generate a conserving integral equation for the proper electron–hole polarization propagator $\Lambda_{sc}[G]$, whose trace over momentum and spin variables is the proper polarization function $\chi_{sc}(\mathbf{q},\omega)$ (in the definition (2.18) denoted as $\chi_0(\mathbf{q},\omega)$), which is intimately related to the dynamical structure factor $S(\mathbf{q},\omega)$ via (2.17 and 2.20). The equation for $\Lambda_{sc}[G]$ contains an effective two-body interaction $\Xi_{sc}[G]$, so that $\Lambda_{sc}[G]$ can be expanded in powers of $\Xi_{sc}[G]$. Then we obtain a series of terms $\chi_0, \chi_1, \chi_2, \ldots$ of increasing order in the effective interaction. We will come back to this point later. Already this short summary of how the corrections to the RPA result from basic relations will make clear that it is two effects that matter: On the one hand, the self-energy effects via the fully dressed Green's function, and on the other hand, the two-body interaction described by $\Xi_{sc}[G]$. Of course, these two effects must not be considered as being independent one from the other. On the contrary, there exists the following interrelationship

$$\Xi_{sc}[G] = i\left(\frac{\delta \Sigma[G]}{\delta G}\right) - V, \qquad (2.81)$$

with V the Coulomb potential, a relation, which guarantees, along with the space–time symmetry of the diagrammatic expansion of the self-energy $\Sigma[G]$, that the resulting polarization function $\chi_{sc}(\mathbf{q},\omega)$ is conserving.

2.3.3.1 *Self-energy effects* More intuitively speaking, the self-energy correction consists in taking into account that the excited particle or hole can polarize the electron fluid surrounding it, so that this polarization can act back on the particle or hole thus modifying the particle/hole propagator. Formally this type of reaction is expressed for the jellium case in the so-called self-energy, a complex function of momentum $\mathbf{k}$ and energy E, which is given in the lowest order approximation (Hedin 1965) as

$$\Sigma(\mathbf{k},E) = \left[\frac{i}{(2\pi)^4}\right]\int v(\mathbf{k}')d^3k'\int \exp(-i\eta E')\left[\frac{1}{\varepsilon(\mathbf{k}',E')}\right]_{RPA}$$
$$\times (E-E'-\varepsilon_{\mathbf{k}-\mathbf{k}'})^{-1}dE'; \quad \eta \to +0, \qquad (2.82)$$

where the dynamic screening of the Coulomb interaction $v(\mathbf{k}')$ (see (equation 2.16 for its definition) is performed via the RPA response function. $\varepsilon_{\mathbf{k}}$ is the kinetic energy of a free particle with momentum $\mathbf{k}$. The first-order self-energy correction of a particle (hole) propagator (Green's function), writes

$$G(\mathbf{k},\varepsilon) = G^0(\mathbf{k},\varepsilon) + G^0(\mathbf{k},\varepsilon)\Sigma(\mathbf{k},\varepsilon)G^0(\mathbf{k},\varepsilon). \qquad (2.83)$$

Allowing for the self-energy correction to infinite order, according to the so-called GW approximation, as shown in Fig. 2.36, the simple one-particle picture, where the free particle is represented by its momentum and the corresponding kinetic energy $\varepsilon_{\mathbf{k}}$, must be modified into a quasiparticle picture, dominated by the so-called spectral density function $A(\mathbf{k}, E)$, which expresses for positive (negative) energies E the relative probability per energy unit for the system to be in a state with an energy $E + \mu$ ($-E - \mu$) above the ground state just after the injection of one electron (hole) of momentum $\mathbf{k}$ (μ is the chemical potential) (Schriefer 1964). The spectral density function is connected to the self-energy by

$$A(\mathbf{k}, E) = -\left(\frac{1}{\pi}\right) \frac{\mathrm{Im}\Sigma(\mathbf{k}, E)}{[E - \varepsilon_{\mathbf{k}} + \mu - \mathrm{Re}\Sigma(\mathbf{k}, E)]^2 + [\mathrm{Im}\Sigma(\mathbf{k}, E)]^2}, \qquad (2.84)$$

so that the imaginary part of the self-energy stands for a broadening of the spectral density function, which is connected with the finite lifetime of the quasiparticle, and the real part of the self-energy characterizes the energy shift relative

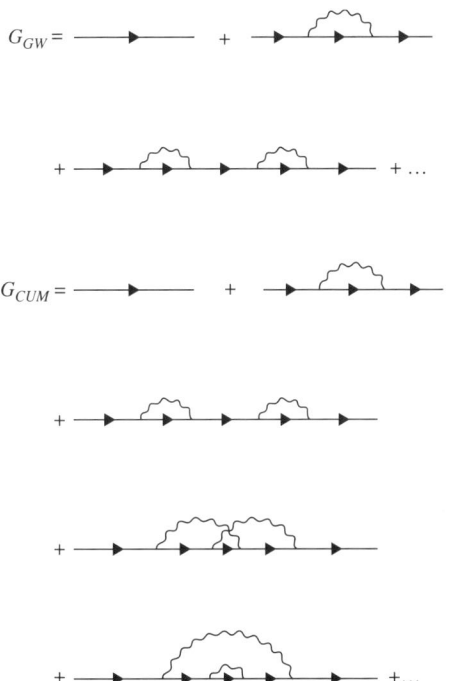

Fig. 2.36. Diagrammatic expansion for the Green's function to second order in the GWA and the cumulant expansion, respectively. The solid lines represent the noninteracting Green's functions G^0, the wiggly lines the screened Coulomb interactions. (Originally published by Holm and Aryasetiawan (1997); copyright (1997) by the American Physical Soiciety.)

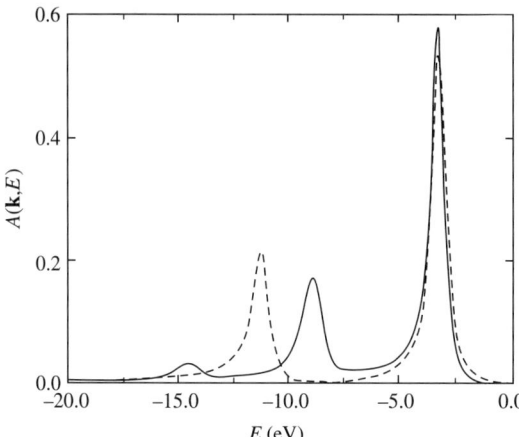

Fig. 2.37. Self-consistent spectral density function $A(\mathbf{k}, E)$ for $k = 0$, $r_s = 4$; energy-zero is the Fermi-energy. Solid curve: cumulant expansion; dashed curve: jellium GW approximation. (Originally published by Holm and Aryasetiawan (1997); copyright (1997) by the American Physical Society.)

to the free particle kinetic energy $\varepsilon_\mathbf{k}$. It was first shown by Lundqvist (1967) that the jellium spectral density function (2.84) is not simply a one-peak function but exhibits a much more complicated structure, depicted in Fig. 2.37, with two prominent peaks for $k < k_\mathrm{F}$, one peak near $\varepsilon_\mathbf{k} - \mu$, and a second one at lower energies roughly $-1.5\hbar\omega_\mathrm{p}$ (ω_p = free electron plasmon frequency) apart from the first one, known as the plasmaron peak, since it stands for a bound plasmon–hole state. Also for $k > k_\mathrm{F}$ the spectral density function is at least double peaked. If one allows for self-energy corrections as shown in the lower panel of Fig. 2.36, the so-called cumulant expansion correction (Holm and Aryasetiawan 1997), the double-peak structure of the spectral density function changes into a three-peak structure, as shown in Fig. 2.37, so that the two additional peaks, for $k < k_\mathrm{F}$, are by $-\hbar\omega_\mathrm{p}$ and $-2\hbar\omega_\mathrm{p}$ apart from $\varepsilon_\mathbf{k} - \mu$, respectively.

Of course, also the definition of the Green's function given in (6.77) must be changed accordingly into

$$G(\mathbf{k}, \varepsilon) = \left(\frac{1}{2\pi}\right)\left(\int_{-\infty}^{0} dE \frac{A(\mathbf{k}, E)}{\varepsilon - E - \mu - i\eta} + \int_{0}^{\infty} dE \frac{A(\mathbf{k}, E)}{\varepsilon - E - \mu + i\eta}\right). \quad (2.85)$$

The chemical potential μ is given by the renormalized quasiparticle energy evaluated at the Fermi momentum.

If we replace the "undressed" Green's function in the expression (6.76) for the Lindhard polarization function by the self-energy corrected Green's function (2.85) (often designated as the "fully dressed" propagator) we will speak of the fully self-energy corrected polarization function $\chi_{\mathrm{SE},0}(\mathbf{q}, \omega)$.

One can easily show that the imaginary part of $\chi_{\text{SE},0}(\mathbf{q},\omega)$ is then given by convoluting the spectral density function of the particle with that of the hole left behind:

$$\text{Im}\chi_{\text{SE},0}(\mathbf{q},\omega) = -\int \frac{\mathrm{d}^3 k}{8\pi^3} \int_{-\omega+E_\text{F}}^{E_\text{F}} \frac{\mathrm{d}E}{2\pi} A(\mathbf{k},E)\, A(\mathbf{k}+\mathbf{q}, E+\omega). \qquad (2.86)$$

If one neglects the real part of the self-energy and interprets the "on-shell" imaginary part of the self-energy, $\Gamma(\mathbf{k}) \equiv \text{Im}\Sigma(\mathbf{k},\varepsilon_\mathbf{k})$, as the inverse lifetime of the particle with momentum $\mathbf{k}$, one can insert $i[\Gamma(\mathbf{k}) + i\Gamma(\mathbf{k}+\mathbf{q})]$ into the energy denominator $(E_{\mathbf{k}+\mathbf{q}} - E_\mathbf{k} + i\varepsilon)$ appearing in the relation for the Lindhard polarization function as given explicitly in (6.74) (Mukhopadhyay et al. 1975). This approximation is called the quasiparticle lifetime model (Green et al. 1985b; Green et al. 1987a) and, in spite of the rather crude approximation, it has been shown in a number of studies that it can bring the calculation nearer to experiment, when compared with the pure RPA. This has been demonstrated for Al (Mukhopadhyay et al. 1975, Rahman and Vignale 1984, Schülke et al. 1993), for Li (Schülke et al. 1986), for Be (Schülke et al. 1989), and for liquid Al (Sternemann et al. 1998). Even a double-peak structure or at least a peak–shoulder structure of $S(\mathbf{q},\omega)$ could be reproduced, and was physically traced back to the characteristic $\mathbf{k}$-dependence of the inverse lifetime: $\Gamma(\mathbf{k})$ undergoes a steep rise at a certain value k_0, where a particle with the kinetic energy $k_0^2/2m$ can excite a plasmon. Thus the spectral density functions of the excited quasiparticles contributing to the low-ω part of $S(\mathbf{q},\omega)$ are much less broadened than those of excited quasiparticles constituting the high-ω part. In this way, a dip between both contributions can arise, as first demonstrated by Niklasson et al. (1983). We show exemplarily in Fig. 2.38 the result of corresponding calculations of Rahman and Vignale (1984) confronted with an Al experiment of Platzman and Eisenberger (1974). Nevertheless the quasiparticle lifetime model was criticized for various reasons. Niklasson (1985) has stressed the fact that this model assumes a symmetric spectral density function of the quasiparticle, whereas this function is, according to the results of Lundqvist (1967) everything else but symmetric. He has shown how the double-peak structure disappears even using a simple model for the asymmetric spectral density function. Ng and Dabrowski (1985) came to similar conclusions by calculating separately the contributions to the dynamical structure factor of the different components of the realistic spectral density function. Of much more fundamental nature was the criticism of Green et al. (1987a) who stated that the quasiparticle lifetime model possesses a divergent third-moment sum rule. This is apparently due to the failure of this model to incorporate the well-known cancellation between the self-energy insertions and dynamical vertex corrections within the polarization function. The dynamic vertex corrections are totally lost in the static local-field approximation, as used by the authors of the quasiparticle lifetime model. Thus it was not surprising that the application of the full self-energy corrected polarization function $\chi_{\text{SE},0}(\mathbf{q},\omega)$

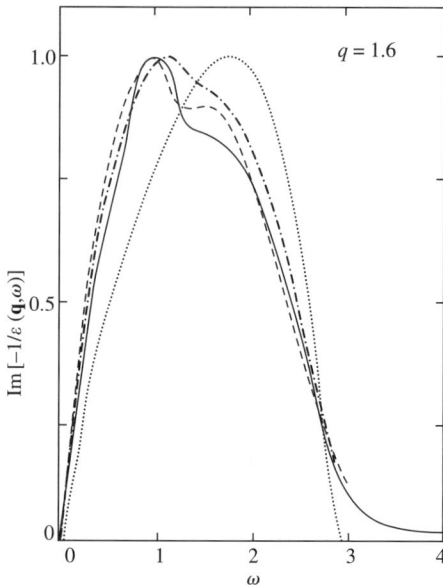

Fig. 2.38. Calculated $\mathrm{Im}(-1/\varepsilon(\mathbf{q},\omega))$ for $r_s = 2$ and $q/k_\mathrm{F} = 1.6$ in the quasiparticle lifetime model: Solid line: plasmon-pole model for the self-energy, local-field correction (see Section 2.3.3.3) included (Singwi et al. 1970). Dashed-dotted line: RPA self-energy with vertex correction (see Section 2.3.3.5), local-field correction included. Dotted line: RPA. ω in units of $2E_\mathrm{F}$; dashed line: experimental $S(\mathbf{q},\omega)$ of Al (Platzman and Eisenberger 1974). (Originally published by Rahman and Vignale (1984); copyright (1984) by The American Physical Society.)

of (2.86) to the computing of the dynamic structure factor does not lead to a better agreement with experiment compared with RPA, when all dynamical vertex corrections were neglected. This is shown in Fig. 2.41 for the case of Al (Schülke 2004), and must be traced back to the two-peak structure of the hole spectral density function with a so-called plasmaron peak shifted to lower energies against the quasihole peak by more than the plasmon energy, as shown in Fig. 2.36. Therefore, it is high time to discuss now in more detail the second source of deviations from the RPA, already mentioned above, namely the two-body interaction described by $\Xi_\mathrm{sc}[G]$.

2.3.3.2 *Vertex corrections* We will follow again the presentation of Green et al. (1985b) and Green et al. (1987a), since they are to the belief of the author, the most thorough ones in this special aspect. Thus we start with the following approximation for the proper polarization function $\chi_\mathrm{sc}(\mathbf{q},\omega)$, the result of expanding the electron–hole polarization propagator $\Lambda_\mathrm{sc}[G]$ in powers of the effective two-body interaction $\Xi_{\mathrm{sc}}[G]$. Then we have to first order in $\Xi_{\mathrm{sc}}[G]$ ($\hbar = 1$)

$$\chi_{\mathrm{sc}}(\mathbf{q}, \omega) \approx \chi_{\mathrm{SE},0}(\mathbf{q}, \omega) + \chi_{\mathrm{SE},1}(\mathbf{q}, \omega) \tag{2.87}$$

$$\chi_{\mathrm{SE},0}(\mathbf{q}, \omega) = 2 \sum_{\mathbf{k}_1} \int \frac{\mathrm{d}\omega_1}{2\pi \mathrm{i}} G(\mathbf{k}_1, \omega_1) G(\mathbf{q} + \mathbf{k}_1, \omega + \omega_1) \tag{2.88}$$

$$\chi_{\mathrm{SE},1}(\mathbf{q}, \omega) = 4 \sum_{\mathbf{k}_1} \sum_{\mathbf{k}_2} \int \frac{\mathrm{d}\omega_1}{2\pi \mathrm{i}} \int \frac{\mathrm{d}\omega_2}{2\pi \mathrm{i}} G(\mathbf{k}_1, \omega_1) G(\mathbf{q} + \mathbf{k}_1, \omega + \omega_1)$$

$$\times \left(\frac{1}{4} \sum_{\sigma_1 \sigma_2} \{\Xi_{\mathrm{sc}}[G]_{\sigma_1 \sigma_2}(\mathbf{k}_1, \omega_1; \mathbf{k}_2, \omega_2; \mathbf{q}, \omega)\} \right)$$

$$\times G(\mathbf{k}_2, \omega_2) G(\mathbf{q} + \mathbf{k}_2, \omega + \omega_2). \tag{2.89}$$

Here $(\mathbf{k}_1, \varepsilon_1, \sigma_1)$ and $(\mathbf{k}_2, \varepsilon_2, \sigma_2)$ are single-particle momentum, energy (frequency), and spin variables. G is the "fully dressed", this means self-energy corrected, Green's function. Therefore, the zero-order polarization function (2.88) is identical with that, whose imaginary part is defined in (2.86). Let us now consider the first-order contribution to the polarization function $\chi_{\mathrm{sc}}(\mathbf{q}, \omega)$.

After the frequency integrals in (2.89) have been performed, the polarization function $\chi_{\mathrm{SE},1}$ becomes the sum of all possible time-ordered Goldstone diagrams (Brown 1972). Thus $\Xi_{\mathrm{sc}}[G]$ in (2.89) separates out into a set of scattering matrix elements linking the initial and the final particle–hole pairs. We can now expand these matrix elements in terms of the components of either the T-matrix or the shielded potential V_{sc}. Without going further into details, we follow Green et al. (1985b) and distinguish between the following three classes of contributions to $\chi_{\mathrm{SE},1}$:

Class A are contribution in which the initial and the final electron–hole pairs both carry the external momentum and energy $(\mathbf{q}, \omega)$ in the same time direction, so that we have either electron–electron or hole–hole correlations. It was stressed by Green et al. (1985b) that only contributions of this class A, although of rather small amplitude, have a structure which leads to a ω-dependence varying rapidly enough to possibly account for fine structure in $S(\mathbf{q}, \omega)$, as found experimentally (e.g. the double-peak structure). Figure 2.39, taken from Schüke et al. (1985), of course, proves that this class A induced fine structure cannot explain, at least for the case of Li, the fine structure found experimentally and must be traced back, as shown in Section 2.3.2.2, to band structure effects. However, it is not the fine structure but the overall shape of $S(\mathbf{q}, \omega)$ and its behavior for large ω, which makes up the big difference between RPA and experiment, and which can satisfactorily be decribed by the above scheme, by the class B and, above all, the class C contributions.

Class B are contributions in which the initial and final electron–hole pairs propagate in opposite time directions, so that there is no overall energy transfer. Therefore they have to be considered as purely static.

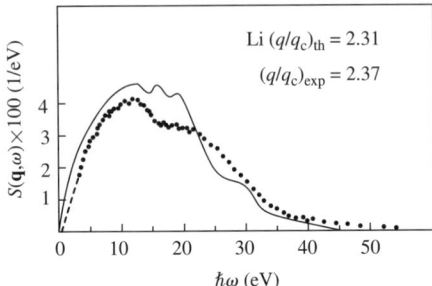

Fig. 2.39. The dynamic structure factor $S(\mathbf{q},\omega)$ of Li ($r_s = 3.2$). Points: experimental values for $\mathbf{q} \parallel [111]$, $q/q_c = 2.37$ (Schülke *et al.* 1986); solid curve: jellium calculation on the basis of the conserving dynamic theory of Green *et al.* (1985a) including class A induced fine structure (Green *et al.* 1985b,c); $q/q_c = 2.31$. (Reprinted with permission from Schülke *et al.* (1986); copyright (1986) by the American Physical Society.)

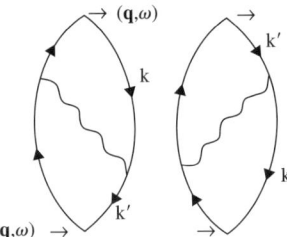

Fig. 2.40. Class C contributions to $\chi_{\text{SE},1}(\mathbf{q},\omega)$. Solid line: full self-energy corrected Green's function; wiggly line: screened particle–hole interaction.

Class C contributions, which are by far the most prominent ones, are those in which the initial and final pairs propagate in the same time direction but such that the correlations link electron and hole. The dominant class C contribution, where the single electron–hole is correlated through the fully dynamically screened Coulomb interactions, is shown diagrammatically in Fig. 2.40. In other words, electron and hole are interacting by the exchange of a plasmon. The corresponding expression of the first-order correction to $\chi_{\text{sc}}(\mathbf{q},\omega)$, often denoted in the literature as the first-order vertex correction, reads as follows ($\hbar = 1$):

$$\chi_{\text{SE},1}(\mathbf{q},\omega) = -2 \int \frac{d^3k_1}{8\pi^3} \int \frac{d\omega_1}{2\pi i} G(\mathbf{k}_1,\omega_1)\, G(\mathbf{q}+\mathbf{k}_1,\omega_1+\omega)$$
$$\times \int \frac{d^3k_2}{8\pi^3} \int \frac{d\omega_2}{2\pi i} \left[\frac{v(\mathbf{k}_2-\mathbf{k}_1)}{\varepsilon_{\text{RPA}}(\mathbf{k}_2-\mathbf{k}_1,\omega_2-\omega_1)} \right]$$
$$\times G(\mathbf{k}_2,\omega_2)\, G(\mathbf{q}+\mathbf{k}_2,\omega_2+\omega), \qquad (2.90)$$

where the dynamical screening of the Coulomb potential is performed by means of the RPA dielectric function. Green *et al.* (1987a) have demonstrated that this dynamic vertex correction brings about not only a shift to smaller ω of the $S(\mathbf{q},\omega)$ peak with respect to its RPA counterpart, as claimed by the experiment, but cancels also the long ω-tail of the purely self-energy corrected dynamic structure factor to such an extent that the third-moment sum rule can be fulfilled. Unfortunately, the only experimental $S(\mathbf{q},\omega)$ data the above authors have compared with their $r_s = 3.2$ calculations (Green *et al.* 1985c) are those of Priftis *et al.* (1978) on Li, still taken with conventional sources and therefore of poor statistics. Moreover these calculations are based on a combination of approximately local (Hartree–Fock) contributions to $S(\mathbf{q},\omega)$ (see below) with the contributions of the class A terms, rather than on a full first–order vertex correction according to (2.90). These computations are in sufficient agreement with the rather noisy data, as far as peak position and the overall shape are concerned. A similar statement was made by Schülke *et al.* (1993) when they compared their experimental $S(\mathbf{q},\omega)$ peak positions of Al with center-of-gravity extrapolated peak positions calculated by Green *et al.* (1985c) for $r_s = 2$ (see Fig. 2.15). With the Fig. 2.41, Schülke (2004) has brought the comparison between state-of-the-art $S(\mathbf{q},\omega)$ experiments and computations going beyond the RPA on a firm basis. $S(\mathbf{q},\omega)$ measurements with $q = 1.5$ a.u. on liquid Al (Sternemann *et al.* 1998) are compared with jellium calculations going beyond RPA. The discrepancy between

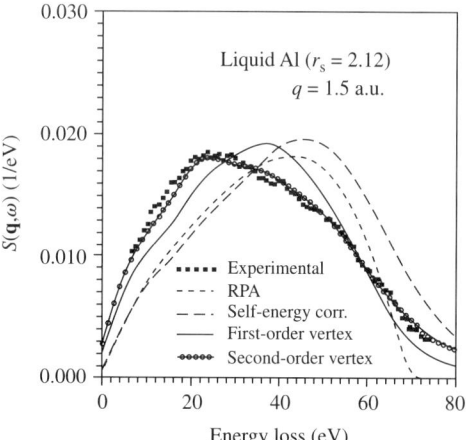

Fig. 2.41. Dynamic structure factor of liquid Al ($r_s = 2.12$), $q = 1.5$ a.u. Filled squares: experimental result (Sternemann *et al.* 1998). Short-dashed line: jellium RPA calculated; long dashed line: calculated fully self-energy (cumulant expansion) corrected; solid line: first-order in the screened particle–hole interaction vertex corrected. Solid line, connected open circles: second-order vertex corrected.

experiment and RPA as seen in Fig. 2.41 is evident. But it is also clear that taking into account the full self-energy correction (here by using the cumulant expansion scheme of Holm and Aryasetiawan (1997)) does not improve but worsens the situation, as already mentioned above. The peak positions shifts to still larger values of ω compared with RPA, and the large-ω tail is overestimated compared with experiment. Only when the vertex corrections (here performed by using exclusively the dominant class C contributions according to (2.90)) are added, the corrections of the RPA go into the right direction bringing the peak position in agreement with the measured one and the high ω-tail in quantitative agreement with the experiment, provided one allows for vertex corrections to second order in the screened Coulomb potential, simply by adding a second virtual plasmon exchange channel into the (self-energy corrected) Lindhard polarization bubble. The remaining marginal deviation of the liquid Al experiment from the computations can be ascribed to short-range order effects, which should resemble lattice effects in a reduced manner. This apparent systematic cancellation between self-energy insertion $\Sigma[G]$ and particle–hole interaction $\Xi_{sc}[G]$ has been discussed by Green et al. (1985b). Beeferman and Ehrenreich (1970) who predicted such a cancellation also for the case of optical absorption gave an intuitive explanation for this phenomenon by saying that the plasmons emitted and recaptured by a particle, when the self-energy effect is adequately described, must have a phase shift of π with respect to the plasmon emitted by the particle in order to act on the hole in the course of the particle–hole interaction. It is remarkable that this mechanism applies only to the off-shell self-energy insertions, since only these have a **particle** intermediate state between plasmon emission and recapture when considering the self-energy diagram of a particle (Green et al. 1987a). On-shell self-energy insertions are characterized by a **hole** intermediate state between plasmon emission and recapture. Thus it is only the off-shell self-energy effect (multipeak structure of the spectral density function) which will be cancelled by the destructive plasmon interference. This may explain the rather big success of the quasiparticle lifetime model, mentioned above, since it is just the on-shell self-energy insertions which determine the lifetime in this model, and which survives the cancellation.

Another diagrammatical approach to the jellium proper polarization function of Sturm and Gusarov (2000) needs to be mentioned, since their evaluation of the leading corrections to RPA have provided possibilities for comparing experimental $S(\mathbf{q},\omega)$ results outside the single particle–hole excitation spectrum with experiment. These authors have evaluated diagrams, partly analytically, which are the lowest-order corrections to the RPA "bubble", and found by Geldart and Vosko (1966) to be crucial to satisfy the compressibility sum rule. This subset of diagrams consists of the first-order self-energy correction for particles and holes, as well as the first-order vertex correction. In addition, Geldart and Vosko (1966) have included a pair of diagrams, a class A and a class B diagram, respectively, in the notation of Green et al. (1985b), which both, as discussed by Sturm and Gusarov (2000), enable double-plasmon excitations. Unfortunately, the summing up of these diagrams produces a logarithmic singularity at the boundaries of the (single) particle–hole excitations, so that calculation within and at the boundaries of this regime are not applicable for comparison with experiment. On the contrary, those comparisons are feasible in the tail regions of the dynamic structure factor, outside the (single) particle–hole excitation regime, and were performed with Al measurements of Schülke et al. (1993) and of Platzman et al. (1992), where the agreement was not so bad. Of special importance was the inclusion of band structure effects on the basis of a pseudopotential scheme into the computation, so that one can estimate to what extent band structure is present in these $S(\mathbf{q},\omega)$ tails of simple metals. We will come later in Section 2.3.3.6 to an experimental proof of double-plasmon excitation.

2.3.3.3 *Local-field corrections*   There exists a direct relationship between the vertex-correction as described above and the so-called dynamic local-field correction, whose static approximation was first introduced by Hubbard (1958). The corresponding so-called (complex) dynamic local-field correction factor $\widetilde{G}(\mathbf{q},\omega)$ is defined by writing the relation for the macroscopic dielectric function $\varepsilon_\mathrm{M}(\mathbf{q},\omega)$ in terms of the Lindhard polarization function $\chi_0(\mathbf{q},\omega)$ (see 6.74) as follows:

$$\varepsilon_\mathrm{M}(\mathbf{q},\omega) = \frac{1 - v(\mathbf{q})(1 - \widetilde{G}(\mathbf{q},\omega)\chi_0(\mathbf{q},\omega)}{1 + \widetilde{G}(\mathbf{q},\omega)v(\mathbf{q})\chi_0(\mathbf{q},\omega)}. \qquad (2.91)$$

Note that $\varepsilon_\mathrm{M}(\mathbf{q},\omega) \to \varepsilon_\mathrm{RPA}(\mathbf{q},\omega)$ (see 6.90) for $\widetilde{G}(\mathbf{q},\omega) \to 0$. One can arrive at (2.91), if one replaces in (2.90) the screened interaction $v(\mathbf{k}_2 - \mathbf{k}_1)/\varepsilon_\mathrm{RPA}(\mathbf{k}_2 - \mathbf{k}_1, \omega_2 - \omega_1)$ between the particle and its hole formally by $v(\mathbf{q})\widetilde{G}(\mathbf{q},\omega)$. Then one can sum up the series, which is represented in (2.87) by its first two terms, to infinite order of the effective two-body interaction.

A great deal of theoretical work, which cannot be presented by describing the calculation methods in any detail, has been invested to find expressions of the static approximation $\widetilde{G}(\mathbf{q},0)$ for homogeneous systems, which satisfy simultaneously the compressibility sum rule, all frequency moment sum rules and the nonnegative condition of the pair-correlation function $g(r)$ at the origin. Nevertheless, the expressions found are rather different, and one finds only a few common aspects. All authors agree with a quadratic dependence of $\widetilde{G}(\mathbf{q},0)$ on q for smaller q, which is necessary to satisfy the compressibility sum rule. This dependence holds for the whole q-range, if we introduce the static local-field correction on the basis of the time-dependent local density approximation (TDLDA) (Fleszar et al. 1995), as sketched in Section 6.5. Brosens et al. (1980) and Holas et al. (1979) discovered a more or less pronounced peak of $\widetilde{G}(\mathbf{q},0)$ around $2k_\mathrm{F}$ with values clearly above one, whereas Geldart and Taylor (1970) failed to see this peak in their calculations. Also Utsumi and Ichimaru (1980), whose analytical expression for $\widetilde{G}(\mathbf{q},0)$ was often used in the literature, found this peak but substantially moderated compared with that of Brosens et al. (1980). Their analytical expression for $\widetilde{G}(\mathbf{q},0)$ exhibits the same limiting behavior for short wavelengths, first introduced by Niklasson (1974), namely

$$\lim_{q\to\infty} \widetilde{G}(q,\omega) = \tfrac{2}{3}[1 - g(0)]. \tag{2.92}$$

But it has been pointed out by Holas (1987) that this limiting behavior is only valid if the Lindhard polarization function $\chi_0(\mathbf{q},\omega)$, used in the defining equation (2.90), is a modified one with the momentum distribution function of non-interacting electrons, $n_0(k) = \theta(1 - k/k_\mathrm{F})$, replaced by that of interacting ones. It has been shown by Holas (1987) that using the nonmodified Lindhard polarization function leads to an expansion of $\widetilde{G}(\mathbf{q},0)$ for larger q, which contains, besides the right-hand side of (2.92), also contributions quadratic in q, so that values larger than one can be reached for sufficiently large q's. Farid et al. (1993) came to a similar result. Their expression for $\widetilde{G}(\mathbf{q},0)$ contains, as leading parameter, the relative changes of, respectively, the mean kinetic energy and the mean-squared kinetic energy that occur when going from a noninteracting to an interacting electron system, so that $\widetilde{G}(\mathbf{q},0)$ is directly related to the momentum distribution function of an interacting electron system. The result of Hong and Shim (1993) looks completely different, although these authors, as also Utsumi and Ichimaru (1980), Holas (1987) and Farid et al. (1993), claim to satisfy the first and third frequency moment sum rule, the compressibility sum rule, and the nonnegative condition of $g(0)$. In order to demonstrate the very different results we have plotted the Utsumi and Ichimaru (1980) result, the local-field correction factor as obtained within the LDA by using (6.87) and (6.88), and $\widetilde{G}(\mathbf{q},0)$ of Farid et al. (1993) in Fig. 2.42. Nevertheless, utilizing the static approximation of $\widetilde{G}(\mathbf{q},\omega)$, when the interpretation of $S(\mathbf{q},\omega)$ measurements needs to go beyond the RPA, turned out to be rather fruitful, as demonstrated in Fig. 2.43 for polycrystalline Be, taken from Schülke et al. (1989). It is evident how only the combination of the quasiparticle lifetime correction (self-energies according to (2.82) with a static local-field correction (Utsumui and Ichimaru 1980) brings the experiment near enough to the calculations, at least as far as the overall shape and the peak position is concerned, and $q < 2k_\mathrm{F}$. Such rather good agreement between computation and experiment was achieved despite the criticism of the quasiparticle lifetime concept (Green et al. 1987a) and rather crudely approximating the particle–hole interaction by a static local-field correction. As already mentioned above, this may be due to a (partial) cancellation between the off-shell self-energy correction and the dynamically screened part of the particle–hole interaction.

There exists also a few approaches to calculate the jellium full dynamic (complex) local-field correction function $\widetilde{G}(\mathbf{q},\omega)$. The author of the first one is Kugler (1975). But his scheme is, with some justification, only applicable for small $q (q < k_\mathrm{F})$. Arawind et al. (1982) found a solution, which satisfies the first and third frequency moment sum rule, but exhibits a singularity at $\hbar\omega = \hbar^2 q^2/2m + \hbar^2 q k_\mathrm{F}$. The same singularity was found by Dellafiore and Matera (1988). Holas and Rahman (1987) gave a thorough analysis of their dynamic

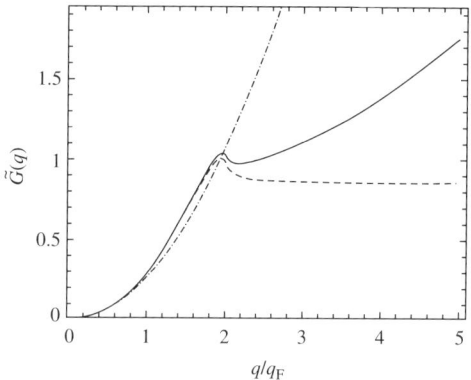

Fig. 2.42. Static local-field correction factor $\widetilde{G}(q)$. Solid curve: Farid et al. (1993); dashed curve: Utsumi and Ichimaru (1980); dashed-dotted curve: LDA.

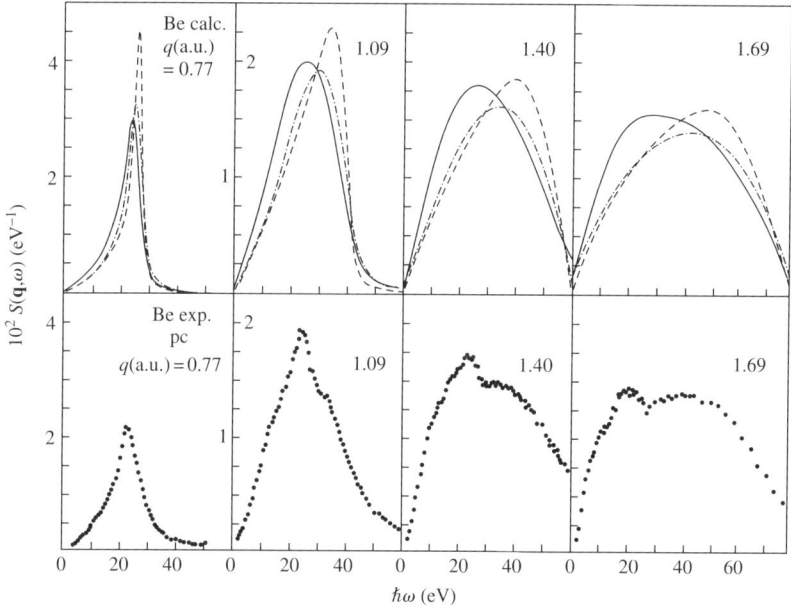

Fig. 2.43. Upper panel: jellium calculated $S(\mathbf{q},\omega)$ for $r_s = 1.87$ (Be) for various values of q as indicated. Dashed curve: RPA; dashed-dotted curve: self-energy corrected within the quasiparticle lifetime model; solid curve: lifetime and local-field corrected (Utsumi and Ichimaru 1980) Lower panel: experimental $S(\mathbf{q},\omega)$ of polycrystalline beryllium metal. (Reprinted with permission from Schülke et al. (1989); copyright (1989) by the American Physical Society.)

local-field correction function, which possesses a singularity of its first derivative at the boundaries of the particle–hole continuum. The overall dependence of $\widetilde{G}(\mathbf{q},\omega)$ on ω is rather weak, much weaker than in the computations of Hasegawa and Shimizu (1975), so that the static approximation seems to be justified. The only calculation that has been used to make comparison with experiment (Green et al. 1985c) is that of Dharma-wardana and Taylor (1980), based on a summation of the Hartree–Fock ladder series, and which is approximately related to the Lindhard polarization function $\chi_0(\mathbf{q},\omega)$ via

$$\widetilde{G}(\mathbf{q},\omega) = v^{-1}(\mathbf{q})[\chi_0(\mathbf{q},\omega)^{-1} - \chi_{\mathrm{HF}}(\mathbf{q},\omega)^{-1}] \tag{2.93}$$

$$\mathrm{Re}\chi_{\mathrm{HF}}(\mathbf{q},\omega) \cong \frac{\mathrm{Re}\chi_0(\mathbf{q},\omega)}{1 + v(k_{\mathrm{F}})\mathrm{Re}\chi_0(\mathbf{q},\omega)/4}$$

$$+ \frac{v(k_{\mathrm{F}})}{4}\left(\frac{\mathrm{Im}\chi_0(\mathbf{q},\omega)}{1 + v(k_{\mathrm{F}})\mathrm{Re}\chi_0(\mathbf{q},\omega)/4}\right)^2 \tag{2.94}$$

$$\mathrm{Im}\chi_{\mathrm{HF}}(\mathbf{q},\omega) \cong \frac{\mathrm{Im}\chi_0(\mathbf{q},\omega)}{[1 + v(k_{\mathrm{F}})\mathrm{Re}\chi_0(\mathbf{q},\omega)]^2} \tag{2.95}$$

However, like in the case of the static local-field correction, also a first-principle investigation of the frequency dependence of the local-field correction factor remains a theoretical challenge.

Therefore, faced with this unsatisfying situation at the theoretical front it does not seem to be misguided to rely upon a comparison with experiment, here $S(\mathbf{q},\omega)$ measurements, in order to come to an evaluation of what the adequate computational basis for calculating $\widetilde{G}(\mathbf{q},\omega)$ should be.

Although equation (2.91) defining $\widetilde{G}(\mathbf{q},\omega)$ can be solved, so that the corresponding relation for $\widetilde{G}(\mathbf{q},\omega)$ contains $\varepsilon_{\mathrm{M}}(\mathbf{q},\omega)$ as the only unknown quantity. In principle, $\varepsilon_{\mathrm{M}}(\mathbf{q},\omega)$ can be obtained from experimental values of $S(\mathbf{q},\omega)$, brought to an absolute scale by using the f-sum-rule (2.21), via the Kramers–Kronig relations (2.22). But the application of (2.21) and (2.22) presumes that the experimental values of $S(\mathbf{q},\omega)$ are known for the valence electron system over the whole ω-range. But, at least for larger q's, the onset of the core contributions (for Li e.g at $\hbar\omega \approx 51$ eV) makes the valence contributions above these frequency very uncertain. Moreover, there are band-structure related fine structures in the experimental $S(\mathbf{q},\omega)$'s at larger ω's. Therefore, fitting of the complex $\widetilde{G}(\mathbf{q},\omega)$ to the experimental structure factor seems to be the right way, only if one fits within a certain limited frequency range, where overlap with the core contribution and strong band structure effects can be excluded. Such a fitting procedure was performed by Schülke et al. (1996) on Li for q between k_{F} and $2.5k_{\mathrm{F}}$ within an energy loss interval of 3–4 eV in the rising part of $S(\mathbf{q},\omega)$ around 5 eV energy loss. The mean value of energy loss within this interval is designated with $\bar{\omega}$ The result of this fitting procedure is shown in Fig. 2.44. The real part of the semiempirical $\widetilde{G}(\mathbf{q},\bar{\omega})$, obtained this way, is compared with various theoretical $\widetilde{G}(\mathbf{q},0)$'s, a comparison, which seems to be justified, since $\bar{\omega}$ is near $\omega = 0$, and

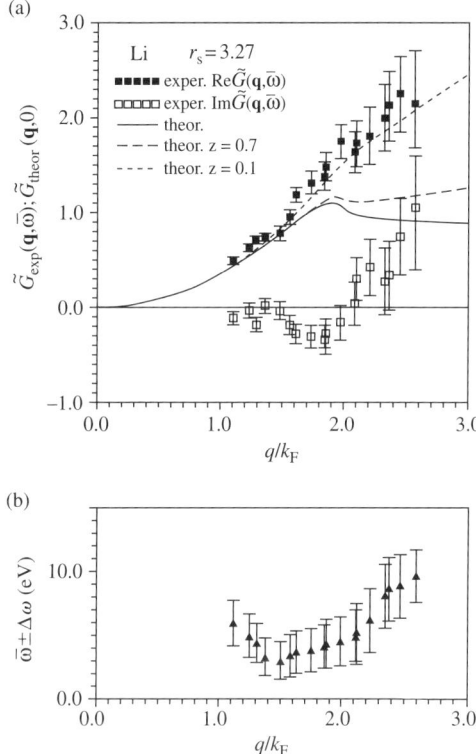

Fig. 2.44. (a) Real (filled squares) and imaginary part (open squares) of the local-field correction factor $\widetilde{G}(\mathbf{q},\bar{\omega})$ as obtained by fitting Li [111] $S(\mathbf{q},\omega)$ measurements within the frequency range $\bar{\omega} \pm \Delta\omega$ to equation (2.91). Solid curve: theory of Utsumi and Ichimaru (1980); long-dashed curve: theory of Farid et al. (1993) with $z = 0.7$; short-dashed curve: theory of Farid et al. (1993) for $z = 0.1$. (b) Frequency range, within which the fitting was performed. (Reprinted with permission from Schülke et al. (1996); copyright (1996) by the American Physical Society.)

for general reasons the ω-dependence of $\widetilde{G}(\mathbf{q},\omega)$ is weak for small ω (Holas and Rahman 1987). The most remarkable result of this fitting procedure is the perfect agreement between experiment and theory in the range $q < 1.5k_{\rm f}$ and the steep rise of the semiempirical Re$\widetilde{G}(\mathbf{q},\bar{\omega})$, beginning at $q = 1.5k_{\rm F}$ when compared with various theories. Only the theory of Farid et al. (1993) brings experiment and theory to a satisfactory agreement, certainly only for an unrealistic value of the renormalization constant z, determining the momentum distribution of the interacting electron system, namely $z = 0.1$ (for a discussion of the renormalization constant see Chapter 4). The imaginary part of the semiempirical $\widetilde{G}(\mathbf{q},\bar{\omega})$

exhibits a change of its sign at $q = 2k_\text{F}$, as predicted by Holas and Rahman (1987). For $q > 1.5k_\text{F}$ the same steep rise of $\text{Re}\widetilde{G}(\mathbf{q},\omega)$, fitted to experimental Al $S(\mathbf{q},\omega)$, was found by Larson et al. (1996), who performed $S(\mathbf{q},\omega)$ measurements using a sagittal focusing pyrolytic graphite analyzer in conjunction with a position-sensitive detector (see Section 2.2.5.3). Their fitting procedure was based on a somewhat different definition of the dynamic local-field function, where the Lindhard jellium $\chi_0(\mathbf{q},\omega)$ in (2.90) was replaced by the proper polarizability of noninteracting electron–hole pairs as calculated for the Al lattice within the limits of the LDA. Of course, these band structure based calculation of the polarizability include exchange and correlation within the LDA. But the authors have found, at least for values of q for which $\text{Re}\widetilde{G}(\mathbf{q},\omega) \approx 1$, that their calculated polarizability was indistinguishable from that calculated from eigensolution of the ground-state problem in the Hartree approximation. Moreover, Larson et al. (1996) have fitted calculation and experiment over the whole range of ω as accessible to experiment. They interpret the likewise good agreement of their empirical $\widetilde{G}(\mathbf{q},\omega)$ with computations of the static local-field factor for $q < 1.5k_\text{F}$ to be an indication of an only weak dependence of the dynamic local-field function on ω thus confirming the theoretical findings of Holas and Rahman (1987). Among the large number of calculations of the static local-field correction function, Larson et al. (1996) have found one, namely that of Brosens et al. (1980), which fits their measurements quite well until $q \approx 2k_\text{F}$. But likewise as with the fitting results of Schülke et al. (1996) and contrary to the findings of Brosens et al. (1980), their semiempirical $\widetilde{G}(\mathbf{q},\omega)$ remains at rather high values up to $q \approx 2.5k_\text{F}$. Therefore, it continues to be a challenge for theoreticians to find a basis for calculating the dynamic local-field correction functions in agreement with experiment.

2.3.3.4 Effective Hamiltonian for attractive electron–hole interaction

Both schemes discussed above, the vertex correction as well as the local-field correction, have taken into account the electron–hole interaction by correcting supplementarily a zero-approach of the proper polarization function. Only recently Calibe et al. (2000) have applied a first-principle technique introduced by Rohlfing and Louie (1998a, 1998b), Benedict et al. (1998a, 1998b) and Benedict and Shirley (1999) to the calculation of the dynamic structure factor, where an effective Hamiltonian obtained from the Bethe–Salpeter equation is used. This effective Hamiltonian includes a one-particle term, additionally a term, which describes the attractive interaction between the electron and the hole, and finally the corresponding repulsive exchange term. The electron–hole pair wavefunction uses a basis of one-particle eigenstates calculated within the limits of the LDA. The single-particle energies were corrected by means of a GW calculation (Hybertsen and Louie 1985). As shown by Caliebe et al. (2000), $S(\mathbf{q},\omega)$ measurements on diamond and LiF (see Section 2.4) could be brought into agreement with calculations only by applying this scheme, demonstrated in Fig. 2.45. It is the attractive interaction between the electron and the hole left behind in

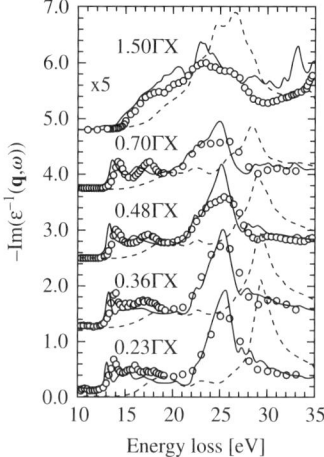

Fig. 2.45. $-\mathrm{Im}[\varepsilon^{-1}(\mathbf{q},\omega)]$ of LiF as a function of momentum transfer in the ΓX direction. Open circles: experimental data; solid lines: theoretical data including electron–hole interaction in the final state; dashed lines: theoretical data without electron–hole interaction. A constant background was added to the theoretical data. (Reprinted with permission from Caliebe et al. (2000); copyright (2000) by the American Physical Society.)

the course of a particle–hole excitation, which shifts the calculated peaks to lower energy losses, so that they reach their experimental positions.

Galambosi et al. (2001) came to a similar conclusion, when applying the above scheme to an IXS study on cBN. Again the results of the calculations could only be brought into agreement with experiment, if the electron–hole interaction was taken into account otherwise the calculations exhibited a 2 eV shift to higher energies.

2.3.3.5 Self-consistent self-energy and vertex function The scheme, represented in terms of imaginary Matsubara frequencies, which we will discuss now, and which goes back to Takada (1995), is based on an exact functional relation between the self-energy Σ and the vertex function Γ, where Γ represents, on the one hand, the vertex correction of the polarization function to infinite order in the irreducible particle–hole interaction which can formally be written ($\hbar = 1$)

$$\chi(q) = -T \sum_{\omega_n} \sum_{\mathbf{p},\sigma} G(p)\, G(p+q)\, \Gamma(p, p+q), \qquad (2.96)$$

where $q \equiv (\mathbf{q}, i\omega_l)$, $i\omega_l \equiv 2\pi i T_l$ – Matsubara frequency, $p \equiv (\mathbf{p}, i\omega_n)$; $i\omega_n = \pi i T(2n+1)$, and $G(p)$ is the self-energy corrected Green's function

$$G(p)^{-1} = G^0(p)^{-1} - \Sigma(p) \qquad (2.97)$$

with $G^0(p)^{-1} = i\omega_n - \mathbf{p}^2/2m$, so that (2.87)–(2.89) can be considered to be the first-order expansion of (2.96), when $\chi_{\rm sc}(\mathbf{q},\omega)$ is understood as the analytic continuation of $\chi(q)$. On the other hand, Γ also takes care of vertex corrections to the self-energy by writing

$$\Sigma(p) = -T \sum_{\omega_l} \sum_{\mathbf{q}} \left[\frac{v(\mathbf{q})}{\varepsilon(q)}\right] G(p+q)\Gamma(p,p+q) \qquad (2.98)$$

where $\varepsilon(q) = 1 + v(\mathbf{q})\chi(q)$, so that equation (2.82) for the self-energy is the zero-order expansion of (2.98), when $\Sigma(\mathbf{p},\varepsilon)$ is understood as the analytic continuation of $\Sigma(p)$. If one has available an appropriate expression for the vertex function as a functional of the dressed Green's function G, one can implement a self-consistent iteration loop using (2.98) together with the relation (2.97) between the self-energy and Green's function starting from the noninteracting solution and ending the loop if the relative difference in the self-energy becomes less than a given value. Takada (2001) has offered an easily to handle expression for the vertex function Γ as the product of two factors $\Gamma_{\rm LFC}(q) = 1 - f_{\rm xc}^{\rm hom}(q)\chi(q)$ and

$$\Gamma_{\rm WI}(p,p+q) = \frac{G(p)^{-1} - G(p+q)^{-1}}{G^0(p)^{-1} - G^0(p+q)^{-1}}, \qquad (2.99)$$

where $f_{\rm xc}^{\rm hom}(q)$ is the frequency (energy) dependent exchange correlation local-field correction appearing in the time-dependent density functional theory (Gross et al. 1996). The factor $\Gamma_{\rm WI}$ guarantees the observance of the Ward identity (WI).

Having thus found a self-consistent self-energy, one obtains a self-consistent Green's function (2.97), a self-consistent vertex function and a self-consistent polarization function. The authors of this approach are claiming that the effect of a strong periodic potential from ion cores is much easier to be added than with the scheme of Green et al. (1985a,b,c). Takada and Yasuhara (2002) have used the self-consistent approach to calculate the dynamic structure factor of jellium for several values of $r_{\rm s}$ and a large range of q values using

$$S(\mathbf{q},\omega) = \left[\frac{-1}{\pi(1-\exp(-\omega/k_{\rm B}T))}\right] {\rm Im}\left(\frac{-\chi_{\rm sc}(\mathbf{q},\omega)}{1+v(\mathbf{q})\chi_{\rm sc}(\mathbf{q},\omega)}\right), \qquad (2.100)$$

where $\chi_{\rm sc}(\mathbf{q},\omega)$ is the analytic continuation of the self-consistent $\chi(q)$. They found in Fig. 2.46, where the case for Al, $r_{\rm s} = 2.08$ is shown, for $q < q_{\rm c}$, a one-peak structure, which has to be attributed to plasmon excitation, but contrary to the RPA result, a strong damping was visible, which was attributed to coupling of the plasmon with multipair excitations. This was in good agreement with experimental results of Schülke et al. (1993) and Larson et al. (1996). For $q > q_{\rm c}$, and especially for $1.2 p_{\rm F} < q < 2.4 p_{\rm F}$ (here: $p_{\rm F} \equiv k_{\rm F} = $ Fermi wavevector), a shoulder emerges, indicated by a' in Fig. 2.46, which is not present in the RPA result. This structure is ascribed to the electron–hole multipair scattering represented by $\Gamma_{\rm LFC}$ in (2.98). The attractive screened potential between the pairs reduces the excitation energy and leads to a pile-up of transition probability at

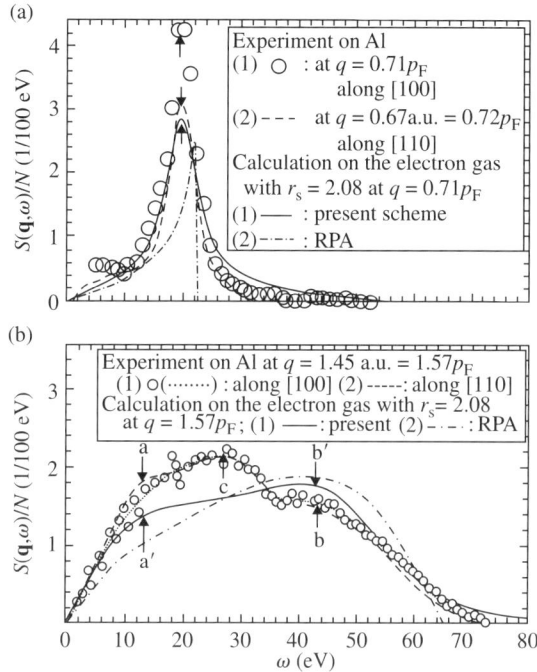

Fig. 2.46. Comparison of the Takada and Yasuhara calculation scheme with experiment. (a) Experimental $S(\mathbf{q},\omega)$ of Al: open circles: $q = 0.71 p_F$ ($p_F = k_F$ = Fermi wavevector) along [100] (Larson et al. 1996); dashed curve: $q = 0.72 p_F$ along [110] (Schülke et al. 1993). Jellium calculation of $S(\mathbf{q},\omega)$, $r_s = 2.08$; $q = 0.71 p_F$: Solid curve: Takada and Yasuhara (2002); dashed-dotted curve: RPA. (b) Experimental $S(\mathbf{q},\omega)$ of Al $q = 1.57 p_F$: points and open circles: along [100], long-dashed curve: along [110] (Schülke et al. 1993). Jellium calculation of $S(\mathbf{q},\omega)$ $r_s = 2.08$; $q = 1.57 p_F$: Solid curve: Takada and Yasuhara (2002); dashed-dotted curve: RPA. (Reprinted with permission from Takada and Yasuhara (2002); copyright (2002) by the American Physical Society.)

smaller energy transfer. The energy position b' of the main peak of $S(\mathbf{q},\omega)$ is not far from the RPA position and is also distinguishable in the experiment (Schülke et al. 1993) as a shoulder designated b, whereas the structure a' is represented only in the [110] experiment. Contrary to RPA, the calculated dynamic structure factor exhibits a tail, in good agreement with experiment. This tail is attributed to multipair excitations outside the one-pair region and would not be present without Γ_{WI}. Nevertheless, a rather big discrepancy between these calculations and experiment remains, as visible in Fig. 2.46; especially the experimental peak c is by no means present in the computations. The authors hold lattice effects, which have already been discussed by Schülke et al. (1993), responsible for the dip between c and b. Above all, it is the pseudogap between the unoccupied 3d and 4f bands as demonstrated by Takada and Yasuhara (2002), which produces the corresponding excitation gap. The authors of the new approach claim to confirm the correctness and the good self-consistency of their scheme by showing that their calculated static polarization function $\chi(\mathbf{q}, 0)$ coincides well with the corresponding quantum Monte Carlo calculated one (Moroni et al. 1995).

2.3.3.6 Correlation induced plasmon–plasmon excitations

It has already been mentioned in Section 2.3.3.2 that the inclusion of a certain subset of diagrams for the perturbation expansion of the proper polarizability, as shown in Fig. 2.47, is necessary in order to guarantee the correct r_s dependence of the compressibility sum rule. These diagrams are second order in the dynamically screened Coulomb interaction and signify the interaction between the initial and the final particle–hole pair, where both carry the external momentum $\mathbf{q}$ and energy $\hbar\omega$ in the same time direction (class A and class C contributions in the classification scheme of Green et al. 1985b). This interaction is mediated by the polarization of the electron system via the RPA response function $1/\varepsilon^{\mathrm{RPA}}(\mathbf{k}, \omega')$ and $1/\varepsilon^{\mathrm{RPA}}(\mathbf{k}+\mathbf{q}, \omega'+\omega)$ (see (6.90), where $\mathbf{k}$ and ω' are internal momentum and frequency variables, respectively). As predicted by Sturm and Gusarov (2000), the plasmon poles of these two RPA response functions will give rise to two simultaneous plasmon excitations at an energy $\hbar\omega = \hbar\omega_{\mathrm{p}}(k)+\hbar\omega_{\mathrm{p}}(|\mathbf{k}+\mathbf{q}|)$, provided q and ω are large enough. Since these plasmons can exist only for $k \leq q_c$ (q_c = plasmon cut-off vector) and $|\mathbf{k}+\mathbf{q}| \leq q_c$, this plasmon–plasmon (pl–pl) excitation can occur only within a limited range on the energy-loss scale, namely between $2\hbar\omega_{\mathrm{p}}(q/2)$ and $2\hbar\omega_{\mathrm{p}}(q_c)$. The peak-like structure in the tails of $S(\mathbf{q}, \omega)$ which is produced by this pl–pl excitation, has its peak roughly at $\hbar\omega_{\mathrm{pl-pl}}^{\mathrm{peak}} = \hbar[2\omega_{\mathrm{p}}(0) + \alpha q^2]$, where α is the plasmon dispersion coefficient, defined by $\omega_{\mathrm{p}}(q) = \omega_{\mathrm{p}}(0) + \alpha q^2$, and can thus easily be distinguished and subtracted from the continuously varying contribution of both the particle–hole pair/particle–hole pair (ph–ph) and the particle–hole-pair/plasmon (ph–pl) excitation. The diagrams of Fig. 2.47 were

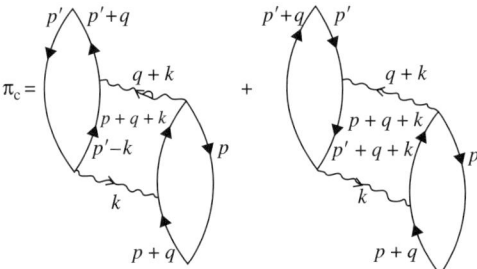

Fig. 2.47. Diagrams for the perturbation expansion of the proper polarizability, second order in the screened Coulomb interaction signified by the wavy lines. The italic letters are four-vectors including frequencies.

calculated by Sturm and Gusarov (2000), where the corresponding expressions could be reduced analytically to three-dimensional integrals, which were then evaluated numerically.

Measurements of the dynamic structure factor on Al single crystals with $\mathbf{q} \parallel [100]$ and $\mathbf{q} \parallel [110]$ and on polycrystalline Na were performed by Sternemann et al. (2005) at the ID16 beamline of the ESRF utilizing the setup with intermediate energy resolution of 1.5 eV at 8994 eV incident photon energy. Additional scans have been performed within the energy-loss range of the pl-pl excitation in order to increase the statistical accuracy of the rather faint signals. After having normalized the $S(\mathbf{q}, \omega)$ spectra according to the f-sum rule, the $S(\mathbf{q}, \omega)$ tails outside the range of the pl-pl excitations were fitted to the sum of a Pearson VII (Prevey 1986) and an exponential function, so that the pl-pl contribution could be separated. Figure 2.48(a) shows the experimental pl-pl contribution of Al for a set of q-values compared with the jellium calculated ones of Fig. 2.48(b) convoluted with the experimental resolution. An excellent agreement was achieved with respect to the shape, the q-dependence of its integrated intensity which follows $q^{5.3}$, and the dispersion coefficient $\alpha = 0.27 \pm 0.02$ of the peak position $\hbar\omega_{\text{pl-pl}}^{\text{peak}}(q)$. The extrapolated peak position at $q = 0$ was by 2.10 eV smaller in the experiment than in the jellium computation. Apparently this shift has to be traced back to a lattice effect. Within the limits of statistical accuracy no significant $\mathbf{q}$-orientation dependence of the pl-pl peak could be found. Of course, the experimental integrated spectral intensity was roughly only 50% of the jellium calculated ones, over the whole q-range measured. This might be due to the so-called exchange counterparts of Fig. 2.47, which are obtained from the latter by the exchange of two particles entering one of the interactions. These exchange diagrams, also second order in the screened Coulomb interaction, do not contribute to the compressibility sum rule (Geldart and Vosko 1966) but can diminish the structures attributed to the pl-pl excitations by up to 50% (Bachlechner et al. 1996). The authors stress the fact that these pl-pl excitations are purely correlation induced, whereas the remaining tail of $S(\mathbf{q}, \omega)$ beyond the particle–hole excitation continuum is widely influenced by interband transitions. Since there exist diagrams other than the Fig. 2.47 diagrams of second order in the screened Coulomb interaction, the good agreement found above between jellium calculation and experiment can be considered as an indication of the special importance which the corresponding diagrams, together with their exchange counterparts, occupy in the hierarchy of diagrammatic perturbation expansion of the polarizability, as already stressed by Geldart and Vosko (1966).

2.4 Study of specific solid state properties

2.4.1 *Study of band gaps and combined density of states*

Writing down the expression for the dynamic structure factor in terms of Bloch-type one-electron states $|\mathbf{k}\nu\rangle$ ($\mathbf{k}$ = Bloch wavevector; ν = band index) as derived

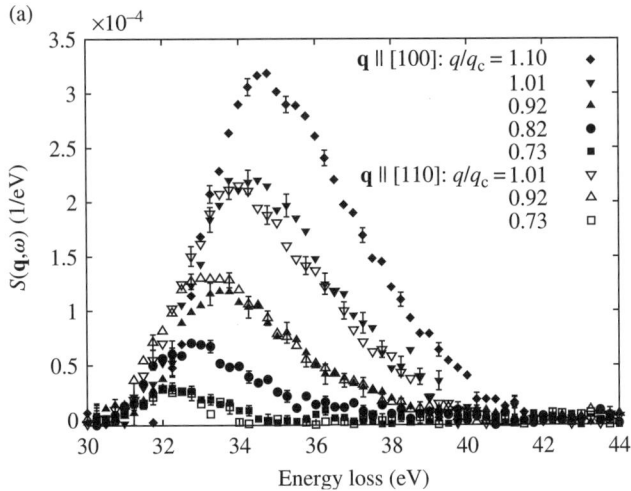

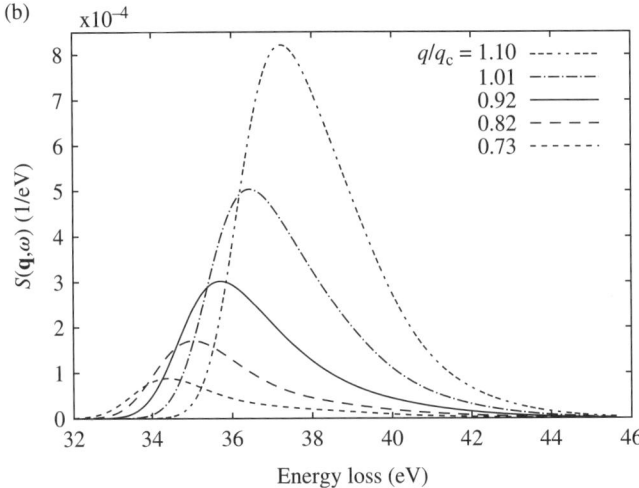

Fig. 2.48. (a) Contribution of plasmon–plasmon (pl-pl) excitations to the dynamic structure factor outside the particle–hole continuum for single-crystal Al sodium for various values of the momentum transfer q and **q**-orientation as indicated. ($q_c = 0.73$ a.u.). (b) Jellium calculated contributions of pl-pl excitations to the dynamic structure factor outside the particle–hole continuum for the electron density of Al ($r_s = 2.07$), convoluted with the experimental energy resolution for various values of the momentum transfer q as indicated. (Reprinted with permission from Sternemann *et al.* (2005); copyright (2005) by the American Physical Society.)

in Chapter 6 by inserting the equation for the Ehrenreich-Cohen (1959) self-consistent dielectric function (see 6.95)

$$\varepsilon^{\text{SCF}}(\mathbf{q},\omega) = 1 - v(\mathbf{q}) \lim_{\varepsilon \to 0_+} \left(\frac{2}{V}\right) {\sum_{\substack{\mathbf{k}\nu \\ \mathbf{k}'\nu'}}}' |\langle \mathbf{k}'\nu'| \exp(i\mathbf{q}\cdot\mathbf{r})|\mathbf{k}\nu\rangle|^2$$

$$\times \frac{f(E(\mathbf{k},\nu)) - f(E(\mathbf{k}',\nu'))}{\hbar\omega + E(\mathbf{k},\nu) - E(\mathbf{k}',\nu') + i\varepsilon} \qquad (2.101)$$

together with its imaginary part

$$\text{Im}\varepsilon^{\text{SCF}}(\mathbf{q},\omega) = v(\mathbf{q}) \left(\frac{2\pi}{V}\right) {\sum_{\substack{\mathbf{k}\nu \\ \mathbf{k}'\nu'}}}' |\langle \mathbf{k}'\nu'| \exp(i\mathbf{q}\cdot\mathbf{r})|\mathbf{k}\nu\rangle|^2$$

$$\times [f(E(\mathbf{k},\nu)) - f(E(\mathbf{k}',\nu'))]\delta[\hbar\omega + E(\mathbf{k},\nu) - E(\mathbf{k}',\nu')] \qquad (2.102)$$

into the relation for the RPA(SCF) dynamic structure factor (see 6.92)

$$S^{\text{SCF}}(\mathbf{q},\omega) = \left(\frac{q^2}{4\pi^2 e^2}\right) \text{Im} \frac{[\varepsilon^{\text{SCF}}(\mathbf{q},\omega)]}{|\varepsilon^{\text{SCF}}(\mathbf{q},\omega)|^2}, \qquad (2.103)$$

one can easily verify that each measurement of the dynamic structure factor within the range of particle–hole excitations should give information about the combined density of states according to the δ-function in (2.102), of course weighted by the corresponding matrix elements. Thus the dynamic structure factor is determined by interband transitions. Therefore, a band gap between the highest occupied and the lowest unoccupied band should be directly reflected as a zero of $S(\mathbf{q},\omega)$ between $\omega = 0$ and the smallest value $\omega_{\min}(\mathbf{q})$, at which, for a distinct momentum transfer $\mathbf{q}$, $S(\mathbf{q},\omega)$ starts to increase from zero. This special momentum transfer $\mathbf{q} = \Delta\mathbf{k}_{\text{gap}} + \mathbf{g}$ characterizes the band gap, so that $\Delta\mathbf{k}_{\text{gap}}$ is the difference between the Bloch wavevector corresponding to the valence band maximum and that of the conduction band minimum. A value of $|\Delta\mathbf{k}_{\text{gap}}| \neq 0$ is an indication for an indirect band gap. Thus inelastic X-ray scattering can determine the energy width of the gap, and, in the case of an indirect gap, also its relative momentum $\Delta\mathbf{k}_{\text{gap}}$, which is not possible with the conventional optical absorption or reflectivity measurements, since those experiments cannot determine the momentum of the phonon, necessary to observe the indirect gap.

The feasibility of investigating the properties of an indirect band gap was demonstrated, for the case of diamond, with an experiment performed by Caliebe *et al.* (2000) at the NSLS, using, in inverse geometry, the combination of a focusing cylindrically bent Si(200) monochromator crystal and a spherically bent analyzer at a fixed Bragg angle. The total energy resolution was 0.6 eV, the momentum resolution, given by the diameter of the analyzer crystal, 0.1 Å^{-1}. Figure 2.49 presents experimental diamond data of $-\text{Im}[1/\varepsilon(\mathbf{q},\omega)]$ near the band gap as a function of momentum transfer in the ΓX direction along with corresponding calculations, where the solid curves take into account the electron–hole

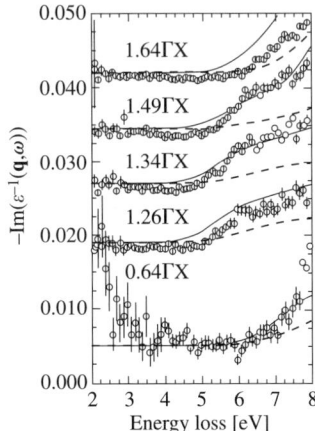

Fig. 2.49. $-\text{Im}[\varepsilon^{-1}(\mathbf{q},\omega)]$ of diamond near the band gap as a function of momentum transfer in the ΓX direction. Open circles: experimental data; solid lines: theoretical data including electron–hole interaction in the final state; dashed lines: theoretical data without electron–hole interaction. A constant background was added to the theoretical data. (Reprinted with permission from Caliebe *et al.* (2000); copyright (2002) by the American Physical Society.)

interaction (see Section 2.3.3.4) the dashed curves don't. One can clearly see that the gap of 5.5 eV (the minimum of the first rise) is not observed at $q = 0.64\Gamma$X, as predicted by theory (0.75ΓX), but at $1.34\ \Gamma$X, which is $q \approx 0.64\Gamma$X + 2ΓX, where 2ΓX is the corresponding reciprocal lattice vector. This behavior can be explained by taking into account the symmetry of the wavefunction at the valence band maximum and at the conduction band minimum, when calculating the matrix elements in (2.102). The authors of this study stressed the point that with the availability of synchrotron radiation of fourth generation sources also the full information (energy and momentum) about much smaller gaps, as for instance in high-T_c superconductors, should be feasible although not easy.

As already mentioned, measurements of $S(\mathbf{q},\omega)$, performed within the range of particle–hole excitation, are directly related to the combined density of occupied and unoccupied states (δ-function in 2.102). But this general property leads only then to pronounced structures of $S(\mathbf{q},\omega)$, when the energy difference between the conduction band $E(\mathbf{k}',\nu')$ and the valence band $E(\mathbf{k},\nu)$ in the δ-function of (2.102) is nearly constant over a wide range of $\mathbf{k}$, where $\mathbf{k}' = \mathbf{k} + \mathbf{q} + \mathbf{g}$ and $\mathbf{g}$ reduces $\mathbf{k} + \mathbf{q}$ into the first Brillouin zone. A classical case is offered by the band structure of graphite as shown partially in Fig. 2.50. The occupied σ_2-bands run for $\mathbf{k} \parallel \Gamma$K and $\mathbf{k} \parallel \Gamma$M nearly parallel to the unoccupied π_2-bands, just as the occupied π_1-band and the unoccupied σ-type so-called interlayer band (Posternak *et al.* 1983). The latter band has its name from the

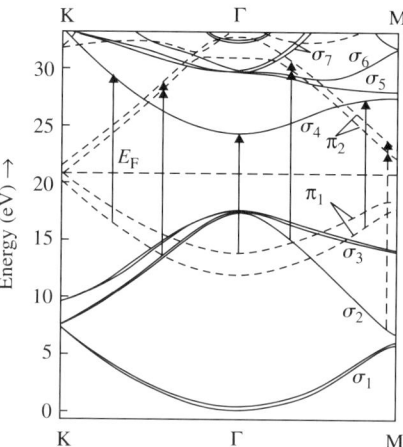

Fig. 2.50. Partial band structure of graphite (Holzwarth *et al.* 1982). Solid curves: σ-bands; dashed curves: π-bands. The arrows indicate some dipole-allowed vertical transitions for $\mathbf{q} \| c$-axis. (Reprinted with permission from Schülke *et al.* (1988); copyright (1988) by the American Physical Society.)

fact that, according to band structure calculations, the wavefunctions of these states are centered between the graphite layers. Interband transitions between the above nearly parallel bands are allowed by dipole selection rules, if one chooses a momentum transfer with $\mathbf{q} \| c$-axis. Since the dispersion of the graphite bands, except that of the interlayer bands, is rather weak for $\mathbf{k} \| c$-axis, the above transitions, in the section of the band structure shown as arrows in Fig. 2.50, can be considered as being nearly perpendicular. Of course, this is not exactly true for the transitions into the unoccupied interlayer bands, so that the transitions for one k-value always cover a certain energy range, which is, according to band structure calculations (Tatar and Rabii 1982), of the order of less than 2 eV ($E_A - E_\Gamma$). This special parallel band case was utilized by Schülke *et al.* (1988a) to get information about the interlayer conduction bands of pristine graphite, the existence of which was difficult to prove with other spectroscopies and whose energy position was controversial among theoreticians (Willis *et al.* 1974, Tatar and Rabii 1982, Holzwarth *et al.* 1982) at the time of the experiment. Moreover, also the shift of the interlayer state upon lithium intercalation was demonstrated this way. The upper part of Fig. 2.51 shows the dynamic structure factor of highly oriented pyrolytic graphite (HOPG) with $\mathbf{q} \| c$-axis, as obtained by Schülke *et al.* (1988a) using a focusing dispersion compensating analyzer at the HARWI/HASYLAB beamline, together with the corresponding real, $\varepsilon_1(\mathbf{q}, \omega)$, and imaginary part, $\varepsilon_2(\mathbf{q}, \omega)$, of the dielectric function as obtained by Kramers–Kronig analysis of the experimental data. A detailed analysis of the spectra by approximating the squared matrix elements of (2.102) for the $\pi \to$ interlayer transition by the Compton profiles of the occupied π-states

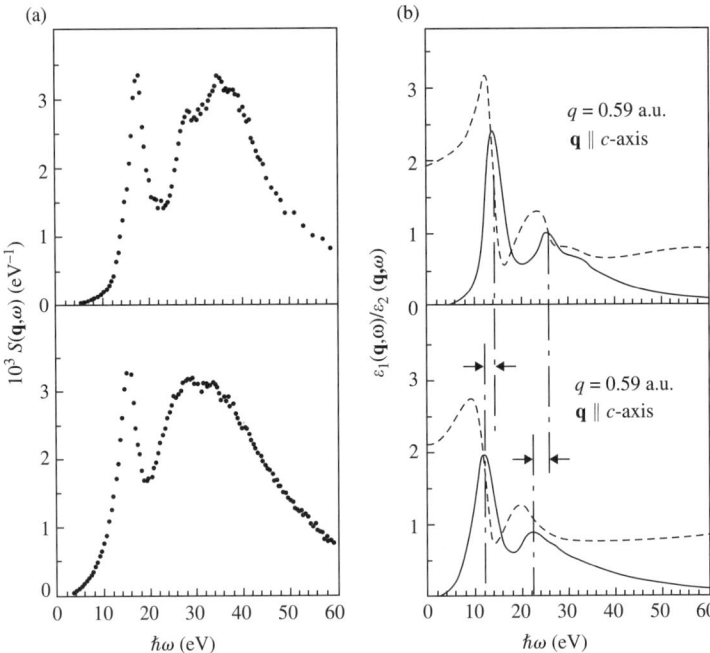

Fig. 2.51. (a) Measured $S(\mathbf{q},\omega)$ with $\mathbf{q}$ along c-axis ($q = 0.59$ a.u.) of HOPG (upper panel) and LiC_6 (lower panel). (b) $\varepsilon_1(\mathbf{q},\omega)$ (dashed curve) and $\varepsilon_2(\mathbf{q},\omega)$ (solid curves) of HOPG (upper panel) and LiC_6 (lower panel) according to Kramers–Kronig analysis of the experimental data. Peak shift upon Li intercalation are indicated. (Reprinted with permission from Schülke et al. (1988); copyright (1988) by the American Physical Society.)

(Chou et al. 1986) have established that part of the first peak in the spectra can be attributed to perpendicular transitions between the π-type valence bands and the σ-type interlayer conduction band, which are over a large range of $\mathbf{k} \perp c$-axis nearly parallel (see Fig. 2.50). The other part of this peak structure is due to the transition between the nearly parallel σ-type valence and the π-type conduction band. As a result of this analysis the energy position of the interlayer conduction band could be determined in rather good agreement with band structure calculations of Willis et al. (1974) and of Tatar and Rabii (1982), whereas the calculations of Holzwarth et al. (1982) could not be confirmed. In another study Schülke et al. (1988b) utilized this well-established origin of the first peak of $\varepsilon_2(\mathbf{q},\omega)$ of graphite to investigate how the first stage intercalation of Li into graphite (LiC_6) may influence the interlayer states, which have a big overlap with the Li ion cores. What is seen in Fig. 2.51 as a result of intercalation is a shift of the first peak together with a broadening. This shift is interpreted as being due to a corresponding nonrigid band shift of the interlayer state to lower energies as a consequence of strong hybridization with the Li metal 2s band in good agreement with predictions of Holzwarth *et al.* (1984) and calculations of Chen and Rabii (1985). The broadening is the result of the rigid band behavior of the $\sigma \to \pi$ transition, responsible for one part of this peak structure, upon intercalation, so that this transition remains at the pristine graphite position. A similar study was performed on potassium intercalated graphite KC_8 (Schülke *et al.* 1991).

It should be mentioned that the dynamic structure factor $S(\mathbf{q}, \omega)$ measured on HOPG for $\mathbf{q} \perp c$-axis by Schülke *et al.* (1988) look very different from that with $\mathbf{q} \| c$-axis, as can be seen in Fig. 2.52. This behavior reflects the fact that for $\mathbf{q} \perp c$-axis, within the limits of the 2D band structure, the dipole selection rules allow exclusively $\sigma \to \sigma^*$ and $\pi \to \pi^*$ transitions. Moreover, these transitions cannot be considered as being vertical for finite q. The resulting $S(\mathbf{q}, \omega)$ exhibits, for smaller values of q ($q < 0.6$ a.u.), two peaks, one, a rather sharp one, around 8 eV and a broader one at 30 eV. The 8 eV peak must be attributed to $\pi \to \pi^*$ interband transitions, distributed over the whole Brillouin zone, and is consequently found in the corresponding $\mathrm{Im}\varepsilon(\mathbf{q}, \omega)$ at the same energy position. The second peak, on the contrary, must be considered as a consequence of a nearly zero of $\mathrm{Re}\varepsilon(\mathbf{q}, \omega)$ and rather weak values of $\mathrm{Im}\varepsilon(\mathbf{q}, \omega)$ at its energy position. Therefore, we can take this peak as being due to a plasmon, of course displaced on the energy scale against its free electron value of 21.5 eV, as must be ascribed to interband transitions. The low-energy peak of $S(\mathbf{q}, \omega)$ remains at its energy position, as shown in Fig. 2.52, with tilting $\mathbf{q}$ by more than 60° into the direction of the c-axis. Also the second peak does not change its shape and energy position appreciably, at least till to a tilting of 45°. Thus it is not surprising that Isaacs *et al.* (1992) found, in a study on single-crystal C_{60}, the same two-peak structure of $S(\mathbf{q}, \omega)$, if one takes into account the rotational degree of freedom of the C_{60}-molecules in the crystal at room temperature. It is the predominance of transitions induced by a momentum transfer $\mathbf{q}$, oriented by more than, let's say 15°, away from the direction of the p_z-orbitals, which are perpendicular on the C_{60}-sphere. Of course, the energy position of the two peaks is somewhat different from that of HOPG, namely 4.9 eV for the first and 27 eV for the second peak. This is a consequence of the specific molecular structure (angles between σ-bonds) and the different electron density, which results in a free-electron plasmon energy of 21.5 eV for C_{60}. In a very recent IXS study on single-crystal graphite, performed at the IXS beamline (BL12XU) of SPring-6, Hiraoka *et al.* (2005) were able to distinguish between the $\mathbf{q} \| \Gamma M$ and $\mathbf{q} \| \Gamma K$ excitations and have compared the resulting $\mathrm{Im}\varepsilon(\mathbf{q}, \omega)$ with the predictions of band structure calculations.

Transitions between occupied and unoccupied d-states in transition metals should also be candidates for interesting peak-like structures in $S(\mathbf{q}, \omega)$ because of the rather small dispersion of the corresponding d-bands. Indead, Gurtubay *et al.* (2005) have found, in an experiment on the UNICAT beamline at the APS with 1.1 eV energy resolution, peaks around 4 eV in $S(\mathbf{q}, \omega)$ spectra of scandium

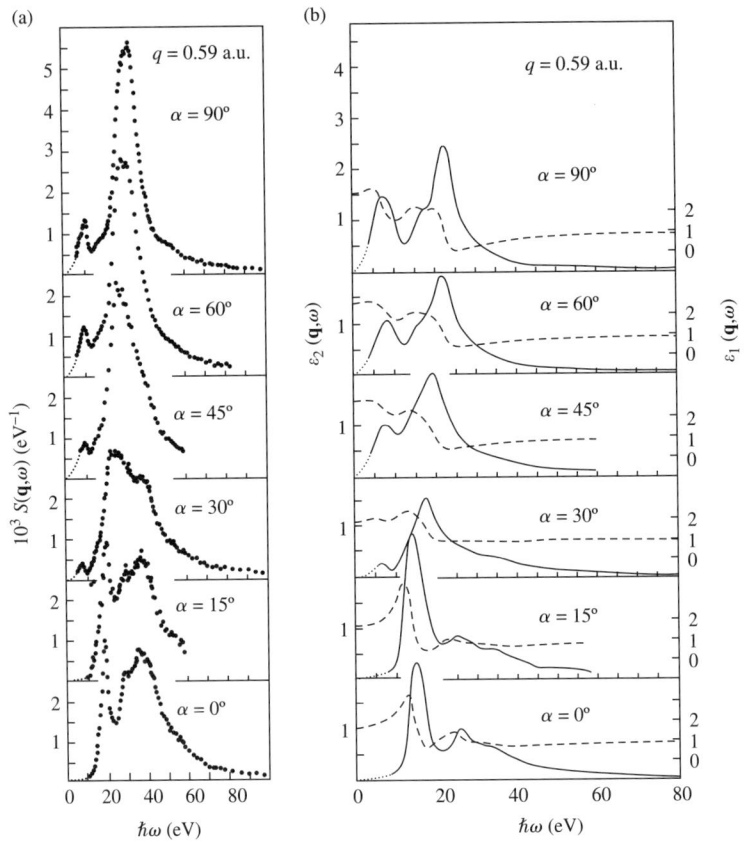

Fig. 2.52. (a) Dynamic structure factor $S(\mathbf{q},\omega)$ of HOPG, measured with $q = 0.59$ a.u.; α is the angle between $\mathbf{q}$ and the c-axis. (b) Dashed line: real part, $\varepsilon_1(\mathbf{q},\omega)$; solid line: imaginary part, $\varepsilon_2(\mathbf{q},\omega)$, of the dielectric function, as obtained by Kramers–Kronig analysis of the experimental $S(\mathbf{q},\omega)$ data. (Reprinted with permission from Schülke et al. (1988); copyright (1988) by the American Physical Society.)

and chromium that must be attributed, according to their calculations, to d-d transitions. Of course, these peaks appear only for larger q-values, as shown in Fig. 2.53 for the case of chromium with $\mathbf{q}$ along [100] and [111] direction. This q-value dependence can easily be understood as a consequence of the matrix elements $\langle f|\exp(i\mathbf{q}\cdot\mathbf{r})|i\rangle$ in (2.3). If one expands the exponential, the first term of this expansion gives a vanishing result since the states $|i\rangle$ and $|f\rangle$ are orthogonal. The second term linear in q (dipole term), gives a nonvanishing result only for $\Delta l = \pm 1$, if Δl is the difference in the angular quantum number between initial and final state, so that d-d transitions are not represented by this term. Only

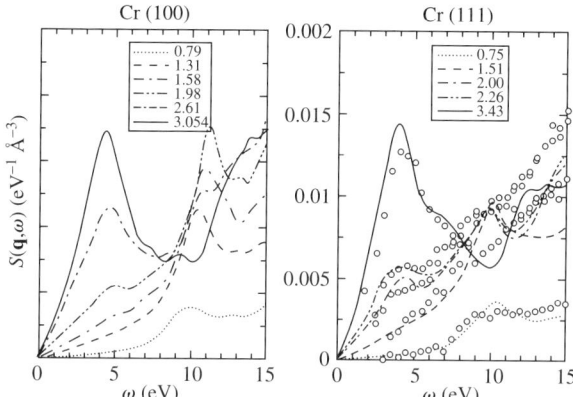

Fig. 2.53. Formation of the d-d transition peak in the dynamic structure factor per unit volume of Cr along the (100) and (111) directions for different values of the wave vector (in units of Å^{-1}). Lines: calculated; open circles: measured. (Reprinted with permission from Gurtubay *et al.* (2005); copyright (2005) by the American Physical Society.)

terms of higher order in q will contribute to the matrix element, provided that $qa > 1$, where a stands for the orbital radius of the participating states. Thus these d-d transitions can be made visible only with spectroscopies like IXS which provide a sufficiently large momentum transfer. Optical spectroscopy must fail and also electron energy loss spectroscopy (EELS) can hardly handle those large q's due to the difficulties with multiple losses. It must be stressed that the rather good agreement between calculation and experiment could only be obtained by the authors when crystal local-field effects are included; this means that the dynamical structure factor is calculated by inverting a complete dielectric matrix till to a maximum reciprocal lattice vector $\mathbf{g}_{max}$ in (6.103a), where $R_{MT} \times g_{max}$ was chosen to be 8 (R_{MT} is the radius of the smallest muffin-tin sphere in the unit cell). Moreover, exchange and correlation effects were included within the limits of the adiabatic local density approximation (ALDA) defined in equation (6.87). Similar strong crystal local-field effects were found by Gurtubay *et al.* (2004) in an IXS study TiO$_2$ on the UNICAT APS beamline.

2.4.2 *Metal–antiferromagnetic insulator transition*

V$_2$O$_3$ is a strongly renormalized Fermi liquid at room temperature. It can be transformed into an antiferromagnetic insulator by cooling, or into a paramagnetic insulator by doping with Cr (McWhan and Remeika 1970). It is well known that this behavior cannot be understood in terms of one-electron band theory, which predicts that these systems should be metallic. V$_2$O$_3$ has been considered to be a classic example of a Mott–Hubbard system in which an insulating gap is formed due to the short-range Coulomb interaction between electrons on the same atomic site, even if this simple picture must be revised because of the many degrees of freedom of the atomic V($3d^3 4s^1$) and O($1s^2 sp^4$) orbitals near the Fermi surface (Mattheis 1994). It must be expected that the excitation spectrum, as obtained by IXS, should exhibit interesting differences between the metallic and insulating phases of $V_{2-x}Cr_xO_3$ ($x = 0.022$). The metal–insulator transition occurred at $T_{AFI} = 173$ K. The inelastic scattering measurements on this system were performed by Isaacs et al. (1996) at the X21 beam line at the NSLS, equipped with a bent Si(220) monochromator and a spherically bent Si(444) analyzer. The sample was mounted inside a Be cell on the cold finger of a closed-cycle He refrigerator. The overall energy resolution was 1.1 eV. In Fig. 2.54 the inelastic X-ray scattering spectra with $q = 2.1$ Å^{-1}, $\mathbf{q} \| c$-axis, for the metallic (200 K) and the antiferromagnetic insulator (AFI) phase (10 K) are shown. The single peak at 12 eV energy loss in the metallic phase was interpreted by the authors to be due to a plasmon-like excitation on the basis of particle–hole transitions O(2p) → V(3d) and a zero-crossing of the real part of the diagonal dielectric function at an energy where the imaginary part of the dielectric function is locally small. This assignment is confirmed by a q^2-type dispersion of the peak position. The plasmon energy of 12 eV is consistent with the density of O(2p) valence electrons. In the AFI phase a second peak appears at 9.2 eV, whereas the plasmon-like peak shifts to 12.9 eV. It must be noted that the plasmon-like peak is also observed optically ($q \approx 0$), the lower energy peak, however, is only observed in the finite-q IXS measurement. The authors assign this lower energy peak to an exciton-like charge transfer excitation, where the electron–hole pair is formed by transferring a g angular momentum electron

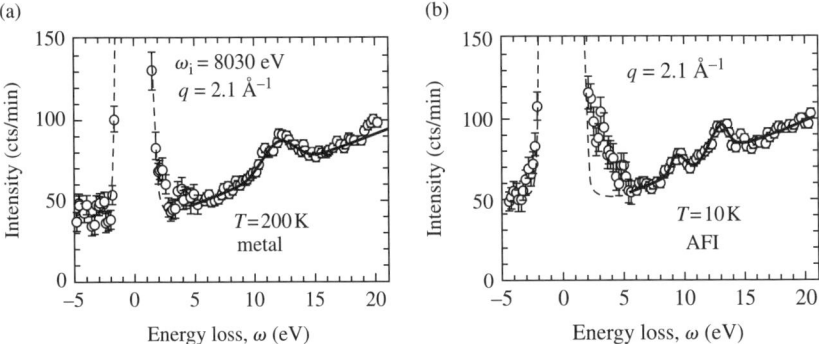

Fig. 2.54. Inelastic X-ray scattering spectrum $S(\mathbf{q}, \omega)$ (open circles) in $V_{1.978}Cr_{0.022}O_3$ with a momentum transfer, $q = 2.1$ Å^{-1}, $\mathbf{q} \| c$-axis. Solid line: fit of a single Gaussian in the metallic phase (a) ($T = 200$ K) and of two Gaussians in the antiferromagnetic insulating (AFI) phase (b) ($T = 10$ K). Dashed line: fit to the elastic line. (Reprinted with permission from Isaacs et al. (1996); copyright (1996) by the American Physical Society.)

from the neighboring oxygen to the empty V(3d) E_{2g} states, so that an angular momentum change of two takes place. This makes the transition dipole forbidden but quadrupole allowed, with the consequence (Platzman and Wolff 1973) that the corresponding dynamic structure factor becomes q^4 dependent. Such a q^4 dependence was indeed found in the dispersion measurements. The disappearance of this excitonic excitation in the metallic phase was traced back to the metallic screening. A careful examination of the AFI results in Fig. 2.54 reveals a shoulder in the elastic line near 2 eV energy loss. This feature must most probably be attributed the Mott–Hubbard gap, since it is not present in the metallic phase. We will come to this point later, when we discuss resonant IXS in Chapter 5.

2.5 Coherent inelastic X-ray scattering

2.5.1 *Access to off-diagonal response*

According to the definition, given in greater detail in Section 6.11.1, we will denote an experiment to be a coherent inelastic X-ray scattering experiment, whenever the initial photon state is built up by two coherently coupled plane waves. We can prepare these two plane waves in such a way that the resulting spatial modulation of the excitation is commensurable with the intrinsic modulation of the inhomogeneous system under investigation. This is achieved by using two incident plane waves whose wavevectors $\mathbf{K}_{10}$ and $\mathbf{K}_{1h}$ differs exactly by a reciprocal lattice vector $\mathbf{g}$, with other words, if one utilizes, in terms of the dynamical theory of X-ray diffraction (von Laue 1960), the incident and the Bragg diffracted wave, which are propagating within the crystal under the Bragg condition for $\mathbf{g}$, to be the initial photon state of an inelastic scattering experiment. The principle setup of such an experiment is sketched in Fig. 2.55 (Schülke *et al.* 1991). One is using the second crystal of a double-crystal monochromator together with the single crystalline scattering sample to form a nondispersive two-crystal arrangement set to meet the condition for Bragg diffraction in the so-called Bragg case. Within the limits of the two-beam approximation (von Laue 1960) the amplitude ratio A_{1h}/A_{10} of the the $\mathbf{K}_{1h}$ and $\mathbf{K}_{10}$ wave within the sample crystal as well as their mutual phase shift both as a function of the incidence angle, can be read by the intensity of the Bragg-reflected beam, monitored by detector D1. Especially the mutual phase shift of the two plane wave components changes drastically from 0 to π from one bound of the so-called total Bragg reflection range to the other one, an incident angular range which counts only a few seconds of arc. We will see that it is this mutual phase shift which is crucial for the application of coherent inelastic scattering. If one observes the scattered radiation with wavevector $\mathbf{K}_2$ and frequency ω_2 one has two transferred momenta $\mathbf{q}_0 = \mathbf{K}_{10} - \mathbf{K}_2$ and $\mathbf{q}_h = \mathbf{K}_{1h} - \mathbf{K}_2 = \mathbf{q}_0 + \mathbf{g}$. Likewise the wave field within the sample crystal, also the response of the sample electron system, consists of a coherent superposition of two transition amplitudes, each brought about by one of the two components $\mathbf{K}_{10}$ and $\mathbf{K}_{1h}$ of the initial photon

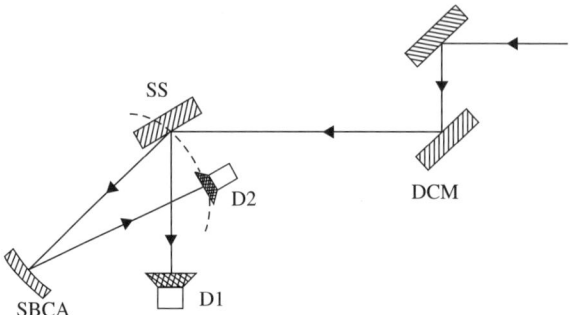

Fig. 2.55. Experimental setup for coherent inelastic X-ray scattering: DCM = double-crystal monochromator; SS = scattering sample (constitutes a nondispersive double-crystal setting with the second crystal of the DCM); SBCA = spherically bent crystal analyzer; D1 = monitor detector; D2 = detector of the analyzer. (Reprinted with permission from Schülke and Kaprolat (1991); copyright (1991) by the American Physical Society.)

wave field. The detailed derivation is found in Section 6.11.1, which ends up with the following expression for the double differential scattering cross-section of a coherent inelastic X-ray scattering experiment:

$$\frac{d^2\sigma}{d\Omega_2 d\hbar\omega_2} = \left[\frac{r_0^2}{A_0^2 + A_h^2}\right] \left(\frac{\omega_2}{\omega_1}\right)$$
$$\times (A_0^2 (\mathbf{e}_{10} \cdot \mathbf{e}_2)^2 S(\mathbf{q}_0,\omega) + A_h^2 (\mathbf{e}_{1h} \cdot \mathbf{e}_2)^2 S(\mathbf{q}_h,\omega) + 2A_0 A_h (\mathbf{e}_{10} \cdot \mathbf{e}_2^*)$$
$$\times (\mathbf{e}_{1h}^* \cdot \mathbf{e}_2)\{\mathrm{Re}[S(\mathbf{q}_0,\mathbf{g},\omega)] \cos(\Delta\phi) - \mathrm{Im}[S(\mathbf{q}_0,\mathbf{g},\omega)] \sin(\Delta\phi)\}), \tag{2.104}$$

where $\mathbf{e}_{10}$ and $\mathbf{e}_{1h}$ are the polarization vectors of the component photons with wavevectors $\mathbf{K}_{10}$ and $\mathbf{K}_{1h}$, respectively. The diagonal dynamic structure factor $S(\mathbf{q},\omega)$ is given by (2.6), whereas the non-diagonal dynamic structure factor $S(\mathbf{q}_0,\mathbf{g},\omega)$ reads, for $T=0$,

$$S(\mathbf{q}_0,\mathbf{g},\omega) \equiv \sum_{\mathrm{f}} \langle \mathrm{i}| \sum_j \exp(-i\mathbf{q}_0 \cdot \mathbf{r}_j)|\mathrm{f}\rangle$$
$$\times \langle \mathrm{f}| \sum_j \exp[i(\mathbf{q}_0 + \mathbf{g}) \cdot \mathbf{r}_j]|\mathrm{i}\rangle \, \delta(E_\mathrm{f} - E_\mathrm{i} - \hbar\omega). \tag{2.105}$$

This way the coherent superposition of two transition amplitudes belonging to the two momentum transfers $\mathbf{q}_0$ and $\mathbf{q}_h$ becomes transparent. Additionally one realizes easily that $S(\mathbf{q}_0,\mathbf{g},\omega)$ is real for the case of a centrosymmetric crystal structure, so that the second term in the { } brackets of (2.104) vanishes. If, additionally the two momentum transfers $\mathbf{q}_0$ and $\mathbf{q}_h$ are of equal modulus and are chosen to be symmetry equivalent with respect to the crystal structure of the sample, we have for a given triplet $(\mathbf{q}, \mathbf{g}, \omega)$ only two unknown quantities, namely $S(\mathbf{q}, \omega)$ and $S(\mathbf{q}, \mathbf{g}, \omega)$, so that these can be determined by two measurements of the double differential scattering cross-section at two positions of the sample crystal at the bounds of the total Bragg-reflection range. As already mentioned, this position can be read from the so-called rocking curve (Bragg-reflected intensity measured at the detector D1 as a function of the angular position of the sample crystal relative to the second monochromator crystal), since both crystals constitute a nondispersive double crystal setting. Then the corresponding values of A_0, A_h and $\Delta\phi$ can be determined from relations of the dynamical theory of X-ray diffraction (Schülke and Mourikis 1986). Of course, the measurements must be brought onto an absolute scale, which can be achieved by utilizing the generalized f-sum rule (Johnson 1974)

$$\int_0^\infty S(\mathbf{q}, \mathbf{g}, \omega)\omega\, d\omega = \frac{[\hbar\mathbf{q}\cdot(\mathbf{q}+\mathbf{g})/2m]\rho_\mathbf{g}}{\rho_0}, \qquad (2.106)$$

where $\rho_\mathbf{g}$ is the $\mathbf{g}$th Fourier component of the electron density ρ. The f-sum rule (2.21) for the diagonal dynamic structure factor is a special case of (2.106). One may object that this whole procedure developed to obtain information about off-diagonal response is of rather limited applicability. Indeed, the sample and double crystal monochromator must be of the same crystal material and this must be nearly perfect, in order to apply the formalism of the dynamical theory of X-ray diffraction. But the idea behind this procedure can also be realized otherwise. Modern insertion devices of third generation synchrotron radiation sources provide beams with extremely small vertical divergence, which can be further reduced, so that it becomes smaller than the Darwin width of low-indexed Bragg reflections. This means that one is no longer dependent on the nondispersive double crystal setting. Moreover, it has been demonstrated by Spiertz (1996) that one can determine separately diagonal and off-diagonal structure factors on the basis of (2.104) without explicitly knowing A_0, A_h and $\Delta\phi$ for the different positions of the scattering sample within the Bragg-reflection range, even for a mosaic crystal. The only requirement is that the spectra measured for different positions must not be linearly dependent, they must look really different. This is, of course, only possible if the mosaic spread is not much larger than the Darwin width. The above approach utilizes, in addition to the sum rule (2.106) for the diagonal and off-diagonal dynamic structure factor, spectra measured far away from the Bragg position, where the off-diagonal contribution vanishes. Moreover, one needs knowledge of the integral

$$\mathbf{I}(\omega_1, \omega_2) \equiv \int_{\omega_1}^{\omega_2} S(\mathbf{q}, \omega)\, d\omega \qquad (2.107)$$

between two frequencies ω_1, $\omega_2 \gg E_B/\hbar$, where E_B is the binding energy of the shallowest core level. If there is no significant overlap between the valence and the core part of $S(\mathbf{q}, \omega)$, then $\mathbf{I}(\omega_1, \omega_2)$ can be calculated by using free atomic eigenstates (Nagasawa et al. 1998). By solving a corresponding set of equations one obtains a complete diagonal and off-diagonal spectrum for every angular position of the sample crystal. This redundancy can be exploited for the appropriate handling of the background, which is of course of importance because of the rather small core contributions to $S(\mathbf{q}, \omega)$.

Let us now express the diagonal and the off-diagonal dynamic structure factor in terms of the dielectric matrix $1 + \mathbf{T}$, as defined in (6.103a), and which we will use for sake of simplicity in its RPA approximation, as given in (6.366) for the off-diagonal dynamic structure factor:

$$[S(\mathbf{q}_{0r} + \mathbf{g}_0, \mathbf{g}, \omega)]_{\text{RPA}} = -\left(\frac{|\mathbf{q}_{0r} + \mathbf{g}||\mathbf{q}_{0r} + \mathbf{g}_0|}{4\pi e^2}\right) \text{Im}[(1 + \mathbf{T})^{-1}]_{\mathbf{g}_0 \mathbf{g}}. \quad (2.108)$$

Its diagonal counterparts reads

$$[S(\mathbf{q}_{0r} + \mathbf{g}_0, \omega)]_{\text{RPA}} = -\left(\frac{|\mathbf{q}_{0r} + \mathbf{g}_0|^2}{4\pi e^2}\right) \text{Im}[(1 + \mathbf{T})^{-1}]_{\mathbf{g}_0 \mathbf{g}_0}, \quad (2.109)$$

where $\mathbf{g}_0$ reduces $\mathbf{q}$ into the first Brillouin zone. The dielectric matrix with its components $\varepsilon_{\mathbf{g}\mathbf{g}'}$ can be written (as shown in more detail in Chapter 6) in terms of the Bloch states $|\mathbf{k}, \nu\rangle$ with energy eigenvalues $E(\mathbf{k}, \nu)$ ($\mathbf{k}$ reduced Bloch wavevector, ν band index)

$$\varepsilon_{\mathbf{g}\mathbf{g}'} \equiv \delta_{\mathbf{g}\mathbf{g}'} - [v(\mathbf{q}_{0r} + \mathbf{g})]\left(2 \lim_{\eta \to 0_+} \sum_{\substack{\mathbf{k}\nu \\ \mathbf{k}'\nu'}}{}' \langle \mathbf{k}\nu| \exp[-i(\mathbf{q}_{0r} + \mathbf{g}) \cdot \mathbf{r}]|\mathbf{k}'\nu'\rangle \right.$$
$$\left. \times \langle \mathbf{k}'\nu'| \exp[i(\mathbf{q}_{0r} + \mathbf{g}') \cdot \mathbf{r}]|\mathbf{k}\nu\rangle \frac{f(E(\mathbf{k}, \nu)) - f(E(\mathbf{k}', \nu'))}{\hbar\omega + E(\mathbf{k}, \nu) - E(\mathbf{k}', \nu') + i\eta}\right). \quad (2.110)$$

The collective behavior of such an inhomogeneous electron system is determined by the plasmon band structure, a continuation of the plasmon dispersion relation known for the homogeneous electron system. As pointed out by Saslow and Reiter (1973) the plasmon band structure arises out of the zeros of the determinant of the dielectric matrix, that is

$$\det[\varepsilon_{\mathbf{g}\mathbf{g}'}(\mathbf{q}, \omega)] = 0. \quad (2.111)$$

In order to make transparent the physics behind this rather bulky formalism and also to interpret the experiments devoted to the plasmon band structure, we will introduce a rigorous reduction by switching over to the so-called two-plasmon band model, already used in Section 2.3.2.3. This means nothing else than to reduce the dielectric matrix into a 2×2 matrix. According to Cohen and Heine (1970) one can formally reduce the dielectric matrix into a 2×2 matrix with the components $\tilde{\varepsilon}_{\mathbf{g}\mathbf{g}'}$ by writing (Oliveira and Sturm 1980)

$$\tilde{\varepsilon}_{\mathbf{g}\mathbf{g}'} = \varepsilon_{\mathbf{g}\mathbf{g}'} - \sum_{\mathbf{g}''}{}'' \frac{\varepsilon_{\mathbf{g}\mathbf{g}''}(\mathbf{q}, \omega)\varepsilon_{\mathbf{g}''\mathbf{g}'}(\mathbf{q}, \omega)}{\varepsilon_{\text{RPA}}(|\mathbf{q} + \mathbf{g}''|, \omega)}, \quad (2.112)$$

where $\sum''$ implies that the sum extends over all reciprocal lattice vectors $\mathbf{g}''$ except $\mathbf{g}$ and $\mathbf{g}'$. As shown by Sturm and Oliveira (1980), one can calculate the matrix elements $\tilde{\varepsilon}_{\mathbf{gg}'}$ to second order in the Fourier components U_g of the pseudopotential analytically, proposed $\mathbf{g} = \mathbf{0}$, and $\mathbf{q}$ lies within the first Brillouin zone and is parallel to $\mathbf{g}'$, a situation, which we will discuss in the following. Having thus folded down the dielectric matrix to

$$\begin{pmatrix} \tilde{\varepsilon}_{00} & \tilde{\varepsilon}_{0\mathbf{g}} \\ \tilde{\varepsilon}_{\mathbf{g}0} & \tilde{\varepsilon}_{\mathbf{gg}} \end{pmatrix} \qquad (2.113)$$

the solution of the corresponding determinant (2.111) leads to a two-plasmon band structure as plotted in Fig. 2.32. The plasmon band gap at the zone boundary turns out to be proportional to $\rho_\mathbf{g}$ (Saslow and Reiter 1973).

2.5.2 Shape and gap of plasmon bands

Schülke and Kaprolat (1991) and Sturm and Schülke (1992) have applied coherent inelastic X-ray scattering to the problem of getting information about the plasmon band gap of Si, as predicted by Saslow and Reiter (1973), especially about the plasmon band gap of Si at the L-point of the $\mathbf{g}_{111} = (2\pi/a)(1,1,1)$ Brillouin zone boundary (BZB). This is feasible, since in the case of Si $|\mathbf{g}_{111}/2| < q_c$ (q_c = plasmon cut-off vector). The result of a coherent inelastic X-ray scattering experiment, performed at the DORIS storage ring by means of the set-up shown in Fig. 2.55 is presented in Fig. 2.56. The wave field of the symmetrical $\mathbf{g}_{111}$ Bragg case was utilized as the initial photon state. The radiation inelastically scattered with momentum transfer $|\mathbf{q}_0| = |\mathbf{q}_h| = |\mathbf{g}_{111}/2|$ was analyzed by means of a spherically bent Si(777) analyzer in inverse geometry at different angular positions of the sample crystal, as indicated at the rocking curve in the inset. The imaginary part of the diagonal and the off-diagonal inverted dielectric matrix, proportional to the diagonal and off-diagonal dynamic structure factor, respectively, are plotted, separated as described above. The diagonal part is single-peaked, the off-diagonal part exhibits a peak–valley structure. We will see that the shape of the two plasmon bands at the L-point can be deduced together with the energy gap between them.

If the plasmon band structure at the BZB looks as shown in Fig. 2.32 the diagonal dynamic structure factor, as can be measured for $q = g_{111}/2$ in a coherent scattering experiment, reads, within the limits of the above described two-plasmon band model, according to (2.109, 2.113) ($\mathbf{g}_0 = \mathbf{0}$) as

$$[S(\mathbf{q},\omega)]_{\text{RPA}} = \left(\frac{\mathbf{q}^2}{4\pi e^2}\right) \text{Im}\left(\frac{-\tilde{\varepsilon}_{\mathbf{gg}}(\mathbf{q},\omega)}{\tilde{\varepsilon}_{00}(\mathbf{q},\omega)\,\tilde{\varepsilon}_{\mathbf{gg}}(\mathbf{q},\omega) - \tilde{\varepsilon}_{0\mathbf{g}}(\mathbf{q},\omega)\,\tilde{\varepsilon}_{\mathbf{g}0}(\mathbf{q},\omega)}\right), \quad (2.114)$$

and the off-diagonal structure factor as

$$[S(\mathbf{q},\mathbf{g},\omega)]_{\text{RPA}} = \left(\frac{|\mathbf{q}+\mathbf{g}||\mathbf{q}|}{4\pi e^2}\right) \text{Im}\left(\frac{-\tilde{\varepsilon}_{0\mathbf{g}}(\mathbf{q},\omega)}{\tilde{\varepsilon}_{00}(\mathbf{q},\omega)\,\tilde{\varepsilon}_{\mathbf{gg}}(\mathbf{q},\omega) - \tilde{\varepsilon}_{0\mathbf{g}}(\mathbf{q},\omega)\,\tilde{\varepsilon}_{\mathbf{g}0}(\mathbf{q},\omega)}\right)$$
$$(2.115)$$

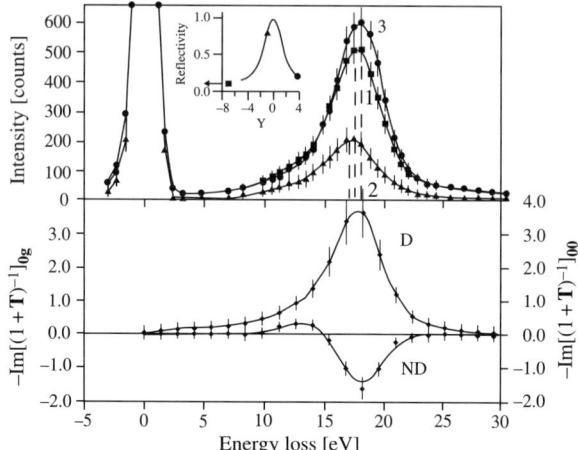

Fig. 2.56. Experimental result of a coherent inelastic X-ray scattering experiment. Upper panel: measured data for three different angular positions of the scattering sample crystal relative to the second crystal of the monochromator, as indicated by the points on the rocking curve (reflectivity of the sample crystal as a function of the reduced angular variable y) in the inset. $\mathbf{q} \parallel [111]$; $q = 0.51$ a.u. Lower panel: diagonal (D) contribution $-\text{Im}[(1+\mathbf{T})^{-1}]_{00}$ and nondiagonal (ND) contribution, $-\text{Im}[(1+\mathbf{T})^{-1}]_{0\mathbf{g}}$, to the inverted dielectric matrix, obtained from the above data by means of the separation procedure described in the text. (Reprinted with permission from Schülke and Kaprolat (1991); copyright (1991) by the American Physical Society.)

Because of the inversion symmetry of the Si-lattice we have with $\mathbf{g} \equiv -\mathbf{g}_{111}$

$$\tilde{\varepsilon}_{\mathbf{gg}}(\mathbf{g}_{111}/2, \omega) = \tilde{\varepsilon}_{\mathbf{00}}(\mathbf{g}_{111}/2, \omega) \qquad (2.116)$$

and

$$\tilde{\varepsilon}_{\mathbf{0g}}(\mathbf{g}_{111}/2, \omega) = \tilde{\varepsilon}_{\mathbf{g0}}(\mathbf{g}_{111}/2, \omega), \qquad (2.117)$$

so that (2.114) and (2.115) can be written as (we suppress the arguments $\mathbf{q} = \mathbf{g}_{111}/2, \omega$)

$$[S(\mathbf{q}, \omega)]_{\text{RPA}} = \left(\frac{(\mathbf{g}_{111}/2)^2}{8\pi e^2}\right) \text{Im}\left(\frac{-1}{\tilde{\varepsilon}_{\mathbf{00}} + \tilde{\varepsilon}_{\mathbf{0g}}}\right) + \text{Im}\left(\frac{-1}{\tilde{\varepsilon}_{\mathbf{00}} - \tilde{\varepsilon}_{\mathbf{0g}}}\right) \qquad (2.118)$$

$$[S(\mathbf{q}, -\mathbf{g}_{111}, \omega)]_{\text{RPA}} = \left(\frac{(\mathbf{g}_{111}/2)^2}{8\pi e^2}\right) \text{Im}\left(\frac{-1}{\tilde{\varepsilon}_{\mathbf{00}} + \tilde{\varepsilon}_{\mathbf{0g}}}\right) - \text{Im}\left(\frac{-1}{\tilde{\varepsilon}_{\mathbf{00}} - \tilde{\varepsilon}_{\mathbf{0g}}}\right). \qquad (2.119)$$

Remembering that the energy position of the two plasmon bands ω_l and ω_u, separated by a gap at the L-point is the result of setting the corresponding

determinant
$$(\tilde{\varepsilon}_{00} + \tilde{\varepsilon}_{0g})(\tilde{\varepsilon}_{00} - \tilde{\varepsilon}_{0g}) \qquad (2.120)$$
of the dielectric matrix (2.113) equal to zero, we have
$$(\tilde{\varepsilon}_{00}(\mathbf{g}_{111}/2, \omega_u) + \tilde{\varepsilon}_{0g}(\mathbf{g}_{111}/2, \omega_u)) = 0 \qquad (2.121)$$
$$(\tilde{\varepsilon}_{00}(\mathbf{g}_{111}/2, \omega_l) - \tilde{\varepsilon}_{0g}(\mathbf{g}_{111}/2, \omega_l)) = 0. \qquad (2.122)$$

Therefore, $\mathrm{Im}(-1/[\tilde{\varepsilon}_{00} \pm \tilde{\varepsilon}_{0g}])$ represent the shape of the upper(u) and the lower(l) plasmon band, respectively. Thus we conclude that the diagonal dynamic structure factor is made up by the sum of the two plasmon lines, so that the plasmon band gap should be visible as a double-peak structure, provided the width of both bands is small enough. The single-peak structure of the diagonal dynamic structure factor manifests that this is not the case. The plasmon bands are strongly lifetime broadened due to interband transitions. On the other hand, the off-diagonal dynamic structure factor is determined by the difference of the two plasmon lines, so that the two components become visible as a peak–valley structure. Of course, it is easy to reconstruct the shape of the two plasmon lines, as shown in Fig. 2.57. One can read from Fig. 2.57 both a plasmon band gap of $\sim 1\,\mathrm{eV}$ and the interesting fact that the lower plasmon band is broader than the upper plasmon band, opposite to what one might expect in terms of plasmon

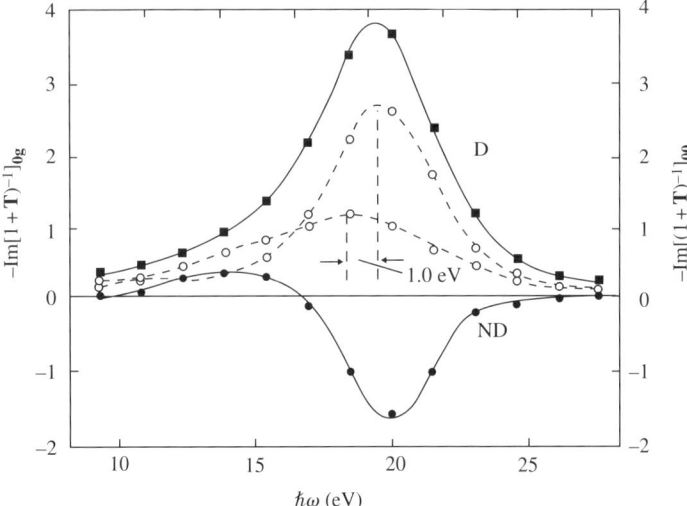

Fig. 2.57. Upper and lower plasmon band (open circles), shifted by 1 eV as obtained from the diagonal (D), $-\mathrm{Im}[(1+\mathbf{T})^{-1}]_{00}$ (filled squares) and the nondiagonal (ND) contribution, $-\mathrm{Im}[(1+\mathbf{T})^{-1}]_{0g}$, (filled circles) to the inverted dielectric matrix. (Reprinted with permission from Sturm and Schülke (1992a); copyright (1992) by the American Physical Society.)

damping, but in good agreement with predictions of Daling et al. (1991) and pseudopotential calculations of Sturm and Schülke (1992). It should be stressed that signatures of this plasmon band gap couldn't be found either in an EELS experiment, also due to the strong damping, which makes the width of the plasmon bands of the order of 5 eV. Therefore, this coherent inelastic scattering experiment on Si can be considered to be the only direct proof for a plasmon band structure in bulk crystals.

2.6 Inverting $S(\mathbf{q}, \omega)$ measurements into time and space. Imaging electron dynamics

All that we have learnt so far by inelastic X-ray scattering about electron dynamics has been examined in the frequency or momentum representation, where the spectra may be compared to theoretically calculable response functions for the electron density. However, the fundamental relations (2.12) between the dynamic structure factor $S(\mathbf{q}, \omega)$ and the van Hove space time correlation function $G_c(\mathbf{r}, t)$ as defined in (2.11) should enable, at least in principle, converting experimental results from the momentum–frequency into the space–time domain. This possibility is the more interesting, since it would mean experimental access to correlated movements on the attosecond time scale (1 as = 10^{-18} s) and probing correlation distances of one Å and less. Abbamonte et al. (2004) were the first to utilize the momentum flexibility of inelastic X-ray scattering to invert its loss function with the aim of imaging density disturbances in a medium. Their starting point was equation (6.44) where the response $\delta\langle\rho(\mathbf{q})\rangle(t)$ of the time-dependent **q**th Fourier component of the density fluctuation is linearly dependent on the Fourier components $\phi_{\text{ext}}(\mathbf{q}, \omega)$ of the disturbing potential:

$$\delta\langle\rho(\mathbf{q})\rangle(t) = (-e) \lim_{\varepsilon\to 0_+} \chi(\mathbf{q}, \mathbf{q}, \omega)\, \phi_{\text{ext}}(\mathbf{q}, \omega) \exp[-i(\omega + i\varepsilon)t], \qquad (2.123)$$

here written for the case of a homogeneous electron system, which reduces the density response function $\chi(\mathbf{q}', \mathbf{q}, \omega)$ to terms diagonal in **q**. $\phi_{\text{ext}}(\mathbf{q}, \omega)$ is the time–space Fourier transform of the disturbing potential $\phi_{\text{ext}}(\mathbf{r}, t)$. In the following equation (2.124)

$$\chi(\mathbf{q}', \mathbf{q}, \omega) = \lim_{\varepsilon\to 0_+} \left(\frac{i}{V\hbar}\right) \int d^3 r \int d^3 r'\, \exp(-i\mathbf{q}' \cdot \mathbf{r})$$
$$\times \int_0^\infty d\tau \exp[i(\omega + i\varepsilon)\tau]\, \chi(\mathbf{r}, \mathbf{r}', \tau) \exp(i\mathbf{q} \cdot \mathbf{r}) \qquad (2.124)$$

the density response function $\chi(\mathbf{q}', \mathbf{q}, \omega)$ is expressed in terms of the density–density response function $\chi(\mathbf{r}, \mathbf{r}', t - t')$, defined by

$$\chi(\mathbf{r}, \mathbf{r}', t' - t) \equiv \left(\frac{-i}{\hbar}\right) \theta(t - t') \langle[\rho(\mathbf{r}, t), \rho(\mathbf{r}', t')]\rangle, \qquad (2.125)$$

where $\theta(\tau) = 0, \tau < 0$; $\theta(\tau) = 1$, $\tau > 0$ guarantees that the response at time t is due only to disturbances at time $t' \leq t$ (causality). Then the response $\delta\langle\rho(\mathbf{r})\rangle$

of the density fluctuation is given by

$$\delta\langle\rho(\mathbf{r})\rangle(t) = (-e)\int d^3r' \int_0^\infty dt'\chi(\mathbf{r},\mathbf{r}',t'-t)\,\phi_{\text{ext}}(\mathbf{r}',t') \qquad (2.126)$$

so that a δ-like external disturbance

$$\phi_{\text{ext}}(\mathbf{r}',t') = \delta(\mathbf{r}')\delta(t') \qquad (2.127)$$

gives rise to a density fluctuation

$$\delta\langle\rho(\mathbf{r})\rangle(t) = (-e)\chi(\mathbf{r},\mathbf{0},t), \qquad (2.128)$$

or, in other words, $\chi(\mathbf{r},\mathbf{0},t)$ is the density disturbance produced by a delta function source at the origin at $t=0$. Abbamonte et al. (2004) have found a way to extract exactly this quantity $\chi(\mathbf{r},\mathbf{0},t)$ by inverse Fourier transform out from inelastic X-ray scattering data. First they got $S(\mathbf{q},\omega)$ on an absolute scale (f-sum rule) from measurements on water within a sufficient wide range q and ω values at the C-1 station of CHESS and at CMC-CAT of APS. Then they utilized equation (2.14) to obtain $\text{Im}\,\chi(\mathbf{q},\omega)$ from $S(\mathbf{q},\omega)$. The Kramers–Kronig relation (6.44) provided access to $\text{Re}\,\chi(\mathbf{q},\omega)$, certainly after extrapolating $\text{Im}\,\chi(\mathbf{q},\omega)$ to sufficiently large values of ω. $\chi(t)$ can then be evaluated with the time transform

$$\chi(t) = \pi^{-1}\int[\sin(\omega t)\,\text{Im}\,\chi(\omega) + \cos(\omega t)\,\text{Re}\,\chi(\omega)]d\omega, \qquad (2.129)$$

where beforehand the discrete spectra were analytically continued onto a continuous frequency interval in order to prevent $\chi(t)$ becoming periodic, a property incompatible with the constraint that $\chi(t)$ must vanish for all $t < 0$. Finally, the spatial inversion $\chi(\mathbf{q},\mathbf{q},t) \to \chi(\mathbf{r},\mathbf{0},t)$ was performed by means of a spherical Fourier integral. According to the energy and momentum resolution of $\Delta\omega = 0.30\,\text{eV}$ and $\delta_q = 0.146\,\text{Å}^{-1}$, respectively, the field of view was $T = 13.8\,\text{fs}$ and $R = 54.3\,\text{Å}$. The space and time resolution are determined by the scan ranges to be $\Delta t = 2\pi\hbar/100\,\text{eV} = 41.3$ as and $\Delta r = 2\pi/4.95\text{Å}^{-1} = 1.27\,\text{Å}$. Figure 2.58 shows the time frames of $\chi(\mathbf{r},\mathbf{0},t)$ in units of Å^{-6}. (b) Shortly after the impulse, a large (off-scale) negative recoil at $r = 0$ is visible, surrounded by a compensating positive buildup. In (c)–(g) the further evolution of this disturbance is depicted. (h)–(i) on an expanded scale, show distant Friedel-like oscillations.

Abbamonte et al. (2004) have used these $\chi(\mathbf{r},\mathbf{0},t)$ data to determine, by means of superposition, the effect of extended sources, e.g. showing the wake produced in water by a 9 MeV gold ion. It is stressed by the authors that the tunability of X-rays over broad ranges of energy makes accessible time-scales that are currently out of the reach of laser-based techniques. This way inelastic X-ray scattering can provide an alternative window on attosecond phenomena.

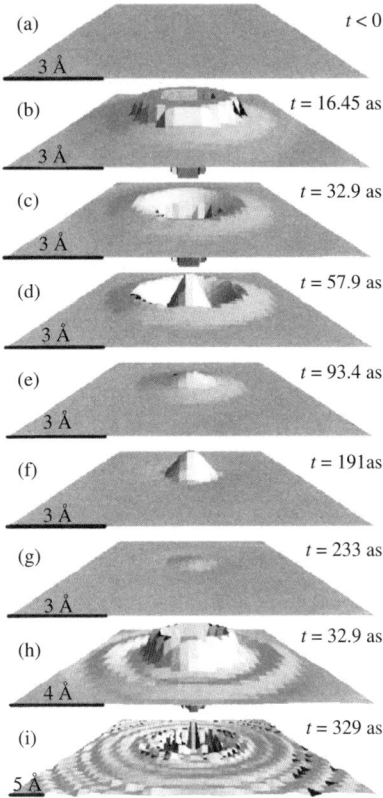

Fig. 2.58. Time frames of $\chi(\mathbf{r}, \mathbf{0}, t)$ in units of Å^{-6}. The vertical scale has been clipped at $1\,\text{Å}^{-6}$ in frames (a)–(g), $0.1\,\text{Å}^{-6}$ in (h) and $0.005\,\text{Å}^{-6}$ in (i). Distances are indicated with scale bars. (a) At $t < 0$, before the perturbation the system is "placid". (b)–(g) Evolution of the disturbance at selected times after the perturbation (h) same as (b) but on an expanded scale, showing distant Friedel-like oscillations. (i) After the disturbance has damped, scale expanded to show the experimental noise level. (Reprinted with permission from Abbamonte *et al.* (2004); copyright (2004) by the American Physical Society.)

2.7 Static structure factor $S(\mathbf{q})$

2.7.1 $S(\mathbf{q})$, *pair-correlation function, exchange-correlation potential and correlation energy*

Because of its very near relationship to the dynamical structure factor we will devote an extra section to the static structure factor, even if the book is mainly dealing with the dynamic properties of matter as seen by inelastic X-ray scattering.

As shown in Chapter 6, equation (6.37), the static structure factor $S(\mathbf{q})$ is, within the limits of the so-called quasielastic approximation ($\omega_2 \approx \omega_1$), nothing else than the frequency integral of the dynamic structure factor:

$$S(\mathbf{q}) = \hbar \int_{-\infty}^{\infty} S(\mathbf{q},\omega)\,d\omega. \tag{2.130}$$

Of course, the dynamic structure factor contains contributions from both electrons and nuclei, where the latter have not been considered so far, since the corresponding excitation energies are far too small to be detectable in experiments with 0.1–1 eV resolution. These contributions are hidden in the quasielastic line. But now the frequency integration make us take them into account. However, because of the large energy difference between the electron and the nuclei contribution, which allows only negligible coupling between the two, they can be separated writing

$$S(\mathbf{q}) = S_{\text{TDS}}(\mathbf{q}) + S_{\text{ee}}(\mathbf{q}), \tag{2.131}$$

where the origin of the nuclei contribution is traced back to the thermal diffuse scattering (TDS).

The most important static quantity, which is connected intimately with the electron part $S_{\text{ee}}(\mathbf{q})$ of the static structure factor, is the pair correlation function $g(\mathbf{r}', \mathbf{r} - \mathbf{r}')$, the probability density of finding two electrons $\mathbf{r}$ apart, the first being at $\mathbf{r}'$:

$$S_{\text{ee}}(\mathbf{q}) - Z = \int d^3r' \int d^3r\, n(\mathbf{r}')\, n(\mathbf{r}' + \mathbf{r})\, [g(\mathbf{r}'_1, \mathbf{r}' + \mathbf{r}) - 1]\, \exp(-i\mathbf{q}\cdot\mathbf{r}) \tag{2.132}$$

if $S_{\text{ee}}(\mathbf{q})$ is normalized to the number Z of electrons per atom, and $n(\mathbf{r})$ is the number density of the electrons.

Note that equation (6.41) is equivalent to (2.132) when written for a homogeneous electron system of number density $n(r)$, since in the sense of the local density approximation

$$Z S_{\text{ee}}(\mathbf{q}) \approx 4\pi \int_{\Omega_0} n(\mathbf{r}) S_{\text{h}}[\mathbf{q}; n(\mathbf{r})]\, r^2 dr \tag{2.133}$$

if $S_{\text{h}}(\mathbf{q}; n)$ refers to the homogeneous electron gas of density n. Ω_0 is the atomic sphere.

A second important quantity, which can directly be related to the static structure factor, is the exchange-correlation interaction energy E_{xc} per atom (Mazzone and Sacchetti 1984); (see also the definition of the static structure factor in terms of the electron density operators $\rho(\mathbf{q})$ (6.40) together with

(6.28 and 6.29))

$$E_{xc} \equiv (e^2/4) \int d^3r \int d^3r' |\mathbf{r}-\mathbf{r}'|^{-1} [\langle \rho^+(\mathbf{r})\rho(\mathbf{r}')\rangle - n(\mathbf{r})n(\mathbf{r}')]$$
$$= (e^2/8\pi^2) \int d^3q ([(S(\mathbf{q})-Z)/q^2]), \qquad (2.134)$$

where $\rho(\mathbf{r})$ is the electron number density operator in position space and $n(\mathbf{r})$ the electronic ground-state number density.

Finally the quantity

$$\gamma(r_s) \equiv -(2k_F)^{-1} \int_0^\infty [S(q)-1] dq \qquad (2.135)$$

defined for the homogeneous electron system is simply related to the ground-state correlation energy $\varepsilon_c(r_s)$ (in Ryd) by the expression (Singwi et al. 1968)

$$\varepsilon_c(r_s) = r_s^{-2} \int_0^{r_s} [-(4/\pi)(9\pi/4)^{1/3}\gamma(r'_s) + 0.9163] dr'_s. \qquad (2.136)$$

Since these three quantities, $g(\mathbf{r},\mathbf{r}')$, E_{xc} and ε_c, are of special interest for the valence electron system of condensed matter, it is advisable to write the static structure factor as the sum of valence and core electron contributions:

$$S(\mathbf{q}) = S_{val}(\mathbf{q}) + S_{core}(\mathbf{q}). \qquad (2.137)$$

In this equation the core-orthogonalization contribution is neglected considering that it is, in any case, fairly small.

Equation (2.132) demonstrates that the pair-correlation function of an inhomogeneous system, which depends on the individual electron coordinates $\mathbf{r}',\mathbf{r}$ and not only on $|\mathbf{r}-\mathbf{r}'|$, and where also the electron number density is a varying function of the position, cannot derived directly from measurements of the static structure factor $S(\mathbf{q})$. Therefore, we define an average pair-correlation function, $\bar{g}(\mathbf{r})$, via:

$$\bar{g}(\mathbf{r}) = [(2\pi)^3 \bar{n} Z]^{-1} \int d^3q [S(\mathbf{q})-Z] \exp(i\mathbf{q}\cdot\mathbf{r})$$
$$= (\bar{n})^{-1} \int d^3r'\, n(\mathbf{r}')n(\mathbf{r}'+\mathbf{r})[g(\mathbf{r}',\mathbf{r}'+\mathbf{r})-1], \qquad (2.138)$$

where $\bar{n}$ is the average number density. But even if one wants to utilize (2.138) in order to derive $\bar{g}(\mathbf{r})$, the knowledge of $S(\mathbf{q})$ throughout the whole reciprocal space would be necessary. If this knowledge is restricted to only a few directions of $\mathbf{q}$, one either needs an appropriate interpolation scheme or one is using a simple spherical average $S(q)$, so that one ends up with a spherical average of the pair correlation function $g(r)$ via:

$$g(r) = 1 + (2\pi^2 \bar{n} r)^{-1} \int_0^\infty [S(q)-1] \sin(qr) q\, dq. \qquad (2.139)$$

2.7.2 *Experimental access to* $S(\mathbf{q})$

Until now three different experimental methods have been used to get access to the static structure factor:

(i) The most direct one is suggested by (2.130) and consists of measuring the dynamic structure factor $S(\mathbf{q},\omega)$ for a larger number of $\mathbf{q}'$s (Schülke et al. 1969, Schülke et al. 1995, Watanabe et al. 1997, Watanabe et al. 1998). After removing the quasielastic line and the core contribution, one brings the measurement on to an absolute scale by using the f-sum rule (2.21). These absolute values of $S(\mathbf{q},\omega)$ have to be integrated with respect to ω to end up with the valence part, $S_{\text{val}}(\mathbf{q})$, of the static structure factor. This method suffers from two drawbacks: (1) The removing of the quasi elastic line is always connected with uncertainties, since the form of the quasielastic line is in general unknown. (2) Whenever the valence and the core contributions are overlapping, the removing of the core contribution needs an extrapolation of the valence part, which is again connected with appreciable uncertainties. Therefore, this method is mainly restricted to the range of smaller q's, where this overlap is not so important.

(ii) A synchrotron radiation related method has been introduced by Eisenberger et al. (1980). These authors have measured the intensity of photons scattered from the sample after having traveled through a foil of a certain element (here platinum) positioned between sample and detector, and transferring a certain momentum $\mathbf{q}$ for two different incident photon energies, as available from a monochromatized synchrotron X-ray. The two photon energies were chosen in such a way that one coincides with the photon energy, which determines the top of the absorption curve of the foil ("white line"), whereas the other photon energy is connected with the lower (long-wavelength) part of the absorption edge. This way the elastic and quasielastic component of $S(\mathbf{q})$ (the latter is denoted in (2.131) by $S_{\text{TDS}}(\mathbf{q})$) are attenuated by a factor $\exp(-3.4)$ for the former photon energy, but only by $\exp(-1.4)$ for the latter one. Assuming that the electron part $S_{\text{ee}}(\mathbf{q})$ is attenuated by nearly the same factor as given by the long-wavelength part of the absorption curve (small differences can be corrected for), one can remove the elastic and quasielastic contribution, ending up with relative values of $S_{\text{ee}}(\mathbf{q})$. These can be brought to an absolute scale by using

$$S_{\text{ee}}(q \to q_{\text{max}}) = 1. \tag{2.140}$$

Moreover, one can also remove the core contribution to S_{ee} by using atomic calculations, e.g. for two 1s electrons (Li, Be) by using (Eisenberger et al. 1980)

$$S_{\text{core}}(q) = 1 - \left(1 + \frac{0.264q}{Z - (5/16)}\right). \tag{2.141}$$

(iii) The third method (Mazzone et al. 1983, Petrillo and Sacchetti 1995, Calzuola et al. 1999) works with conventional X-ray sources and is using a standard X-ray diffractometer, properly adapted to measure diffuse scattering with sufficient q-resolution. Characteristic $K\alpha$ radiation is applied monochromatized by a pyrolytic graphite crystal in order to prevent $K\beta$ contamination of the incident radiation. The single-crystal sample orientation and the scattering angles are adjusted in such a manner that contact of the Ewald sphere with reciprocal lattice points is avoided. Nevertheless, the measured intensity as a function of scattering angle is strongly structured due to thermal diffuse scattering, which has to be subtracted. The TDS contribution was calculated, within the limits of the harmonic approximation, by employing the dynamical matrix as deduced from the neutron measured phonon dispersion relation. Also multiphonon contributions were taken into account using the same dynamical matrix. Moreover, Monte Carlo simulations of the multiple scattering contributions were performed using the free atom scattering factors and static structure factors. The experimental intensity, after background correction, was normalized to the sum of the free-atom theoretical structure factor, TDS, and multiple-scattering contribution at high momentum transfer. This normalization procedure is believed to be very reliable, since at high momentum transfer the static structure factor must converge to the number of electrons, and solid state effects are minimized in this region.

2.7.3 *Correlation effects studied by $S(\mathbf{q})$ measurements*

It has been recognized rather early that the measurements of the static structure factor can offer valuable insight into electron–electron correlation in solids. We will exemplify the chances and problems connected with the evaluation of measured static structure factor in terms of correlation effects on the simple metal beryllium.

It was found by Schülke et al. (1969) using the method (i) with a conventional X-ray source (W Lβ_1 radiation) that the static structure factor of Be metal is much better described by the jellium RPA than by the jellium Hartree–Fock approximation of $S(\mathbf{q})$, thus giving rather direct evidence for Coulomb correlation in simple metals. It was the reduction of $S(\mathbf{q})$ against its Hartee–Fock value for small q which indicated the strong screening due to plasmon excitation. Later Eisenberger et al. (1980) studied again the static structure factor of Be together with graphite by using the method (ii) with synchrotron radiation from SPEAR. The core contribution was subtracted by means of equation (2.141). The resulting valence part of $S(\mathbf{q})$ exhibited rather strong positive deviation from the jellium RPA around $q = 2k_\mathrm{F}$ for graphite and even stronger for Be. These deviations were connected with corresponding non-RPA fine structure of the dynamic structure factor (double-peak structure, see Section 2.3.1). The authors of this study have then used the relation (2.139) to get the valence electron pair correlation function, where they allow $S(\mathbf{q})$ to "approach unity with the asymptotic RPA form such that $g(0)$ was equal to zero". They obtained, as already predicted by Platzman and Eisenberger (1974), a pair correlation function both for graphite and beryllium with a very pronounced peak at $rk_F \approx \pi/5$, which they attribute to the deviations from RPA around $q = 2k_F$. Note that r_{peak} is smaller than r_s. Since the peak-like features in $g(r)$ look very similar when scaled to electron gas parameter, it was concluded that they have to be interpreted as being due to electron–electron interactions at distances shorter than r_s. Nevertheless, these findings of the Eisenberger group were criticized by Mazzone et al. (1983), who also performed a study of the Be $S(\mathbf{q})$, this time by using the method (iii) with conventional X-ray sources. The measurements were performed on single crystals, so that $\mathbf{q}$ was set by several degrees off a high-symmetry direction in order to prevent low-indexed Bragg reflections. Their valence electron part of $S(\mathbf{q})$ (denoted in this study as the band-electron static structure factor) obtained after subtraction of the core part, as calculated by Thakkar and Smith (1978), averaged with respect to a small anisotropy not larger than the other estimated uncertainties, is shown in Fig. 2.59. Also in this study, a distinct deviation from the jellium Hartree–Fock $S(\mathbf{q})$ is found, whereas the experiment agrees much better with calculations of Singh and Pathak (1973) which are founded on moment sum rules. Moreover, a two-parameter fit of the experiment to a "model" $S(\mathbf{q})$ based on a "model" pair-correlation function of Rajagopal et al. (1978) shows

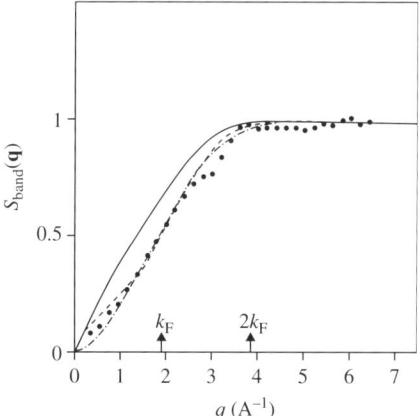

Fig. 2.59. Static structure factor $S_{\text{band}}(\mathbf{q})$ of the valence electrons of Be metal. Solid circles: experimental data (conventional X-ray source); solid line: Hartree–Fock (HF) jellium calculation; dashed-dotted line: calculation of Singh and Pathak (1973); Dashed line: best fit to $S(\mathbf{q})$ calculated from a parametrized $g(r)$-function according to Rajagopal et al. (1978). (Reprinted with permission from Mazzone et al. (1983); copyright (1983) by the American Physical Society.)

fairly good agreement with the measurement. In order to get the pair-correlation function $g(r)$, according to (2.139), directly from the experiment, a polynomial fit of the experimental data up to $4\,\text{Å}^{-1}$ has been used, while from that point on, $S_{\text{band}}(\mathbf{q})$ was considered to have reached its asymptotic value. The result of this inversion procedure is shown in Fig. 2.60. Plotted also is the well-known jellium Hartree–Fock result, the "model" pair-correlation function of Rajagopal et al. (1978) as obtained by the above mentioned two-parameter fit, and a curve, which is the result of a direct Monte Carlo calculation of Ceperly (1978). Note that the pair correlation function as deduced from the experiment does not exhibit the oscillatory structure of the Eisenberger result, but shows unphysical negative values for $r < 0.5\,\text{Å}$, as is well known for the jellium RPA case. This negative $g(r)$ must not speak against the reliability of the experiment, since the large-q extrapolation to the asymptotic value of $S(\mathbf{q})$ amounts to losing most of the relevant information for $r < 2\pi/4\,\text{Å}$. The large sensitivity of the inversion procedure for $g(r)$, according to (2.139) in the range of smaller r, against fine details of $S(\mathbf{q})$ for larger q becomes evident if one sees the large discrepancies between the "experimental" $g(r)$ and the Rajagopal-type $g(r)$ in Fig. 2.60 on the one hand, and the rather good agreement between the corresponding $S(\mathbf{q})$ in Fig. 2.59. At this point Mazzone et al. (1983) have fixed their criticism of the Eisenberger $g(r)$ result. Since they did not find the oscillatory structure when they applied their inversion scheme to the Eisenberger data with setting $S_{\text{band}}(\mathbf{q}) = 1$ for $q > 8\,\text{Å}^{-1}$, they assume that it was the RPA-like extrapolation of the data with the constraint $g(0) = 0$, which has produced the oscillations as artifacts. (By the way, one can find serious physical reasons that $g(0)$ must

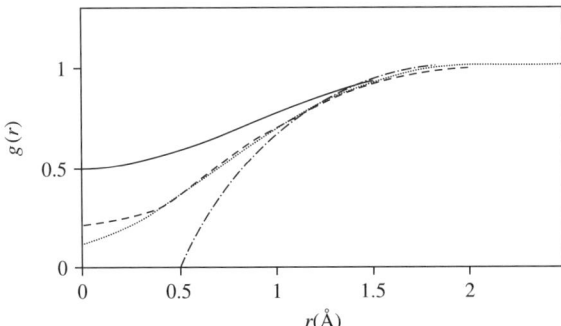

Fig. 2.60. Pair-correlation function $g(r)$ of Be valence electrons. Dashed-dotted line: experimental $g(r)$ as obtained from a polynomial fit to the experimental $S_{\text{band}}(\mathbf{q})$ of Fig. 2.59; solid line: Hartree–Fock (HF) jellium calculation; dashed line: calculation of Zabolitzki (1980); dotted line: Rajagopal's $g(r)$ with parameters, which best fit the experimental $S_{\text{band}}(\mathbf{q})$ of Fig. 2.59. (Reprinted with permission from Mazzone et al. (1983); copyright (1983) by the American Physical Society.)

be > 0 and will approach zero only for $r_s \to \infty$; see, e.g. Gori-Giorgi and Perdew 2002.) Concluding, one needs experimental data of $S(q)$ up to very high values of q with high statistical accuracy in order to get reliable information about the electronic pair-correlation function for the physically so interesting range $r \to 0$, in other words, information about the exchange and the Coulomb hole in electron systems. These experimental problems remain a challenge for the future.

Much more favorable is the situation with extracting information about the correlation energy ε_c related quantity $\gamma(r_s)$ out of experimental data of $S_{ee}(\mathbf{q})$ as being made possible by (2.135). Here it is the q-integral of $(S(q) - 1)$ which allows rather accurate values of $\gamma(r_s)$, of course, only for one value of r_s. Contini and Sacchetti (1981) have used a linear interpolation between their Be value ($r_s = 2.072$) and the high-density (exact) value of Gell-Mann and Brueckner (1957) in order to employ the relation (2.136) for calculating the correlation energy as a function of r_s. The result is seen in Fig. 2.61 and compared with an interpolation scheme of Hedin and Lundqvist (1971). The departure between both is rather small, being 0.02 Ryd at $r_s = 2$.

Similar accuracy can be obtained with calculating the exchange-correlation energy per atom, E_{xc}, using (2.134), since in this case the **q**-integral over $(S(\mathbf{q}) - Z)/q^2$ matters. But it is more the contribution of exchange and correlation to the cohesive energy, which is highly interesting. Thus one has

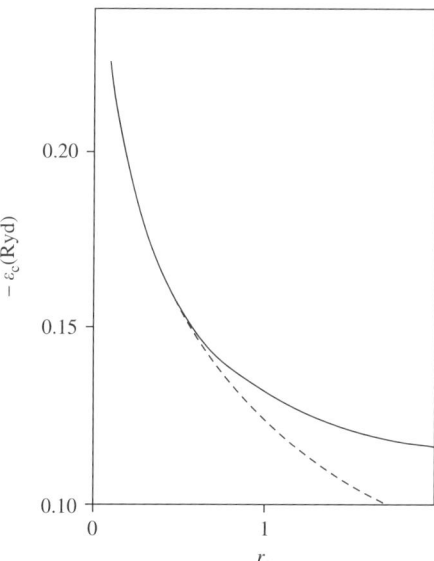

Fig. 2.61. Jellium correlation energy as a function of r_s. The full curve is deduced from the experimental $\gamma(r_s)$. The dashed curve is the interpolation scheme of Hedin and Lundqvist (1971). (Reprinted with permission from Contini and Sacchetti (1981); copyright (1981) by the Institute of Physics.)

to build the difference between the experimental value of E_{xc} measured on solids and the theoretical value as calculated for a free atom (in the case of Be: Benesh and Smith, Jr. 1970), a small difference between large numbers, which requires high accuracy both with the measurement and with the calculations. This exchange-correlation induced contribution to the cohesive energy is $-3.8\,\text{eV/atom}$ for Be metal and must be compared with the static contribution of $-0.7\,\text{eV/atom}$ for Be, being the difference between measured structure factors (Be: Dovesi et al. 1982) and their free atomic counterparts (Be: Benesh and Smith Jr. 1970). The sum of both contributions is $-4.5\,\text{eV/atom}$ and compares well with the thermochemical value of $-3.34\,\text{eV/atom}$. Thus one ends up with the remarkable result that exchange and correlation, in other words dynamical interactions, contribute much more than the static interaction to the cohesive energy. Petrillo and Sacchetti (1995) came to a similar conclusion in their $S(\mathbf{q})$ study on diamond, where the evaluation of orientation averaged $S(\mathbf{q})$ data together with free-atom calculations (Brown 1972) yield an exchange-correlation contribution to the cohesive energy of $4.35\,\text{eV/atom}$, whereas the static contribution as deduced from free atomic scattering factors (Brown 1972) and X-ray scattering data of Spackman (1991) amounts to $1.63\,\text{eV/atom}$, so that also in the diamond case the exchange-correlation interaction is superior to the static one. The sum of both experimentally found contributions is not too far from both the thermodynamic measurement ($7.34\,\text{eV/atom}$) and the Monte Carlo calculation ($7.48\,\text{eV/atom}$) (Fahy et al. 1990).

The averaged pair-correlation function was obtained in a similar manner as in the Be case by fitting the orientation averaged $S(\mathbf{q})$ to a "model" $S(\mathbf{q})$ deduced from a two(four)-parameter "model" pair-correlation function (Rajagopal et al. 1978). The $g(0)$ value was found to be 0.285, which compares well with Monte Carlo calculations of Fahy et al. (1990), who got a value 0.35 when the first electron is at the bond center and 0.25 when the first electron is at the tetrahedral interstitial site.

Such an indirect way to obtain the pair correlation function from measured static structure factor was also used by Schülke et al. (1995) with a study on single crystals of Si following the method (i). The authors found clear deviations of the orientation average $S(q)$ from RPA but much better agreement with $S(q)$ deduced from local-field corrected (Ichimaru and Utsumi 1981) jellium dynamic structure factors. Therefore, it might be justified to invert these local-field corrected $S(q)$, especially because experimental data are missing for $q > 2k_\text{F}$. The $g(0)$ value obtained this way was slightly positive, contrary to the corresponding unphysical negative RPA value.

An $S(\mathbf{q})$ study on LiF, performed by Calzuola et al. (1999) using method (iii) has helped to elucidate how electron correlation works in a strongly inhomogeneous electron system of an ionic solid. Figure 2.62 depicts the valence part $S_\text{val}(\mathbf{q})$ after subtraction of both the core part as calculated by means of configuration interaction (CI) calculations for free ions by Thakkar and Smith (1978) and the core-orthogonalization part according to Hartree–Fock (HF) calculations

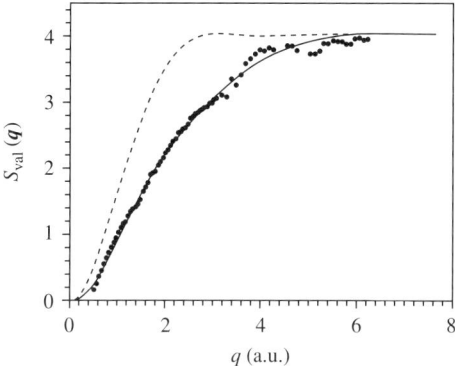

Fig. 2.62. Dots: experimental valence static structure factor $S_{val}(q)$ of LiF. Solid line: fit to the experimental data according to equation (2.135) using quantum Monte Carlo calculations of Ortiz and Ballone (1994). Dashed line: quantum Monte Carlo result of Ortiz and Ballone for jellium with $r_s = 1.48$. (Reprinted with permission from Calzuola et al. (1999); copyright (1999) by the American Physical Society.)

of Shukla (1999). The solid line, which accounts extremely well for the experimental $S_{val}(\mathbf{q})$, apart from the region where the subtraction of the $S_{TDS}(\mathbf{q})$ is less satisfactory, is a fit to the following partition of the valence structure factor

$$S_{val}(\mathbf{q}) = Z_1 S_h(q, n_1) + Z_2 S_h(q, n_2), \qquad (2.142)$$

where $S_h(q,n)$ is the static structure factor of the homogeneous interacting electron gas with density n as obtained by quantum Monte Carlo simulation of Ortiz and Ballone (1994). n_1, n_2, Z_1, and Z_2 are left as free parameters with the additional condition $Z_1 + Z_2 = 4$. On the contrary, the homogeneous interacting electron gas model, namely the Monte Carlo simulation for $r_s = 1.48$, corresponding to the average electron density in LiF, remarkably fails in reproducing the experimental data, thus suggesting that the pair-correlation function is defined by the local density. The best fit parameters were found to be: $Z_1 = 0.91 \pm 0.02$; $r_{s1} = 1.87 \pm 0.2$; $Z_2 = 3.09 \pm 0.02$, and $r_{s2} = 0.708 \pm 0.01$, so that a small number of low-density electrons together with a larger number of high-density electrons are necessary to describe the correlation function in LiF. This behavior is believed to be direct experimental evidence of the local-density approximation (LDA) used so successfully to describe the electron states in solids.

A very interesting $S(\mathbf{q})$ study on molecular liquids using the method (i) has been published by Watanabe et al. (1997) (H_2O) and by Watanabe et al. (1998) (CH_3OH, CH_3CN, C_6H_6 and C_6H_{12}), performed at BL-16A of the Photon Factory, Tsukuba, Japan. The authors were using a double-crystal monochromator with a sagittaly focusing second crystal and a vertically focusing mirror. Energy analysis is done by combining a cylindrically bent Ge crystal and a position sensitive detector, ending up with 2 eV total energy resolution. Since the 1s contributions were outside the experimental energy-loss range, calculated values from Thakkar and Smith (1978) were added in order to make comparison with all-electron theoretical calculations. Figure 2.63 shows the static structure factor of four organic carbon compounds, confronted with calculations of different levels of approximation. The largest discrepancy with experiment is found in the

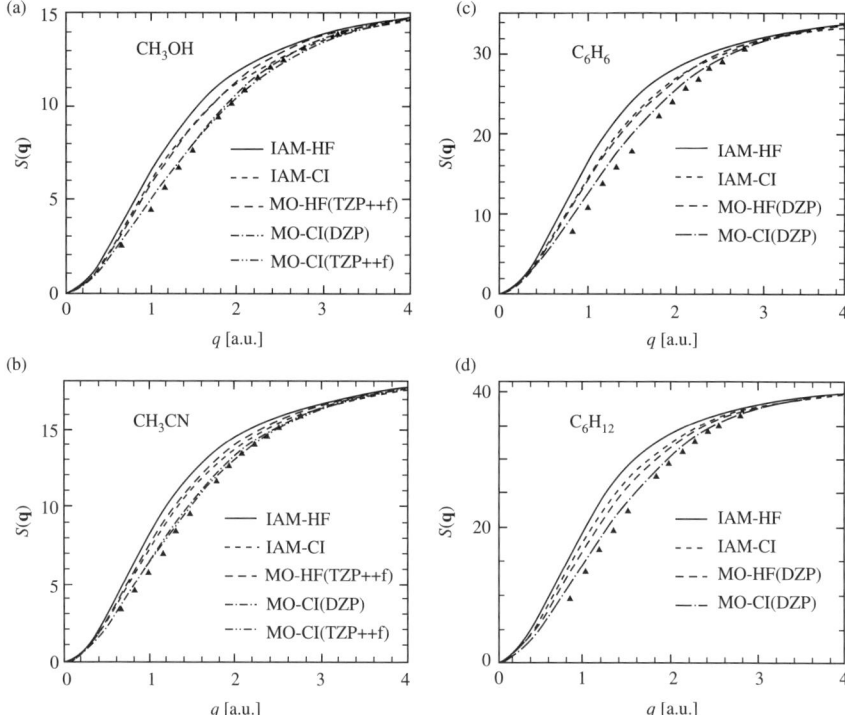

Fig. 2.63. Dots: experimental static structure factor $S(\mathbf{q})$ of varoius liquid carbon compounds as indicated, confronted with calculations of increasing quality of approximation. IAM-HF: Hartree–Fock independent atomic model; IAM-CI: independent atomic model with configuration interaction; MO-HF(TZP++f): Molecular orbitals with Hartee–Fock wavefunctions, triple-zeta plus polarization basis set augmented with diffuse and f-polarization functions; MO-CI(TZP++f): the same as the previous one but with configuration interaction; MO-CI(DZP): molecular orbitals with configuration interaction, double-zeta plus polarization basis set; MO-HF(DZP): The same as the previous one but with Hartree–Fock wavefunctions. (Reprinted with permission from Watanabe et al. (1998); copyright (1999) by the American Institute of Physics.)

independent atom model (IAM) on the basis of Hartree–Fock (HF) atomic wave-functions. A little bit better agreement is obtained with IAM, but on the basis of atomic configuration interaction (CI). Of similar deviation from the experiment is the result of a molecular orbital (MO) calculation in Hartree–Fock approximation using a "triple zeta plus polarization" basis set augmented with diffuse and f-polarization functions (TZP++f). Finally, a MO-CI calculation both with a "double zeta plus polarization basis" set and those with a TZP++f basis set exhibits the best approach to the experiment. This synopsis furnishes evidence for the high sensitivity of $S(\mathbf{q})$ measurements to electron correlation effects. The same is true with the water measurements, where the MO-CI calculation yields nearly complete agreement with experiment.

2.7.4 *The role of multiple scattering*

The evaluation of $S(\mathbf{q})$ measurements in terms of pair-correlation presupposes that only single-scattering events contribute to the final result. The same is true with the judgement of $S(\mathbf{q},\omega)$ measurements, especially in the case that multiple scattering would give rise to additional fine structure of the spectra. Felsteiner and Schülke (1997) have performed Monte Carlo calculation of the double-scattering contributions to the $S(\mathbf{q},\omega)$ spectra of Li, Be and Al using the RPA result for $S(\mathbf{q})$ of Utsumi and Ichimaru (1980), the local-field corrected jellium $S(\mathbf{q},\omega)$, thermal diffuse scattering according to the approximation of independently moving atoms and taking into account Bragg reflection in the secondary scattering process. These Monte Carlo calculations have shown that the total double-scattering contribution for $q \approx 1$ a.u. did not exceed 4%, even in the worst case of transmission geometry with $d\mu > 1$ (d = sample thickness, μ = linear absorption coefficient), and that this contribution is decreasing with increasing atomic number. This means that $S(\mathbf{q})$ measurements on low-Z materials in cases where they claim an accuracy of 1% or better must be multiple scattering corrected. On the other hand, the Monte Carlo calculations of Felsteiner and Schülke (1997) have clearly demonstrated that multiple scattering does not produce additional fine structure in the $S(\mathbf{q},\omega)$ spectra. The energy distribution of the double-scattered intensity exhibits in all cases treated so far a steep rise till to the plasmon energy and then a slowly decreasing behavior far beyond the range of one-particle–hole excitation.

2.7.5 *Feasibility of measuring the spin-dependent static structure factor*

In Section 6.7 we have demonstrated that, in principle, the magnetic dynamic structure factor, that is the contribution of unpaired spins to the density–density correlations can be measured by utilizing the coupling of the electromagnetic field to the magnetic moments of the spins. The magnetic (spin-only) contribution to the double differential scattering cross-section was expressed (see 6.124) as the Fourier transform in time and space of the time-dependent ground state density correlation of electrons with unpaired spins

$$\left(\frac{d^2\sigma}{d\Omega_2 d\hbar\omega_2}\right)_m = \left(\frac{r_0^2}{2\pi}\right)\left(\frac{\hbar}{mc^2}\right)^2\left(\frac{\omega_2}{\omega_1}\right)\int_{-\infty}^{\infty}dt\,\exp(-i\omega t)|D_\zeta|^2$$
$$\times \int d^3r\,d^3r'\,\exp(-i\mathbf{q}\cdot\mathbf{r})$$
$$\times \sum_a g_a\langle a|[\rho_\uparrow(\mathbf{r}',0)-\rho_\downarrow(\mathbf{r}',0)][\rho_\uparrow(\mathbf{r}'+\mathbf{r},t)-\rho_\downarrow(\mathbf{r}'+\mathbf{r},t)]|a\rangle,$$
(2.143)

where $|a\rangle$ is a spin-free ground state vector of the electron system, and g_a its probability. We can introduce into this relation the dynamic structure factor of the electrons with unpaired spins by writing (see 6.125)

$$\left(\frac{d^2\sigma}{d\Omega_2 d\hbar\omega_2}\right)_m = \left(\frac{r_0^2}{2\pi}\right)\left(\frac{\hbar}{mc^2}\right)^2\left(\frac{\omega_2}{\omega_1}\right)|D_\zeta|^2 S_{\text{unpaired}}(\mathbf{q},\omega). \qquad (2.144)$$

Here D_ζ is the component of the vector $\mathbf{D}$ in the direction of the quantization axis, where $\mathbf{D}$ is given by

$$\mathbf{D} = \left(\frac{1}{2}\right)(\omega_1+\omega_2)[\mathbf{e}_2^*\times\mathbf{e}_1 - (\hat{\mathbf{K}}_2\times\mathbf{e}_2^*)\times(\hat{\mathbf{K}}_1\times\mathbf{e}_1)]$$
$$+ [-\omega_1(\mathbf{e}_2^*\cdot\hat{\mathbf{K}}_1)(\hat{\mathbf{K}}_1\times\mathbf{e}_1) + \omega_2(\mathbf{e}_1\cdot\hat{\mathbf{K}}_2)(\hat{\mathbf{K}}_2\times\mathbf{e}_2^*) \qquad (2.145)$$

in terms both of the polarization unit vectors $\mathbf{e}_1(\mathbf{e}_2)$ of the incident (scattered) wave and of the unit vectors $\hat{\mathbf{K}}_1(\hat{\mathbf{K}}_2)$ in the direction of the incident (scattered) wavevectors. Two attributes are characteristic for the magnetic contribution to the double differential scattering cross-section, when compared with the contributions due to charge fluctuations (Thomson scattering) (see 2.3): (i) the double differential cross-section of magnetic (spin-only) scattering is by $(\hbar\omega_1/mc^2)^2$ smaller than its charge counterpart; (ii) the Thomson scattering part of the double differential scattering cross-section is determined by the polarization factor $|\mathbf{e}_2^*\cdot\mathbf{e}_1|^2$, so that a scattered wave polarization perpendicular to the incident polarization is excluded. On the other hand, the polarization factor $|D_\zeta|^2$ makes possible a scattered wave with polarization perpendicular to the incident polarization. We will see that this behavior has been utilized by Petrillo et al. (1998) to offer experimental evidence for inelastic magnetic scattering in a simple metal (Be). A few years before, Petrillo and Sacchetti (1994) had shown the feasibility of getting information about spin-dependent correlation in electron liquids by measuring the magnetic static structure factor $S_m(\mathbf{q})$. The way to reach this goal is to use a linearly polarized incident beam as provided in good "polarization" quality by a storage ring together with an appropriate monochromator, which suppresses unwanted polarization components. Let say the incident polarization is $\mathbf{e}_1 = (0,1,0)$. Then, if one has available a polarization filter in the scattered beam, which transmits only the polarization $\mathbf{e}_2 = (0,0,1)$, the Thomson scattering will completely vanish, whereas the magnetic scattering is determined by the nonvanishing vector $\mathbf{D}$

$$\mathbf{D} = \omega_1(\sin^2\theta + \cos\theta - 1, \ \sin\theta - \sin\theta\cos\theta, \ 0), \tag{2.146}$$

where θ is the scattering angle, and where we have already anticipated the quasielastic limit, $\omega_1 = \omega_2$. Assuming a cubic paramagnetic scattering electron system, the three Cartesian components of $\mathbf{D}$ in (2.146) have equal probability to act as D_ζ in (2.144), so that one obtains for the magnetic (spin-only) contribution to the double differential scattering cross-section

$$\left(\frac{d^2\sigma}{d\Omega_2 d\hbar\omega_2}\right)_m = \left(\frac{r_0^2}{2\pi}\right)\left(\frac{\hbar\omega_1}{mc^2}\right)^2 [(1-\cos\theta)^2/2]\, S_m(\mathbf{q},\omega) \tag{2.147}$$

with

$$S_m(\mathbf{q},\omega) \equiv S_{\text{unpaired}}(\mathbf{q},\omega) = [S_{\uparrow\uparrow}(\mathbf{q},\omega) - S_{\uparrow\downarrow}(\mathbf{q},\omega)]. \tag{2.148}$$

Petrillo and Sacchetti (1994) have now used (2.147 and 2.148) to make an estimate of how the magnetic static structure factor $S_m(\mathbf{q})$, deduced from (2.147) via

$$\left(\frac{d\sigma}{d\Omega_2}\right)_m = \int d\omega \left(\frac{d^2\sigma}{d\Omega_2 d\hbar\omega_2}\right) = \left(\frac{r_0^2}{2\pi}\right)\left(\frac{\hbar\omega_1}{mc^2}\right)^2 [(1-\cos\theta)^2/2]\, S_m(\mathbf{q}) \tag{2.149}$$

can be victorious over the predominance of the charge-fluctuation induced static structure factor $S_c(\mathbf{q})$ by utilizing an incident polarization $\mathbf{e}_1 = (0,1,0)$ together with a polarization filter in the scattered beam, so that $\mathbf{e}_2 = (0,0,1)$. In order to have an easily to handle system they assumed the scattering electron system to be paramagnetic, jellium treated within the RPA. In this case the magnetic (spin-only) dynamic structure factor is simply given by the dynamical spin misceptibility χ_s (Zhu and Overhauser 1986)

$$S_m(\mathbf{q},\omega) = -\left(\frac{1}{\pi}\right) \text{Im}\chi_s(\mathbf{q},\omega) \tag{2.150}$$

whereas the dynamic structure factor due to charge fluctuations is (see 2.19)

$$S_c(\mathbf{q},\omega) = -\left(\frac{1}{\pi}\right) \text{Im}\frac{[\chi_L(\mathbf{q},\omega)]}{1 - v(\mathbf{q})\chi_L(\mathbf{q},\omega)}, \tag{2.151}$$

where $\chi_L(\mathbf{q},\omega)$ is the Lindhard proper polarization function.

On this basis, Petrillo and Sacchetti (1994) have calculated $S_m(\mathbf{q})$ and $S_c(\mathbf{q})$ in relative units, which they would obtain by exploiting the experimental setup sketched in Fig. 2.64. The incident monochromatic X-ray with wavevector $\mathbf{K}_1$ might be linearly polarized in the scattering plane, so that the Thomson-scattered part with wavevector $\mathbf{K}_2$ is also polarized in the scattering plane. The

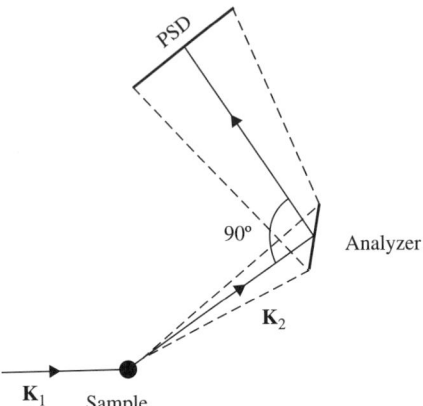

Fig. 2.64. Proposed experimental setup for measuring the magnetic contribution to the static structure factor (PSD = position-sensitive detector). (Reprinted with permission from Petrillo and Sacchetti (1994); copyright (1994) by Elsevier Science Ltd.)

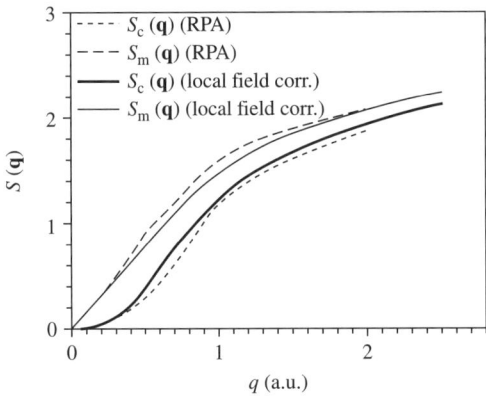

Fig. 2.65. Long-dashed line: magnetic static structure factor of Li in relative units; RPA treatment; short dashed line: residual charge scattering static structure factor; RPA treatment; continuous lines: local-field corrected (Zhu and Overhauser 1986). (Reprinted with permission from Petrillo and Sacchetti (1994); copyright (1994) by Elsevier Science Ltd.)

incident wavelength should be chosen such that the elastically scattered radiation ought to be Bragg-reflected at the analyzer crystal exactly under a Bragg angle of 45°. Thus the polarization of the elastically scattered radiation completely prevents Bragg-reflection into the position-sensitive detector. But even radiation, which is inelastically Thomson scattered, and therefore Bragg-reflected not exactly under 45° is to such an extent retained by the Thomson polarization factor $\cos^2 2\theta_B$ that mainly radiation magnetically scattered at the sample will, despite the extremely small magnetic cross-section, reach the detector being Bragg-reflected at the analyzer crystal, since this radiation contains also polarization components perpendicular to the scattering plane. If the experiment with keV incident photon energy and Li metal as scatterer is arranged such that the energy range Bragg-reflected at the analyzer spans 50 eV, the relative values of $S_m(\mathbf{q})$ and $S_c(\mathbf{q})$ are of comparable magnitude as shown in Fig. 2.65. Surprisingly no experiment has been done on the lines of this proposal, probably since it is rather difficult to obtain a synchrotron radiation beam which exhibits, to an extent necessary for those experiments, a clean linear polarization in the scattering plane. It was just this difficulty Petrillo *et al.* (1998) had to fight with, when they tried successfully to do the first step of such a synchrotron radiation based experiment, namely to identify in the X-radiation scattered on Be the polarization component perpendicular to the scattering plane as being really due to magnetic scattering and not caused by the pollution of the incident beam with this polarization component. They had to perform a rather sophisticated polarization analysis, which cannot be described in detail. The reader is referred to the original literature. Nevertheless, the rich information about spin-dependent correlation, offered by those magnetic $S_m(\mathbf{q})$ measurements, justifies further efforts in this direction.

2.8 References

Abbamonte, P., K.D. Finkelstein, M.D. Collins, and S.M. Gruner (2004). *Phys. Rev. Lett.* **92** 237401

Aiyama, T., and K. Yada (1974). *J. Phys. Soc. Japan* **36** 1554

Alexandropoulos, N.G. (1971). *J. Phys. Soc. Japan* **31** 1790

Aravind, P.K., A. Holas, and K.S. Singwi (1982). *Phys. Rev. B* **25** 561

Bachlechner, M.E., A. Holas, H.M. Böhm, and A. Schinner (1996). *Phys. Rev. B* **54** 2360

Batterman, B.W., and L. Berman (1983). *Nucl. Instrum. & Methods* **208** 327

Batterman, B.W., and D.H. Bilderback (1991). *Handbook on Synchrotron Radiation Vol. 3*, ed. G.S. Brown and D.E. Moncton (North-Holland, Amsterdam) p. 105

Beeferman, L.W., and H. Ehrenreich (1970). *Phys. Rev. B* **2** 364

Benedict, L., E.L. Shirley, and R.B. Bohn (1998a). *Phys. Rev. B* **57** 9385

Benedict, L., E.L. Shirley, and R.B. Bohn (1998b). *Phys. Rev. Lett.* **80** 4514

Benesh, R., and V.H. Smith, Jr (1970). *Acta Crystallogr. Sect A* **26** 586

Berthold, A., S. Mourikis, J.R. Schmitz, W. Schülke, and H. Schulte-Schrepping (1992). *Nucl. Instrum & Methods A* **317** 373

Bilderback, D.H. 1986. *Nucl. Instrum. & Methods A* **246** 434

Blanke, W. (1989). *Thermophysikalische Stoffgrößen* (Springer, Berlin)

Bonse, U., and M. Hart (1965). *Appl. Phys. Lett.* **7** 151

Bonse, U., G. Materlik, and W. Schröder (1976). *J. Appl. Crystallogr.* **9** 223
Brosens, F., J.T. Devreese, and L.F. Lemmens (1980). *Phys. Rev. B* **21** 1363
Brown, G.E. (1972). *Many-Body Problems* (North-Holland, Amsterdam)
Brown, R.T. (1972). *Phys. Rev. A* **5** 2141
Bross, H., and G. Bohn (1975). *Z. Phys. B* **20** 261
Bross, H. (1978). *J. Phys. F: Metal Phys.* **8** 2631
Bross, H., and M. Ehrnsperger (1995). *Z. Phys. B* **97** 17
Buras, B., and S. Tazzari (1984). *European Synchrotron Radiation Facility Report of the ESRP* (Geneva) pp. 2.1–2.6
Burkel, E. (1991). *Inelastic Scattering of X-rays with Very High Energy Resolution* (Berlin: Springer) pp. 85–89
Burkel, E., J. Peisl, and B. Dorner (1987). *Europhys. Lett.* **3** 957
Burns, C.A., P. Abbamonte, E.D. Isaacs, and P.M. Platzman (1999). *Phys. Rev. Lett.* **83** 2390
Burns, C.A., P. Giura, A. Said, A. Shukla, G. Vanko, M. Tuel-Benckendorf, E.D. Isaacs, and P.M. Platzman (2002). *Phys. Rev. Lett.* **89** 236404
Cai, Y.Q., P.C. Chow, O.D. Restrepo, Y. Takano, K. Togano, H. Kito, H. Ishii, C.C. Chen, K.S. Liang, C.T. Chen, S.Tsuda, S. Shin, C.-C. Kao, W. Ku, and A.G. Eguiluz (2006) Cond-mat/0605320
Caliebe, W.A., J.A. Soininen, E.L. Shirley, C.-C. Kao, and K. Hämäläinen (2000). *Phys. Rev. Lett.* **84** 3907
Calzuola, G., C. Petrillo, and F. Sacchetti (1999). *Phys. Rev. B* **59** 12853
Ceperley, D. (1978). *Phys. Rev. B* **18** 3126
Chen, N., and S. Rabii (1985). *Phys. Rev. B* **31** 4784
Ching, W.Y., and J. Callaway (1974). *Phys. Rev. B* **9** 5115
Chou, M.Y., M.L. Cohen, and St. G. Louie (1986). *Phys. Rev. B* **33** 6619
Cohen, M.L., and V. Heine (1970). *Solid State Physics*, ed. by H. Ehrenreich, F. Seitz and D. Turnbull (Academic, New York) Vol **24** p. 34
Contini, V., and F. Sacchetti (1981). *J. Phys. F: Metal Phys.* **11** L1
Daling, R., W. van Haeringen, and B. Farid (1991). *Phys. Rev. B* **44** 2952
Dellafiore, A., and F. Matera (1988). *Phys. Rev. B* **37** 8588
Dorner, B., anf J. Peisl (1983). *Nucl. Instrum & Methods* **208** 587
Dovesi, R., C. Pisani, F. Ricca, and C. Roetti (1982). *Phys. Rev. B* **25** 3731
Eguiluz, A.G., A. Fleszar, and J. Gaspar (1995). *Nucl. Instrum & Methods B* **96** 550
Ehrenreich, H., M.H. Cohen (1959). *Phys. Rev.* **115** 786
Eisenberger, P., P.M. Platzman, and K.C. Pandy (1973). *Phys. Rev. Lett.* **31** 311
Eisenberger, P., W.C. Marra, and G.S. Brown (1980). *Phys. Rev. Lett.* **45** 1439
Fahy, S., X.W. Wang, and S.G. Louie (1990). *Phys. Rev. B* **42** 3503
Fano, U. (1961). *Phys. Rev.* **124** 1866
Farid, B., V. Heine, G.E. Engel, and I.J. Robertson (1993). *Phys. Rev. B* **48** 11602
Felsteiner, J., and W. Schülke. *Nucl. Instrum & Methods in Phys. Res. B* **132** 1
Fleszar, A., A.A. Quong, and A.G. Eguiluz (1995). *Phys. Rev. Lett.* **74** 590

References

Fujii, Y., J.B. Hastings, S.C. Ulc and D.E. Moncton (1982). SSRL Activity Report p. VIII–95

Galambosi, S., J.A. Soininen, K. Hämäläinen, E.L. Shirley, and C.-C. Kao (2001). *Phys. Rev. B* **64** 024102

Galambosi, S., J.A. Soininen, A. Mattila, S. Huotari, S. Manninen, Gy. Vanko, N.D. Zhigadlo, J. Karpinski, and K. Hämäläinen (2005). *Phys. Rev. B* **71** 060504(R)

Geldart, D.J.W., and S.H. Vosko (1966). *Can. J. Phys.* **44** 2137

Geldart, D.J.W., and R. Taylor (1970). *Can. J. Phys.* **48** 167

Gell-Mann, M., and K.A. Brueckner (1957). *Phys. Rev.* **106** 364

Gori-Giorgi, P., and J.P. Perdew (2002). *Phys. Rev. B* **66** 165118

Green, F., D. Neilson, and J. Szymanski (1985a). *Phys. Rev. B* **31** 2779; (1985b). *Phys. Rev. B* **31** 2796; (1985c). *Phys. Rev. B* **31** 5837; (1987a). *Phys. Rev. B* **35** 124

Green, F., D. Neilson, D. Pines, and J. Szymanski (1987b). *Phys. Rev. B* **35** 133

Gross, E.K.U., et al. (1996). in *Density Functional Theory II*, ed. R.F. Nalewajski (Springer, Berlin) Chap. 2, p. 81

Gurtubay, I.G., W. Ku, J.M. Pitarke, A.G. Eguiluz, B.C. Larson, J. Tischler, and P. Zschak (2004). *Phys. Rev. B* **70** 201201(R)

Gurtubay, I.G., J.M. Pitarke, W. Ku, A.G. Eguiluz, B.C. Larson, J. Tischler, P. Zschak, and K.D. Finkelstein (2005). *Phys. Rev. B* **72** 125117

v. Hamos, L. (1938). *J. Sci. Instrum* **15** 87

Hedin, L. (1965). *Phys. Rev.* **139** A796

Hedin, L., and B.I. Lundqvist (1971). *J. Phys. C: Solid St. Phys.* **4** 2064

Hill, J.P., C.-C. Kao, W.A.C. Caliebe, D. Gibbs, and J.B. Hastings (1996). *Phys. Rev. Lett.* **77** 3665

Hiraoka, N., H. Ischi, J. Jarrige, and Y.Q. Cai (2005). *Phys. Rev. B* **72** 075103

Höppner, K., A. Kaprolat, and W. Schülke (1998). *Eur. Phys. J. B* **5** 53

Holas, A., (1987). in *Strongly Coupled Plasma Physics*, ed. by F.J. Rogers and H.E. Dewitt (Plenum, New York) pp. 463–482

Holas, A., and S. Rahman (1987). *Phys. Rev. B* **35** (1987) 2720

Holas, A., P.K. Aravind, and K.S. Singwi (1979). *Phys. Rev. B* **20** 4912

Holm, B., and F. Aryasetiawan (1997). *Phys. Rev. B* **56** 12825

Holzwarth, N., A.W., St. G. Louie, and S. Rabii (1982). *Phys. Rev. B* **26** 5382

Holzwarth, N.a., W., St. G. Louie, and S. Rabii (1984). *Phys. Rev. B* **30** 2219

Hubbard, J., (1958). *Proc. R. Soc. A* **243** 336

Hybertsen, M.S. and S.G. Louie (1985). *Phys. Rev. Lett.* **55** 1418

Ice, G.E., C.J. Sparks, Jr. (1990). *Nucl. Instrum. & Methods A* **291** 110

Isaacs, E.D., P.M. Platzman, P. Metcalf, and J.M. Honig (1996). *Phys. Rev. Lett.* **76** 4211

Isaacs, E.D., P.M. Platzman, P. Zschack, K. Hämäläinen, A.R. Kortan (1992). *Phys. Rev. B* **46** 12910

Ishikawa, T., Y. Yoda, K. Izumi, C.K. Suzuki, W. Zhang, M. Ando, S. Kikuta (1992). *Rev. Sci. Instrum.* **63** 1015

Johnson D.L. (1974). *Phys. Rev. B* **9** 4475
Joksch, S., A. Freund, M. Krisch, and G. Marot (1991). *Nucl. Instrum. & Methods A* **306** 386
Kerker, G.P., (1980). *J. Phys. C* **13** L189
Krishnan, S., and P.C. Nordine (1993). *Phys. Rev. B* **47** 11780
v. Laue, M. (1960). *Röntgenstrahl-Interferenzen* (Akademische Verlagsgesellschaft, Frankfurt)
Ku, W., W.E. Pickett, R.T. Scalettar, and A.G. Eguiluz (2002). *Phys. Rev. Lett.* **88** 057001
Kugler, A.A. (1975). *J. Stat. Phys.* **12** 35
Larson, B.C., J.T. Tischler, E.D. Isaacs, P. Zschack, A. Fleszar, and A.G. Eguiluz (1996). *Phys. Rev. Lett.* **77** 1346
Lundqvist, B.I. (1967). *Phys. Kondens. Materie* **6** 193
Macrander, A.T., V.I. Kushnir, and R.C. Blasdell (1995). *Rev. Sci.Instrum.* **66** 1546
Macrander, A.T., P.A. Montano, D.L. Price, V.I. Kushnir, R.C. Blasdell, C.-C. Kao, and B.R. Cooper (1996). *Phys. Rev. B* **54** 305
Maddocks, N.E., R.W. Godby, and R.J. Needs (1994a). *Phys. Rev. B* **49** 8502
Maddocks, N.E., R.W. Godby, and R.J. Needs (1994b). *Europhys. Lett.* **27** 681
Mahan, G.D. (1981). Many Particle Physics (New York: Plenum)
Manzke, R., J. (1978). *Phys. C: Solid State Phys.* **11** L349
Masciovecchio, C., U. Bergmann, M. Krisch, G. Ruocco, F. Sette, and R. Verbeni (1996). *Nucl. Instrum. & Methods B* **111** 181
Matsushita, T., and H. Hashizume (1983). *Handbook on Synchrotron Radiation, Vol. 1a*, ed. E.E. Koch (North-Holland, Amsterdam) p. 261
Mattheiss, L. (1994). *J. Phys. Condens. Matter* **6** 6477
Mazzone, G., F. Sacchetti, and V. Contini (1983). *Phys. Rev. B* **28** 1772
McWhan, D.B. and J.P. Remeika (1970). *Phys. Rev. B* **2** 3734
Montano, P.A., D.L. Price, A.T. Macrander, and B.R. Cooper (2002). *Phys. Rev. B* **66** 165218
Moroni, S. et al. (1995). *Phys. Rev. Lett.* **75** 689
Mukhopadhyay, G., R.K. Kalia, and K.S. Singwi (1974). *Phys. Rev. Lett.* **34** 950
Nagasawa, H., S. Mourikis, and W. Schülke (1997). *J. Phys. Soc. Japan* **66** 3139
Ng, T.K., and B. Dabrowski (1986). *Phys. Rev B* **33** 5358
Niklasson, G. (1974). *Phys. Rev. B* **10** 3052
Niklasson, G. (1985). *Solid State Commun.* **54** 665
Niklasson, G., Sjölander, A., and F. Yoshida (1983). *J. Phys. Soc. Japan* **52** 2140
Ojala, E. (1983). *Phys. Stat. Solidi, B* **119** 269
Oliveira, L.E., and K. Sturm (1980). *Phys. Rev. B* **22** 6283
Ortiz, G., and P. Ballone (1994). *Phys. Rev. B* **50** 1391
Petrillo, C., and F. Sacchetti (1994). *Solid State Commun.* **91** 895

Petrillo, C., and F. Sacchetti (1995). *Phys. Rev. B* **51** 4755
Petrillo C., F. Sacchetti, E. Bubucci, M. Colapietro, A. Pifferi (1998). *Europhys. Lett.* **42** 179
Platzman, P.M. and P.A. Wolff (1973). *Waves and Interactions in Solid State Plasmas* (Academic Press, New York)
Platzman, P., and P. Eisenberger (1974). *Phys. Rev. Lett.* **33** 152
Platzman, P.M., E.D. Isaacs, H. Williams, P. Zschak. and G.E Ice (1992). *Phys. Rev. B* **46** 12943
Posternak, M., A. Baldereschi, A.J. Freeman, E. Wimmer, and M. Weinert (1983). *Phys. Rev. Lett.* **50** 761
Prevey, P.S. (1986). *Adv. in X-ray Anal.* **29** 103
Priftis, G.D., J. Boviatsis, and A. Vradis (1978). *Phys. Lett.* **68A** 482
Raether H., (1980). *Excitation of Plasmons and Interband Transitions by Electrons* (Berlin: Springer)
Rahman, S., and G. Vignale (1984). *Phys. Rev. B* **30** 6951
Rajagopal, A.K., J.C. Kimball, and M. Banerjee (1978). *Phys. Rev. B* **18** 2339
Rohlfing, M., and S.G. Louie (1998a). *Phys. Rev. Lett.* **80** 3320
Rohlfing, M., and S.G. Louie (1998b). *Phys. Rev. Lett.* **81** 2312
Russel T.P. (1991). *Handbook on Synchrotron Radiation* Vol. 3 ed. by G. Brown and D.E. Moncton (Elsevier Science Publishers B.V.)
Saslow, W.M., and G.F. Reiter (1973). *Phys. Rev. B* **7** 2995
Schriefer, J.R. (1964). Theory of Superconductivity (New York: Benjamain)
Schülke, W., (1986). Nucl. Instrum. & Methods A **246** 491
Schülke, W., (1991). *Handbook on Synchrotron Radiation* Vol. 3 ed. by G. Brown and D.E. Moncton (Elsevier Science Publishers B.V.)
Schülke, W., (2004). unpublished
Schülke, W., and A. Kaprolat (1991). *Phys. Rev. Lett.* **67** 879
Schülke, W., and S. Mourikis (1986). *Acta Cryst A* **42** 86
Schülke, W., U. Berg, and O. Brümmer (1969). *Phys. Stat. Sol.* **35** 227
Schülke, W., H. Nagasawa, and S. Mourikis (1984). *Phys. Rev. Lett.* **52** 2065
Schülke, W., H. Nagasawa, S. Mourikis, and P. Lanzki (1986). *Phys. Rev. B* **33** 6744
Schülke, W., U. Bonse, H. Nagasawa, S. Mourikis, and A. Kaprolat (1987). *Phys. Rev. Lett.* **59** 1361
Schülke, W., U. Bonse, H. Nagasawa, A. Kaprolat, and A. Berthold (1988a). *Phys. Rev. B* **38** 2112
Schülke, W., A. Berthold, A. Kaprolat, and H.-J. Güntherodt (1988b). *Phys. Rev. Lett.* **60** 2217
Schülke, W., H. Nagasawa, S. Mourikis, and A. Kaprolat (1989). *Phys. Rev. B* **40** 12215
Schülke, W., A. Berthold, H. Schulte-Schrepping, V. Thommes-Geiser, and H.-J. Güntherodt (1991). *Solid State Commun.* **79** 661
Schülke, W., H. Schulte-Schrepping, and J.R. Schmitz (1993). *Phys. Rev. B* **47** 12426

Schülke, W., J.R. Schmitz, H. Schulte-Schrepping, and A. Kaprolat (1995). Phys. Rev. B **52** 11721
Schülke, W., K. Höppner, and A. Kaprolat (1996). Phys. Rev. B **54** 17464
Shukla, A., (1999). Phys. Rev. B **60** 4539
Singh, H.B., and K.N. Pathak (1973). Phys. Rev. B **8** 6035
Singwi, K.S., M.P. Tosi, R.H. Land, and A. Sjölander (1968). Phys. Rev. **176** 589
Singwi, K.S., A. Sjölander, M.P. Tosi, and R.H. Land (1970). Phys. Rev. B **1** 1044
Smither, R., (1989). Rev. Sc. Instrum. **60** 2044
Spackman, M.A. (1991). Acta Crystallogr. Sec. A **47** 420
Sparks, C.J., G.E. Ice, J. Wong and B.W. Batterman (1982). Nucl. Instrum. & Methods **195** 73
Spiertz, A. (1996). Thesis, Dortmund
Sprösser-Prou, J., A. vom Felde, and J. Fink (1989). Phys. Rev. B **40** 5799
Sternemann, C., A. Kaprolat, and W. Schülke (1998). Phys. Rev. B **57** 622
Sternemann, C., S. Huotari, G. Vanko, M. Volmer, G. Monaco, and W. Schülke (2005). Phys. Rev. Lett. **95** 157401
Sturm, K., and A. Gusarov (2000). Phys. Rev. B **62** 16474
Sturm, K., and L.E. Oliveira (1980). Phys. Rev. B **22** 6268
Sturm K., and L.E. Oliveira (1981). Phys. Rev. B **24** 3054
Sturm, K., and L.E. Oliveira (1984). Phys. Rev. B **30** 4351
Sturm, K., and L.E. Oliveira (1989). Phys. Rev. B **40** 3672
Sturm, K., and W. Schülke (1992a). Phys. Rev. B **46** 7193
Sturm, K., W. Schülke, and J.R. Schmitz (1992b). Phys. Rev. Lett. **68** 228
Takada, Y., (1995). Phys. Rev. B **52** 12708
Takada, Y., (2001). Phys. Rev. Lett. **87** 226402
Takada, Y., and H. Yasuhara (2002). Phys. Rev. Lett. **89** 216402
Tatar, R.C. and S. Rabii (1982). Phys. Rev. B **26** 5382
Taut, , M. and W. Hanke (1976). Phys. Status Solidi, B **77** 543
Thakkar, A.J., and V.H. Smith (1978). J. Phys. B **11** 3803
Thompson, J.C. (1976). Electrons in Liquid Ammonia (Oxford University Press, New York)
Tischler J.Z., B.C. Larson, P. Zschack, A. Fleszar and A.G. Eguiluz (2003). phys. stat. sol. (b) 280
Utsumi, K., and S. Ichimaru (1980). Phys. Rev. B **22** 5203
Watanabe, N., H. Hayashi and Y. Udagawa (1997). Bull. Chem Soc. Jpn. **70** 719
Watanabe, N., H. Hayashi, Y. Udagawa, S. Ten-no, and S. Iwata (1998). J. Chem. Phys. **108** 4545
Willis R.F., B. Fitton, and G.S. Painter (1974). Phys. Rev. B **9** 1926
Wohlert, F., (2001). Thesis, Dortmund
Zachariasen, W.H. (1967). Theory of X-ray diffraction in Crystals (Dover Publications, Inc, New York)
Zabolitzki, J.G. (1980). Phys. Rev. B **22** 2353

Zaeper, R., M. Richwin, D. Lützenkirchen-Hecht, and R. Frahm (2002). *Rev. Sci. Instrum.* **73** 1564

Zeppenfeld, K., (1969). *Z. Physik* **223** 32

Zhu, X., and A.W. Overhauser (1986). *Phys. Rev. B* **33** 925

Zhukov, V.P., V.M. Silkin, E.V. Chulkov, and P.M. Echenique (2001). *Phys. Rev. B* **64** 180507

3

Nonresonant inelastic X-ray scattering: Regime of core-electron excitation (X-ray Raman scattering)

3.1 Basic relations; the role of momentum transfer q in X-ray Raman scattering

In Chapter 2 we utilized the basic relation between the double differential scattering cross-section and excitations in the scattering system, as given in equation (2.3) only by considering excitations of the valence electron systems of condensed matter. We will now examine inelastic scattering processes connected with excitations of core-electrons from closed shells into unoccupied states about which we want to learn by this type of spectroscopy. In this case we will speak, for more historical reasons, of non-resonant X-ray Raman scattering or spectroscopy, or still more briefly of X-ray Raman scattering or spectroscopy (XRS), if mixing up with resonant X-ray Raman scattering is not to be suspected. We start again with the following expression for the double differential scattering cross-section derived in a first-order perturbation treatment of a scattering event connected with the transition of the electron system from its ground state $|i\rangle$ with energy E_i into a final state $|f\rangle$ for $T=0$:

$$\frac{d^2\sigma}{d\Omega_2 d\hbar\omega_2} = \left(\frac{d\sigma}{d\Omega_2}\right)_{Th} S(\mathbf{q},\omega) \tag{3.1}$$

with the Thomson scattering cross-section

$$\left(\frac{d\sigma}{d\Omega_2}\right)_{Th} = r_0^2 \left(\frac{\omega_2}{\omega_1}\right) |\mathbf{e}_1 \cdot \mathbf{e}_2^*|^2 \tag{3.2}$$

and the dynamic structure factor

$$S(\mathbf{q},\omega) = \sum_f \left|\langle i|\sum_j \exp(-i\mathbf{q}\cdot\mathbf{r}_j)|f\rangle\right|^2 \delta(E_i - E_f + \hbar\omega), \tag{3.3}$$

where $\hbar\mathbf{q} = \hbar(\mathbf{K}_1 - \mathbf{K}_2)$ is the momentum transfer, $\hbar\omega = \hbar(\omega_1 - \omega_2)$ the energy transfer (energy loss), $\mathbf{K}_1(\mathbf{K}_2)$, $\mathbf{e}_1(\mathbf{e}_2)$ and $\omega_1(\omega_2)$ are the wavevector, the polarization (unit) vector and the frequency of the incident (scattered) radiation, respectively. The summation j is over all electrons of the system.

Following Mizuno and Ohmura (1967) we will express (3.3) for the case of core excitation within the limits of the one-electron approximation neglecting, to begin with, both correlation and electron–hole interaction. Then the initial state wavefunction $|i\rangle$ in (3.3) is approximated by a Slater determinant composed of the one-electron orbitals ϕ_i of the core electrons and the other occupied orbitals ψ_j, where every different orbital is orthogonal to each other. In the final state, one of the core electrons, say ϕ_n with spin α, jumps to an unoccupied Bloch state with the same spin, say $\psi_{\mathbf{k}\alpha}$, so that $\psi_{\mathbf{k}}$ replaces ϕ_n in the Slater determinant representing the final state wave function $|\text{f}\rangle$. Then the matrix element $\langle \text{i}| \sum \exp(-i\mathbf{q}\cdot\mathbf{r}_j)|\text{f}\rangle$ in (3.3) is reduced to

$$\langle \text{i}| \sum_j \exp(-i\mathbf{q}\cdot\mathbf{r}_j)|\text{f}\rangle = \int \psi_{\mathbf{k}}^*(\mathbf{r}) \exp(-i\mathbf{q}\cdot\mathbf{r})\phi_n(\mathbf{r}) \, d^3r, \quad (3.4)$$

where $\phi_n(\mathbf{r})$ represents the core orbital localized around the nth lattice site $\mathbf{R}_n$. Writing

$$\mathbf{r} = \mathbf{R}_n + \mathbf{r}' \quad (3.5)$$

and making use of the periodicity of the Bloch orbital $\psi_{\mathbf{k}}$, one obtains

$$\langle \text{i}| \sum_j \exp(-i\mathbf{q}\cdot\mathbf{r}_j)|\text{f}\rangle = \exp[-i(\mathbf{q}+\mathbf{k})\cdot\mathbf{R}_n] \int \psi_{\mathbf{k}}^*(\mathbf{r}') \exp(-i\mathbf{q}\cdot\mathbf{r}') \phi(\mathbf{r}') \, d^3r', \quad (3.6)$$

where $\phi(\mathbf{r}') = \phi_n(\mathbf{r})$, and $\mathbf{k}$ being the Bloch wavevector associated with $\psi_{\mathbf{k}}$.

The momentum transfer dependence of XRS can most easily be understood by expanding the exponential in the matrix element on the right-hand side of (3.6) as

$$\exp(i\mathbf{q}\cdot\mathbf{r}) = 1 + i\mathbf{q}\cdot\mathbf{r} + (i\mathbf{q}\cdot\mathbf{r})^2/2 + \cdots \quad (3.7)$$

For low momentum transfer $(q\cdot a) \ll 1$, where a is the radius of the core orbital, the second term dominates (the first one does not contribute because of the orthogonality of the participating orbitals in equation (3.6)), and mostly dipole-allowed transition are probed. This can be seen, when we insert the first two terms of (3.7) into (3.6):

$$\exp[-i(\mathbf{q}+\mathbf{k})\cdot\mathbf{R}_n] \int \psi_{\mathbf{k}}^*(\mathbf{r}) \exp(-i\mathbf{q}\cdot\mathbf{r}) \phi(\mathbf{r}) \, d^3r$$

$$\approx i \exp[-i(\mathbf{q}+\mathbf{k})\cdot\mathbf{R}_n] \left(\int \psi_{\mathbf{k}}^*(\mathbf{r})\,\mathbf{r}\,\phi(\mathbf{r}) \, d^3r \right) \cdot \mathbf{q}, \quad (3.8)$$

so that the double differential scattering cross-section can be written, according to (3.1)–(3.3),

$$\frac{d^2\sigma}{d\Omega_2 d\hbar\omega_2} = \left(\frac{d\sigma}{d\Omega_2}\right)_{\text{Th}} \mathbf{q}\cdot\mathbf{T}(\omega)\cdot\mathbf{q}, \quad (3.9)$$

where the tensor $\mathbf{T}(\omega)$ is given by

$$\mathbf{T}(\omega) = \sum_{\mathbf{k}(\text{unoccupied})} \ll 0|\mathbf{r}|\mathbf{k} \gg \ll \mathbf{k}|\mathbf{r}|0 \gg \delta(\hbar\omega - E_{\mathbf{k}} - E_0) \quad (3.10)$$

and $\ll \mathbf{k}|\mathbf{r}|0 \gg$ being shorthand for $\int \psi_\mathbf{k}^*(\mathbf{r})\, \mathbf{r}\, \phi(\mathbf{r})\, d^3r$. $E_\mathbf{k}$ and E_0 are the energies of the orbitals $\psi_\mathbf{k}$ and ϕ, respectively. This means that the double differential scattering cross-section is determined by the density of those unoccupied states which are accessible by the excited core electron according to the dipole matrix elements in (3.10). Additionally the symmetry-selective role of $\mathbf{q}$ in (3.9) restricts the possible transitions probed by the inelastic experiment.

It will be shown in what follows that a similar restriction is found in the absorption spectroscopy, where the direction of the incident electric field vector takes over the role of $\mathbf{q}$. To this end we start with the interaction Hamiltonian of (6.6), which governs the absorption of the photon field $\mathbf{A}(\mathbf{r}_j)$ in a system of electrons at position $\mathbf{r}_j$

$$H_{i2} = -\left(\frac{e}{mc}\right) \sum_j \mathbf{A}(\mathbf{r}_j) \cdot \mathbf{p}_j, \qquad (3.11)$$

where the vector potential operator $\mathbf{A}(\mathbf{r}_j)$ is given by

$$\mathbf{A}(\mathbf{r}) = \sum_{\mathbf{K}\lambda} \left(\frac{2\pi\hbar c^2}{V\omega_\mathbf{K}}\right)^{1/2} [\mathbf{e}(\mathbf{K}\lambda)c(\mathbf{K}\lambda)\exp(i\mathbf{K}\cdot\mathbf{r} - i\omega_\mathbf{K}t)$$
$$+ \mathbf{e}^*(\mathbf{K}\lambda)c^+(\mathbf{K}\lambda)\exp(i\mathbf{K}\cdot\mathbf{r} + \omega_\mathbf{K}t)]. \qquad (3.12)$$

Here the summation $\mathbf{K}\lambda$ is over all modes of the photon field with frequency $\omega_\mathbf{K}$, where λ counts the two orthogonal polarization states of the photon field corresponding to the wavevector $\mathbf{K}$ with the polarization unit vector $\mathbf{e}(\mathbf{K}\lambda)$. $c^+(\mathbf{K}\lambda)$ and $c(\mathbf{K}\lambda)$ are the photon creation and annihilation operators, respectively. By using

$$\mathbf{p}_j = \left(\frac{m}{i\hbar}\right)[\mathbf{r}_j, H_0] \qquad (3.13)$$

where H_0 is the ground-state Hamiltonian of the system, one obtains for the matrix element describing the transition of the electron system to an excited state with a core hole, as induced by the absorption of a photon $\mathbf{e}(\mathbf{K})$,

$$\langle f|H_{i2}|i\rangle = \left(\frac{e}{i}\right)\left(\frac{2\pi}{V\omega_\mathbf{K}}\right)^{1/2}(E_i - E_f)\langle f|\sum_j \exp(i\mathbf{K}\cdot\mathbf{r}_j)[\mathbf{e}(\mathbf{K}\lambda)\cdot\mathbf{r}_j]|i\rangle. \qquad (3.14)$$

Assuming the one-electron approximation as above, the matrix element on the right-hand side of (3.14) can be written as

$$\langle f|\sum_j \exp(i\mathbf{K}\cdot\mathbf{r}_j)[\mathbf{e}(\mathbf{K}\lambda)\cdot\mathbf{r}_j]|i\rangle = \int \psi_\mathbf{k}^*(\mathbf{r})\,(\mathbf{e}(\mathbf{K}\lambda)\cdot\mathbf{r})\,\exp(i\mathbf{K}\cdot\mathbf{r})\,\phi_n(\mathbf{r})\,d^3r. \qquad (3.15)$$

If one can assume that the orbital radius of the core-electron is much smaller than the wavelength K^{-1} of the absorbed radiation, we can make the so-called dipole approximation by setting $\exp(i\mathbf{K}\cdot\mathbf{r}) \approx 1$ (3.15) thus ending up, by using the same notation as above, with the following expression for the transition probability for absorption:

$$W_{abs}(\omega_\mathbf{K}) \sim (\hbar\omega_\mathbf{k})^2 \; \mathbf{e}(\mathbf{K}\lambda) \cdot \mathbf{T}(\omega_\mathbf{K}) \cdot \mathbf{e}(\mathbf{K}\lambda) \tag{3.16}$$

where $\mathbf{T}(\omega_\mathbf{K})$ writes, in anlogy to (3.10),

$$\mathbf{T}(\omega_\mathbf{K}) = \sum_{\mathbf{k}(\text{unoccupied})} \ll 0|\mathbf{r}|\mathbf{k} \gg \ll \mathbf{k}|\mathbf{r}|0 \gg \delta(\hbar\omega_\mathbf{K} - E_\mathbf{k} - E_0). \tag{3.17}$$

Thus we can state that XRS, connected with the excitation of a core electron into an unoccupied level, delivers formally the same resulting information as X-ray absorption spectroscopy (XAS), when the absorption process induces the same transition of a core electron, provided both spectroscopies are considered within the limits of the dipole approximation. Then $\mathbf{q}$, ω in the former spectroscopy plays the same role as $\mathbf{e}(\mathbf{K}\lambda)$, $\omega_\mathbf{K}$ in the latter.

But one must not forget that the inelastic scattering spectroscopy has two decisive advantages over soft X-ray absorption spectroscopy, as far as low-Z elements are involved.

(i) Since in the case XAS, $\hbar\omega_\mathbf{K}$ is only between 10 and 300 eV the penetration depth of the radiation is so small that surface effects will play an important role. Moreover, the sample must be held under vacuum conditions, so that investigation of liquids or of samples under inconvenient ambient conditions (e.g. reactive gases) are very difficult or have to put up with disturbing effects. On the other hand, inelastic X-ray scattering spectroscopy can be performed with hard X-rays, so that the above limitations will not happen. One gets really bulk information of the sample, and difficult ambient conditions will not hamper investigations that much. Of course, one needs a rather good spectral resolution, in order to reach results, comparable with soft X-ray absorption spectroscopy.

(ii) We have seen that it is rather easy to reach the range $qa \geq 1$ in inelastic X-ray scattering spectroscopy thus enabling dipole forbidden transitions with high probability, simply by increasing the scattering angle θ and shifting this way the value of $q \cong 2K \sin\theta$ into the desired range. This can be done without changing the incident wavelength. On the other hand, in the case of soft X-ray absorption spectroscopy, the δ-function in (3.17) restricts $\hbar\omega_\mathbf{K}$ to values near the core binding energy, so that there is no free play for K.

Another method competing with inelastic X-ray scattering spectroscopy is electron energy loss spectroscopy (EELS). Two disadvantages of EELS make this spectroscopy less versatile compared with IXS spectroscopy: (i) On the one hand, the cross-section of electron scattering is much larger than the Thomson scattering cross-section. This might be considered as an advantage. On the other hand, the scattering cross-section of EELS goes as q^{-2}. Both properties of EELS together have the consequence that EEL spectra for larger q are strongly contaminated with multiple scattering contribution, mainly composed of small-q spectra. Therefore, EEL spectra for larger q, as desirable to yield non-dipole contributions, need rather sumptuous correction procedures. (ii) The penetration depth of electrons with energies used for EELS are of the order of a few nm, so that surface effects contributing to the final result cannot excluded. The desired bulk information, provided by IXS, must be considered as doubtful in the EELS case.

3.2 Core-hole–electron interaction in X-ray Raman scattering

Until now we have considered XRS only in terms of the one-electron approximation. But it turns out that we must not neglect the strong influence of the core hole in the final state, even if screened to a certain extent by the valence electrons and by the other core electrons. Especially the core hole in the final state must be made responsible for two effects, namely the edge singularity for simple metals and an exciton-like pre-peak in the case of semiconductors and insulators. Both effects have found a rather different theoretical treatment. The effect of core- hole interaction with conduction electrons in inelastic X-ray scattering was first investigated by Doniach et al. (1971) on the basis of the so-called Mahan–Nozières–de Dominicis (MND) theory of edge singularities (Mahan 1967, Nozières and de Dominicis 1969), which predicts a spectral shape $I(\hbar\omega)$ near an X-ray threshold $\hbar\omega_0$ of the following form:

$$I(\hbar\omega) \sim \sum_l |W_l(\hbar\omega)|^2 \left(\xi/[\hbar(\omega-\omega_0)]\right)^{\alpha_l} \qquad (3.18)$$

due to the screening of the core-hole potential by the conduction electrons close to the Fermi level, which induces anomalies in the absorption and emission threshold. Here ξ is a range parameter of the order of the kinetic energy of the electrons at the Fermi level. $|W_l|^2$ is proportional to the transition density of states between the core level and the conduction electron states with the relative angular quantum number l. The many-body effect is condensed in the threshold exponent α_l, which is connected with the phase-shift δ_l characterizing the scattering of conduction electrons at the Fermi level from a core hole in a state of relative angular momentum l, via

$$\alpha_l = 2\delta_l/\pi - \alpha; \quad \alpha = 2\sum_l (2l+1)(\delta_l/\pi)^2. \qquad (3.19)$$

These relations, often called sum rules, are derived neglecting exchange, so that the threshold exponents do not depend either on the orientation of the spin nor on the orientation of the orbital angular momenta. An extension of the MND theory by including exchange has been worked out by Girvin and Hopfield (1976).

Doniach et al. (1971) start their application of the MND theory to inelastic X-ray scattering by using the dynamic structure factor as written in (6.34) in terms of time-dependent density–density correlations:

$$S(\mathbf{q},\omega) = \left(\frac{1}{2\pi\hbar}\right) \int_{-\infty}^{\infty} dt \exp(-i\omega t) \langle \rho(\mathbf{q},0)\, \rho^+(\mathbf{q},t)\rangle, \qquad (3.20)$$

where the density operator writes

$$\rho(\mathbf{q}) \equiv \int d^3r \exp(-i\mathbf{q}\cdot\mathbf{r})\, \rho(\mathbf{r}); \quad \rho(\mathbf{r}) \equiv \sum_j \delta(\mathbf{r}-\mathbf{r}_j). \qquad (3.21)$$

At the energy transfer of interest, the core conduction electron transition, the density operator can be written in second quantized form

$$\rho(\mathbf{r}) = \sum_n \sum_\mathbf{k} \psi_\mathbf{k}^*(\mathbf{r})\, \phi_n(\mathbf{r})\, c_\mathbf{k}^+ a_n + \text{H.C.}, \qquad (3.22)$$

where $\psi_\mathbf{k}$ are the conduction band Bloch functions, ϕ_n the core states and $c_\mathbf{k}^+$ and a_n the corresponding electron and hole creation operators. Assuming the core states to be tightly bound we can write (3.20) as

$$S(\mathbf{q},\omega) = \sum_\mathbf{k} \sum_{\mathbf{k}'} M_\mathbf{k}(\mathbf{q}) M_{\mathbf{k}'}(\mathbf{q})\, R(\mathbf{k},\mathbf{k}',\omega) \qquad (3.23)$$

where

$$R(\mathbf{k},\mathbf{k}',\omega) = \int_{-\infty}^{\infty} dt \exp(i\omega t) \langle c_\mathbf{k}^+(t)\, a(t)\, a^+(0)\, c_{\mathbf{k}'}(0)\rangle \qquad (3.24)$$

stands for the many-body effects, and the matrix element

$$M_\mathbf{k}(\mathbf{q}) = \int d^3r\, \psi_\mathbf{k}^*(\mathbf{r})\, \phi_n(\mathbf{r}) \exp(i\mathbf{q}\cdot\mathbf{r}) \qquad (3.25)$$

determines the scattering process from a single atom.

The correlation function $R(\mathbf{k},\mathbf{k}',\omega)$ is now evaluated according to the assumptions of the MND theory: the electrons are noninteracting with one another and with the periodic lattice potential; the core hole is infinitely heavy; the interaction between the electron and the hole can be described by a central potential. The conduction electrons are represented by orthogonalized plane waves (OPW's). Then one ends up with the following expression for the dynamic structure factor

$$S(\mathbf{q},\omega) = \sum_l A_l(\mathbf{q})\, \overline{R}_l(\omega), \qquad (3.26)$$

where, for ω near the threshold ω_0, the quantity $\overline{R}_l(\omega)$ is given by

$$\overline{R}_l(\omega) = \frac{1}{[\hbar(\omega-\omega_0)]^{\alpha_l}}, \qquad (3.27)$$

with α_l of equation (3.19).

The generalized matrix element $A_l(\mathbf{q})$ is plotted in Fig. 3.1 for Li with q in units of $2\pi/a$. a is the Li lattice constant (3.50 Å). The 1s core wavefunction of Li was taken to be hydrogen-like

$$\phi_{\text{core}}(r) = (\lambda^3/\pi)^{1/2} \exp(-\lambda r); \quad \lambda = 1/a_{\text{Li}}; \quad a_{\text{Li}} = 0.371 \text{ a.u.} \quad (3.28)$$

It becomes evident from Fig. 3.1 that, as the scattering angle and with it also q is increased, the threshold behavior of the double differential scattering cross-section should change over from that predicted for the $l=1$ phase shift in (3.19) to that predicted for the $l=0$ phase shift. The $l=2$ contribution can be neglected. We will come back to this threshold characteristic later, when we discuss an actual experiment. For the present, the anomalous shape of the spectra at the threshold depends both on the weight of the l-contribution and on the corresponding α_l, and can extend from inhibition to enhancement. Of course, this anomaly has only a finite range on the energy scale, determined by the validity of (3.27), where the validity can be limited by introducing a range parameter ξ in (3.27) as in (3.18). ξ is of the order of the electron kinetic energy at the Fermi level.

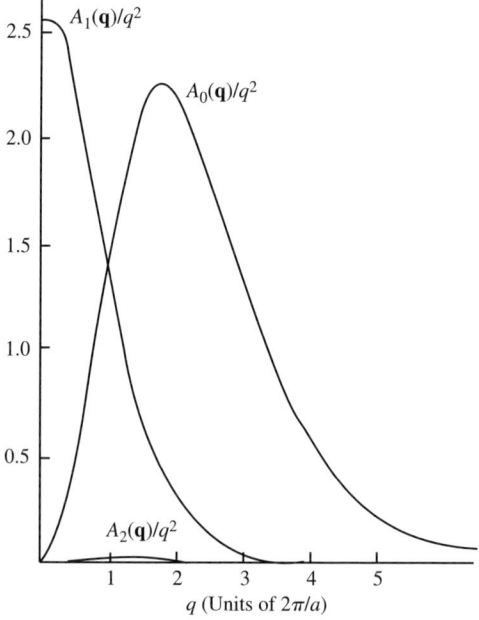

Fig. 3.1. The matrix element $A_l(\mathbf{q})/q^2$ as a function of the momentum transfer $\mathbf{q}$ in units of $2\pi/a$. a is the lattice parameter of Li (3.50 Å). (Originally published by Doniach et al. (1971); copyright (1971) by the American Physical Society.)

The core-hole–electron interaction has a much more far-reaching effect in the case of semiconductors and insulators and needs therefore another theoretical treatment. Following Soininen and Shirley (2001), we start again with the expression (3.3) for the dynamic structure factor. Using the completeness of the final states and taking the ground state $|0\rangle$ as the initial state one can approximate $S(\mathbf{q}, \omega)$, when ω is close to a core binding energy, as

$$S(\mathbf{q}, \omega) = -\pi^{-1}\mathrm{Im}\, \langle 0|\rho(\mathbf{q})[\hbar\omega - H_{\mathrm{eff}} + i\Gamma(\omega)]^{-1}\rho(\mathbf{q})^+|0\rangle. \quad (3.29)$$

H_{eff} is an approximate Hamiltonian of the excited system and $\Gamma(\omega)$ accounts for the lifetime broadening of the spectrum. The state $|0\rangle$ denotes a Slater determinant type of ground-state wavefunction. $\rho(\mathbf{q})^+$ is the Fourier component of the density operator which we will express in second-quantized form

$$\rho(\mathbf{q})^+ = \sum_{j,j',k,k'} \langle \psi_{jk}|\exp(i\mathbf{q}\cdot\mathbf{r})|\psi_{j'k'}\rangle\, a_{jk}^+ a_{j',k'} + \mathrm{H.C.}, \quad (3.30)$$

where ψ_{jk} and $\psi_{j'k'}$ are unoccupied and occupied single-particle states, respectively, a_{jk+} and $a_{j'k'}$ the corresponding creation and annihilation operators, respectively. The core-excited-state wavefunction is now approximated with an electron–core-hole pair wavefunction:

$$\Phi(\mathbf{r}_e, \mathbf{r}_h) = \sum_{\nu\mathbf{k}} C_{\nu\mathbf{k}}\,\psi_{\nu\mathbf{k}}(\mathbf{r}_e)\,[\psi_{\mathbf{k}-\mathbf{q}}^{\mathrm{TB}}\,\alpha(\mathbf{r}_h)]^*, \quad (3.31)$$

where $\psi_{\nu\mathbf{k}}$ is a conduction band state (ν = band index, $\mathbf{k}$ = Bloch wavevector), $\psi_{\mathbf{k}-\mathbf{q}}^{\mathrm{TB}}\alpha$ is a tight binding state and $C_{\nu\mathbf{k}}$ are the expansion coefficients. The atomic state $n\,l\,m$ of the core hole and its position $\boldsymbol{\tau}$ in the unit cell shall be represented by the parameter α:

$$\psi_{\mathbf{k}-\mathbf{q}}^{\mathrm{TB}}\,\alpha(\mathbf{r}_h) = N^{-1/2}\sum_{\mathbf{R}} \exp[i(\mathbf{k}-\mathbf{q})\cdot(\mathbf{R}+\boldsymbol{\tau})]\,\phi_{nlm}(\mathbf{r}_h - \mathbf{R} - \boldsymbol{\tau}), \quad (3.32)$$

where N is the number of $\mathbf{k}$-points, and ϕ_{nlm} the atomic wavefunction. We will denote the electron–core hole wavefunction by $|\nu\mathbf{k}\alpha\rangle$ in what follows. The approximate Hamiltonian in that scheme is composed of a single-particle part H_0, a direct (V_{D}) and an exchange (V_{X}) interaction of the electron with the core hole

$$H_{\mathrm{eff}} = H_0 + V_{\mathrm{D}} + V_{\mathrm{X}}, \quad (3.33)$$

so that the matrix element of H_{eff} can be written

$$\langle \nu\mathbf{k}\alpha|H_{\mathrm{eff}}|\nu'\mathbf{k}'\alpha'\rangle = \delta_{\alpha\alpha'}[(\varepsilon_{\nu\mathbf{k}} - E_\alpha)\delta_{\nu\nu'}\delta_{\mathbf{k}\mathbf{k}'} + \langle\psi_{\nu\mathbf{k}}|V_{\mathrm{D}}(\alpha) + V_{\mathrm{X}}(\alpha)|\psi_{\nu'\mathbf{k}'}\rangle], \quad (3.34)$$

where $\varepsilon_{\nu\mathbf{k}}$ is the single-particle energy of the conduction band electron and E_α is the core-hole energy. V_{D} must be divided into a bare core-hole potential and two correction terms taking into account the screening by the core electrons and the valence electrons, respectively. Thus the expression (3.29) for the dynamic structure factor can now be written

$$S(\mathbf{q},\omega) = -\pi^{-1}\,\mathrm{Im}\sum_{\nu\mathbf{k}}\sum_{\nu'\mathbf{k}'}\langle 0|\rho(\mathbf{q})|\nu\mathbf{k}\alpha\rangle\langle\nu\mathbf{k}\alpha|(\hbar\omega - H_{\mathrm{eff}} + i\Gamma(\omega))^{-1}|\nu'\mathbf{k}'\alpha\rangle$$
$$\times \langle\nu'\mathbf{k}'\alpha'|\rho^{+}(\mathbf{q})|0\rangle. \tag{3.35}$$

We will see in the following sections that this scheme seems to be adequate to account for the core-hole–electron interaction when making a reasonable ansatz for the quantities to be inserted into (3.34).

3.3 Special instrumentation with synchrotron radiation. Data processing

In principle nonresonant XRS experiments can be performed by utilizing the various elements of instrumentation, which has already been described in detail in Section 2.2. The choice of a specific instrumentation depends mainly on the desired total energy resolution, the sample size, and the momentum-transfer resolution one wants to achieve. Of course, often the high statistical accuracy one would like to obtain forces oneself to make compromises with respect to energy and, especially, to momentum transfer resolution. An experimental setup, which is the result of such a compromise, and which we have not yet mentioned in Section 2.2, should be shown in some detail, especially since a number of very nice and successful experiments have been performed at it. The instrument is installed at the APS undulator beamline 18ID and is described in detail by Bergmann and Cramer (1998). The monochromator is a Si(400) double-crystal monochromator and works in focusing mode. The analyzer consists of eight spherically bent Si/Ge crystals of 9 cm diameter on Rowland circles (see Section 2.2.5.1), which are intersecting in one point on the sample and one point on the detector, so that they are cooperating as shown in Fig. 3.2. The inelastic scattered spectrum is swept by scanning the beamline monochromator energy at fixed analyzer setting (inverse geometry). At an analyzer Bragg angle of approximately 88° the spectrometer captures a solid angle of 0.5% of 4π sr. The resolution is 0.3 eV at 6.46 keV using Si(440) reflections or 0.5 eV at 9.7 keV using Si(660). Together with the monochromator a total energy resolution of better than 1 eV is obtained. The price for the rather high angular acceptance of this multicrystal spectrometer is its nonuniform momentum transfer both with respect to the value of q, and with respect to the orientation of $\mathbf{q}$ relative to the sample. Faced with the momentum transfer dependence of the core-excitation spectra, as discussed in Section 3.1, this can be a disadvantage for special applications.

Inelastic X-ray scattering spectra due to core excitation are in most cases superimposed by spectra originating from valence electron excitations, as treated in Chapters 2 and 4. For small q ($\hbar^2 q^2/2m < E_{\mathrm{B}}$; E_{B} = core-electron binding energy) it is mainly the high-ω tail of the valence electron $S(\mathbf{q},\omega)$ spectrum, as shown, for example, in Fig. 3.3 for Li ($E_{\mathrm{B}} \cong 55$ eV). For larger q ($\hbar^2 q^2/2m > E_{\mathrm{B}}$)

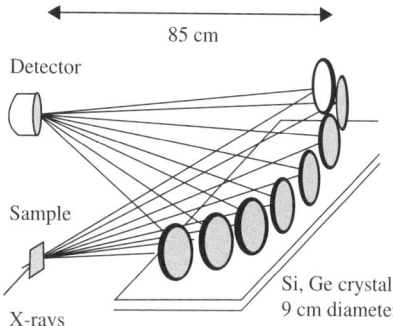

Fig. 3.2. Multicrystal analyzer system in Rowland geometry. (Reprinted with permission from Bergmann *et al.* (2002); copyright (2002) by Elsevier Science B. V.

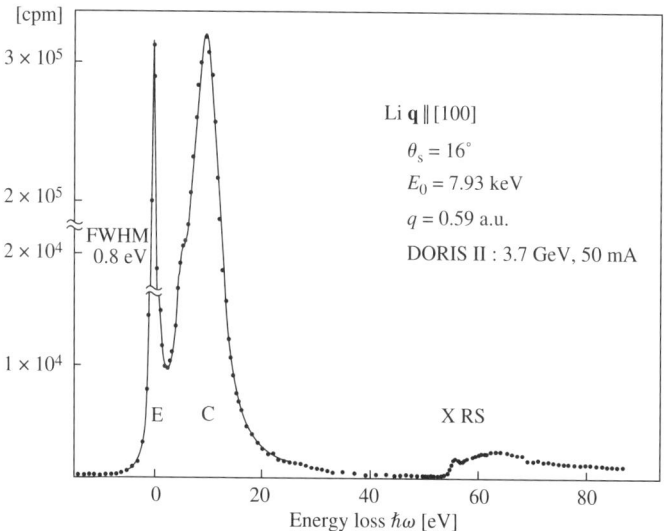

Fig. 3.3. Raw data of the experimental inelastic scattering spectrum of a Li single crystal. The XRS part has an edge-like onset on the high-energy-loss tail of the valence electron $S(\mathbf{q},\omega)$ spectrum (denoted C) at the binding energy of the Li K electron of about 55 eV. (Reprinted with permission from Nagasawa *et al.* (1989); copyright (1989) by the Physical Society of Japan.)

it is the low-ω tail of the valence electron $S(\mathbf{q},\omega)$ spectrum, which is already approaching the Compton profile (see Chapter 4), as shown in Fig. 3.4. In both cases this valence electron contributions needs to be subtracted, in order to get the pure core-electron excitation part. As long as it is mainly the low- or

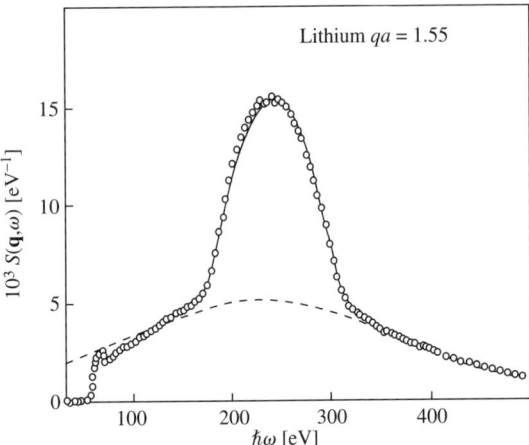

Fig. 3.4. Raw data of the experimental inelastic scattering spectrum of Li with $qa = 1.55$ (a is the radius of the K-shell core orbital). The XRS part has an edge-like onset on the low-energy-loss tail of the $S(\mathbf{q},\omega)$ Compton-like spectrum. The dashed curve is the calculated core contribution to the Compton profile in the impulse approximation. (Reprinted with permission from Nagasawa *et al.* (1997); copyright (1997) by the Physical Society of Japan.)

high-ω tail of the valence electron spectrum, which is in most cases a rather smooth function of ω, one can fit a corresponding smooth function to the low-ω behavior of the valence contribution for $\hbar\omega < E_B$, in order to extrapolate this function into the range $\hbar\omega > E_B$. It has been shown that the character of this smooth function, whether an exponential or a hyperbolic, is quite irrelevant, since essentially the same result was found for the corrected spectrum in the near-edge region (Bergmann *et al.* 2002). The range $\hbar^2 q^2/2m \cong E_B$ should be prevented, when planning an experiment, since the valence contribution dominates the core-electron part, and exhibits a peak- and possibly a fine-structure, which is difficult to be simulated by simple smooth functions. If, together with the X-ray Raman spectrum, also the whole valence part of the IXS spectrum has been measured, so that this valence part (after subtraction of the XRS) is completely represented by experimental data and can therefore be brought to an absolute scale by means of the f-sum rule (see 2.21), also the X-ray Raman spectrum can be reduced to an absolute scale and related to one K electron by multiplication of the reduced data with a factor $\alpha = V/K$, where V and K are the number of valence and core electrons in the sample, respectively. In cases where, for large q-values, the IXS spectrum is a Compton profile (CP) with the near-edge contribution on its low-ω tail, so that the CP with its core contribution extends to rather large values of ω, it is rather time-consuming to measure the whole CP. In those cases one can use free atomic CP's as available in the literature (Hartree–Fock CP's of Biggs *et al.* 1975) and which can be brought to the energy-loss scale $\hbar\omega$ thus resulting in $S_{CP}(\mathbf{q},\omega)$. Then the application of the f-sum rule yields absolute values of $S(\mathbf{q},\omega)$ via

$$S(\mathbf{q},\omega) = \left(\frac{2\hbar q^2}{2m} - \int_{\omega_m}^{\omega_l} S_{CP}(\mathbf{q},\omega')\omega'\,d\omega'\right) I(\mathbf{q},\omega) \left(\int_0^{\omega_m} I(\mathbf{q},\omega')\omega'\,d\omega'\right)^{-1}, \quad (3.36)$$

where $I(\mathbf{q},\omega)$ is the measured spectrum in arbitrary units, $(0,\omega_m)$ is the frequency range of the measurement, and ω_l corresponds to the p_z, where the calculated core contribution is fading out.

3.4 Early XRS experiments; EXAFS; comparison with soft-X-ray absorption spectroscopy

The first experiments to investigate systematically core excitation induced inelastic X-ray scattering were still performed using the characteristic radiation of conventional X-ray sources. Special attention should be called to the measurements of Suzuki and Nagasawa (1975) on polycrystalline beryllium (using CrKβ radiation) and polycrystalline graphite (using CuKα radiation), where the scattering angle was varied between 39° and 140° in order to scan a wide range of aq between 0.24 and 0.81. The authors have done their measurements by utilizing an analyzer composed of a flat crystal [(303) topaz single crystal] and a Sollar slit to reduce the divergence to 0.15°. This setup leads to an energy resolution not better than 10 eV. Nevertheless, in all spectra obtained (an example is shown in Fig. 3.5) the threshold of the core excitation contribution was clearly visible and distinguishable from the $S(\mathbf{q},\omega)$ part of the spectrum, which has been subtracted in order to get the pure core excitation contribution. The peak position on the ω-scale of this core excitation contribution, in Fig. 3.5 denoted by R, turned out to be independent of the scattering angle θ. The dependence on the scattering angle of the peak intensity of this contribution was compared with the theoretical result of Mizuno and Ohmura (1967) as given in (3.9) and (3.10) and which reads, specialized to the case of a polycrystalline sample and unpolarized incident radiation, as

$$\left(\frac{d^2\sigma}{d\Omega_2 d\omega_2}\right)_{\omega_{peak}} \sim I_{\omega_{peak}}(\theta) \sim (1+\cos^2\theta)\sin^2(\theta/2)\,t(\omega_{peak}), \quad (3.37)$$

with

$$t(\omega) = (1/3)\,\text{trace}\,\mathbf{T}(\omega). \quad (3.38)$$

By plotting $y \equiv I_{\omega_{peak}}(\theta)/(1+\cos^2\theta)$ as a function of $x \equiv (qa)^2 = (4\pi a/\lambda_1)^2 \sin^2(\theta/2)$, where a is the radius of the K-shell core orbital, and λ_1 the incident wavelength, the experiments on Be showed, beginning at $q^2 \cong 0.2$, distinct deviation from a linear relationship, as predicted by the Mizuno and Ohmura theory, depicted in Fig. 3.6. Having in mind that this theory is based

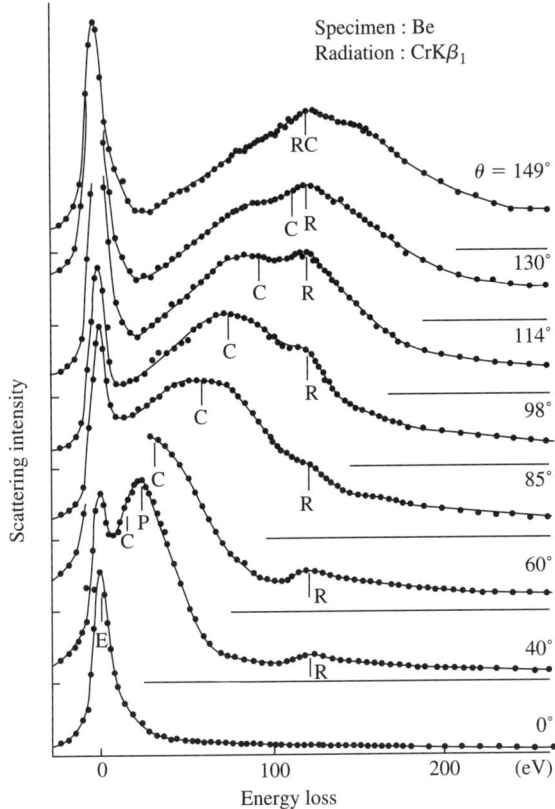

Fig. 3.5. Inelastic scattering spectra of beryllium at different scattering angles. Primary radiation is CrKβ1. The symbols E, R, C, and P indicate the Rayleigh, Raman, Compton and plasmon scattering peak, respectively. The peak of the Raman band does not shift with scattering angle. (Originally published by Suzuki and Nagasawa (1975); copyright (1975) by the Physical Society of Japan.)

on the dipole approximation, this deviation can be attributed to increasing contributions of nondipolar excitations with increasing q, the first time detected in experiment. The authors of this experiment have also tried to quantify these contributions by adding two more terms to the expansion of $\exp(i\mathbf{q}\cdot\mathbf{r})$ than taken into account in the dipole approximation. A theoretical curve, which is parametrizing the q^2 dependence of this extension of the Mizuno-Ohmura model, is also shown in Fig. 3.6.

Indications of well-resolved fine structure (0.9 eV) in the core excitation part of inelastic X-ray scattering as obtained with synchrotron radiation on Li metal was reported by Schülke et al. (1986). One year later, Tohji and Udagawa (1987)

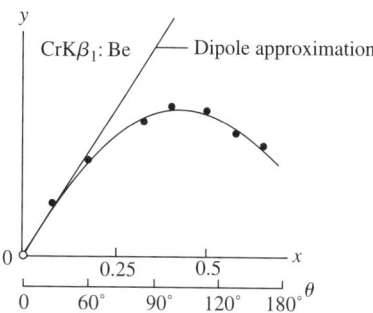

Fig. 3.6. Relation for Be between $y \equiv I_{\omega_{\text{peak}}}(\theta)/(1+\cos^2\theta)$ and $x \equiv (qa)^2 = (4\pi a/\lambda_1)^2 \sin^2(\theta/2)$, where a is the radius of the K-shell core orbital, θ the scattering angle and λ_1 the incident wavelength (Cr K$_{\beta 1}$). (Originally published by Suzuki and Nagasawa (1975); copyright (1975) by the Physical Society of Japan.)

reported on a synchrotron radiation based measurement on graphite, which extends over an energy range of somewhat more than 300 eV beyond the K threshold of carbon. Monochromatized [Si(111) double crystal monochromator] and double mirror-focused synchrotron radiation hits the sample and was energy-analyzed by the combined action of a curved Johansson-type Ge(111) crystal and a position-sensitive detector (see Section 2.2.5.3), where the sample was positioned between the Rowland circle and analyzer crystal. The total energy resolution was 6 eV read from the FWHM of the Raleigh line. The large energy range beyond threshold and the high statistical accuracy enabled the authors to make the first complete EXAFS (extended **X**-ray **a**bsorption **f**ine **s**tructure) analysis of the X-ray data. This means that the observed oscillations $\chi(k)$ of the scattered spectrum brought to a k-scale, where k is the photoelectron wave vector as calculated from $k = \hbar^{-1}[2m(\hbar\omega - E_\text{B})]^{1/2}$, can be interpreted as being due to interferences of the outgoing photoelectron wave with partial waves scattered at the atoms nearest to the photoelectron emitting reference atom, so that

$$\chi(k) = \sum_j (N_j/kr_j^2)\, F_j(k) \exp(-2k^2\sigma_j^2 - 2r_j/\lambda)\, \sin[2kr_j + \alpha_j(k)], \quad (3.39)$$

where N_j is the number of atoms in the jth shell, F_j is the scattering amplitude, σ_j is the Debye–Waller factor, and α_j the phase shift. r_j is the distance from the central photoelectron emitting atom to the atom in the jth shell. λ is the mean free path of the photoelectron in the solid. Thus one can extract structure information by Fourier transforming $\chi(k)$, more precisely, information about the r_j, proposed the phase shifts α_j are known or can theoretically be calculated (Teo and Lee 1979). The result of such Fourier transforming of the graphite $\chi(k)$ data are depicted in Fig. 3.7. The position of the first two peaks of the Fourier transform magnitude at 1.43 and 2.49 Å is not far from the first two

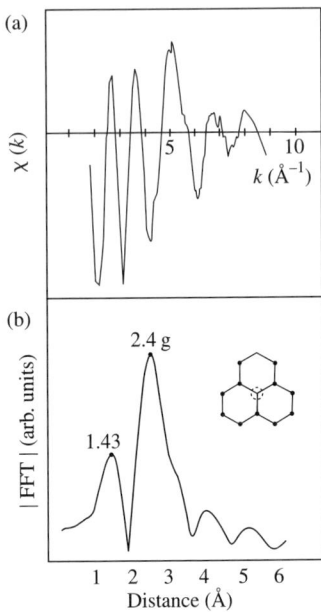

Fig. 3.7. (a) Extended oscillations of the XRS spectrum of graphite as a function of k. (b) Associated Fourier transform (radial structure function). (Originally published by Tohji and Udagawa (1987); copyright (1987) by the American Physical Society.)

carbon–carbon distances in graphite, namely 1.42 and 2.46 Å. Application of the so-called FEFF code (Rehr *et al.* 1992), as now routinely used in the evaluation of EXAFS data could improve the agreement.

The first systematic synchrotron radiation based high-resolution study of X-ray Raman scattering has been performed by Nagasawa *et al.* (1989) at HASY-LAB, DESY Hamburg on single crystals of Li and Be and on highly oriented pyrolytic graphite (HOPG). The experimental setup was the same as used by Schülke *et al.* (1986) in their IXS study on Li. An energy resolution of 0.8 eV was achieved. One aim of these experiments was to look for a **q**-orientation dependence of the X-ray Raman spectra in the near-threshold range (denoted as the **n**ear-edge **X**-ray **a**bsorption **f**ine **s**tructure (NEXAFS) in the case of X-ray absorption spectroscopy). Whereas the Raman spectra of cubic Li were found to be independent of the **q**-direction, the Be and especially the graphite spectra exhibited a pronounced **q**-orientation dependence, shown in Fig. 3.8 for graphite. This **q**-orientation sensitivity can be traced back to the selective action of **q** on the dipole matrix element, as is made clear in (3.8). In graphite the orientation **q** $\|$ c-axis allows only transition from the 1s core state to unoccupied π-states (p_z-orbitals, where the z-axis coincides with the crystallographic c-axis), whereas the orientation **q** $\perp$ c-axis prohibits these transitions and enables transitions to

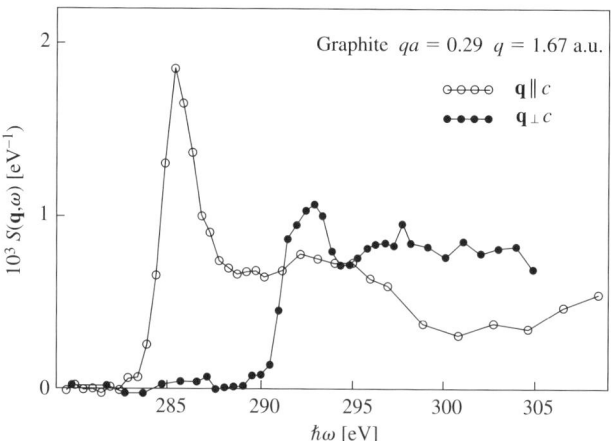

Fig. 3.8. The anisotropic XRS spectrum of graphite for $\mathbf{q} \parallel c$ and $\mathbf{q} \perp c$-axis reflect the density of unoccupied carbon π-states and σ-states, respectively. (Reprinted with permission from Nagasawa et al. (1989); copyright (1989) by the Physical Society of Japan.)

unoccupied σ-states (linear combinations of s, p_x and p_y orbitals). Thus the energy distance of about 6 eV between the two edges in Fig. 3.8 can be interpreted as the energy distance between the bottom of the π-and the σ-bands in graphite, and was found in excellent agreement with 3D band structure calculations of Tatar and Rabii (1982). A complete group theoretical inspection of the symmetry selectivity of X-ray Raman scattering can be found in a paper of Schülke et al. (1988).

The study of Nagasawa et al. (1989) is especially noteworthy, since it contains an explicit experimental verification of the equivalence of soft X-ray absorption and X-ray Raman spectroscopy, as predicted in Section 3.1, provided the assumptions made to allow for the dipole approximation of X-ray Raman scattering are valid. Already Fig. 1.7 has shown the X-ray Raman spectrum of Li for two different values of qa, where a is the orbital radius of the 1s electrons of Li ($a = 0.376$ a.u.). The X-ray Raman spectrum is confronted with the soft X-ray absorption data, where these data were derived from the published $\mu(\omega)$ data of Haensel et al. (1970) ($\mu(\omega)$ is the linear absorption coefficient) by the relation

$$S_a(\omega) = (k/\omega)[\mu(\omega) - \mu_v(\omega)] \quad (3.40)$$

Here $\mu_v(\omega)$ is the contribution of the valence electrons to the absorption coefficient extrapolated beyond the core threshold by assuming $\mu_v(\omega) \sim \omega^{-3}$, $\hbar\omega$ is the absorbed photon energy and k is the normalization constant, giving the same area of the absorption curve as the X-ray Raman spectra over the energy range up to 57.5 eV. The XRS spectrum for $qa = 0.22$ is in fairly good agreement with the absorption curve, as predicted for XRS experiments, when performed within

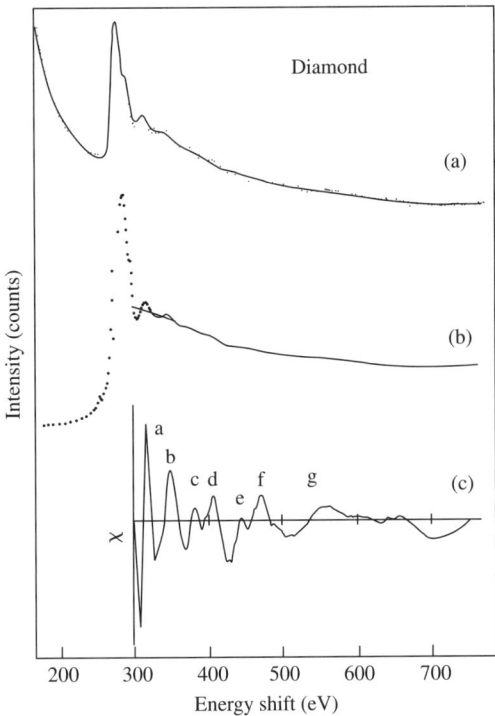

Fig. 3.9. (a) Part of the raw inelastic scattering spectrum from diamond (dots) and the smoothed spectrum (solid line) observed at 60°. (b) XRS spectrum obtained from (a) by removing the tail of the valence electron $S(\mathbf{q},\omega)$, together with the assumed smooth contribution. (c) Extracted oscillations from (b). Features (a) to (g) correspond to those in Fig. 3.10. (Originally published by Tohji and Udagawa (1989); copyright (1989) by the American Physical Society.)

a q-range valid for the dipole approximation. Contrary, the XRS spectrum for $q = 0.38$ exhibits already a certain deviation from the absorption curve, an indication for contributions to the spectrum other than the dipole allowed ones. We will specify this aspect of XRS in what follows. Similar conclusions were drawn by Nagasawa et al. (1989) by comparing XRS data on Be with corresponding soft-X-ray measurements of Haensel et al. (1970). The equivalence between XRS and soft-X-ray absorption spectroscopy refers not only to the near-edge part of the spectra but holds also for the extended fine structure as demonstrated by Tohji and Udagawa (1989) for the case of diamond. In Fig. 3.9 the XRS spectrum of diamond is shown in different stages of data evaluation. Figure 3.10 confronts this with the corresponding soft-X-ray absorption spectrum. One finds in the figures a one-to-one correspondence of the peaks (denoted with letters a to g) in the oscillatory part.

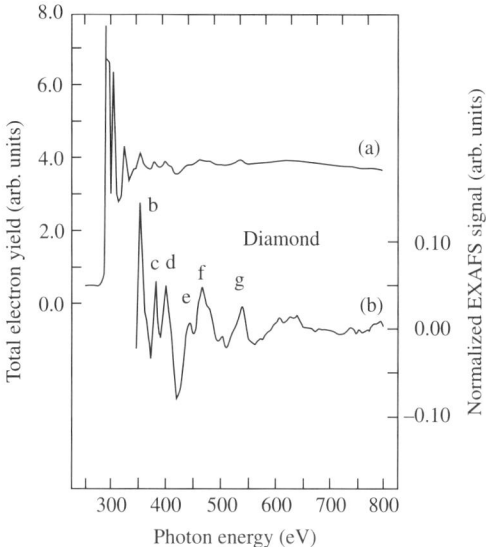

Fig. 3.10. (a) Soft-X-ray absorption spectrum of diamond. (b) The extracted oscillation (Comelli et al. 1988). (Originally published by Tohji and Udagawa (1989); copyright (1989) by the American Physical Society.)

Very similar XRS spectral characteristics as those seen in the case of graphite were found by Watanabe et al. (1996) in XRS experiments on layered hexagonal BN performed at the multipole wiggler beamline BL16X of the Photon Factory, Japan equipped with a Si(111) double-crystal monochromator and cylindrically bent Ge(440) analyzer working according to the van Hamos principle (see Section 2.2.5.3) in connection with a position-sensitive detector. The spectra were taken on single crystals both at the N-K and at the B-K edge with $\mathbf{q} \parallel c$-axis and $\mathbf{q} \perp c$-axis. Both the $\mathbf{q} \parallel c$-axis spectra start with a very strong exciton-like peak, which must be attributed, as in the case of graphite, to transition into the lowest lying π^*-states. No contribution above noise is visible in this range of energy loss in the $\mathbf{q} \perp c$-axis spectra. The corresponding part of the unoccupied band structure consists entirely of π^*-states. The $\mathbf{q} \perp c$-axis spectra (to be attributed to transitions into σ^*-states) have their first prominent peak shifted by 7 eV to larger energy loss with respect to the first peak of the $\mathbf{q} \parallel c$-axis ones. At this position a range with strong overlap of π^*- and σ^*-type bands starts.

3.5 Momentum transfer dependence in XRS experiments

It has been recognized above to be an important advantage of XRS over other spectroscopies that simply by varying the value of momentum transfer q transitions other than the dipole-allowed ones can be "switched on". This means that XRS offers a specific sensitivity to angular momentum projections of the density of states (DOS). Further symmetry selectivity is achieved by utilizing the q-orientation dependence as based on (3.9) and (3.10). In order to verify these theoretical predictions by selected experiments, Sternemann et al. (2003) have performed a systematic study of both momentum transfer dependencies on Be by comparing synchrotron radiation-based measurements on polycrystalline and single crystal Be with calculations, which were using the scheme of Soininen et al. (2001), so that also interactions between the excited electron and the core hole as presented in Section 3.2 were taken into account. The experiments were done at the SAW2-beamline of the DELTA storage ring. The radiation was monochromatized by means of a Si(311) double crystal setup with a saggitally focussing second crystal. The analyzer consisted of a spherically bent Si(800) crystal run in inverse geometry. The overall energy resolution was 1.4 and 1.2 eV for the single crystal and the poycrystalline measurement, respectively. The spectra taken with qa values ranging from 0.17 to 1.29 ($a = 0.198$ Å) and with $\mathbf{q} \parallel c$-axis and $\mathbf{q} \perp c$-axis, respectively, were normalized by using the f-sum rule (see Section 3.3).

The dependence of the XRS spectra on the q-values was tested on polycrystalline samples. Figure 3.11 shows the experimental and theoretical Be XRS spectra for different values of qa as indicated. In the left-hand panel (a) the absolute values of $S(\mathbf{q}, \omega)$ are plotted, whereas in the right-hand panel (b) the spectra are scaled to their value at 116 eV energy loss to allow a detailed discussion of the shape of the spectra, the main features of which are denoted from A to D*. The spectra measured at low momentum transfer $qa = 0.17$ and 0.34, for which the dipole approximation can be assumed to be valid, exhibit two main peaks A and C and a broad shoulder between B and C with a small peak-like structure at B*. The tail behind C changes its slope at D*. The XRS measurements with $qa = 1.18$ and 1.29 show clear differences when compared with those of smaller momentum transfer. The structures A and C are more pronounced, and an additional peak structure B appears. Furthermore, the peak at C is shifted to lower energy losses. The slope behind C changes now at D. It can clearly be demonstrated how these changes are connected with deviation from the dipole approximation. While, within the limits of the dipole approximation, only excitations into p-type final states are allowed, the larger values of the momentum transfer enables also monopolar transitions into s-type and quadrupolar transitions into d-type final states. To make this transparent, Fig. 3.12 shows in addition to the total theoretical XRS spectrum of Be for $qa = 1.29$ the p- and s-contributions (the d-contributions turned out to be negligible small). The dependence of the XRS spectra on the q-orientation and on the q-value which was tested on Be single crystals. Figure 3.13 shows in the upper panel (a) theoretical and experimental XRS spectra with $\mathbf{q} \perp c$-axis for $qa = 0.17$ and 1.29, respectively, in the lower panel spectra with $\mathbf{q} \parallel c$-axis. For the smaller momentum transfer, $qa = 0.17$, we find a distinct difference between the two q-orientations. The $\mathbf{q} \parallel c$-axis result exhibits two main peaks and a shoulder at

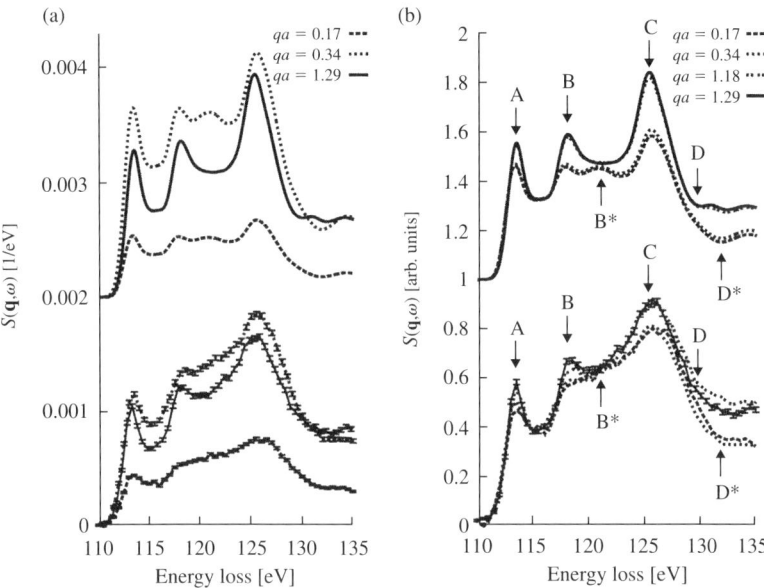

Fig. 3.11. (a) q-dependence of experimental (lower part) and theoretical (upper part) K-edge XRS spectra of polycrystalline Be on an absolute scale. The theoretical spectra are shifted vertically by 0.002 eV^{-1}. (b) Experimental (lower part) and theoretical (upper part) K-edge spectra of polycrystalline Be on an arbitrary scale, so that they agree at 116 eV energy loss, in order to allow detailed discussion of the spectral shape. The calculated spectra are shifted vertically by 1. (Reprinted with permission from Sternemann et al. (2003); copyright (2003) by the American Physical Society.)

B, whereas the $\mathbf{q}\perp c$-axis spectrum displays much less pronounced features at A, B, B* and C. At larger momentum transfer, $qa = 1.29$, nearly no **q**-orientation dependence can be detected. This behavior must be ascribed to the increasing role of the s-contributions with increasing q, where these s-type contributions turn out to be isotropic contrary to the p-type ones. In conclusion, it was clearly demonstrated how the XRS spectroscopy provides, with its momentum transfer dependence, a high degree of selectivity when investigations of unoccupied states are concerned.

An earlier synchrotron radiation based study of the momentum transfer dependence of XRS spectroscopy was published by Nagasawa et al. (1997) for the case of Li, performed at HASYLAB/DESY using the same setup as Nagasawa et al. (1989). The authors have measured the XRS spectra of Li for $qa = 0.22$; 0.38, 1.17 and 1.55. The near-edge range of these spectra were compared with calculations according three different schemes: (i) The so-called on-site approximation (OSA) of Arimitsu, Kobayashi and Mizuno (1987), where initial and

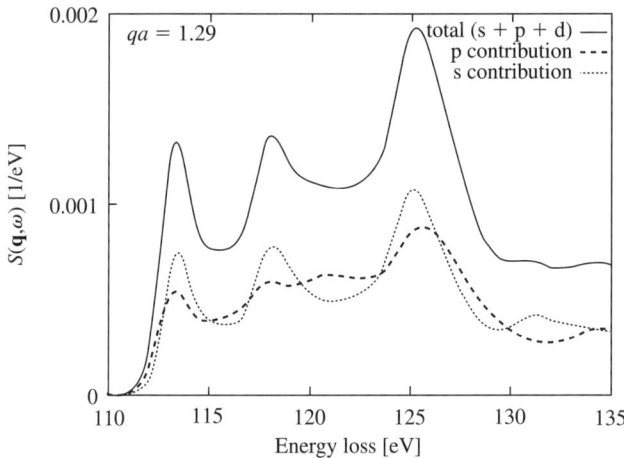

Fig. 3.12. Symmetry projected calculation of the K-edge XRS spectra of polycrystalline Be for $qa = 1.29$. Solid line: full calculation including s-, p-, and d-type final states; short dashed: s-type contribution; long dashed: p-type contribution. The d-type contribution is negligibly small within this energy-loss range. (Reprinted with permission from Sternemann et al. (2003); copyright (2003) by the American Physical Society.)

final state wavefunctions in (3.25) are deduced from the same simplified potential describing the atomic configuration around the scattering atom, namely a spherical symmetric potential within an inscribed sphere embedded in a constant potential. In the two other schemes the initial state wavefunction is the 1s hydrogenic ground state wavefunction given in (3.28), whereas the excited state wavefunctions are (ii) hydrogenic continuous waves (HCW), as used already by Eisenberger and Platzman (1970) for $S(\mathbf{q},\omega)$ calculations and (iii) a single orthogonal plane wave (OPW), respectively, as employed by Doniach et al. (1971), who obtained the results of Fig. 3.1 by means of this scheme. It turned out that the OSA and the HCW calculations are in reasonable agreement with the overall shape of the spectra in a range 50 eV above threshold. On the contrary, the OPW approximation failed totally to describe the experiment. This result is all the more important, since the HCW on the one hand and the OPW scheme on the other hand exhibit a very different q-dependence of the $l = 0$ (transitions to s-type excited state wavefunctions) and $l = 1$ (transitions to p-type) partial-wave contributions to $S(\mathbf{q},\omega)$, respectively, as shown in Fig. 3.14 for $\hbar\omega = 55$ eV. The OPW result of Fig. 3.14 resembles that of Fig. 3.1: The s-type and p-type contributions become equal size at $q \approx 1$ a.u. The inadequacy of the OPW approximation doubts the applicability of Fig. 3.1 for further discussions. The HCW approximation, however, yields equal values of s- and p-contributions not before $q = 3.5$ a.u.

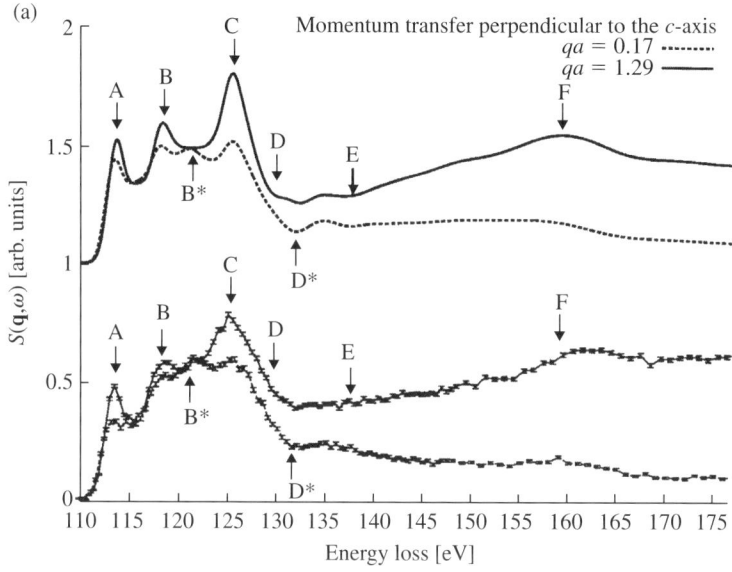
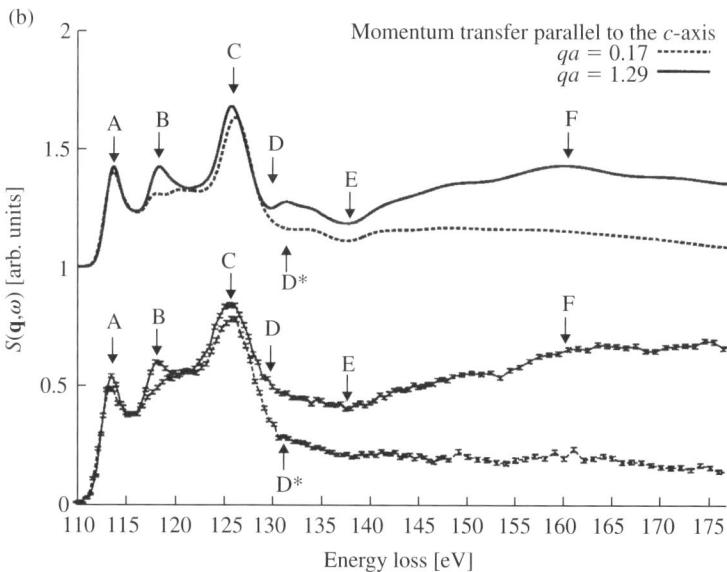

Fig. 3.13. (a) Experimental (lower part) and theoretical (upper part) Be K-edge spectra for $\mathbf{q} \perp c$-axis on an arbitrary scale, so that they agree at 116 eV energy loss. The calculated spectra are shifted vertically by 1. (b) Same as (a), but with momentum transfer $\mathbf{q} \parallel c$-axis. (Reprinted with permission from Sternemann et al. (2003); copyright (2003) by the American Physical Society).

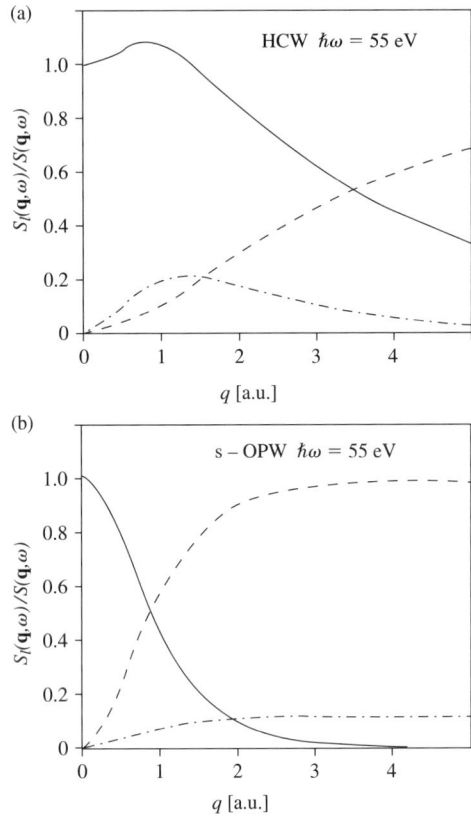

Fig. 3.14. (a) Partial-wave contributions to $S(\mathbf{q},\omega)$ for $\hbar\omega = 55$ eV, calculated in HCW approximation. Solid line: p-contribution ($l = 1$); dashed line: s-contribution ($l = 0$); dashed dotted line: d-contribution ($l = 2$). (b) The same as in (a), but calculated in OPW approximation. (Reprinted with permission from Nagasawa et al. (1997); copyright (1997) by the Physical Society of Japan.)

The result of this Nagasawa study, as far as the fine structure of the near-edge spectrum is concerned, is shown in Fig. 3.15, where the normalized modulation

$$N(\mathbf{q},\omega) \equiv [S(\mathbf{q},\omega) - T(\mathbf{q},\omega)]/T(\mathbf{q},\omega) \tag{3.41}$$

is plotted for different values of qa, as indicated. Here $T(\mathbf{q},\omega)$ is the calculated smooth HCW spectrum. $S(\mathbf{q},\omega)$ is the experimental spectrum brought to an absolute scale by means of the f-sum rule. One sees distinct differences in Fig. 3.15 between the low-q and the high-q $N(\mathbf{q},\omega)'s$ (denoted NONES). These differences again can be attributed to increasing contributions of monopolar transitions with increasing q. According to the HCW computations of q-dependence

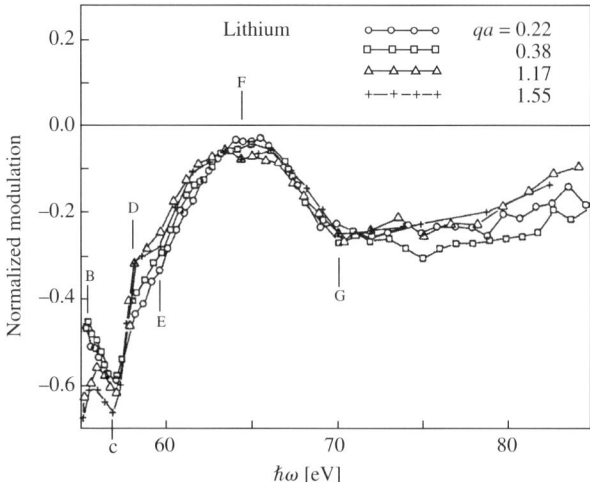

Fig. 3.15. Normalized modulation of the near K-edge fine structure of Li metal (NONES, see text) for different values of qa as indicated. (Reprinted with permission from Nagasawa *et al.* (1997); copyright (1997) by the Physical Society of Japan.)

of the partial-wave contributions to $T(\mathbf{q},\omega)$, as shown in Fig. 3.14, the s-type contributions become equal to the p-type ones at $qa \approx 1.3$, so that the $qa = 1.29$ spectra should contain s- and p-contribution to equal parts. This is visualized in Fig. 3.16, where the NONES are confronted with a difference between the p-partial density of states profile and the half-and-half mixed profile of the s- and p-partial densities of states as calculated by Ojala and Lähdeniemi (1982). Details of the experimental fine structure denoted C to F are in good agreement with the theoretical differences, again indicating that the half-and-half mix of s- and p-contribution for $q \approx 3.5$, as predicted by the HCW-scheme, is in much better agreement with experiment than the OPW treatment, which predicts for that q-value negligible p-contributions.

Two more studies on the momentum transfer dependence of XRS must be mentioned, without going into details. Both are devoted to the pre-edge structures, which are attributed to excitonic like core excitations. The first one has been performed by Hämäläinen *et al.* (2002) on the F-K edge of LiF, accompanied by first-principles calculations taking into account the core-hole–electron interaction. A pre-edge peak, which emerges with increasing q-values, was ascribed to an s-type core exciton. This conclusion was also confirmed by theoretical computations. The second study by Feng *et al.* (2004) on the q-dependence of the B-K edge XRS spectra of icosahedral B_4C point at a p-type core exciton at the so-called B3 site of the B_{12}-CBC structure, where the contributions of the different B sites to the XRS spectra could be separated by means of site-specific *ab initio* calculations going with the experiment.

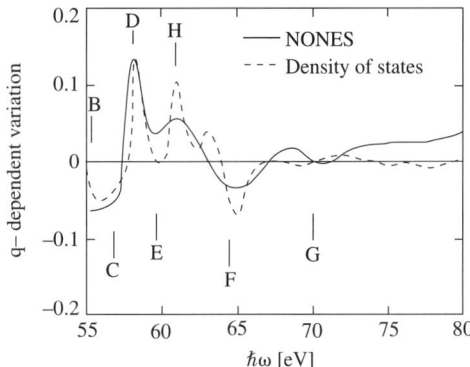

Fig. 3.16. Difference between the p-partial density of state profile and the half-and-half mixed profile of the s- and p-partial densities of states (Ojala and Lähdeniemi 1982). (Reprinted with permission from Nagasawa et al. (1997); copyright (1997) by the Physical Society of Japan.)

The interpretation of XAS spectra both in the so-called NEXAFS (near-edge X-ray absorption fine structure) and the EXAFS (extended X-ray absorption fine structure) regime is very often performed on the basis of the multiple scattering approach (see, for a recent review, Rehr and Albers 2000), which can be regarded as the real space analog of the KKR band structure method by calculating the one-electron Green's functions for the excited state. But this scheme cannot be simply transferred from XAS to XRS. As already noted above, the main difference between inelastic scattering and X-ray absorption lies in the transition matrix elements. The operator mediating the transitions in inelastic scattering is $\exp(i\mathbf{q}\cdot\mathbf{r})$, as compared to the dipole operator $(\mathbf{e}_1\cdot\mathbf{r})$ (and in some cases also the quadrupole operator) in X-ray absorption. Without going into details, we point to a very recent paper of Soininen et al. (2005), who applied the real-space multiple-scattering approach to calculate the core-excited states in XRS over a wide energy range. The momentum transfer dependence of both the near-edge spectra and the extended fine structure in the inelastic loss spectra are analyzed in detail and compared with experimental results. The final expression for the dynamical structure factor reads

$$S(\mathbf{q},\omega) = \sum_{L\,L'} M_L(-\mathbf{q},E)\,\rho_{L0,L'0}(E)\,M_{L'}(\mathbf{q},E), \tag{3.42}$$

where $M_L(\mathbf{q},E) = \langle R_L(E)|\exp(i\mathbf{q}\cdot\mathbf{r})|i\rangle$ is the embedded atomic matrix element with scattering states $R_L(\mathbf{r},E)$ at energy $E = E_i + \hbar\omega$ which are regular at the origin (L stands for (l,m)). $\rho_{L\mathbf{R},L'\mathbf{R}'}(E) = (-1/\pi)\mathrm{Im}G_{L\mathbf{R},L'\mathbf{R}'}(E)$ denote matrix elements of the final state density matrix, including the effect of the core-hole potential. $G_{L\mathbf{R},L'\mathbf{R}'}(E)$ is the one-particle Green's function represented in an angular momentum L and site $\mathbf{R}$ basis $|L, \mathbf{R}\rangle$. Thus the calculation of the dynamical structure factor separates into two parts: (i) the $\mathbf{q}$-dependent transition matrix element, which describes the production of photoelectrons into various final states; and (ii) a propagator matrix $G_{LL'}$ which governs the scattering of the photoelectrons within the system at a given excitation energy.

3.6 Core-hole–electron interaction in XRS experiments

As shown in Section 3.2 on the basis of theoretical considerations, the interaction of the core hole with the excited electron in the final state of XRS process needs special attention. Even in metals, where one expects the core-hole potential to be strongly screened by the nearly free electrons, the special situation at the Fermi edge leads to singularities at the threshold of the XRS spectra, the shape of which depends, according to (3.27), on the exponent α_l. Inserting the phase shifts δ_l, as calculated by Ausman and Glick (1969), into the sum rules of (3.19), Doniach et al. (1971) have predicted a very weak singularity for small q's, where the $l = 1$ (p-type) contribution dominates. On the contrary, one expects a spectrum strongly enhanced at the threshold, when, for larger q, the $l = 0$ (s-type) transitions are superior. Of course, one should have in mind, as shown above, that the theoretically predicted partition of s- and p-contributions as a function of q depends strongly on the approximation scheme used.

It suggests itself to use an XRS experiment in order to get direct information about the threshold exponent α_l, since one can switch, as shown above, from dipolar to preferentially monopolar transitions simply by changing the q-value. Such an experiment was performed by Krisch et al. (1997) on Li, using the inelastic scattering beamline ID16 at the ESRF. They applied a cryogenically cooled Si(111) premonochromator followed by a four-bounce Si(444) main monochromator (see Section 2.2.2.2). The scattered radiation was analyzed in the horizontal plane in full Rowland geometry by means of a segmented Si crystal (see Section 2.2.5.4), bent to a radius of 2 m, set to a fixed energy of 9885 eV (inverse geometry). The overall experimental resolution was 80 or 210 meV, depending on the monochromator configuration. The sample temperature was 30 K. Spectra were recorded at scattering angles 10° and 150° corresponding to a momentum transfer of $qr = 0.176$ and $qr = 1.95$ (r is here the core radius). The low-q spectrum taken with 80 meV resolution resembled quite well the p-partial density of unoccupied states (Papaconstantopoulos 1986), whereas the high-q spectrum exhibits much more similarity with the s-partial density of states, thus confirming the increasing weight of monopolar transitions with increasing q.

In order to extract information about the threshold exponent α_1 in the low-q and about α_0 in the high-q experiment, the incident flux was increased by a factor of 5 by removing the second crystal pair of the four-bounce monochromator, so that the overall resolution was 210 meV. Figure 3.17 shows the inelastic X-ray scattering spectra in the immediate vicinity of the Li K-edge for the two q-values. The fitting procedure starts with the assumption that the low-q

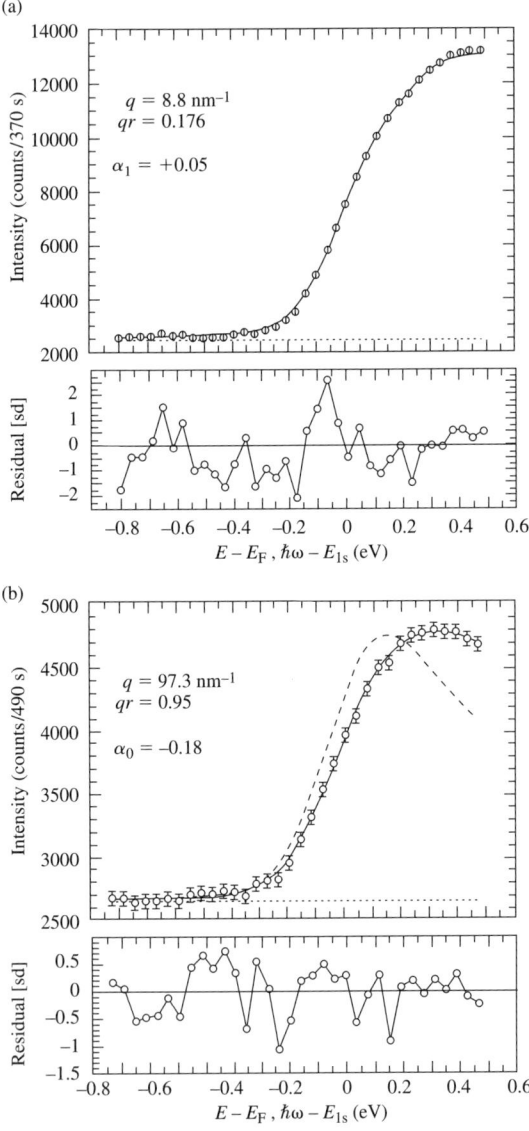

Fig. 3.17. Inelastic X-ray scattering spectra in the immediate vicinity of the Li K-edge at $q = 8.8\,\text{nm}^{-1}$ (a) and $q = 97.3\,\text{nm}^{-1}$ (b). Open circles: data; solid line: fit obtained as described in the text with $\alpha_1 = +0.05$ (a) and $\alpha_0 = -0.18$ (b). The dashed line in (b) has been obtained by keeping fixed in the fit the values of α_0 and the Gaussian broadening, respectively, to 0.274 and 290 meV. The background (dotted line) was determined from data below the edge. The bottom panels give the difference between the data and the fit in standard deviations (sd). (Reprinted with permission from Krisch *et al.* (1997); copyright (1997) by the American Physical Society.)

spectrum reflects the dipolar and the high-q spectrum exclusively the monopolar transitions. Therefore the increasing part of the spectrum was fitted to a function, modeled by the product of the partial unoccupied density of states (p DOS for low-q and s DOS for high-q) times the energy-dependent term $\xi/[\hbar(\omega-\omega_0)]^{\alpha_l}$ of (3.27). Broadening is introduced by the convolution with a Lorentzian of 40 meV FWHM accounting for the lifetime broadening (Citrin *et al.* 1977), and with a Gaussian of adjustable width allowing for both the phonon broadening, estimated to be 200 meV and the instrumental resolution of 210 meV. Background subtraction is based on the data of the pre-edge region. The Gaussian width, the intensity (including the range parameter ξ) and the threshold exponent α_l were left as free parameters. The solid line represents the best fit which gives $\alpha_0 = -0.18 \pm 0.07$ and $\alpha_1 = 0.05 \pm 0.03$. The quality of the fit can be read from the bottom panels, where the difference between experiment and fit is given in standard deviation units. The value of α_1 is not far from that obtained by soft X-ray absorption (Citrin *et al.* 1979) and by EELS measurements of Ritsko *et al.* (1974). On the contrary, the α_0 value is in disagreement with the prediction that transitions to final states of s-symmetry should give a large positive α_0. In order to demonstrate how far the experiment is from a larger positive value of α_0, the dashed line in Fig. 3.17(b) is the fit with α_0 fixed to 0.274 (a value obtained by using the fit result of Fig. 3.17(a) and the sum rules (3.19)) and the same Gaussian width as obtained in the best fit of Fig. 3.17(b). But it must be kept in mind that the sum rules (3.19) do not include exchange. Indeed the values of α_0 and α_1 found in this study are in good agreement with those obtained in a theoretical study of Girvin and Hopfield (1976), who included exchange when calculating the phase shifts δ_l. This result must be considered as a rather direct hint that exchange interactions play a very important role in the screening of a 1s core hole, at least in the case of Li. Nevertheless, it must be stressed that this result is based on the assumption that in the low-q experiment only dipole transitions are involved, in the high-q measurement exclusively monopolar transition. As shown by Nagasawa *et al.* (1998), this can be stated only if one represents the final state wavefunctions by OPW's, a simplification which found no backing by the Nagasawa experiment.

It was predicted in Section 3.2 that the core-hole–electron interaction should affect the results of XRS experiments on insulators and semiconductors in a much more pronounced manner than in metals, where it is only the peculiarity of the core hole screening near the Fermi-level which matters. In the case of insulators and semiconductors the attractive core-hole potential will influence more or less the whole near-edge spectrum, presumably by shifting the DOS induced fine structures of the XRS spectra to smaller energy losses. Moreover, core excitons should be expected. We have presented in Section 3.2 a formal background for those effects of the core-hole–electron interaction. The computational scheme presented there needs confirmation by experiments. These have been performed by Soininen *et al.* (2001) on LiF, BeO and diamond, strictly attended by theory. The experimental setup used at the beamline X21 of the NSLS was a combination of a horizontally bent Si(220) monochromator (see Section 2.2.2.3) with a Rowland circle spectrometer utilizing a spherically bent Si(111) analyzer in nearly back-scattering geometry with bending radius of 1m. The experiment was run in inverse geometry. The total energy resolution was between 0.7 and 1 eV depending on the sample geometry. Figure 3.18 compares the experimental XRS spectrum of BeO ($q = 1.77\,\text{Å}^{-1}$) (open circles with error bars) with the calculated ones. The solid line is the calculated spectrum with and the dashed line that without the core-hole–electron interaction. The most pronounced feature of the calculated spectrum is the strong core-exciton peak around 119 eV that is about 0.74 eV below the edge of the conduction band, indicated by the edge of the calculated spectrum without core-hole–electron interaction. The agreement between theory (with core-hole–electron interaction) and experiment is rather good as far as the peak positions are concerned. However, the calculation slightly overestimates the spectral weight of the core excitons. On the other hand, the near-edge structure and the overall agreement are completely lost if the interaction is neglected. Similar good agreement between calculation including interaction and experiment was achieved with LiF, where the relative weight of the even- and odd-parity core exciton could be changed experimentally by varying the q-value. The measured energy splitting between these two core excitons

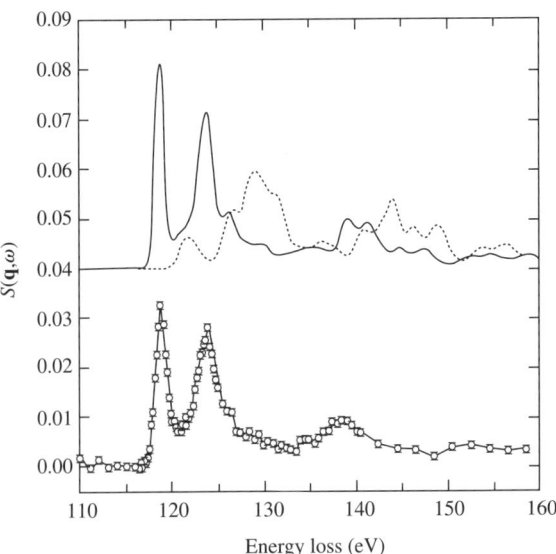

Fig. 3.18. Calculated (offset vertically) and measured XRS spectra at the Be K-edge in BeO; $q = 1.77\,\text{Å}^{-1}$. Calculated spectra: solid line: calculated with core-hole–electron interaction; dashed line: calculated without core-hole–electron interaction. (Reprinted with permission from Soininen *et al.* (2001); copyright (2001) by IOP Publishing Ltd.)

was somewhat larger than the calculated ones. The authors make the neglect of correlation between the excited electron and the remaining Li 1s electron responsible for this deviation. Also in the case of diamond the calculated result with including the core-hole–electron interaction predicts the weight and shape of most of the various structures adequately, especially the energy position of the shallow core exciton, whereas the calculations without including this interaction failed totally to achieve the experiment.

One must assume that also the sharp peak at the threshold, seen in the $\mathbf{q} \parallel c$-axis spectrum of graphite (Fig. 3.8), has to be traced back to a shallow core exciton as discussed by Veenendaal and Carra (1997).

3.7 XRS studies on molecular gases, liquids and solids

We will summarize in what follows the most recent applications of XRS on problems which are most suited to this method, namely those, where one can make full use of the bulk sensitivity as well as of the insensibility against surface faults and gaseous environment of the sample, heavy obstacles for methods such as soft X-ray absorption spectroscopy (SXAS) or electron energy loss spectroscopy (EELS). Liquid samples will therefore play an important role. Knowledge collected in experiments described above might be helpful for the interpretation of the results.

3.7.1 *Hydrogen bonds and O-O-correlation in water (O K-edge)*

Despite its rather simple elemental composition water has a complicated local structure. Especially it is the hydrogen bond (H-bond) in liquid water which holds the key to its unique behavior, being decisive for chemical, biological and geological processes. The dynamical motion of the atoms in liquid water at the picosecond time-scale let the H-bonds break and then re-form, so that a statistical distribution of different coordinations of the water molecules is constituted. Typical local structures are shown in Fig. 3.19 (Myneni *et al.* 2002). Denoted with SYM is the fully coordinated and symmetric H-bonded species, one central water molecule is coordinated by four neighboring waters through H-bonds, two H-bonds to the oxygen atom (donor bonds), and one to each hydrogen (acceptor bonds). This coordination is found in ice. A-ASYM is characterized by a broken H-bond on the oxygen site, D-ASYM on the hydrogen site. What remained an unanswered question is the portion of these types of coordination in liquid water, and whether also types of coordination exist, where both an oxygen related and a hydrogen related H-bond is broken.

If we study in SXAS or XRS the excitation of O 1s electrons, the dipole selection rule allows the measurement of the local atomic p-contribution, and these are primarily levels derived from the antibonding O-H molecular orbitals The oxygen p-character of these orbitals must be strongly influenced by the local geometrical arrangement, i.e. the H-bonding around the probed O atom. This probing takes place on the time-scale of the core-hole lifetime, much shorter

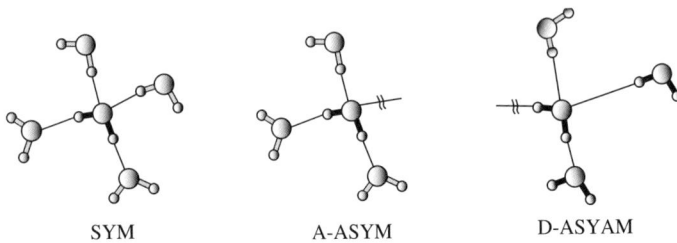

Fig. 3.19. Typical local structures of hydrogen bonding in liquid water. SYM: fully coordinated and symmetric H-bonded species. A-ASYM: broken asymmetric H-bond network on the oxygen site. D-ASYM: broken asymmetric H-bond network on the hydrogen site. (Reprinted with permission from Myneni et al. (2002); copyright (2001) by IOP Publishing Ltd.)

than the picosecond time-scale on which the changes of the local coordination occur. Thus it is a frozen-in structure studied in core-excitation spectroscopies. Along with the dependence of the 1s electron excitations on the local coordination of the water molecules also the O-O-coordination must be influenced by varying this local coordination. Therefore, also EXAFS studies at the O K-edge, especially those performed on the basis of XRS, can help to shed light on the O-O-coordination in liquid water.

The first to attack these problems by using XRS were Bowron et al. (2000); interestingly their results were published before the first conventional SXAS on liquid water microjets (Wilson et al. 2001) and on water flowing over a metal plate (Myneni et al. 2002). This might indicate the experimental difficulties conventional SXAS is confronted with in such work. The measurements were performed on the inelastic scattering beamline II (ID28) of the ESRF using a cryogenically cooled Si(111) double-crystal monochromator and a spherically bent Si(660) crystal analyzer in nearly back-scattering mode, set to a fixed photon energy (inverse geometry). The overall energy resolution was 2 eV. A scattering angle of 50° guarantees with $qa = 0.29$ (a = radial extent of the O 1s wavefunction) that the excitation of the O 1s electrons are dominated by electric dipole transitions. The near-edge spectra of polycrystalline ice Ih and of liquid water thus obtained exhibited distinct differences. The liquid water spectrum showed a shoulder in the first rising, not present in the ice data, and the ice spectrum had a second peak structure 8 eV above the edge not present in the liquid water spectrum. We will come back to these differences when discussing the better resolved spectra of Bergmann et al. (2002). Since the spectra were taken over a range of 200 eV, the oscillatory part $\chi(k)$ over a sufficiently large domain of k, with $k = \hbar^{-1}[2m(\hbar\omega - E_\mathrm{B}]^{1/2}$, could be extracted by subtracting the free atomic contribution. Utilizing modern EXAFS spectroscopy analytic methods (Filipponi and Cicco 1995), developed to handle especially the liquid state analysis by introducing known physical constraints of the bulk density and isothermal compressibility of water (Filipponi 1994) by using the ice data, i.e.

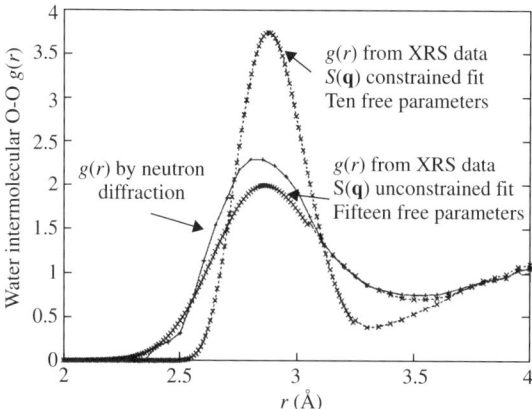

Fig. 3.20. The oxygen–oxygen partial pair distribution function as determined by neutron scattering (+++) and XRS (×××). The two XRS curves demonstrate the limiting structural models consistent with the data. (Originally published by Bowron et al. (2000); copyright (2000) by the American Physical Society.)

the data of a well-characterized system as a reference, the authors were able to obtain the oxygen–oxygen partial pair distribution function $g(r)$, at least as lower and upper limit as shown in Fig. 3.20 together with the $g(r)$ obtained by neutron diffraction. The lower limit is defined by an unconstrained fit, whereas the upper limit is based on a highly constrained model. The first neighbor oxygen–oxygen distance is found consistently at a value of 2.87 Å. The coordination number ranges between 4 (unconstrained) and 7 (constrained), consistent with the neutron scattering result. Improvements of the counting statistics as possible with the current experimental state of the art could reduce this uncertainty.

A later XRS experiment on ice and liquid water was performed by Bergmann et al. (2002) at the multicrystal analyzer instrument, described in Section 3.3. The total energy resolution was 1 eV. The near-edge part of the XRS spectra of ice and liquid water are shown in Fig. 3.21. The spectra exhibit three features, a pre-edge structure at 535 eV, the main absorption edge (537 eV) and a broad continuum structure around 541 eV. The intensity of these three features is clearly different when comparing ice with liquid water. The pre-edge structure is more emphasized in the liquid water, the broad continuum around 541 eV much stronger in the case of ice. Density functional calculations of Myneni et al. (2002) done in order to interpret their SXAS measurements on ice and liquid water have clearly demonstrated how these differences can be traced back to different contributions of the various types of local structures of hydrogen bonding as shown in Fig. 3.19. The result of these calculations on many different structures as generated in the molecular dynamic simulations are shown in Fig. 3.22, here confronted with the SXAS data, which are taken with a much

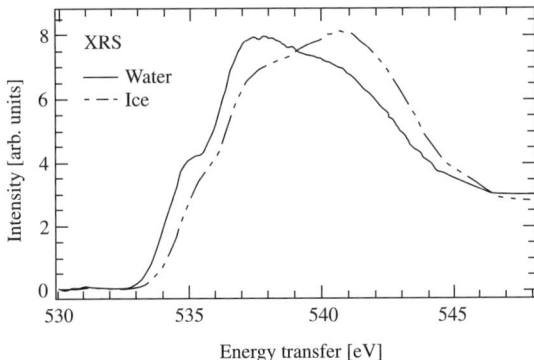

Fig. 3.21. XRS spectra of the oxygen K-edge of liquid water (solid line) and ice (dashed line). (Reprinted with permission from Bergmann *et al.* (2002); copyright (2002) by the American Physical Society.)

better energy resolution of 0.1 eV, so that the pre-edge structure is more pronounced. It becomes transparent that it is the D-ASYM type of broken hydrogen bond which dominates the spectral shape of liquid water with the strong pre-edge structure and a more accentuated main edge. On the contrary, the contribution of the A-ASYM type differs hardly from that of the SYM type, which constitutes crystalline ice. Thus by removing the ice features around 541 eV from the liquid spectrum without introducing any strong negative features in the difference spectrum a limit of 20–40% is found of how much of the ice spectrum can be subtracted from that of the liquid. The remaining 60–80% can then be assigned to the D-ASYM species. Assuming that the number of broken H-bonds at the donor and acceptor sites should be equal, we end up with a number of 1.2–1.4 broken H-bonds per molecule, or an average number of H-bonds per molecule as 2.4–2.8, much lower than found in previous molecular dynamic simulations (Luzar and Chandler 1996, Soper *et al.* 1997). Bergmann *et al.* (2002) arrive at the same estimate with their XRS measurements. Thus one may ask which additional information XRS can contribute. In this case it is mainly the control to what extent SXAS is able to yield bulk information and not to be influenced by other limitations of this method. Bergmann *et al.* (2002) could verify, within the limits of their reduced energy resolution, good agreement between the results of both methods, certainly with one exception. The liquid water spectra of SXAS, after being convoluted with the XRS energy resolution, differ from the XRS spectra above 230 eV. It is probably a saturation effect connected with the fluorescence yield mode of the SXAS liquid water measurement, which has produced this difference. This difference does not appear with the ice measurement, where partial electron yield mode of SXAS is used. This XRS study on water has revealed the great potential of XRS to determine the structure of water under extreme condition such as high temperature and high pressure up to the supercritical regime.

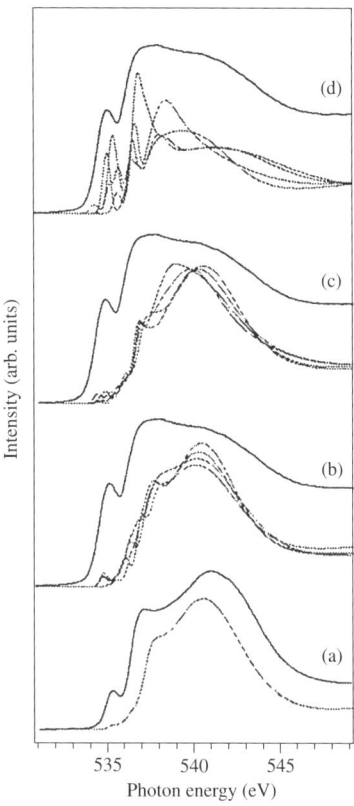

Fig. 3.22. Calculated soft X-ray absorption spectra of (a) ice and of the three typical hydrogen-bond structures as shown in Fig. 3.19 (b) SYM, (c) A-ASYM, and (d) D-ASYM. For each of the liquid species, four spectra corresponding to typical variations of distances and angles are shown. The experimental spectra of water and ice, respectively, are shown as solid lines above each set of calculated spectra. (Reprinted with permission from Myneni *et al.* (2002); copyright (2001) by IOP Publishing Ltd.)

An extension of the above studies on the H-bonds in water was provided by Wernet *et al.* (2004) where again both SXAS and XRS measurements were used. Besides near O-K-edge spectra of bulk ice Ih (SXAS) and of bulk liquid water at 25°C and 90°C (XRS), also ice Ih topmost surface (Auger) spectra (XAS) and spectra of an NH3-terminated first half bilayer of an ice Ih surface are compared, as shown in Figure 3.23. The difference between the bulk liquid spectra of 90°C water minus 25°C water (magnified by a factor of 10), depicted in Fig. 3.23(f), exhibits the same principal shape with the same isosbestic point at 538.8 eV as the difference between the spectra of 25°C water minus bulk ice, indicating that the heating of liquid water causes similar types of changes as

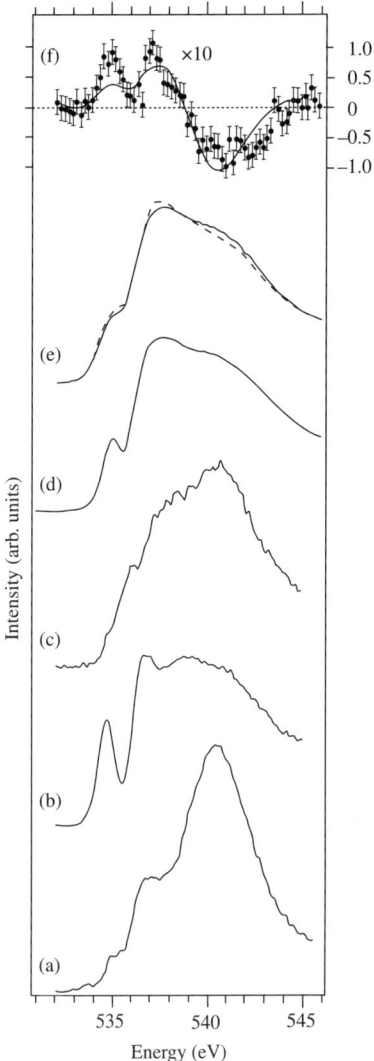

Fig. 3.23. Experimental near-edge spectra (a) bulk ice Ih (SXAS secondary electron yield); (b) ice Ih surface (topmost surface layer, SXAS Auger electron yield); (c) NH$_3$-terminated first half bilayer of ice Ih surface (SXAS Auger electron yield); (d) liquid water at ambient condition (SXAS fluorescence yield); (e) bulk liquid water at 25°C (solid line), and 90°C (dashed line) (XRS spectra normalized to the same area); (f) difference spectra: 25°C water minus bulk ice (solid curve) and 95°C water minus 25°C water (circles with error bars). The latter difference spectrum is multiplied by a factor of 10. (Reprinted with permission from Wernet *et al.* (2004); copyright (2004) by the American Association for the Advancement of Science.)

those connected with the ice–liquid phase transition, but of about one-tenth the magnitude, namely: increase of configurations with one uncoordinated or weakly H-bonded O-H-group replacing tetrahedral ones. A detailed cluster calculation based on density functional theory supported these phenomenological findings. A cluster of 11 water molecules is considered, with a central molecule surrounded by the first coordination shell (four molecules) plus part of the second shell (six molecules saturating the dangling O-H groups of the first shell). In order to simulate "broken" or weakened donor H-bonds, distortions of the O-O distances r_i (in Å) and angles θ_i (in degrees) of the nearest neighbors ($i = 1, 2$) on the H-sides have been systematically varied as given for each spectrum in Fig. 3.24a. It turned out that donor H-bond distortions of a certain degree manifest as a distinct pre-edge peak and an intense main edge in the oxygen K-edge spectra (see zone B in Fig. 3.24b), as found experimentally in the cases of liquid water and ice surfaces (b, d and e in Fig. 3.23). It is interesting that varying r at constant θ and varying θ at constant r yield similar spectra, so that it seems to be justified to speak of broken H-bonds when crossing the boundary between A and B zones in Fig. 3.24b. A more detailed analysis shows that breaking one donor H-bond while keeping the other intact entails s-p rehybridization in the orbitals close to 535 eV (Cavalleri *et al.* 2002). This leads to an increase of the p-character and translates intensity into the pre-edge region of the spectra due to the dipole selection rule.

Cai *et al.* (2005) were the first to extend XRS investigations on the ordering of hydrogen bonds to much better energy resolution of 170 meV by using th IXS installation of the BL 12XU at SPring-8, however, with rather low statistical accuracy. Moreover, these investigations on water, ice II, ice III, and ice IX were performed both at ambient pressure and 0.25 GPa, and in the case of the ice modifications down to 4 K. The behavior of the pre-edge, the main edge and the post edge of the O K XRS spectra of water and the various ice modifications corresponds, in agreement with the above results, to the expected variation of the hydrogen bond ordering. New is the statement of the authors that the spectral details coupled with density functional theory calculations indicate that the Madelung potential of the proton ordered lattice in ice II and ice IX contributes also to the pre-edge intensity.

Näslund *et al.* (2005) have studied in a combined SXAS and XRS investigation the ion solvation effect on the bulk hydrogen bonding structure of water. They showed that the hydrogen bond network in bulk water, in terms of forming and breaking hydrogen bonds, remains unchanged, and only the water molecules in the close vicinity to the ions are affected.

3.7.2 *Characterization of carbon species in fused-ring hydrocarbons and polymerized fullerenes (C K-edge)*

Whereas the carbon sites in graphite, diamond, benzene and C_{60} fullerenes are all equivalent, fused-ring hydrocarbons as well as polymerized fullerenes

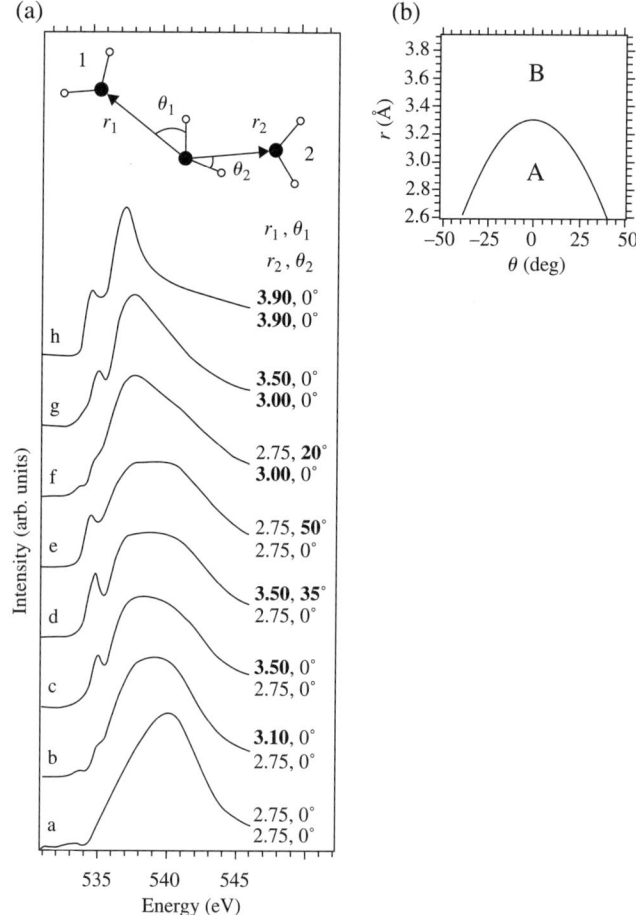

Fig. 3.24. (a) Calculated spectra for a cluster of 11 water molecules with O-O distances r_i (in Å) and angles θ_i (in degrees) of the nearest neighbors $i = 1, 2$ on the H-sides systematically varied as given for each spectrum. (b) The systematic spectral changes allow for a definition of zones A and B to denote the positions of the two nearest-neighbor H-bond acceptor molecules. Spectra show similar features for positions within one zone. Compare, e.g. (a) and (b) or (c) to (e) (Reprinted with permission from Wernet et al. (2004); copyright (2004) by the American Association for the Advancement of Science.)

exhibit inequivalent carbon sites, which must leave behind footprints in the near-edge region of carbon K-edge XRS spectra. This was demonstrated in an XRS study on polymerized C_{60} fullerenes by Rueff et al. (2002) at the ESRF inelastic scattering beamline ID16/ID28, the setup of which has already been described in Section 3.6. The experimental result of this study is shown in Figure 3.25. Along with the XRS spectra of a 1D and a 2D polymerized C_{60} also the spectrum of a pure C_{60}, the spectrum of graphite with $\mathbf{q} \parallel c$ and $\mathbf{q} \perp c$, and of diamond are presented. One must have in mind that the first strong peak of the $\mathbf{q} \parallel c$ graphite spectrum must be attributed to transitions into unoccupied π^*-states in a sp^2-carbon network, whereas the first peak of the diamond spectrum has to be ascribed to transitions into σ^* states in a network of sp^3-hybridized bonds. Then a preliminary conclusion can be drawn: a C_{60} molecule must essentially be composed of an sp^2-carbon network comparable with that of graphite leading to the structures A, B and C in the C_{60} spectrum of Fig. 3.25, although the cage bending results in a certain admixture of sp^3-character, represented by the features D and C. Going from pure C_{60} to the polymerized C_{60} fullerenes, the intensity of the structures A and B are clearly decreasing, the structure C vanishes completely. The features denoted

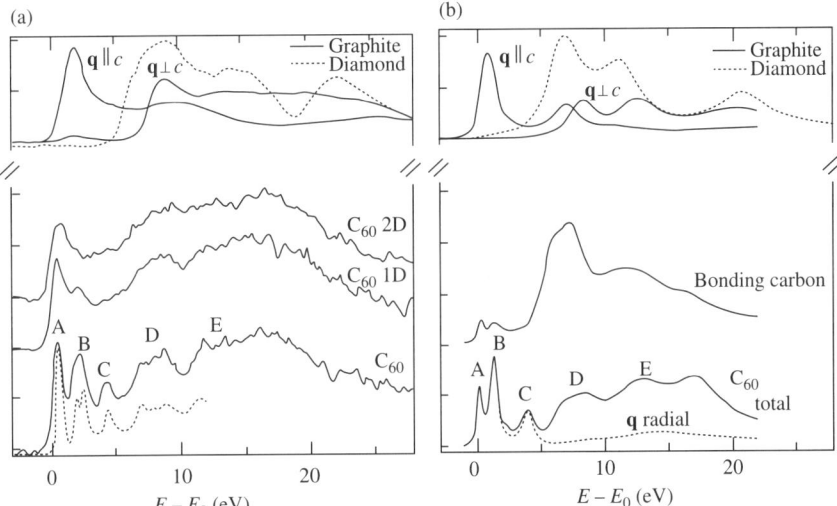

Fig. 3.25. (a) Lower panel: experimental XRS spectra of pure C_{60}, 1D polymerized C_{60}, and 2D polymerized C_{60}. Dotted line: soft X-ray absorption spectrum. Upper panel: for comparison: Solid lines: XRS spectra of graphite with $\mathbf{q} \parallel c$-axis and $\mathbf{q} \perp c$-axis; dotted line: XRS spectra of diamond. (b) Lower panel: solid lines: cluster calculated XRS spectrum of C_{60} (18 C atoms), and cluster calculated XRS spectrum to simulate a C_{60} dimer by choosing the most distorted C atom (denoted C* in Fig. 3.26), which makes the bond in the dimer, to lodge the core-hole. Dotted line: the radial $\mathbf{q}$-component of the total C_{60} spectrum. Upper panel: for comparison: solid lines: calculated XRS spectra of graphite wit $\mathbf{q} \parallel c$-axis and $\mathbf{q} \perp c$-axis; dotted line: XRS spectra of diamond. (Reprinted with permission from Rueff et al. (2002); copyright (2002) by IOP Publishing Ltd.)

D and E, however, remain with the same intensity. This can be understood on the basis of inequivalent C sites in the polymers as shown in the right-hand panel of Fig. 3.25. Here calculations are presented performed on the basis of a finite-difference method (Joly 2001), which allows us to use a non-muffin-tin potential, a key point when the near-edge fine structure in systems of low symmetry is calculated. The influence of the core hole is taken into account by complete screening of the hole in the $Z+1$ approximation. The simulation of the C_{60} molecule was performed on a cluster of 4 Å radius containing 18 atoms. The calculated spectrum for the pure C_{60} is in fairly good agreement with experiment. So do the calculated graphite and diamond spectra. The changes due to polymerization have been simulated by using a C_{60} dimer molecule as depicted in Fig. 3.26. The most distorted atom C^*, which makes the bond to the adjacent C_{60} molecule was chosen to lodge the core hole. Note that both molecules are distorted. The spectrum calculated for this atom with the same 4 Å cluster radius is also shown in Fig. 3.25. The features A and B are strongly suppressed, the feature C is completely absent. In general this spectrum resembles that of diamond. This tells us that the atom participating in the dimerization is approaching a sp^3-configuration. Of course the spectrum retains also some aspects of the original C_{60} molecule. The atoms not participating in the interconnection will be intermediate between the C^* one and that of the original C_{60}, so that one can understand, on this basis, the changes occuring upon polymerization.

Another example of the role of inequivalent C sites has been presented by Gordon *et al.* (2003) in an XRS investigation on fused-ring aromatic molecules performed at the APS 18ID beamline using the multicrystal analyzer as already described in Section 3.3. Figure 3.27 presents the inelastic scattering spectra (XRS) near the C K-edge of benzene, naphtalene, anthrazene, tiphenylene, and 1,2-benzanthracene taken at the solid phase of these fused-ring aromatic compounds, together with the corresponding inner-shell electron energy loss

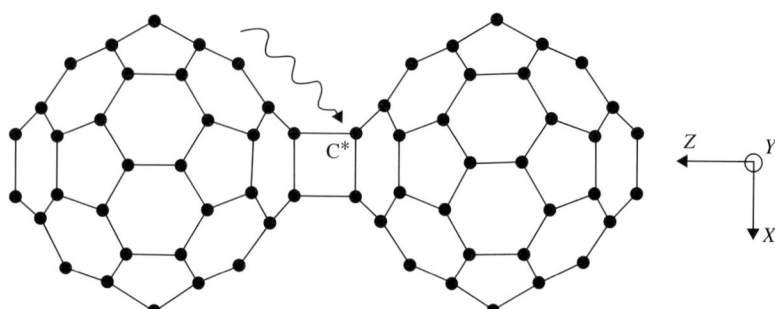

Fig. 3.26. C_{60} dimer. The bonding C atom is marked C^*. (Reprinted with permission from Rueff *et al.* (2002); copyright (2002) by IOP Publishing Ltd.)

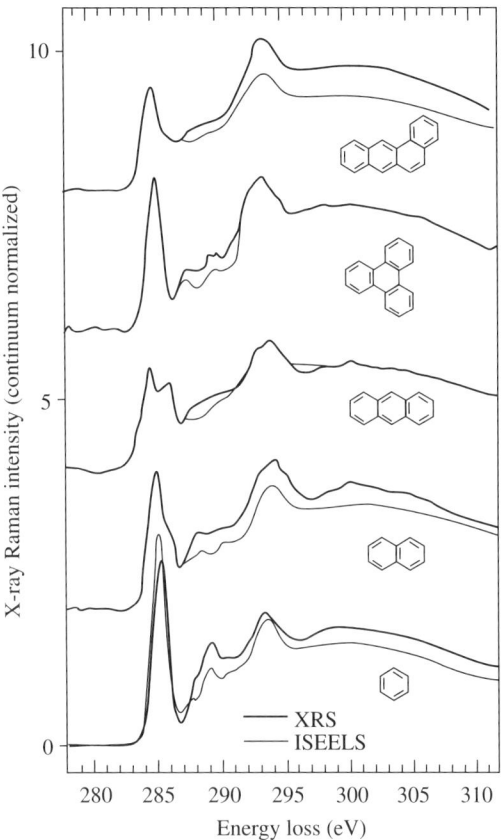

Fig. 3.27. Thick lines: XRS spectra at the C K-edge of benzene, naphthalene, anthracene, triphenyl, and 1,2 benzanthrazene. Thin lines: inner-shell EELS spectra of the same species, broadened to match the width of the lowest-energy π^*-band. (Reprinted with permission from Gordon et al. (2003); copyright (2003) by the American Institute of Physics.)

spectra (ISEELS) measured in the gas phase, broadened by smoothing to match the lowest energy π^*-band. Apart from slight differences in the range between 287 and 290 eV, possibly due to quenching of Rydberg transitions in the condensed phase, the spectra are in good agreement with respect to the clear differences between the various compounds. It was shown by Gordon et al. (2003) on the basis of the GSCF3 (Gaussian self-consistent field version 3) code (Kosugi 1987), which is designed specifically for inner-shell excitation calculations, that these differences can be traced back to the effect of different C sites in these fused-ring aromatic molecules. They calculated separately the contribution of the different sites to the total spectrum and were able to assign specific structures of the spectra to the symmetry of differing C sites.

3.7.3 Positional selective investigation of the LiC_6 band structure (Li K-edge)

A special possibility for investigating band structure peculiarities positionally selective is offered by XRS, when a low-Z atomic species is embedded into a lattice consisting of other atoms, without disturbing too much the original coordination of bounds. This can be the case with the so-called intercalation compounds, whenever the intercalation of the guest atom occurs between only loosely bound atomic layers, for instance with the alkali intercalation compounds of graphite. This way the near-edge RXS spectrum of the intercalated alkali atom, for instance Li, will act as a probe of unoccupied states centered at the Li position, since these have a maximum overlap with the Li core states. Indeed those unoccupied states centered between the van der Waals bound graphite layers, the so-called interlayer states, were a big matter of debate in the mid-1980s. Their energy position and dispersion could not unambiguously be determined by those methods such as inverse photoemission (Fauster et al. 1983, Schäfer et al. 1987), constant final state spectroscopy (Law et al. 1985) and angular resolved secondary-electron emission spectroscopy (Maeda et al. 1988). Moreover, also theoretical calculations of Tatar and Rabii (1982) on the one hand and those of Holzwarth et al. (1978) on the other found different energy positions for the interlayer bands.

Schülke et al. (1991) have measured the XRS spectrum near the Li K-edge on LiC_6 (prepared on the basis of HOPG) with $\mathbf{q} \parallel c$ and $\mathbf{q} \perp c$-axis employing, at HASYLAB/DESY, a Ge (311) double-crystal monochromator and a Si(444) spherically bent analyzer, utilizing dispersion compensation and running the spectrometer in inverse geometry. Additionally a pure HOPG XRS spectrum was measured for the two $\mathbf{q}$ orientations. In order to isolate the Li-core part of the spectrum, the valence contributions of C and Li must be subtracted. This could be achieved by fitting an exponential of the form

$$f(\omega) = A \exp(B\omega^2 + C\omega + D) \tag{3.43}$$

to the pure HOPG spectrum between 53 and 90 eV. Then the LiC_6 spectrum was scaled to this exponential between 53 and 55 eV, i.e. in the pre-edge range, in order to obtain the Li core contribution by subtracting the exponential from the LiC_6 spectrum between 53 and 90. The XRS spectra of the Li core contribution in LiC_6 for $\mathbf{q} \parallel c$ and $\mathbf{q} \perp c$-axis, respectively, are shown in the upper panel of Fig. 3.28 together with a spectrum of Li metal, for sake of comparison. The relation to special features of the LiC_6 band structure, shown in the lower panel is marked by encircled numbers. A section of the LiC_6 band structure calculated by Holzwarth et al. (1978) is presented in the lower panel, where the energy scale was fitted to the energy-loss scale of the measurement by fixing the Fermi energy E_F at 57.1 eV, according to the photoemission result of Wertheim et al. (1980). π-type bands are dashed, σ-type bands (including the interlayer states) are solid. Bands of special importance for the comparison with the experiment are set off. Thus features in the spectra can be correlated with characteristics of the band structure: The small peak, marked 1 in Fig. 3.28 can be attributed to the high

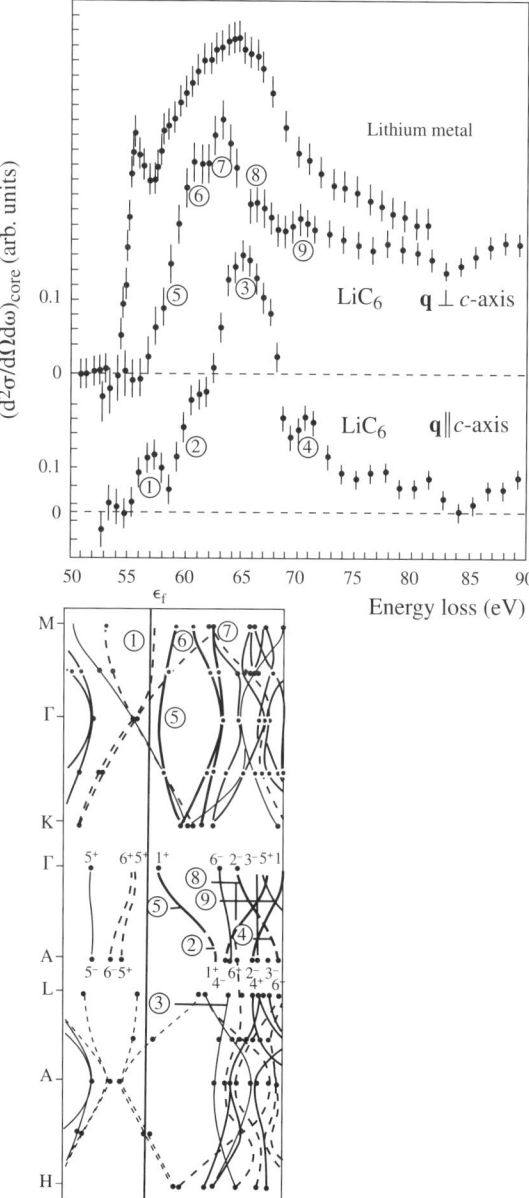

Fig. 3.28. Upper panel: XRS spectra (after subtraction of the valence contributions) at the Li K-edge of Li metal, LiC$_6$ with $\mathbf{q} \perp c$-axis and LiC$_6$ with $\mathbf{q} \parallel c$-axis. The relation to special features of the LiC$_6$ band structure shown in the lower panel is marked by encircled numbers. Lower panel: section of the calculated band structure of LiC$_6$ (Holzwarth *et al.* 1983). The energy scale of the band structure calculation has been brought to coincide with the experimental scale by fixing the Fermi energy at 57.1 eV. (Reprinted with permission from Schülke *et al.* (1991); copyright (1991) by Pergamon Press Ltd.)

DOS of bands, derived from the graphite π-states near the M-point just above the Fermi energy ($\mathbf{q} \parallel c$-axis scattering is projecting π-symmetry). The relative small intensity of this peak in spite of the very high DOS of the π-band at M is due to the rather small overlap of these states with the Li core. A much stronger overlap can be predicted for the interlayer states which first come into play with a steep rise, marked 5 in the $\mathbf{q} \perp c$-axis spectra, and which can be linked to the σ-type ($\mathbf{q} \perp c$-axis) interlayer band starting at the Γ-point, and extending its contribution also through half of the strongly dispersing ΓA band. This band is then switching to π-symmetry and therefore contributing to the strong shoulder marked 2 in the $\mathbf{q} \parallel c$-axis spectrum. The structure marked 3 has to be attributed to the high DOS of π-type interlayer states near the L-point, again emphasized by the strong spatial overlap with the Li core. For the same reason the double peak marked 6 and 7 in the $\mathbf{q} \perp c$-axis spectra must be attributed to the 2 eV splitting of the high DOS part of the σ-type interlayer band at the M-point. The dip between the peak marked 7 and the shoulder marked 8 reflects the 2 eV wide gap of the backfolded σ-type interlayer band at the Γ-point. It must be stressed that in the energy loss range 5 eV above the Fermi level, the energy position of the calculated bands agrees within 0.5 eV with the position of the corresponding structures of the spectra. Above that range, the experimental structures are always at somewhat higher energies with increasing tendency. Possibly this can be traced back to correlation effects. Nevertheless, this study can be regarded as a convincing example of how XRS can help with both its site and symmetry selectivity to test band structure calculations in a very specific manner.

3.8 High-pressure XRS experiments

As already emphasized several times, XRS gains its superiority over many other methods as SXAS, EELS, inverse photoemission, etc. by the high penetration capability of the hard X-rays used. This characteristic property of XRS is decisive, when investigation of electronic excitations on samples under high pressure, more precisely, on those encapsulated in a high-pressure cell are demanded. Such a situation arises in the case of solid hcp ^{4}He, since the remarkable quantum nature of the atomic constituents allows solidification only under externally applied pressure, even at very low temperatures. Thus, at the time, when the first XRS experiment on solid hcp ^{4}He was performed by Schell et al. (1995), no experimental data on the optical and electronic properties of solid helium were available. The only method successfully applied to obtain information about the dynamics of solid helium in its various modifications has been neutron scattering. The lack of experimental data about electronic properties must be considered to be regrettable, because solid helium is a case amenable to essentially first-principle theoretical studies. Indirect knowledge of electronic properties of solid helium could be obtained by extrapolating the results of measurements on atoms utilizing the relations between atomic and bulk properties in other noble gases, where one presumes that these relations are transferable to solid helium. Those investigations have shown that solid state effects are noticeable at the onset of transitions from inner shells, and as weak structures superimposed upon the absorption continuum.

The first XRS experiment on solid hcp ^{4}He (Schell *et al.* 1995) was performed at the HASYLB/DESY inelastic scattering beamline equipped with a Si(511) double crystal sagittally focusing monochromator and a Si(12 0 0) spherically bent analyzer utilizing dispersion compensating. The spectrometer was working in inverse geometry. A liquid helium flow cryostat with a high-pressure polycrystalline Be cell (10 mm height, 2.1 mm outer and 1.2 mm inner diameter) inside was installed at the sample position. The helium crystal was grown within this Be cell by pressurizing it through a stainless steel capillary to 61.5 MPa, where the pressure was generated by a thermal pump. A temperature gradient along the cell axis could be maintained. When the whole cell was filled with solid, the cell was sealed, so that constant volume conditions could be kept. At 61.5 MPa, ^{4}He solidifies to the hcp modification with the *c*-axis perpendicular to the long axis of the cell, which was controlled by the (002) Bragg reflection. The XRS measurements were performed at 4.2 K. The molar volume at the freezing temperature of 10.2 ± 0.1 K was 13.5 ± 0.1 cm^3/mol. Three different *q*-values were chosen, $q = 0.45$, 0.97 and 1.24 a.u. with **q** in the direction of the *c*-axis. In order to obtain the pure XRS spectrum of solid helium, one has to measure the contribution of the empty cell. This contribution was scaled between 5 and 15 eV energy loss to the measurement with the helium crystal. Since the lowest free atomic excitation $1^1S_0 \rightarrow 2^3S_1$ is at 20.6 eV, one can be sure that it is extremely unlikely to have contributions to the solid He spectrum at energy losses smaller than 15 eV. Integrated over the whole measured energy loss range, the contribution of the He crystal to the whole scattering spectrum was less than 2%. Nevertheless, a dominant structure was clearly visible in all three pure He spectra, shown exemplarily for $q = 0.45$ a.u. in Fig. 3.29, namely a sharp peak, the average position of which at 21.9 ± 0.3 eV is, within the experimental error, independent of the *q*-value. Other oscillations are more or less within the statistical noise. The authors assign the sharp peak to a Frenkel-type exciton of atomic parentage related to the free atomic $1^1S_0 \rightarrow 2^1P_1$ transition, which is at 21.2 eV and has been found in liquid helium by Surko *et al.* (1969) at 21.6 eV. This shift to higher energies was interpreted to be due to the large extension of the excited-state wavefunction, which gives rise to a significant overlap with the ground-state wavefunction of the nearest neighbors. This interpretation found confirmation by fluorescence excitation spectroscopy of Joppien *et al.* (1993) on He clusters and/or droplets, where the shift to higher excitation energies was found with increasing number of atoms in the cluster.

A second XRS measurement on solid helium was performed by Arms *et al.* (2001) at the 15ID beamline of the APS employing a high heat load (111) diamond double-crystal monochromator and a 1 m spherically bent Si(555) analyzer. The overall energy resolution was 1.1 eV. The hcp ^{4}He crystal was again grown within a Be cell, but now under a pressure of 211.15 MPa and a freezing

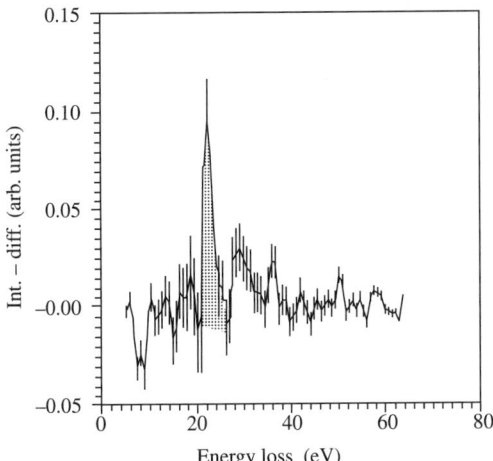

Fig. 3.29. Experimental XRS at the He K-edge of solid hcp ^{4}He at 61.5 MPa and 4.2 K with $q = 0.45$ a.u. after subtraction of the Be-cell contribution. (Reprinted with permission from Schell *et al.* (1995); copyright (1995) by the American Physical Society.)

temperature of 22.335 ± 0.005 K, so that the molar volume was significantly smaller than in the first measurement, namely 10.720 cm^3/mol. The measuring temperature was 14.0 K. Again the XRS measurements was done with $\mathbf{q} \parallel c$-axis, the q-values were 1.01, 1.37 and 1.59 a.u., corresponding to 1.5, 2.1 and 2.3 in units of $2\pi/c$ (c = lattice parameter in c-direction). The other steps of data processing were very similar to those of the first measurement. As in this first measurement, the most prominent structure of the measured XRS spectra was an excitonic peak, now found at an energy loss ~ 1 eV larger than in the former experiment. This must be traced back to the significantly higher density. Due to the much better statistical accuracy attainable with a third generation synchrotron radiation source as the APS, a clear indication for dispersion of the excitonic peak was found, shown in Fig. 3.30. Also plotted are the band bending of the valence band and of the conduction band in the (002) periodic zone, as calculated by the authors, where the minimum band separation at the Γ-point was 31 eV. This would mean that the exciton binding energy is 8.4 eV at Γ.

Another high-pressure application of XRS was reported by Meng *et al.* (2004), who have investigated the boron and nitrogen near K-edge of BN under high pressure. It is well known that hexagonal graphite-like boron nitride (h-BN) can, under high pressure, directly be transformed to a hexagonal close-packed polymorph (w-BN) that can be partially quenched after releasing pressure. The spectra of the high-pressure XRS study on w-BN demonstrate clearly the conversion process of boron and nitrogen sp(2)- and p-bonding to sp(3) bonding with its specific directional nature. In combination with *in situ* X-ray diffraction,

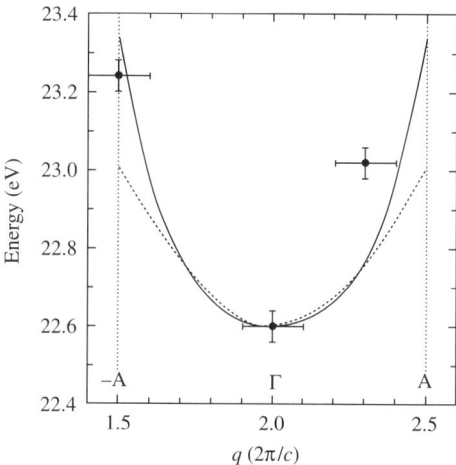

Fig. 3.30. The measured dispersion of the excitonic excitation in solid ^{4}He together with the calculated band bending for the valence band (dashed line) and the conduction band (solid line) in the periodic zone scheme. (Reprinted with permission from Arms et al. (2001); copyright (2001) by the American Physical Society.)

the mechanism of the structure transformation could be clarified. This archetypal example opens a wide field of research on high-pressure bonding evaluation of boron and nitrogen, where each of these two elements and their respective compounds have shown a wealth of fascinating pressure induced phenomena.

3.9 Coincidence experiments

In all cases presented so far, the core excitation spectra were obtained in pure form only after subtraction of the valence part, or of the contribution of other closed shells. We have seen that such a "purification" can be connected with rather risky extrapolation. But it was first shown by Fukamachi and Hosoya (1972) and later by Namikawa and Hosoya (1984) that, at least in principle, the contribution to the XRS spectrum of a deep lying core state can be obtained in pure form by utilizing coincidence techniques, more precisely, by measuring and energy analyzing the photons scattered as a result of exciting a core electron in coincidence with the fluorescence photon emitted subsequently, when the core hole is filled. The first measurements of Namikawa and Hosoya (1984) performed on Cu still using a ^{241}Am γ-source were suffering from false coincidences, which created structures in the Cu K-edge XRS, partly misinterpreted by the authors as discussed by Manninen (1986). Later investigations on Nb (Hämäläinen et al. 1990), on Cu (Laukkanen et al. 1996), and on Ag (Laukkanen et al. 1998) have largely avoided false coincidences and have carefully corrected the spectra with respect to random coincidences. The latter appear when an inelastically scattered photon and a fluorescence photon is simultaneously but accidentally observed, if the inelastic scattering event and a fluorescence detection following (i) an independent photoabsorption process or (ii) an independent inelastic scattering process takes place within the time window of the coincidence setup. The arrangement of the scattering sample, of the Ge solid state detector energy-analyzing the scattered photons, and of the NaJ detector counting the fluorescence photons is shown in Fig. 3.31. Note the special shielding of both detectors in order to avoid false coincidence events, which can appear if the scattering in the sample is quasielastic and inelastic scattering takes place in the NaJ detector. If direct scattering from one detector to another were not prevented by appropriate shielding, this would produce a final photon spectrum centered at the strong valence electron Compton peak belonging to the scattering angle of this cross-talk event and recoil electrons counted by the fluorescence detector. The primary photons were either taken from monochromatized synchrotron radiation (Hämäläinen et al 1990, Laukkanen et al. 1998) or from monochromatized characteristic W $K\alpha$ radiation (Laukkanen et al. 1996). A typical raw coincidence spectrum scattered at 90° on Cu is shown in Fig. 3.32. The spectrum comprises both true and random coincidences. Additionally a spectrum of exclusively random coincidences is shown, which has been produced in the following way. The signal of one counting chain is delayed in such a way that the true coincidences no longer give any contribution. By using two separate multichannel analyzers one can measure

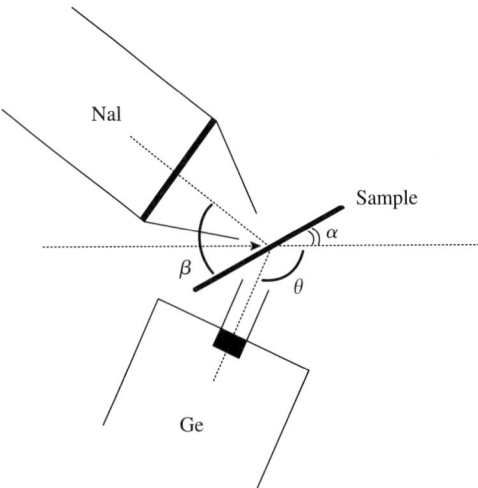

Fig. 3.31. Arrangement of the scattering sample, of the Ge solid state detector, and of the NaJ detector in a core-excitation coincidence experiment. (Reprinted with permission from Laukkannen et al. (1996); copyright (1996) by IOP Publishing Ltd.)

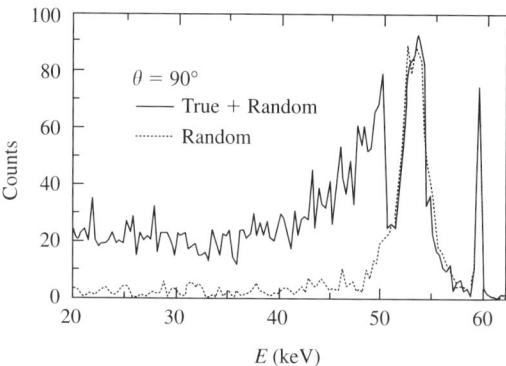

Fig. 3.32. Solid line: raw coincidence spectrum scattered at 90° on Cu. The spectrum comprises both true and random coincidences. Dotted line: spectrum of exclusively random coincidences. (Reprinted with permission from Laukkannen *et al.* (1996); copyright (1996) by IOP Publishing Ltd.)

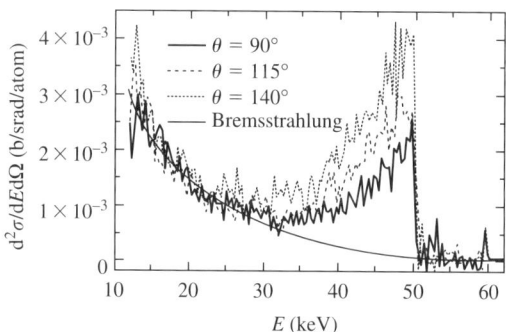

Fig. 3.33. Contribution of the Cu 1s electrons to the double differential scattering cross-section in absolute units as obtained by means of core excitation coincidence measurements for three different values of the scattering angle. The thin solid line is the contributions of bremsstrahlung produced by photoelectrons. (Reprinted with permission from Laukkannen *et al.* (1996); copyright (1996) by IOP Publishing Ltd.)

the true+random and the random coincidences simultaneously. Of course, events which lead to random coincidences of the type described above will happen most frequently in the energy-loss range, where the valence electron Compton profile peaks. Due to the rather small scattering angle of $\theta = 90°$, this range stood clearly out against the K-shell contribution. It is not the aim of those XRS measurements with coincidence technique to get information about the near-edge fine structure. For that reason the energy resolution of a solid state detector, which is a few hundred eV, is much too bad. Measurements of absolute values of the energy-dependent double differential scattering cross-section, however, are a worthwhile object, and Fig. 3.33 presents the result of such a measurement in absolute units on Cu by Laukkanen et al. (1996) after subtraction of the random coincidences and the absorption correction. Nevertheless, there is still a contribution in the spectra, which is clearly not due to inelastic scattering from 1s electrons. The origin of this low-energy tail is the bremsstrahlung produced by the photoelectrons in the sample. These events are detected as true coincidences, since the K-shell fluorescence is created after the photoabsorption. A computational fit taking into account all aspects of this bremsstrahlung component was calculated and subtracted from the experimental data as seen in Fig. 3.33. The motivation for those measurements can be understood in the context of the question of how good the so-called impulse approximation works, which is the basis for interpreting Compton data in terms of electron momentum distribution (See Chapter 4).

3.10 References

Arimitsu, N., Y. Kobayashi, and Y. Mizuno (1987). *J. Phys. Soc. Jpn.* **56** 2940

Arms, D.A., R.O. Simmons, M. Schwoerer-Böhning, A.T. Macrander, and T.J. Graber (2001). *Phys. Rev. Lett.* **87** 156402

Ausmann, G.A. and A.J. Glick (1969). *Phys. Rev.* **183** 687

Bergmann, U. and S.P. Cramer (1998). *SPIE-The International Society for Optical Engineering* (Society of Photo-Optical Instrumentation Engineers, San Diego, California) Vol 3448, p. 198

Bergmann, U., P. Glatzel, and S. P. Cramer (2002). *Microchem. J.* **71** 221

Bergmann, U., Ph. Wernet, P. Glatzel, M. Cavalleri, L.G.M. Petterson, A. Nilsson, and S.P. Cramer (2002). *Phys. Rev. B* **66** 092107

Biggs, F., L.B. Mendelsohn, and J.B. Mann (1975). *At. Data Nucl. Data Tables* **16** 201

Bowron, D.T., H.M. Krisch, A.C. Barnes, J.L. Finney, A. Kaprolat, and M. Lorenzen (2000). *Phys. Rev. B* **62** R9223

Cai, Y.Q., H.-K. Mao, P.C. Chow, J.S. Tse, Y. Ma, S. Patchkovskii, J.F. Shu, V. Stuzhkin, R.J. Hemley, H. Ishii, C.C. Chen, I. Jarrige, C.T. Chen, S.R. Shieh, E.P. Huang, and C.-C. Kao (2005). *Phys. Rev. Lett.* **94** 025502

Cavalleri, M., H. Ogasawara, L.G.M. Pettersson, and A. Nilsson (2002). *Chem. Phys. Lett.* **364** 363

Citrin, P.H., G.K. Wertheim, and Y. Bear (1977). *Phys. Rev. B* **20** 3067

Citrin, P.H., G.K. Wertheim, and M. Schlüter (1979). *Phys. Rev. B* **20** 3067

Comelli, G., J. Stohr, W. Jark, and B.B. Pate (1988). *Phys. Rev. B* **37** 4383

Doniach, S., P.M. Platzman, and J.T. Yue (1971). *Phys. Rev. B* **4** 3345

Eisenberger, P. and P.M. Platzman (1970). *Phys. Rev. A* **2** 415

Fauster, Th., F.J. Himpsel, J.E. Fischer, and E.W. Plummer (1983). *Phys. Rev. Lett.* **51** 430

References

Feng, Y., G.T. Seidler, J.O. Cross, A.T. Macrander, and J.J. Rehr (2004). *Phys. Rev. B* **69** 125402

Filipponi, A. (1994). *J. Phys.: Condens. Matter* **6** 8415

Filipponi, A. and A. Di Cicco (1995). *Phys. Rev. B* **52** 15135

Fukamachi, T. and S. Hosoya (1972). *Phys. Lett.* **38A** 341

Girvin, S.M. and J.J. Hopfield (1976). *Phys. Rev. Lett.* **37** 1091

Gordon, M.L., D. Tulumello, G. Cooper, A.P. Hitchcock, P. Glatzel, O.C. Mullins, S.P. Cramer, and U. Bergmann (2003). *J. Phys. Chem. A* **107** 8512

Haensel, R., G. Keitel, B. Sonntag, C. Kunz, and P. Schreiber (1970). *Phys. Sat. Sol. (a)* **2** 85

Hämäläinen, K., S. Manninen, and J.R. Schneider (1990). *Nucl. Instrum. and Methods, A* **297** 526

Hämäläinen, K., S. Galambosi, J.A. Soininen, E. L. Shirley, J.-P. Rueff, and A. Shukla (2002). *Phys. Rev. B* **65** 155111

Holzwarth, N.A.W., S. Rabii and L.A. Girifalco (1978). *Phys. Rev. B* **18** 5190

Joly, Y. (2001). *Phys. Rev. B* **63** 125120

Joppien, M., R. Karnbach, and T. Möller (1993). *Phys. Rev. Lett.* **71** 2654

Kosugi, N. (1987). *Theor. Chim. Acta* **74** 149

Krisch, M., F. Sette, C. Masciovecchio, and R. Verbeni (1997). *Phys. Rev. Lett.* **78** 2843

Law, A.R., M.T. Johnson, H.P. Hughes, and H.A. Padmore (1985). *J. Phys. C.: Solid State Phys.* **18** L297

Laukkanen, J., K. Hämäläinen, and S. Manninen (1996). *J. Phys.: Condens. Matter* **8** 2153

Laukkanen, K. Hämäläinen, S. Manninen, and V. Honkimäki (1998). *Nucl. Instrum. and Methods A* **416** 475

Luzar, A. and D. Chandler (1996). *Phys. Rev. Lett.* **76** 928

Maeda, F., T. Taahashi, H. Ohsawa, S. Suzuki, and H. Suematsu (1988). *Phys. Rev. B* **37** 4482

Mahan, G.D. (1967). *Phys. Rev.* **163** 612

Manninen, S. (1986). *Phys. Rev. Lett.* **57** 1500

Meng, Y., H.K. Mao, P.J. Eng, T.P. Trainor, M. Newville, M.Y. Hu, C.C. Kao, J.F. Shu, D. Häusermann, and R.J. Hemley (2004). *Nature Mat.* **3** 111

Mizuno, Y. and Y. Ohmura (1967). *J. Phys. Soc. Japan*, **22** 445

Myneni, S., Y. Luo, L.-A. Näslund, M. Cavalleri, L. Ojamäe, H. Ogasawara, A. Pelmenschikov, Ph. Wernet, P. Väterlein, C. Heske, Z. Hussain, L.G.M. Pettersson, and A. Nilsson (2002). *J. Phys.: Condens. Matter* **14** L213

Näslund, L.-A., D.C. Edwards, Ph. Wernet, U. Bergmann, H. Ogasawara, L.G.M. Pettersson, S. Myneni, and A. Nilsson (2005). *J. Phys. Chem. A* **109** 5995

Nagasawa, H., S. Mourikis, and W. Schülke (1989). *J. Phys. Soc. Japan* **58** 710

Nagasawa, H., S. Mourikis, and W. Schülke (1997). *J. Phys. Soc. Japan* **66** 3139

Namikawa, K. and S. Hosoya (1984). *Phys. Rev. Lett.* **53** 1606
Nozières, P. and C. de Dominicis (1969). *Phys. Rev.* **178** 1097
Ojala, E. and M Lähdeniemi (1982). *Phys. Scr.* **25** 703
Papaconstantopoulos, D.A. (1986) *Handbook of the Bands of Elemental Solids* (Plenum Press, New York and London)
Rehr, J.J. and R.C. Albers (2000). *Rev. Mod. Phys.* **72** 621
Rehr, J.J., R.C. Albers, and S.I. Zabinsky (1992). *Phys. Rev. Lett.* **69** 3397
Ritsko, J.J., S.E. Schnatterly, and P.C. Gibbons (1974). *Phys. Rev. B* **10** 5017
Rueff, J.P., Y. Joly, F. Bartolome, M. Krisch, J.L. Hodeau, L. Marques, M. Mezouar, A. Kaprolat, M. Lorenzen, and F. Sette (2002). *J. Phys.: Condens. Matter* **14** 11635
Schell, N., R.O. Simmons, A. Kaprolat, W. Schülke, and E. Burkel (1995). *Phys. Rev. Lett.* **74** 2535
Schülke, W., H. Nagasawa, S. Mourikis, and P. Lanzki (1986). *Phys. Rev. B* **33** 6744
Schülke, W., U. Bonse, H. Nagasawa, A. Kaprolat, and A. Berthold (1988). *Phys. Rev. B* **38** 2112
Schülke, W., K.-J. Gabriel, A. Berthold, and H. Schulte-Schrepping (1991). *Solid State Commun.* **79** 657
Soininen, J.A. and E.L. Shirley (2001). *Phys. Rev. B* **54** 165112
Soininen, J.A., K. Hämäläinen, W.A. Caliebe, C.C. Kao, and E. L. Shirley (2001). *J. Phys.: Condens. Matter* **13** 8039
Soininen, J.A., A.L. Ankudinov, and J.J. Rehr (2005). *Phys. Rev. B* **72** 045136
Soper, A.K., F. Bruni, and M.A. Ricci (1997). *J. Chem. Phys.* **106** 247
Sternemann, C., M. Volmer, J.A. Soininen, H. Nagasawa, M. Paulus, H. Enkisch, G. Schmidt, M. Tolan, and W. Schülke (2003). *Phys. Rev. B* **68** 035111
Surko, C.M., G.J. Dick, F. Reif, and W.C. Walker (1969). *Phys. Rev. Lett.* **23** 842
Suzuki, T. and H. Nagasawa (1975). *J. Phys. Soc. Japan* **39** 1579
Tatar, R.C. and S. Rabi (1982). *Phys. Rev. B* **25** 4126
Teo, Boon-Teng and P.A. Lee (1979). *J. Am. Chem. Soc.* **101** 2815
Tohji, K. and Y. Udagawa (1987). *Phys. Rev. B* **36** 9410
Tohji, K. and Y. Udagawa (1989), *Phys. Rev. B* **39** 7590
Venkataraman, C. T. (private communication)
Watanabe, N., H. Hayashi, Y. Udegawa, K. Takeshita, and H. Kawata (1996). *Appl. Phys. Lett.* **69** 1370
Wernet, Ph, D. Nordlund, U. Bergmann, M. Cavalleri, M. Odelius, H. Ogasawara, L.A. Näslund, T.K. Hirsch. L. Ojamäe, P. Glatzel, L.G.M. Pettersson, and A. Nilsson (2004). *Science* **304** 995
Wertheim, G.K., P.M.Th.M. Vanattekum, and S. Basu (1980). *Solid State Commun.* **33** 1127
Wilson, K.R., B.S. Rude, and T. Catalano (2001). *J. Phys. Chem. B* **105** 3346

4
The Compton scattering regime

4.1 The impulse approximation; access to electron density in momentum space

It has already been pointed out in Chapter 1, on the basis of more intuitive arguments, and will be derived in greater detail in Chapter 6, how the measurement of the double differential inelastic X-ray scattering cross-section can yield information about a fundamental ground-state quantity, namely the electron density in momentum space. We will summarize the relevant relations, stressing the prerequisites, on which these relations are applicable, and will then deduce practicable rules for extracting the desired information on momentum densities out of the experimental data. In a first step we will do this in the nonrelativistic limit, and in a second approach relativistically. In every case the reader is referred to Chapter 6 of this book for further and more thorough derivation of the fundamental relations.

4.1.1 *The nonrelativistic limit*

The fundamental step towards the Compton regime with its chance to get ground state information starts with the general relation for the double differential scattering cross-section (DDSCS) in the nonrelativistic limit, as already given, in the so-called excitation picture, with equation (1.9), and which may once more be repeated here:

$$\frac{d^2\sigma}{d\Omega_2 d\hbar\omega_2} = r_0^2 |\mathbf{e}_1 \cdot \mathbf{e}_2^*|^2 \left(\frac{\omega_2}{\omega_1}\right) \sum_{i,f} p_i |\langle f| \sum_j e^{i\mathbf{q}\cdot\mathbf{r}_j} |i\rangle|^2 \delta(E_f - E_i - \hbar\omega), \quad (4.1)$$

where p_i is the probability of the initial state $|i\rangle$ with energy E_i, $|f\rangle$ the many-electron final state with energy E_f, $\mathbf{q}$ the momentum transfer

$$\mathbf{q} = \mathbf{K}_1 - \mathbf{K}_2 \quad (4.2)$$

$$q \approx 2K_1 \sin\left(\frac{\theta}{2}\right) \quad (4.3)$$

and $\mathbf{K}_1$, $\mathbf{e}_1$ and ω_1 the wavevector, the polarization (unit) vector and the frequency of the incident beam, respectively, and $\mathbf{K}_2$, $\mathbf{e}_2$ and ω_2 the corresponding quantities of the scattered beam, r_0 is the so-called classical electron radius, and θ the scattering angle. The summation j is over all electrons of the system. Then the Compton regime of inelastic scattering can be defined over the scattering vector **q**. If q^{-1} is much smaller then the interparticle distances, we can neglect interference between different particles contributing to the DDSCS, so that only single-particle properties, and not the collective behavior of the scattering system matters. More precisely it is the correlation between the position of the same particle at different time, which determines the squared matrix element of (4.1). Of course, this can be made more transparent when going from the excitation to the correlation picture of the DDSCS, as will be done in Chapter 6, where this step will be executed by going from (6.25) to (6.34). In order to probe single-particle **ground-state** properties a second condition must be fulfilled. The scattering process should be so fast, this means the transferred frequency $\omega = \omega_1 - \omega_2$ should be so high, that the remaining system cannot become rearranged. This is the case if $\hbar\omega$ is much larger than characteristic energies of the scattering system, i.e. the Fermi energy for solid state valence electrons or the binding energy of electrons in core states. Mathematically this last condition means that the contribution of the potential energy V to E_i and E_f must remain the same, so that V cancels in the δ-function of (4.1). The difference $E_f - E_i$ in the argument of the δ-function in (4.1) can thus be expressed in terms of the electron momentum **p** of the initial one-electron state, and of the final electron momentum after having incorporated the transferred momentum $\hbar\mathbf{q}$

$$E_f - E_i = \varepsilon(\mathbf{p} + \hbar\mathbf{q}) - \varepsilon(\mathbf{p}) = \hbar^2 q^2/2m + \hbar\mathbf{p}\cdot\mathbf{q}/m. \tag{4.4}$$

$\varepsilon(\mathbf{p})$ is the kinetic energy of an electron with momentum **p**. In Chapter 6 we will show in greater detail that the above-mentioned two conditions are the basis for what is called the impulse approximation of the DDSCS of (4.1) (Eisenberger and Platzman 1970), which reads with the Thomson cross-section

$$\left(\frac{d\sigma}{d\Omega_2}\right)_{Th} \equiv r_0^2 \left(\frac{\omega_2}{\omega_1}\right) |\mathbf{e}_1 \cdot \mathbf{e}_2^*|^2 \tag{4.5}$$

$$\frac{d^2\sigma}{d\Omega_2 d\hbar\omega_2} = \left(\frac{d\sigma}{d\Omega_2}\right)_{Th} \left(\frac{m}{\hbar q}\right) \int\int \rho(p_x, p_y, p_z = p_q) dp_x\, dp_y$$

$$= \left(\frac{d\sigma}{d\Omega_2}\right)_{Th} \left(\frac{m}{\hbar q}\right) J(p_q), \tag{4.6}$$

where we have introduced the so-called (directional) Compton profile

$$J(p_q) \equiv \int\int \rho(p_x, p_y, p_z = p_q)\, dp_x\, dp_y, \tag{4.7}$$

which we interpret as the projection of the momentum space density $\rho(\mathbf{p})$ along the direction of the scattering vector **q**, which might coincide with the z-direction of the sample. The component $p_q \equiv \mathbf{p}\cdot\mathbf{q}/q$ of the electron momentum in the direction of the momentum transfer **q**, which corresponds to the energy transfer $\hbar\omega$ according to the impulse approximated argument of the δ-function in (4.1), is given by:

$$p_q = \omega m/q - \hbar q/2 \approx mc \left(\frac{\hbar\omega_1 - \hbar\omega_2 - (\hbar^2\omega_1\omega_2/mc^2)(1-\cos\theta)}{[\hbar^2\omega_1^2 + \hbar^2\omega_2^2 - 2\hbar^2\omega_1\omega_2\cos\theta]^{1/2}} \right), \quad (4.8)$$

where $\hbar^2(\omega_1 - \omega_2)^2/2mc^2$ has been omitted.

4.1.2 *Relativistic treatment of the Compton limit*

The relativistic treatment of the Compton limit as will be used in this chapter is based on a heuristic approach as first introduced by Eisenberger and Reed (1974) restricted in its applicability to scattering angles very near to 180°, and then extended to the more general cases by Ribberfors (1975a,b). The adequacy of this approach found its confirmation by a more rigorous treatment by Holm (1988). For details of the derivation the reader again is referred to Chapter 6. Both heuristic treatments start with an expression of Jauch and Rohrlich (1955) for the total relativistic scattering cross-section σ for colliding beams, an initial photon beam, characterized by the four-vector $\kappa_1 = (\mathbf{K}_1, i\omega_1)$ and a scattering free electron beam described by the four-vector $\pi_1 = (\mathbf{p}_1, iE_1)$. After the scattering process the electrons are in the state $\pi_2 = (\mathbf{p}_2, iE_2)$ and the photons in $\kappa_2 = (\mathbf{K}_2, i\omega_2)$. Then σ reads ($\hbar = c = 1$) for the case of polarized photons

$$\sigma = m^2 r_0^2 \int d\mathbf{K}_2 d\mathbf{p}_2 \left(\frac{1}{2k_1 E_2 \omega_2} \right) X(k_1, k_2)\, \delta(\pi_1 + \kappa_1 - \pi_2 - \kappa_2), \quad (4.9)$$

where

$$k_1 = E_1\omega_1 - \mathbf{p}_1 \cdot \mathbf{K}_1 \quad (4.10)$$

$$k_2 = E_1\omega_2 - \mathbf{p}_1 \cdot \mathbf{K}_2 = k_1 + \kappa_1\kappa_2 = k_1 - \omega_1\omega_2(1-\cos\theta), \quad (4.11)$$

and

$$X(k_1, k_2) = \tfrac{1}{2}[(k_1/k_2) + (k_2/k_1)] - 1 \\ + 2(\mathbf{e}_1 \cdot \mathbf{e}_2 + ((\mathbf{e}_1 \cdot \mathbf{p}_1)(\mathbf{e}_2 \cdot \mathbf{p}_2)/k_1) - ((\mathbf{e}_2 \cdot \mathbf{p}_1)(\mathbf{e}_1 \cdot \mathbf{p}_2)/k_2))^2. \quad (4.12)$$

The key points of using (4.9) for the calculation of the relativistic DDSCS are: (i) to apply this expression for the case of scattering photons against electrons in bound states, characterized, in the spirit of the impulse approximation, by the momentum distribution $\rho(\mathbf{p}_1)$, which means that the flux factor $k_1/E_1\omega_1$ in (4.9) has to be replaced by one ($c = 1$); (ii) to assume the scattering electron system at rest, so that $\langle \rho(\mathbf{p}_1) \rangle = 0$; (iii) to introduce $d\mathbf{K}_2 = \omega_2^2 d\hbar\omega_2 d\Omega_2$ in (4.9); (iv) and to execute the integration over $\mathbf{p}_2$. If one limits oneself to isotropic momentum distributions one ends up, after a huge amount of algebra and neglecting terms of the order p_{1z}/mc, with the following relation that connects the DDSCS with the Compton profile (these relations have been written down in a manner independent of any unit system)

$$\frac{d^2\sigma}{d\hbar\omega_2 d\Omega_2} = \left(\frac{mr_0^2\omega_2}{2\omega_1 q}\right) J(p_{1z})X, \qquad (4.13)$$

where

$$X = \frac{1}{2}\left(\frac{R_1}{R_2} + \frac{R_2}{R_1}\right) - 1 + 2(\varepsilon_1 \cdot \varepsilon_2)^2, \qquad (4.14)$$

and

$$\mathbf{p}_1 \cdot \mathbf{q}/q \equiv p_{1z} = \frac{mc[(\omega_1 - \omega_2) - \hbar\omega_1\omega_2(1 - \cos\theta)/mc^2]}{[\omega_1^2 + \omega_2^2 - 2\omega_1\omega_2\cos\theta]^{1/2}} \qquad (4.15)$$

$$R_1 = \hbar\omega_1[mc^2 - (\omega_1 - \omega_2\cos\theta)p_{1z}/q] \qquad (4.16)$$

$$R_2 = R_1 - \hbar^2\omega_1\omega_2(1 - \cos\theta). \qquad (4.17)$$

It is of importance to know (Ribberfors 1975b) that the X-factor of (4.12) applies also to anisotropic systems, if the very small $1/k_1$ and $1/k_2$ terms of (4.12) are neglected.

Thus one realizes that, conceding the approximations done, the Compton profile can directly be extracted from the measured DDSCS, as in the nonrelativistic limit. One has only to replace the Thomson scattering cross-section $(d\sigma/d\Omega_2)_{\text{Th}}$ by the somewhat more complicated expression

$$\left(\frac{d\sigma}{d\Omega_2}\right)_{\text{rel}} \equiv \frac{1}{2}r_0^2\left(\frac{\omega_2}{\omega_1}\right)\left[\frac{1}{2}\left(\frac{R_1}{R_2} + \frac{R_2}{R_1}\right) - 1 + 2|\mathbf{e}_1 \cdot \mathbf{e}_2^*|^2\right], \qquad (4.18)$$

which, in the isotropic limit goes to (one has to average with respect to the two incident polarization directions and to summarize over the two polarization directions of the scattered beam)

$$\left(\frac{d\sigma}{d\Omega_2}\right)_{\text{rel}} \equiv \frac{1}{2}r_0^2\left(\frac{\omega_2}{\omega_1}\right)\left[\left(\frac{R_1}{R_2} + \frac{R_2}{R_1}\right) - \sin^2\theta\right] \qquad (4.19)$$

and approaches, in the nonrelativistic limit, for $\theta \to \pi$, the Thomson scattering cross-section.

If the incident beam is linearly polarized, one has to summarize over the two polarization directions of the scattered beam, so that one obtains

$$\left(\frac{d\sigma}{d\Omega_2}\right)_{\text{rel}} \equiv \frac{1}{2}r_0^2\left(\frac{\omega_2}{\omega_1}\right)\left[\left(\frac{R_1}{R_2} + \frac{R_2}{R_1}\right) - 2\sin^2\theta\cos^2\beta\right], \qquad (4.20)$$

where β is the angle between the incident polarization vector and the scattering plane.

4.2 The reciprocal form factor, properties and relations to ground-state quantities

When discussing the information which can be extracted from measurements of Compton profiles, very often the 3D Fourier transform of the momentum space density, the so-called reciprocal form factor (Pattison and Williams 1976, Pattison et al. 1977) is introduced:

$$B(\mathbf{r}) \equiv \int \rho(\mathbf{p}) \exp(-i\mathbf{r} \cdot \mathbf{p}/\hbar) \mathrm{d}\mathbf{p}. \tag{4.21}$$

Using (4.7) one sees immediately that the following relationship exists between the Compton profile and the reciprocal form factor, if one lets the z-axis in real space coincide with the **q**-direction:

$$B(0,0,z) = \int J(p_z) \exp(-izp_z/\hbar) \mathrm{d}p_z. \tag{4.22}$$

The z-component of the reciprocal form factor is directly related to the 1D Fourier transform of the Compton profile.

We will see later that this relation builds the basis for 3D reconstruction of the momentum space density $\rho(\mathbf{p})$ using measurements of Compton profiles with different directions of the momentum transfer **q**.

If one expresses the electron momentum density $\rho(\mathbf{p})$, within the limits of the one-electron approximation, in terms of a one-electron wavefunction $\psi(\mathbf{r})$, as

$$\rho(\mathbf{p}) = \left(\frac{1}{2\pi\hbar}\right)^3 \left| \int \psi(\mathbf{r}) \exp(-i\mathbf{p} \cdot \mathbf{r}/\hbar) \mathrm{d}\mathbf{r} \right|^2 \tag{4.23}$$

then the application of the convolution theorem of the Fourier transform results in the following expression for the reciprocal form factor (for details see Section 6.8.2)

$$B(\mathbf{r}) = \int \psi(\mathbf{r}') \psi^*(\mathbf{r}' + \mathbf{r}) \mathrm{d}\mathbf{r}', \tag{4.24}$$

which says that the reciprocal form factor is the autocorrelation of the position space wavefunction. Furthermore, if we describe one-electron properties of a many-electron system in terms of the one-particle density matrix $\Gamma_1(\mathbf{r}|\mathbf{r}')$ in position space (for details see Section 6.8.2), then the relation between the momentum space density and the one-electron density matrix reads

$$\rho(\mathbf{p}) = \left(\frac{1}{2\pi\hbar}\right)^3 \int \Gamma_1(\mathbf{r}|\mathbf{r}') \exp[-i(\mathbf{p} \cdot (\mathbf{r} - \mathbf{r}'))/\hbar] \mathrm{d}\mathbf{r} \, \mathrm{d}\mathbf{r}', \tag{4.25}$$

so that we get the following expression for the reciprocal form factor

$$B(\mathbf{r}) = \int \Gamma_1(\mathbf{r}'|\mathbf{r}' + \mathbf{r}) \mathrm{d}\mathbf{r}', \tag{4.26}$$

which is nothing else than the position space average of the corresponding nondiagonal element of the one-particle density matrix.

Two important properties of the reciprocal form factor $B(\mathbf{r})$ (Schülke 1976) should be added. They are based on the representation of a solid state valence electron system by one-electron Bloch functions $\phi_{\mathbf{k},\nu}(\mathbf{r})$ expanded into plane waves as

$$\phi_{\mathbf{k},\nu}(\mathbf{r}) = \left(\frac{1}{2\pi\hbar}\right)^{3/2} \sum_{\mathbf{g}} a_\nu(\mathbf{k}+\mathbf{g}) \exp[i(\mathbf{k}+\mathbf{g}) \cdot \mathbf{r}]. \tag{4.27}$$

($\mathbf{k}$ = Bloch wavevector, ν = band index, $\mathbf{g}$ = reciprocal lattice vector), and the corresponding diagonal occupation number density $n_{\nu\nu}(\mathbf{k})$. The momentum space density reads then (for details see Chapter 6)

$$\rho(\mathbf{p}) = \sum_\nu \sum_\mathbf{k} \sum_\mathbf{g} n_{\nu\nu}(\mathbf{k}) \, a_\nu^*(\mathbf{k}+\mathbf{g}) \, a_\nu(\mathbf{k}+\mathbf{g}) \, \delta(\mathbf{k}+\mathbf{g}-\mathbf{p}/\hbar), \tag{4.28}$$

where we have neglected the nondiagonal elements $n_{\nu\nu'}(\mathbf{k})$ of the occupation number density, which are due to mixing between different bands caused by the electron–electron interaction. The diagonal elements describe the shape of the Fermi surfaces in the case of metals.

Inserting (4.28) into (4.21) yields

$$B(\mathbf{r}) = \sum_\nu \sum_\mathbf{g} \sum_\mathbf{k} n_{\nu\nu}(\mathbf{k}) \, a_\nu^*(\mathbf{k}+\mathbf{g}) \, a_\nu(\mathbf{k}+\mathbf{g}) \exp[i(\mathbf{k}+\mathbf{g}) \cdot \mathbf{r}], \tag{4.29}$$

and by looking at the values of $B(\mathbf{r})$ at lattice translation vectors $\mathbf{R}$, one finds easily as a consequence of the normalization of the Bloch waves

$$B(\mathbf{R}) = \sum_\nu \sum_\mathbf{k} n_{\nu\nu}(\mathbf{k}) \exp(i\mathbf{k} \cdot \mathbf{R}). \tag{4.30}$$

Since for insulators $n_{\nu\nu}(\mathbf{k})$ is either 1 or 0 throughout the whole Brillouin zone, $B(\mathbf{R})$ vanishes for all $\mathbf{R}$, a very good test for the reliability of theoretical calculations of Compton profiles for insulators. In the case of metals the sum over the bands of the occupation number density in the repeated zone scheme

$$N(\mathbf{k}) \equiv \sum_\nu n_{\nu\nu}(\mathbf{k}) \tag{4.31}$$

can be reconstructed by the following Fourier series:

$$N(\mathbf{k}) = \frac{1}{N_\mathbf{R}} \sum_\mathbf{R} B(\mathbf{R}) \exp(-i\mathbf{k} \cdot \mathbf{R}), \tag{4.32}$$

where $N_\mathbf{R}$ is the number of $\mathbf{R}$-values taken into account in the Fourier series of (4.32). In the case of monovalent metals, $N(\mathbf{k})$ represents the shape of the Fermi surface, so that $B(\mathbf{R})$ plays the same role for reconstructing the occupation number density as $B(\mathbf{r})$ for the reconstruction of the momentum space density $\rho(\mathbf{p})$.

4.3 Electron exchange and correlation in Compton scattering

Until now the treatment of Compton scattering was based on the independent particle approximation, where the wavefunctions used to derive the momentum space density are solution of one-electron Schroëdinger-type equations, in most actual band structure calculations, solutions of the local-density approximation (LDA) on the basis of the Kohn–Sham equation (Kohn and Sham 1965). In the latter case exchange and correlation is introduced by means of the nonlocal exchange-correlation potential $\delta E_{xc}[\rho]/\delta\rho(\mathbf{r})$, where the LDA approximation of the exchange-correlation energy functional reads (see the detailed derivation in Section 6.8.4)

$$E_{xc}^{LDA}[\rho] = \int \rho(\mathbf{r})(\varepsilon_0^h[\rho(\mathbf{r})] - \varepsilon_0^f[\rho(\mathbf{r})])d\mathbf{r}. \quad (4.33)$$

Hereby $\varepsilon_0^h[\rho(\mathbf{r})]$ is the ground-state energy per electron of a gas of interacting electrons of density $\rho(\mathbf{r})$, and $\varepsilon_0^f[\rho(\mathbf{r})]$ the ground-state energy of the electron gas of density $\rho(\mathbf{r})$ constituted of noninteracting particles subjected only to the Pauli principle. Both quantities are available in analytical form (Hedin and Lundquist 1971).

Applying the LD approximated Feynman's theorem (Feynman 1939) to the momentum space density operator, its ground-state expectation value is composed of the expectation value, as calculated using a Slater determinant built by solutions the LDA Kohn–Sham equation, and a correction term, which is given in LDA by (Lam and Platzman 1974)

$$\Delta\rho^{LDA}(p) = \int \rho(\mathbf{r})(n_0^h(p)[\rho(\mathbf{r})] - n_0^f(p)[\rho(\mathbf{r})])d\mathbf{r}. \quad (4.34)$$

Here $n_0^h(p)$ and $n_0^f(p)$ are the momentum density of an interacting and a noninteracting homogeneous electron system. These quantities have been calculated within different approaches by a large number of authors (see e.g. Lundqvist, 1967a,b, Farid et al. 1993, Barbiellini and Bansil 2001). In Fig. 4.1 $\rho(p) \equiv n_0^h(p)$ has been plotted for different electron density parameters $r_s \equiv (3/4\pi\rho)^{1/3}$, where ρ is the average electron density in atomic units ($\hbar = e = m = 1$). One can immediately see that, allowing for exchange and correlation in an electron gas, the momentum space density is shifted to momenta greater than the Fermi momentum p_F, so that a tail of $\rho(p)$ exists for $p > p_F$. Certainly a step at $p = p_F$ persists, although smaller than in the case of a noninteracting electron gas. Equation (4.34) is the so-called Lam–Platzman correction, which has the drawback to be isotropic according to its very nature as derived from the behavior of an isotropic free electron system. An empirical correlation correction scheme, which can also adapt the anisoptropic case, has been developed by Wakoh and Matsumoto (1989). Also the heuristic scheme of Lundqvist and Leyden (1971), which will be discussed in greater detail in connection with experiment, can be considered as an approach to correlation correction of a monovalent metal with

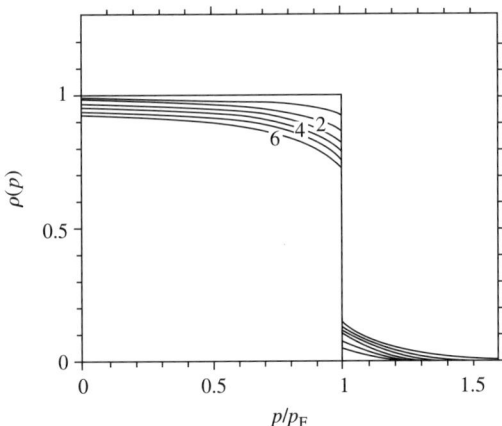

Fig. 4.1. Electron momentum densities of an interacting homogeneous electron system for different density parameters r_s as indicated. (Originally published by Lundquist (1967); copyright (1967) by Springer Verlag, Heidelberg).

anisotropic momentum density. An interesting step forward has been done by Kubo (1997), who has implemented the GW approximation into the calculation of solid state electron momentum densities, although the plasmon pole model he has used lead to some misleading results (Schülke 1999). We will come to this point later in Section 4.8.1.

A complete new approach to treat correlation in Compton scattering has been introduced by Barbiellini and Bansil (2001) based on a BCS-like approach in which they start with a singlet pair wavefunction (so-called geminal)

$$C(\mathbf{r}_1\uparrow, \mathbf{r}_2\downarrow) = \sum_i h_i \psi_i^*(\mathbf{r}_1)\psi_i(\mathbf{r}_2)[|\uparrow\rangle|\downarrow\rangle - |\downarrow\rangle|\uparrow\rangle], \qquad (4.35)$$

where the $\psi_i(\mathbf{r}_{1/2})$ are natural orbitals, the eigenfunctions of the one-particle density matrix defined in equation (6.157), so that the electron momentum density is given by

$$\rho(\mathbf{p}) = \sum_i n_i |\langle \mathbf{p}|\psi_i\rangle|^2, \qquad (4.36)$$

where n_i are the occupation numbers associated to ψ_i. The many-body wavefunction is then constructed by taking an antisymmetrized geminal product (AGP) Ψ, which can be cast into an $N/2 \times N/2$ determinant. The normalization of the AGP's requires the following relationship between h_i and n_i

$$h_i = \pm[n_i(1-n_i)]^{1/2}. \qquad (4.37)$$

The change of sign has to occur at the "pseudo-Fermi surface", which corresponds to the Hartree–Fock solution.

The BCS solution means to determine the ψ_i's and h_i's of equation (4.35) by minimizing the total energy. The non-Hartree–Fock-like term of the total energy may then be written as

$$E[h_i, \psi_i] = E_{\mathrm{HF}}[\rho] + E_{\mathrm{BCS}}[h_i, \psi_i] + O(1/N). \qquad (4.38)$$

$E_{\mathrm{HF}}[\rho]$ is the Hartee–Fock functional of the density operator

$$\rho = \sum_i n_i |\psi_i\rangle\langle\psi_i|, \qquad (4.39)$$

and E_{BCS}, the BCS-type functional, is given by

$$E_{\mathrm{BCS}} = \frac{1}{2} \langle C | V_{12} | C \rangle, \qquad (4.40)$$

where the pair potential V_{12} is repulsive, so that energy will be gained only through the exchange part of the Hartee–Fock functional, E_{HF}.

The BCS solution along the lines sketched above should be considered by approximating the natural orbitals by the Kohn–Sham orbital together with the corresponding eigenvealues ε_i, where $\varepsilon_i = 0$ defines the Fermi level. A solution within the limits of this approximation (Blatt 1964) implies occupation numbers given by

$$n_i = \frac{1}{2}\left(\frac{1 - \varepsilon_i}{(\varepsilon_i^2 + \Delta_i^2)^{1/2}}\right). \qquad (4.41)$$

n_i is greater than $1/2$ for $\varepsilon_i < 0$ and less than $1/2$ for $\varepsilon_i > 0$, where Δ_i determines the mixing of states above and below the Fermi level. In order to get an estimate of Δ_i it is not misguided to proceed from the assumption that the total energy of the AGP wavefunction should not be far from that of the Kohn–Sham Hamiltonian. Therefore, Barbiellini and Bansil made the assumption that the energy cost of electron pairing will be roughly compensated by the exchange energy, which implies

$$|\Delta_i| \sim I_i, \qquad (4.42)$$

where the exchange-type integral I_i reads within the limits of the LDA

$$I_i = \frac{1}{3} \int d\mathbf{r}\, |\psi_i(\mathbf{r})|^4 \, v_x(\mathbf{r})/\rho(\mathbf{r}). \qquad (4.43)$$

$v_x(\mathbf{r})$ is the Kohn–Sham exchange potential (Kohn and Sham 1965)

$$v_x(\mathbf{r}) = \frac{2}{\pi}[3\pi^2 \rho(\mathbf{r})^{1/3}]. \qquad (4.44)$$

Additionally the n_i's have to be renormalized according to the sum rule:

$$N = \sum_i n_i \qquad (4.45)$$

The correlation effect is controlled by the parameter

$$\alpha = \frac{\Delta_i}{w} \sim \frac{I_i}{w}, \tag{4.46}$$

where w is the valence electron bandwidth. If $\alpha \approx 1$ the spectral weight deep in the Fermi sea is shifted above $\varepsilon_i = 0$, whereas for $I_i \ll w$ only states near the Fermi momentum are redistributed. If Δ_i goes through zero at the Fermi level, this model admits discontinuities of the momentum distribution at the Fermi momentum. We will see later in Section 4.8.1 that α varies between 0.4 for Li and 0.05 for Al with dramatic consequences for the behavior of the Fermi surface discontinuity of these simple metals. Moreover, this scheme can also account for the anisotropy of the correlation correction. Apparently, the I_i's and hence the Δ_i's are energy and also Bloch-**k** dependent, so that anisotropies of the Fermi surface will reappear in the correlation correction according to (4.41).

4.4 Compton scattering versus positron annihilation

Inelastic X-ray scattering in the Compton regime provides information about the electron momentum density of the scattering system. Very similar information can be obtained by measuring the angular correlation of the two γ-photons emitted, when a thermalized positron is annihilated by an electron in a condensed matter system (for a review see Mijnarends 1979). Figure 4.2 shows a schematic diagram of a so-called long-slit device for an angular correlation measurement on the two γ-photons, where θ_y and θ_z represent the angular deviation from exact collinearity induced by the momentum of the electron positron pair. (If this pair momentum was zero, exact collinearity will be enforced by momentum conservation.) Because the angular deviations θ_y and θ_z are rather small (a few mrad) they are connected with the y- and z-component of the pair

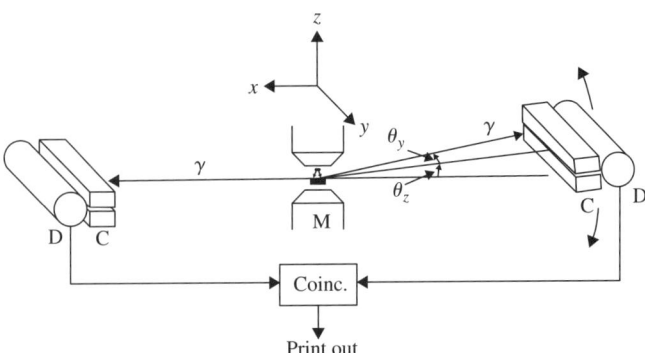

Fig. 4.2. Schematic diagram of an ACAR long-slit geometry. C = collimator, D = detector, M = magnet. (Originally published by Mijnarends (1979); copyright (1979) by Springer Verlag, Berlin.)

momentum **p** by

$$\theta_y \cong p_y/m_0 c \qquad \theta_z \cong p_z/m_0 c, \tag{4.47}$$

where m_0 is the electron rest mass. The longitudinal component of the pair momentum (here p_x) produces a Doppler shift ΔE at the energy $m_0 c^2$ of the γ-photon given by

$$\Delta E = c p_x / 2, \tag{4.48}$$

so that also the lineshape of the annihilation radiation reflects the pair momentum distribution.

Treating the positron and the electron as independent particles, the coincidence countrate $N(p_z)$ in the so-called long-slit geometry, as sketched in Fig. 4.2, is given by (DeBenedetti et al. 1950)

$$N(p_z)\Delta p_z \sim \Delta p_z \int dp_x \int dp_y \, \rho_p(\mathbf{p}), \tag{4.49}$$

where the pair momentum density $\rho_p(\mathbf{p})$ reads

$$\rho_p(\mathbf{p}) = \left(\frac{1}{2\pi\hbar}\right)^3 \sum_{\text{occ}} \left| \int \exp(-i\mathbf{p}\cdot\mathbf{r}/\hbar) \, \psi(\mathbf{r}) \, \psi_+^*(\mathbf{r}) d\mathbf{r} \right|^2. \tag{4.50}$$

$\psi(\mathbf{r})$ and $\psi_+^*(\mathbf{r})$ represent the electron and positron wavefunction. The summation extends over all occupied electron and positron states. Since the number of thermalized positrons present in the sample at any one time is of the order of one, their momentum distribution has the Boltzmann form

$$f_+(p_+) = (2m^*\pi k_B T)^{-3/2} \, \exp(-p_+^2/2m^* k_B T), \tag{4.51}$$

where m^* is the positron effective mass and k_B the Boltzmann constant. The distribution of the electrons over the available states $|\mathbf{k}\rangle$ is given by the Fermi–Dirac distribution function

$$f[E(\mathbf{k})] = (\exp\{[E(\mathbf{k}) - E_F]/k_B T\} + 1)^{-1}, \tag{4.52}$$

where E_F is the Fermi energy. Therefore, the pair momentum density $\rho_p(\mathbf{p})$ at finite temperature is given by the convolution of (4.51) and (4.52). At $T = 0$ the positron is in its ground state $\mathbf{k}_+ = 0$, so that (4.51) reduces to a delta function at $p_+ = 0$. With electrons in Bloch states $\phi_{\mathbf{k},\nu}(\mathbf{r})$, the Fermi–Dirac function is identical with the diagonal elements $n_{\nu\nu}(\mathbf{k})$ of the occupation number density introduced in Section 4.2, and the pair momentum density can be written as

$$\rho_p(\mathbf{p}) = \sum_{\mathbf{k},\nu} n_{\nu\nu}(\mathbf{k}) \left(\frac{1}{2\pi\hbar}\right)^3 \left| \int \exp(-i\mathbf{p}\cdot\mathbf{r}/\hbar) \, \phi_{\mathbf{k},\nu}(\mathbf{r}) \, \psi_+^*(\mathbf{r}) d\mathbf{r} \right|^2 \tag{4.53}$$

or by expanding the electron Bloch wavefunctions into plane waves according to (4.27) and likewise the wavefunction of the positron in its ground state $\mathbf{k}_+ = 0$ with coefficients $\beta(\mathbf{g}'')$, one ends up with (Berko and Plaskett 1958)

$$\rho_p(\mathbf{p}) = \sum_{\mathbf{k},\nu} n_{\nu\nu}(\mathbf{k}) \sum_{\mathbf{g},\mathbf{g}'} |a_\nu(\mathbf{k}+\mathbf{g})\,\beta(\mathbf{g}-\mathbf{g}')|^2 \,\delta(\mathbf{k}+\mathbf{g}'-\mathbf{p}/\hbar), \qquad (4.54)$$

where $\mathbf{g}' = \mathbf{g}'' + \mathbf{g}$. This equation might be compared with the expression (6.206), setting there $\nu' = \nu$. One sees that the pair momentum density $\rho_p(\mathbf{p})$, which determines the angular correlation of the two γ-photons produced when a positron is annihilated in a system of Bloch electrons, is identical with the momentum space density $\rho(\mathbf{p})$ of these electrons, provided

$$\beta(\mathbf{g}-\mathbf{g}') = \delta_{\mathbf{gg}'}, \qquad (4.55)$$

i.e. the positron can be represented by a single plane wave. In that case, the long-slit geometry as shown in Fig. 4.2 provides the same information about the electron momentum density as Compton profile measurements. Obviously, (4.55) is a crude simplification, so that the information about electron momentum density provided by **a**ngular **c**orrelation of **a**nnihilation **r**adiation (ACAR) is not a very direct one as in the case of Compton scattering. Nevertheless, the identical dependence of $\rho_p(\mathbf{p})$ and $\rho(\mathbf{p})$ on the occupation number density $n_{\nu\nu}(\mathbf{k})$ produces, due to Fermi surfaces, the same discontinuities in the first derivative of ACAR and Compton profiles, as will be shown later. Additionally it must be stressed that by using point detectors instead of long-slit geometry much more detailed information can be obtained, since then two of the three momentum components can be resolved according to

$$N(p_y, p_z)\Delta p_y \Delta p_z \sim \Delta p_y \Delta p_z \int \mathrm{d}p_x\, \rho_p(\mathbf{p}). \qquad (4.56)$$

Information about more than one electron momentum component can also be achieved by means of the γ-eγ Compton scattering technique as will be shown in Section 4.12.

It should be mentioned that also the Doppler broadening of the annihilation radiation due to the p_x component of the pair momentum makes feasible the measurement of this component by utilizing the energy resolution of a solid state detector (Hotz et al. 1968). Certainly, the momentum space resolution attainable with the Doppler technique is nearly one order of magnitude lower than with ACAR (an energy resolution of 1.0–1.5 keV is equivalent to 4–6 mrad angular resolution). The combination of the Doppler technique with point-slit ACAR is an attractive variant (Singru 1973), although overloaded with countrate problems.

All relations concerning ACAR and the Doppler technique have been derived within the independent particle model (IPM) without taking into account positron–electron (e$^+$–e$^-$) and electron–electron (e$^-$–e$^-$) correlation effects. The most important many-body effect is the attractive interaction of the positron with the nearly free electrons in condensed matter, which makes the electron density at the position of the positron higher than predicted by the IPM, so that the annihilation rate is enhanced, and the positron lifetime shortened by nearly one order of magnitude compared with the predictions of the IPM. On the other hand, the experimental ACAR results could always be described reasonably well within the limits of the IPM. Kahana (1960, 1963) was the first to solve this apparent contradiction by using a Green's function technique for the positron–electron pair. He showed that the enhancement of the annihilation rate, described by an enhancement factor $\varepsilon(p)$, is only weakly dependent on the momentum of the electron–positron pair, and can be parametrized in the following way

$$\epsilon(\mathbf{p}) = a + b\gamma^2 + c\gamma^4; \quad \gamma = p/p_F \leq 1. \tag{4.57}$$

Thus the electron momentum density $\rho(\mathbf{p})$ deduced from an ACAR measurement should be enhanced near the Fermi momentum as shown in Fig. 4.3, when compared both with the electron momentum density of the free-electron gas and with that found in Compton measurements on an interacting electron system according to Fig. 4.1. All three curves of Fig. 4.3 are normalized to the same total density. Whereas the discontinuity at the Fermi momentum is reduced as compared with the independent particle case when taking into account (e^--e^-) correlations, this step is enhanced with (e^--e^-) interactions, since, for momenta $p > p_F$, the (e^--e^-) interactions cancel the tails resulting from the (e^--e^-) interactions nearly completely (Carbotte and Kahana 1965). Therefore, discontinuities found in the first derivative of ACAR profiles as a consequence of Fermi surfaces are much stronger than those in Compton profiles. Of course, these correlation effects were derived only for the jellium case, so that they can change, as shown in Section 4.8 for metals which are far from being well approximated by the jellium model.

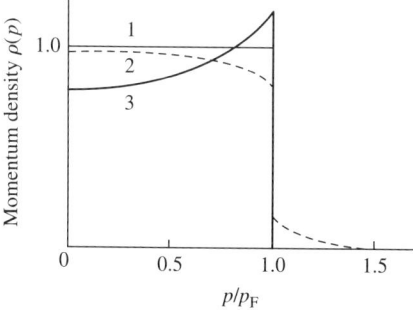

Fig. 4.3. Momentum density $\rho(\mathbf{p})$ for $r_s = 4$. 1: Noninteracting free electrons; 2: interacting electron gas; 3: with e^+-e^- interactions. Curves are normalized to the same density. (Originally published by Mijnarends (1979); copyright (1979) by Springer Verlag, Berlin.)

4.5 Special instrumentation for the Compton scattering regime

4.5.1 General considerations

The Compton scattering regime makes special demands upon the instrumentation, which result from

  (i) the validity of the impulse approximation;
  (ii) the low Compton scattering cross-section, when compared with the photoabsorption cross-section and its photon energy dependence;
  (iii) the details of the valence electron momentum density, which should be resolved adequately;
  (iv) the rather small details of a Compton profile reflecting bond- and Fermi surface-induced deviations from free atomic and free electron gas behavior, respectively. The two-dimensional momentum space integration in (4.7) makes it difficult to reveal finer details.

Let us summarize how these special demands determine the parameter of experiments in the Compton scattering regime.

(i) In order to guarantee the validity of the impulse approximation, the primary photon energy together with the scattering angle should be chosen such that the transferred energy $\hbar\omega$, according to (4.4), is much greater than the Fermi energy we can ascribe the valence electron system. Moreover, the binding energy of the core-shells must not be located within the $\hbar\omega$-range of the valence electron Compton profile, so that we can subtract the free atomic core contribution as calculated within the limits of the impulse approximation, in order to get the valence electron Compton profile we are interested in. The latter specification is of importance, since the deviation of a closed-shell DDSCS from the result of the impulse approximation is greatest for energy transfers a little bit above the core-shell binding energy. (The core-shell contribution to the total Compton profile becomes zero for energy transfers smaller than the core-shell binding energy.)

(ii) The total Compton scattering cross-section per electron is, in the nonrelativistic limit and averaged over the polarization (total Thomson scattering cross section) independent of the incident photon energy:

$$\lim_{\omega_1 \to 0} \sigma = \sigma_{\text{Th}} = \left(\frac{8\pi}{3}\right) r_0^2. \qquad (4.58)$$

This value is not far from the total Klein–Nishina scattering cross-section in the limit of very high photon energy.

On the contrary, the total photoabsorption cross-section per electron depends strongly on the incident photon energy according to the interpolation formula of Victoreen (1948)

$$\sigma_{\text{ph}} = \frac{(C\lambda_1^3 - D\lambda_1^4)M}{N_A Z}, \qquad (4.59)$$

where λ_1 is the wavelength (in Å) of the incident photon, M and Z are the atomic weight and the atomic number, respectively, of the target element, and N_A is the Avogadro's number. The constants C and D in (4.59) are specific for elements of the target and are limited to a certain energy range enclosed by absorption edges (for tabulated values see *International Tables of Crystallography*, Vol. III, 1968).

In order to have available the largest possible effective scattering volume, σ_{Th} and σ_{ph} should be of comparable order of magnitude. According to Fig. 4.4, this means that for elements of intermediate atomic number $10 < Z < 50$, photon energies between 50 and 200 keV are well matched. Only for light elements with $Z < 10$ photons with energies commonly used in X-ray crystallography are suited.

(iii) One scales the fine structure of the valence electron momentum density as a certain fraction of the Fermi momentum k_F, where k_F can be calculated assuming the valence electrons to be a free-electron gas. A momentum space resolution of $0.1 k_F$ can be considered to be sufficient to resolve the most important details of the electron momentum density. Of course, many older Compton experiments must be content with much lower resolution. If Δp_z is the momentum space resolution one is aspiring to, the necessary energy resolution $\Delta \hbar \omega_2$ of the scattering experiment can be calculated via the first derivative of p_z in (4.8) with respect to ω_2 at $p_q \equiv p_z = 0$:

$$\Delta \hbar \omega_2 = \left(\frac{[(\hbar \omega_1)^2 + (\hbar \omega_2)^2 - 2\hbar^2 \omega_1 \omega_2 \cos \theta]^{1/2}}{mc[1 + \hbar \omega_1 (1 - \cos \theta)/mc^2]} \right) \Delta p_z, \qquad (4.60)$$

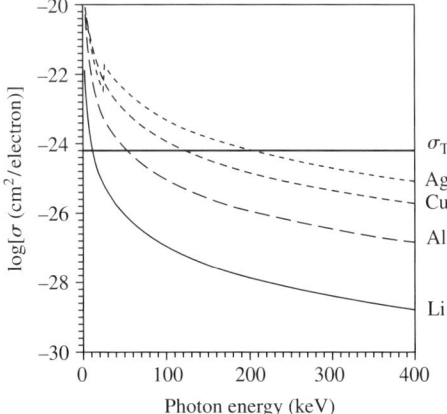

Fig. 4.4. The total photoabsorption cross-section σ_{ph} (thin curves) for different elements as indicated, and the total Thomson scattering cross-section σ_{Th} (thick curve) as a function of photon energy in the logarithmic representation.

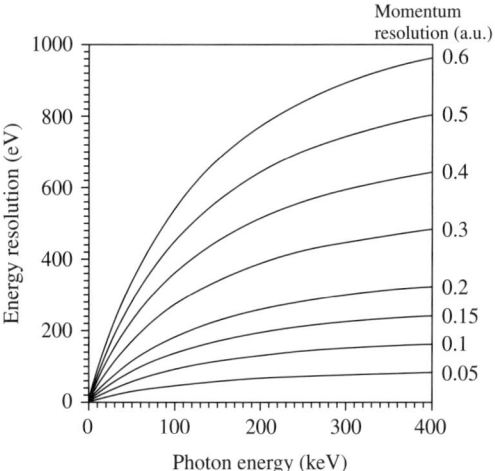

Fig. 4.5. Necessary energy resolution for different desired momentum space resolutions as a function of incident photon energy.

where $\hbar\omega_2$ for $p_z = 0$ is given by

$$\hbar\omega_2 = \frac{\hbar\omega_1}{1 + \hbar\omega_1(1 - \cos\theta)/mc^2}. \tag{4.61}$$

Figure 4.5 shows the necessary energy resolution for different desired momentum space resolutions plotted as a function of the incident photon energy.

In (4.60) we have assumed that the energy resolution is the only quantity which determines the momentum space resolution; this means that the influence of the beam divergence $\Delta\theta$ can be neglected. How $\Delta\theta$ determines the momentum space resolution Δp_z may be calculated via the first derivative of p_z in (4.8) with respect to θ again taken at $p_z = 0$:

$$\Delta\theta = \frac{\Delta p_z[(\hbar\omega_1)^2 + (\hbar\omega_2)^2 - 2\hbar^2\omega_1\omega_2\cos\theta]^{1/2}}{(\hbar^2\omega_1\omega_2)\sin\theta/c}. \tag{4.62}$$

In Fig. 4.6 $\log(\Delta\theta)$ is plotted, for different momentum space resolutions, as a function of the incident photon energy. One realizes that the requirements upon beam collimation increase drastically with increasing incident energy. This means that the benefit offered by the increasing scattering volume, when going to higher incident energies, can be eaten up by the necessary higher beam collimation, which costs flux. Only if, for a given incident energy and a desired momentum space resolution, the experimentally attainable beam divergence is considerable smaller than that given by (4.62) and Fig. 4.6, its influence on the momentum space resolution can be neglected. Otherwise both energy resolution and beam divergence must be taken into account when planning a Compton experiment

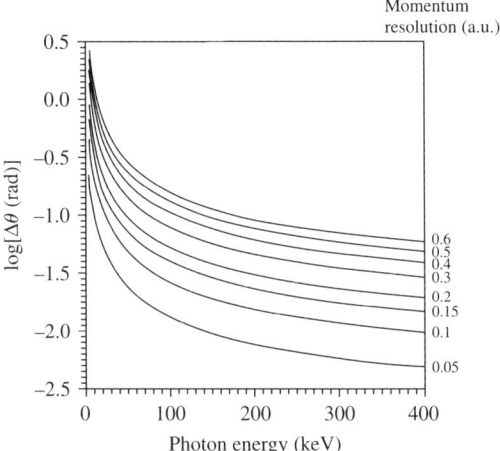

Fig. 4.6. Maximum tolerable beam divergence $\Delta\theta$ for different desired momentum space resolutions as a function of incident photon energy in the logarithmic representation. Here the influence of the instrumental energy resolution on the momentum resolution is neglected.

with a desired momentum space resolution. We will see later in Section 4.10.4 that possibly final state effects (lifetime of the recoil electrons) can additionally contribute to the momentum space resolution.

(iv) Detecting finer details of the two-dimensional integrated momentum density requires not only higher momentum space resolution but also appreciable statistical accuracy, where it is the number of counts accumulated in the total Compton profile, which matters. The interplay of both is best represented by the reciprocal form factor $B(z)$, as introduced in Section 4.2. If the experimental p_z-resolution function is a Gaussian, so that the theoretical Compton profile must be considered to be convoluted with this Gaussian, the corresponding experimental $B(z)$ function is the product of the theoretical $B(z)$ function with the Fourier-transformed p_z resolution function, again a Gaussian. Thus the experimental $B(z)$ function exhibits oscillations which are fading out with increasing z. On the other hand, the statistical error of the $B(z)$ function is in good approximation equal to the $ZN^{-1/2}$, where N is the total number of counts in the Compton profile, when the latter is normalized to the atomic number Z of the target atoms. Therefore, in order to match statistical accuracy and momentum space resolution, one should look for a simple model of the momentum space density of the sample under consideration. (e.g. the corresponding free-electron Fermi surface or the so-called Jones zone in the case of an insulator). Moreover one should estimate the total number of counts, which can be obtained using the given instrumentation within a reasonable time. One calculates the $B(z)$ function for the above model momentum density. There will be a certain $z = z_0$, where the oscillations of the $B(z)$ function become smaller than the statistical error. A well-matched p_z resolution function is that, the Fourier transform of which, multiplied with the model $B(z)$, will truncate $B(z)$ around that z_0.

4.5.2 Conventional X-ray sources

Early Compton scattering experiment used conventional X-ray sources which were also employed in X-ray diffraction, in most cases the $K\alpha$ doublet of molybdenum or in exceptional cases also the $K\beta$ line. The energy analysis was performed, to begin with, by means of a single plane crystal analyzer, where the divergence of the Compton scattered radiation was limited using a Sollar-slit collimator (Cooper et al. 1965). The energy resolution of such a device is mainly determined by the angular divergence σ transmitted by the Sollar slit via

$$\Delta E_\mathrm{A}/E = \sigma \cot\theta_\mathrm{A}, \tag{4.63}$$

where θ_A is the Bragg angle of the analyzer crystal. The analyzer transmission, defined in Section 2.2.5, is given by

$$T_\mathrm{A} = R_\mathrm{iA}/2.35\sigma. \tag{4.64}$$

R_iA is the integrated reflectivity of the analyzer as can be calculated using (2.32). These considerations apply to the situation in the mid-1960s, when the revival of Compton scattering took place, and registration by scintillation counters was already in use. In the late 1920s and early 1930s only photographic registration was available. But nevertheless rather sophisticated analyzer systems were developed (DuMond and Kirkpatrick 1930), which employed a large number of plane crystals working together in a way shown in Fig. 4.7 so that one scattered wavelength is focused by all crystals to a certain point on a circle, on which also the reflecting crystals are positioned.

Moreover, by using a point source and extended scatterer on another circle uniformity of the scattering angle could be achieved. Such an analyzer must be considered as a precursor of systems utilizing focusing bent crystals as first installed for Compton scattering by Schülke et al. (1969), and whose characteristics are already described in Section 2.2.5.1. Since unmonochromatized characteristic radiation of conventional sources was used, the spectra exhibited an unwanted background due to bremsstrahlung. Moreover in the case of K_α-radiation the α_2-component must be separated.

4.5.3 γ-ray sources

As can be read from Fig. 4.4, conventional X-ray sources with photon energies not above 20 keV could effectively be employed only for scattering from elements with atomic numbers smaller than 10. This was the reason for using γ-ray sources with photon energies up to 500 keV. A high specific activity and low self-absorption, in line with monochromaticity within the energy range utilized are

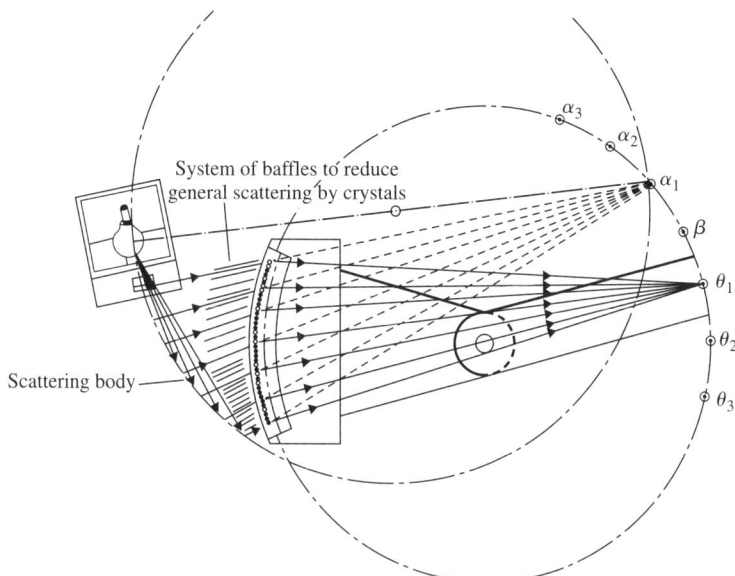

Fig. 4.7. Schematic diagram of DuMond's multicrystal spectrograph. The 50 calcite analyzing crystals lie on the circumference of the right-hand circle. X-rays of a given wavelength are focused to a well-defined point on that circle. (Originally published by DuMond and Kirkpatrick (1930); copyright (1930) by the American Institute of Physics.)

the most important criteria to be satisfied by those γ-ray sources. Either long-lived isotopes like ^{241}Am ($E_\gamma = 60$ keV) (Bonse et al. 1979) or isotopes with a half-life of only a few days like ^{198}Au ($E_\gamma = 412$ keV) (Schneider 1974) came into use, where the latter can only be employed when the γ-spectrometer is sited near a neutron source in order to produce the radioactive isotope by an (n, γ)-reaction. A detailed review of γ-ray sources in Compton spectroscopy has been given by Cooper (1979). Of course, the application of γ-ray sources in Compton spectroscopy was not possible without the timely developement of solid state detectors with an energy resolution $\Delta E/E$ of better than 10^{-3}, so that, according to Fig. 4.5, momentum space resolutions between 0.6 and 0.4 a.u. became attainable. The application of crystal-dispersive analyzers together with γ-ray sources was excluded because of the low flux of these sources. Since the solid state detectors obey the following equation for the instrumental broadening ΔE

$$(\Delta E)^2 = \alpha^2 + \beta E, \tag{4.65}$$

where the first term is a noise contribution and the second term is due to the statistics of electron–hole pair production, in principle, γ-ray sources with higher photon energy are to be preferred. But also the problems with shielding the detector against the γ-radiation, which finds its way directly to the detector,

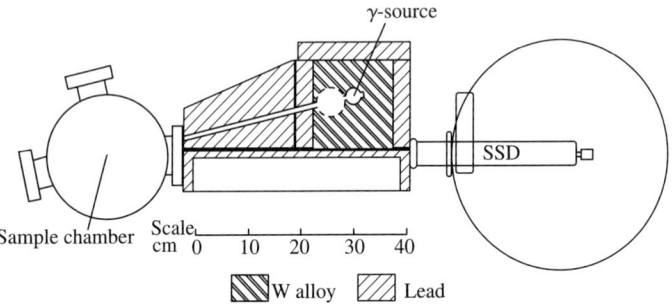

Fig. 4.8. Schematic diagram of a typical γ-ray Compton spectrometer. The region immediately surrounding the γ-source is made of tungsten alloy replacing lead for better shielding. The sample chamber is to the left and the solid state detector (SSD) to the right. (Reprinted with permission from Cooper (1985); copyright (1985) by IOP Publishing Ltd.)

especially for the case of large scattering angles, increases with increasing photon energy. This can be understood when looking at Fig. 4.8, which shows a typical γ-Compton spectrometer. An additional complication is the shape of the detector resolution function, which exhibits, in addition to the Gaussian shaped main line, a low-energy tail due to incomplete charge collection within the detector and so-called escape events, where characteristic K-radiation of the detector material excited by the incident radiation can partly escape, taking away its energy so that it does not contribute to the electron–hole pair production. Knowing the shape of this resolution function as a result of an independent measurement, one can correct for its influence by fast Fourier deconvolution of the experimental Compton profile with respect to this resolution function and subsequent convolution with the Gaussian shaped main line (Bauer 1984).

4.5.4 *Synchrotron radiation sources*

The advent of dedicated synchrotron radiation sources with critical photon energies larger than 50 keV (especially so-called wavelength-shifter) has radically changed the situation with Compton scattering. Now crystal-dispersive analysis of scattered radiation to primary photon energies above 60 keV becomes feasible, where the momentum space resolution could be held near 0.1 a.u. and the statistical accuracy in the range of 0.1% for each data point. We will describe in what follows the main lines of Compton instrumentation with synchrotron radiation often having reference to Sections 2.2.2–2.2.5, where already synchrotron radiation based IXS instrumentation was presented.

The very first application of synchrotron radiation for Compton profile measurement was composed of a double-crystal monochromator as described in Section 2.2.2.1 with an energy dispersive solid state detector as an analyzer, first introduced by Holt *et al.* (1978). Here it was the improvement in statistical accuracy due to the much higher flux of the synchrotron radiation source compared with a γ-ray source, which counted. The momentum space resolution was limited by the finite energy resolution of the solid state detector as in the γ-ray case. Even in these days, this technique is used to complement a Compton spectrometer with crystal dispersive analysis, when an extremely high statistical accuracy at the expense of resolution is aspired to. In those cases often segmented solid state detectors are in use, each segment with its own chain of amplifiers in order to increase the accepted countrate and to diminish deadtime losses.

The first synchrotron radiation based setup with a crystal dispersive analyzer in connection with a double-crystal monochromator was introduced by Loupias and Petiau (1980). The analyzer consisted of a cylindrically bent curved Cauchois-type focusing crystal (cylinder axis in the horizontal plane) in connection with a gas filled position-sensitive detector, as sketched in Fig. 4.9, so that the whole Compton spectrum of width $\delta\lambda$ could be detected simultaneously.

Similar setups at HASYLAB (Hamburg) were described by Berthold *et al.* (1992b), where the position-sensitive detector was a 200-strip Ge solid state detector (Berthold *et al.* 1992b) and by Sakurai *et al.* (1992a) (KEK-accumulation ring, Photon Factory, Japan), who applied image plate detection. The specialty of the latter spectrometer was the simultaneous use of four Cauchois-type bent crystal analyzers and image plates, arranged on the surface

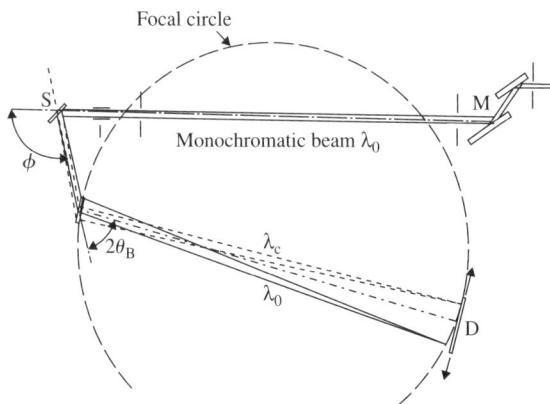

Fig. 4.9. Schematic diagram of the focusing Compton spectrometer installed at LURE. The white beam is monochromatized by a channel-cut double-crystal monochromator M, passes through an ionization chamber monitor I, and hits the sample S. The curved Cauchois-type focusing analyzer generates the whole Compton spectrum at the position sensitive detector D. (Reprinted with permission from Loupias and Petiau (1980); copyright (1980) by Editions de Physique, Paris.)

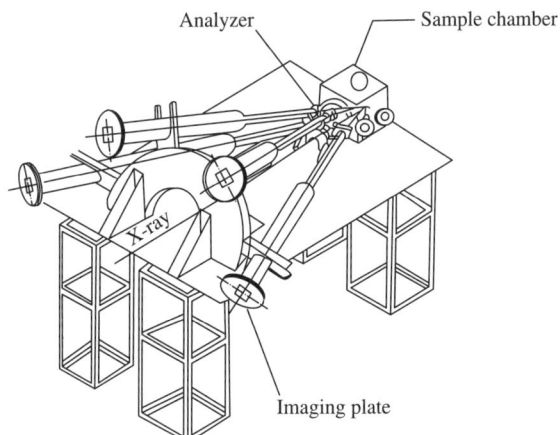

Fig. 4.10. The KEK-PF-AR Compton spectrometer. The four sets of analyzer and image plates are mounted on the surface of a cone, whose axis defines a scattering angle of 160°. (Reprinted with permission from Sakurai *et al.* (1992a); copyright (1992) by the American Institute of Physics.)

of a cone sharing the same scattering angle, as depicted in Fig. 4.10. Thus four Compton profiles with different orientation of the scattering vector **q** can be recorded. A Compton spectrometer for 90–120 keV synchrotrons radiation using a Cauchois-type analyzer and a solid state position sensitive detectors was installed at SPring-8 by Hiraoka *et al.* (2001)

If ε is the efficiency of the position-sensitive detector/image plate the analyzer transmission T_A can be approximated by

$$T_A = R_{iA} \varepsilon / \Delta \psi_v, \tag{4.66}$$

where $\Delta\psi_v$ is the vertical divergence of the scattered radiation as determined by the most restrictive aperture of the analyzer. Due to both the change in lattice parameter and crystal orientation of the bent crystal along the direction of the incident beam, the reflection width of the analyzer crystal can become larger than the Darwin width. Therefore, the analyzer's integral reflectivity R_{iA} can approach, in the limit, the value of an ideal mosaic crystal. The energy resolution of the Cauchois-type analyzer is mainly determined by the analyzer crystal thickness D, the spatial resolution Δx of the position sensitive detector and R_{iA} via

$$\left(\frac{\Delta E}{E}\right)_A = \frac{[D^2 \sin^2 \theta_A + (\Delta x \cot \theta_A)^2 + R_{iA}^2 R^2 \cos^2 \theta_A]^{1/2}}{R \sin \theta_A}. \tag{4.67}$$

R is the bending radius of the crystal.

Another type of Compton spectrometer as used at the ESRF (Suortti *et al.* 1999) is a scanning spectrometer composed of a focusing cylindrically bent

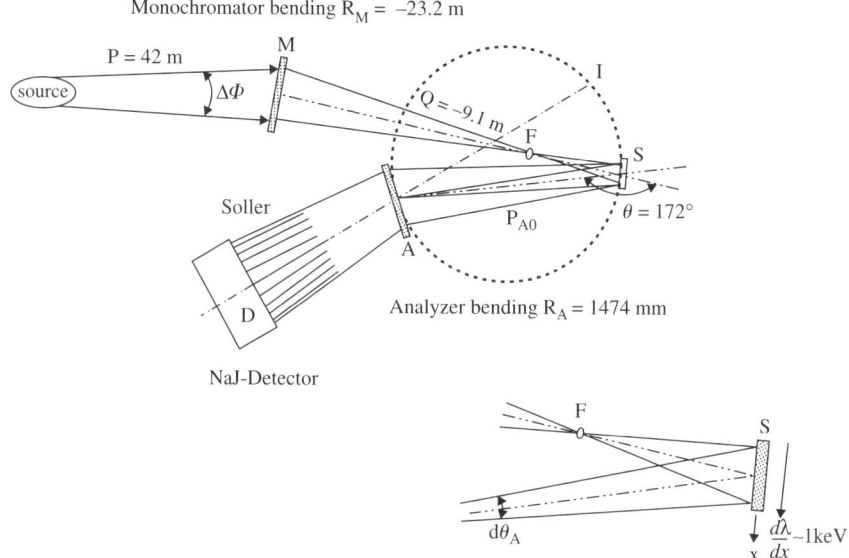

Fig. 4.11. Principle of the dispersion compensating Compton spectrometer at the ESRF. The incident beam is focused at the entrance slit F, and it diverges on the sample S. The Rowland circle of the analyzer crystal A is indicated by the broken line. The tapered Sollar slits in front of the detector D point to the virtual source of scattering I. (Reprinted with permission from Suortti *et al.* (2001); copyright (2001) Elsevier Science Publisher B.V.)

monochromator described in Section 2.2.2.3 and a focusing cylindrically bent analyzer in full Rowland geometry, discussed in Section 2.2.5.1., where the analyzer crystal and the NaJ detector are moving on the focusing circle by synchronized translations and rotations.

A further setup used for Compton spectroscopy with primary photon energies in the 100 keV range has also been introduced at the ESRF (Suortti *et al.* 2001a) and is a combination of a cylindrically bent monochromator crystal in transmission geometry and a cylindrically bent Laue analyzator crystal, as depicted in Fig. 4.11. The special feature of this spectrometer consists in working according to the principle of dispersion compensating as described in Section 2.2.5.2 in connection with a slightly different analyzer geometry. This means that the beam incident on the sample has an energy gradient equal to the gradient of energy reflected by the analyzer. This way the same energy loss is observed, at a given analyzer Bragg angle, for the entire incident beam. This principle is realized in the following way: A bent transmission crystal focuses the beam to a narrow line, then the beam diverges, so that the energy gradient at the sample meets the condition of dispersion compensation. The focal distances P and Q obey the following equation (lens equation):

$$\frac{P_0}{P} + \frac{Q_0}{Q} = 2, \quad (4.68)$$

where

$$P_0 = R_M \cos(\alpha_M + \theta_M)$$
$$Q_0 = R_M \cos(\alpha_M - \theta_M). \quad (4.69)$$

Here R_M is the bending radius, α_M the asymmetric cut angle and θ_M the Bragg angle of the monochromator. The energy bandwidth due to the angular width $\Delta\phi$ of the beam incident on the monochromator is

$$\frac{\Delta E}{E} = \cot\theta_M \left(\frac{P_0}{P} - 1\right)\Delta\phi. \quad (4.70)$$

The transmission T_M of the monochromator can be estimated to

$$T_M = \frac{R_{iM} P}{2.35\sigma_x}, \quad (4.71)$$

provided that $\sigma_x/P < (\sigma_x'^2 + \sigma_R'^2)^{1/2}$. Here R_{iM} is the integrated reflectivity of the monochromator, which is in the present Laue case one-half of that given by (2.32) for the Bragg case. σ_x is the source dimension in the reflecting plane, σ_x' the source divergence and σ_R' the natural divergence of the synchrotron radiation in x-direction.

With a proper choice of α, the polychromatic focus coincides with the geometrical focus, a necessary condition for a well-defined energy gradient on the sample. The wavelength gradient fitting the Bragg-angle change of the analyzer is

$$\frac{d\lambda}{dx} = \frac{2d_A}{R_A \cos\alpha_A \cos\theta_A (1 - \tan\alpha_A \tan\theta_A)}, \quad (4.72)$$

where d_A is the lattice spacing and θ_A the Bragg angle of the analyzer. The correction term $\tan\alpha_A \tan\theta_A$ is, for incident photon energies in the range of 100 keV, rather small and can therefore be neglected, when formulating the condition

$$\frac{d\lambda}{dx} = \frac{2d_A}{R_A \cos\theta_A} = \frac{2d_A}{P_{0A}}, \quad (4.73)$$

for dispersion compensation with $\alpha_A = 0$, which has to be maintained during the scan. The transmission of the analyzer is given by

$$T_A = \frac{R_{iA}\varepsilon}{\Delta\psi_x},$$

where R_{iA} is the integrated reflectivity of the analyzer crystal and $\Delta\psi_x$ the divergence of the scattered radiation in the x-direction limited by the tapered Sollar slits in front of the NaJ detector with efficiency ε.

The total energy resolution of this Compton spectrometer is composed of the monochromator and the analyzer resolution and can be estimated to

$$\left(\frac{\Delta E}{E}\right)_{\text{tot}} = \left[\left(\frac{\Delta E}{E}\right)_{\text{M}}^2 + \left(\frac{\Delta E}{E}\right)_{\text{A}}^2\right]^{1/2} \quad (4.74)$$

with

$$\left(\frac{\Delta E}{E}\right)_{\text{M}} = \left(\frac{2.35\sigma_x}{P}\right)\cot\theta_{\text{M}} + \frac{R_{\text{iM}}\sqrt{|b|}\sin 2\theta_{\text{M}}}{\sin^2\theta_{\text{M}}}, \quad (4.75)$$

where b is the asymmetry factor given in (2.30), and

$$\left(\frac{\Delta E}{E}\right)_{\text{A}} = \frac{[D^2\sin^2\theta_{\text{A}} + (\Delta x)^2 + R_{\text{iA}}^2 R_{\text{A}}^2 \cos^2\theta_{\text{A}}]^{1/2}}{R_{\text{A}}\sin\theta_{\text{A}}}. \quad (4.76)$$

Here D is the thickness and R_{A} the bending radius of the analyzer crystal. Δx is the width in the x-direction of the scattering volume of the sample, if one takes into account the finite penetration depth t of the monochromatic incident radiation:

$$\Delta x = 2t\tan\left(\frac{\pi}{4} - \frac{\theta}{2}\right). \quad (4.77)$$

An interesting possibility to utilize the chemical shift of the X-ray absorption edge for the energy analysis of inelastically scattered X-radiation was proposed by Schülke et al. (1983, 1984). As shown in Fig. 4.12 (Schülke et al. 1984), a high-resolution X-ray passband filter was built by combining a high-pass filter provided by the fluorescence yield $I_e(E)$ of Ge, and a low-pass filter, the transmission $T_L(E)$ of a GeO_2 foil, positioned in front of the Ge block. The X-ray absorption edge of GeO_2 is shifted by 4.0 eV to higher photon energies compared with the Ge edge, so that the resulting bandpass $T(E)$ has a width of the same order of magnitude, namely 5 eV. The peculiarity of the bandpass combination shown in Fig. 4.12 lies in the fact that the Ge K-fluorescence recorded by the NaI crystal is emitted by a Ge solid state detector (Ge SSD). Thus each fluorescence photon leaving the Ge SSD contributes to the so-called escape peak in the SSD spectral distribution. This way one can discriminate fluorescence events from other counts (scattered radiation) recorded by the NaI detector by looking

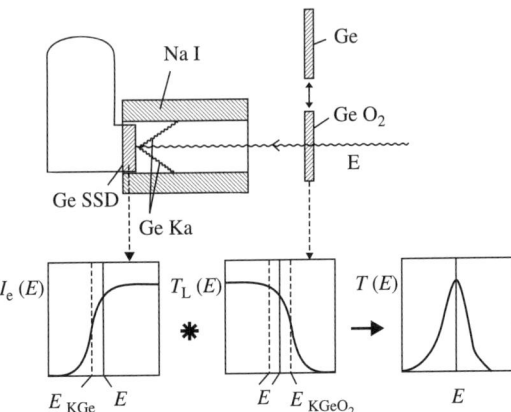

Fig. 4.12. Principle of a GeO_2–Ge X-ray bandpass filter. (Reprinted with permission from Schülke et al. (1984); copyright (1984) by Elsevier Science Publisher B.V.)

at coincidences of escape peak counts of the SSD with counts of the NaI detector. Due to the resonant Raman effect the transmission curve $T(E)$ exhibits a certain asymmetry with a tail on the low-energy side. This asymmetry can be prevented if one measures the transmission twice: first with the GeO_2 foil and then with a regular Ge foil, and subtracting the latter transmission curve from the former. The thickness of both foils has to be adjusted so that they have equal transmission at the low-energy side of their K-edges far enough from the edges.

Finally, quite another way of energy-analyzing the Compton scattered radiation must be mentioned, although only tested once in connection with monochromatized incident synchrotron X-rays, namely the application of a microcalorimeter to measure directional silicon Compton profiles by Stahle et al. (1992). In principle, these systems exhibit high potentiality for Compton scattering, since they are superior to conventional solid state detectors as far as energy resolution is concerned, although, in this first application, only a momentum space resolution of 0.4 a.u., corresponding to an energy resolution of 180 eV could be achieved.

4.5.5 Data processing; from raw data to the Compton profile

It is the goal of an experiment in the Compton scattering regime to obtain finally a well-normalized (directional) Compton profile of the electron system under investigation. On the other hand, the basic experimental data consist of the scattered countrate as a function of the energy loss, including the inelastically scattered contribution, which comprises both the result of single-scattering and multiscattering events, the quasielastic part of the spectrum, and a background.

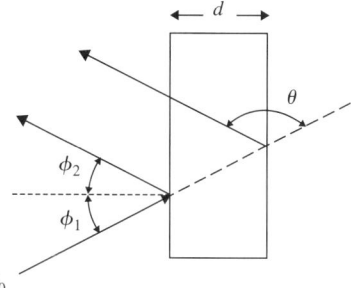

Fig. 4.13. Schematic diagram of a back-scattering experiment. I_0 is the intensity of the incident beam, ϕ_1 and ϕ_2 is the incident and the exit angle, respectively; θ is the scattering angle, and d the sample thickness.

Possibly the spectrum of the inelastically scattered events is not completely recorded, but truncated at a certain energy loss corresponding to the low-energy tail of the core Compton profile. Figure 4.13 is the scheme of a back-scattering experiment. The number of photons scattered under an angle θ into a solid-angle element $\Delta\Omega_2$, transmitted by the energy analyzer at its position between x and $x + \Delta x$, and recorded by the detector, is given by

$$I(x) = \int_0^d \mathrm{d}z \, I_0 \exp\left(-\left[\frac{\mu(E_1)}{\cos\phi_1} + \frac{\mu(E_2)}{\cos\phi_2}\right] z\right)$$
$$\times \left(\frac{\mathrm{d}^2\sigma}{\mathrm{d}\Omega_2 \mathrm{d}x}\right) \Delta x \, \Delta\Omega_2 \, \rho T(E_2) E(E_2) A(E_2), \quad (4.78)$$

where I_0 is the number of photons hitting the sample. The exponential takes into account the photoabsorption within the sample with the linear absorption coefficient $\mu(E)$. ϕ_1 and E_1 (ϕ_2, E_2) are the incidence angle and the energy of the primary (scattered) photons. (For sake of convenience we are using in this subsection, contrary to our remaining convention, $E_{1,2}$ to indicate photon energies.) ρ is the electron density of the sample. The DDSCS is related to position x of the energy analyzer, so that the following relation holds to the DDSCS used so far:

$$\frac{\mathrm{d}^2\sigma}{\mathrm{d}E_2\mathrm{d}x} = \left(\frac{\mathrm{d}E_2}{\mathrm{d}x}\right)\left(\frac{\mathrm{d}^2\sigma}{\mathrm{d}\Omega_2\mathrm{d}E_2}\right). \quad (4.79)$$

$T(E_2)$ stands for the transmission of the analyzer system (see the definition in Section 2.2.5), $E(E_2)$ for the efficiency of the detector, and $A(E_2)$ takes into account the weakening of the scattered photon beam by windows or air between the sample and analyzer system.

As found out in Section 4.1.2, the DDSCS is connected with the Compton profile via

$$\frac{\mathrm{d}^2\sigma}{\mathrm{d}\Omega_2 \mathrm{d}E_2} = \left(\frac{\mathrm{d}\sigma}{\mathrm{d}\Omega_2}\right)_{\mathrm{rel}} \left(\frac{m}{\hbar q}\right) J(p_z) \equiv F(E_2)\, J(p_z). \qquad (4.80)$$

If we insert (4.79) and (4.80) into (4.78) and execute the integration, we end up with

$$I(x) = C\, f(E_2)\, J(p_z). \qquad (4.81)$$

The constant $C = I_0 \Delta\Omega_2 \Delta x \rho$ loses its significance in the course of the normalization of the Compton profile to its theoretical value. The quantity $f(E_2)$, on the contrary, comprises all energy-dependent factors and reads:

$$f(E_2) = \left(\frac{1 - \exp[-\{\mu(E_1)/\cos\phi_1 + \mu(E_2)/\cos\phi_2\}d]}{\{\mu(E_1)/\cos\phi_1 + \mu(E_2)/\cos\phi_2\}d}\right)$$
$$\times \left(\frac{\mathrm{d}E_2}{\mathrm{d}x}\right) F(E_2)\, T(E_2)\, A(E_2)\, E(E_2). \qquad (4.82)$$

As already said, the measured intensity $I(x)$ comprises along with the number of scattered photons also a background, which has to be subtracted. Moreover, the direct relation between the Compton profile and the measured intensity is only valid if the photon has been scattered within the sample under the angle θ only once. But one must be aware that also multiple scattering processes can occur. Those photons, which are multiply scattered and leave the sample exactly in the scattering direction, will also be energy-analyzed and registered. Let $U(x)$ be the background and $M(p_z)$ the multiple scattering spectrum. Then the following relation

$$I(x) - U(x) = C f(E_2)[J(p_z) + M(p_z)], \qquad (4.83)$$

will build the basis for the evaluation of the measured data.

According to (4.82) we have to consider five energy-dependent corrections of the measured data $I(x)$, expressed in the corresponding correction factors:

(i) for absorption in the sample

$$K_1(E_2) = \left(\frac{1 - \exp[-\{\mu(E_1)/\cos\phi_1 + \mu(E_2)/\cos\phi_2\}d]}{\{\mu(E_1)/\cos\phi_1 + \mu(E_2)/\cos\phi_2\}d}\right)^{-1}; \qquad (4.84)$$

(ii) for absorption in the windows and air

$$K_2(E_2) = \exp\left[\sum_n \mu_{w_n}(E_2) d_{w_n}\right] \exp[\mu_{\mathrm{air}}(E_2) d_{\mathrm{air}}]; \qquad (4.85)$$

(the summation is over the different windows);

(iii) for scale correction

$$K_3(E_2) = \left(\frac{dE}{dx}\right)^{-1} \qquad (4.86)$$

on the basis of an experimental energy calibration of the analyzer system, performed by means of X-ray fluorescence lines;

(iv) for relativistic scattering cross-section correction

$$K_4(E_2) = \left(\left(\frac{d\sigma}{d\Omega_2}\right)_{\text{rel}}(E_2)\left[\frac{m}{\hbar q(E_2)}\right]\right)^{-1}; \qquad (4.87)$$

and

(v) for analyzer transmission correction

$$K_5(E_2) = T(E_2)^{-1}. \qquad (4.88)$$

Appropriate expressions for $T(E_2)$ are found in Sections 4.5.2 and 4.5.4. If a position-sensitive detector is used along with a crystal dispersive analyzer, also the efficiency $\varepsilon(x)$ of the positions on the detector can vary with x, and have to be calibrated experimentally. Using the position–energy calibration, this leads to an additional correction factor

$$K_6(E_2) = \varepsilon(E_2)^{-1}. \qquad (4.89)$$

The multiple scattering contribution $M(p_z)$ in (4.82) has been a matter of long debate and effort during the early period of Compton scattering in the 1970s. The contribution of multiple scattering was first evaluated by measuring the Compton scattering on samples with varying thickness and by extrapolating to the zero thickness, where the multiple scattering part is assumed to vanish (Reed and Eisenberger 1972, Tanner and Epstein 1974). The breakthrough was managed by Felsteiner et al. (1974). They simulated the spectral distribution of the multiple scattered photons by means of a Monte Carlo code, where the crystallographic anisotropy was neglected, the simulation was performed on the basis of theoretical atomic isotropic Compton profiles (see Chapter 6), unpolarized incident radiation and energy dispersive analysis was assumed. Later on Chomilier et al. (1985) extended these simulations to the cases where synchrotron radiation sources with linear polarization of the incident radiation and crystal dispersive analysis were taken into account. Such a simulation program for calculating the amount of double scattering comprises the following steps:

1. Fixing the point of impact of the photon on the sample surface.
2. Determining the penetration depth within the sample; this means fixing the place of the first scattering event.
3. Deciding whether an elastic or an inelastic scattering event should be simulated, or the photon should be absorbed.

4. Fixing the momentum component p_z of the scattering electron. Fixing the scattering angle θ. Fixing the angle η between the polarization plane of the incident radiation and the plane spanned by incident and scattered wave vector. Fixing the angle β between the plane spanned by the incident polarization vector and the scattered wavevector, and the polarization plane of the scattered radiation. The probability for a certain combination of θ, η and β corresponds to the intensity of the experimental spectrum at p_z.

5. Determination of the pathlength of the photon till the next scattering event.

6. Repetition of steps 3–5.

7. Determination of the exit parameter of the photon: exit point on the surface, energy, scattering angle. The decisions in the steps 1–5 were taken randomly (random number generator) using probability distributions which are based on the scattering laws.

The measured spectral distribution of the scattered photons contains a background, which has to be subtracted before the energy-dependent corrections can be applied and the normalization can be performed. The background has different sources, which depend strongly on the method used for measuring the spectral distribution. Often a linear background is assumed, which is determined by a linear interpolation between experimental values on the high-energy side (energy higher than the quasielastic line) and on the low-energy side of the measured spectrum, where the long extension of the core contribution on the low-energy side must be taken into account. We will discuss here the more general case of how to handle an unknown nonlinear background. If one subdivides the whole spectrum into small sections, enumerated with i, (4.83) can be written accordingly as

$$J(p_z(i)) - M(p_z(i)) = C\,K(E)[I(i) - U(i)], \qquad (4.90)$$

where $K(E_i) = K_1(E_i)K_2(E_i)K_3(E_i)K_4(E_i)K_5(E_i)K_6(E_i)$ describes the total energy-dependent correction, and C is a normalization constant. Outside of the central section ($|p_z| > p_0$) of the Compton profile, where only the core electrons are contributing to the Compton profile, the background can be determined by

$$U(i) = I(i) - (C\,K(E_i))^{-1}[J_{\text{core}}(p_z(i)) + M(p_z(i))] \quad \text{for } |p_z| > p_0, \qquad (4.91)$$

where we assume that the core Compton profile can be represented by a theoretical free atomic profile as calculated by Biggs et al. (1975). The still unknown normalization constant C can be obtained by using the normalization for $J(p_z)$, namely

$$\int_{-\infty}^{+\infty} J(p_z)\,\mathrm{d}p_z = Z. \qquad (4.92)$$

Here Z is the number of electrons per atom. Within the interval $[-p_0; p_0]$ the normalization integral reads

$$\int_{-p_0}^{p_0} J(p_z) \mathrm{d}p_z = Z_{\mathrm{val}} + \int_{-p_0}^{p_0} J_{\mathrm{core}}(p_z) \mathrm{d}p_z, \qquad (4.93)$$

where Z_{val} is the number of valence electrons per atom. Using (4.90) and (4.82) the following normalization constant is obtained:

$$C = \frac{Z_{\mathrm{val}} + \int_{-p_0}^{p_0} J_{\mathrm{core}}(p_z) \mathrm{d}p_z}{\int_{-p_0}^{p_0} K(E)[I(i) - U(i)] \mathrm{d}p_z}. \qquad (4.94)$$

Equations (4.91) and (4.94) represent a system of coupled equations, which can be solved iteratively, where a polynomial is least-square fitted to $U(i)$ within the intervals $[-p_{\min}; -p_0]$ and $[p_0; p_{\max}]$. This procedure is repeated till self-consistency is reached for C. Since often the scattering angle is difficult to be determined with sufficient accuracy, it is introduced as a free parameter into the above procedure. After the iteration, which might be performed assuming a starting value for the scattering angle θ, the position of the Compton profile peak on the energy scale is determined, so that, together with the well-known incident photon energy, the scattering angle can be calculated according to (4.8) with $p_q = p_z = 0$. With this new value of the scattering angle the whole iterative procedure is repeated till the change in the Compton profile peak is smaller than a small fraction ($\sim 10^{-3}$) of the momentum space resolution.

4.6 Compton scattering of atoms and molecules

Before starting the next sections, all devoted to the application of Compton scattering to various problems in condensed matter physics, it must be stressed that the author had to restrict this part of the book to exemplary cases of special importance for the further development of the method, where those utilizing synchrotron radiation had the highest priority. Otherwise taking into account the full amount of applications found in the literature since the revival of Compton scattering in the mid 1960s would blow up the size of the book in an unwanted manner.

Already in the early 1970s Compton profiles of atomic and diatomic molecular gases (liquids) were investigated using both conventional X-ray sources (Mo K and Ag K) and γ-ray sources ($^{123\mathrm{m}}$Te 160 keV). As pointed out by Eisenberger (1970) these investigations may provide a check on the appropriate choice between different variational forms of wavefunctions This is of special importance, since it is well known that, even though one may approach very closely the lowest energy of the system, one may not have a wavefunction which can predict observables other than energy with any accuracy. Eisenberger (1970) presented X-ray Compton profiles of He and H_2 in liquid and compressed gaseous form. The He Compton profiles agreed perfectly with calculated ones using nonrelativistic Hartree–Fock wavefunctions (Biggs *et al.* 1975). Huotari (2002) came to the same conclusion in a more recent synchrotron radiation based measurement on He, performed with much higher statistical accuracy than the former one. His calculation used the LCAO method with the nonrelativistic quantum chemistry program DALTON (Helgaker et al. 2001). Moreover, Huotari (2002) could show that, including correlation by utilizing the Möller–Plesset second-order perturbation theory (Möller and Plesset 1934, Binkley and Pople 1975) no change of the theoretical result occurred, where this change was comparable with the statistical accuracy of the experiment. As far as Eisenberger's Compton measurement on H_2 molecules are concerned, the effect of chemical binding was clearly visible, since the measured Compton profile was more than 10% smaller at $p_z = 0$ than the sum of two Hartree–Fock free atomic H-profiles. Both Hartree–Fock type self-consistent field (HF-SCF) and multiconfigurational self-consistent field (MC-SCF) calculations (Wahl and Das 1966), where the latter include correlation to a certain extent, exhibit Compton profile values at $p_z = 0$ in the average 3.5% above the experiment. It is worth mentioning that deviation of the MC scheme from experiment was even larger than that of the HF calculation, even though the former produces a lower energy state, an illustration of the weakness of the variational approach mentioned above. Later calculations by Brown and Smith (1972) accounting for 98% of the correlation energy have removed this discrepancy to some extent, but the theoretical Compton profile at $p_z = 0$ remained 2% higher than the experiment. Compton profile measurements on H_2 by means of high-energy electron impact spectroscopy (HEEIS) performed by Lee (1977) saw the profile 2% higher at $p_z = 0$ than the X-ray experiments, so that they were in good agreement with the Brown and Smith (1972) computations. The latter were also confirmed by Jeziorski and Szalewicz (1979) who used an explicitly correlated Gaussian wavefunction accounting for 99.9% of the correlation energy.

In another X-ray Compton study Eisenberger (1972) has presented Compton profiles of N_2, O_2 and Ne gases, compressed in a pressure cell. Also in this case the role of chemical bonding in the diatomic molecules is evident in so far as two free atomic HF profiles of N and O, respectively are at $p_z = 0$, 13% larger than the corresponding experimental profiles of the diatomic molecules. This could also be confirmed by Huotari (2002) in a more recent synchrotron radiation based study on N_2 with much higher statistical accuracy, by comparing, as shown in Fig. 4.14, the experimental profile with two-times a free atomic profile, calculated in a nonrelativistic HF scheme. Huotari (2002) has also compared his experimental result with two LCAO calculations on N_2 molecules using the DALTON program and a Gaussian basis set first introduced by Dunning (1989), where in one of the computations correlation effects were taken into account, again on the basis of the second-order perturbation treatment of Müller and Plesset (1934). In this case, as shown in Fig. 4.15, the inclusion of correlation improves the agreement with experiment, even though a sizeable discrepancy between theory and experiment remains. Also the calculations of Thakkar et al. (1990) (shown in Fig. 4.15), though performed by using still more elaborate basis sets, could not lead to full agreement with experiment, so that one must assume that in all cases the correlation effects are not completely accounted for.

The first application of a γ-ray source in Compton scattering was also devoted to free atomic and molecular systems. Eisenberger and Reed (1972) have investigated He, N_2, Ar and Kr using a ^{123m}Te 160 keV γ-ray source. With respect to

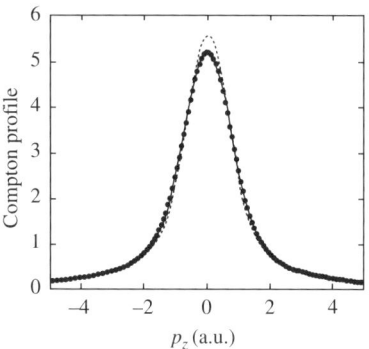

Fig. 4.14. The experimental Compton profile of N_2 molecules (dots). LCAO calculated Compton profile of N_2 including correlation (solid line). The dashed line is the corresponding free-atom nitrogen Compton profile multiplied by two. (Reprinted with permission from Huotari (2002).)

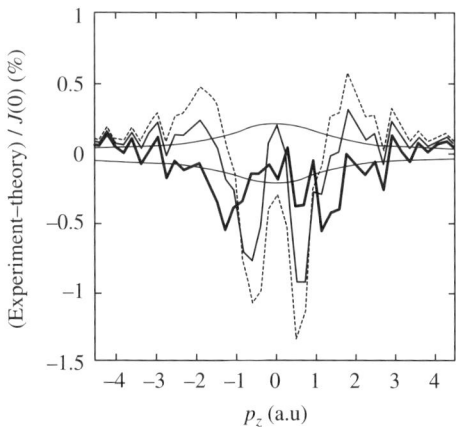

Fig. 4.15. The difference between the experimental and various theoretical N_2 Compton profiles. Dashed line: LCAO calculation without correlation; thin solid line: LCAO calculation including correlation; thick solid line: calculation by Thakkar *et al.* (1990). Note the confidence margins of the experiment. (Reprinted with permission from Huotari (2002).)

He and N_2 former results were confirmed. The Ar and Kr results were in excellent agreement with HF calculations. Indeed later calculations by Tong and Lam (1978) using a correlation correction scheme proposed by Lam and Platzman (sketched in Section 6.8.4) showed that even for heavier closed-shell atoms the correlation correction is rather small (less than 2% of the Compton profile at $p_z = 0$). These γ-ray experiments were at that time the first Compton measurement on higher-Z elements and can be considered as a breakthrough of Compton scattering techniques.

Another problem connected with the Compton profiles of closed shells is the core contribution to the Compton profiles of otherwise open systems. As shown in Section 4.5.5 these core contributions must be taken for granted in order to come to an appropriate normalization of the profiles. Especially the K- and L-shell contributions of heavier elements offers some problems, since they have to be calculated relativistically. Therefore, experimental tests on the validity of relativistic Hartree–Fock schemes as used by Biggs *et al.* (1975) were of importance for the Compton community. Figure 4.16 shows the result of such a test, namely the high-momentum tail of the Compton profile of gold as measured by Pattison and Schneider (1979) using 412 keV γ-radiation. The sharp cut-off in the high-energy profile tail when the energy transfer drops below the K or L binding energy is well pronounced and fits quite well the relativistic calculations of Biggs *et al.* (1975).

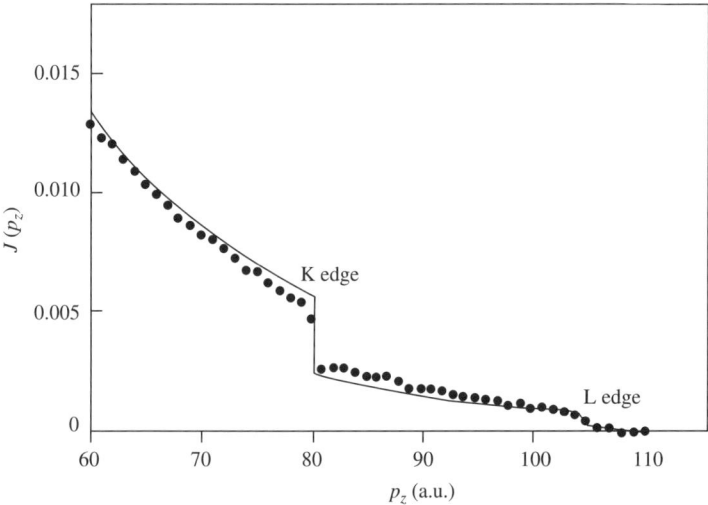

Fig. 4.16. The high-momentum tail of the Compton profile of gold (filled circles) measured with 412 keV radiation by Pattison and Schneider (1979). The solid curve is the relativistic HF profile of Biggs *et al.* (1975). (Reprinted with permission from Cooper (1985); copyright (1985) by IOP Publishing Ltd.)

4.7 Compton scattering and chemical bonds

4.7.1 *Momentum space and position space representations of chemical bonds*

If one describes chemical bonds in terms of single-particle wavefunctions, either in the form of the molecular orbital (MO) model (see 6.194), or within the so-called valence bond (VB) model (see 6.199), it is the phase correlation (Löwdin 1967) which counts. Without interaction between the atoms, the phases of their orbitals remain completely uncorrelated and all quantum mechanical observables are simply superpositions of the pure atomic contributions. The phase correlation due to the covalent interaction leads to interference phenomena, which are visible in the additional terms of the physical observables. It is these interference terms by which covalent bonding is recognizable.

Within the limits of the MO model with a linear combination of two (real) atomic orbitals $\psi_a(\mathbf{r})$ and $\psi_b(\mathbf{r})$, the momentum space density $\rho(\mathbf{p})$ writes (see 6.197)

$$\rho(\mathbf{p}) = [2(1 \pm S)]^{-1}(|\chi_a(\mathbf{p})|^2 + |\chi_b(\mathbf{p})|^2 \pm 2\chi_a(\mathbf{p})\chi_b(\mathbf{p})\cos(\mathbf{p}\cdot\mathbf{R}/\hbar)), \quad (4.95)$$

where $\chi_{a,b}(\mathbf{p})$ is the Fourier transform of $\psi_{a,b}(\mathbf{r})$, $\mathbf{R}$ the internuclear separation and S the overlap integral defined by (6.195). If S is positive, the positive sign in (4.94) stands for a binding, the negative sign for an antibonding orbital. For negative S the opposite signs are valid. Thus a directional Compton profile of a bonding MO, derived from (4.95), exhibits oscillations along the bonding direction with a period equal to $2\pi/R$; an antibonding MO will exhibit similar oscillations 90° out of phase with those of the bonding orbital. Moreover, as shown in (6.196), the charge density of a MO possesses a peak midway between the nuclear positions, causing the wavefunction to vary less abruptly along the internuclear axis than it did in the absence of the bond. Since the momentum is the gradient of the wavefunction, the component of momentum along the bond is decreased. Therefore, one can state generally that the momentum of an electron in a chemical bond is more likely directed perpendicular to that along the bond axis.

Of course, in reality, bonds are either isotropically distributed as for instance in diatomic gases or share their role with other bonds in a solid, so that the signature for a specific bond in a Compton profile is not easily revealed. One possibility is to look at differences between an isotropically averaged profile (polycrystalline sample) and a directional profile with a scattering vector in the direction of the bond axis. Then the difference profile should exhibit signatures of the oscillations given by (4.95), which could be visualized by calculating the so-called power spectrum (modulus squared of the Fourier transform) of the difference profile. By measuring a number of directional Compton profiles with different directions of the scattering vector the momentum space density $\rho(p)$ can be reconstructed, as shown in Section 4.7.2. In that case the signatures of different bonds in various directions can be observed.

A wealth of information about chemical bonds offers the reciprocal form factor $B(\mathbf{r})$, as first pointed out by Weyrich et al. (1979). $B(z)$ is the 1D Fourier transform of a Compton profile (see Section 4.2), where the z-directions coincides with the direction of the scattering vector q, so that, as will be shown in Section 4.7.2, the complete function $B(\mathbf{r})$ can be reconstructed out of a set of Compton profile measurements with different directions of $\mathbf{q}$ with respect to the sample. For a diatomic molecule built by two identical atoms in a distance $\mathbf{R}$ and described within the limits of the MO model, the reciprocal form factor $B(\mathbf{r})$ writes (see Section 6.8.3.2)

$$B(\mathbf{r}) = F_1(B_\mathrm{a}(\mathbf{r}) + B_\mathrm{b}(\mathbf{r}) \pm F_2[B_\mathrm{ab}(\mathbf{r}+\mathbf{R}) + B_\mathrm{ba}(\mathbf{r}-\mathbf{R})]);$$
$$F_1 = [2(1\pm S)]^{-1}; \quad F_2 = 1 \qquad (4.96)$$

where the same sign convention is valid as for (4.95). $B_{\mathrm{a,b}}(\mathbf{r})$ is the reciprocal form factor of the atom with orbital wave function $\psi_{\mathrm{a,b}}(\mathbf{r})$.

$$B_\mathrm{ab}(\mathbf{r}\pm\mathbf{R}) = \int \psi_\mathrm{a}(\mathbf{s}+\mathbf{r}\pm\mathbf{R})\,\psi_\mathrm{b}(\mathbf{s})\,\mathrm{d}\mathbf{s}. \qquad (4.97)$$

The valence bond (VB) model leads to the same expression, only the prefactors are different, namely $F_1 = [2(1\pm S^2)]^{-1}$ and $F_2 = S$. Since $B_{\mathrm{a,b}}(\mathbf{R})$ and $B_\mathrm{ab}(2\mathbf{R})$ are small compared to $B_\mathrm{ab}(0)$, $B(\mathbf{r})$ shows a secondary extremum at $\mathbf{r} = \mathbf{R}$.

4.7.2 3D reconstruction of $B(\mathbf{r})$ and $\rho(\mathbf{p})$

As already stressed, the signatures of chemical bonds in crystalline matter, both in real and in momentum space representation, can be found, when the complete $B(\mathbf{r})$ and $\rho(\mathbf{p})$ functions are reconstructed by using a larger number of directional Compton profiles with different q-orientation. Since we will discuss in what follows characteristic chemical bonds based on those reconstructions, a short description of reconstruction methods is necessary at this point. This presentation is restricted to the so-called Fourier–Bessel method, as introduced by Hansen (1980).

If the momentum density of a system under investigation is anisotropic, $\rho(\mathbf{p})$ and $B(\mathbf{r})$ are functions both of the value and of the direction of the $\mathbf{p}$- and the $\mathbf{r}$-vector, respectively. In order to decouple the radial from the angular part, we expand the $B(\mathbf{r})$ function into lattice harmonics $F_l(\theta,\phi)$:

$$B(\mathbf{r}) = \sum_l b_l(r)\, F_l(\theta,\phi), \qquad (4.98)$$

where θ and ϕ are the angular variables of $\mathbf{r}$. For reconstruction we need the coefficients $b_l(r)$ for as many lattice harmonics as can be extracted from experiment. Let their number be N. If we have measured a finite set of directional Compton profiles of dimension M, N must obey $N \leq M$, since the $b_l(r)$ are obtained by solving the following system of linear equations:

$$B(\mathbf{r}_i) = \sum_l b_l(r) F_l(\theta_i, \phi_i), \quad i = 1, \ldots, M, \qquad (4.99)$$

where the sum is over N different values of l, according to the symmetry of the lattice under investigation. Because of (4.22) θ_i and ϕ_i are the angular variables of the M different $\mathbf{q}$'s of the directional Compton profiles. If we write (4.99) in the form of a matrix equation

$$\mathbf{B} = \mathbf{F}\mathbf{b} \qquad (4.100)$$

one obtains, for $M > N$,

$$\mathbf{b} = (\mathbf{F}^t \mathbf{F})^{-1} \mathbf{F}^t \mathbf{B} \qquad (4.101)$$

and, for $M = N$

$$\mathbf{b} = \mathbf{F}^{-1} \mathbf{B}. \qquad (4.102)$$

Using this set of $b_l(r)$, we obtain the 3D reconstructed $B(\mathbf{r})$ according to (4.98), and the 3D reconstructed momentum space density $\rho(\mathbf{p})$ by means of the inverse Fourier transform of $B(\mathbf{r})$:

$$\rho(\mathbf{p}) = (2\pi\hbar)^{-3} \sum_l \int d^3r\, b_l(r) F_l(\theta_{\mathbf{r}}, \phi_{\mathbf{r}}) \exp(i\mathbf{p}\cdot\mathbf{r}/\hbar)$$

$$= \sum_l \rho_l(p) F_l(\theta_{\mathbf{p}}, \phi_{\mathbf{p}}), \qquad (4.103)$$

with ($\hbar = 1$)

$$\rho_l(p) = (i^l/2\pi^2) \int_0^\infty b_l(r) j_l(pr)\, r^2 dr, \qquad (4.104)$$

where $j_l(pr)$ are spherical Bessel functions. As pointed out by Hansen (1980), the quality of the reconstruction depends not only on the number of contributing Compton profiles but also on how the corresponding directions of $\mathbf{q}$ are distributed within the irreducible triangle of their stereographic projections. By means of a covariance analysis of error propagation one can calculate error maps of the 3D reconstructed distributions (for details see Schülke et al. 1996). Alternatively to the Fourier–Bessel method, the direct Fourier transform scheme of Mueller (1977) has been applied by several authors, where $\rho(\mathbf{p})$ is obtained directly, by inverting (4.21), from $B(\mathbf{r})$'s, which are deduced from the Compton profiles and interpolated on an appropriate $\mathbf{r}$-mesh. Moreover, other reconstruction schemes for $\rho(\mathbf{p})$ has been proposed, which are not based on the preceding reconstruction of the $B(\mathbf{r})$ function. This is the so-called Fourier–Hankel method, first introduced by Mijnarends (1967, 1969) and the application of Cormack's method (1963, 1964) by Kontrym-Sznajd et al. (1997). Gillet et al. (1995) have gone along quite another path by refining models of the momentum densities using the data of anisotropic Compton profiles. Coupling coefficients between various atomic functions and constants describing the radial part of the model wavefunctions were used as adjustable parameters. One step further has been done by Gillet *et al.* (2001), who carried on an idea of Schülke and Kramer (1979), when they applied a joint refinement of a local wavefunction model from Compton and Bragg scattering data of MgO. Finally, the maximum entropy method (MEM), which was already an established method in the field of charge density reconstruction, was adapted to momentum density reconstruction by Dobrzynski and Holas (1996).

It should be mentioned at this place that the $B(\mathbf{r})$ function, reconstructed according to the above formalism, can also be used to reconstruct, by means of (4.32), the occupation number function $N(\mathbf{k})$ which was defined in (4.31).

4.7.3 *Covalent bonding, semiconductors*

As pointed out in Section 4.7.1, the basic phenomenon of binding is the phase correlation of the participating orbitals, which leads to interference, detectable by additional terms of the observables. It is these interference terms that indicate experimentally covalent bonding. Especially the $B(\mathbf{r})$ function is, in principle, capable of distinguishing between different types of covalent bonding, as shown by Weyrich *et al.* (1979) on the basis of simple calculations for homonuclear MO's, depicted in Fig. 4.17.

A σ-bond between two 1s orbitals, likewise a π-bond between two 2p orbitals, leads to a pure positive $B(\mathbf{r})$ function for $\mathbf{r}$ in the bond direction. On the contrary, an antibonding σ-type phase-correlated superposition of two 1s orbitals, as well as a σ-bond between a 1s and a 2p orbital results in one zero-passage of $B(\mathbf{r})$. The same is found for an antibonding σ-type superposition of two 2p orbitals, provided the binding distance R is small, and for a π-type antibonding phase-correlated superposition of two 2p orbitals for $\mathbf{r}$ in the bond direction. Of course, as shown exemplarily in the 2D representation of Fig. 4.18, the $B(\mathbf{r})$ function for two π-bounded 2p orbitals exhibits zero-passages for $\mathbf{r}$ in the direction of the atomic orbital axes (x-axis in Fig. 4.18).

The behavior of the $B(\mathbf{r})$ function ($\mathbf{r}$ in the bond direction) for other bonding and antibonding configurations can be read from Fig. 4.17, where it can easily be recognized that the sequence of signs in $B(\mathbf{r})$ allows one to distinguish between binding and antibinding orbitals, provided the symmetry of the contributing orbitals is known.

A very intuitive example for the application of the above sign rules of $B(\mathbf{r})$ is provided by investigations of Pattison *et al.* (1981) on the classical covalent-bonded solid, diamond. They used the Compton profile calculations of Seth and Ellis (1977) to compute the anisotropy of the $B(\mathbf{r})$ function of diamond by omitting the first term in the harmonic expansion of $B(\mathbf{r})$ (4.98). In the bonding direction [111], they found, as shown in Fig. 4.19, a strong negative depression along [111] at a separation of 1.54, which is exactly the nearest neighbor

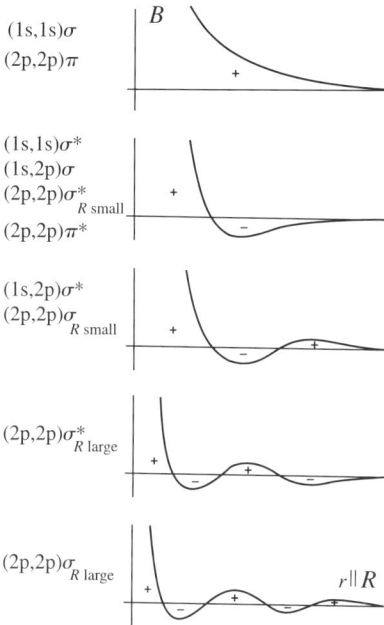

Fig. 4.17. Sequence of signs in $B(r)$ for various molecular orbitals along the direction of the bond. (Reprinted with permission from Weyrich et al. (1979); copyright (1979) by North-Holland Publishing Company.)

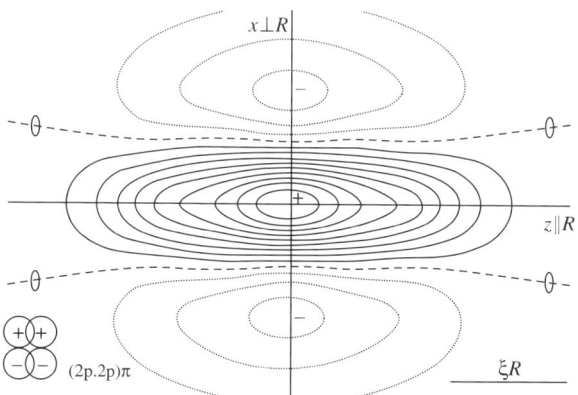

Fig. 4.18. $B(\mathbf{r})$ function for a homonuclear molecular orbital of two π-bounded 2p orbitals. Contours of $B(\mathbf{r})$ are shown in a sectional plane, the orientation of which is defined in a schematic orbital picture. The interval between contours is 0.1, starting from 1.0 at the origin. (Reprinted with permission from Weyrich et al. (1979); copyright (1979) by North-Holland Publishing Company.)

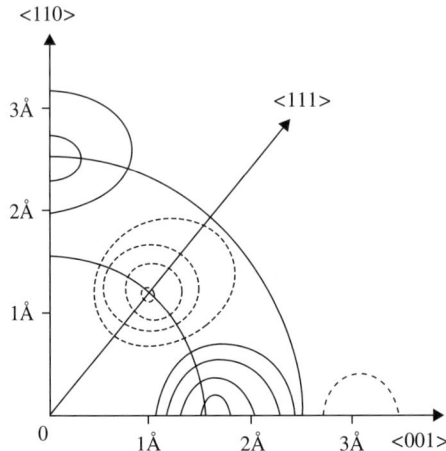

Fig. 4.19. The anisotropic part of $B(\mathbf{r})$ for diamond derived from Compton profile calculations of Seth and Ellis (1977). The contours are given in intervals of 0.05 electrons. The radial distances corresponding to the bond length (1.54 Å) and the second neighbor bond distance are indicated. (Reprinted with permission from Pattison *et al.* (1981); copyright (1981) by North-Holland Publishing Company.)

distance. Such a negative depression was also present in the $B(\mathbf{r})$ reconstructions of experimental data of Si, though not that pronounced as in the case of diamond. This was interpreted as a clear indication of the strong covalent (2p,2p) σ-bond in the [111] direction, where, according to (4.96), $B(\mathbf{r})$ will exhibit a negative extremum at $\mathbf{r} = \mathbf{R}$ ($\mathbf{R}$ is the bonding distance), since $S < 0$. Of course, the anisotropic part of a $B(\mathbf{r})$ reconstruction is not free from artifacts, which cannot be directly related to bond features. Thus a strong feature of the $B(\mathbf{r})$ function in some direction will also contribute to the spherical average. When this is subtracted from the total $B(\mathbf{r})$, information can be transferred to an other angular position. This is clearly the case with the strong positive enhancement of $B(\mathbf{r})$ in the [001] direction just in the vicinity of the bonding distance in Fig. 4.19. Therefore, the theoretical difference between the diamond $B(\mathbf{r})$ along [111] and [001] exhibits still much more pronounced the deep depression at the bonding distance R_b in very good agreement with the experimental results of Reed and Eisenberger (1972). The same feature at the bonding distance of the ([111]-[100])-$B(\mathbf{r})$ difference has also been found in the Hartree–Fock calculations on SiC-3C of Ayma *et al.* (1998b) and on cubic BN of Ayma *et al.* (1998a). The positive contours in the [110] direction of Fig. 4.19, which is close to the second neighbor distance along this direction, arises from an orthogonalization correction between different bonds, i.e. a nonbonding interaction. This interpretation is supported by a later 3D reconstruction of $B(\mathbf{r})$ on the basis of six directional Compton profiles of Si (Hansen *et al.* 1987) and a comparison with calculations of Kane and Kane (1978), who employed a localized model for bond Wannier functions, where they varied this model by increasing the sphere around the reference bound that contains neighboring bounds taken into account in the process of orthogonalization. This way the above positive contour in the [110] direction, as seen in the experiment, appeared in the calculations only after the overlapping of all the simple two-centre bonds in the crystal had been considered using Löwdin's symmetrical orthogonalization procedure (Löwdin 1956). The 3D reconstruction of $\rho(\mathbf{p})$, simultaneously performed, does not allow such an intuitive interpretation on the basis of extended orthogonalization. Thus this example along with all other above presentations may show that the Fourier transform of the electron momentum density is a useful representation for identifying the bond information in Compton data, since it makes possible the position space separation between the various types of electronic interaction in crystals.

The lattice structure of covalent-bond semiconductors is represented in the zero-passages of $B(\mathbf{r})$ at the lattice translation vectors $\mathbf{r} = \mathbf{R_n}$, as already discussed in Section 4.2 (see equation 4.30). These zero-passages are the consequence of the orthogonality of the corresponding Wannier functions, and can be used to test both the reliability of theoretical calculations and of Compton measurements on single crystals. They were discussed, e.g. by Krusius (1977) and Krusius *et al.* (1982) in connection with Compton measurements on Se single crystals, and by Pattison and Schneider (1978) along with measurements on Si and Ge. But it turned out, already in the earliest experimental investigation of the behavior of $B(\mathbf{r})$ of insulators or semiconductors, that the first zero-passage is very near to the nearest neighbour distance, but a little above it (see Kramer *et al.* (1977) for the case of trigonal Se; and Schülke (1977) for the case of Si). In a general study on the interpretation of the Compton profiles of diamond structure semiconductors, MacKinnon and Kramer (1980) have shown that this first zero, near the binding distance, is intimately connected with the need to orthogonalize the ansatz functions of their model for different atoms. Without orthogonalization the first zero-passage is found, in the case of a Gaussian basis, at the next-nearest neighbor in [110] direction, rather than at the binding distance. On the other hand, the first zero is rather insensitive against the choice of the basis functions and independent of the details of the Hamiltonian, whereas these parameters determine strongly the overall form of the density of states. This leads the authors of the study to the conclusion that information which can be obtained from Compton scattering is largely complementary to that obtained from the density of states. Krusius *et al.* (1979) have studied the behavior of $B(\mathbf{r})$, especially the zero-passages, of ordered and disordered one-dimensional potential arrays, where the situation of filled and partly filled bands could be simulated. They found, like in the experiment, that the zeros of the $B(\mathbf{r})$ function of an ordered elemental insulator or semiconductor are related to characteristic separations, such as the atomic separations, of the system. For ordered systems with partly filled bands (metallic systems), the zeros are related to the Fermi momentum.

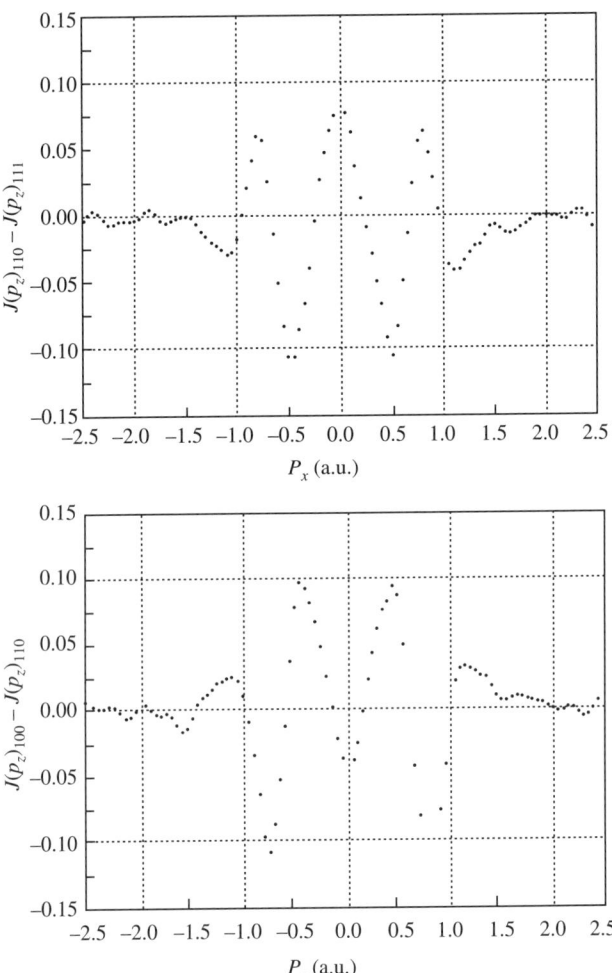

Fig. 4.20. Compton profile differences of Si (a) [110] − [111]; (b) [100] − [110]. (Reprinted with permission from Berthold *et al.* (1992a); copyright (1992) by Elsevier Science Publisher B.V.)

The influence of pure structural or compositional disorder on $B(\mathbf{r})$ turned out to be negligible. This was verified by a Compton experiment on Se, where the $B(\mathbf{r})$ function of polycrystalline and amorphous Se was nearly indistinguishable. Preceding investigations (Kramer *et al.* 1977), which found considerable differences in $B(\mathbf{r})$ between polycrystalline and amorphous Se, showed afterwards incorrect evaluation of the amorphous data.

Until now we have discussed the covalent bonding effects in semiconductors and insulators mainly in the position space representation of the Compton

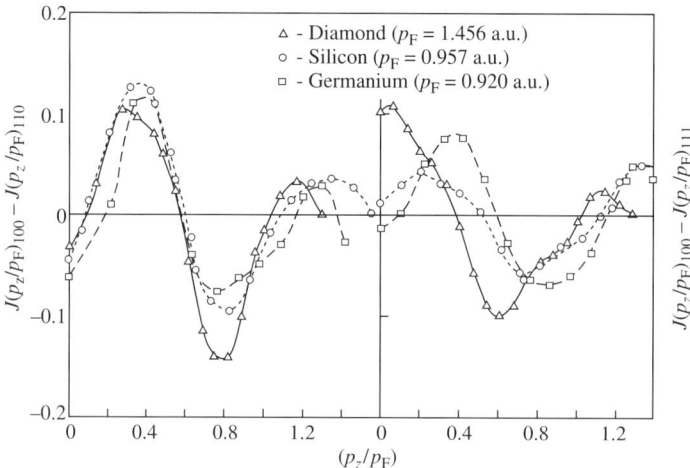

Fig. 4.21. Anisotropies of the Compton profiles at equal electron density for diamond (triangles), Si (open circles), and Ge (squares). Left panel: [100] − [110]. Right panel: [100] − [111]. (Originally published by Reed and Eisenberger (1972); copyright (1972) by the American Physical Society.)

data. However, most Compton experiments and calculations on these systems are represented in momentum space. We will show some characteristics of this representation, where again only the most prominent examples, mainly those obtained with synchrotron radiation, shall be singled out. Until now Si, the tetrahedrally coordinated covalent bound semiconductor, has been investigated most frequently, using conventional X-rays (Schülke 1974), γ-ray sources (Reed and Eisenberger 1972, together with investigations on Ge and diamond; Hansen et al. 1987, Schülke and Kramer 1979, together with investigation of Ge), and synchrotron radiation (Sakai et al. 1989; Berthold et al. 1992a, Kubo et al. 1997). In all cases the measurements of directional Compton profiles are presented in the form of directional differences, where the differences [100] − [110], and [110] − [111] exhibit the strongest amplitudes (see Fig. 4.20 from Berthold et al. 1992a). These directional differences are best suited for comparison with theory, since systematic errors of the experiment arising from multiple scattering corrections, background subtraction, and normalization are widely suppressed by building the differences. Nevertheless, the hope that these differences might help to test sensitively band structure calculations, more precisely, to investigate the quality of wavefunctions obtained for different potentials, was, to some extent, disappointed. Already Reed and Eisenberger (1972) have stressed that the [100] − [110] difference profiles of diamond, Si and Ge look nearly the same, when plotted against p_z/p_F (p_F is the Fermi momentum according to the corresponding valence electron density), as shown in Fig. 4.21.

This scaling behavior can easily be understood on the basis of the "modified nearly free electron model", where the momentum density $\rho(\mathbf{p})$ has the form of a step function, that is $\rho(\mathbf{p})$ has a constant value in a closed zone of reciprocal space and is zero elsewhere. The conventional choice for a diamond-structure crystal is the so-called "Jones zone" (Mott and Jones 1936), where the 12 {220}-type Bragg planes form the boundary of that zone. Within the limits of this "modified free electron model" the directional profile differences of all diamond-structure crystals must look the same, when scaled to their characteristic length in momentum space. Of course, this simple model, which would provide amplitudes of the [100] – [110]-difference up to five times larger than in experiment, must be revised according to the real Bloch character of the occupied states, which we know give rise to higher momentum components, so that the characteristics of the Jones-zone model are widely smeared out. Nevertheless, a theoretical study of Nara *et al.* (1984) of the directional Compton profile differences of a number of IV, III-V and II-VI tetrahedrally bonded semiconductors on the basis of a local pseudopotential scheme has demonstrated that indeed the Compton profile directional differences of Si, Ge and Sn (IV-semiconductors), when taken relative to the [100] direction, have all the same shape and even the same amplitude, with the exception of the range for small p_z. The same is true with the series of III-V semiconductors GaAs and InSb, but with smaller amplitudes. Still smaller amplitude, but again equal shape, is found for the II-VI semiconductors ZnSe and CdTe. This behavior tells us that the very different potential used in the band structure calculation, e.g. for Si and Ge, has only a minor influence on the anisotropy of the Compton profiles. However, it is worthy to note that the heights or depths of the extrema of the Ge-GaAs-ZnSe series, alongside the corresponding value of Ge, are reasonably close to the values of the covalency parameters of Philips (1973). The amplitude of the directional Compton profile differences decreases with increasing ionicity. A somewhat more pronounced sensitivity to changes of the local pseudopotential are found, when the profile in the [111] direction is chosen as a reference. In Fig. 4.22 experimental Si [100] – [111] directional differences of Kubo *et al.* (1997) are confronted with the (resolution function convoluted) pseudopotential calculations of Delaney *et al.* (1998). One finds good qualitative agreement as far as the position of the extrema and the zero passages are concerned, though quantitatvely there are differences between the heights of the peaks. The dashed curve in Fig. 4.22 belongs to *ab initio* pseudopotential calculations, using the norm-conserving Hamann–Schlüter–Chiang (1979) pseudopotential. The pseudowavefunctions are constructed so as to have a smooth behavior near to the atomic cores, and so they lack some of the high momentum Fourier components that the all-electron valence wavefunctions have. Therefore, the all-electron valence wavefunctions were reconstructed by changing the pseudowave functions inside a reconstruction sphere around the atoms by re-solving there the Kohn-Sham equation self-consistently using an all-electron potential. The result of this reconstruction procedure is depicted in Fig. 4.22 by the solid line, which brings the calculated and the experimental profile differences

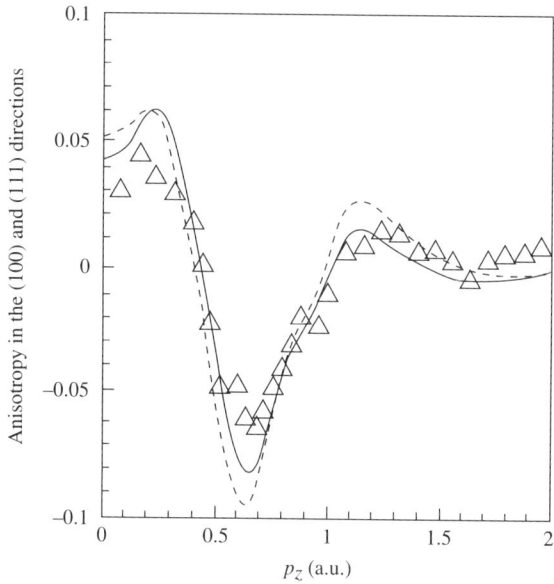

Fig. 4.22. Directional [100] − [111] Compton profile differences of Si. Triangles: experimental data of Kubo *et al.* (1997). Dashed line: pseudopotential calculation of Delaney *et al.* (1998). Solid line: reconstruction calculation of Delaney *et al.* (1998). (Originally published by Delaney *et al.* (1998); copyright (1998) by the American Physical Society.)

closer. The remaining discrepancies can probably be traced back to correlation effects, although the directional differences of corresponding correlation corrections on the basis of an *ab initio* nonlocal pseudopotential variational quantum Monte Carlo calculations (Kralik *et al.* 1998) exhibit much smaller values. On the other hand, the directional averages of these Monte Carlo computations agree quite well with the Lam–Platzman correction (Lam and Platzman 1974) according to (4.34), which, added to the calculated total Si valence Compton profile of Delaney *et al.* (1998), brings the calculations into excellent agreement with more recent measurements of three groups: Dortmund (synchrotron radiation), ESRF (conventional source) and Konstanz (γ-source), as shown in the informal *Proceedings of the Second International Workshop on Compton Scattering and Fermiology*, Tokyo, 1995. Corresponding experimental total Si profiles of the KEK (Japan, synchrotron radiation) and the Bialystok (Poland, γ-source) group, which show a more than 5% smaller value at $p_z = 0$, were apparently suffering from some inadequacies of data processing, which lead to an (unphysical) constant offset of the measured core contribution compared to the free atomic one, over a large range of p_z. It might be mentioned that, in the case of Si, a Compton profile calculation (Kubo *et al.* 1997) using the self-interaction correction scheme of Perdue and Zunger (1981) gave only a minor correction to the result of the FLAPW (full-potential linearized augmented plane wave) method.

So far we have exemplarily discussed the Compton profiles and $B(\mathbf{r})$ functions of covalent bound tetrahedrally coordinated IV semiconductors in order to provide proof of how well the experimental valence electron momentum density of these systems is understood and can be represented, in its general properties, by actual theoretical approaches. Therefore, other systems, at least semicovalent bonded, will be treated only in a short survey. Directional γ-Compton profiles of GaAs, a prototype of III-V semiconductors, have been investigated by Panda et al. (1990). The results are compared with local and nonlocal pseudopotential calculations. The measured Compton profile anisotropy $J_{100}(p_z) - J_{111}(p_z)$ exhibits an oscillatory period distinctly longer than the theoretical one. The same was observed with GaP (Panda et al. 1992). This discrepancy might be due to the partial ionicity, which is not well represented in a pseudopotential scheme. But a still larger discrepancy between experimental and calculated directional Compton profile differences with an antiphase behavior of the oscillations was found in a γ-Compton investigation on the wurtzite type CdS by Perkkiö et al. (1989), although one might think that an LCAO (linear combination of atomic orbitals) scheme on the basis of Cd^{2+} and S^{2-} ionic wavefunctions (the S^{2-} wavefunction was not a free atomic one, but calculated in a potential describing the crystal surrounding) should be much better suited than a pseudopotential scheme with a limited number of plane waves. In this case it was concluded that it was the inadequate consideration of the partial covalency, which has to be made responsible for that discrepancy. On the other hand, Lichanot et al. (1996) demonstrated, for the case of hexagonal BN, that self-consistent field (SCF) LCAO schemes, both at the Hartree–Fock (HF) and at the LDA level (by solving the Kohn–Sham (KS) equation after each SCF cycle) can be brought to excellent agreement with experiment (Loupias et al. 1994), as shown in Fig. 4.23, provided an all-electron basis set including d-like polarization functions is used rather than a minimal basis set. Calculations on a similar footing were also performed for SiC-3C (space group F-43m) by Ayma et al. (1998) and were compared with rather old γ-Compton measurements of Mahapatra and Padhi (1982) on polycrystalline material. It should be mentioned that the $J_{100}(p_z) - J_{111}(p_z)$ directional difference of these calculations are in rather severe disagreement with those of Seth and Ellis (1977) but in good qualitative agreement with calculations and measurements of tetrahedrally coordinated semiconductors. This might be traced back that the basis set (Slater type orbitals) of Seth and Ellis (1977) was of poorer quality.

Chemists are used to thinking in terms of the transferability of bonds from molecule to molecule when correlating, predicting and considering a multitude of molecular properties, especially in larger molecules. The reason for the success of this way of thinking is that many properties appear to depend almost completely on the type of orbital and much less on the environment in which this orbital is found. Thus it was obvious to ask whether or not such a localized

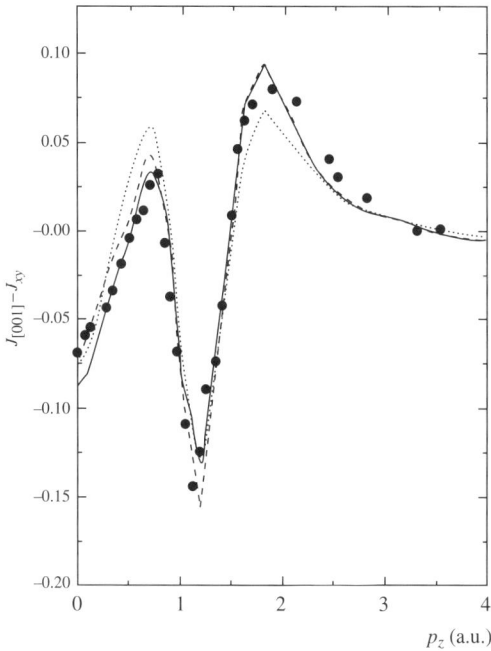

Fig. 4.23. Hexagonal BN valence Compton profile anisotropy $J_{[001]} - J_{xy}$, where J_{xy} is the basal-plane Compton profile defined by $(1/2)(J_{[110]} - J_{[1\bar{1}0]})$. The experimental data are from Loupias *et al.* (1994). Full, dashed and dotted lines correspond to the calculations done at the LDA level with the use of a basis set B_2, B_1, and at the HF level using basis set B_2, respectively. (Originally published by Lichanot *et al.* (1996); copyright (1996) by IOP Publishing Ltd.)

orbital picture can also be of use in understanding momentum space properties, or in other words, to what extent momentum density or Compton profiles can be partitioned into contributions from bonding and other localized orbitals, whether these contributions can be transferred from molecule to molecule, and how sensitive these contributions are to the molecular environment. The most conclusive evidence for the transferability of the bond picture in the Compton profile analysis was already provided in the very early days of Compton scattering, namely by the investigations of Eisenberger and Marra (1971). They carried out Compton profile measurements of hydrocarbons. By assuming that C–C and C=C bonds have distinct profiles, they calculated these profiles from their measured methane, ethane and ethylene Compton profiles. Then they used these data to calculate the profiles of larger molecules, which proved to be in excellent agreement with their experiments. The same idea was followed up, on the theoretical side, by Epstein (1970) and Epstein and Lipscomb (1970), investigating hydrocarbons and boron hydrides. But rather than decomposing the Compton profiles with respect to single and double carbon bindings, these authors used localized molecular orbitals (LMO's) for decomposition. They found inner shells almost perfectly transferable, B—H and C—H bonds a bit less, and B—B and C—C bonds showing the greatest variation from molecule to molecule.

4.7.4 Ionic crystals

Since the point charge model of ionic crystals reproduces with high accuracy the lattice energy (Madelung energy), one could believe that the electronic structure of ionic crystals is nothing than the superimposing contributions of free ions. However, two facts enforce a change of the electronic structure: The virial theorem and orthogonalization. In order to diminish the total energy by the value of the Madelung energy, the virial theorem lets the potential energy decrease by twice this value, and increases the kinetic energy by this value. This can be achieved by shrinking of the ions, and we can read the lattice energy from the second moment of the Compton profiles, see equations (6.185) and (6.187). The claim for orthogonality forces an admixture of orbitals of adjacent ions. This becomes transparent, when looking at the $B(\mathbf{r})$ function. Following Pattison and

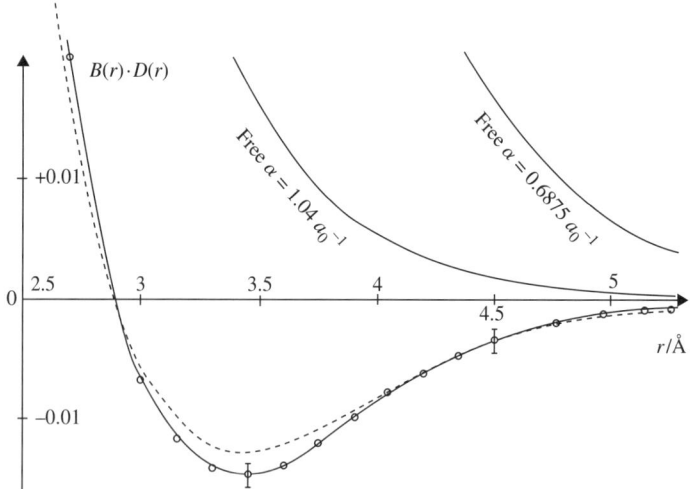

Fig. 4.24. Comparison between the experimental $B(r) \cdot D(r)$ ($D(r)$ is the Fourier transformed resolution function) for LiH powder (full line) and the model of an antibonding-type hydride cluster orbital (dashed curve) as described in the text. The results of free hydride ions with the same orbital exponent $\alpha = 1.04$ as well as with $\alpha = 0.6875$ are also given. (Reprinted with permission from Pattison and Weyrich (1979); copyright (1979) by Pergamon Press.)

Weyrich (1979), let us take LiH as the simplest example of an ionic solid. Because of the pure 1s character of the wavefunctions, the $B(\mathbf{r})$ function of the free ions is completely positive, as shown in Fig. 4.24 for different orbital exponents of the H$^-$ ions. On the contrary, both the γ-Compton experiment on a polycrystalline sample, and the results of a calculations including orthogonalization of the ionic orbitals exhibit, as shown in Fig. 4.24, a negative minimum of $B(\mathbf{r})$. According to the rules fixed in Section 4.7.3, the admixture of orbitals from the neighbouring ions must therefore be antibonding. In the position space density, the antibonding admixture leads to a decreasing of $\rho(\mathbf{r})$ in between the ions, energetically this admixture results in a repulsion. Without this repulsion, the ions would collapse due to the electrostatic attractions. Because of the small spatial extent of the Li$^+$ ion, it is exclusively the H$^-$ contribution which determines the free ion $B(\mathbf{r})$ in Fig. 4.24. Therefore, only the orthogonalization of the H$^-$ ions must be made responsible for the negative minimum. The crystal orbital ψ_μ obtained by symmetrical orthogonalization from the ion orbitals ϕ_α are, according to Löwdin (1970), given by

$$\psi_\mu = \phi_\mu - \sum_\alpha \phi_\alpha S_{\alpha\mu} + \frac{3}{8} \sum_\alpha \sum_\beta \phi_\alpha S_{\alpha\beta} S_{\beta\mu}$$

$$- \frac{5}{16} \sum_\alpha \sum_\beta \sum_\gamma \phi_\alpha S_{\alpha\beta} S_{\beta\gamma} S_{\gamma\mu} + \cdots \qquad (4.105)$$

where $S_{\kappa\lambda}$ are the overlap matrix elements between the orbitals ϕ_κ and ϕ_λ. In the case of LiH with s-type basis functions, the terms with negative sign provide the antibonding contributions. If the modulus of $S_{\kappa\lambda}$ is not close to unity, the second term of (4.105) is the dominant admixture to ϕ_μ. It is typical for ionic bonding that a trial function of the form

$$\psi(\mathbf{r}) = N \left[\phi(\mathbf{r}) + c \sum_{j=1}^{12} \phi(\mathbf{r} - \mathbf{R}_j) \right] \qquad (4.106)$$

is able to describe any covalent interaction between a central hydride ion and its 12 nearest neighbor hydride ions located with the internuclear vector $\mathbf{R}_j$. Choosing by trial and error $\alpha = 1.04$ and $c = -0.01685$ (the constant c quantifies the admixture from the neighbors at $\mathbf{R}_j$), the experimental zero-passage of $B(\mathbf{r})$ were reproduced. The result is shown in Fig. 4.24 by the dashed curve. Thus the importance of antibonding interaction is clearly demonstrated. In a later γ-Compton investigation of Asthalter and Weyrich (1993) on single crystal LiH, the experimental $B(\mathbf{r})$ in [100], [110] and [111] are compared with different free ion and crystal basis sets. It turned out that only crystal wavefunctions using the Hartree–Fock program for periodic systems of Pisani et al. (1988) based on an extended basis set, which possesses p-type polarization functions on both ions (Dovesi et al. 1984), were able to fit the experimental data well. The most negative orbital autocorrelation occurred in the [110] direction, where the overlap between the diffuse charge distribution of adjacent hydride ions reaches a maximum. More recent high-resolution synchrotron radiation based Compton profile data on LiH, collected by Loupias and Mergy (1980) at LURE DCI, consisting of 12 directional profiles, were used by Gillet et al. (1995) to fit a rather simple covalent model to the momentum space density as reconstructed using these Compton data. This simple model of a central hydrogen surrounded by six octahedrally coordinated Li atoms had only five adjustable parameters, namely the exponents z_H and z_{Li} of the Slater-type orbitals, a covalent parameter λ, which measures the overlap of the H 1s with the Li 2sp orbitals, and μ, the amount of polarization (2p admixture) of the Li atom. Surprisingly, the optimized fit of this model leads to an agreement with the experimental directional Compton profile differences comparable in quality with the calculations of Dovesi et al. (1984). The advantage of such a model fit is the transparent physical interpretation of its parameters. Thus the above fit exhibits a partial (antibonding) covalency with $\lambda = -0.318$, where the directional differences prove to be extremely sensitive to this covalence parameter.

Like LiH, also Li_3N proves to be an exemplary ionic solid, whose electronic structure can be understood, in a first approximation, as being composed of Li^+ and N^{3-} ions. γ-Compton profile measurements on single crystalline Li_3N (Pattison and Schneider, 1980, Pattison et al. 1984) have revealed how orthogonalization changes this simple picture, and what the most important ionic interactions are, to be responsible for the anisotropy found both in the directional Compton profiles and in the directional reciprocal form factor $B(\mathbf{r})$. Self-consistent pseudopotential band structure calculations of the Li_3N reciprocal form factor of Kerker (1981) displayed a pronounced directional dependence, which pointed to a remarkable deviation from the simple ionic picture. A combined analysis of X-ray diffraction data of Schulz and Schwarz (1978) and of NMR studies of Differt and Messer (1980) by Lewis and Schwarzenbach (1981) indicated that there must be a contraction of the N^{3-} charge density distribution along the c-direction of hexagonal Li_3N, which was not observed in the difference charge density (crystal calculation minus superposition of Li^+ and N^{3-} ionic charge densities determined by the Wattson sphere (Watson 1958) model) as predicted by Kerker (1981). An LCAO analysis of the $B(\mathbf{r})$ anisotropy, as performed by Pattison et al. (1984) for eight $\mathbf{q}$-directions within the basal plane and already shown in the form of a contour diagram in Fig. 1.8, has attributed the remarkable depression around 3 Å in the N–N direction to the most significant antibonding orthogonalization term of the N–N interaction, where the sp^2 hybrid orbital on the two nitrogen sites is oriented with its lobe pointing towards its neighbor. The long-range peak in the experimental $B(\mathbf{r})$ around 4 Å can also be explained from this N–N interaction. The Li 1s and nitrogen L-shell orbital overlaps are of the same order as the N–N overlaps. Therefore, also the influence of the N–Li othogonalization term must be considered. A peak of $B(\mathbf{r})$ is found between 2 and 4 Å in the N–Li [120] direction of Fig. 1.8 due to the interaction of the sp^2 orbital on the N with its lobe directed towards the 1s orbital of the neighboring Li. This example of analyzing experimental results in terms of MO overlap might be indicative of the amount of chemical information in the reciprocal form factor $B(\mathbf{r})$, especially, if a 2D mapping of the $B(\mathbf{r})$ distribution is relied upon.

MgO is a third example of an ionic crystal, which exhibits severe deviations from a simple isotropic free-ion model. The O^{2-}-ion must be stabilized by the crystalline field. In the simplest case, this field can be simulated by a hollow charged sphere, the Watson sphere, with its radius equal to the nearest-neighbor separation. A set of directional γ-Compton profiles of MgO from Aikala et al. (1982) served for many years as a reference for theoretical studies. Thus Aikala et al. (1982) found that their theoretical results on the basis of cluster calculations, which were either utilizing varying degrees of orthogonalization or the LCAO band structure calculations of Pantelides et al. (1974), reproduced the main features of the experiment, but definite disagreement remained in all cases, especially at low momenta, in particular along the [100] nearest neighbor direction. The best agreement was attained with the wavefunctions of Pantelides et al. (1974). A later theoretical study of Podloucky and Redinger (1984) on the basis of augmented plane-wave band structure calculation found qualitative agreement with respect to the oscillations of the directional differences, but in all cases the amplitude of these oscillations are overestimated. An analysis of the directional reciprocal form factors exhibits a strong deviation from experiment in [100] directions, shown in Fig. 4.25, found both for the exchange correlation potential of Hedin and Lundquist (1972) as well as using the Xα-method. Similar deviations of theoretical (LCAO and APW) Compton profile and $B(\mathbf{r})$ anisotropies from experimental ones which include the [100] direction have been found in a study on NaF and CaO of Redinger et al. (1989). Causa et al. (1986b) have tried to find an explanation for this discrepancy by means of Hartree–Fock LCAO "cohesive-energy" calculations (Causa et al. 1986a). Performing a decomposition of the various pair contributions to the reciprocal form factor in [100]-direction the authors found out that, within the relevant r-range between 3 and 4 a.u., the O(0)O(1) and the O(0)Mg(0) terms vary very rapidly and largely compensate each other, so this range of $B(\mathbf{r})$ is critically dependent on the relative importance of these two contribution. (Here the number in parentheses identifies the cell.) Compared with the Podloucky and Redinger (1984) result, the big discrepancy of $B(\mathbf{r})$ in the [100] direction is somewhat lifted, although still rather far from experiment, and also the amplitudes of the directional differences are nearer to the experiment. A new set of eight high-resolution synchrotron radiation based directional Compton profiles of MgO from the ESRF, taken by Gillet et al. (2001), was used by the above authors to perform a joint refinement of a local wavefunction model from X-ray diffraction and Compton scattering data. Apparently, the flexibility of the model wavefunctions was not extensive enough to bring the directional differences of the Compton profile together with the diffraction data to an agreement with the experiment that was comparably good as the excellent agreement between calculations using the

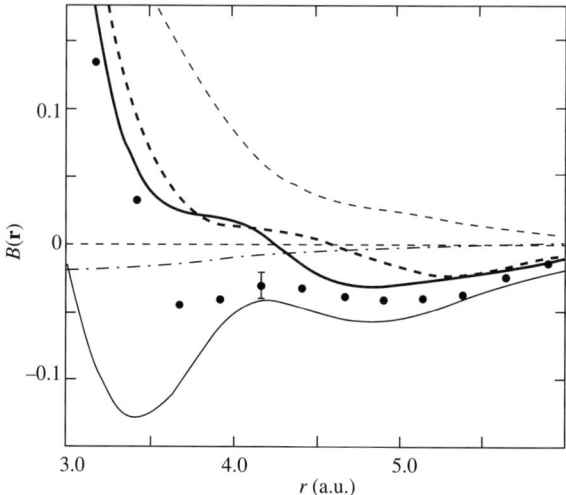

Fig. 4.25. $B(\mathbf{r})$ function along [100] of MgO. Experiment (Aikala *et al.* 1982): full circles. Theoretical results (Podloucky and Redinger 1984) multiplied with experimental resolution function: thick full curve is the total $B(\mathbf{r})$ function; the O s band contribution is the broken curve; the O p band contribution is the thinner full curve; the chain curve is the core contribution. Also shown is the total B_{100} for a momentum cut-off at p = 3.4 a.u. as the thick broken curve. (Originally published by Podloucky and Redinger (1984); copyright (1983) by IOP Publishing Ltd.)

CRYSTAL92 code of Causa *et al.* (1986) with the new high-resolution Compton data, shown in Fig. 4.26. So one might conclude that the above-mentioned rather strong deviations between calculations and low-resolution γ-Compton data are due to systematic errors of the latter. Interestingly the joint refinement from reconstructed electron momentum densities of MgO and structure factors have indicated a remarkable sensitivity of Compton profile directional differences on a parameter of the model wavefunctions which measures the degree of covalency and came out to be around 3%.

There exists a larger number of other Compton sudies on ionic crystals (e.g. Loupias and Petiau 1980 (LiF), Das and Padhi 1986 (NaCl), Das and Padhi 1987 (KCl), Das and Padhi 1988 (MgF), Joshi *et al.* 1999 (BeO)). These studies are either less conclusive as far as the theoretical interpretation is concerned, or performed with low momentum space resolution (γ-Compton), so that the reader is referred to the above publications.

4.7.5 *Hydrogen bonding*

It was the first measurement of the Compton profile anisotropy if ice Ih by Isaacs *et al.* (1999) at the Compton beamline ID15b of the ESRF which has again

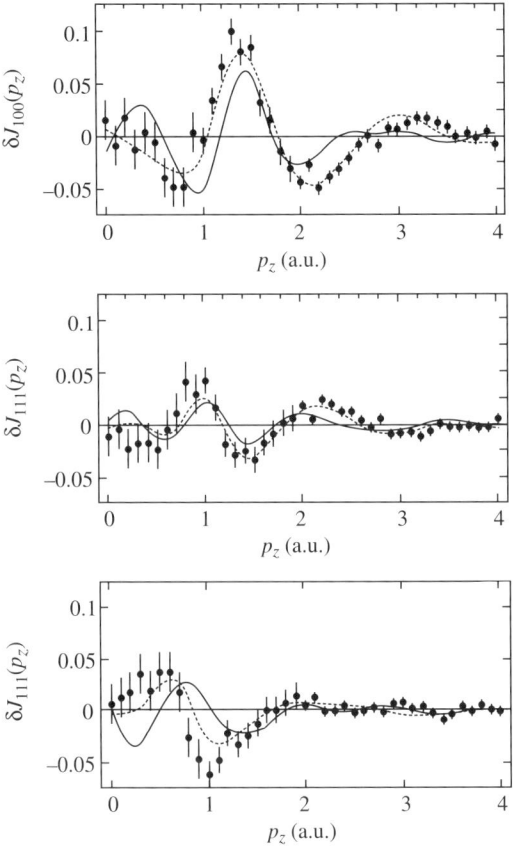

Fig. 4.26. MgO Compton profile anisotropies $\delta J_{hkl}(p_z) \equiv J_{hkl}(p_z) - J_{\text{iso},3}(p_z)$, where $J_{\text{iso},3}$ is the weighted average over the three main directions [100],[110],[111]. Experiment: full circles with error bars; solid line: joint refinement of the local model, here with a localized wave functions of a density functional theory of Cortona (1991), from a set of six structure factors and three directional Compton profiles; Dashed line: CRYSTAL92 calculation. (Reprinted with permission from Gillet *et al.* (2001); copyright (2001) by the American Physical Society.)

excited the debate about the signature of hydrogen bonding in Compton profile anisotropies and about its physical interpretation. Figure 4.27 shows the experimental anisotropy ($J_{c\text{-axis}} - J_{a\text{-}b\ \text{plane}}$). The c-direction contains one quarter of the hydrogen bonds whereas the other are distributed over directions close to the a-b plane. The oscillations of the Compton profile anisotropy exhibit two prominent frequencies, which manifest themselves as peaks at 1.72 Å and 2.85 Å, depicted in the inset of Fig. 4.27. Since the first peak position was near to the hydrogen

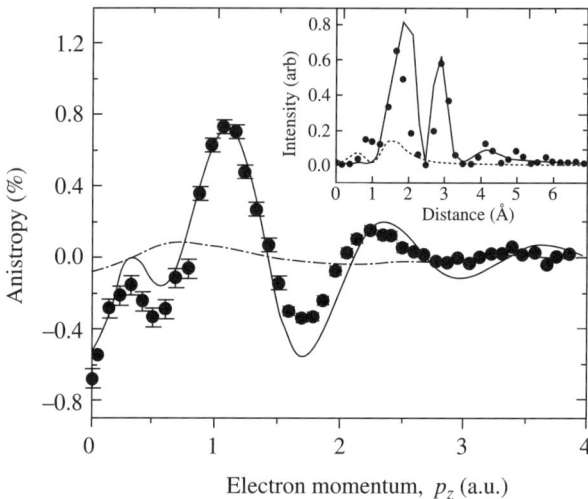

Fig. 4.27. Anisotropy of the experimental Compton profile defined by ($J_{c\text{-axis}}-J_{a\text{-}b\ \text{plane}}$) of ice Ih plotted as a percentage of the Compton profile as a function of p_z in a.u. Filled circles: experiment; solid line: fully quantum mechanical bonding model; dashed-dotted line: a purely electrostatic bonding model. Inset: The power spectrum of the data and the theory. (Reprinted with permission from Isaacs *et al.* (1999); copyright (1999) by the American Physical Society.)

bond length of 1.75 Å, the corresponding oscillations were interpreted as being due to the phase coherence of the neighboring water monomer wavefunctions, thus indicating the covalency of the hydrogen bond in ice. This interpretation found support by a full quantum mechanical calculation in a 12-molecule supercell using pseudopotential methods with a gradient density functional formalism (Hamann 1997). The result of this calculation is plotted as a solid line in Fig. 4.27, although reduced by 40%. The dotted-dashed line, however, is simply the sum of the contribution from individual water molecules, oriented according to the Bernal–Fowler ice rules (Bernal and Fowler 1935), so that the hydrogen bonds are treated as purely electrostatic. This way the quantum mechanical origin of these oscillations becomes evident, and again the high sensitivity of Compton profile anisotropies to the coherent superposition of wavefunctions at different positions in a crystal is evident.

Nevertheless, one very important aspect of this interpretation was criticized by several authors, namely the statement that the nature of this coherent superposition is **bonding**. Thus Ghanty *et al.* (2000) have demonstrated that not only a Hartree–Fock calculation of an ice-like water dimer but already the antisymmetrized product of the isolated monomer wavefunctions gives rise to the Compton profile directional anisotropy as measured by Isaacs *et al.* (1999).

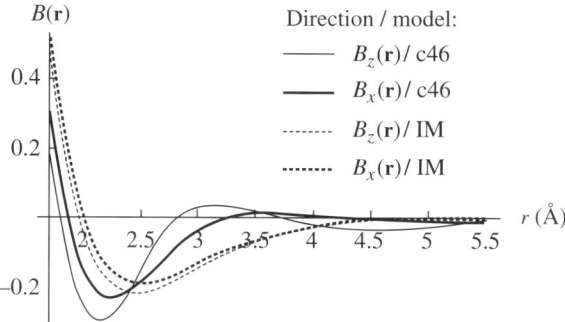

Fig. 4.28. Calculated reciprocal form factor $B(\mathbf{r})$ of ice Ih for the x- and z-directions and different models as described in the text. (Reprinted with permission from Ragot et al. (2002); copyright (2002) by the American Physical Society.)

Ganthy et al. (2000) argue that antisymmetrization of the product of monomer wavefunctions gives a large exchange repulsion, reduces the charge density between the monomers and is therefore net **antibonding**. A still more detailed study on that question has been presented by Ragot et al. (2002) by using a cluster partitioning method, where the first-order reduced density matrix (IRDM) is decomposed into one- and two-center terms, so that the latter stand for the overlap between orbitals at different positions. Calculation were performed for two different clusters of 22 (c22) and 46 (c46) molecules, obeying the Bernal–Fowler rules. These authors extend the discussion on the directional reciprocal form factors, which have already been used by Weyrich (1978) to qualify the covalent interaction between water monomers in liquid water to be antibonding. Weyrich (1978) has found the experimental reciprocal form factor of liquid water above the calculated one of Smith et al. (1975) for an isolated water molecule for $r > 2.7$ Å, which is, according to the rules given in Section 4.7.1, an indication for antibonding interaction between the water molecules. Since the attraction between the water molecules is of electrostatic origin, amplified by the polarization of the O—H bond and possibly a charge transfer component, this antibonding repulsive interaction is necessary, similar as in the case of ionic crystals, in order to equilibrate the electrostatic attraction. Ragot et al. (2002) has followed this interpretation line as documented in Fig. 4.28, where the reciprocal form factor for the z- and x-directions of those model calculations ($Bz/$c46 and $Bx/$c46) are shown, which brought good agreement with the experimental Compton profile anisotropies: $B(\mathbf{r})$ is, for $r > 2.5$ Å, above the reciprocal form factor of the independent water molecule model ($Bz/$IM and $Bx/$IM)), thus offering proof for the antibonding nature of the water molecule interaction as seen in the $B(\mathbf{r})$ function. It should be stressed that the conclusive findings about the antibonding nature of the coherent superposition of molecular wavefunction have been deduced from the **signs** of reciprocal form factor differences, whereas the covalent bonding character of the hydrogen bond was concluded by Isaacs et al. (1999) from the modulus squared of those differences, the so-called power spectrum. It is interesting that the c22 calculated $(B_c(\mathbf{r}) - B_{a/b}(\mathbf{r}))$ exhibits the same structure, although with much smaller amplitude, when the water molecules are replaced by the isoelectronic Ne atoms. This way it could be excluded that the 1.72 Å peak in the power spectrum of the directional difference of Isaacs et al. (1999) has to do with the hydrogen bond length.

Another compound with directed hydrogen bonds is urea. Shukla et al. (2001) have measured directional high-resolution Compton profiles at the ID15b beamline of the ESRF, which were evaluated in the same manner as the ice measurements of Isaacs et al. (1999). The power spectrum of those directional differences, which contains the Compton profiles with **q** in the [001] direction, exhibited a tiny peak at 1.9 Å. This is near 1.8 Å, the projected length of the hydrogen bonds in [001] direction. The peak is also visible in a quantum mechanical GGA (generalized gradient approximation) based calculations of the wavefunctions for the urea crystal, but absent in a computation using isolated urea molecule wave functions obtained by precisely the same calculation procedure as for the solid. It is concluded that for these reasons the peak is a direct indication of the intermolecular interaction arising from the linear hydrogen bonds, so that Compton scattering spectroscopy can be used to study chemical bonding in relative complex molecular crystals.

Two further theoretical works should be mentioned, which came, with respect to the hydrogen bonding, to the same conclusions. One is from Barbiellini and Shukla (2002), who predicted on the basis of an *ab initio* simulation using the CRYSTAL98 program (Saunders et al. 1998) a remarkable sensitivity of the negative radial derivative of Compton profiles to hydrogen bonds by comparing this quantity for an isolated water molecule, a dimer, a water molecule cluster and an ice crystal. They deduced a correlation between the full width at half maximum of the radial derivative and an angle, which describes the mixing of the highest occupied molecular orbital (HOMO) of a proton acceptor monomer and the lowest unoccupied molecular orbital (LUMO) of the donor neighbor molecule, and consequently the charge transfer connected with this mixing.

The most important aspect of the second theoretical work published by Hakala et al. (2004) is a constrained space orbital variation (CSOV) analysis (Bagus et al. 1984) of a water dimer with an O-O distance of 2.75 Å (ice) as shown in Fig. 4.29, where the difference between the isotropic Compton profile of the dimer and two free water molecules is shown for three different CSOV steps: (i) Frozen means that only the electrostatic and the exchange interaction between the molecules is taken into account. One sees that this interaction provides the biggest contribution to the Compton profile differences. (ii) The contribution of the polarization of both molecules is only marginal, whereas (iii) The charge transfer between the molecules (here denoted as "relaxed") is of more importance as already stated by Barbiellini and Shukla (2002).

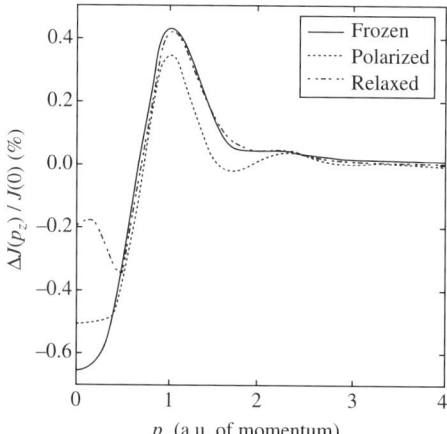

Fig. 4.29. Compton profile differences at the various stages of a constrained space orbital variation (CSOV) analysis between a water dimer with R_{OO} distance of 2.75 Å and two free water molecules. (Published with permission from Hakala et al. (2004); copyright (2004); by the American Physical Society.)

Summarizing, these findings will stimulate further experiments on ice and water under different conditions as well as experiments on free water clusters and ionic solvations.

4.8 Compton scattering of metals

4.8.1 Simple metals and alloys, reconstructed occupation number density, Fermiology and electron correlation

Nearly free-electron metals were the first samples when Compton scattering was applied by DuMond (1929) to a solid state problem, namely to the question of what the momentum distribution of the valence electrons in a metal might look like. The result of this experiment on Be metal provided the first direct experimental evidence for the Fermi–Dirac statistics. Thus one can consider this Compton experiment as the most important one ever done. After the revival of Compton scattering in the mid 1960s by Cooper et al. (1965), lithium metal was the first sample. We will give in what follows a survey of the more recent synchrotron radiation based high-resolution Compton experiments on simple metals and their alloys, where it should not be forgotten that many prevailing experiments have been performed using conventional X-ray and γ-ray sources. To give some examples, we refer to experiments by Phillip and Weiss (1968) on Li, Be, Na, Mg and Al, by Currat et al. (1971) on Be, by Eisenberger et al. (1972) on Li and Na, by Manninen et al. (1974) on Al, by Wachtel et al. (1975) on Li, by Hansen et al. (1979) on Be, by Berndt and Brümmer (1976) on LiMg alloys, by

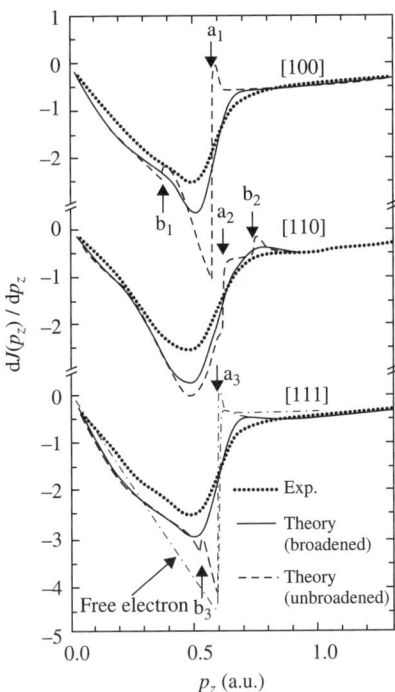

Fig. 4.30. The first derivative of the Compton profiles of Li along the three principal directions. Full circles: experiment; solid curve: theory broadened with the experimental resolution; dashed curve: theory unbroadened, a_1-a_3 and b_1-b_3 denote Fermi surface breaks, dotted-dashed curve shown on the [111] curve: The free electron prediction. The error bars (not shown) are approximately of the size of the data points. (Reprinted with permission from Sakurai et al. (1995); copyright (1995) by the American Physical Society.)

Manninen and Paakkari (1981) and by Sakai and Sekizawa (1981) on Mg, and by Cardwell and Cooper (1986) on Al.

The synchrotron radiation based Compton experiments on simple metals aimed at two main goals: (i) to quantify the influence of the lattice potential on the total profile, the directional differences, the reconstructed 3D valence electron momentum density and the Fermi surface; (ii) to investigate the influence of electron–electron correlation on the valence electron momentum density, in order to test different theoretical models. The Compton scattering experiments on single-crystal Li metal of Sakurai et al. (1995), performed at the NE1 beamline of the KEK accumulation ring, Japan, with a momentum space resolution of 0.12 a.u., were prevailing for further investigations of this kind. Fermi surface radii anisotropy of Li, defined as $\delta = (k_{110} - k_{100})/k_F$ (k_F = free electron Fermi wavenumber), was deduced from the peaks of the second derivative of the profiles (after some corrections due to the asymmetry of the underlying distribution) to be $\delta = (4.6 \pm 1.0)\%$ and compared with the LDA theoretical value of $\delta = 5.9\%$. The most remarkable result of this study was the rather large discrepancy between the LDA calculations and the experiments both with respect to the total profile, where $J(0)$ was found too small in experiment, and with respect to the Fermi surface break, which was detected with the first derivative of the Compton profiles, shown in Fig. 4.30, to be much sharper in the (resolution convoluted) theory. Finally, both discrepancies were traced back to an inadequate treatment of electron–electron correlation in the theoretical computations. Schülke et al. (1996) came to a similar, although more quantitative, conclusion in their evaluation of 11 directional Compton profiles of Li, measured with 0.14 a.u. resolution at the Compton beamline of HASYLAB, Hamburg. The authors have reconstructed the occupation number density $N(\mathbf{k})$ (defined by 4.31) by using (4.32) and the Fourier–Bessel method to obtain $B(\mathbf{r})$. They compared the occupation number density $N(\mathbf{k})$, reconstructed from the 11 measured Compton profiles, with a model $N(\mathbf{k})$, reconstructed the same way on the basis of an empirical interpolation formula for $N(\mathbf{k})$ of a correlated electron liquid (fitted with the renormalization constant Z_F as the only parameter to data of Takada and Yasuhara 1991). This way they found surprisingly for Li metal a Z_F-value as small as $Z_F = 0.1 \pm 0.1$. It must be mentioned that already the differences between LDA calculated and measured directional Compton profiles had shown values which could be reduced only by 30%, when the Lam-Platzman correction (see 4.34) was applied to the LDA theory. Moreover, these difference plots exhibit fine structures which can be understood as indicating an overestimation of the contribution from secondary Fermi surfaces (see Fig. 6.7 in Section 6.8.3.3) due to correlation effects as first discussed by Bauer and Schneider (1984). An experimental contour diagram of reconstructed $N(\mathbf{k})$ in the ΓNPH plane of the repeated zone scheme is shown in Fig. 4.31, together with the corresponding error map. The trace of the Fermi surface is indicated by the dotted line, obtained by determining that $N(\mathbf{k})$ = constant surface which encloses the volume of the corresponding free electron Fermi sphere. The Fermi surface anisotropy δ, thus found, was, with $\delta = 3.6 \pm 1.1\%$, smaller than that obtained in the Sakurai et al. (1995) study. Later Tanaka et al. (2001a) performed a reconstruction of the Li momentum density by means of the direct Fourier method using 12 directional Compton profiles, measured at the NE1 beamline of the KEK-accumulation ring, Japan. Their results were compared with momentum densities reconstructed using LDA calculated Compton profiles (including the Lam–Platzman correction) of the same $\mathbf{q}$-orientations, so that systematic errors and additional broadening due to the reconstruction procedure were the same in both the experimental and the calculated densities. As in the previous Li $\rho(\mathbf{p})$ reconstruction of Schülke et al. (1996), also in this study the break at the Fermi momentum appears in the experiment much more smeared than in the reconstructed theoretical $\rho(\mathbf{p})$, clearly indicating the importance of including electron correlation effects beyond the LDA. The Fermi surface asphericity shown

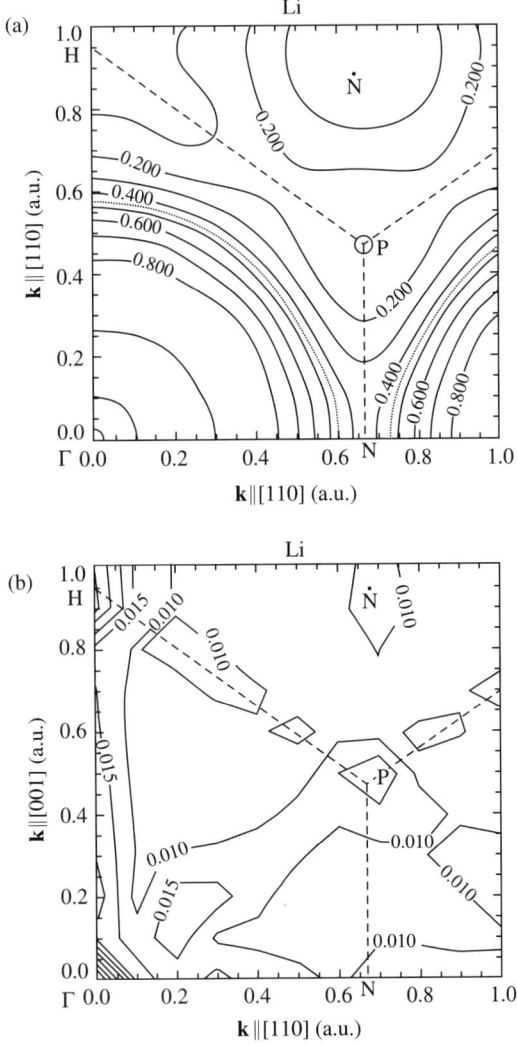

Fig. 4.31. (a) Level diagram of the occupation number density $N(\mathbf{k})$ of the repeated zone scheme of Li in the ΓNPH plane reconstructed from 11 directional Compton profiles. The level-line distance is 0.1 a.u.$^{-3}$. The trace of the Fermi surface is the dotted line. The dashed lines are the Brillouin zone boundaries. (b) Error map corresponding to (a); the level line distance is 0.005 a.u.$^{-3}$. (Reprinted with permission from Schülke et al. (1996); copyright (1996) by the American Physical Society.)

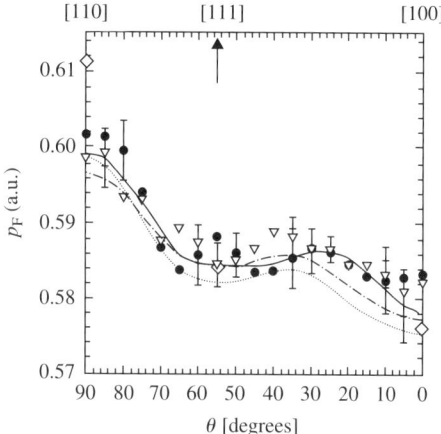

Fig. 4.32. Fermi radii in the (110) plane in Li. Filled circles are the experimental values obtained from the inflection points of the reconstructed momentum density corrected for the influence of finite momentum resolution (see Tanaka *et al.* 2001 for details). Chains (dots) are the theoretical values obtained the same way from the momentum density reconstructed from resolution folded (non-resolution folded) theoretical Compton profiles. The open diamonds give the radii along the three principal symmetry directions obtained via LDA band structure calculations which do not involve a reconstruction. Results of Schülke *et al.* (1996) (solid line) and of Oberli *et al.* (1985) (open triangles) are added for comparison. (Reprinted with permission from Tanaka *et al.* (2001a); copyright (2001) by the American Physical Society.)

in Fig. 4.32 was in good agreement with that obtained by Schülke *et al.* (1996) but in disagreement with the LDA calculations as far as the rather local protrusion of the theoretical Fermi surface in the [110] direction is concerned. It was argued that apparently the reconstruction on the basis of only 12 directional Compton profiles is unable to reproduce faithfully this feature. More directions should be chosen around the [110] axis. Two further Compton experiments by Sternemann *et al.* (2000) at beamline G3 of DESY/HASYLAB and at beamline X21 of NSLS on Li singles crystals aiming at a resolution between 0.02 and 0.03 a.u., found also a large smearing of the Fermi break (seen as a broadening of the peak in the second derivative of the Compton profiles at the Fermi momentum), when compared with the Lam–Platzmann corrected LDA calculations of Sakurai *et al.* (1995). This smearing could, of course, be partly attributed to the folding with the particle spectral density function (for details see Section 4.10.4), but there remained an additional smearing, which again must be ascribed to correlation effects, not considered in the LDA computations.

In a rather extensive Compton study on $Li_{100-x}Mg_x$ alloys ($0 \leq x \leq 40$), Stutz *et al.* (1999) have proved the ability of Compton scattering to yield detailed

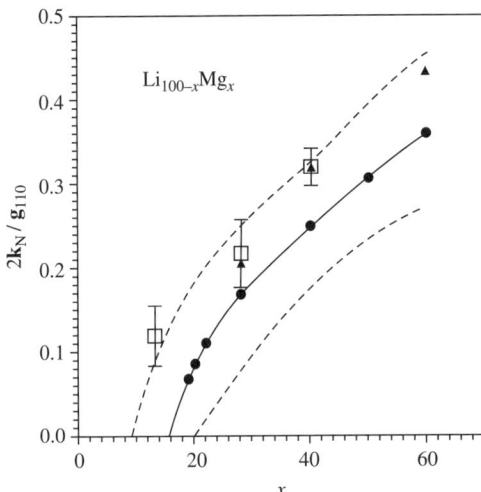

Fig. 4.33. Fermi surface neck diameter of $Li_{100-x}Mg_x$ on the 110 Brillouin zone boundary in units of the reciprocal lattice vector $\mathbf{g}_{110}$ as a function of x. Open squares are the experiment of Stutz et al. (1999). Triangles are the 2D-ACAR experiment of Rajput et al. (1993). Filled circles connected by an interpolating line are the KKR-CPA calculations. Dashed lines mark the range of smearing of the theoretical neck radius due to the disorder scattering. (Reprinted with permission from Stutz et al. (1999); copyright (1999) by the American Physical Society.)

information about the behavior of the Fermi surface of a monovalent metal upon alloying of a divalent metal, which cannot be obtained with transport methods due to the chemical disorder. The Compton profiles, twelve for each of the four Mg concentrations $x = 0, 13, 28, 40$, were measured at the NE1 beamline of the KEK-accumulation ring (Japan) with a resolution of 0.11 a.u. at a primary energy of 59.38 keV. Reconstruction of the momentum density $\rho(\mathbf{p})$ and of the occupation number density $N(\mathbf{k})$ were performed by means of the Fourier–Bessel method. The $N(\mathbf{k})$ reconstruction provides, among other things, information about the Fermi surface neck diameter on the 110 Brillouin zone boundary as shown in Fig. 4.33, where the neck diameter in units of the $\mathbf{g}_{110}$ reciprocal lattice vector is shown as a function of x together with the result of 2D-ACAR measurement (Rajput et al. 1993) and KKR-CPA calculations. Apparently the calculations systematically underestimate the neck diameter.

Compton measurements on polycrystalline sodium has been reported by Sakurai et al. (1992b). The profile coincides nicely with a theoretical profile calculated on the basis on the interacting electron gas model of Daniel and Vosko (1969).

The first specimen of synchrotron radiation based Compton scattering was a Be single crystal, investigated by Loupias et al. (1980) at LURE-DCI using 10 keV

X-rays and a Cauchois-type analyzer in connection with a position-sensitive detector, achieving 0.15 a.u. momentum space resolution. Chou et al. (1982) compared their results with pseudopotential calculations and found orientation-dependent differences between experiment and theory, which could only be partly removed by introducing a free electron correlation correction given by Rennert (1981). Finer details of the Fermi surface topology of Be were expected from Compton profile measurements of Hämäläinen et al. (1996), performed at the NSLS X21 beamline using a focusing spectrometer in Rowland geometry (see Chapter 2) with X-rays of 8 keV, and a momentum resolution between 0.02 and 0.03. However the comparison between LDA calculated and experimental first derivative of directional Compton profiles exhibited a much stronger smearing of Fermi surface induced details in the experiment than in the theory convoluted with the resolution function. Surprisingly Itou et al. (1998) found in their Compton study on Be fine structures in the first derivative of directional Compton profiles, which were predicted by theory but either not visible or strongly smeared in the previous experiment, although their resolution was with 0.08 a.u. much poorer, certainly their incident photon energy with 59.38 keV much higher. But also in this experiment a distinct discrepancy between experiment and convoluted theory remained, as far as the Fermi surface signatures are concerned, shown exemplarily in Fig. 4.34. The apparent contradiction that a better instrumental resolution yields a poorer resolution of structural details was

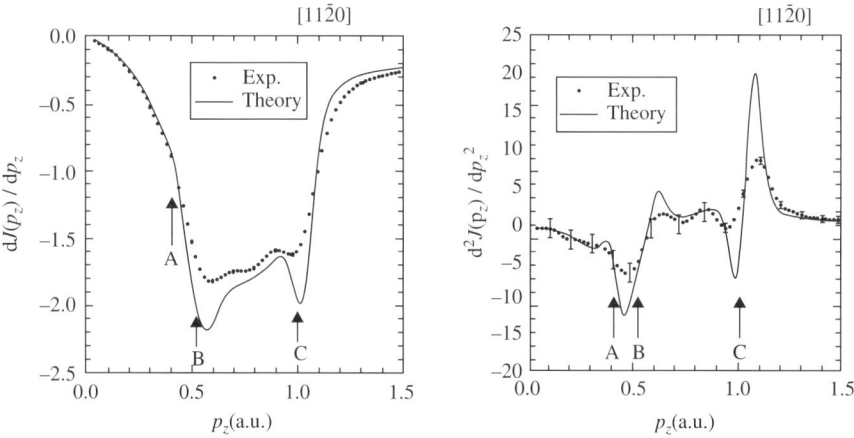

Fig. 4.34. First (left-hand panel) and second (right-hand panel) derivatives of the experimental and theoretical Compton profiles of Be along the [1120] direction. The Fermi surface feature A arises from the "cigar". The broad feature B is due to a combination of the "coronet" and the "cigars". The sharp dip C is due to the surface of the "cigar" in the third Brillouin zone around the K-point. (Reprinted with permission from Itou et al. (1998); copyright (1997) by Elsevier Science Ltd.)

solved by investigations of Huotari et al. (2000) on Be performed at the beamlines ID15b and ID16 of the ESRF, who came to the same result as Sternemann et al. (2000) in their above-mentioned high-resolution Compton study on Li: it is the additional convolution of the experiment with the spectral density function of the recoil electron, which is of importance for low energy Compton measurements, as will be shown in some detail in Section 4.10.4. Another result of the Be study of Huotari et al. (2000) is shown in Fig. 4.35, where

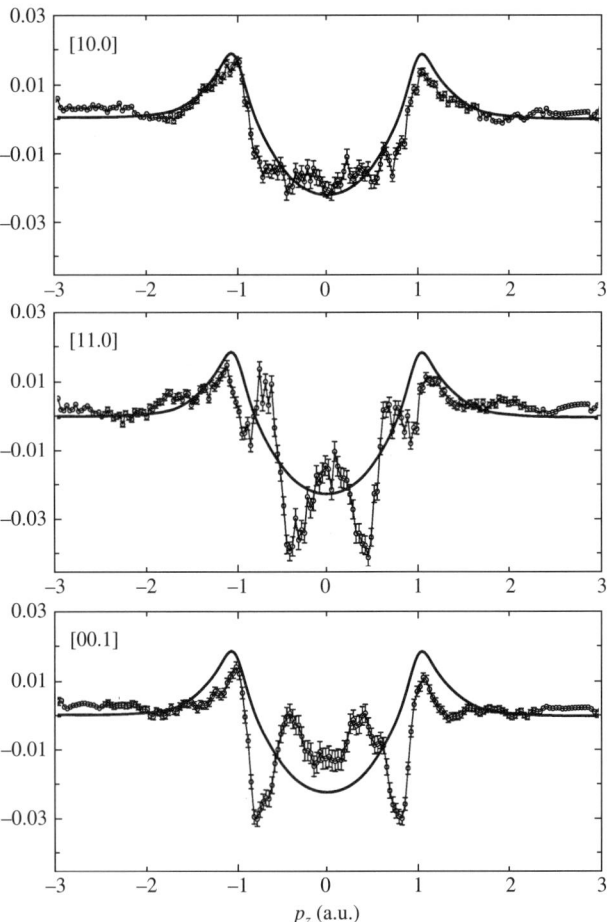

Fig. 4.35. The difference between experimental and Lam–Platzman corrected theoretical Compton profiles of Be with incident energy of 59 keV along the [10.0], [11.0] and [00.1] directions. The thick solid line represents the Lam–Platzman correction. The thin solid line is a guide to the eye. (Reprinted with permission from Huotari et al. (2000); copyright (2000) by the American Physical Society.)

the strongly orientation-dependent difference between the experimental and the Lam–Platzman corrected theoretical Compton profiles are shown. Additionally the Lam–Platzman correction is plotted, thus showing that, on the average, the Lam–Platzman correction can only account for one-half of the discrepancy between experiment and calculation.

The first synchrotron radiation based Compton profile measurements with 0.084 a.u. resolution on Al were reported by Shiotani *et al.* (1989). The profile in the [111] direction was compared with a 1D ACAR curve of Okada *et al.* (1976) measured with a similar resolution. This way the enhancement due to electron–positron interaction could be demonstrated. Fermi surface induced features both in the first derivative of directional Compton profiles and in the measured anisotropies were found in good qualitative agreement with corresponding LDA-KKR calculations in a Compton study on Al of Ohada *et al.* (2000) using 60 keV synchrotron X-rays and an overall momentum resolution of 0.12 a.u.. Suortti *et al.* (2000) have analyzed the Al Compton profiles, measured at the ID15B beamline of the ESRF in terms of a one-parameter model of the correlation correction. They concluded that in contrast to the case of Li and Be the LDA predictions were met, by using in the above model a renormalization constant Z between 0.7 and 0.8, a value which agrees quite reasonably with free-electron calculations. A combined theoretical and experimental (ESRF) Compton investigation on Al and a $Al_{0.97}Li_{0.03}$ alloy of Suortti *et al.* (2001b) with 29 keV (0.08 a.u. resolution) and 58 keV (0.16 a.u. resolution) has demonstrated how also minute anisotropies and differences between Al and the alloy, both of the order of a few tenths of a percent of $J(0)$, can be measured with high statistical accuracy. It is typical for Al that the amplitudes of the directional differences as well as the first and second derivatives of the Compton profiles were found in good agreement with the KKR-CPA calculations, differently from the Li and Be Compton measurements. Another Compton study on Al and an Al-3at%Li disordered alloy by Matsumoto *et al.* (2001) must be mentioned, since in that case a variant of the Lock–Chrisp–West (LCW) folding theorem (Lock *et al.* 1973) was used to visualize the Fermi surface projection on the (110) plane. The LCW theorem tells us how to get the band sum of the occupation number density, $N(\mathbf{k})$, in the repeated zone scheme, defined in equation (4.31), by backfolding the momentum density $\rho(\mathbf{p})$ into the first Brillouin zone:

$$N(\mathbf{k}) \equiv \sum_{\nu} n_{\nu\nu}(\mathbf{k}) = \sum_{\mathbf{g}} \rho(\mathbf{k} + \mathbf{g}). \qquad (4.107)$$

Here ν is the band index and $\mathbf{g}$ a reciprocal lattice vector. One can easily show that (4.107) is equivalent to equation (4.32), which relates the occupation number density $N(\mathbf{k})$ in the repeated zone scheme to the values of the reciprocal form factor $B(\mathbf{R})$ at the lattice translation vectors $\mathbf{R}$. The variant of the LCW scheme introduced by Matsumoto *et al.* (2001b) is characterized by rewriting this theorem to relate the two-dimensional projection of the occupation number density on

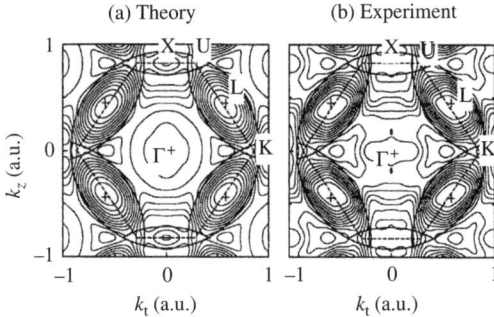

Fig. 4.36. Contour maps of the theoretical and experimental 2D occupation number densities (after projection along [110]) $N(k_z, k_t)$ in the (110) plane based on the partial LCW folding scheme in Al-3at%Li. Resolution broadening is included in the theory. The Brillouin zone boundary and the projected free electron sphere are shown. + signs indicate high-density regions. (Reprinted with permission from Matsumoto et al. (2001b); copyright (2001) by the American Physical Society.)

a certain plane to the corresponding two-dimensional projection of the momentum density. To apply these relations one needs only directional Compton profiles with **q**-vectors lying in this plane. Moreover, the sum over the reciprocal lattice vectors in (4.107) is restricted to a certain set of **g**'s, so that the strong overlap with Fermi surface images obtained through projections from higher Brillouin zones could be prevented. Figure 4.36 depicts a contour map of the theoretical and experimental 2D occupation number densities based on the partial LCW folding scheme in the Al-3at%Li alloy.

Of course, the deficits of the LDA treatment, as found in connection with the Compton measurements on Li and Be, have attracted the attention of theoreticians. Kubo (1997) has applied the so-called GW approximation (Hedin 1965) to solve this problem. Within the limits of this approximation, the diagonal elements of the occupation number density in (4.28) no longer have only the values one or zero but are given by

$$n_{\nu\nu}(\mathbf{k}) = \pi^{-1} \int_{-\infty}^{\mu} \operatorname{Im} A_{\nu\nu}(\mathbf{k}, E) \, dE, \qquad (4.108)$$

where μ is the chemical potential. The spectral density function $A_{\nu\nu}(\mathbf{k}, E)$ writes

$$A_{\nu\nu}(\mathbf{k}, E) = [E - E_\nu(\mathbf{k}) - \Sigma(\nu, \mathbf{k}, E)]^{-1}_{\nu\nu}, \qquad (4.109)$$

where $E_\nu(\mathbf{k})$ are the eigenenergies of the Bloch states $|\mathbf{k},\nu\rangle$. The self-energy $\Sigma(\nu,\mathbf{k},E)$ has to be calculated as follows:

$$\Sigma(\nu,\mathbf{k},E) = i(2\pi)^{-1} \int dE' \exp(-i\eta E')$$
$$\times \sum_{\nu',\mathbf{k}',\mathbf{g},\mathbf{g}'} D_{\mathbf{g},\mathbf{g}'}(\nu',\nu,\mathbf{k}',\mathbf{k}) W_{\mathbf{g},\mathbf{g}'}(\mathbf{k}',E')$$
$$\times [E - E' - E_{\nu'}(\mathbf{k}+\mathbf{k}') - i\eta]^{-1}; \quad \eta \to +0 \quad (4.110)$$

$$D_{\mathbf{g},\mathbf{g}'}(\nu',\nu,\mathbf{k}',\mathbf{k}) = \langle \nu,\mathbf{k}|\exp[-i(\mathbf{k}'+\mathbf{g}')\cdot\mathbf{r}]|\nu',\mathbf{k}+\mathbf{k}'\rangle$$
$$\times \langle \nu',\mathbf{k}+\mathbf{k}'|\exp[i(\mathbf{k}'+\mathbf{g})\cdot\mathbf{r}|\nu,\mathbf{k}\rangle, \quad (4.111)$$

$$W_{\mathbf{g},\mathbf{g}'}(\mathbf{k}',E') = 4\pi \frac{\left([1+\mathbf{T}(\mathbf{k}',E)]^{-1}\right)_{\mathbf{g},\mathbf{g}'}}{|\mathbf{k}'+\mathbf{g}|^2}, \quad (4.112)$$

where $([1+\mathbf{T}(\mathbf{k}',E)]^{-1})_{\mathbf{g},\mathbf{g}'}$ is the $\mathbf{g},\mathbf{g}'$th element of the inverse dielectric matrix as defined in (6.105), which was calculated in the random phase approximation (RPA) and by using the plasmon pole model proposed by Hamada *et al.* (1990). The occupation number density for the one occupied band of Li exhibited both a strong orientation dependence and a remarkable reduction of the renormalization constant Z_F when compared with the free-electron RPA value, $Z_F \approx 0.7$, of Lundqvist (1967b) used frequently in connection with the Lam–Platzman correction of LDA calculations, namely $Z_F = 0.35, 0.15$ and 0.25 for the three directions [100], [110] and [111], respectively. This reduction of Z_F brought the theoretical directional GWA Compton profiles very close to the experimental one of Sakurai *et al.* (1995) and of Schülke *et al.* (1996). But the discrepancy between calculated and measured Compton profiles of Li was solved only apparently. Schülke (1999) has substantiated that the strong reduction of Z_F in Kubo's calculation must be traced back to an unphysical behaviour of the imaginary part of the self-energy induced by the imaginary part of the plasmon-pole frequency: the spectral density function did not vanish at the Fermi momentum. By calculating, within Kubo's scheme, Z_F using the real part of the self-energy rather than the imaginary part, according to

$$Z_F = \frac{1}{[1 - \partial \operatorname{Re} \Sigma(\mathbf{k},E)/\partial E]_{E=E_F}} \quad (4.113)$$

one ends up with Z_F values around 0.75. Another application of the GWA approach to Compton scattering has been proposed by Eguiluz *et al.* (2000). By treating the spectral density function, the self-energy, the dielectric function, and the screened Coulomb interaction self-consistently they ended up with much smaller corrections to the independent occupation number density than Kubo (1997).

Quite another approach to solving the problem of discrepancies between calculated and measured Compton profiles of Li has been suggested by Dugdale and Jarlborg (1998). They ascribed this discrepancies to thermal disorder, which they treat within the limits of a static model. The mean displacements at a given temperature were introduced into an eight-atom supercell for Li and Na, averaging calculations with many configurations. As a result, the disordered Compton profiles became more delocalized in momentum space, when compared with the ordered ones, thus explaining the above discrepancies, which exhibit the same tendency. This delocalization is attributed to an increase of the higher momentum components due to the higher "roughness" of the disordered structure. It should be stressed that changes of the lattice constant with temperature were neglected. In order to test these predictions, Sternemann et al. (2001) have performed high-resolution Compton profile measurements at the ID15B beamline of the ESRF on Li at 95 and 295 K and on Al at 15 and 560 K with a resolution of 0.1 a.u. They found in both cases, contrary to the predictions of Dugdale and Jarlborg (1998), that the Compton profiles became more localized in momentum space with increasing temperature. This temperature dependence could be completely explained as being due to both the change in the lattice constant resulting in a change of the Fermi momentum, and the decrease of the higher-momentum components as a consequence of the attenuation of the periodic lattice potential described by a Debye–Waller factor. Thus no place is left for disorder effects as predicted by Dugdale and Jarlborg (1998).

Quantum Monte–Carlo (QMC) calculations are known to offer very direct access to electron–electron correlation in condensed matter. Therefore, a QMC approach to Compton profiles of Li, as performed by Filippi and Ceperley (1999), gain our attention. Possibly they can solve the correlation problem. However, without going into details of this computation, it should be stressed that the many-body wavefunctions used are built from Slater determinants of single-particle orbitals with an implicit nodal structure and plasmon-type correlation treatment, so that they have a similar basis as the GW approximation. Therefore, it might not be surprising that the correlation correction of the LDA profiles resulting from the QMC computations are not far from the Lam–Platzman corrections.

A completely new ansatz to solve the correlation problem was attempted by Barbiellini and Bansil (2001) utilizing the antisymmetrized geminal products (AGP), an ansatz which has already been sketched in Section 4.3, and whose most important results can be comprehended as follows. The occupation number n_i of a Kohn–Sham orbital $\psi_i(\mathbf{r})$ with energy eigenvalues ε_i is given by

$$n_i = \frac{1}{2}\left[\frac{1-\varepsilon_i}{(\varepsilon_i^2 + \Delta_i^2)^{1/2}}\right], \tag{4.114}$$

where Δ_i determines the mixing of states above and below the Fermi level. Barbiellini and Bansil (2001) have argued that $|\Delta_i|$ can be approximated, within the LDA, by the exchange type integral I_i of equation (4.43), where I_i is large when $\psi_i(\mathbf{r})$ is confined to a small spatial region.

This means that the correlation effect is controlled by the parameter

$$\alpha = \frac{\Delta_i}{w} \sim \frac{I_i}{w}, \qquad (4.115)$$

where w is the valence electron bandwidth. If $\alpha \approx 1$, the renormalization takes place deep in the Fermi sea shifting spectral weight from below to above $\varepsilon_i = 0$, whereas for $I_i \ll w$ only states near the Fermi momentum are concerned. Thus Barbiellini and Bansil (2001) obtained $\alpha \approx 0.4$ for Li, which means that the occupation number density of the independent-particle model will be modified strongly by correlation, so that the cusp of the Compton profile at the Fermi momentum is smeared, just as found in the experiment (e.g. Schülke et al. 1996). For Be a value of $\alpha = 0.14$ is estimated. Thus the correlation effect is somewhat smaller than in the case of Li. But, more importantly, Δ_i of equation (4.43) becomes directional dependent for Bloch-states, so that also the difference between the AGP and the LDA Compton profile (where the latter includes the Lam–Platzman correction), shown in Fig. 4.37, is **q**-direction dependent and exhibits fine structures, which resembles quite nicely those of Huotari et al. (2000), depicted in Fig. 4.35.

Finally, in the case of Al, $\alpha \approx 0.05$ is estimated. This rather low value is in accord with the smaller correlation effects of Al as stated above. Summarizing,

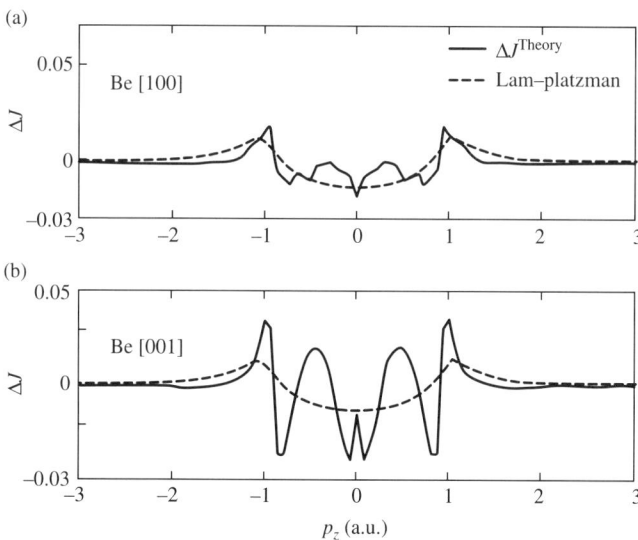

Fig. 4.37. Effect of correlation on the Compton profile along two different directions in Be. Solid curve: residual difference, $\Delta J_{\text{theory}} = J_{\text{AGP}} - J_{\text{LDA}}$, between the LDA and the AGP profiles. The LDA profiles include the Lam–Platzman correction, which is shown separately (dashed line). (Reprinted with permission from Barbiellini and Bansil (2001); copyright (2001) by Elsevier Science Ltd.)

306 The Compton scattering regime

the AGP's seem to be an efficient ansatz for treating electron–electron correlation in momentum space quantities.

4.8.2 *Transition and noble metals, alloys and compounds*

Like simple metals, also 3d transition metals (including copper), their alloys and compounds, are building a good testing ground for investigating how well momentum space related ground-state quantities, especially electron correlation in narrow bands, are understood by state-of-the-art theoretical approaches. Also in this survey of Compton data, we will concentrate our attention on most recent high-resolution results based on instrumentation using synchrotron radiation. Only when problems of general interest are concerned, also reference will be made to low-resolution investigations.

The most frequently investigated 3d transition metal was vanadium. Rollason et al. (1983) have measured low-resolution γ-Compton profiles with a 412 keV

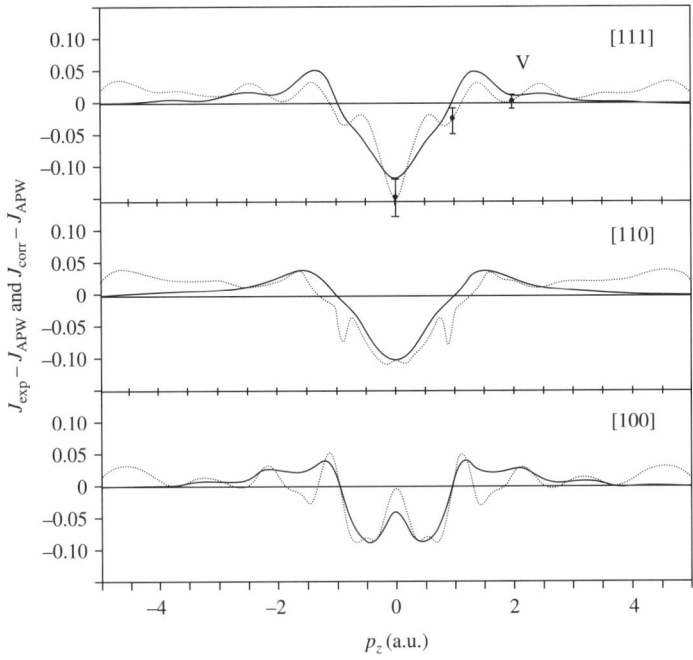

Fig. 4.38. Dotted lines: differences between experimental Compton profiles of vanadium, J_{exp}, and the corresponding APW profiles, J_{APW}, convoluted with the experimental resolution. Solid lines: differences between the correlation corrected APW profiles and the noncorrected J_{APW}. (Reprinted with permission from Wakoh and Matsumoto (1990); copyright (1990) by IOP Publishing Ltd.)

^{198}Au γ-ray source achieving 0.4 a.u. momentum resolution. The experimental profiles (after subtraction of a 12% multiple scattering contribution) were compared with LCAO (Laurent et al. 1978) and APW (Wakoh et al. 1976) calculation. The Compton profile differences (γ-ray experiment minus APW theory), depicted in Fig. 4.38, show, as a general tendency, that the theory overestimates the lower and underestimate the higher momentum range. This is already well known from corresponding investigations of simple metals, but the strong orientation dependency of the difference profiles of Fig. 4.38 could not be explained by an isotropic Lam–Platzman-type corrections. In order to allow for this orientation dependence, Wakoh and Matsumoto have introduced an empirical energy-dependent occupation function $n(E)$ defined in analogy with an empirical momentum distribution function of Cardwell and Cooper (1989) as follows:

$$n(E) = \begin{cases} 1 - A(E - E_1)/(E_F - E_1) - D & \text{for } E < E_F \\ B((E - E_b)^2/(E_F - E_b)^2) & \text{for } E_F < E < E_b. \end{cases} \quad (4.116)$$

Here the coefficients A, B, D have to be determined empirically. E_1 is a reference energy like that of the Γ_1 point, and E_b is the upper bound energy of the distribution. This energy occupation function $n(E)$ is implemented into the band structure calculation of the directional Compton profiles, which have to be corrected by

$$\Delta J_\mathbf{q}^{\text{corr}}(p_z) = - \sum_{E=\Gamma_1}^{E_F} (1 - n(E))\, J_\mathbf{q}(p_z, E) \Delta E + \int_{E_F}^{\infty} n(E)\, J_\mathbf{q}(p_z, E) \Delta E, \quad (4.117)$$

where $J_\mathbf{q}(p_z, E)$ is a partial directional Compton profile (scattering vector $\mathbf{q}$) contributed from states whose energy is between E and $E + \Delta E$. These anisotropic correlation corrections for V, calculated on the basis of an APW scheme and convoluted with the experimental resolution, are shown in Fig. 4.38. The main anisotropic features of the experimental difference curve are well reproduced, of course using optimized coefficients A, B, C. About 12% of the electron density from below the Fermi level is shifted to energy states higher than the Fermi level. A high-resolution Compton study on V has been performed by Shiotani et al. (1993) at the KEK accumulation ring Japan, using 59.38 keV synchrotron radiation with an energy resolution of 0.12 keV. Figure 4.39 shows the directional Compton profiles compared with APW calculations of Wakoh and Matsumoto (1990). The orientation-dependent differences between experiment and theory show the same tendency as in the case of the γ-Compton measurement, but the amplitudes are much larger. This unexpected result might be due to a systematic error in the data processing, either with the synchrotron or the γ-data. The directional differences of the synchrotron radiation experiment, as shown in Fig. 4.40, exhibit the same structure as the γ-study, but with larger amplitudes, as it should be because of the higher resolution. But nevertheless, the amplitudes are smaller than those of the

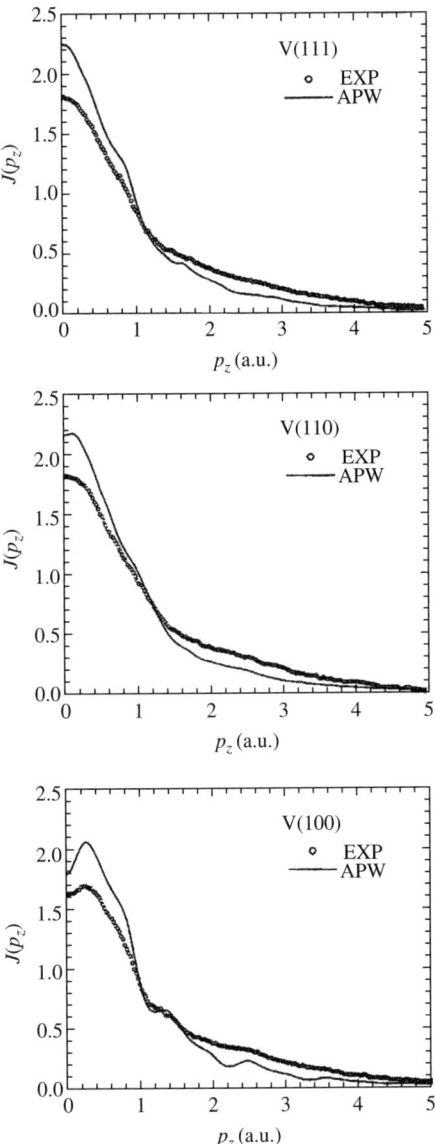

Fig. 4.39. Compton profiles of the valence electrons of vanadium along the [111], [110] and [100] direction. Circles: experiment; full lines: APW calculations of Wakoh and Matsumoto (1990). The size of the error bars for the data points near $p_z = 0$ is about twice the diameter of the circles and that near $p_z = 1.0$ is about the same as that of the circles. (Reprinted with permission from Shiotani et al. (1993); copyright (1993) by the Physical Society of Japan.)

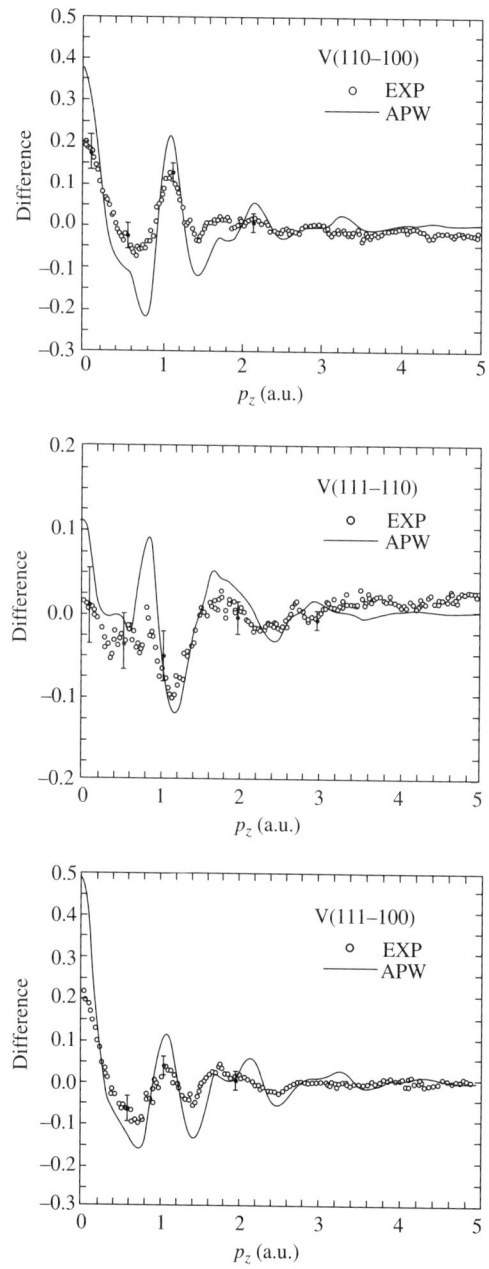

Fig. 4.40. Differences between pairs of Compton profiles of vanadium along different directions. Circles: experiment. Solid lines: APW calculations by Wakoh and Matsumoto (1990). (Reprinted with permission from Shiotani *et al.* (1993); copyright (1993) by the Physical Society of Japan.)

APW calculations. This has been attributed by the authors to directional dependent correlation effects, which, in principle, could be treated empirically by means of the energy occupation function of Wakoh and Matsumoto (1990), but they criticize this scheme for accommodating the correlation effects into the band theory by altering only the occupation probability in **k**-space and leaving the electron wavefunctions unaffected. The orientation differences can be understood qualitatively in terms of the geometry of the Fermi surface. The most important feature of the Fermi surface is the lattice of interconnecting holes running along [100] directions in reciprocal space. Thus the [100] profile has a reduced contribution from the conduction band at $p_z = 0, 2\pi/a, 4p/a, \ldots$, where the planes of integration intersect the lattice of the holes.

Chromium exhibits a similar behavior as vanadium as far as the orientation dependent difference between APW calculated (Wakoh et al. 1976) and experimental 412 keV ^{198}Au γ-Compton profiles (Cardwell et al. 1989) are concerned. The directional difference between LDA theory and low-resolution experiment can also be fitted to a Compton profile correction based on the energy occupation function of Wakoh and Matsumoto (1990). More recent higher resolved (0.18 a.u.) directional Compton profiles of Cr, measured at beamline ID15b of the ESRF, were used by Kubo (2001) to show that the GWA (see Section 4.8.1) can account for the difference between experiment and LDA calculations. The same seven Cr Compton profiles all measured with **q**-vectors lying within the same (110) plane were employed to reconstruct the 2D projection of the electron momentum density on the (110)-plane by means of the direct Fourier method (Tanaka et al. 2000) and by Dugdale et al. (2000) using the Cormack's method, as developed by Kontrym-Sznajd (1990). In both cases the projected momentum density was backfolded into the projected occupation number density utilizing the LCW theorem. The result of Tanaka et al. (2000) is delineated in Fig. 4.41 in form of a contour map and compared with KKR computations convoluted with the resolution of the experiment. Most of the details of the Fermi surface can be seen in experiment and theory, the electron jack with two knobs at Γ, a hole octahedron at H and hole ellipsoids at the N-point, as predicted by Mattheiss (1965). But there are some differences: the shape of the H-holes are different, the hole pockets at the N-point are much less prominent in the experiment, and finally the details of the electron jack at the Γ-point are somewhat different. Nevertheless, this experiment together with the reconstruction procedure has proved the ability of high-resolution Compton scattering to provide information about details of the Fermi surface of more complicated systems.

Nonspin-selective γ-Compton measurements (^{198}Au, 412 keV) on iron single crystals were reported by Rollason et al. (1983). The comparison of their data (directional differences, individual profiles, directional reciprocal form factors) with APW calculations of Wakoh and Kubo (1977), and with LCAO computations of Rath et al. (1973) was not very conclusive. The theories differ substantially with respect to the individual profiles, but agree as far as the directional differences are concerned. The experiment is in various aspects

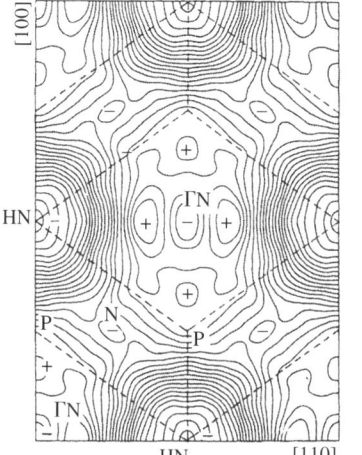 

Fig. 4.41. Left-hand panel: contour map of the experimental occupation number density of chromium projected on the (110) plane. Dashed lines represent the projected image of the Brillouin zone. Right-hand panel: contour map of the theoretical occupation number density projected on the (110) plane. The density is convoluted with the experimental resolution. The contour intervals are the same as in the left-hand panel. (Reprinted with permission from Tanaka *et al.* (2000); copyright (2000) by Elsevier Science Ltd.)

in disagreement with one or the other of the theories, so that it is difficult to say something about the reasons for these disagreements. Later Compton experiments on iron have been made spin selectively, and will be reviewed in Section 4.13.

A very instructive example of a high-resolution Compton study of Fermi surface details on an ordered alloy, namely FeAl, was provided by Blaas *et al.* (1995). It was the first experiment at the ESRF Compton beamline ID15b with 49.6 keV and a resolution of 0.15 a.u. The measurements were compared and interpreted by means of LDA-type FLAPW calculations. Frame (a) of Fig. 4.42 shows the experimental [100]–[111] difference Compton profile as obtained in a low-resolution (conventional W $K\alpha_1$ source, 0.55 a.u. resolution) preliminary experiment together with the theory convoluted with the experimental resolution. Frame (b), again the same difference profile (experiment and theory), demonstrates the progress which is achieved with the higher resolution. The theory tells us that the 11 valence electrons of FeAl which require six bands are occupying four bands completely. The fourth, fifth, and sixth bands (all Fe d-like) are filled partly with 1.95, 1.80, and 1.25 electrons, respectively. Frame (c) makes transparent to what extent Fermi surface features are contributing to the oscillating fine structure of the difference profile: the solid line is again the

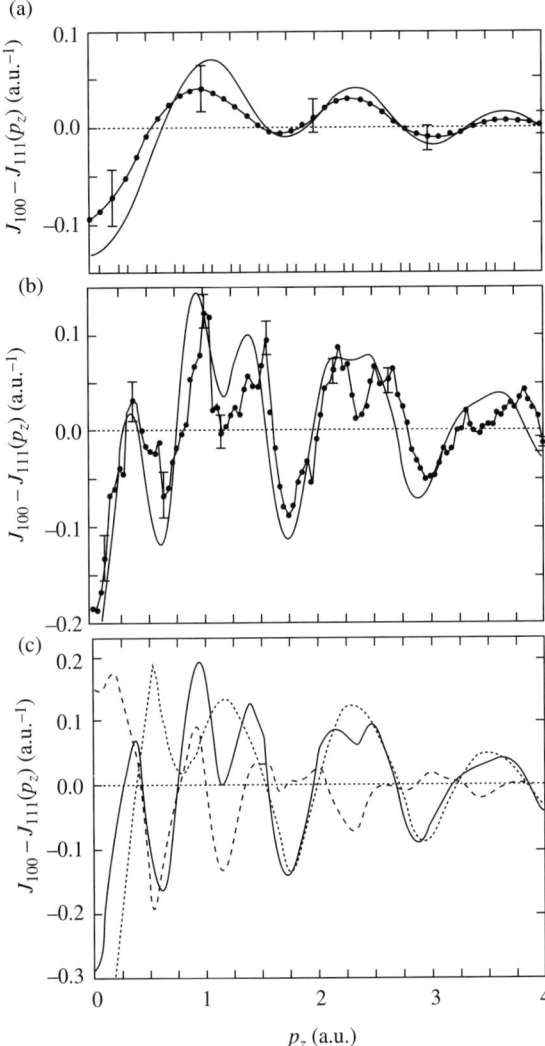

Fig. 4.42. Experimental and theoretical difference Compton profile [100]–[111] for FeAl. (a) Low-resolution experiment (bullets); theory convoluted with the resolution function (solid line). (b) High-resolution experiment (bullets); theory convoluted with the resolution function. (c) Fermi-surface induced variations (unconvoluted: difference Compton profile (solid); difference Compton profile with six bands assumed to be filled completely (dotted); the difference between a partly filled and a completely filled sixth band (dashed). (Reprinted with permission from Blaas *et al.* (1995); copyright (1995) by the American Physical Society.)

(unconvoluted) theoretical directional difference. The dotted line is the directional difference assuming the sixth band to be completely filled. Finally, the dashed line emphasizes the influence of the sixth band Fermi surface by showing the difference between the partly filled and completely filled sixth band. It becomes transparent that it is the Fermi surface of the sixth band which determines the main features of the oscillation of these directional differences.

Copper metal, a so-called noble metal, is assumed to have fully occupied 3d states, so that its Fermi surface should be simpler than that of the real 3d transition metals. Thus special features of the directional Compton profiles should be less influenced by the Fermi surface than by the shape of the occupied region of the extended zone scheme. This is one reason why Cu is especially suited to study anisotropic correlation effects, as we will see in what follows. Leading in this respect was a γ-Compton investigation (412 keV, ^{198}Au) on Cu of Pattison *et al.* (1982) followed by a study of Bauer and Schneider (1984b, 1985). First of all these authors have found that the directional differences of their experimental Cu Compton profiles are well reproduced by corresponding (resolution convoluted) LDA computations of Bagayoko *et al.* (1980) as far as the frequencies and phases of the oscillation are concerned, whereas their amplitudes are up to two times larger in the calculations. This must be seen as an indication of an anisotropic correlation effect. Indeed, even after performing the isotropic Lam–Platzmann correction to the theoretical Compton profile in [110] directions, there remains an oscillating behavior of the difference between the LDA theory and experiment, which is plotted in Fig. 4.43. The XY plane of the Brillouin zone and the Fermi surface of copper in the repeated zone scheme, as depicted in the lower part of Fig. 4.43, can give an explanation for these oscillations. The peaks of the oscillations are, as indicated in Fig. 4.43, connected with integration planes that constitute the Compton profiles and are perpendicular to the **q**-vector in [110] direction, relative to which all secondary Fermi surfaces are well aligned (according to the 3D structure of the fcc unit cell, the [110] direction is the only one which fulfills the above requirement). As the correlation effect in the homogeneous electron gas corresponds to a transfer of momentum density from the occupied Fermi surface into the unoccupied region, also in the inhomogeneous electron gas the same effect is observable not only from the primary Fermi surface, but also from its images centered at reciprocal lattice vectors. A transfer of momentum density into the interstitial regions with their small momentum density takes place, thus washing out the contrast predicted by the one-electron model, the difference between one-electron theory and experiment peaks on the planes mentioned above. The necessary d-character of states contributing to the secondary Fermi surfaces can be understood on the basis of the hybridization between the plane-wave like states and the atomic-like d-states as proposed in a tight-binding model of Bauer and Schneider (1984a), so that the images of the Fermi surface are projected onto the momentum density, not only at the origin, but at all reciprocal lattice vectors. The net transfer of momentum density from inside the Fermi surface to outside has been estimated by Bauer and Schneider

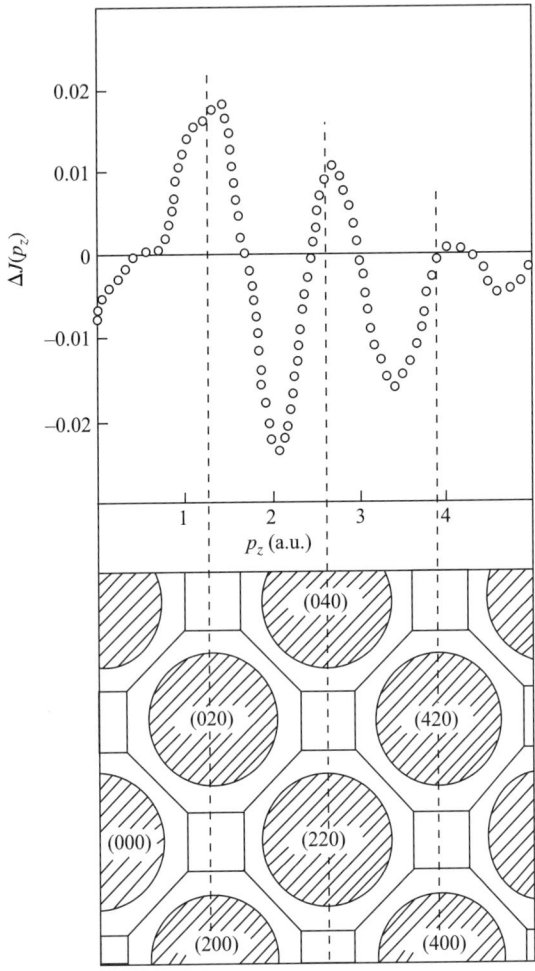

Fig. 4.43. Upper part: difference between the LDA calculated and Lam–Platzman corrected Compton profile of Cu in the [110] direction (Bagayoko et al. 1980) and the experimental one. Lower part: XY plane of the Brillouin zone and the Fermi surface of copper in the repeated zone scheme. (Reprinted with permission from Bauer and Schneider (1984b); copyright (1984) by the American Physical Society.)

(1985) in the following way: They considered the experimental reciprocal form factor $B(\mathbf{R})$ at the lattice vector $\mathbf{R} = (1,1,0)a/\sqrt{2}$, which is, according to equation (4.30), directly related to a model occupation number function $N(\mathbf{k})$. $N(\mathbf{k})$ is reduced by correlation effects relative to the independent particle $N_0(\mathbf{k})$ by $2z$ inside the Fermi surface and exhibits a constant increase of $2z$ outside the

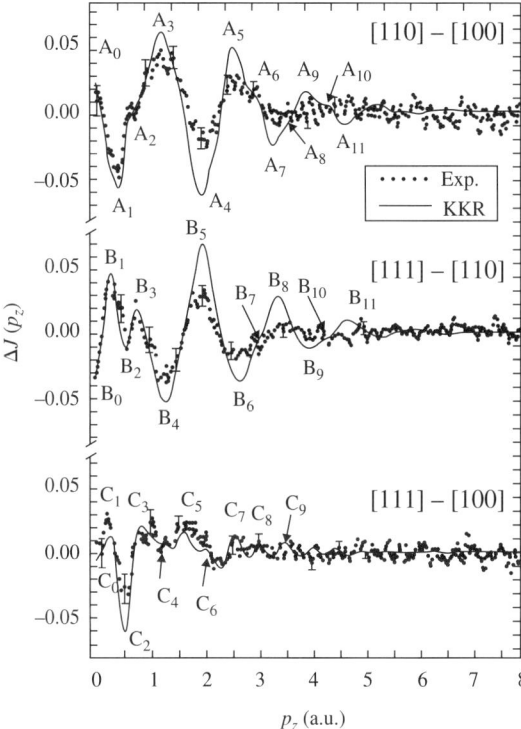

Fig. 4.44. Anisotropic Compton profiles in Cu. Experimental (full circled) and theoretical (solid lines) Compton profile differences along the three pairs of high-symmetry directions were compared. (Reprinted with permission from Sakurai *et al.* (1999); copyright (1999) by Elsevier Science Ltd.)

Fermi surface. (The factor 2 is included to take account of the spin degeneracy.) Taking into account the damping of $B(\mathbf{r})$ due to the finite resolution, z was determined to be 0.12 which is roughly two times the value for a homogeneous electron gas of the valence electron density of copper. All these deviations of the correlation effects from the LDA picture were attributed by Bauer and Schneider to nonlocal exchange-correlation due to the satellite structure of the spectral density function in the momentum space representation (see e.g. Almbladh and Hedin 1983).

More recent high-resolution Compton profile measurements on Cu using synchrotron radiation of the KEK accumulation ring at 59.38 keV with 0.12 a.u. momentum resolution (Sakurai *et al.* 1999) show finer details of the directional differences. As depicted in Fig. 4.44, the oscillations of these difference curves agree well, with respect to their frequencies and phase, with the LDA KKR calculations, but, as in the low-resolution case, their amplitudes are significantly smaller than in the theory. The difference between theoretical and experimental directional profiles is positive between $p_z = 0$ and 1 a.u., negative between 1 and 6 and slightly positive between 6 and 8 a.u. The amplitudes of these differences are much larger than in the low-resolution case, at $p_z = 0$, namely 9% of the experimental Compton profile peak, whereas the corresponding Lam–Platzman correction, which was sufficient to account for this difference at $p_z = 0$ in the low-resolution case, is less than 2%. As with Si and vanadium this discrepancy between different experiments on the same sample must be left unsolved and points to systematic errors in the course of data processing. Therefore, it seems to be premature to discuss the overall shape of these difference between theory and experiment in terms of correlation effects. On the contrary, the oscillating structure in the [110] difference, shown in Fig. 4.43, and thoroughly discussed above as being correlation induced, is also visible in the high-resolution experiment, even though superimposed by the much larger low-frequency oscillations of unknown origin. Kubo *et al.* (1999) have corrected the full-potential linearized APW (FLAPW) LAD-type computations of copper Compton profiles by taking into account self-interaction (SI) first introduced by Perdew and Zunger (1981). This leads to the following consequences for the band structure. The relative position of the d-bands with respect to the Fermi energy is lowered by 2 eV, and the width of the d-band is reduced by 15%. As a result the electrons in the d-bands are more localized and the sp-d hybridization near the Fermi energy is much reduced. Thus the sp part of the Compton profile (between 0 and 1 a.u.) became sharper (more free-electron like) and reduced in height, and the d part (between 1 and 6 a.u.) became a little bit broader, so that the overall shape of the SI-corrected copper Compton profile came nearer to the high-resolution experiment of Sakurai *et al.* (1999), a new discrepancy between theory and experiment around 0.8 a.u. notwithstanding. The amplitudes of the directional differences were reduced by the SI correction, thus achieving a better agreement with experiment.

The first comprehensive Compton study of an alloy system was done by Benedek *et al.* (1985) on a series of Cu_xNi_{1-x} alloys still performed with a 60 keV ^{241}Am γ-source on seven polycrystalline samples with $x = 1.0$, 0.65, 0.6, 0.55, 0.50, 0.40, and 0.0. The experimental results were compared with KKR-CPA calculations. For some values of p_z the theoretical Compton profiles exhibited a nonlinear behavior on x, which could qualitatively be verified by the experiment. On the other hand, lack of resolution and statistics prevented extraction of Fermi surface dimensions out of the polycrystalline data. This points to the necessity of higher resolution and **q**-orientation dependent measurements on single crystals, when conclusive data on Fermi surface dimensions should be extracted from Compton profiles. The synchrotron radiation based Compton measurements of Matsumoto *et al.* (2001a) on a Cu-27.5at% Pd disordered alloy performed at the PF-AR NE1A1 beamline of the KEK (Japan) is a good example of a successful approach to the Fermiology of an alloy. This alloy was of special interest since it was predicted by Moss (1969) and later by Gyorffy and

Stocks (1983) that flat segments of the Fermi surface, connected by $\mathbf{k}_0$, can be made responsible for short-range order seen by means of diffuse scattering. These flat segments of the Fermi surface could be identified by means of the $\rho(\mathbf{p})$ reconstruction (using the direct Fourier method) on the basis of 28 directional Compton profiles and were compared with those of pure copper measured the same way. The Fermi radii were determined from the position of a peak in the first derivatives of the reconstructed momentum density and plotted in Fig. 4.45 together with the Fermi radii of pure copper normalized to the Brillouin zone dimensions. The disappearance of the neck at the L-point due the alloying and the flattening in the [110] direction forming a set of parallel sheets can clearly be seen. The k_F in the [110] direction is 0.59 ± 0.01 a.u. in the alloy while 0.69 ± 0.01 a.u. in pure copper. This is in excellent agreement with the alloy value of 0.586 ± 0.001 as deduced from diffuse scattering measurements of Ohshima

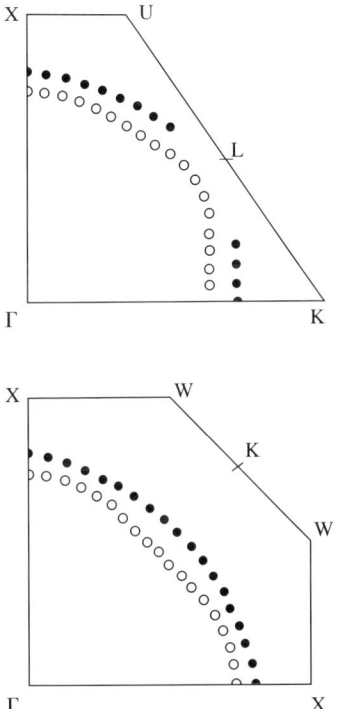

Fig. 4.45. The Fermi radii of pure Cu (full circles) and Cu-27.5at% Pd (open circles) in the disordered phase on the ΓXULK plane (upper panel) and the ΓXWK plane (lower panel). The Fermi radii are normalized by the Brillouin zone dimension. The size of the circles is the error involved in the radial direction. (Reprinted with permission from Matsumoto et al. (2001a); copyright (2001) by the American Physical Society.)

and Watanabe (1973). It is this kind of study which proves the ability of high-resolution Compton scattering to extract Fermiology parameters in disordered systems, where conventional methods on the basis of transport properties must fail due to the short mean free path of the electrons. Even 2D-ACAR can yield misleading data as a consequence of inhomogeneities of the positron wavefunction (see Section 4.4).

In a similar study, in that case on a disordered $Cu_{0.842}Al_{0.158}$ alloy, again the nesting properties of the Fermi surface has been investigated by Kwiatkowska et al. (2004), utilizing 28 directional Compton profiles from the same beamline as for the CuPd alloy and confronted with KKR-CPA calculations.

The Fermi surface nesting problem is also the matter of a high-resolution (0.13 a.u.) synchrotron radiation based Compton study on the shape memory alloy $Ti_{48.5}Ni_{51.5}$ by Shiotani et al. (2004) at the PF-AR NE1A1 beamline of the KEK. The momentum density $\rho(\mathbf{p})$ reconstructed from 28 directional profiles was LCW backfolded to obtain a contour map of the occupation number density as shown for the XMR plane of the repeated zone scheme in Fig. 4.46. Here the arrow marked b indicates a possible nesting vector in [110] direction. Its length $(2/3)[1,1,0]2\pi/a$ (a is the lattice constant) can be reduced, by subtracting the reciprocal lattice vector [110], to $(1/3)[1,1,0]2\pi/a$, where the $TA_2[110]$ phonon branch in the β-phase becomes soft, according to neutron scattering data of Moine et al. (1984), where the origin of the phonon softening has long been speculated to be Fermi surface nesting.

The oscillatory structure of the Compton profile difference between LDA calculations and experiment of Cu, especially in the [110] direction, which has

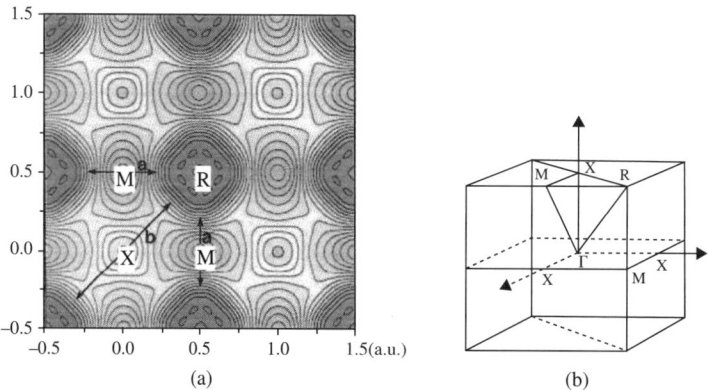

Fig. 4.46. Contour map of the occupation number density of $Ti_{48.5}Ni_{51.5}$ on the X-M-R (a) plane in the repeated zone scheme. The first Brillouin zone with high- symmetry points is depicted in (b). The arrows a and b indicate possible nesting vectors. (Reprinted with permission from Shiotani et al. (2004); copyright (2004) by the Physical Society of Japan.)

been attributed by Bauer and Schneider (1985) to nonlocal correlation effects, has also been found by Rollason *et al.* (1987), and by Anastassopoulos *et al.* (1991) in γ- Compton investigations of Ni, where the theoretical reference profiles of the former experiment is that of Wang and Callaway (1975). The latter study utilizes own APW computations.

A high-resolution study of directional Compton profiles of $CoSi_2$ of Bellin *et al.* (1995) brought a good overall agreement between experimental and theoretical profile anisotropies, where the calculations are based on self-consistent linear-muffin- tin orbital band structure, again with a slight overestimation of the experimental anisotropies by the band structure result, which were attributed to the specific properties of the d-type bonds in [110] direction.

Charge transfer upon alloying in amorphous alloys has been investigated by means of γ-Compton scattering by Zukowski *et al.* (1990) on $Co_{70-x}Ni_xFe_5Si_{15}B_{10}$ and by Anderejczuk *et al.* (1992) on $Fe_{82-x}Ni_xB_{18}$ both with increasing varying x. The authors have tried to shed light on the mechanism of magnetic moment formation in amorphous metals. By representing the difference profiles with different x, after subtracting the core contributions, by (positive) free atomic 3d functions and (negative) free electron 4s contributions, it was concluded that the replacement of cobalt and iron, respectively, by nickel leads to a charge transfer from the 4s to the 3d band, which is consistent with their magnetic behavior.

Among the 4d-transition metals, yttrium is by far the most thoroughly investigated one both by high-resolution Compton scattering and by 2D-ACAR. In a paper by Kontrym-Sznajd *et al.* (2002) a reconstruction of the electron momentum density $\rho(\mathbf{p})$ from 12 directional Compton profiles measured with a resolution of 0.16 a.u. at the ID15b beamline of the ESRF was performed using the Cormack method. Additionally, a reconstruction of the electron–positron pair momentum density $\rho_p(\mathbf{p})$ from five 2D-ACAR spectra with projection directions between ΓM and ΓK and with a momentum space resolution comparable with that of the Compton profiles (Dugdale *et al.* 1997) was obtained. This reconstruction was either done with the raw positron data or with those deconvoluted with respect to the instrumental broadening by a maximum entropy formalism. In order to facilitate a realistic confrontation of Compton and positron data, also the 1D projection of the reconstructed electron–positron pair momentum density was calculated. The experimental results were compared with fully relativistic APW calculations, where for the positron spectra the independent particle model (IPM) was used. The most striking result of comparing Compton and positron data was the distinct similarity of the directional differences of the 1D profiles for both techniques, and the fact that both are roughly by a factor 2 smaller than the corresponding theoretical ones. Since this discrepancy between theory and experiment has been found in most of the transition metal Compton data, and has been ascribed to correlation effects, it was concluded that the e–e correlation in the positron data is not compensated by the electron–positron correlation, as predicted for the jellium case by Carbotte

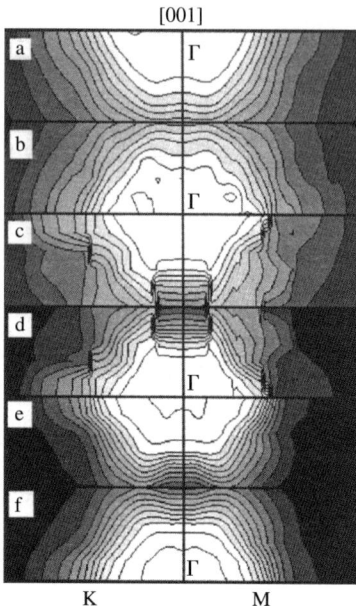

Fig. 4.47. Momentum densities in Y, containing the core contributions, along ΓK (left side) and ΓM (right side) and parallel directions, for momenta up to 1.37 a.u., on six planes up to 0.73 a.u.: (a) $\rho(\mathbf{p})$ reconstructed from Compton profiles, 36 Jacobi polynomials; (b) $\rho(\mathbf{p})$ reconstructed from Compton profiles, 60 Jacobi polynomials; (c) APW calculation $\rho(\mathbf{p})$; (d) calculated electron–positron momentum density (IPM); (e) reconstructed electron–positron momentum density (deconvoluted); (f) reconstructed electron–positron momentum density (raw data). (Reprinted with permission from Kontrym-Sznajd et al. (2002); copyright (2002) by the American Physical Society.)

and Kahana (1965). The difference between the isotropic averaged Compton profile and the APW calculated counterpart is not adequately described by the Lam–Platzman correction as in many other cases, but exhibits a characteristic oscillatory structure which points to directional dependent correlation effects. In Fig. 4.47 the momentum densities of Y for the two directions ΓK and ΓM are shown. Panels (a) and (b) demonstrate how a larger number of Jacobi polynomials (36 and 60) used in the course of the reconstruction improved the detail resolution; (c) and (d) substantiate the similarity between theoretical $\rho(\mathbf{p})$ and (IPM)-$\rho_p(\mathbf{p})$; and (e) and (f) compared with (a) and (b) illustrate the strong similarity between momentum density reconstructed from positron and Compton data.

Another member of the noble metals, silver, has gained considerable attention. Exemplarily the γ-Compton study of Andrejczuk et al. (1993) should be mentioned, especially since it contains a comparison between the results from a ^{198}Au (412 keV) and a ^{137}Cs (662 keV) experiment. Such a comparison is of importance because the bremsstrahlung background, a source of a big systematic error in the case of high γ-energy Compton investigations, has a strong dependence on the incident photon energy. Since the measured Ag Compton profile exhibited only a small left–right asymmetry it was assumed that the additional background cannot possess a substantial energy-loss dependence in the region of the Compton peak. Therefore, this background was modeled by a constant term and its magnitude used to fit the theory. Thus the rather small and nearly orientation independent difference between the theory and the directional Compton profiles might not be too surprising. But contrary to the other noble metal copper, the directional differences (as an example the [111]–[110] difference is shown in Fig. 4.48) are in rather good agreement (with respect to shape and amplitude) both with the relativistic RKKR computations (Lam–Platzman correction included) and with the results of a nonrelativistic LCGO (linear combination of Gaussian orbitals) scheme of Fuster et al. (1990).

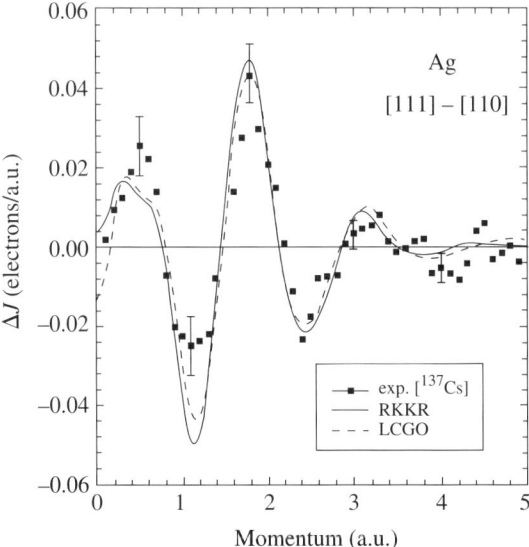

Fig. 4.48. Directional difference Compton profile [111]–[110] of silver: experiment (filled squares); theory (RKKR) (solid line); theory (LCGO) (dashed line). Both are convoluted with the experimental resolution (Gaussian of 0.4 a.u. FWHM). (Reprinted with permission from Andrejczuk et al. (1993); copyright (1993) by the American Physical Society.)

4.9 Compton scattering of special systems

4.9.1 *Intercalation compounds*

Graphite intercalation compounds (GIC's) are synthesized by inserting foreign atoms or molecules (intercalants) between carbon layers in graphite. Since graphite is a semimetal, the electron donated or accepted by the intercalant modifies the electronic properties of graphite and can result in metallic behavior of the final material. It was the goal of Compton investigations on this class of compounds to find out how the electronic structure of graphite is changing upon intercalation. Let us start with the prototype of those GIC's, LiC_6, where Li is intercalated between the graphite layers, so that one Li atom sits on top of each third hexagonal C ring constituting the graphite layer. Additionally intercalation of Li changes the stacking sequence of the graphite layers from ABAB in graphite to AαAα in LiC_6, where the B configuration means a C atom on top of the hexagonal C ring of the A layer underneath. α indicates the alkali metal layer in between the AA stacking. The interlayer spacing increases by 10% upon Li intercalation. The AαAα stacking is also called a stage I intercalation compound. Compton scattering should decide between different models for the intercalation process: the simplest model is the rigid band model, where it is assumed that the Li atom hands over its 2s electron to the unoccupied π^* states of graphite. These states and all other states of graphite remain unchanged. This results in an increased Fermi level. Alternative models are those where no or only a part of the 2s charge of Li is transferred to graphite states. Finally, it may happen that the Li ion interact with the neighboring C atoms, the carbon charge distribution becomes polarized, and the carbon layer distance increases, so that the whole electron structure of the host will be changed. Each model will leave behind specific traces in the Compton profile differences between the intercalation compound and the pure host. Such a difference profile for $\mathbf{q} \parallel c$-axis, measured first by Loupias *et al.* (1984) in a synchrotron radiation based experiment at LURE-DCI with 12.85 keV photons and 0.15 a.u. resolution is shown in Fig. 4.49. On the basis of this result, two models can immediately be excluded: (i) the rigid band model, since in this case the difference profile must be a pure π-type Compton profile with zero value at $p_z = 0$ for the above $\mathbf{q}$-orientation, as shown with the dashed curve (Berko *et al.* 1957); (ii) the model with zero charge transfer, since in this case the difference profile should resemble a free-electron-like inverted parabola for the Li 2s electron with a peak value at $p_z = 0$, as shown with the dashed-dotted curve. The local minimum of the experimental curve tells us that a certain charge transfer to the unoccupied π^*-states must have taken place, but that additionally changes of the graphite wavefunctions have to be made responsible for the finite value at $p_z = 0$, and the much smaller width of the experimental difference profile compared with that of the graphite π-Compton profile (dashed curve). A more detailed analysis of the difference Compton profile for $\mathbf{q} \parallel c$-axis is provided by a theoretical calculation done by Chou *et al.* (1984), and by Chou *et al.* (1986) using a pseudopotential technique

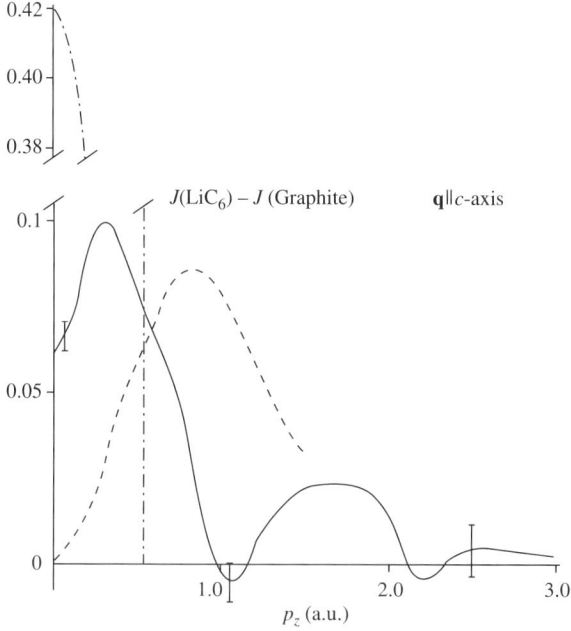

Fig. 4.49. Experimental Compton profile difference for $\mathbf{q} \parallel c$, $J(\mathrm{LiC}_6) - J(\mathrm{graphite})$ (solid line). The experiment is compared with a graphite π-electron Compton profile with $\mathbf{q} \parallel c$-axis (dashed line) and a free-electron-like Compton profile. All profiles are normalized to 1/6 electron. (Reprinted with permission from Loupias et al. (1984); copyright (1984) by Editions de Physique.)

on the basis of LDA. They separated the difference profile into two parts. One is the contribution from the difference between LiC_6 and graphite in the profiles of the filled σ and π bonding states as shown in Fig. 4.50(a). This part indicates that the electron charge in LiC_6 is more spread out into the interlayer region due to the polarization effect of the Li ions. The other part is the contribution from the conduction electrons in LiC_6, the partly filled π antibonding states, shown in Fig. 4.50(b). The structure of this contribution, periodic with g_3 the shortest reciprocal lattice vector in the c-direction, is a consequence of the fact that the states near the L-point ($k_z = g_3/2$) are occupied, while the states near the M-point ($k_z = 0$) are not occupied. Both parts together amount to a shape of the difference profile which is in good agreement with experiment as demonstrated in Fig. 4.51 (here the experimental difference profiles are from Berthold 1993). It was shown by Rabii et al. (1989) that the two contributions to the difference profile are changing their weight, when going from a stage I (LiC_6) to a stage II (LiC_{12}; AαAAαA stacking) Li-intercalated graphite, so that the first part dominates. This leads to an experimental difference profile of LiC_{12} which is negative

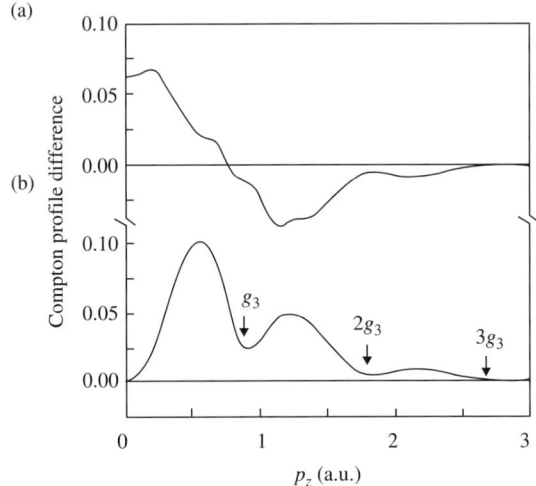

Fig. 4.50. (a) Contribution from the difference between LiC$_6$ and graphite in the Compton profiles of valence bands (filled σ and π bonding states). (b) is the contribution from the conduction electrons in LiC$_6$ (partly filled π antibonding states). (Originally published by Chou *et al.* (1984); copyright (1984) by the American Physical Society.)

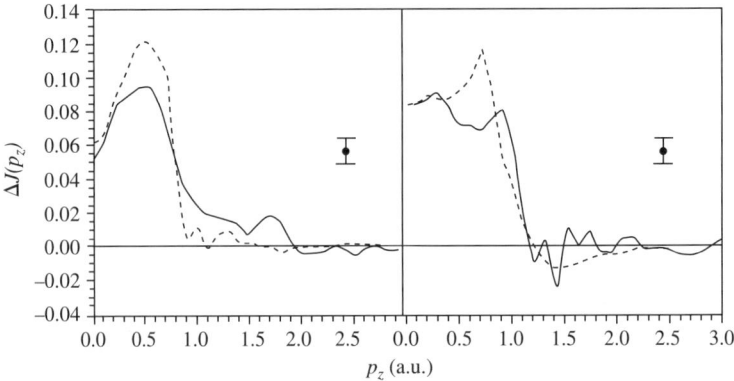

Fig. 4.51. Difference between the Compton profiles of LiC$_6$ and graphite. Experiment (Berthold 1993): solid line; calculations (Chou *et al.* 1986): dashed line. Left-hand panel: $\mathbf{q} \parallel c$-axis; right-hand panel: $\mathbf{q} \perp c$-axis. (Reprinted with permission from Berthold (1993).)

between 1 and 2.5 a.u. Synchrotron radiation based Compton profile measurements at the Compton beamline of HASYLAB of Berthold (1993) on potassium intercalated graphite KC_8 yielded difference profiles between KC_8 and graphite similar to those of LiC_6, but with a less distinct minimum at $p_z = 0$. This could mean that the charge transfer from K to the π^*-states of graphite upon intercalation is not complete. Loupias et al. (1990) came to a similar conclusion in a study of the near-edge photoabsorption measurement on KC_8.

Another class of intercalation compounds systematically investigated by Compton scattering are the C_{60}-fullerenes intercalated with heavy alkali atoms (K, Rb and Cs). Marangolo et al. (1998) have measured Compton profiles of polycrystalline C_{60} and K_6C_{60} using 16.3 keV synchrotron radiation at LURE achieving 0.15 a.u. resolution by means of a Cauchois-type analyzing system. The measurements were compared with LCAO-LDA computations. Both from the theoretical and the experimental difference profiles $K_6C_{60} - C_{60}$ the K 3p pseudocore contribution, obtained in the course of the computations, was subtracted out. The agreement between theory and experiment shown in Fig. 4.52 is satisfactory. Especially, the shoulder around 0.6 and 1.0 a.u. can be attributed to that part of the difference profile which is due to the distortion of the C_{60}-charge density by the K ions. The other part has to be ascribed to the charge transfer from potassium to the t_{1u} band of K_6C_{60}, which is completely filled this way.

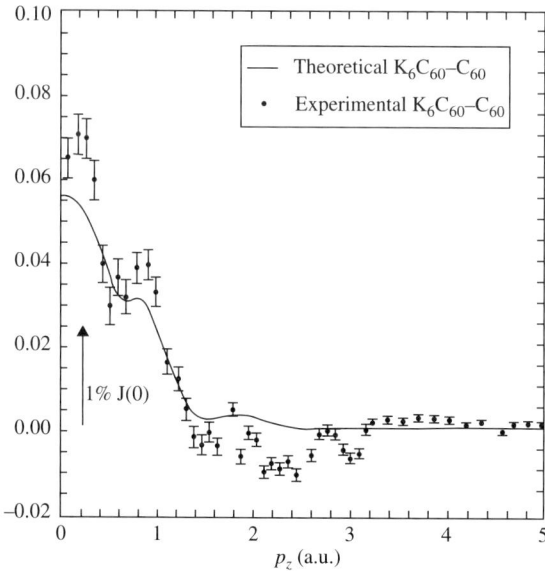

Fig. 4.52. Difference between valence Compton profiles (CP's) of K_6C_{60} and C_{60}. The potassium 3p band-like contribution to the CP has been subtracted. (Reprinted with permission from Marangolo et al. (1998); copyright (1998) by the American Physical Society.)

Similar conclusions were drawn on the basis of further Compton experiments and calculations on K_3C_{60}, K_4C_{60}, $CsRb_2C_{60}$, and Rb_4C_{60} performed by Marangolo et al. (1999). In all cases a distinct distortion of the C_{60} charge density could be established. It was shown that the Compton profile differences between intercalation compound and host depend on the number of ions and not on their nature.

Li intercalated NiO_2, $LiNiO_2$, possesses a layered rock-salt structure consisting of a stacking of an O plane, and Ni plane, an O plane and a Li plane, and so on. A high-resolution Compton scattering study on this intercalation compound has been performed by Chabaud et al. (2004) at the Compton beamline of the ESRF with 55.8 keV photon energy and 0.22 a.u. momentum resolution. The difference Compton profile between $LiNiO_2$ and NiO_2 is in good agreement with a theoretical one calculated on the basis of an LDA-type *ab initio* pseudopotential scheme. One quarter of this calculated difference profile could be attributed to the 2s electrons transferred from Li into the e_g^* states of the host. The rest of the difference profile, the so-called total distortion profile, must be ascribed to deformation of the deep oxygen 2s states, to the Jahn–Teller effect (distortion of the NiO_6 octahedron), change of the lattice constant, and polarization of the host electron density brought about by Li ions.

4.9.2 *Hydrides*

Even in the early days of Compton scattering metal hydrides were considered to be an ideal testing ground for using Compton profiles as sensitive indicators for changes in the valence electron distribution upon hydrogenating. Actually, there are three models for incorporating hydrogen into metals the effect of which on the momentum density was used to let Compton profiles decide on one of them. The first one is the so-called anionic model which assumes that the incorporated hydrogen takes electrons from the conduction band to form hydrogen anions with bound 1s states so that the electronic wavefunction can be approximated by a hydrogenic 1s wavefunction with a screened nuclear charge of $Ze = (11/16)e$ (Bethe and Salpeter 1957). The valence Compton profile of the MH_x metal hydride then consists of the valence profile of the host reduced by the contribution of x electrons and a profile of $2x$ electrons in a screened hydrogen bound state. The second model is the neutral atomic model, where the hydrogen exists as a neutral atom in the metal lattice, so that the difference Compton profile between the metal hydride and the pure host metal resembles simply a free atomic hydrogen Compton profile. The third model for a MH_x metal hydride is the so-called protonic model, which, in its simplest form, the rigid band model, assumes that the electrons from the hydrogen atoms are donated to the unoccupied states in the conduction band of the metal without changing the wavefunctions of the host valence band. The charge of the resulting bare proton is screened by the conduction electrons. Thus the Fermi level is shifted to higher energies in the band structure of the host metal often simply simulated by increasing the valence electron part of the host Compton profile by a factor of x/N_v, where N_v is the number of valence electrons of the host metal, this way simulating the Compton profile of the metal hydride. A more sophisticated version of the protonic model leaves the assumptions of the rigid band model assuming that new states are created upon hydrogenation lying below the Fermi level and being occupied by the donated electrons. The new band can correspond to the bonding or antibonding hybrid of the hydrogen 1s orbitals on certain interstitial positions in the metal lattice. Moreover, also bands due to H-metal interactions have been discussed. The employment of those variations of the protonic model are often called band structure models, since their utilization either need calculations of band structure or even momentum densities of the stochiometric counterparts of the hydrides or more elaborate model calculations to simulate the hydrogen hybrids. If the incorporation of hydrogen takes place in an amorphous phase of a tetrahedrally coordinated covalent bounded solid, this hydrogenation is described in terms of binding (part of the) hydrogen to the dangling bounds assumed to exist in the disordered phase.

We will briefly summarize some of the Compton studies, in most cases γ-Compton experiments, on hydrides published during the last 30 years by saying, what conclusions with respect to the above models were drawn by the authors. γ-Compton measurements of Pattison *et al.* (1976) on polycrystalline $NbD_{0.6}$, of Pattison *et al.* (1977) on single crystalline $NbH_{0.76}$, and of Alexandropoulos and Reed (1977) on $NbH_{0.29}$ were interpreted to prove the protonic model with filling conduction band states observed as a reduction of the V Compton profile anisotropy, even though their conclusions were only of a qualitative nature. The results of a Compton study on $PdH(D)_{0.71}$, $VD_{0.77}$ and $VH_{0.71}$ by Lässer and Lengeler (1978) were successfully interpreted on the basis of a modified protonic model, where, according to band structure calculations of Switendick (1972, 1978), hydrogen–palladium and hydrogen–vanadium bonding states below the Fermi level created by the introduction of hydrogen are filled with the donated electrons. These more qualitative arguments found their confirmation in a theoretical study of Harmalkar *et al.* (1985) on Pd and stoichometric PdH. γ-Compton profile measurements on lutetium and lutetium hydride of Lässer *et al.* (1979), and on FeTi and FeTi hydride by Lässer *et al.* (1980) also turned out to be in agreement with a modified protonic model. Following Switendick (1971) they introduced bands below the Fermi level corresponding to antibonding hybrids of the two hydrogen 1s orbitals on the two tetrahedral sites in the unit cell of Lu. Theoretical Compton profiles of Ti and TiH2, APW-calculated by Bacalis *et al.* (1986), found experimental support by Boulakis (1986), who stressed that it is the appearance of Ti-H bonding states, and reordering of energy levels which clear the way for incorporating the donated electrons. Again a modified protonic model, the difference profile of which is confronted with the experiment together with the difference profiles of other models, is shown in Fig. 4.53. A γ-Compton investigation of Theodoridou and Alexandropoulos (1984) has shown that, under special circumstances, the atomic

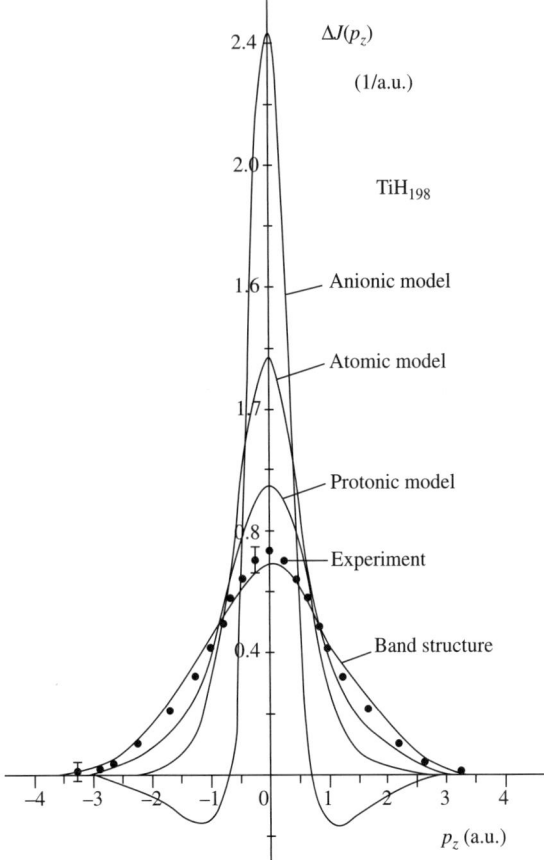

Fig. 4.53. Difference Compton profiles between polycrystalline TiH$_{1.98}$ and Ti. Experimental and theoretical results. Experiment and model calculations (Boulakis 1986). Band structure calculations (Bacalis et al. 1986). (Originally published by Boulakis (1986); copyright (1986) by Editions de Physique, Paris.)

model can fit experimental data. The authors measured the Compton profile of polycrystalline NbH$_{1.2}$ and the spherical average of three directional Compton profiles of NbH$_{0.3}$. The difference of these profiles bears a striking similarity to the Compton profiles of atomic hydrogen. This can be understood assuming that a large part of the dilute hydrogen forms atomic clusters at the grain boundaries.

Amorphous hydrogenated silicon, a-Si:H, has been investigated by Bonse et al. (1980) using γ-Compton scattering looking at the difference Compton profiles between a-Si:H and polycrystalline Si. In a similar study, Bellin et al. (1997) have utilized synchrotron radiation based Compton profile measurements studying the difference profiles between a-Si:H and both dehydrogenated and recrystallized Si. Based on models, where either hydrogen is bound to Si by occupying the dangling bonds of amorphous Si or two hydrogen atoms are inserted into a broken Si—Si bond, the latter authors extracted Si—H bond and H—H bond profiles out of these difference profile. The lack of theoretical calculation prevented a quantitative analysis of these results.

4.9.3 *Highly correlated superconducting systems*

High T_c ceramic superconductors have, very soon after their detection, attracted the attention of those who hoped to find any relationship between the transition to superconductivity and momentum space properties, as Compton profiles. Aleandropoulos *et al.* (1988) determined the Compton profiles of polycrystalline $YBa_2Cu_3O_{7-\delta}$ samples below and above the transition temperature of 90 K using 60 keV γ-rays. No difference was found within the experimental error. On the contrary Priftis *et al.* (1994) found surprisingly a large temperature effect in $Bi_{1.6}Pb_{0.4}Sr_2Ca_2Cu_3O_{10-\delta}$ polycrystalline samples. Of course, the observed effect of $\Delta J(0)/J(0) \approx -0.01$ in the difference Compton profiles between the normal (295 K) and superconducting state (83 K) cannot be explained by the subtle difference close to the Fermi momentum taking place in the superconducting transition. The effect was explained with changes of the electron density in such a way that the number of d-electrons in Cu—O bands is diminished below the transition temperature due to the increased coupling p-type states in Bi—O bands. This would lead to the observed narrowing of the Compton profiles at lower temperature, so that this effect should be seen in other types of experiments as well. The above authors have found a similar effect also in $YBa_2Cu_3O_7$ (Priftis *et al.* 1994). However Manninen *et al.* (1999) failed in a synchrotron radiation based Compton experiment at the beamline ID15B of the ESRF to reproduce the above findings. Their difference profiles between the superconducting (40 K) and normal state (100 K) both of polycrystalline $YBa_2Cu_3O_{7-\delta}$ and polycrystalline $Bi_2Sr_2CaCu_2O_8$ showed no structure beyond statistical error, neither in an experiment, where the energy analysis of the scattered radiation was performed by means of a solid state detector (statistical accuracy $\Delta J(0)/J(0) = \pm 0.002$), nor by utilizing the high-resolution crystal spectrometer (statistical accuracy between 0.5 and 0.7%).

A Compton scattering study on twinned single crystalline $YBa_2Cu_3O_{7-\delta}$ and twinned single crystalline nonsuperconducting $PrBa_2Cu_3O_{7-\delta}$ performed at the beamline ID15b of the ESRF by Shukla *et al.* (1999) has shed light on a long standing question of why the preparation conditions of $PrBa_2Cu_3O_{7-\delta}$ decide whether this compound is superconducting or not. The upper panel of Fig. 4.54 shows the experimental Compton profile anisotropy ($J_{[100/010]} - J_{[110]}$) of $YBa_2Cu_3O_{7-\delta}$ together with the LDA calculation for $YBa_2Cu_3O_{7-\delta}$ (solid line), and for $PrBa_2Cu_3O_{7-\delta}$, where the calculations are scaled down by 0.71. Good agreement is found between the experiment and the computation.

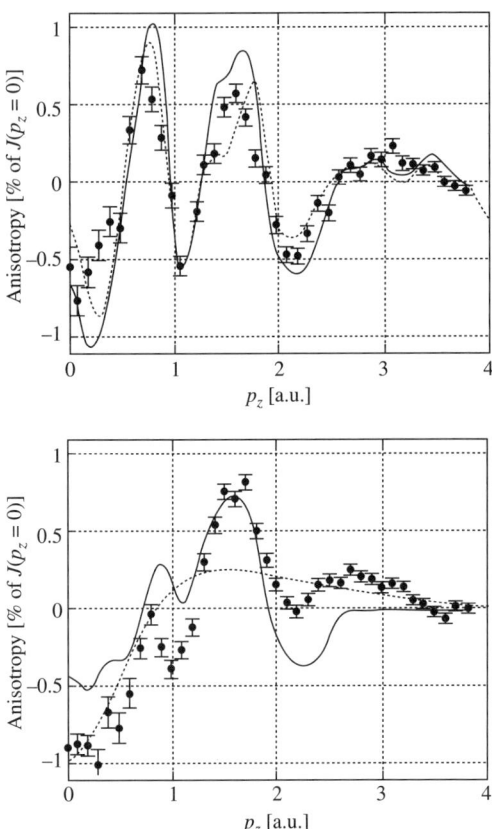

Fig. 4.54. Upper panel: anisotropy ($J_{[100/010]} - J_{[110]}$) in the Compton profiles. Filled circles: experiment on $YBa_2Cu_3O_{7-\delta}$; Solid line: LDA calculation for $YBa_2Cu_3O_{7-\delta}$; dashed line: LDA calculation for $PrBa_2Cu_3O_{7-\delta}$. The calculations are scaled down by 0.71. Lower panel: anisotropy ($J_{[100/010]} - J_{[110]}$) in the Compton profiles of $PrBa_2Cu_3O_{7-\delta}$. Filled circles: experiment. Note the remarkable difference between experiment and LDA calculations in the upper panel. Solid line: theoretical model accounting for Pr-Ba disorder. Dashed line: theoretical model with undoped Cu—O planes. (Reprinted with permission from Shukla *et al.* (1999); copyright (1999) by the American Physical Society.)

Moreover, one sees only small differences between the LDA calculations for the two materials. However, the experimental anisotropy of $PrBa_2Cu_3O_{7-\delta}$, shown in the lower panel of Fig. 4.54, has no similarity with the LDA calculated one in the upper panel. This discrepancy is explained by the authors to be due to the preparation conditions which give rise to a nonsuperconducting compound, and which produce a certain degree of substitutional disorder concerning the Pr and

Ba sites. If such disorder were present, it would diminish the contribution of the Pr and Ba—O planes to the total anisotropy. This was successfully simulated by means of a linear muffin-tin orbital (LMTO) calculation in which the contribution of Pr and Ba—O planes were neglected. One sees with the solid line of the lower panel of Fig. 4.54 that the suppression of the first peak in the anisotropy is well reproduced by the model calculation. Moreover, it is assumed that the hybridization of Pr f-electrons with O 2p-electrons produces hole depletion in the Cu—O layers (this means metal–insulator transition of the Cu—O layers), which again could be simulated by a model LMTO calculation shown as the dashed line in the lower panel of Fig. 4.54. Thus it is the presence of some Pr on the Ba site, probably in a 4+ oxidation state, while maintaining the carrier concentration in the Cu chains, which inhibits hole doping in the Cu—O planes and/or localizes doped carriers. Superconductivity appears in $PrBa_2Cu_3O_{7-\delta}$, if this substitution is suppressed.

This is a comprehensible example of how the cooperation of directional Compton profile measurements with intelligent model calculation can help to solve a very complex problem in the field of highly correlated materials.

4.9.4 *Quasicrystals*

Understanding of the cohesion mechanism of quasicrystals (QCs) is one of the outstanding problems one can hope to solve by means of Compton scattering. The crucial quantity, the origin of which should be clarified, is the pseudogap in the density of states across the Fermi level. Above all, the so-called Hume–Rothery mechanism is a matter of debate: The quasi-Brillouin zone (q-BZ), built by strong Bragg reflections of electron wavefunctions causes a gap at the Fermi energy in the reciprocal space when the Fermi sphere (FS) touches the q-BZ, so that possibly the electronic energy is lowered in order to stabilize the quasicrystals. This mechanism might work most effectively in icosahedral materials, since the q-BZs possess icosahedral symmetry, which is nearly spherical. The FS can touch the q-BZ at many places. Moreover, the sp-d hybridization mechanism can possibly further enhance the depth and width of the pseudogap.

Tanaka *et al.* (1994) have carried out measurements of high-resolution Compton profiles of Al-Li-Cu icosahedral quasicrystals (i-QCs) and have shown that the nearly spherical Fermi surface touches the q-BZ boundary thus supporting the Hume–Rothery mechanism. A similar study was performed by Okada *et al.* (2002) on a single grain of decagonal $Al_{72}Ni_{12}Co_{16}$ using the BL08W beam line at SPring-8 with 115 keV incident energy and 0.14 a.u. momentum resolution. The valence Compton profile obtained was decomposed into two parts, an inverted parabola with a tail in the region of $p_z < 1.5$ a.u., and a broad Gaussian-like profile extended to p_z above 5.0 a.u. The latter can be fitted to the sum of $Ni(3d^8)$ and $Co(3d^7)$ Compton profiles multiplied by 1.29, as shown in Fig. 4.55 for two directions. The area under the d-electron profile gives 9.58 electrons. This value suggests that a sizeable number of s,p electrons are transferred into the

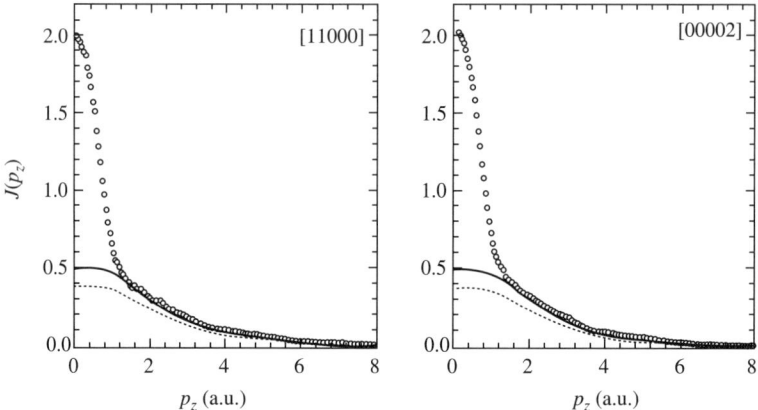

Fig. 4.55. Experimental valence electron Compton profile of decagonal $Al_{72}Ni_{12}Co_{16}$ for two directions as indicated (open circles). The dashed curve represents the nominal d-electron profile given by the concentration-weighted sum of $Ni(3d^8)$ and $Co(3d^7)$ free atom profiles. The thick full curve is the fitted d-electron profile. (Reprinted with permission from Okada et al. (2002); copyright (2002) by IOP Publishing Ltd.)

3d-electron states. Subtracting the 3d-electron part from the Compton profile, one gets the partial profile of a free-electron-like inverted parabola with a tail, where the latter can be attributed to electron–electron and electron–lattice interactions. The area under the free-electron-like profiles is 2.105 on the average. This corresponds to $p_F = 0.86$ a.u., whereas the peak of the second derivative of the partial free-electron-like profile is around 1.0 a.u., so that one can assume that p_F is between these two values. However, it turns out that even with $p_F = 0.86$ a.u. the Fermi sphere is larger than the q-BZ built by the strongest Bragg-reflection planes. Therefore, it was concluded that the Hume–Rothery mechanism does not play an important role in explaining the stability of the decagonal quasi crystals.

On the contrary, a high-resolution Compton study of Okada et al. (2003) on icosahedral $Cd_{84}Yb_{16}$ resulted in a strong indication for the important role of the Hume–Rothery mechanism in stabilizing the icosahedral phase in Cd-Yb binary alloys. Decomposing again the valence Compton profile into an inverted parabola-like part and into a Gaussian-like part they found the area under the inverted parabola-like part to be consistent with an electron per atom (e/a) of 1.26, so that 0.74 electrons are transferred from the free-electron-like s-p bands into the Cd 5d states. The corresponding radius of the Fermi sphere is 0.63 ± 0.02 a.u. coinciding nicely with the q-BZ formed by the strong Bragg diffraction planes (211111) and (221001). Therefore the Hume–Rothery mechanism seems to work very effectively in this compound. Moreover, the Gaussian-like part of the valence profile resembles a Compton profile of Yb 5d states lowered below

E_F forming the theoretically predicted (Ishii and Fujiwara 2001) bonding states by strong hybridization with the Cd 5p states. Thus both the Hume–Rothery mechanism and the sp-d hybridization play important roles in forming a deep pseudogap, stabilizing the icosahedral phase $Cd_{84}Yb_{16}$.

4.9.5 Systems under high pressure

There exists only a few Compton studies on systems under high pressure. But it can be presumed that these applications will increase in the near future. Therefore, the few existing ones should be briefly described in what follows. A first high-pressure study was carried out by Oomi et al. (1998) with 1.5 GP on Li metal, resulting in an increase of the Fermi momentum. Hämäläinen et al. (2000) have reported Compton profile measurements on sodium under high pressure, carried out at the Compton beamline ID15b of the ESRF with 55.8 keV incident photon energy. The scattered spectrum was analyzed by means of a Ge solid state detector. The momentum space resolution was 0.55 a.u. The high

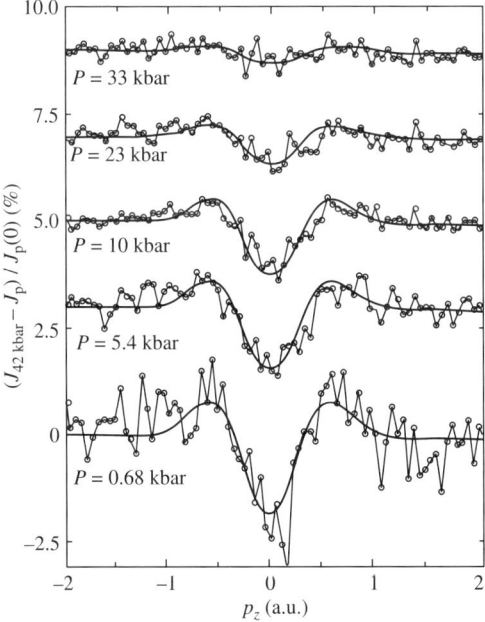

Fig. 4.56. The difference between the Compton profile of sodium at 42 kbar and the profile at various pressures as indicated. The experiment is compared with RPA calculations including correlation effects. (Reprinted with permission from Hämäläinen et al. (2000); copyright (2000) by the American Physical Society.)

pressure of 0.68, 5.4, 10, 23, 33, and 42 kbar was applied by a mechanical "Paris–Edinburgh" press with hard metal anvils, where the special geometrical design guaranteed a uniform hydrostatic pressure over the full sample volume. Small beam and slit sizes together with a careful alignment made it possible to discriminate almost completely the scattering from the pressure cell. Nevertheless, a small boron nitride (gasket) Compton profile had to be subtracted when taking the differences $(J_{42\ \text{kbar}} - J_\text{p})/J_\text{p}(0)$, shown in Fig. 4.56. The experimental results are confronted with free-electron RPA calculations including correlation effects according to Daniel and Vosko (1960), which agree well with the measurements. It is the increase of the Fermi momentum with pressure, due to the decreasing lattice constant, together with an increase of the renormalization constant Z, which have to be made responsible for these differences, even if these two effects cannot be distinguished experimentally, because of the poor statistics and the low resolution. Nevertheless, this study has demonstrated the feasibility of high-pressure Compton profile measurements. It is stressed by the authors that with 1–2 orders of magnitude increase in photon flux which is obtainable with focusing optics such an experiment can be performed with almost one order of magnitude better momentum resolution. This would open a great potential for studying ground-state electronic properties by means of high-pressure Compton scattering.

4.10 Compton scattering beyond the impulse approximation

In presenting the impulse approximation in Section 6.8 we started with the following expression for the nonrelativistic DDSCS (see equation 6.21), obtained within the limits of the Born approximation (first-order perturbation treatment in Section 6.2), and written in terms of one-electron wavefunctions ψ_i and ψ_f, which, in the sense of the independent particle model, are inserted for the state vectors $|\text{i}\rangle$ and $|\text{f}\rangle$:

$$\frac{d^2\sigma}{d\Omega_2 d\hbar\omega_2} = \left(\frac{d\sigma}{d\Omega_2}\right)_\text{Th} \sum_\text{f} |\langle \psi_\text{f}| \exp(i\mathbf{q}\cdot\mathbf{r})|\psi_\text{i}\rangle|^2 \delta(E_\text{f} - E_\text{i} - \hbar\omega), \qquad (4.118)$$

which could be transformed into

$$\frac{d^2\sigma}{d\Omega_2 d\hbar\omega_2} = \left(\frac{\hbar}{2\pi}\right)\left(\frac{d\sigma}{d\Omega_2}\right)_\text{Th} \int_{-\infty}^{\infty} dt \exp(-i\omega t) \sum_\text{f} \langle \psi_\text{i}| \exp(-i\mathbf{q}\cdot\mathbf{r})|\psi_\text{f}\rangle$$
$$\times \langle \psi_\text{f}| \exp(iHt/\hbar) \exp(i\mathbf{q}\cdot\mathbf{r}) \exp(-iHt/\hbar)|\psi_\text{i}\rangle, \qquad (4.119)$$

when the integral representation of the δ-function in equation (4.118) is utilized, E_i and E_f are understood as energy eigenvalues of the Hamiltonian H, and the averaging over the initial states is dropped. Finally we can get rid of the final states $|\psi_\text{f}\rangle$ by closure, since the continuum states form a complete set together with the bound states:

$$\frac{d^2\sigma}{d\Omega_2 d\hbar\omega_2} = \left(\frac{\hbar}{2\pi}\right)\left(\frac{d\sigma}{d\Omega_2}\right)_{Th} \int_{-\infty}^{\infty} dt \exp(-i\omega t)\, \langle\psi_i|\exp(-i\mathbf{q}\cdot\mathbf{r})$$
$$\times \exp(iHt/\hbar)\exp(i\mathbf{q}\cdot\mathbf{r})\exp(-iHt/\hbar)|\psi_i\rangle$$
$$- \left(\frac{d\sigma}{d\Omega_2}\right)_{Th} \sum_j |\langle\psi_j|\exp(i\mathbf{q}\cdot\mathbf{r})|\psi_i\rangle|^2\, \delta(E_j - E_i - \hbar\omega), \quad (4.120)$$

where $|\psi_j\rangle$ stands for a bound state with energy E_j, so that the last term in (4.120) can be removed, because, in the case of Compton scattering, $E_j - E_i - \hbar\omega < 0$.

As shown in Section 6.8, the essence of the impulse approximation consists in replacing $\exp(iHt/\hbar)$ by its approximation $\exp(iH_0 t/\hbar)\exp(iVt/\hbar)$ and $\exp(-iHt/\hbar)$ by $\exp(-iVt/\hbar)\exp(-iH_0 t/\hbar)$, so that the potential energy operator cancels, since $\exp(iVt/\hbar)$ commutes with $\exp(i\mathbf{q}\cdot\mathbf{r})$. H_0 is the kinetic part of the Hamiltonian.

This was the reason that a complete set of eigenfunction of H_0, that is plane waves, could be inserted into (4.120) ending up with

$$\frac{d^2\sigma}{d\Omega_2 d\hbar\omega_2} = \left(\frac{1}{2\pi}\right)^3 \left(\frac{d\sigma}{d\Omega_2}\right)_{Th} \int |\chi_i(\mathbf{p})|^2\, \delta\left(\frac{\hbar\omega - \hbar^2 q^2}{2m} - \frac{\hbar\mathbf{p}\cdot\mathbf{q}}{m}\right) d\mathbf{p}, \quad (4.121)$$

the basic equation for using Compton scattering to get information about electron momentum densities.

We shall now briefly answer the question, how will the double differential Compton scattering cross-section deviate from the Compton limit, whenever one goes beyond the impulse approximation. Quantifying those deviations can help to correct experimental results in the process of extracting Compton profiles from measured differential scattering cross-sections. We shall summarize:

(1) how the insertion of real continuum states as final states in (4.118) may influence the DDSCS;

(2) how an appropriate expansion of the operator, whose ground-state expectation value determines the DDSCS in (4.120), gives rise to deviations from the impulse approximation, in other words, the role the incomplete cancellation of the potential energy operator plays;

(3) to what extent the interaction of the recoil electron with the many-particle system (final state interaction) of valence electrons in a solid brings about effects that are neglected in the limits of the impulse approximation, namely the self- energy corrections; and

(4) how the interaction of the recoil electron with the hole in the Fermi sea left behind (vertex correction) may influence the DDSCS along with the self-energy correction.

4.10.1 The Born approximation

Eisenberger and Platzman (1970), Bloch and Mendelsohn (1974), Mendelsohn and Bloch (1975), and Issolah et al. (1988) have demonstrated how the DDSCS is modified compared with the impulse approximation deduced Compton limit, when hydrogenic Slater type wavefunctions for the initial state and hydrogenic continuum wavefunctions are inserted for the final state in the expression of the Born approximation (4.118)

In order to demonstrate exemplarily how a computation within the limits of the Born approximation will change the results of the impulse approximation a "Compton profile" is defined as

$$J = \frac{\mathrm{d}^2\sigma/\mathrm{d}\Omega_2 \mathrm{d}\hbar\omega_2}{(\omega_2/\omega_1)(\mathrm{d}\sigma/\mathrm{d}\Omega_2)_{\mathrm{Th}}}. \quad (4.122)$$

Inserting a continuum hydrogenic-like state of an electron in the Coulomb potential with effective charge Z_f^* (Joachain 1975) as final state $|\psi_f\rangle$ in (4.118), and an initial 1s-state hydrogenic wavefunction ψ_i according to equations (6.191) and (6.192), where the screening constant γ_{1s} is given by (a_0 = Bohr radius)

$$\gamma_{1s} = 2^{-2/3} Z_i^* / a_0 \pi^{1/3} \quad (4.123)$$

one obtains, assuming $Z_i^* = Z_f^* = Z^*$, with κ the wavevector of the final electron,

$$J(\kappa) = \left(\frac{256 m a_0 K^2}{3\hbar Z^{*2}}\right)[1 - \exp(-2\pi/P)]^{-1} \boldsymbol{g}((1 + 3K^2 + P^2)/[1 + K^2 - P^2)^2$$
$$+ 4P^2]^3 \boldsymbol{g}) \exp(-(2/P)) \tan^{-1}[2P/(1 + K^2 - P^2)], \quad (4.124)$$

where

$$K = \frac{q a_0}{Z^*} \quad (4.125)$$

and

$$P = \frac{\kappa a_0}{Z^*}. \quad (4.126)$$

This "Compton profile" $J(\kappa)$ can also be expressed as a function of p_z by using the definition of p_z, as given in equation (4.8), and the relation

$$\kappa = [(2m/\hbar^2)(\varepsilon_i + \hbar\omega)]^{1/2}. \quad (4.127)$$

This way one can compare $J(p_z)$ as deduced from equation (4.124) with a Compton profile, as obtained by using equations (4.7), where the momentum density $\rho(\mathbf{p})$ is derived from the Fourier transform of the hydrogenic wavefunction of equation (6.192) together with (4.127). The "Compton profile" of the Born approximation becomes flattened and asymmetrical with a peak shift to negative p_z (larger scattered photon energy) for 1s-hydrogenic wavefunctions, when compared with the Compton profile of the impulse approximation. This tendency increases rapidly with decreasing incident photon energy. As shown by

Mendelsohn and Bloch, the sign of the asymmetry can change when going from s to p states. Issolah *et al.* (1991) have proposed, again within the limits of the Born approximation, an alternative choice of final-state wavefunction, which they call the quasi-self-consistent field (QSCF) for a spherical symmetric potential. It turned out that the corrections to the impulse Compton profiles, as obtained by calculating the DDSCS in the Born approximation, depend sensitively on the choice of the final-state wavefunctions. Therefore, Holm and Ribberfors (1989) have looked for corrections, which are independent of final-state wavefunctions. Their procedure shall be outlined in the next section.

4.10.2 *First-order correction to the impulse approximation*

The Holm–Ribberfors procedure starts from equation (4.120), more precisely, tries to find an approximation of

$$\exp(-i\mathbf{q}\cdot\mathbf{r})\exp(iHt/\hbar)\exp(i\mathbf{q}\cdot\mathbf{r})\exp(-Ht/\hbar), \tag{4.128}$$

going beyond the impulse approximation, the main approach of which is

$$\exp(iHt/\hbar) = \exp(iH_0 t/\hbar)\exp(iVt/\hbar) \tag{4.129}$$

$$\exp(-iHt/\hbar) = \exp(-iVt/\hbar)\exp(-iH_0 t/\hbar). \tag{4.130}$$

Using the operator relation

$$H\exp(i\mathbf{q}\cdot\mathbf{r}) = \exp(i\mathbf{q}\cdot\mathbf{r})(\hbar\mathbf{q}\cdot\mathbf{p}/m + \hbar^2 q^2/2m + H) \tag{4.131}$$

we can write the expression (4.128) in the following way

$$\exp(-i\mathbf{q}\cdot\mathbf{r})\exp(iHt/\hbar)\exp(i\mathbf{q}\cdot\mathbf{r})\exp(-Ht/\hbar) = \exp(i[\mathbf{q}\cdot\mathbf{p}/m + \hbar q^2/2m]t)X, \tag{4.132}$$

where

$$X = \exp(i\mathbf{q}\cdot\mathbf{p}t/m)\exp(i[\hbar\mathbf{q}\cdot\mathbf{p}/m + H]t/\hbar)\exp(-iHt/\hbar). \tag{4.133}$$

Inserting these expressions into equation (4.119) yields

$$\frac{d^2\sigma}{d\hbar\omega_2 d\Omega_2} = \left(\frac{d\sigma}{d\Omega_2}\right)_{\text{Th}}\left(\frac{1}{2\pi}\right)^4 \int d\mathbf{p} \int d\mathbf{p}' \chi_i^+(\mathbf{p})\chi_i(\mathbf{p}')$$

$$\times \int_{-\infty}^{\infty} dt \langle \exp(i\mathbf{p}\cdot\mathbf{r})|X|\exp(i\mathbf{p}'\cdot\mathbf{r})\rangle$$

$$\times \exp[i(\hbar\mathbf{q}\cdot\mathbf{p}/m + \hbar^2 q^2/2m - \hbar\omega)t], \tag{4.134}$$

where, for $X = 1$, this relation reduces to the impulse approximation as given by equation (4.121). Therefore, in order to obtain a first-order correction to the impulse approximation we have to expand X, where the zeroth order term is 1, the first term contains the factor $1/q$, the second $(1/q)^2$ and so on, since we assume the impulse approximation to work well, so that q is large. Without presenting details of the derivations it was shown by Holm and Ribberfors (1989) that the contribution of the first-order term to the total Compton profile, $J_1(p_z)$, reads, for closed shell Slater-type hydrogenic wavefunctions (see 6.192),

$$J_1^{1s}(p_z) = J_0^{1s}(p_z)(1/q)[-2\gamma_{1s}\arctan(p_z/\gamma_{1s}) + (3p_z/2)] \quad (4.135)$$

$$J_1^{2p}(p_z) = J_0^{2p}(p_z)(1/q)(-4\gamma_{2p}\arctan(p_z/\gamma_{2p})$$
$$+ p_z[80\gamma_{2p}^2 + 120p_z^2]/[12\gamma_{2p}^2 + 60p_z^2]). \quad (4.136)$$

where the zeroth order terms are connected with the impulse approximation profiles by

$$J_0^{1s}(p_z) = J_{1s}(p_z) \quad \text{and} \quad J_0^{2p}(p_z) = 3J_{2p}(p_z). \quad (4.137)$$

One realizes immediately that the expressions (4.135) and (4.136) are asymmetric with respect to p_z, so that also the total Compton profile becomes asymmetric.

4.10.3 *Core Compton profile asymmetry in experiment*

In order to test the theoretical results of Sections 4.10.1 and 4.10.2, Huotari et al. (2001) have presented a detailed study of the asymmetry of Compton profiles. They first looked for external sources of asymmetry such as asymmetries of the instrumental function, asymmetries caused by asymmetric distribution of the scattering angle and multiple scattering. Under the conditions in which they have performed their test experiments these experimental artifacts were much smaller than the asymmetry

$$\Delta J(p_z) = [J(p_z) - J(-p_z)]/J(0) \quad (4.138)$$

they found, e.g. in an experiment on sodium as shown in Fig. 4.57. Qualitatively the asymmetry is in agreement with the Holm–Ribberfors (HR) calculations. Note by looking at Fig. 4.58 that the sign of the asymmetry belongs to the 2p contributions, whereas the 1s, 2s contributions have opposite sign. The agreement with the QSCF computation according to Issolah is nearly perfect. Also in this case the 1s, 2s contributions cancel partly those from the 2p states (see Fig. 4.58).

Based on these experimental results, and taking into consideration the simple additive nature of the correction to the impulse approximation, Huotari et al. (2001) have proposed to correct experimental Compton profiles with respect to the impulse approximation failure in the following way:

$$J_{\text{corr}}(p_z) = J_{\exp}(p_z) - B(p_z); \quad B(p_z) = \Delta J(p_z)J(0)/2. \quad (4.139)$$

The correction formalism proposed in Sections 4.10.1 and 4.10.2 applies only to closed-shell core states which can be represented by Slater-type wavefunctions, with other words to the core states. This limitation compels us to consider deviations from the impulse approximation in the following section also for the valence electron system.

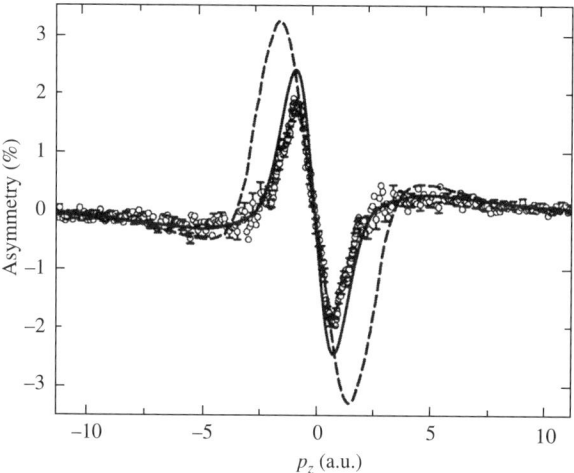

Fig. 4.57. Asymmetry of an experimental Na Compton profile, incident X-ray energy 56 keV and scattering angle of 173°, compared with corresponding predictions of the first-order correction of Holm and Ribberfors (1989) (dashed line), and with QSCF theory (solid line). The data are presented as a percentage of the peak value $J(0) = 3.7$. (Reprinted with permission from Huotari et al. (2001); copyright (2001) by Esevier Science Ltd.)

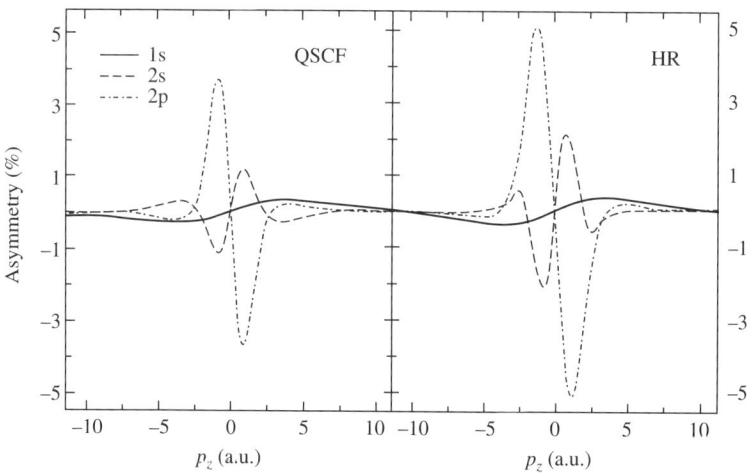

Fig. 4.58. Asymmetry of the Compton profile of Na core electrons (incident photon energy 56 keV; $\theta = 173°$), as calculated by QSCF method (left-hand panel) and by using the Holm–Ribberfors (1989) correction (right-hand panel). (Reprinted with permission from Huotari et al. (2001); copyright (2001) by Elsevier Science Ltd.)

4.10.4 Final-state self-energy effects and particle–hole interaction in Compton scattering from valence states. Theory and experiment

Even if we can assume that the final state electron in an inelastic scattering process does not experience the crystal potential, e.g. since we are considering the scattering valence electron system as being adequately described within the limits of the jellium model, it is part of this interacting electron system. This means that the recoil electron can polarize this system, and this polarization can act back on the final-state electron, in other words, the recoil electron is subjected to self-energy effects. It is well known (see, e.g. Lundqvist 1967a,b) that these self-energy effects can be treated in terms of the so-called spectral density function $A(\mathbf{p}, E)$, as defined in Section 2.3.3.1 being determined by the real and imaginary part of the self-energy (see 2.84), so that the width of the spectral density function is directly related to the imaginary part of the self-energy, which determines the reciprocal lifetime of the quasiparticle represented by $A(\mathbf{p}, E)$. The real part of the self-energy indicates how far the energy position of the center of $A(\mathbf{p}, E)$ is from the free-particle energy.

According to Ng and Dabrowski (1986) the electronic dynamic structure factor $S(\mathbf{q}, \omega)$ is, for $\hbar q \gg p_F$, directly related to the spectral density function:

$$S(\mathbf{q}, \omega) = \frac{1}{\pi}(2\pi)^{-3} \int d\mathbf{p} \int_{-\hbar\omega+E_F}^{E_F} \left(\frac{dE}{2\pi}\right) A(\mathbf{p}, E) A(\mathbf{p} + \hbar\mathbf{q}, E + \hbar\omega), \quad (4.140)$$

where E_F is the Fermi energy of the homogeneous system.

We will now use the fact that, according to the derivation between (6.136) and (6.141), the Compton profile is, apart from a prefactor, nothing else than the impulse approximation of the dynamic structure factor.

Based on the definition of the spectral density function (see 2.84), the electron density in momentum space must be obtainable by integrating $A(\mathbf{p}, E)$ between appropriate limits with respect to E:

$$\rho(\mathbf{p}) = \int_{-\infty}^{E_F} \left(\frac{dE}{2\pi}\right) A(\mathbf{p}, E). \quad (4.141)$$

Therefore, neglecting $\mathrm{Re}\Sigma(\mathbf{p} + \hbar\mathbf{q}, E)$ and replacing, in the sense of the impulse approximation, E in the spectral density function $A(\mathbf{p}+\hbar\mathbf{q}, E+\hbar\omega)$ of the recoil electron by the free-particle energy $\varepsilon(\mathbf{p})$, we end up with

$$S(\mathbf{q}, \omega) = \frac{1}{\pi}(2\pi)^{-3} \int d\mathbf{p}\, \rho(\mathbf{p}) A(\mathbf{p} + \hbar\mathbf{q}, \varepsilon(\mathbf{p}) + \hbar\omega), \quad (4.142)$$

which means that the effect of the imaginary part of the recoil electron self-energy consists in replacing the $\delta(\hbar^2 q^2/2m + \hbar\mathbf{p}\cdot\mathbf{q}/m - \hbar\omega)$ function in equation (6.146), which constitutes the Compton profile, by the Lorentzian shape spectral density function $A(\mathbf{p} + \hbar\mathbf{q}, e(\mathbf{p}) + \hbar\omega)$ of the recoil electron, the width of which in momentum space is given, according to equation (2.84), by the imaginary part of the self-energy $\Sigma(\mathbf{p}+\hbar\mathbf{q},\varepsilon(\mathbf{p})+\hbar\omega)$. In other words, the Compton profile appears convoluted with the spectral density function of the recoil electron, first noted by Sternemann et al. (2000). As pointed out in an ultrahigh-resolution Compton experiment on Li by Sternemann et al. (2000) and another one on beryllium by Huotari et al. (2000), this effect has to be taken into account, whenever the width of the momentum space related resolution function of the experiment falls short of that of the spectral density function of the recoil electron, and sharp features of the momentum density and of its derivative, for instance Fermi breaks, are the object of investigations.

Since the numerical computation of self-energies, even when using the GW approximation, is rather cumbersome, Bell (2003) has proposed a simplified scheme to calculate the inverse lifetime $\Gamma(p) = -\mathrm{Im}\Sigma[p,\varepsilon(p)]$, which determines the width of the Lorentzian shape spectral density function, at least for the case of the recoil electrons with $p > p_\mathrm{F}$.

We have seen in Sections 4.10.1 and 4.10.2 that the deviation from the impulse approximation can be traced back to changes the potential undergoes when, on the one hand, seen by the electrons in the ground state, and, on the other hand, by the recoil electrons after the scattering process. Allowing for these effects leads to corresponding corrections of the impulse approximation, which were either restricted to a spherical symmetric potential for the final states or to electrons in closed shells, so that these schemes are limited to electrons in core states. Let us now have a look at electrons in valence states, where we will utilize the jellium model to describe these, as far as corrections to the impulse approximation are concerned. Two effects, which give rise to deviations from the impulse approximation, have to be considered. One is connected with the interaction of the holes left behind in the scattering process with the surrounding electron system. As we have already stated above, the hole polarizes the electron system, so that the polarization can act back on the holes. Theses many-particle effects lead to a characteristic double-peak structure of the corresponding hole spectral density function $A(\mathbf{p},E)$ in equation (4.142) with a so-called plasmaron peak shifted to lower energies against the quasihole peak by more than the plasmon energy as first shown by Lundqvist (1967a). According to equation (4.142) this double-peak structure should give rise to an asymmetry of the corresponding Compton profile as shown for the case of Li by Sternemann et al. (2000). But this asymmetry is opposite in sign to the asymmetry experimentally found for Li metal (Sternemann et al. 2000) and Be metal (Huotari et al. 2001). The reason for this apparent discrepancy is another interaction, which counteracts the aforementioned one, namely screened Coulomb interaction between the excited quasiparticle (recoil electron) and the quasihole left behind. This interaction gives rise to the so-called vertex correction, and can adequately be allowed for by the first-order (in the screened Coulomb potential) diagrams (shown in Fig. 2.40) belonging to an expansion of the polarization function $\chi(\mathbf{q},\omega)$ into irreducible diagrams (see Nolting 1991), where $\chi(\mathbf{q},\omega)$ constitutes the dynamic structure factor according to

$$S(\mathbf{q},\omega) = -(\pi)^{-1}\mathrm{Im}\chi(\mathbf{q},\omega). \quad (4.143)$$

It was demonstrated by Sternemann *et al.* (2000) in an ultrahigh-resolution Compton scattering experiment on Li metal, performed at HASYLAB (NSLS) with 8.7 (7.9) keV incident photon energy and 0.02 (0.03) a.u. momentum space resolution, the results of which are shown in Fig. 4.59, that both many-particle effects together produce an asymmetry in good agreement with experiment, so that they can be considered as the main contributions for deviations from the impulse approximation in the case of valence electron systems, at least for simple metals.

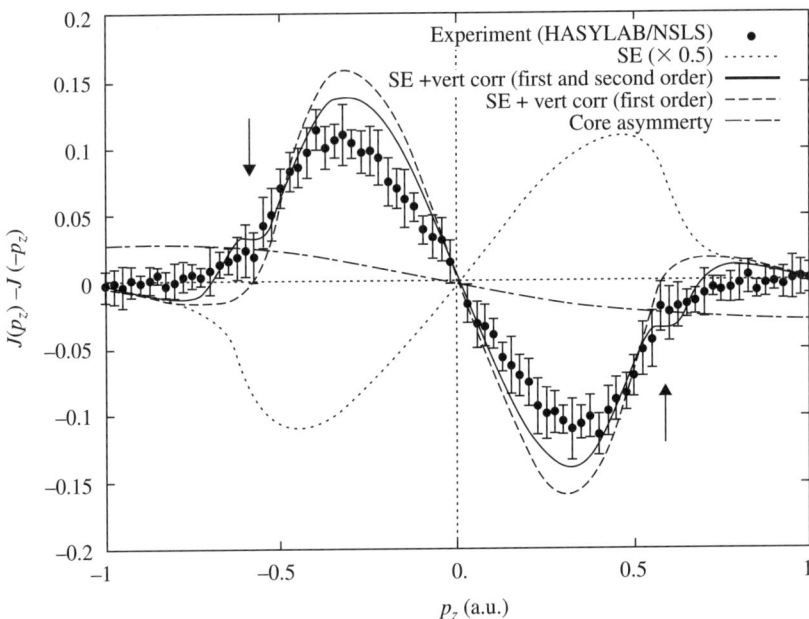

Fig. 4.59. Valence Compton profile asymmetry compared to the calculation taking into account the self-energy effect, i.e. the double-peak structure of the spectral density function (short-dashed line), and the screened Coulomb interaction between recoil electron and hole left behind (long-dashed line first order, solid line second order in the screened Coulomb interaction). The jellium p_F is indicated by the arrows, and the Holm–Ribberfors core asymmetry is the dashed-dotted line. (Reprinted with permission from Sternemann *et al.* (2000); copyright (2000) by the American Physical Society.)

4.11 Coherent Compton scattering

It has already been shown in Section 2.5 that, by utilizing the coherent superposition of two plane waves with wavevectors $\mathbf{K}_{10}$ and $\mathbf{K}_{1h}$, and with $\mathbf{K}_{1h} - \mathbf{K}_{10} = \mathbf{g}$, to play the role of the initial photon state of an inelastic X-ray scattering experiment, its information content is substantially extended, when such a coherent inelastic scattering experiment is compared with a conventional one. In Section 6.11.1 it will be shown in detail that it is the so-called nondiagonal structure factor $S(\mathbf{q}, \mathbf{g}, \omega)$, the double spatial Fourier transform with respect to $\mathbf{r}$ and the reference coordinate $\mathbf{r}'$ and time Fourier transform of the time-dependent two-particle density correlation function $n_2(\mathbf{r}', \mathbf{r}, t)$ (see equation 6.365) one can get information about, by coherent inelastic scattering. Moreover, it is pointed out in Section 6.11.2 that, by applying the formalism of the impulse approximation, the information about the nondiagonal structure factor changes over into information about nondiagonal elements $\Gamma_1(\mathbf{p} + \mathbf{g}|\mathbf{p})$ of the one-particle density matrix in momentum space, provided momentum and energy transfer are large enough to meet the requirements of the impulse approximation.

If one designs a Compton scattering experiment such that the sample crystal is in the Bragg position, a wave field consisting of the coherently coupled incident wave $\mathbf{K}_{10}$ with amplitude A_0 and polarization $\mathbf{e}_0$, and the Bragg-reflected wave $\mathbf{K}_{1h}$ with amplitude A_h and polarization $\mathbf{e}_h$ exists within the crystal. Both waves exhibit a phase difference $\Delta\phi$ which depends on the angular position of the sample crystal relative to the incident wavevector $\mathbf{K}_{10}$ and changes, according to the dynamical theory of X-ray diffraction (von. Laue 1960) from 0 to π within the Bragg-reflection range (Darwin width). The wave $\mathbf{K}_2$, Compton scattered off this wave field, makes a momentum transfer $\mathbf{q}_0 = \mathbf{K}_2 - \mathbf{K}_{10}$ and a momentum transfer $\mathbf{q}_1 = \mathbf{K}_2 - \mathbf{K}_{1h}$ to the crystal. Then the double differential scattering cross-section writes, for the special case that $\mathbf{q}_0$ and $\mathbf{q}_h$ are symmetry equivalent, $q_0 = q_h = q$, and the crystal under investigation is centrosymmetric:

$$\frac{d^2\sigma}{d\Omega_2 d\hbar\omega_2} = \left[\frac{r_0^2}{A_0^2 + A_h^2}\right]\left(\frac{\omega_2}{\omega_1}\right)([A_0^2(\mathbf{e}_{10}\cdot\mathbf{e}_2)^2 + A_h^2(\mathbf{e}_{1h}\cdot\mathbf{e}_2)^2]$$

$$\times \int \Gamma_1(\mathbf{p}|\mathbf{p})\,\delta(\hbar\omega - \hbar^2 q^2/2m - \hbar\mathbf{p}\cdot\mathbf{q}/m)\,d\mathbf{p}$$

$$+ 2A_0 A_h (\mathbf{e}_{10}\cdot\mathbf{e}_2)(\mathbf{e}_{1h}\cdot\mathbf{e}_2)\cos(\Delta\phi)$$

$$\times \int \Gamma_1(\mathbf{p}+\mathbf{g}|\mathbf{p})\,\delta(\hbar\omega - \hbar^2 q^2/2m - \hbar\mathbf{p}\cdot\mathbf{q}/m)\,d\mathbf{p}, \qquad (4.144)$$

where r_0 is the classical electron radius, ω_1 and ω_2 is the incident and Compton scattered frequency, respectively, and $\omega = \omega_1 - \omega_2$ (Schülke and Mourikis 1986).

In this way we obtain direct experimental access to the projections of the nondiagonal elements of the momentum space one-particle density matrix

$\Gamma(\mathbf{p}+\mathbf{g}|\mathbf{p})$ on the scattering vector $\mathbf{q}$, rather than only the momentum space average of $\Gamma(\mathbf{p}+\mathbf{g}|\mathbf{p})$ as obtainable in a diffraction experiment (see equation 6.177). By means of a reconstruction procedure one can assemble the full nondiagonal matrix element out of their projections as shown in Section 4.7.2. Decisive for a successful coherent Compton scattering experiment is the separation of the diagonal from the nondiagonal term in (4.144). This can be achieved if one changes the phase difference $\Delta\phi$ by rotating the sample crystal within the Bragg reflection range, which is best done when the sample is the second crystal of a nondispersive double-crystal setting, as realized by Schülke et al. (1981) and Schülke and Mourikis (1986) with conventional X-ray sources. In this case $\Delta\phi$ can be read from the so-called rocking curve which is the Bragg-reflected intensity as a function of the incidence angle α within the Bragg-reflection range, provided the sample crystal is perfect enough to meet the requirements of the dynamical theory. But one has to be aware that also the amplitude A_h of the Bragg-reflected wave depends on α, so that the separation of the nondiagonal from the diagonal contribution in (4.143) is rather complicated.

In order to make coherent Compton scattering also applicable to less perfect crystals, Spiertz (1996) has developed an alternative method to separate diagonal and nondiagonal contributions. He performed a synchrotron radiation based coherent Compton scattering experiment at the G3 beamline of HASYLAB. The experimental setup consisted of a double-crystal Ge(511) monochromator for the 50 keV incident radiation. The monochromatized radiation hits a Si crystal which forms, together with the Si sample crystal, a double-crystal setting, where (111), (220), (311) and (400) pairs could be installed. The crystal pairs were mounted on a rotating stage with 1/12 arcsec step width, which made it possible to rock through the Bragg-reflection range. The Bragg reflected intensity could be measured by means of an ionization chamber. Since the application of (4.144) implies that the Compton scattered beam has to leave the crystal parallel to the Bragg-reflecting lattice plane, the surface of the scattering sample was tilted with an axis in the Bragg-diffraction plane. The Compton scattered beam was collimated on a Ge solid state detector which carried out the energy analysis. A momentum space resolution of 0.65 a.u. was obtained. The separation of the diagonal and nondiagonal contributions in (4.143) took place without making reference to the dynamic theory of X-ray diffraction and is therefore also applicable to less perfect crystals. The only preposition is that the second term on the right-hand side of (4.143) really changes with angular positions within the Bragg-reflection range. The separation was achieved by solving a system of equations as set up in what follows: Let us denote the projection of $\Gamma_1(\mathbf{p}|\mathbf{p})$ on $\mathbf{q}$ the diagonal Compton profile $J_{\mathbf{q},\mathbf{0}}(p_z)$, the projection of $\Gamma_1(\mathbf{p}+\mathbf{g}|\mathbf{p})$ on $\mathbf{q}$ the nondiagonal Compton profile $J_{\mathbf{q},\mathbf{g}}(p_z)$, then we can write, according to (4.144), the measured intensity I_i of the whole spectrum (after subtracting a Monte Carlo calculated multiple scattering contribution) for an angular position i

$$x_i\, J_{\mathbf{q},\mathbf{0}}(p_z) + y_i\, J_{\mathbf{q},\mathbf{g}}(p_z) = I_i(p_z). \tag{4.145}$$

If we integrate (4.145) over p_z, we get the second equation

$$x_i z + y_i \, \mathbf{F}(\mathbf{g}) = \int_{-\infty}^{+\infty} I_i(p_z) \mathrm{d}p_z \qquad (4.146)$$

where the normalization of the diagonal Compton profile to the number of electrons z per atom, together with equation (6.177), have been utilized. The pure diagonal Compton profile can be obtained by measuring at an angular position far beyond the Bragg-reflection range, and by normalization to the number of electrons per atom. Finally we exploit the fact that both the diagonal and the nondiagonal Compton profile can be calculated by using well-known free atomic wavefunctions for a p_z-range Δp_z, where only the core electrons are contributing. Thus one get a third equation

$$x_i \int_{\Delta p_z}^{\text{free atom}} J_{\mathbf{q},\mathbf{0}}(p_z)\, \mathrm{d}p_z + y_i \int_{\Delta p_z}^{\text{free atom}} J_{\mathbf{q},\mathbf{g}}(p_z)\, \mathrm{d}p_z = \int_{\Delta p_z} I_i(p_z)\, \mathrm{d}p_z. \qquad (4.147)$$

For the Si measurements Δp_z was chosen between 4.5 and 9.5 a.u. (4.146) and (4.147) together yield a pair x_i, y_i for each angular position, which inserted into (4.145) results in the nondiagonal Compton profile, since the diagonal Compton profile has been measured independently. Of course, the shape of the nondiagonal Compton profiles should be independent of the angular position i. Deviations from this demand were attributed to a nonlinear background which therefore could be determined empirically by optimizing the mutual conformity of the nondiagonal Compton profiles belonging to different angular positions. Finally the nondiagonal core Compton profiles were subtracted.

In Fig. 4.60 the final experimental nondiagonal valence Compton profiles for four different reciprocal lattice vectors are shown, where in each case also the momentum transfer vector $\mathbf{q}$ of the experiment is indicated. The error bars are the result of a covariance analysis. Two theoretical curves are plotted in this figure. One (the solid line) is the result of a local pseudopotential calculation with potential coefficients of Harrison (1986). The other one (dashed line) is a free atomic calculation. Unfortunately no first principle calculation was at hand. The pseudopotential computation is in every case below the experiment. Having in mind that the nondiagonal Compton profiles are determined to a large extent by the higher momentum components (constituting the tails of the diagonal Compton profiles) this is not surprising. The pseudopotential calculation takes into consideration only higher momentum components due to lattice induced "Umklapp" processes. Contributions from the orthogonalization of the valence electron wavefunctions to the core states are neglected. On the other hand, these contributions are the only ones of the free atomic wavefunctions. Therefore, it is reasonable that the sum of the two theoretical curves coincide approximately with experiment, at least for the last three reciprocal lattice vectors. The 111 nondiagonal profile is an exceptional case since $|\mathbf{g}_{111}|$ is smaller than the diameter of the Jones zone, so that the nondiagonal Compton profile is less determined by

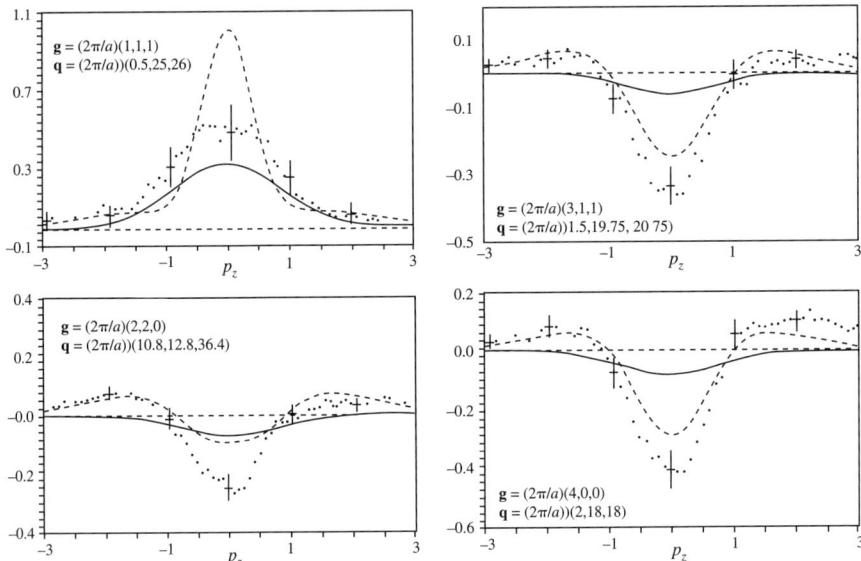

Fig. 4.60. Measured (points, partly with error bars) and calculated nondiagonal valence Compton profiles of Si, dashed line: free atomic wavefunctions; solid lines: pseudopotential calculations. The corresponding **g** and **q** vectors are indicated. (Reprinted with permission from Spiertz 1996.)

the higher momentum components. In that case the pseudopotential calculation is somewhat nearer in the experiment.

4.12 γ-eγ experiments

γ-eγ experiments can be understood as an extension of conventional Compton scattering. The twofold integration over the electron momentum distribution which determines the Compton profile, the outcome of this conventional technique, results from the lack of information about the momentum distribution of the recoiling electrons. Since integration averages over large volumes in momentum space, detailed information about solid state effects might be difficult to obtain. Therefore, it is desirable to measure the electron momentum density directly, based on the complete scattering kinematics: if the momenta of the primary ($\mathbf{K}_1$) and scattered photon ($\mathbf{K}_2$) in addition to that of the recoil electron ($\mathbf{p}_2$) are measured simultaneously, this means in coincidence, the momentum of the electron in its initial state ($\mathbf{p}_1$) is determined in a unique manner, via the momentum conservation law

$$\mathbf{p}_1 = \hbar \mathbf{K}_2 + \mathbf{p}_2 - \hbar \mathbf{K}_1. \qquad (4.148)$$

The corresponding triple-differential scattering cross-section turns out to be proportional to the electron momentum density $\rho(\mathbf{p}_1)$ itself. This relation can be derived by starting with the Jauch–Rohrlich (1955) expression for the total relativistic cross-section ($\hbar = 1; c = 1$) (see 6.222–6.227):

$$\sigma = m^2 r_0^2 \int d\mathbf{K}_2 d\mathbf{p}_2 \left(\frac{1}{2K_1 E_2 \omega_2}\right) X(k_1, k_2) \, \delta(\mathbf{p}_1 + \mathbf{K}_1 - \mathbf{p}_2 - \mathbf{K}_2)$$
$$\times \delta(E_1 + \omega_1 - E_2 - \omega_2) \qquad (4.149)$$

where

$$k_1 = E_1 \omega_1 - \mathbf{p}_1 \cdot \mathbf{K}_1 \qquad (4.150)$$
$$k_2 = E_1 \omega_2 - \mathbf{p}1 \cdot \mathbf{K}_2 \qquad (4.151)$$

and the X-factor for polarized photons

$$X(k_1, k_2) = \frac{1}{2}\left(\frac{k_1}{k_2} + \frac{k_2}{k_1}\right) - 1 + 2\left[\mathbf{e}_1 \cdot \mathbf{e}_2 + \frac{(\mathbf{e}_1 \cdot \mathbf{p}_1 \mathbf{e}_2 \cdot \mathbf{p}_2)}{k_1}\right.$$
$$\left. - \frac{(\mathbf{e}_2 \cdot \mathbf{p}_1 \mathbf{e}_1 \cdot \mathbf{p}_2)}{k_2}\right]^2. \qquad (4.152)$$

Characterizing the electrons in the bound state, in the spirit of the impulse approximation, by the momentum distribution $\rho(\mathbf{p}_1)$, and replacing the flux factor $k_1/E\omega$ by 1, we obtain the triple-differential scattering cross-section utilizing the same physical arguments as in Chapter 6

$$\frac{d^3\sigma}{d\omega_2 d\Omega_{\mathbf{K}_2} d\Omega_{\mathbf{p}_2}} = m^2 r_0^2 \left(\frac{\omega_2}{2\omega_1}\right) \int p_2^2 [X(k_1, k_2)/E_1 E_2] \rho(\mathbf{p}_1)$$
$$\times \delta(E_1 + \omega_1 - E_2 - \omega_2) d\mathbf{p}_2. \qquad (4.153)$$

Using $E_2 = [p_2^2 + m^2]^{1/2}$ and $\int \delta(E_1 + \omega_1 - E_2 - \omega_2) d\mathbf{p}_2 = |\delta E_2/\delta p_2|^{-1} = E_2/p_2$ we end up with

$$\frac{d^3\sigma}{d\omega_2 d\Omega_{\mathbf{K}_2} d\Omega_{\mathbf{p}_2}} = m^2 r_0^2 \left(\frac{\omega_2}{2\omega_1}\right)\left(\frac{p_2}{E_1}\right) X(k_1, k_2) \rho(\mathbf{p}_1). \qquad (4.154)$$

Further reduction of the above relation can be achieved by going with k_1, k_2 and E_1 in (4.153) to the limit $p_1 \to 0$, and by neglecting the very small $1/k_1$ and $1/k_2$ terms in (4.151):

$$\frac{d^3\sigma}{d\omega_2 d\Omega_\gamma d\Omega_e} = m \left(\frac{\omega_1}{\omega_2}\right) p_2 \left(\frac{d\sigma}{d\Omega_\gamma}\right)_{KN} \rho(\mathbf{p}_1), \qquad (4.155)$$

where the Klein–Nishina differential cross-section $(d\sigma/d\Omega_\gamma)$ for polarized photons is given by

$$\left(\frac{d\sigma}{d\Omega_\gamma}\right)_{KN} = \left(\frac{r_0^2}{4}\right)\left(\frac{\omega_2}{\omega_1}\right)^2 \left[\frac{\omega_2}{\omega_1} + \frac{\omega_1}{\omega_2} + 4(\mathbf{e}_1 \cdot \mathbf{e}_2)^2 - 2\right]. \qquad (4.156)$$

The main difficulty of γ-eγ coincidence measurements consists in the strong incoherent elastic scattering of the recoiling electron within the target which prevents the determination of the recoil momentum. Since the mean free path for elastic scattering of electrons with recoil energy is only about 100 nm, the targets should be as thin as possible. To get an impression of how drastic the coincidence rate is influenced by this demand one has to confront the mean free path for Compton scattering which is 3 cm for photon energies of 180 keV in carbon, whereas the target should be as thin as 20 nm. Thus only one out of 10^6 photons will be scattered. Moreover, a solid angle of the photon detector of 10^{-4} srad yields another factor of 10^5. Therefore, a coincidence countrate of a few Hz requires a monochromatized photon flux of 10^{12} photons s^{-1} at the target, which can be delivered only by third-generation synchrotron radiation sources. Rollason and Woolf (1995) have performed a Monte Carlo study of the γ-eγ process. They found out that the controlling parameter in γ-eγ angular correlation measurement is the electron mean free path for scattering at an angle comparable with the desired resolution. Inelastic scattering occurs with a characteristic deviation given by $\Delta E/2E$ which is of the order of 0.1° for scattering 50 keV electrons from Cu (0.03° from Al) and is therefore in general much more forward directed than elastic scattering. On the contrary, the FWHM of the elastic incoherent (atomic) differential scattering cross-section of Cu and Al, when plotted for 50 keV electrons as a function of scattering angle, is of the order of 2°. This corresponds to a momentum transfer of 2 a.u. so that even single scattering events will obscure all details of the Fermi surface of these substances, and the cumulative result of multiple elastic scattering events is virtually independent of the details of the electron momentum density. The Monte Carlo simulations on simple solid state models exhibit a broad, more or less structureless background due to the elastic incoherent scattering events which increases, relative to the electron momentum density signals, with increasing sample thickness up to a certain saturation level. Therefore, it is concluded that it should be possible to obtain useful signals from the momentum density in optically thick targets, providing the background of accidental coincidences can be kept sufficiently low.

A state of the art γ-eγ experiment (Bell and Schneider 2001) is sketched in Fig. 4.61. A monochromatized X-ray beam from the PETRA-undulator (DESY HASYLAB) of 180.3 keV hits the target in a beam spot of $2 \times 2 \, \text{mm}^2$. The photons scattered at an angle of $\theta = 150°$ were detected by a two-dimensional array of 12 intrinsic Ge diodes with an energy resolution $\sigma_{\omega_2} = 0.32$ keV. The electrons were identified by a position-sensitive detector (PSD) consisting of a two-dimensional array of 32 individual PIN diodes, which is set up in such a way that the vector $\mathbf{q}_0 = \mathbf{K}_1 - \mathbf{K}_{20}$, i.e. the momentum transfer to an electron initially at rest, was pointing at the center of the PSD, while $\mathbf{K}_{20}$, the momentum of the corresponding scattered photon, was defined by the center of the Ge diode array. The influence of the electron binding energy ε can be neglected, since it is small (tens of eV) compared to $\omega_{20} = 108.7$ keV (energy of a photon scattered from an electron at rest). With a momentum transfer of $q_0 = 75.0$ a.u.

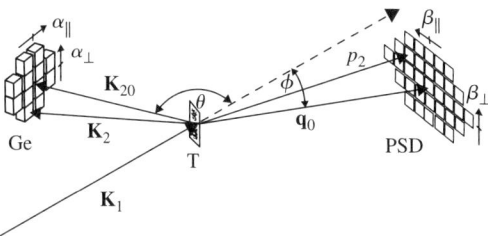

Fig. 4.61. Experimental (γ-eγ) set-up. Ge: 12-pixel Ge diode; T: target; PSD: 32 pixel position-sensitive electron detector. (Reprinted with permission from Bell and Schneider (2001); copyright (2001) by IOP Publishing Ltd.)

the validity of the impulse approximation is guaranteed. A Cartesian coordinate system describes the initial electron momentum **p** (in what follows we skip the index 1), where the p_z-component is parallel to $\mathbf{q}_0$, and the p_x-component is lying in the $(\mathbf{K}_1, \mathbf{K}_{20})$ scattering plane. Then the Cartesian components of **p** read

$$p_x = \mathbf{q}_0 \beta_{||} - ((\omega_{20}/c) \cos \delta) \, \alpha_{||} + ((1/c) \sin \delta) \, \Delta\omega_2 \quad (4.157)$$

$$p_y = \mathbf{q}_0 \beta_\perp + (\omega_{20}/c) \alpha_\perp \quad (4.158)$$

$$p_z = ((\omega_{20}/c) \sin \delta) \, \alpha_{||} - ((c \sin \delta)/(\omega_{20} \sin \theta)) \, \Delta\omega_2, \quad (4.159)$$

where $\alpha_{||,\perp}$ and $\beta_{||,\perp}$ denote the angular deviations of $\mathbf{K}_2$ from $\mathbf{K}_{20}$ and of $\mathbf{p}_2$ from $\mathbf{q}_0$, respectively; $\delta = \theta + \phi$; $\Delta\omega_2 = \omega_2 - \omega_{20}$. The fast signal from the photon detector served as a gate for the coincidence unit. If the coincidence unit detected an electron signal within the open gate, the data set of both the coincidence unit and the ADC for the photon signal were read out.

The momentum resolution $R(p_x, p_y, p_z)$ of the γ-$\varepsilon\gamma$ spectrometer was calculated by means of Monte Carlo simulations which included the triple-differential scattering cross-section (4.155), the solid angles and the energy resolution of the Ge detectors, the energy broadening of the primary photon beam and the extended beam spot size at the target. Fitting of a trivariate Gaussian to this simulation resulted in a covariance matrix the diagonal elements of which delivered standard deviations $(\sigma_{xx}, \sigma_{yy}, \sigma_{zz}) = (0.14, 0.14, 0.27)$ a.u. and small nonvanishing off-diagonal elements leading to some anticorrelations. Theoretical computations which were compared with the γ-eγ experiment were convoluted with the above trivariate Gaussian.

We shall demonstrate the feasibility of the γ-eγ method to reveal the 3D momentum distribution of solids by way of an example, the study of self- supporting carbon foils (Sattler *et al.* 2001). Two different kinds of foils were investigated using the spectrometer described above. One prepared by means of laser plasma ablation (lpa) turned out to possess an isotropic distribution of crystallites. The other one (te) was obtained by evaporation of graphite heated to about 3200 K.

Electron diffraction showed that the basal planes (002) were arranged mostly parallel to the foil surface. The resulting anisotropy of the 3D electron momentum distribution was convincingly demonstrated by plotting in Fig. 4.62 the difference $\Delta\rho$, the electron momentum density of the te foils minus that of the lpa foil as a function of $p_{\|}$ for different values of $p_{\perp}$. The measurements are confronted with three different calculations: (i) an empirical pseudopotential (PP) method (Lou *et al.* 1991) using potential parameter of Reed *et al.* (1974), broken lines in Fig. 4.62; (ii) the full-potential linear muffin-tin orbital method (FP- LMTO) (Kheifets *et al.* 1999), solid lines, and (iii) the modified augmented-plane-wave (MAPW) method of Bross *et al.* (1970), dashed dotted lines. It is evident that

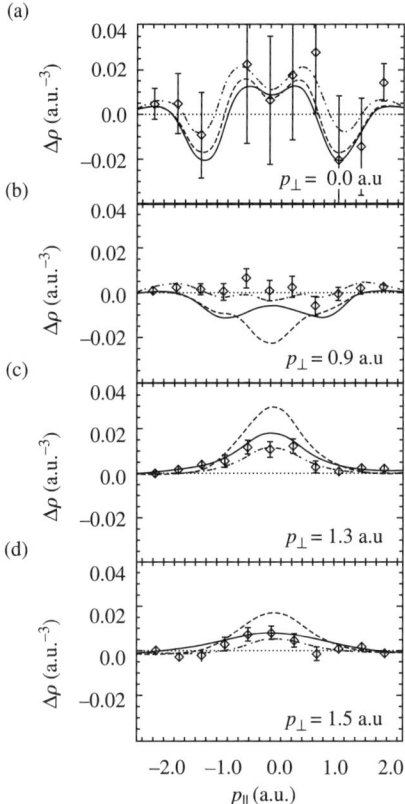

Fig. 4.62. The difference $\Delta\rho$: the electron momentum density of the te foil minus that of the lpa foil, as a function of $p_{\|}$ for different values of $p_{\perp}$. PP calculation: broken line; FP-LMTO calculation: solid line; MAPW calculation: dashed-dotted line. (Reprinted with permission from Bell and Schneider (2001); copyright (2001) by IOP Publishing Ltd.)

the MAPW method is superior to the other ones in providing good agreement with experiment.

When the statistical accuracy of the γ-eγ measurement is limited, one can evaluate the angular correlation density from the γ-eγ data (Metz et al. 1999a):

$$\rho^{2D}(p_x, p_y) = \int \rho(p_x, p_y, p_z) \mathrm{d}p_z, \tag{4.160}$$

which is comparable with the 2D-ACAR data (see Section 4.4). A comparison of angular correlation density deduced for Al from γ-eγ with those of 2D-ACAR has shown remarkable differences which could be traced back to the influence of the positron wavefunction on the positron annihilation data.

If, in a γ-eγ coincidence experiment, all the events for a constant p_z-value are summed up, one obtains a so-called coincident Compton profile J_{coinc}. Due to the limited range of the γ-eγ experiment in the p_x- and p_y-direction, the contributions of the core states to J_{coinc} are strongly reduced if compared with a noncoincident Compton profile. This way Metz et al. (1999a) have evaluated their γ-eγ data of aluminum in order to find out to what extent the Lam–Platzman (1974) correction may improve the agreement between FP-LTMO calculations (Savrasov 1996). Indeed the agreement was slightly improved. Metz et al. (1999b) have investigated the coincident Compton profile of a $Cu_{0.5}Ni_{0.5}$ alloy, where the difference between the profile of a Cu-Ni sandwich and the profile after heat treatment of the sandwich to enforce substitutional alloying is plotted. The comparison with KKR-CPA calculations of Benedeck et al. (1985) shows qualitative agreement. The influence of the substitutional disorder on the electron momentum density becomes evident.

If one writes the energy conservation law of the γ-eγ process in a nonrelativistic manner:

$$\varepsilon = \hbar\omega_1 - \hbar\omega_2 - E_2 \tag{4.161}$$

where a photon with energy $\hbar\omega_1$ is scattered at an electron with binding energy $\varepsilon > 0$, so that the scattered photon has energy $\hbar\omega_2$ and the electron energy E_2 after the interaction, then even the binding energy of the electron state can be determined at which the Compton scattering took place, if also the energy of the recoil electron E_2 can be fixed. One measures the electron momentum density state selectively.

A first step in this direction has taken by Itou et al. (1999). Differently from conventional γ-eγ experimental setups, they have inserted a time-of-flight analysis into the path of the recoil electrons, in order to determine their kinetic energy T. Measurements were performed on a graphite foil. A contour plot of the coincidence events' distribution, electron energy versus photon energy, has already been shown in Fig. 1.13. The center of gravity for events in range I which are mainly due to scattering from the valence electrons lies on the solid line $\hbar\omega_2 + T = \hbar\omega_1$. The center of gravity for events in the ranges II resulting from core-electron scattering follows the dashed line $\hbar\omega_2 + T = \hbar\omega_2 + \varepsilon$, where ε

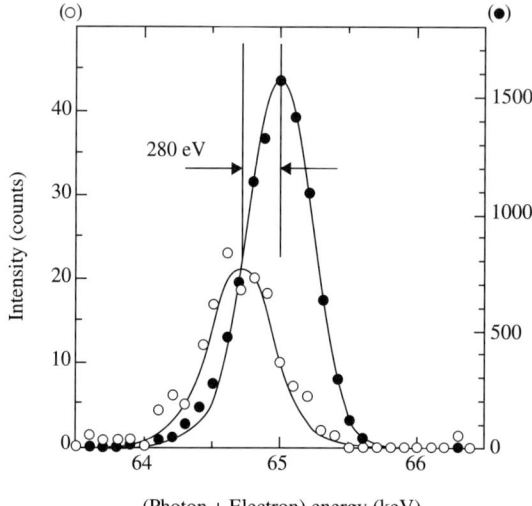

Fig. 4.63. The event distribution on energy, which is the summation of the recoil electron and the Compton scattered photon energy of a coincidence event. Black dots (white circles) correspond to the distribution integrated over a momentum region I(II) in Fig. 1.13. The solid lines are calculations accounting for electron multiple scattering. (Reprinted with permission from Itou et al. (1999); copyright (1999) by the Physical Society of Japan.)

is the binding energy (280 eV) of the 1s electrons of carbon. This is again visualized in Fig. 4.63, which shows the event distribution as a function of the sum of the Compton scattered photon and recoil electron energy, integrated over the range I (black dots) and over the ranges II (white circles) of Fig. 1.13. As shown in Fig. 4.63, the peak of both curves is shifted by 280 eV, the binding energy of the 1s core electrons.

Thus it seems not to be impossible that by utilizing the highly collimated flux and the extremely short photon pulses of fourth generation synchrotron radiation sources together with multiple element photon and recoil electron detection, and a time-of-flight electron energy analysis the electron momentum density can be measured state selectively.

4.13 Magnetic Compton scattering

4.13.1 *Basics, instrumentation and data processing*

We have found in Section 6.7 that the double differential scattering cross-section contains, beside the charge contribution, a magnetic contribution due to both orbital and spin magnetic moments. If one confines the considerations to the spin contribution, and applies the impulse approximation, the charge contribution together with the interference term between the charge and the magnetic scattering reads

$$\frac{d^2\sigma}{d\Omega_2 d\hbar\omega_2} = \left(\frac{r_0^2}{2}\right)\left(\frac{\omega_2}{\omega_1}\right)\int d\mathbf{p}\Big\{(1 + \cos^2\theta + P_3\sin^2\theta)(\rho_\uparrow(\mathbf{p}) + \rho_\downarrow(\mathbf{p}))$$
$$+ \left(2\left(\frac{\hbar\omega_1}{mc^2}\right)P_2(\cos\theta - 1)(\hat{\mathbf{K}}_1\cos\theta + \hat{\mathbf{K}}_2)_z(\rho_\uparrow(\mathbf{p}) - \rho_\downarrow(\mathbf{p}))\right)\Big\}$$
$$\times \delta(\hbar\omega - \hbar^2 q^2/2m - \hbar\mathbf{p}\cdot\mathbf{q}/m) \qquad (4.162)$$

where $\rho_{\downarrow(\uparrow)}(\mathbf{p})$ means the spin-dependent electron momentum density. $\hat{\mathbf{K}}_1$ and $\hat{\mathbf{K}}_2$ are the unit vectors in the direction of the incident and scattered wave, respectively, and θ is the scattering angle. P_2 and P_3 are the second and third Stokes parameters, respectively. According to the definitions given in Section 6.7 $P_2 = +1(-1)$ means an incident wave completely circularly polarized with positive (negative) handedness; $P_3 = +1(-1)$ describes a completely linearly polarized wave with an electric field vector perpendicular to (in) the scattering plane. Therefore, one can get information about the momentum distribution $\rho_\uparrow(\mathbf{p}) - \rho_\downarrow(\mathbf{p})$ of unpaired spin electrons, in other words, information about the spin density in momentum space by using (at least partly) circularly polarized incident X-rays. Then one has either to reverse the magnetic field, which magnetizes a ferromagnetic sample, or to reverse the handedness of the circularly polarized incident photon, so that one obtains, by recording the difference spectrum,

$$\Delta\left[\frac{d^2\sigma}{d\Omega_2 d\hbar\omega_2}\right] = 2r_0^2\left(\frac{\omega_2}{\omega_1}\right)\left(\frac{\hbar\omega_1}{mc^2}\right)P_2\cos\theta(\cos\theta - 1)\int d\mathbf{p}[\rho_\uparrow(\mathbf{p}) - \rho_\downarrow(\mathbf{p})]$$
$$\times \delta\left(\hbar\omega - \frac{\hbar^2 q^2}{2m} - \frac{\hbar\mathbf{p}\cdot\mathbf{q}}{m}\right). \qquad (4.163)$$

The $\mathbf{p}$ integral in this equation is called the magnetic Compton profile, J_{mag}. The equation is already specialized to the case in which $\mathbf{K}_1$ is parallel to the quantization axis (z-axis in (4.162)).

In Section 6.7 we have found arguments that, within the limits of the impulse approximation, the orbital contributions to the magnetic Compton scattering can be neglected.

The first magnetic Compton scattering experiments with the aim to get information about the spin density in momentum space were performed by Sakai and Ono (1976) who used circularly polarized γ-rays from oriented nuclei. Cooper et al. (1986) have utilized, for the first time, off-plane emitted circularly polarized synchrotron radiation for measuring spin-dependent Compton profiles. Later a superconducting wavelength shifter at the ESRF (beamline ID15a) came into use for magnetic Compton scattering, where the circular polarization was produced by selecting a beam below the plane of the electron orbit (McCarthy et al. 1997).

Mills (1987) got the necessary circularly polarized X-rays by using quarter-wave plates. A real breakthrough was achieved by constructing an elliptic multipole wiggler (Yamamoto and Kitamura 1987), the intense elliptically polarized X-rays of which were soon applied for magnetic Compton scattering (Yamamoto et al. 1989). A state-of-the-art experimental station for magnetic Compton scattering, as used at SPring-8 (beamline 08W), is sketched in Fig. 4.64 (Kakutani et al. 2003). The incident X-rays emitted from an elliptical multipole wiggler (EMPW) with a high portion of circular polarization are monochromatized to 175 keV using a bent Si (620) crystal and focused to a spot of 0.5 mm high and 0.3 mm wide at the sample position. The scattered X-rays (scattering angle 178°) are counted by a 10-segmented solid state Ge detector (SSD). The segmentation diminishes dead-time losses. The overall momentum resolution was 0.42 a.u. mainly determined by the energy resolution of the SSD. As already mentioned, there exists two methods to separate the magnetic part from the total Compton profile, which means to get experimental information about

$$M = \frac{I_+ - I_-}{I_+ + I_-} \times 100\%, \qquad (4.164)$$

where I_+ and I_- are the Compton spectra with either opposite magnetization or opposite helicity of the incident radiation. Therefore one method is to change

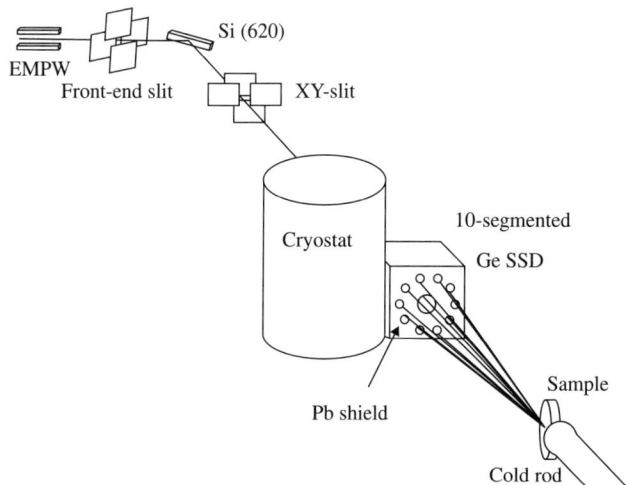

Fig. 4.64. Schematic view of instrumentation for magnetic Compton scattering. The incident X-rays emitted from an elliptical multipole wiggler (EMPW) are monochromatized to 175 keV by a bent Si (620) single crystal. The distance between the sample mounted in a superconducting magnet and the solid state detector (SSD) is 1 m. The scattering angle is 178°. (Reprinted with permission from Kakutani et al. (2003); copyright (2003) by the Physical Society of Japan.)

the direction of the spin magnetization within the sample, and to measure two profiles J_+ and J_- successively in a series $J_+, J_-, J_-, J_+, \ldots$ Each profile is measured a few seconds until a sufficient number of counts is accumulated. This way unwanted fluctuations of the flux intensity, decay of the storage ring or instabilities of the monochromator are cancelled out by subtracting J_- from J_+. The second possibility is to switch the helicity of the incident circularly polarized radiation. In that case the accumulation time must be much longer than the switching time which is several minutes for wigglers composed of permanent magnet arrays, where the phase of the electron circular orbit is changed. Faced with the rather small weight of the magnetic Compton profile relative to the total one, the normalization of the summed up I_+ and I_- profile plays a crucial role. In most cases this is done by summing up either the integrated counts in the quasielastic line each for the plus (A_+) and the minus position (A_-), or to integrate the counts over a momentum range of the core profile which can be assumed not to be influenced by unpaired spins. This way systematic errors due to minute changes of the experimental geometry upon changing from the plus to minus position can be prevented by inserting the quantity $b = A_+/A_-$ into (4.164):

$$M = \frac{I_+ - bI_-}{I_+ + bI_-} \times 100\%. \qquad (4.165)$$

A further important problem is the absolute normalization of the magnetic Compton profile, which is necessary if the value of the total spin magnetic moment is aspired to. This can only be obtained from the integrated appropriately normalized magnetic Compton profile. While the charge Compton profile is normalized to the number of contributing electrons per atom utilizing calculated free-atomic-like core profiles, a comparable procedure is not applicable to magnetic Compton profiles, since the prefactor of the charge and the magnetic Compton profile in (4.161) are completely different containing the Stokes parameter P_2 and P_3 and other geometrical parameters, which are not known with sufficient accuracy. Therefore, the magnetic properties of well-known materials can be used as a normalization standard. Iron with a well-known spin moment is used in most cases. If I_+ and I_- (4.165) are understood to be the integrals over the total Compton profile taken in the plus and the minus position, respectively, then M is proportional to the total spin magnetic moment μ of the specimen under investigation. Thus one obtains the relation between the unknown spin magnetic moment, μ_2, of a specimen and the known reference magnetic moment μ_1:

$$\mu_2 = \left(\frac{M_1}{M_2}\right)\left(\frac{Z_1}{Z_2}\right)\mu_1, \qquad (4.166)$$

where Z_1 and Z_2 are the numbers of electrons contributing to Compton scattering for the two specimens, more precisely the integrals of free atomic profiles in a given momentum range used in the normalization procedures of the corresponding total Compton profiles. Knowing the spin magnetic moment of the specimen under investigation, we can normalize the magnetic Compton profile to the effective number of magnetic electrons per atom in the same manner as the charge Compton profile is normalized to the total number of electrons per atom. Then we can obtain twice the majority-spin Compton profile by adding the normalized magnetic Compton profile to the total charge profile, and twice the minority-spin Compton profile by subtracting the magnetic Compton profile from the total one. This has been exemplified by Sakurai *et al.* (1994) in a high-resolution magnetic Compton profile study on Fe-5.8at%Si, the only one in the literature performed with a momentum resolution of 0.12 a.u.

In a different way from the other subsections of the Compton chapter, we shall not aspire to a more or less complete review of the more recent applications in the different fields of Compton scattering. We shall give, in the next three subsections, in each case only one or two examples of the most important applications of magnetic Compton scattering, namely: (i) the reconstruction of the 3D spin density in momentum space: (ii) the determination of the spin component of the magnetic moment, selectively on the different elements in compounds or selectively on orbitals of different symmetry; and finally (iii) the estimation of the population of 3d orbitals of different symmetry in compounds, where the orbital degree of freedom is crucial for special physical properties.

This procedure seems to be justified faced with the comprehensive review of spin-dependent Compton scattering offered by Sakai (2004).

4.13.2 *Spin density in momentum space*

As demonstrated in Section 4.7.2, the 3D momentum density $\rho(\mathbf{p})$ can be reconstructed using a sufficient number of directional Compton profiles. If the separation of the spin-dependent part of the Compton profile is achieved for different directions of the scattering vector $\mathbf{q}$, (4.163) tells us that the same must also be possible with the magnetic Compton profiles with the aim to obtain $\rho_\uparrow(\mathbf{p}) - \rho_\downarrow(\mathbf{p})$, the spin density in momentum space. Obviously, having in mind that the magnetic Compton profile is only a few percent of the total one, it is a very time-consuming task to accumulate a sufficiently good statistic for that purpose. Therefore, the applications of reconstruction procedures to magnetic Compton profiles are rather scarce.

Tanaka *et al.* (1993) have performed such a 3D reconstruction of the spin density of ferromagnetic Fe+3wt% Si using circularly polarized 60 keV X-rays emitted from an elliptical wiggler at the beamline of the KEK accumulation ring, Japan. Fourteen directional magnetic Compton profiles measured with a statistical accuracy of 1% of the magnetic Compton peak were used for the reconstruction, employing the direct Fourier transform technique of Suzuki and Tanigawa (1989), where the reciprocal form factors $B_\mathrm{m}(\mathbf{r})$, calculated by Fourier transforming the magnetic Compton profiles, were interpolated on a grid which enables the calculation of $\rho_{\text{mag}}(\mathbf{p}) = \rho_\uparrow(\mathbf{p}) - \rho_\downarrow(\mathbf{p})$ by inverse Fourier transformation. Figure 4.65(a) shows a cross-section of the experimental $\rho_{\text{mag}}(\mathbf{p})$ in the (001) plane including the Γ-point. Figure 4.65(b) depicts theoretical $\rho_{\text{mag}}(\mathbf{p})$, according to FLAPW calculations of Kubo and Asano (1990), convoluted with the experimental resolution of 0.76 a.u. Both figures agree with respect to a deep minimum with negative spin polarization around the Γ-point and four peaks of

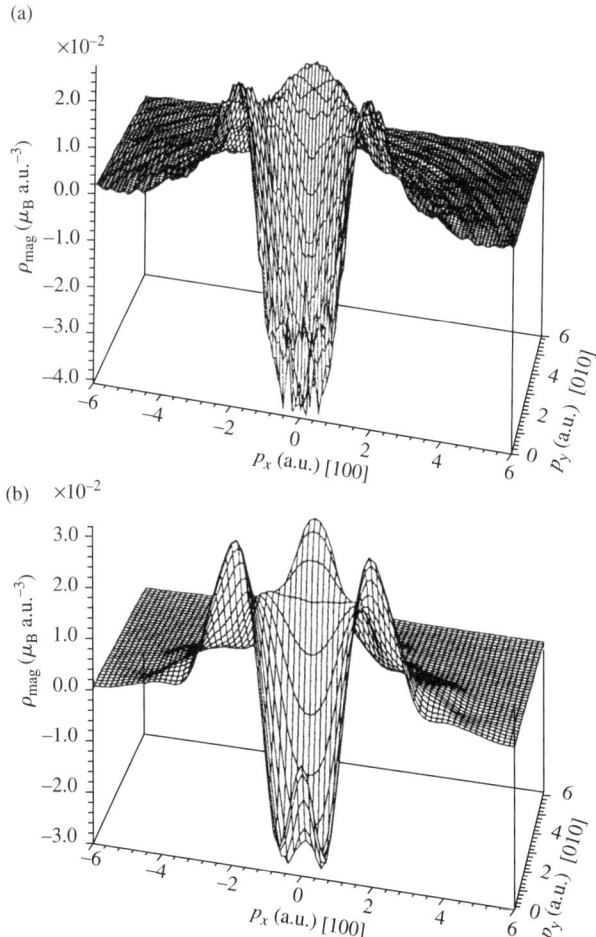

Fig. 4.65. (a) A cross-section of the experimental ρ_{mag} of Fe+3wt% Si in the (001) plane including the Γ-point. (b) A cross-section of the theoretical ρ_{mag} in the (001) plane including the Γ-point. The density is convoluted with the experimental resolution (Gaussian of FWHM 0.76 a.u.). (Reprinted with permission from Tanaka et al. (1993); copyright (1993) by the American Physical Society.)

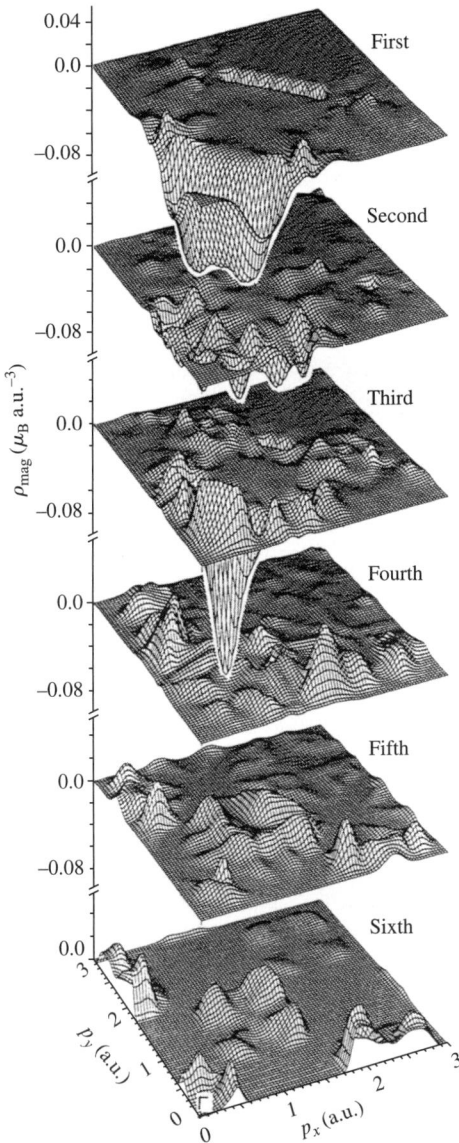

Fig. 4.66. Partial contributions to the theoretical ρ_{mag} of Fe+3wt% Si from the first to the sixth band. Each density is convoluted with a Gaussian of FWHM 0.1 a.u. Each cross-section is of the (001) plane including the Γ-point. (Reprinted with permission from Tanaka *et al.* (1993); copyright (1993) by the American Physical Society.)

positive spin polarization at $p_x = \pm 1.8$ a.u. and $p_y = \pm 1.8$ a.u. The theoretical calculation exhibits no spin polarization at the Γ-point. Due to the failures in the interpolation treatment of the $B(\mathbf{r})$ function, always a sharp spike appears as an artifact at the Γ-point, so that the experimental $\rho_{\text{mag}}(\mathbf{p})$ values in that region are less confident. A band-by-band analysis of the theoretical spin polarization, as shown in Fig. 4.66, reveals negative spin polarization of the first band of s-like electrons, and negative spin polarization of the second and third bands of p-like electrons, where this induced negative spin polarization is restricted to the first Brillouin zone. Positive spin polarization is found in the fourth to the sixth band contributions of d-like electrons.

4.13.3 Spin component of the magnetic moment, element and orbital specific

We have seen how magnetic Compton scattering yields information about the total spin component of the magnetic moment by utilizing a normalization standard as shown above. This information is extracted empirically from the experimental data and can then be used to get more details about magnetic properties of the sample under investigation: (i) about the orbital component of the magnetic moment based on the macroscopically measured total magnetic moment; (ii) about the distribution of the spin magnetic moment on itinerant nonlocal orbitals on the one hand and localized orbitals on the other hand; and (iii) about the distribution of the spin magnetic moment on different elements and orbitals in multicomponent magnetic systems.

(i) If one compares the total spin magnetic moment of a specimen, as obtained by integrating a normalized magnetic Compton profile, with the total magnetic moment determined by means of a magnetometer, the total orbital magnetic moment can be determined in a much more direct way than with magnetic X-ray circular dichroism, where the interpretation of the signal relies on assumptions of the sum rules (Thole et al. 1992). As an example of such an application of magnetic Compton scattering a study of Taylor et al. (2001) on $Pd_{1-x}Co_x$ alloys with $x = 0.6$ and 0.28 may be reported, performed at the bealine ID15a of the ESRF with an incident beam energy of 200 keV reversing the sample magnetization using a 1 T rotating permanent magnet. If one assumes that there is negligible orbital contribution to the induced Pd moment, the orbital contribution per Co atom is 0.08 ± 0.02 μ_B for $x = 0.6$, not that far from calculations of Eriksson et al. (1990), who quoted an orbital moment in metallic fcc Co as 0.12 μ_B using a linear muffin-tin orbital (LMTO) method with spin–orbit coupling and 0.07 μ_B with no spin–orbit coupling. For $x = 0.28$ the orbital contribution was 0.18 ± 0.05 μ_B, in good agreement with that quoted by Miyahaza et al. (1995).

(ii) A classical example for separating the orbital contribution of different symmetry to the spin magnetic moment by means of magnetic Compton scattering is presented by ferromagnetic gadolinium, the simplest ferromagnet among the 4f pure metals. The bulk magnetic moment per atom is found to be 7.63 μ_B,

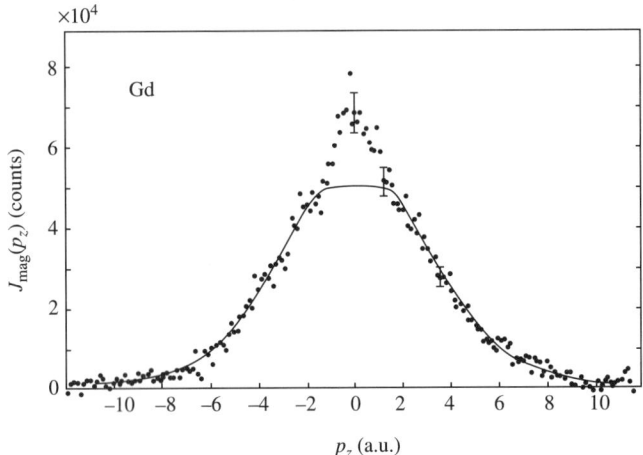

Fig. 4.67. Magnetic Compton profile of polycrystalline gadolinium at 106 K. The dots are the experimental data. The solid line is the theoretical Compton profile from a self-consistent Hartree–Fock calculation for atomic 4f electrons. (Reprinted with permission from Sakai *et al.* (1991); copyright (1991) by the Physical Society of Japan.)

exceeding the localized magnetic moment of 7 μ_B due to the localized $(4f)^7$ electrons. Magnetic Compton profile measurements, performed at the elliptical wiggler beamline installed at the KEK accumulation ring, Japan (Sakai *et al.* 1991), have revealed the presence of spin polarized 5d itinerant electrons, which couple ferromagnetically with the 4f localized electrons. Figure 4.67 shows the experimental result, the magnetic Compton profile of polycrystalline gadolinium. 4f electrons of gadolinium are well localized. Therefore, their Compton profile should reproduce the free atomic 4f Compton profile of gadolinium as calculated in a self-consistent Hartree–Fock computation, and the tails of which can easily be fitted to the gadolinium magnetic Compton profile of Fig. 4.67. The excess above the free atomic 4f profile must then be attributed to the spin magnetic moment of the 5d electrons; its integral corresponds to an excess magnetic moment of $0.53 \pm 0.08 \, \mu_B$.

Another example of separating itinerant and localized contributions to the spin magnetic moment has been submitted by Deb *et al.* (2001) with a study on the Heusler alloy Ni_2MnSn performed with 270 keV incident photon energy at the beamline BL08W of SPring-8. Figure 4.68 shows the magnetic Compton profile of Ni_2MnSn at 10 K, normalized to the saturation moment of 4.05 μ_B for three different orientations of the scattering vector. It turns out that the experimental data are, for $p_z > 1.5$ a.u., well reproduced by free atomic Hartree–Fock calculated Mn 3d Compton profiles convoluted with the experimental resolution of 0.45 a.u. for $p_z > 1.5$ a.u., thus indicating an average Mn 3d contribution of

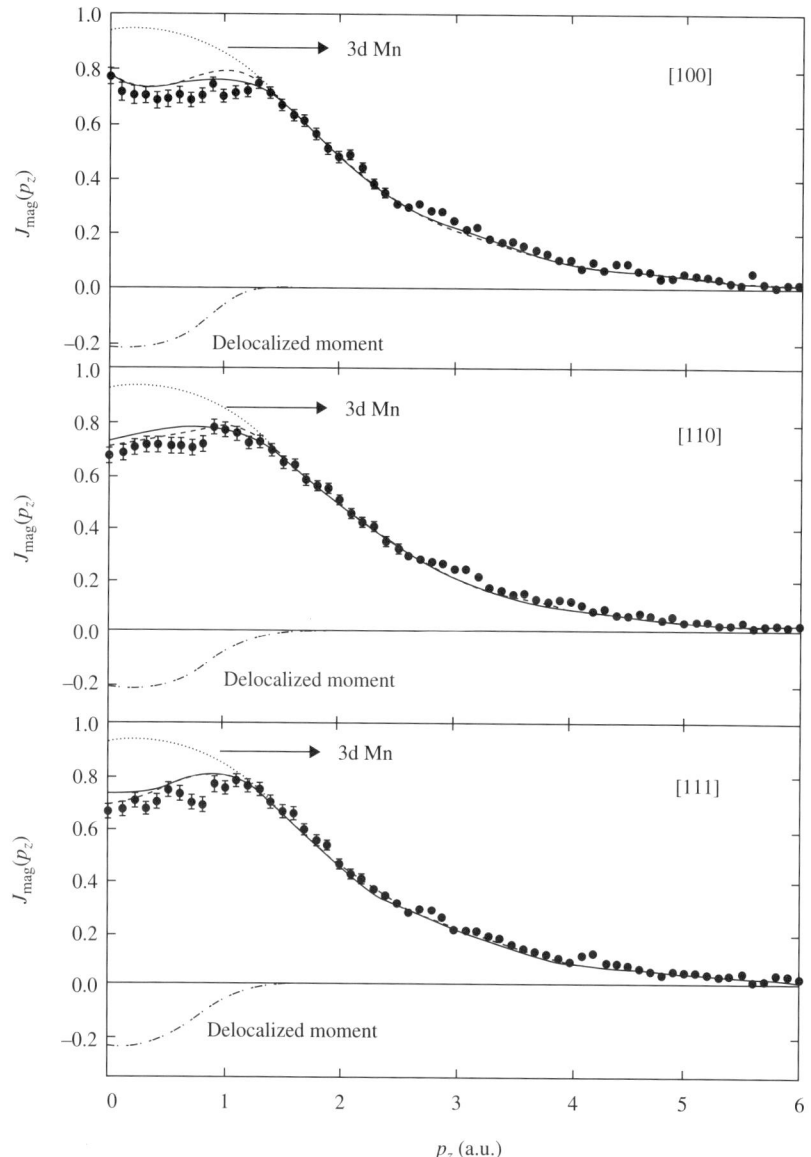

Fig. 4.68. Dots: the experimental magnetic Compton profile of Ni_2MnSn at T = 10 K, normalized to 4.05 μ_B. The theoretical profiles are convoluted with the experimental resolution of 0.48 a.u. Solid line: theoretical FLAPW profile. Decomposition of the experimental data into the Mn 3d free atomic and the free-electron Compton profile is shown by dotted and dashed dotted curves, respectively. The dashed curve is the sum of the 3d and the free-electron profile. (Reprinted with permission from Deb *et al.* (2001); copyright (2001) by the American Physical Society.)

4.39 μ_B to the spin magnetic moment. On the other hand, a central dip around $p_z = 0$ clearly deviates from the 3d Mn profile demonstrating the existence of a diffuse spin magnetic moment opposed to the Mn 3d moment. The three directional measurements predict a negative moment with an average value of 0.34 μ_B antiferromagnetically coupled to the Mn 3d moment.

(iii) We have already seen how the very different shapes of the Compton profiles of different orbitals can help to distinguish between their contributions to the spin magnetic moment. We shall now look at examples where the same philosophy can even help to separate the spin magnetic contribution of different elements in a multielement magnetic system. Let us consider HoFe$_2$, where it can be assumed that both elements bear a spin magnetic moment. Magnetic Compton scattering measurements on that system were performed at the ARNE-1 station of the KEK accumulation ring (Japan) by Cooper et al. (1993) aiming to investigate the temperature dependence between 10 and 300 K of these contributions. Figure 4.69 shows the magnetic Compton profile of HoFe$_2$ at room temperature. The vertical scale is the percentage of magnetic to charge scattering intensities as defined by (4.164). Compared to ferromagnetic iron, which yields nearly 1%

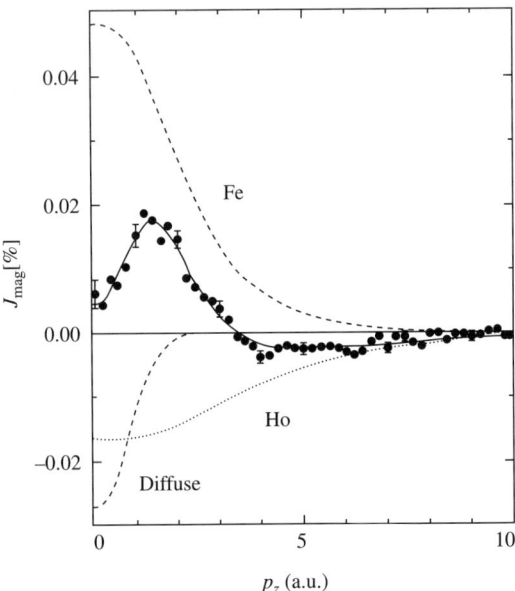

Fig. 4.69. The room-temperature magnetic Compton profile of HoFe$_2$ decomposed into a 3d (Fe site), a 4f (Ho site) and a diffuse component. The vertical scale is the percentage of magnetic to charge scattering intensities as defined by (4.164). (Reprinted with permission from Cooper et al. (1993); copyright (1993) by the American Physical Society.)

of the total scattering for the spin magnetic contribution, here the percentage is only 0.05%, indicating a rather small net spin magnetic moment. However, as shown in Fig. 4.69, one can decompose, in a unique way, this magnetic Compton profile into a free atomic 3d Fe and 4f Ho Compton profile together with a free-electron-like Compton profile. Then it becomes clear that the rather small net moment is the result of a near cancellation of the 3d spin magnetic moment at the iron site on the one hand and of the 4f spin magnetic moment at the holmium site in line with a diffuse moment, both oriented opposite to the iron moment. Detailed measurements at different temperatures between 10 and 300 K (see Fig. 4.70) show that the 4f holmium contribution is increasing with decreasing temperature. This is consistent with the fact that at low temperatures Hund's rule gives a holmium spin magnetic moment of 4 μ_B, whereas room-temperature neutron and magnetization data (cited in Zukowski *et al.* 1993) suggest a softening to 2.52 μ_B. The spin moment at the iron site, however, remains constant at 1.85 μ_B. Also the diffuse moment exhibits only little change between the extreme temperatures. This softening of the holmium spin magnetic moment leads to a reversal of the net spin direction with rising temperature. A similar study of Lawson *et al.* (1995) was devoted to ferrimagnetic DyFe$_2$ and ErFe$_2$, where also the temperature dependence of the individual spin moments was the matter of investigation with the result that a softening of the moment at the rare earth element with rising temperature leads to a characteristic temperature dependence of the net spin magnetic moment.

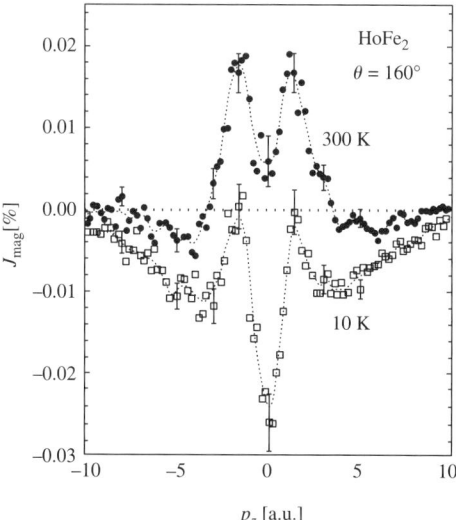

Fig. 4.70. The magnetic Compton profiles of HoFe$_2$ at 300 K (filled circles) and 10 K (open squares). (Reprinted with permission from Cooper *et al.* (1993); copyright (1993) by tThe American Physical Society.)

4.13.4 Population of d-orbitals

The mechanism of the colossal magnetoresistance (CMR) found in manganese oxides is one of the most fascinating problems in solid state physics. It turned out that not only the spin degree of freedom but also the lattice degree of freedom is involved, which is strongly coupled with the orbital degree of freedom through the Jahn–Teller effect (Koizumi et al. 1998). In this context the character of the occupied e_g orbitals strongly affects the rather complicated transport and magnetic properties. Magnetic Compton scattering should be able to extract the character of the occupied e_g and t_{2g} states, because the ferromagnetic moment of a manganese oxide like $La_{2-2x}Sr_{1+2x}Mn_2O_7$, investigated by Koizumi et al. (2001), originates from spins in Mn 3d orbitals. Since the Compton profile of each orbital exhibits a different shape, one can utilize this property to differentiate the population in e_g and t_{2g} orbitals in these manganites. The measurements were

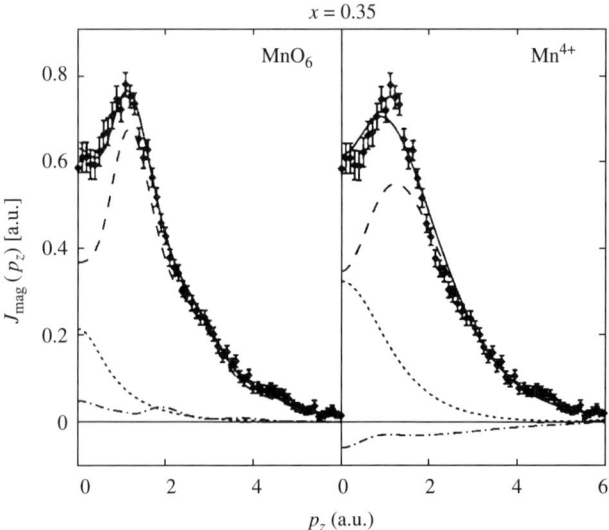

Fig. 4.71. Left panel: diamonds with error bars are the magnetic Compton profile of $La_{2-2x}Sr_{1+2x}Mn_2O_7$. The solid line is the fit using MnO_6 cluster orbitals. The dashed line is the t_{2g} orbital contribution. The dotted line represents the $e_{x^2-y^2}$, the dashed dotted line the $e_{3z^2-r^2}$ orbital contribution. Right panel: the same as the left panel but using the Mn^{4+} orbitals. (Reprinted with permission from Koizumi et al. (2001); copyright (2001) by the American Physical Society.)

performed at the 08W beamline at SPring-8, Japan, using circularly polarized X-rays emitted from a multipole elliptical wiggler, monochromatized to 271 keV, on two single-domain crystals with $x = 0.35$ and $x = 0.42$. The theoretical profiles for each 3d orbital was the result of an *ab initio* molecular orbital calculation for the $(MnO_6)^{8-}$ cluster, which is embedded in point charges (total number 444) that describe the Madelung potential. The same calculation was also performed for an isolated Mn^{4+} to obtain atomic t_{2g} and e_g orbitals. Figure 4.71 presents the experimental magnetic Compton profile along the [001] direction at $x = 0.35$. The left panel shows the fit using the MnO_6 cluster orbitals with a spin density of 3.65 per site. The t_{2g} orbital contribution has been fixed to 3.0 because it is fully occupied. The $e_{x^2-y^2}$ contribution (dotted line) is 0.46, the $e_{3z^2-r^2}$ contribution (dashed-dotted line) is 0.19. The right panel corresponds to a fit using the atomic orbitals for an isolated Mn^{4+} showing poor agreement with an unphysical negative population of the $e_{3z^2-r^2}$ orbital thus indicating the usefulness of the cluster calculation and the significant oxygen hybridization in the e_g orbitals. The fit for the $x = 0.42$ data yields 0.44 and 0.14 for the $e_{x^2-y^2}$ and the $e_{3z^2-r^2}$ orbital population, respectively. Summarizing, the dominant character of the e_g state is the $x^2 - y^2$ type with almost constant occupation number per site. The population in the $3z^2 - r^2$ type orbitals decreases with increasing x. This example demonstrates that magnetic Compton scattering is a very powerful method to study orbitals of spin-polarized electrons in manganites.

4.14 References

Aikala, O., T. Paakkari, and S. Manninen (1982). *Acta Crystallogr. A* **38** 155
Alexandropoulos, N.G. and W.A. Reed (1977). *Phys. Rev. B* **15** 1790
Alexandropoulos, N.G., T. Chatzigeorgiou, and G.E. Giakoumakis (1988). *Phil. Mag. Lett.* **58** 91
Almbladh, C.-O. and L. Hedin (1983). In *Handbook on Synchrotron Radiation*, ed. E.E. Koch (North-Holland, Amsterdam) vol. 1B
Anastassopoulos, D.L., G.D. Priftis, N.I. Papanicolaou, N.C. Bacalis, and D.A. Papaconstantopoulos (1991). *J. Phys.: Condens. Matter* **3** 1099
Andrejczuk, A., L. Dobrzynski, E. Zukowski, M.J. Cooper, S. Hamouda, and J. Latuszkiewicz (1992). *J. Phys.:Condens. Matter* **4** 2735
Andrejczuk, A., L. Dobrzynski, J. Kwiatkowska, F. Maniawski, S. Kaprzyk, A. Bansil, E. Zukowski, and M.J. Cooper (1993). *Phys. Rev. B* **48** 15552
Asthalter, T. and W. Weyrich (1993). *Z. Naturforssch.* **48a** 303
Ayma, D., M. Rerat, and A. Lichanot (1998a). *J. Phys. Condens. Matter* **10** 557
Ayma D., M. Rerat, R. Orlando, and A. Lichanot (1998b). *Acta Cryst. A* **54** 1019
Bacalis, N.C., N.I. Papanicolaou, and D.A. Papaconstantopoulos (1986). *J. Phys. F: Met. Phys.* **16** 1471
Bagayoko, D., D.G. Laurent, S.P. Singhal, and J. Callaway (1980). *Phys. Lett.* **76A** 187

Bagus, P.S., K. Hermann, and C.W. Bauschlicher (1984). *J. Chem. Phys.* **80** 4378

Barbiellini, B. and A. Bansil (2001). *J. Phys. and Chem. Solid,* **62** 2181

Barbiellini, B. and A. Shukla (2002). *Phys. Rev. B* **66** 235101

Bauer, G.E.W. (1984). *Compton profile und elektronische Struktur des Kupfers,* Thesis, Berlin

Bauer, G.E.W. and J.R. Schneider (1984a). *J. Phys. Chem. Phys.* **45** 675

Bauer, G.E.W. and J.R. Schneider (1984b). *Phys. Rev. Lett.* **52** 2061

Bauer, G.E.W. and J.R. Schneider (1985). *Phys. Rev. B* **31** 681

Bell, F. (2003). *Phys. Rev. B* **67** 155110

Bell, F. and J.R. Schneider (2001). *J. Phys.: Condens. Matter* **13** 7905

Bellin, Ch., P. Roca i Cabarrocas, K. Zellama, M.L. Theye, and G. Loupias (1997). *Solid State Commun.* **104** 193

Bellin, Ch., G. Loupias, A.A. Manuel, T. Jarlborg, Y. Sakurai, Y. Tanaka, and N. Shiotani (1995). *Solid State Commun.* **96** 563

Benedek, R., R. Prasad, S. Manninen, B.K. Sharma, A. Bansil, and P.E. Mijnarends (1985). *Phys. Rev. B* **32** 7650

Berko, S. and J.S Plaskett (1958). *Phys. Rev.* **112** 1877

Berko, S., R.E. Kelley, and J.S. Plaskett (1957). *Phys. Rev.* **106** 824

Bernal, J.D. and R.H. Fowler (1935). *J. Chem. Phys.* **1** 515

Berndt, K. and O. Brümmer (1976). *phys. stat. sol.* (b) **78** 659

Berthold, A. (1993). *Unelastische Röntgenstreuung an Kalium-und Lithiuminterkaliertem Graphit,* Thesis, Dortmund

Berthold, A., S. Mourikis, J.R. Schmitz, W. Schülke and H. Schulte-Schrepping (1992a). *Nucl. Instr. and Meth.* **A317** 373

Berthold, A., J. Degenhardt, S. Mourikis, J.R. Schmitz, W. Schülke, H. Schulte-Schrepping, A. Hamacher, D. Protic and G. Riepe (1992b). *Nucl. Instr. and Meth.* **A320** 375

Biggs, F.L., B. Mendelsohn, and J.B. Mann (1975). *At. Data and Nucl. Data Tables* **16** 201

Binkley, J.S. and J.A. Pople (1975). *Int. J. Quantum Chem.* **9** 229

Blaas, C., J. Redinger, S. Manninen, V. Honkimäki, K. Hämäläinen, and P. Suortti (1995). *Phys. Rev. Lett.* **75** 1984

Blatt, J.M. (1964). *"Theory of superconductivity"* (Academic Press, New York)

Bloch, B.J. and L.B. Mendelsohn (1974). *Phys. Rev A* **9** 129

Bonse U., W. Schröder, and W. Schülke (1979). *J. Appl. Crystallogr.* **12** 432

Bonse, U., W. Schülke, and G. Wolf (1980). *Phil. Mag. B* **42** 499

Boulakis, G.D. (1986). *J. Physique* **47** 1523

Bross, H., G. Bohn, G. Meister, W. Schubö, and H. Stöhr (1970). *Phys. Rev. B* **2** 3098

Brown, R.E. and V.H. Smith, Jr (1972). *Phys. Rev. A* **5** 140

Carbotte, J.P. and S. Kahana (1965). *Phys. Rev.* **139** A213

Cardwell, D.A. and M.J. Cooper, (1986). *Phil. Mag. B* **54** 37

Cardwell, D.A. and M.J. Cooper (1989). *J. Phys.: Condens Matter* **1** 9357

Cardwell, D.A., M.J. Cooper, and S. Wakoh (1989). *J. Phys.: Condens. Matter* **1** 541

Chabaud, S., Ch. Bellin, F. Mauri, G. Loupias, S. Rabii, L. Croguennec, C. Pouillerie, C. Delmas, and Th. Buslaps (2004). *J. Phys. Chem Solids* **65** 241

Chomilier, J., G. Loupias, and J. Felsteiner (1985). *Nucl. Instrum & Methods A* **235** 603

Chou, M.Y., M.L. Cohen, and St. G. Louie (1986). *Phys. Rev. B* **33** 6619

Chou, M.Y., P.K. Lam, M.L. Cohen, G. Loupias, J. Chomilier, and J. Petiau (1982). *Phys. Rev. Lett.* **49** 1452

Chou, M.Y., St. G. Louie, M.L. Cohen, and N.A. Holzwarth (1984). *Phys. Rev. B* **30** 1062

Cooper, M.J. (1977). *Compton Scattering* ed. B.C. Williams (New York: McGraw-Hill)

Cooper. M.J. (1979). *Nucl. Instrum. Meth.* **166** 21

Cooper, M.J. (1985). *Rep. Prog. Phys.* **48** 415

Cooper, M.J., J.A. Leake, and R.J. Weiss (1965). *Philos. Mag.* **12** 797

Cooper, M.J., D. Laundy, D.A. Cardwell, D.N. Timms, and R.S. Holt (1986). *Phys. Rev. B* **34** 5984

Cooper, M.J., E. Zukowski, D.N. Timms, R. Armstrong, F. Itoh, Y. Tanaka, M. Ito, H. Kawata, and R. Bateson (1993). *Phys. Rev. Lett.* **71** 1095

Cormack, A.M. (1963, 1964). *J. Appl. Phys.* **34** 2722, **35** 2908.

Cortona, P. (1991). *Phys. Rev. B* **44** 8454

Currat, R., P.D. DeCicco, and R. Kaplow (1971). *Phys. Rev. B* **3** 243

Daniel, E. and S.H. Vosko (1960). *Phys. Rev.* **120** 2041

Das, G. and H.C. Padhi (1986). *Phil. Mag. B* **54** 415

Das, G. and H.C. Padhi (1987). *J. Phys. C: Solid State Phys.* **20** 5253

Das, G. and H.C. Padhi (1988). *Phys. Lett. A* **128** 383

Deb, A., N. Hiaoka, M. Itou, Y. Sakurai, M. Onodera, and N. Sakai (2001). *Phys. Rev. B* **63** 205115

DeBenedetti, S., C.E. Cowan, W.R. Konneker, and H. Primakoff (1950). *Phys. Rev.* **77** 205

Delaney, P., B. Kralik, and S.G. Louie (1998). *Phys. Rev. B* **58** 4320

Differt, K. and R. Messer (1980). *J. Phys. C* **13** 717

Dobrzynski, L. and A. Holas (1996). *Nucl. Instrum. Methods Phys. Res. A* **383** 589

Dovesi, R., C. Ermondi, E. Ferrero, C. Pisani, and C. Roetti (1964). *Phys. Rev. B* **29** 3591

Dugdale, S.B. and T. Jarlborg (1998). *Solid State Commun.* **105** 283

Dugdale, S.B., H.M. Fretwell, M.A. Alam, G. Kontrym-Sznajd, R.N. West, and S. Badrzadeh (1997). *Phys. Rev. Lett.* **79** 941

Dugdale, S.B., H.M. Fretwell, K.J. Chen, Y. Tanaky, A. Shukla, T. Buslaps, Ch. Bellin, G. Loupias, M.A. Alam, A.A. Manuel, P. Suortti, and N. Shiotani (2000). *J. Phys. Chem. Solids* **61** 361

DuMond, J.W.M. (1929). *Phys. Rev.* **33** 643
DuMond, J.W.M. and H.A. Kirkpatrick (1930). *Rev. Sci. Insrum.* **1** 88
Dunning, T.H. (1989). *J. Chem. Phys.* **90** 1007
Eguiluz, A.G., W. Ku, and J.M. Sullivan (2000). *J. Phys. Chem. Solids* **61** 383
Eisenberger, P. (1970). *Phys. Rev. A* **2** 1678
Eisenberger, P. (1972). *Phys. Rev. A* **5** 628
Eisenberger, P. and W.C. Marra (1971). *Phys. Rev. Lett.* **27** 1413
Eisenberger, P. and P.M. Platzman (1970). *Phys. Rev. A* **2** 415
Eisenberger, P. and A.A. Reed (1972). *Phys. Rev. A* **5** 415
Eisenberger, P. and W.A. Reed (1974). *Phys. Rev. B* **9** 3237
Eisenberger, P. L. Lam, P.M. Platzman, and P. Schmidt (1972b). *Phys. Rev. B* **6** 3671
Epstein, I.R. (1970). *J. Chem. Phys.* **53** 4425
Epstein, I.R. and W.N. Lipscomb (1970). *J. Chem. Phys.* **53** 4418
Eriksson, O., B. Johansson, R.C. Albers, A.M. Boring, and M.S.S. Brooks (1990). *Phys. Rev. B* **42** R2707
Farid, B., V. Heine, G.E. Engel, and I.J. Robertson (1993). *Phys. Rev. B* **48** 11602
Felsteiner, J., P. Pattison, and M.J. Cooper (1974). *Phil. Mag.* **30** 537
Feynmann, R.P. (1939). *Phys. Rev.* **56** 340
Filippi, C. and D.M. Ceperley (1999). *Phys. Rev. B* **59** 7907
Fuster, G., J.M. Tyler, N.E. Brener, J. Callaway, and D. Bagayoko (1990). *Phys. Rev. B* **42** 7322
Ghanty, T.K., V.N. Staroverov, P.R. Koren, and E.R. Davidson (2000). *J. Am. Chem. Soc.* **122** 1210
Gillet, J.M., P.J. Becker, and G. Loupias (1995). *Acta Crystallogr.* **A51** 405
Gillet, J.M., P.J. Becker, and P. Cortona (2001). *Phys. Rev. B* **63** 235115
Gyorffy, B.L. and G.M. Stocks (1993). *Phys. Rev. Lett.* **50** 374
Hakala, M., S. Huotari, K. Hämäläinen, S. Manninen, Ph. Wernet, A. Nilsson, L.G.M. Pettersen (2004). *Phys. Rev. B* **70** 125413
Hamada, N., M. Hwang, and A.J. Freeman (1990). *Phys. Rev. B* **41** 3620
Hämäläinen, K., S. Manninen, C.-C. Kao, W. Caliebe, J.B. Hastings, A. Bansil, S. Kaprzyk, and P.M. Platzman (1996). *Phys. Rev. B* **54** 5453
Hämäläinen, K., S. Huotari, J. Lakkanen, A. Soininen, S. Manninen, C.-C. Kao, T. Buslaps, and Mezouar (2000). *Phys. Rev. B* **62** R735
Hamann, D.R. (1997). *Phys. Rev. B* **55** 10157
Hamann, D.R., M. Schlüter, and C. Chiang (1979). *Phys. Rev. Lett.* **43** 1494
Hansen, N.K. (1980). *Reconstruction of the EMD from a set of Compton profiles* (Hahn Meitner Institut, Berlin Rep. HMI B) p. 342
Hansen, N.K., P. Pattison, and J.R. Schneider (1979). *Z. Phys. B* **35** 215
Hansen, N.K., P. Pattison, and J.R. Schneider (1987). *Z. Phys. B-Condensed Matter* **66** 305
Harmalka, A., D.G. Kanhere, and R.M. Singru (1985). *Phys. Rev. B* **31** 6415

Harrison, W.A. (1986). *Pseudopotentials in the Theory of Metals* (W. A. Benjamin, Inc)

Helgaker, T., H.J.A. Jensen, P. Jörgensen, J. Olsen, K. Ruud, H. Agren, A.A. Auer, K.L. Bak, V. Bakken, O. Christiansen, S. Coriani, P. Dahle, E.K. Dalskov, T. Enevoldsen, B. Fernandez, C. Hättig, K. Hald, A. Halkier, H. Heiberg, H. Hettema, D, Jonsson, S. Kirpekar, R. Kobayashi, H. Koch, K.V. Mikkelsen, P. Norman, M.J. Packer, T.B. Pedersen, T.A. Ruden, A. Sanchez, T. Saue, S.P.A. Sauer, B. Schimmelpfennig, K.O. Sylvester-Hvid, P.R. Taylor, and O. Vahtrsa (2001). DALTON, a molecular electronic structure program, release 1.2 http://www.kjemi.uio,no/software/dalton/.

Hedin, L. (1995). *Phys. Rev.* **139** A796

Hedin, L. and B.I. Lundqvist (1971). *J. Phys. C* **4** 2064

Hedin, L. and B.I. Lundqvist (1972). *J. Physique C* **3** 731

Hiraoka, N., M. Itou, T. Ohata, M. Mizumaki, Y. Sakurai, N. Sakai (2001). *J. Synchrotron Rad.* **8** 26

Holm, P. (1988). *Phys. Rev. A* **37** 3706

Holm, P. and R. Ribberfors (1989). *Phys. Rev. A* **40** 6251

Holt, R.S., M. Cooper, and K.R. Lea (1978). *J. Phys. E., Sci. Instr.* **11** 68

Hotz, H.P., J.M. Mathiesen, and J.P. Hurley (1968). *Phys. Rev.* **170** 351

Huotari, S. (2002). *Electronic structure of matter studied by Compton scattering*, Thesis, Helsinki

Huotari, S., K. Hämäläinen, S. Manninen, S. Kaprzyk, A. Bansil, W. Caliebe, T. Buslaps, V. Honkimäki, and P. Suortti (2000). *Phys. Rev. B* **62** 7956

Huotari, S., K. Hämäläinen, S. Manninen, A. Issolah, and M. Marangolo (2001). *J. Phys. Chem. Solids* **62** 2205

Isaacs, E.D., A. Shukla, P.M. Platzman, D.R. Hamann, B. Barbiellini, and C.A. Tulk (1999). *Phys. Rev Lett.* **82** 600

Ishii, Y. and T. Fujiwara (2001). *Phys. Rev. Lett.* **87** 206408

Issolah, A., B. Levi, A. Beswick, and G. Loupias (1988). *Phys. Rev. A* **38** 4509

Itou, M., Y. Sakurai, T. Ohata, A. Bansil, S. Kaprzyk, Y. Tanaka, H. Kawata, and N. Shiotani (1998). *J. Phys. Chem. Solids* **59** 99

Itou, M., S. Kishimoto, H. Kawata, M. Ozaki, H. Sakurai, and F. Itoh (1999). *J. Phys. Soc. Japan* **68** 515

Jauch, J.M. and F. Rohrlich (1955). *The theory of photons and Electrons* (Addison-Wesley, Cambridge, Mass. pp. 163–169 and pp. 228–235

Jeziorski, B. and K. Szalewicz (1979) *Phys. Rev. A* **19** 2360

Joachain, C.J. (1975) *Quantum Collision Theory* (Amsterdam: Elsevier) p. 144

Joshi, K.B., Rajesh Jain, R.K. Pandya, B.L. Ahuja, and B.K. Sharma (1999). *J. Chem. Phys.* **111** 163

Kahana, S. (1960). *Phys. Rev.* **117** 123

Kahana, S. (1963). *Phys. Rev.* **129** 1622

Kakutani, Y., Y. Kubo, A. Koizumi, N. Sakai, B.L. Ahuja, and B.K. Sharma (2003). *J. Phys. Soc. Jpn.* **72** 599

Kane E.O. and A.B. Kane (1978). *Phys. Rev. B* **17** 2691
Kerker, B. (1981). *Phys. Rev. B* **23** 6312
Kheifets, A.S., D.R. Lun, and S. Yu. Savrasov (1999). *J. Phys.: Condens. Matter* **11** 6779
Kohn, W. and L.J. Sham (1965). *Phys. Rev.* **140** A1133
Koizumi, A., S. Miyaki, Y. Kakutani, H. Koizumi, N. Hiraoka, K. Makoshi, and N. Sakai (2001). *Phys. Rev. Lett.* **86** 5589
Koizumi, H., T. Hotta, and Y. Takada (1998). *Phys. Rev. Lett.* **80** 4518
Kontrym-Sznajd, G. (1990). *phys. stat. sol.* (a) **117** 227
Kontrym-Sznajd, G., R.N. West, and S.B. Dugdale (1997). *Mater. Sci. Forum* 255–257 796
Kontrym-Sznajd, G., M. Samsel-Szekala, A. Pietraszko, H. Sormann, S. Manninen, S. Huotari, K. Hämäläinen, R.N. West, and W. Schülke (2002). *Phys. Rev. B* **66** 155110
Kralik, B., P. Delaney, and S.G. Louie (1998). *Phys. Rev. Lett.* **80** 4253
Kramer, B., P. Krusius, W. Schrüder, and W. Schülke (1977). *Phys. Rev. Lett.* **38** 1227
Krusius, P. (1977). *J. Phys. C: Solid St. Phys.* **10** 1875
Krusius, P., H. Isomäki, and B. Kramer (1979). *Phys. Rev.* **B19** 1818
Krusius, P., P. Pattison, B. Kramer, W. Schülke, U. Bonse, and J. Treusch (1982). *Phys. Rev. B* **25** 6393
Kubo, Y. (1997). *J. Phys. Soc. Jpn.* **66** 2236
Kubo, Y. (2001). *J. Phys. Chem. Solids* **62** 2199
Kubo, Y. and S. Asano (1990). *Phys. Rev. B* **42** 4431
Kubo, Y., Y. Sakurai, Y. Tanaka, T. Nakamura, H. Kawata, and N. Shiotani (1997). *J. Phys. Soc. Japan* **66** 2777
Kubo, Y., Y. Sakurai, and N. Shiotani (1999). *J. Phys.:Condens. Matter* **11** 1683
Kwiatkowska, J., F. Maniawski, I. Matsumoto, H. Kawata, N. Shiotani, L. Litynska, S. Kaprzyk, and A. Bansil (2004). *Phys. Rev. B* **70** 075106
Lässer, R. and B. Lengeler (1978). *Phys. Rev. B* **18** 637
Lässer, R., B. Lengeler, K.A. Gschneidner, and P. Palmer (1979). *Phys. Rev. B* **20** 1390
Lässer, R., B. Lengeler, and G. Arnold (1980). *Phys. Rev. B* **22** 663
Lam, L. and P.M. Platzman (1974). *Phys. Rev. B* **9** 5122
Laue, M. von (1960). Röntgenstrahl-Interferenzen, Frankfurt/Main: Akademische Verlagsgesellschaft
Laurent, D.G., C.S. Wang, and J. Callaway (1978). *Phys. Rev. B* **17** 455
Lawson, P.K., J.E. McCarthy, M.J. Cooper, E. Zukowski, D.N. Timms, F. Itoh, H. Sakurai, Y. Tanaka, H. Kawata, and M. Ito (1995). *J. Phys.: Condens. Matter*,**7** 389
Lee, J.S. (1977). *J. Chem. Phys.* **66** 4906
Lewis, J. and D. Schwarzenbach, D. (1981). *Acta Cryst. A* **37** 507
Lichanot, A., M. Rerat, and M. Causa (1996). *J. Phys.:Condens. Matter* **8** 10425
Lock, D.G., V.H.C. Crisp, and R.N. West (1973). *J. Phys. F: Met. Phys.* **3** 561

Löwdin, P.O. (1956). *Adv. Phys.* **5** 1
Löwdin, P.O. (1970). *Adv. Quantum Chem.* **5** 185
Lou, Y., B. Johansson, and R.M. Nieminen (1991). *J. Phys.:Condens. Matter* **3** 1699
Loupias, G. and J. Petiau (1980). *J. Phys. (Paris)* **41** 265
Loupias, G. and J. Mergy (1980). *Etude de la Distribution des Quantites de movemment Electroniques dans l'Hydrure de Lithium par Diffusion Compton*. Raport de L'Ecole Polytechnique X85. (Ecole Polytechnique, Palaisceau France)
Loupias, G., J. Petiau, A. Issolah, and M. Schneider (1980). *phys. stat. sol.* (b) **102** 79
Loupias, G., J. Chomilier, and D. Guerard (1984). *J. Physique Lett.* **45** L-301
Loupias, G., S. Rabii, T. Tarbes, S. Nozieres, and R.C. Tatar (1990). *Phys. Rev. B* **41** 5519
Loupias, G., R. Wentzcovitch, L. Bellaiche, J. Moscovici, and S. Rabii (1994). *Phys. Rev. B* **49** 13342
Lundqvist, B.I. (1967a, 1967b). *Phys. kondens. Materie* **6** 193; **6** 206
MacKinnon, A. and B. Kramer (1980). *J. Phys. C: Solid. State Phys.* **13** 37
Mahapatra, D.P. and H.C. Padhi (1982). *Phil. Mag. B* **46** 607
Manninen, S. and T. Paakkari (1981). *Phil. Mag. B* **44** 127
Manninen, S., T. Paakkari, and K. Kajantie (1974). *Phil. Mag.* **29** 167
Manninen, S., K. Hämäläinen, M.A.G. Dion, M.J. Cooper, D.A. Cardwell, and T. Buslaps (1999). *Physica C* **314** 19
Marangolo, M., J. Moscovici, G. Loupias, S. Rabii, S.C. Erwin, C. Herold, J.F. Mareche, and Ph. Lagrange (1998). *Phys. Rev. B* **58** 7593
Marangolo, M., Ch. Bellin, G. Loupias, S. Rabii, S.C. Erwin, Th. Buslaps (1999). *Phys. Rev. B* **60** 17084
Matsumoto, I., H. Kawata, N. Shiotani (2001a). *Phys. Rev. B* **64** 195132
Matsumoto, I., J. Kwiatkowska, F. Maniawski, M. Itou, H. Kawata, N. Shiotani, S. Kaprzyk, P.E. Mijnarends, B. Barbiellini, and A. Bansil (2001b). *Phys. Rev. B* **64** 045121
Mattheiss, L.F. (1965). *Phys. Rev.* **139** A1893
McCarthy, J.E., M.J. Cooper, P.K. Lawson, D.N. Timms, S.O. Manninen, K. Hämäläinen, and P. Suortti (1997). *J. Synchrotron Radiat.* **4** 102
Mendelsohn, L.B. and B.J. Bloch (1975). *Phys. Rev. A* **12** 551
Metz, C., Th. Tschentscher, P. Suortti, A.S. Kheifets, D.R. Lun, T. Sattler, J.R. Schneider, and F. Bell (1999a). *Phys. Rev. B* **59** 10512
Metz, C., Th. Tschentscher, T. Sattler, K. Höppner, J.R. Schneider, K. Wittmaack, D. Frischke, and F. Bell (1999b). *Phys. Rev. B* **60** 14049
Mijnarends, P.E. (1967, 1969). *Phys. Rev.* **160** 512; **178** 622
Mijnarends, P.E. (1979) in *Positrons in Solids*, ed. by P. Hautojärvi (Springer)
Mills, D.M. (1987). *Phys. Rev. B* **36** 6178
Miyahara, T., S.Y. Park, T. Hanyu, T. Hatano, S. Moto, and Y. Kagoshima (1995). *Rev. Sci. Instrum.* **66** 1558

Müller, C. and M.S. Plesset (1934). *Phys. Rev.* **46** 618
Moss, S.C. (1969). *Phys. Rev. Lett.* **22** 1108
Mott, N.F. and H. Jones (1936). *The Theory of the Properties of Metals and Alloys* (Oxford: Clarendon) p. 159
Mueller, F.M. (1977). *Phys. Rev. B* **15** 3039
Nara, H., T. Kobayasi, and K. Shindo (1984). *J. Phys. C: Solid State Phys.* **17** 3967
Ng, T.K. and B. Dabrowski (1986). *Phys. Rev. B* **33** 5358
Nolting, W. (1991). *Grundkurs: Theoretische Physik 7. Vielteilchentheorie.* (Ulmen: Zimmermann-Neufang) p. 362
Oberli, L., A.A. Manuel, R. Sachot, P. Descouts, and M. Peter (1985). *Phys. Rev. B* **31** 6104
Ohata, T., M. Itou, I. Matsumoto, Y. Sakurai, H. Kawata, N. Shiotani, S. Kaprzyk, P.E. Mijnarends, and A. Bansil (2000). *Phys. Rev. B* **62** 16528
Ohshima, K. and D. Watanabe (1973). *Acta Crystallogr., A* **29** 520
Okada, T., H. Sekizawa, and N. Shiotani (1976). *J. Phys. Soc. Japan* **41** 836
Okada, T., Y. Watanabe, Y. Yokoyama, N. Hiraoka, M. Itou, Y. Sakurai, and S. Nanao (2002). *J. Phys. Condens. Matter* **14** L43
Okada, T., Y. Watanabe, S. Nanao, R. Tamura, S. Takeuchi, Y. Yokoyama, N. Hiraoka, M. Itou, and Y. Sakurai (2003). *Phys. Rev. B* **68** 132204
Oommi, G. and F. Ito (1993). *Jpn J. Appl. Phys. Suppl.* **32**, Supl. 32-1, 352
Panda, B.K., D.P. Mahapatra, and H.C. Padhi (1990). *phys. stat. sol.* (b) **158** 261
Panda, B.K., D. Pal, D.P. Mahapatra, and H.C. Padhi (1992). *J. Phys.: Condens. Matter* **4** 269
Pantelides, S.T.; D.J. Mickish, and A.B. Kunz (1974). *Phys. Rev. B* **10** 5202
Pattison, P. and J.R. Schneider (1978). *Solid State Commun.* **28** 581
Pattison, P. and J.R. Schneider (1979a). *J. Phys. B: Atom. Molec. Phys.* **12** 4013
Pattison, P. and J.R. Schneider (1979b). *Nucl. Instrum. Meth.* **158** 145
Pattison, P. and J.R. Schneider (1980). *Acta Cryst. A* **36** 390
Pattison, P. and W. Weyrich (1979). *J. Phys. Chem. Solids* **40** 213
Pattison, P., M. Cooper, and J.R. Schneider (1976). *Z. Phys. B* **25** 155
Pattison, P., N.K. Hansen, and J. Schneider (1981). *Chem. Phys.* **59** 231
Pattison, P., N.K. Hansen, and J.R. Schneider (1982). *Z. Phys. B-Condens. Matter* **46** 285
Pattison, P., N.K. Hansen, and J.R. Schneider (1984). *Acta Cryst B* **40** 38
Pattison, P., M. Cooper,, R. Holt, J.R. Schneider, and N. Stump (1977). *Z. Phys. B* **27** 205
Perkkiö, S., S. Manninen, and T. Paakkari (1989). *Phys. Rev. B* **40** 8446
Philips, J.C. (1973). *Bonds and Bands in Semiconductors* (New York: Academic Press)
Phillips, W.C. and R.J. Weiss (1968). *Phys. Rev.* **171** 790
Pisani, C., R. Dovesi, and C. Roetti (1988). *Hartree-Fock ab initio Treatment of Crystalline Systems* (Springer, Berlin) p. 15 ff

Podloucky, R. and J. Redinger (1984). *J. Phys. C: Solid State Phys.* **16** 6955
Priftis, G.D., D.L. Anastassopoulos, and A.A. Vradis (1994a). Proc. Sagamore XI Conference, Brest, France, unpublished
Priftis, G.D., D.L. Anastassopoulos, A.A. Vradis, and R. Suryanarayanan (1994b). *Physica C* **223** 106
Ragot, S., J.-M. Gillet, and P.J. Becker (2002). *Phys. Rev. B* **65** 235115
Rabii, S., J. Chomilier, and G. Loupias (1989). *Phys. Rev. B* **40** 10105
Rajput, S.S., R. Prasad, R.M. Singru, W. Triftshäuser, A. Eckert. G. Kögel, S. Kaprzyk, and A. Bansil (1993). *J.Phys.: Condens. Matter* **5** 6419
Rath, J., C.S. Wang, R.A. Tauril, and J. Callway (1973). *Phys. Rev. B* **8** 5139
Redinger. J., R. Podloucky, S. Manninen, T. Pitkänen, and O. Aikala (1989). *Acta Crystallogr.* **A45** (1989) 478
Reed, W.A. and P. Eisenberger (1972). *Phys. Rev B* **6** 4596
Reed, W.A., P. Eisenberger, K.C. Pandey, and L.C. Snyder (1974). *Phys. Rev.* **B10** 1507
Ribberfors, R. (1975a). *Phys. Rev. B* **12** 2067
Ribberfors, R. (1975b). *Phys. Rev. B* **12** 3136
Rollason, A.J. and M.B.J. Woolf (1995). *J. Phys.: Condens. Matter* **7** 7939
Rollason, A.J., R.S. Holt, and M.J. Cooper (1083a). *J. Phys. F: Metal Phys.* **13** 1807
Rollason, A.J., R.S. Holt, and M.J. Cooper (1983b). *Phil. Mag. B* **47** 51
Rollason, A.J., J.R. Schneider, D.S. Laundy, R.S. Holt, and M.J. Cooper (1987). *J. Phys. F: Met. Phys.* **17** 1105
Sakai, N. (2004). In X-*ray Compton scattering* (ed. Cooper M.J., P.E. Mijnarends, N. Shiotani, N. Sakai, and A. Bansil; Oxford University Press) p. 289
Sakai, N. and K. Ono (1976). *Phys. Rev. Lett* **37** 351
Sakai, N. and H. Sekizawa (1981). *J. Phys. Soc. Jpn*, **50** 2606
Sakai, N., N. Shiotani, F. Itoh, O. Mao, M. Ito, H. Kawata, Y. Amemiya, and M. Ando (1989). *J. Phys. Soc. Jpn* **58** 3270
Sakai, N., Y. Tanaka, F. Itoh, H. Sakurai, H. Kawata, and T. Iwazumi (1991). *J. Phys. Soc. Jpn.* **60** 1201
Sakurai, Y., M. Ito, T. Urai, Y. Tanaka, N. Sakai, T. Iwazumi, H. Kawata, M. Ando, N. Shiotani (1992a). *Rev. Sci. Instrum.* **63** 1190
Sakurai, Y., S. Nanao, Y. Nagashima, T. Hyodo, T. Iwazumi, H. Kawata, M. Ito, N. Shiotani, and A.T. Stewart (1992b). *Materials Science Forum* 105–110 803
Sakurai, Y., Y. Tanaka, T. Ohata, Y. Watanabe, S. Nanao, Y. Ushigami, T. Iwazumi, H. Kawata, and N. Shiotani (1994). *J. Phys.: Condens. Matter* **6** 9469
Sakurai, Y., T. Tanaka, A. Bansil, S. Kaprzyk, A.T. Stewart, Y. Nagashima, T. Hyodo, S. Nanao, H. Kawata, N. Shiotani (1995). *Phys. Rev. Lett.* **74** 2252
Sakurai, Y., S. Kaprzyk, A. Bansil, Y. Tanaka, G. Stutz, H. Kawata, and N. Shiotani (1999). *J. Phys. Chem. Solids* **60** 905

Sattler, T., Th. Tschentscher, J.R. Schneider, M. Vos, A.S. Kheifets, D.R. Lun, E. Weigold, G. Dollinger, H. Bross, and F. Bell (2001). *Phys. Rev. B* **63** 155204

Saunders, V.R., R. Dovesi, C. Roetti, M. Causa, N.M. Harrison, R. Orlando, and C.M. Zichovich-Wilson (1998). *CRYSTAL98 User's Manual* (University of Turino, Torino)

Savrasov, S.Y. (1996). *Phys. Rev. B* **54** 16470

Schneider, J.R. (1974). *J. Appl. Crystallogr.* **7** 547

Schülke, W. (1974). *Phys. Stat. Sol.* (b) **62** 453

Schülke, W. (1977). *Phys. Stat. Sol.* (b) **82** 229

Schülke, W. (1999). *J. Phys. Soc. Jpn*, **68** 2470

Schülke, W. and B. Kramer (1979). *Acta Cryst.* **A35** 953

Schülke, W. and S. Mourikis (1983). *Nucl. Instrum. Meth.* **208** 593

Schülke, W. and S. Mourikis (1986). *Acta Cryst.* **A42** 86

Schülke, W., U. Berg, and O. Brümmer (1969). *Phys. Stat. Sol.* **35** 227

Schülke, W., S. Mourikis, and K.D. Liedtke (1984). *Nucl. Instrum. Meth.* **222** 266

Schülke, W., G. Stutz, F. Wohlert, and A. Kaprolat (1996). *Phys. Rev. B* **54** 14381

Schulz, H. and K. Schwarz (1978). *Acta Cryst. A* **34** 999

Seth, A. and D.E. Ellis (1977). *J. Phys. C: Solid State Phys.* **10** 181

Shiotani, N., N. Sakai, M. Ito, O. Mao, F. Itoh, H. Kawata, Y. Amemiya, and M. Ando (1989). *J. Phys.: Condens. Matter* **1** SA27

Shiotani, N., Y. Tanaka, Y. Sakurai, N. Sakai, M. Ito, F. Itoh, T. Iwazumi, and H. Kawata (1993). *J. Phys. Soc. Jpn* **62** 239

Shiotani, N., I. Matsumoto, H. Kawata, J. Katsuyama, M. Mizuno, H. Araki, and Y. Shirai (2004). *J. Phys. Soc. Jpn* **73** 1627

Shukla, A., B. Barbiellini, A. Erb, A. Manuel, Th. Buslaps, V. Honkimäki, and P. Suortti (1999). *Phys. Rev. B* **59** 12127

Shukla, A., E.D. Isaacs, D.R. Hamann, and P.M. Platzman (2001). *Phys. Rev. B* **64** 052101

Singru, R.M. (1973). *Phys. Lett. A* **46** 61

Smith, Jr, V.H., G.H. Diercksen, and W.P. Kraemer (1975). *Phys. Lett.* **54A** 319

Spiertz, A. (1996). *Experimenteller Zugriff auf Nichtdiagonalelemente der Dichtematrix der Valenzelektronen in Silizium*, Thesis, Dortmund

Stahle, C.K., D. Osheroff, L. Kelley, S. Moseley, and A.E. Szymkowiak (1992). *Nucl. Instrum. & Methods* **A319** 393

Sternemann, C., K. Hämäläinen, A. Kaprolat, A. Soininen, G. Döring, C.-C. Kao, S. Manninen, and W. Schülke (2000). *Phys. Rev. B* **62** R7687

Sternemann, C., T. Buslaps, A. Shukla, P. Suortti, G. Döring, and W. Schülke (2001). *Phys. Rev. B* **63** 094301

Stutz, G., F. Wohlert, A. Kaprolat, W. Schülke, Y. Sakurai, V. Tanaka, M. Ito, H. Kawata, N. Shiotani, S. Kaprzyk, and A. Bansil (1999). *Phys. Rev. B* **60** 7099

Suortti, P., T. Buslaps, P. Fajardo, V. Honkimäki, M. Kretzschmer, U. Lienert, J.E. McCarthy, M. Renier, A. Shukla, T. Tschentscher, and T. Meinander (1999). *J. Synchrotron Rad.* 6 69

Suortti, P., T. Buslaps, V. Honkimäki, C. Metz, A. Shukla, Th. Tschentscher, J. Kwiatkowska, F. Maniawski, A. Bansil, S. Kaprzyk, A.S. Kheifets, D.R. Lun, T. Sattler, J.R. Schneider and F. Bell (2000). *J. Phys. Chem. Solids* **61** 397

Suortti, P., T. Buslaps, M. DiMichiel, V. Honkimäki, U. Lienert, J.E. McCarthy, J.M. Merino, and A. Shukla (2001a). *Nucl. Instrum. and Meth. A* 467–468 1541

Suortti, P., T. Buslaps, V. Honkimäki, A. Shukla, J. Kwiatkowska, F. Maniawski, S. Kaprzyk, and A. Bansil (2001b). *J. Phys. Chem. Solids* **62** 2223

Suzuki, R. and S. Tanikawa (1989). In *Positron Annihilation*, ed. L. Dorikens-Vanpraet, M. Dorikens, and D. Segers (World Scientific, Singapore) p. 626

Switendick, A.C. (1971). *Int. J. Quantum Chem.* **5** 459

Switendick, A.C. (1972). *Ber. Bunsenges. phys. Chem.* **76** 535

Switendick, A.C. (1978). In *Hydrogen in Metals* I (ed. G. Alefeld and J. Völkl, Springer, Berlin) p. 101

Takada, Y. and H. Yasuhara (1991). *Phys. Rev. B* **44** 7879

Tanaka, Y., N. Sakai, Y. Kubo, and H. Kawata (1993). *Phys. Rev. Lett.* **70** 1537

Tanaka, Y., Y. Sakurai, S. Nanao, N. Shiotani, M. Ito, N. Sakai, H. Kawata, and T. Iwazumi (1994). *J. Phys. Soc. Jpn* **63** 3349

Tanaka, Y., K.J. Chen, C. Bellin, G. Loupias, H.M. Fretwell, A. Rodrigues-Gonzales, M.A. Alam, S.B. Dugdale, A.A. Manuel, A. Shukla, T. Buslaps, P. Suortti, and N. Shiotani (2000). *J. Phys. Chem. Solids* **61** 365

Tanaka, Y., Y. Sakurai, A.T. Stewart, N. Shiotani, P.E. Mijnarends, S. Kaprzyk, and A. Bansil (2001). *Phys. Rev. B* **63** 045120

Tanner, A.C. and I.R. Epstein (1974). *J. Chem. Phys.* **61** 4251

Thakkar, A.J. and H. Tatewaki (1990). *Phys. Rev. A* **42** 1336

Theodoridou, I. and N.G. Alexandropoulos (1984). *Z. Phys. B.-Condens. Matter* **54** 225

Thole, B.T., P. Carra, F. Sette, and G. van der Laan (1992). *Phys. Rev. Lett.* **68** 1943

Victoreen, J.A. (1948). *J. Appl. Phys.* **19** 855

Wachtel, S., J. Felsteiner, S. Kahane, and R. Opher (1975). *Phys. Rev. B* **12** 1285

Wahl, A.C. and G. Das (1966). *J. Chem. Phys.* **44** 87

Wakoh, S. and Y. Kubo (1977). *Technical Report of ISSP*, A, 807

Wakoh, S. and M. Matsumoto (1990). *J. Phys.: Condens. Matter* **2** 797

Wakoh, S., Y. Kubo, and J. Yamashita (1976). *J. Phys. Soc. Jpn* **40** 1043

Wang, C.S. and J. Callaway (1975). *Phys. Rev. B* **11** 2417.
Watson, R.E. (1958). *Phys. Rev.* **111** 1108
Weyrich, W. (1978). *Einige Beiträge zur Comptonspektroskopie* (Habilitationsschrift, Darmstadt) p. 155
Weyrich, W., P. Pattison, and B.G. Williams (1979). *Chem. Phys.* **41** 271
Yamamoto, S. and H. Kitamura (1987). *Jpn. J. Appl. Phys.* **26** L1613
Yamamoto, S., H. Kawata, H. Kitamura, M. Ando, N. Sakai, and N. Shiotani (1989). *Phys. Rev. Lett.* **62** 2672
Zukowski, E., L. Dobrzynski, M.J. Cooper, D.N. Timms, R.S. Holt, and J. Latuszkiewicz (1990). *J. Phys.: Condens. Matter* **2** 6315
Zukowski, E., S.P. Collins, M.J. Cooper, D.N. Timms, F. Itoh, H. Sakurai, H. Kawata, Y. Tanaka, and A. Malinowski (1993). *J. Phys.: Condens. Matter* **5** 4077

5

Resonant inelastic X-ray scattering (RIXS)

When Sparks jr. (1974), still using conventional X-ray sources, started to investigate resonant emission of X-rays, and Eisenberger *et al.* (1976a,b), using for the first time synchrotron radiation for inelastic X-ray scattering studies, detected the so-called Raman shift and the anomalous width of the resonant emission, nobody would believe that it was just RIXS, which became the most important field of inelastic X-ray scattering, at least for investigating actual materials of high practical importance. Of course, it is mainly the resonantly amplified excitation of the valence electron system at the core excited atom due to the Coulomb interaction of the virtual exciton with this highly correlated valence system, which bears the desired information about dispersion and direction dependence of site-selective excitation across, e.g. the Mott gap. It was Platzman and Isaacs (1998) who stressed the general importance of those shakeup processes. Since then, this type of spectroscopy has experienced a real boom. We will come in detail to these very interesting applications of RIXS later. But also the fact that absorption and reemission processes of RIXS are intimately coupled so that they have to be considered as a unified inelastic scattering event leads to characteristic properties of this spectroscopy which have a wide application potential. Thus momentum conservation, valid for the whole scattering process, has the consequence that conservation of the Bloch-**k** vectors involved in the absorption and re-emission process must be claimed, so that RIXS becomes Boch-**k** selective (Johnson and Ma 1994). Moreover, spin conservation of the whole resonant X-ray scattering process can be utilized to make the absorption process spin selective, if the spin characteristic of the re-emission process is known (Hämäläinen *et al.* 1992). Last but not least, also the so-called Raman shift of the resonant fluorescence emission, which was the special feature of RIXS first detected (Eisenberger *et al.* 1976a,b), must be traced back to the intimate knotting between absorption and re-emission, and delivers a means to suppress the core-hole lifetime broadening of conventional absorption spectroscopy, when the signal is taken from the properly energy-analyzed re-emitted photon of a corresponding RIXS process (Hämäläinen *et al.* 1991). Of course, also the rich variety of coupling polarization states of incident and re-emitted photons, as shown in detail by van Veenendaal (1996) and Luo *et al.* (1994), gains special attention and will find more and more applications, when combined with multipolar expansion of the RIXS cross-section and symmetry assignment of orbitals.

In the following sections of the book we shall deal with all these aspects and applications of RIXS spectroscopy so far as they are related to harder X-rays. Only in some exceptional cases, where the initial experiments were soft X-ray based and later on extended to harder X-rays, will we also refer to these pioneering applications. The reason for this restriction is the bulk sensitivity we claim for all spectroscopic applications in this book.

5.1 Basics of RIXS, the Kramers–Heisenberg formula

We will start our considerations about the basics of RIXS with the generalized Kramers–Heisenberg formula for the double differential scattering cross-section as derived in Section 6.2 within the limits of the second-order perturbation treatment, which was necessary, since the Hamiltonian describing the interaction between the photon field and the scattering electron system contains terms linear and those quadratic in the vector potential $\mathbf{A}$. We will specify the expression (6.21) for the spinless case, thus writing the double differential cross-section for a scattering process connected with the transition of the electron system from the ground state $|i\rangle$ to the final state $|f\rangle$

$$\left(\frac{d^2\sigma}{d\Omega_2 d\hbar\omega_2}\right)_{|i\rangle \to |f\rangle}$$

$$= \left(\frac{e^2}{mc^2}\right)^2 \left(\frac{\omega_2}{\omega_1}\right) \bigg| \langle f| \sum_j \exp(i\mathbf{q}\cdot\mathbf{r}_j)|i\rangle (\mathbf{e}_1\cdot\mathbf{e}_2^*) + \left(\frac{\hbar^2}{m}\right)$$

$$\times \sum_n \sum_{jj'} \bigg\{ \frac{\langle f|(\mathbf{e}_2^*\cdot\mathbf{p}_j/\hbar)\exp(-i\mathbf{K}_2\cdot\mathbf{r}_j)|n\rangle \langle n|(\mathbf{e}_1\cdot\mathbf{p}_{j'}/\hbar)\exp(i\mathbf{K}_1\cdot\mathbf{r}_{j'})|i\rangle}{E_i - E_n + \hbar\omega_1 - i\Gamma_n/2}$$

$$+ \frac{\langle f|(\mathbf{e}_1\cdot\mathbf{p}_j/\hbar)\exp(i\mathbf{K}_1\cdot\mathbf{r}_j)|n\rangle \langle n|(\mathbf{e}_2^*\cdot\mathbf{p}_{j'}/\hbar)\exp(-i\mathbf{K}_2\cdot\mathbf{r}_{j'})|i\rangle}{E_i - E_n - \hbar\omega_2} \bigg\} \bigg|^2$$

$$\times \delta(E_i - E_f + \hbar\omega), \tag{5.1}$$

where $|i\rangle$, $|f\rangle$ and $|n\rangle$ are the initial, final and intermediate states of the scattering electron system with energies E_i, E_f and E_n, respectively. $\mathbf{K}_1$, ω_1 and $\mathbf{e}_1$ are the wavevector, frequency and polarization unit vector of the incident, and $\mathbf{K}_2$, ω_2 and $\mathbf{e}_2$ the wavevector, frequency and polarization unit vector of the scattered photon field, respectively. The summation j, j' is over the electrons of the scattering system. $\mathbf{q} = \mathbf{K}_1 - \mathbf{K}_2$ and $\omega = \omega_1 - \omega_2$. For an adequate handling of the resonance denominator in the second term of (5.1) we have introduced Γ_n which stands for the energy broadening of the resonance due to the finite lifetime of the intermediate state.

The resonant inelastic scattering (RIXS) experiments we will describe in the following section are assumed to be arranged such that $(E_\text{i} - E_n)$ are nearly equal to the incident photon energy $\hbar\omega_1$. In this case the first and the third contribution to the double differential scattering cross-section within $|\ldots|^2$ of (5.1) can be neglected compared with the second, the resonating part. Thus the double differential scattering cross-section for the spin-free case writes:

$$\left(\frac{d^2\sigma}{d\Omega_2 d\hbar\omega_2}\right) = \left(\frac{e^2}{mc^2}\right)^2 \left(\frac{\omega_2}{\omega_1}\right) \left|\sum_\text{f} \left(\frac{\hbar^2}{m}\right) \sum_n \sum_{jj'} \langle \text{f}|\mathbf{e}_2^* \cdot \mathbf{p}_j/\hbar \, \exp(-i\mathbf{K}_2 \cdot \mathbf{r}_j)|n\rangle \right.$$

$$\left. \times \langle n|\mathbf{e}_1 \cdot \mathbf{p}_{j'}/\hbar \, \exp(i\mathbf{K}_1 \cdot \mathbf{r}_{j'})|\text{i}\rangle / (E_\text{i} - E_n + \hbar\omega_1 - i\Gamma_n/2) \right|^2$$

$$\times \delta(E_\text{i} - E_\text{f} + \hbar\omega). \tag{5.2}$$

It must be stressed that, due to the short lifetime of the intermediate state $|n\rangle$, the excitation process $|\text{i}\rangle \to |n\rangle$ is not limited by energy conservation. Only the whole resonant scattering process must conserve energy (see the δ-function in 5.2). Figure 1.4 of Section 1.2 has already shown a resonant scattering process in the single-particle representation. In that case the intermediate state $|n\rangle$ is characterized by a hole in a deeper-lying inner-shell level and an electron in a formerly unoccupied level. Γ_n now stands for the finite lifetime of the core hole. The final state is connected with a hole in a higher-lying level (inner-shell level or a valence electron state) and an electron in a formerly unoccupied level. Therefore, the final state is the same as the one which would be governed by the $\mathbf{A}^2$ term in the nonresonant case. The limiting case of anomalous elastic scattering is reached, when the hole in the deeper-lying core level is filled by the excited electron in the former unoccupied level. Contrary to the nonresonant scattering process, resonant inelastic scattering possesses a much larger variety of combining polarization and level characteristics of the intimately coupled excitation ($|\text{i}\rangle \to |n\rangle$) and de-excitation (re-emission) processes, which are the matter of many applications as shown in the following sections. Of course, the direct connection between the double differential scattering cross-section and correlation functions, which was the main object of studies in Chapter 2, gets lost when switching to the resonant case.

Often (see, e.g. Åberg and Tulkki 1985) one finds equation (5.2) written for the case that the excitation process ejects the electron from the 1s core level into a continuum level with kinetic energy $\hbar\varepsilon$, so that the summation over both the intermediate states and the final states can be replaced by an integral with respect to ε by introducing into (5.2) the density of the continuum states (DOS). Furthermore, the matrix elements in (5.2) are written in terms of the corresponding oscillator strengths. Assuming a final state, which is connected with a hole in the np_j core level, then we end up with

$$\frac{d^2\sigma}{d\Omega_2 d\omega_2} = \frac{(e^2/mc)^2}{2} \int_0^\infty \left(\frac{\omega_2}{\omega_1}\right) \left((\Omega_{1s} - \Omega_{np_j})g_{np_j,1s}(\Omega_{1s} + \varepsilon)\frac{dg_{1s}}{d\varepsilon}\right)$$
$$\times \left(\frac{1}{(\Omega_{1s} + \varepsilon - \omega_1)^2 + \Gamma_{1s}^2/4\hbar^2}\right) \delta(\omega_1 - \Omega_{np_j} - \varepsilon - \omega_2)\, d\varepsilon, \quad (5.3)$$

where Ω_{1s} and Ω_{np_j} are the threshold frequencies of the core levels involved in the intermediate and the final state, respectively. $g_{np_j,1s}$ is the oscillator strength of the transition between the $(1s)^{-1}$ and $(np_j)^{-1}$ hole state, $\hbar\varepsilon$ the kinetic energy of the ejected electron. Γ_{1s} is the lifetime-induced width of the 1s level (FWHM), and $dg_{1s}/d\varepsilon$ is the oscillator density proportional to the density of empty states and to the transition matrix element.

The final-state lifetime can be included by replacing the δ-function in (5.3) by the normalized density function $dN_f/d\omega_2$ with the FWHM of Γ_{np_j}, where

$$\frac{dN_f}{d\omega_2} = \frac{(\Gamma_{np_j}/2\pi\hbar)}{(\omega_1 - \Omega_{np_j} - \varepsilon - \omega_2)^2 + \Gamma_{np_j}^2/4\hbar^2}, \quad (5.4)$$

so that the double differential scattering cross section writes

$$\frac{d^2\sigma}{d\Omega_2 d\omega_2} = \frac{(e^2/mc^2)^2}{2} \int_0^\infty \left(\frac{\omega_2}{\omega_1}\right) \left((\Omega_{1s} - \Omega_{np_j})g_{np_j,1s}(\Omega_{1s} + \varepsilon)\frac{dg_{1s}}{d\varepsilon}\right) \frac{\Gamma_{np_j}}{2\pi\hbar}$$
$$\times ([(\Omega_{1s} + \varepsilon - \omega_1)^2 + \Gamma_{1s}^2/4\hbar^2]$$
$$\times [(\omega_1 - \Omega_{np_j} - \varepsilon - \omega_2)^2 + \Gamma_{np_j}^2/4\hbar^2])^{-1}\, d\varepsilon. \quad (5.5)$$

The content of (5.5) is summarized graphically in Fig. 5.1, which was taken from Hayashi et al. (2003): Fig. 5.1(a) depicts the 1s2p RIXS process. In Fig. 5.1(b) K and L-shell levels have lifetime-induced Lorentzian distributions f_1 and f_2, centered on the ε-axis at $\omega_1 - \Omega_{1s}$ and $\omega_1 - \omega_2 - \Omega_{2p}$, respectively. In a RIXS measurement, ω_1 is fixed and the scattering intensity is monitored as a function of ω_2, which is equivalent to sliding the very sharp Lorentzian f_2 along the ε-axis while keeping f_1 and the oscillator density $dg_{1s}/d\varepsilon$ fixed. The oscillator density in Fig. 5.1(b) is characterized by a narrow discrete band corresponding to vacant 3d states followed by a continuum of empty 4p states, simulating a copper compound. The scattering intensity is governed by the overlap of these three functions f_1, f_2 and $dg_{1s}/d\varepsilon$. Three different cases are possible: $\omega_1 > \Omega_{1s}$ (normal fluorescence), $\omega_1 = \Omega_{1s}$ (transition to RIXS) and $\omega_1 = \Omega_{1s}$ (RIXS). In

Basics of RIXS, the Kramers–Heisenberg formula

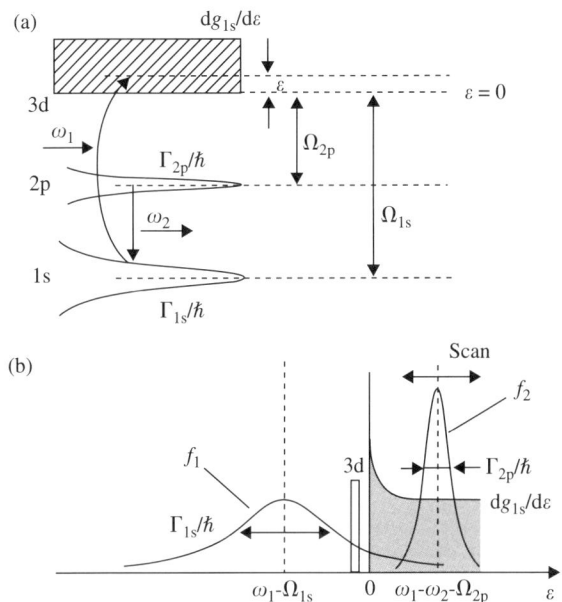

Fig. 5.1. (a) A diagram of the 1s2p resonant inelastic X-ray scattering (RIXS) process. (b) A schematic presentation of a RIXS measurement, where the sharp 2p Lorentzian f_2 is sliding along the ε-axis while keeping the 1s Lorentzian f_1 and the oscillator density $dg_{1s}/d\varepsilon$ fixed. (Reprinted with permission from Hayashi et al. (2003); copyright (2003) by the American Physical Society.)

Fig. 5.1 the RIXS case is depicted. If we consider only the continuous part of the oscillator density, the RIXS line $I(\omega_2)$ corresponding to the (2p,1s) transition, shown schematically in Fig. 5.2, is characterized by a strong asymmetry with a sharp rise at $\hbar\omega_2 = \hbar(\omega_1 - \Omega_{2p})$ followed by a $(\Delta\Omega_{KL} - \omega_2)^{-2}$ fall-off to lower energies, where $\Delta\Omega_{KL} = \Omega_{1s} - \Omega_{2p}$. This corresponds to sharing of energy between the outgoing photon $\hbar\omega_2$ and the electron with energy $\hbar\varepsilon$.

If we adhere to the δ-function in (5.3) we can write this equation for the scattering by $2p_j$ ($j = 1/2, 3/2$) electrons, using $\omega_2 + \varepsilon = \omega_1 - \Omega_{np_j}$; $\Delta\Omega_{KL} \equiv \Omega_{1s} - \Omega_{np_j}$; $\omega_{abs} \equiv \Omega_{1s} + \varepsilon = \Delta\Omega_{KL} + \omega_1 - \omega_2$, and $g_{LK} = g_{np_j,1s}$

$$\frac{d^2\sigma}{d\Omega_2 d\omega_2} = \left(\frac{e^2}{mc^2}\right)^2 \left(\frac{\omega_2}{\omega_1}\right)$$

$$\times \left(\frac{\Delta\Omega_{KL}\, g_{LK}(\Omega_{1s} + \varepsilon)}{2[(\Delta\Omega_{KL} - \omega_2)^2 + \Gamma_{1s}^2/4\hbar^2]}\right) \left(\frac{dg_{1s}}{d\varepsilon}\right)_{\Omega_{1s}+\varepsilon}, \qquad (5.6)$$

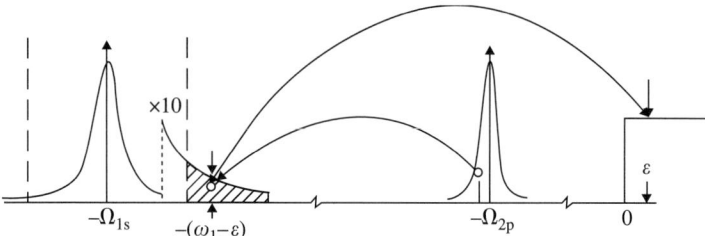

Fig. 5.2. The formation of the RIXS spectrum (hatched) for the case $\omega_1 < \Omega_{1s}$. Only the continuous part of the oscillator density is taken into account. (Reprinted with published by Hämäläinen *et al.* (1989); copyright (1989) by IOP Publishing Ltd.)

where we have assumed that $g_{\mathrm{np}_j,1s}$ is independent of ε, so that it can be taken out of the ε-integration in (5.5).

5.2 Instrumentation for RIXS

In principle the instrumentation for RIXS experiments is composed of the same three main parts as other inelastic X-ray scattering setups described in Chapters 2 and 3, namely monochromator (focusing devices), scattering sample environment, analyzing system, which fixes also the scattering angle. However, since RIXS needs to approach the resonance condition, the incident energy must be tunable to the core binding energy of as many elements as possible. Therefore, all instruments which are exclusively based on the so-called inverse geometry, where the photon energy transmitted by the analyzer is held fixed, whereas the monochromator is tuned, drop out. The desired overall energy resolution depends strongly on the physics under investigation. The resolution described in the literature ranges from 100 meV for the measurements of the dispersion of shakeup processes to one or even more eV for investigations of the resonant fluorescence emission as a probe for core excitation processes. High momentum resolution is required for dispersion measurement of excitations in highly correlated materials as well as for the utilization of Bloch-**k** conservation for band mapping. Special degrees of freedom for the moving parts of an inelastic X-ray scattering device are necessary if the direction of the incident polarization $\mathbf{e}_1$ and of the vector $\mathbf{q}$ of momentum transfer should be kept fixed with respect to (crystal) coordinates of the sample, when the scattering angle is changed. Since the incident synchrotron radiation is mainly polarized in the horizontal plane this requirement can be met only with a vertical scattering geometry, where the sample is rotated through an axis parallel to $\mathbf{e}_1$ and perpendicular to $\mathbf{q}$. The same vertical scattering geometry is required, when, as shown in Fig. 1.18, the polarization $\mathbf{e}_1$ of the incident radiation prefers the excitation of molecules with a certain orientation relative to $\mathbf{e}_1$, which leads to a distinct emission characteristic of aligned molecules. The

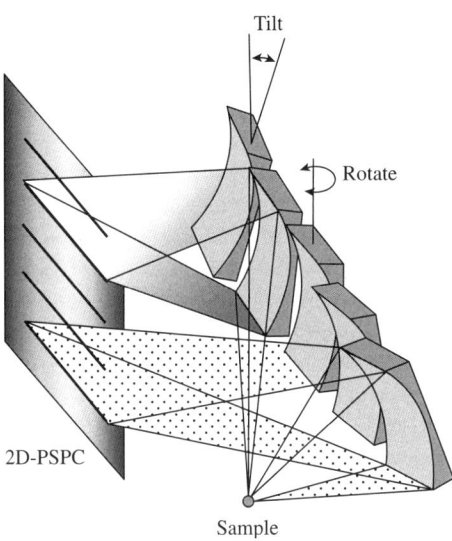

Fig. 5.3. Scheme of the dispersive-type five-crystal spectrometer connected with a two-dimensional detector. (Reprinted with permission from Hayashi *et al.* (2004); copyright (2004) by Elsevier B.V.)

latter can then be observed by means of a polarization analysis of the re-emitted radiation utilizing diffraction under a 45° Bragg angle.

A special setup, particularly devoted to very effective RIXS investigations, is described by Hayashi *et al.* (2004), and is installed at the hard X-ray undulator beamline BL47XU at SPring-8. This dispersive-type five-crystal spectrometer, shown schematically in Fig. 5.3, utilizes the properties of a cylindrically bent analyzer in the so-called von Hamos geometry, as described in Section 2.2.5.3. The five crystals make a captured solid angle of 0.06 sr. The cylindrical axis is in the diffraction plane of the crystals. The crystals are rotated around an axis perpendicular to the cylindrical axis in such a way that each crystal is Bragg reflecting the same spectral range of the radiation scattered by the sample. By tilting the crystals around the cylindrical axis, the spectrum Bragg-reflected by each crystal is exposed as a horizontally dispersed pattern on a 2D position-sensitive multiwire detector so that these pattern are offset vertically. Finally these patterns were converted to a one-dimensional spectrum by adding them together. The gas-filled wire detector has an imaging area of 115 mm diameter and can collect a data frame of 1024×1024 pixels. The detector is sensitive to X-ray photons between 3 and 15 keV with a high quantum efficiency (80% at 8 keV). The dynamic range was $> 10^6$ and the maximum countrate per pixel is about 200 cps. Thus an instrument is available which can satisfy the demands of RIXS upon high statistical accuracy and a rather large ω_1, ω_2 range covered by experiment in sufficiently small steps $d\omega_1, d\omega_2$.

5.3 Investigation of core excitations using RIXS

5.3.1 Determination of core-hole lifetime

We have seen that, within the single-particle limit, the intermediate state $|n\rangle$ of an RIXS process is directly connected with a core hole, so that the energy width Γ_n in (5.2) was determined by the core-hole lifetime, as found by going from (5.2) to (5.5). One can therefore utilize the resonance denominator of (5.6) to get experimental access to the quantity Γ_K. It was first proposed by Suortti (1979) to integrate (5.6) with respect to ω_2 within the limits $0 \leq \omega_2 \leq \omega_1 - \Omega_L$, in order to get the following approximate expression for the ω_1 dependence of the differential scattering cross-section. We perform this integration for $\Omega_{1s} - \omega_1 > \Gamma_{1s}$, so that we are sufficiently far below the threshold. Then $dg_{1s}/d\varepsilon$ can be replaced by the value that corresponds to the average energy of the ejected electron, $\bar{\varepsilon}$, when the cross-section is calculated (constant DOS model). Thus we obtain approximately

$$\left(\frac{d\sigma}{d\Omega_2}\right)_{KL} \approx \frac{(e^2/mc^2)^2}{\omega_1} \left[\frac{2\hbar(\Omega_{1s} + \bar{\varepsilon})(\Delta\Omega_{KL})^2}{\Gamma_{1s}}\right] g_{2p,1s} \left(\frac{dg_K}{d\varepsilon}\right)_{\Omega_{1s}+\bar{\varepsilon}}$$
$$\times \tan^{-1}(\Gamma_{1s}/2\Delta E). \tag{5.7}$$

Here $\hbar\Omega_{1s}$ is the K-shell binding energy, $\hbar\bar{\varepsilon}$ is the average kinetic energy of the ejected electron, ΔE is the incident energy relative to the K-edge ($\Delta E \equiv \hbar(\Omega_{1s} - \omega_1)$), and Γ_{1s} the K-shell hole width, due to its finite lifetime. If we take (5.7) in the limit $(\Omega_{1s} - \omega_1) \to 0$, assuming isotropy of the scattering system, and multiply (5.7) by 4π, we obtain the KL contribution to the total cross-section $\sigma_{1s}(\Omega_{1s} + \bar{\varepsilon})$ for photoelectric absorption:

$$\sigma_{1s}(\Omega_{1s} + \bar{\varepsilon}) = 4\pi^2 \left[\frac{(e^2/mc^2)^2}{\omega_1}\right] \left[\frac{\hbar(\Omega_{1s} + \bar{\varepsilon})(\Delta\Omega_{KL})^2}{\Gamma_{1s}}\right] g_{2p,1s} \left(\frac{dg_{1s}}{d\varepsilon}\right)_{\Omega_{1s}+\bar{\varepsilon}}. \tag{5.8}$$

This expression can be inserted into (5.7) ending up with

$$\left(\frac{d\sigma}{d\Omega_2}\right)_{KL} \approx \left(\frac{1}{4\pi^2}\right) 2\sigma_{1s}(\Omega_{1s} + \bar{\varepsilon}) \tan^{-1}(\Gamma_{1s}/2\Delta E), \tag{5.9}$$

or by introducing the total RIXS cross-section of KL processes, $\sigma_{R,KL}(\omega_1)$, again assuming an isotropic scattering system

$$\sigma_{R,KL}(\omega_1) = 4\pi \left(\frac{d\sigma}{d\Omega_2}\right)_{KL} \tag{5.10}$$

we obtain the following relation

$$\Delta E = \frac{1}{2}\Gamma_{1s} \cot\left[\frac{\pi\sigma_{R,KL}(\omega_1)}{2\sigma_{1s}(\Omega_{1s} + \bar{\varepsilon})}\right] \tag{5.11}$$

which will provide an appropriate plot for determining Γ_{1s}.

Of course, also the RIXS cross-section for other than KL processes can be deduced in the same manner, leading to corresponding expressions, e.g. for LM RIXS, where the lifetime broadening Γ_{2p_j} plays the decisive role.

One has to be aware that the cross-section for the photoelectric absorption as used in (5.11) is the **total** cross-section, which is, due to the Auger effect, different from the corresponding radiative or fluorescence cross-section $\sigma_{1s,f}$. Both are related via the so-called fluorescence yield w_K:

$$\sigma_{1s,f} = w_K \sigma_{1s}. \tag{5.12}$$

Given the similarity to the fluorescence, the RIXS cross-section is also expected to be reduced by the Auger effect, so that we can write for the radiative RIXS cross-section

$$\sigma_{R,KL,r} = w_{RIXS}\, \sigma_{R,KL}, \tag{5.13}$$

where w_{RIXS} is the corresponding RIXS yield. In the following we will assume that it is, within the whole RIXS energy range, equal to w_K, since this must be true at the limit of fluorescence $\omega_1 = \Omega_{1s}$. Experimental values of w_{RIXS} (Suortti 1979, Manninen et al. 1986) are rather scarce, and subject to rather large errors, but show the tendency to be equal to w_K, at least not too far from the threshold.

Hämäläinen et al. (1989) have used (5.11) to determine experimentally the K-shell hole width Γ_{1s} of Cu and Zn, as well as the L-shell hole width $\Gamma_{2p_{3/2}}$ of Ho. In a later experiment (Hämäläinen et al. 1990) the L-shell hole widths in Yb and Ta were measured. These authors employed synchrotron radiation from a bending magnet of the Daresbury storage ring, monochromatized by a

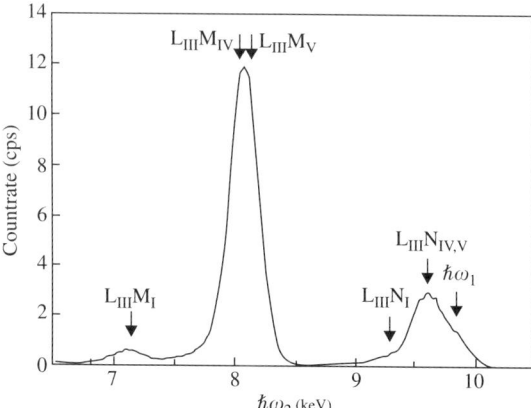

Fig. 5.4. Typical RIXS spectrum of Ta. The incident photon energy is 28 eV below the L_{III} absorption edge. The edge position corresponding to transitions involving the L_{III} hole are indicated by arrows. (Reprinted with permission from Hämäläinen et al. (1990); copyright (1990) by IOP Publishing Ltd.)

Si(111) channel-cut monochromator having a resolution of about 1 eV. The incident energies were tuned from about 1 keV below the desired edge up to about 300 eV above the edge. The scattered radiation was measured at the scattering angle of 90° in the plane of the storage ring. Due to the polarization of the incident radiation, which is directed mainly in the storage ring plane, this scattering angle minimizes the elastic and inelastic scattering arising from the $\mathbf{A}^2$ term which would overlap the RIXS spectrum. The scattered radiation was measured using a Si detector with an energy resolution of 200 eV at 10 keV and a constant efficiency in the range of interest. As an example Fig. 5.4 shows a typical RIXS spectrum from Ta, where the incident radiation is 28 eV below the L_{III} absorption edge at 9.811 keV. The edge positions corresponding to different transitions involving the L_{III}-shell hole are indicated by arrows. A small amount of elastic scattering is visible around 9.8 keV. The radiative RIXS cross-section was deduced from the measured countrate of scattered radiation, integrated over the relevant part of the spectrum; this means including all lines corresponding to transitions under investigation (KL transition, when investigating Γ_{1s} and $L_{III}M$ transitions, when measuring $\Gamma_{2p_{3/2}}$). The countrate was normalized to the current of the ion chamber monitoring the incident photon flux, where the energy dependence of the ion chamber efficiency was taken into account. Additionally the countrates were corrected with respect to the energy dependence of the sample absorption using absolute values of the absorption coefficient from the literature. The RIXS cross-sections were not brought to an absolute scale, since in the analysis according to (5.11), in line with the assumptions concerning the yield factors, only the radiative RIXS cross-section relative to the fluorescence

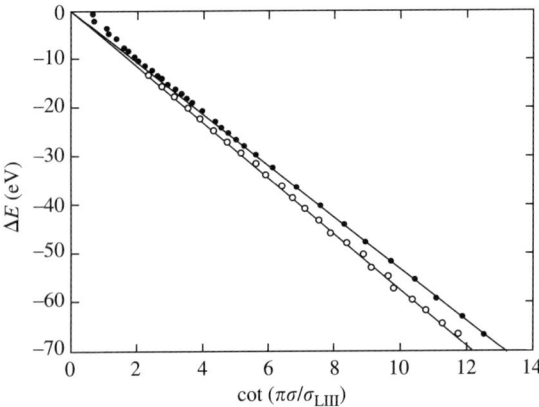

Fig. 5.5. Plot of the experimental cross-section according to equation (5.11). Upper line: Yb; lower line: Ta. The linewidth Γ is read from the slope of the curves. (Originally published by Hämäläinen *et al.* (1990); copyright (1990) by IOP Publishing Ltd.)

Table 5.1. Lifetime core width (eV) (Hämäläinen *et al.* (1989, 1990))

Element	Lifetime core width (eV)
Cu	$\Gamma_{1s} = 1.5 \pm 0.1$ eV
Zn	$\Gamma_{1s} = 1.9 \pm 0.1$ eV
Ho	$\Gamma_{2p_{3/2}} = 4.8 \pm 0.2$ eV
Yb	$\Gamma_{2p_{3/2}} = 5.3 \pm 0.2$ eV
Ta	$\Gamma_{2p_{3/2}} = 5.7 \pm 0.2$ eV

cross-sections is necessary. The determination of the latter was, in principle, performed as that of the radiative RIXS cross-section, but needs special care because the EXAFS oscillations and the near-edge fine structure had to be smoothed by means of a linear fit. How important a reliable value of the fluorescence cross-section is might be elucidated by the fact that a change as small as 3% of the full fluorescence cross-section leads to a ±0.2 eV difference in the Γ value. According to (5.11) the latter was obtained from the slope of a plot of ΔE as a function of $\cot[\pi\sigma_{R,KL,r}(\omega_1)/2\sigma_{1s,f}(\Omega_{1s} + \bar{\varepsilon})]$. A plot of this kind is shown in Fig. 5.5. Table 5.1 offers the results of the above two studies.

5.3.2 The evolution of RIXS spectra with incident photon energy

In order to understand the basic information about electronic excitations offered by RIXS spectra let us go back to (5.6) and Fig. 5.1, where the RIXS process was illustrated for a 2p → 1s re-emission, the oscillator strength g_{LK} of which is assumed to be independent on ε. The more complicated situation, where the final states are characterized by multiplets due to the interaction between the electron excited into an unoccupied level and the final-state core-hole, will be left for the following sections. Here we shall discuss the basic features of RIXS spectra when they are evolving with increasing incident-photon energy. If we rewrite (5.5) in order make transparent how the scattering cross-section depends on ω_2:

$$\frac{d^2\sigma}{d\Omega_2 d\omega_2} = \left(\frac{e^2}{mc^2}\right)^2 \left(\frac{\omega_2}{\omega_1}\right)$$

$$\times \left(\frac{\Delta\Omega_{KL}\, g_{LK}(\Omega_{1s} + \varepsilon)}{2(\Delta\Omega_{KL} - \omega_2)^2 + \Gamma_{1s}^2/4\hbar^2}\right) \left(\frac{dg_{1s}}{d\omega_2}\right)_{\omega_1 - \varepsilon - \Omega_{2p}} \quad (5.14)$$

then (5.14) tells us that the intensity distribution $I(\omega_2)$ of the RIXS spectrum is, apart from the prefactor $(\omega_2/\omega_1)\Delta\Omega_{KL}(\Delta\Omega_{KL} + \omega_1 - \omega_2)g_{LK}$ proportional to $(dg_{1s}/d\omega_2)_{\omega_1-\varepsilon-\Omega_{2p}}$ multiplied by a Lorentzian centered at $\Delta\Omega_{KL}$. Let us start the discussion of (5.14) with a case where the incident photon energy might be $10 \times \Gamma_{1s}$ smaller than the 1s binding energy $\hbar\Omega_{1s}$, so that the oscillator density with its threshold at $\omega_1 - \Omega_{2p}$ on the ω_2 scale is overlapping only the slowly varying tail of the Lorentzian. Since also the prefactor is only a slowly varying function of ω_2, the RIXS spectrum is a nearly perfect image of the oscillator density notwithstanding the small distortion by the Lorentzian tail and the prefactor. In (5.14) we have neglected the influence of the 2p core lifetime. Therefore the oscillator density will appear convoluted with the 2p core lifetime-induced Lorentzian of (5.4). However, the spectral intensity of these RIXS spectra will be rather low. Continuing the discussion of (5.14) we shall assume in the first place that the oscillator density is a continuum with a distinct threshold. Let us now increase ω_1 so that the incident photon energy is approaching $\hbar\Omega_{1s}$. The distortion of the oscillator density by the 1s Lorentzian becomes more important and leads to a RIXS spectrum, which is strongly asymmetric with a peak near the threshold at $\omega_2 = \omega_1 - \Omega_{2p}$. The peak position on the ω_2 scale shifts with ω_1 (so-called Raman shift, see Section 6.10.1) until ω_1 has reached Ω_{1s}. Just at $\omega_1 = \Omega_{1s}$ the shape of the RIXS spectrum is the half of the 1s Lorentzian halved by the threshold, so that its width is only one-half of the width of the fluorescence line. Increasing ω_1 furthermore, the RIXS spectrum is governed more and more by the shape of the 1s Lorentzian and, for $\omega_1 > \Omega_{1s}$, its peak stays, according to (5.14), at $\omega_2 = \Delta\Omega_{KL}$, and the normal fluorescence case is reached. This behavior of the RIXS spectrum was experimentally confirmed for the first time by Eisenberger et al. (1976a) with an experiment on Cu metal. As shown in Fig. 5.6, where the energy distance of the RIXS peak from the normal fluorescence peak, ΔE_0, is plotted as a function of the distance ΔE_R of the incident photon energy from the Cu K edge, the linear relation between both exhibits clearly the Raman-shift characteristic for transitions into a continuum. Moreover, the FWHM of the resonantly excited emission lines follows the predictions of the theory with a minimum width just when the incident photon energy reaches the 1s binding energy. If the continuum of the oscillator density possesses additional thresholds, additional peak structures in the RIXS spectrum can appear.

In those cases where the oscillator density is, in addition to the continuum, characterized by one or more narrow discrete bands, the RIXS spectrum reflects also this discrete nature by exhibiting additional peaks with the following three characteristic properties:

(i) The intensity of these peaks, as a function of the incident photon energy $\hbar\omega_1$, shows a resonance behavior with a maximum at E_d, the energy necessary for exciting a 1s electron into the corresponding discrete level.

(ii) The peak position of maximum intensity is, according to (5.14), always found at $\omega_2 = \Delta\Omega_{KL}$.

(iii) Contrary to the behavior of a continuum, the Raman shift of a discrete level induced peak on the ω_2-scale persists also for $\hbar\omega_1 > E_d$.

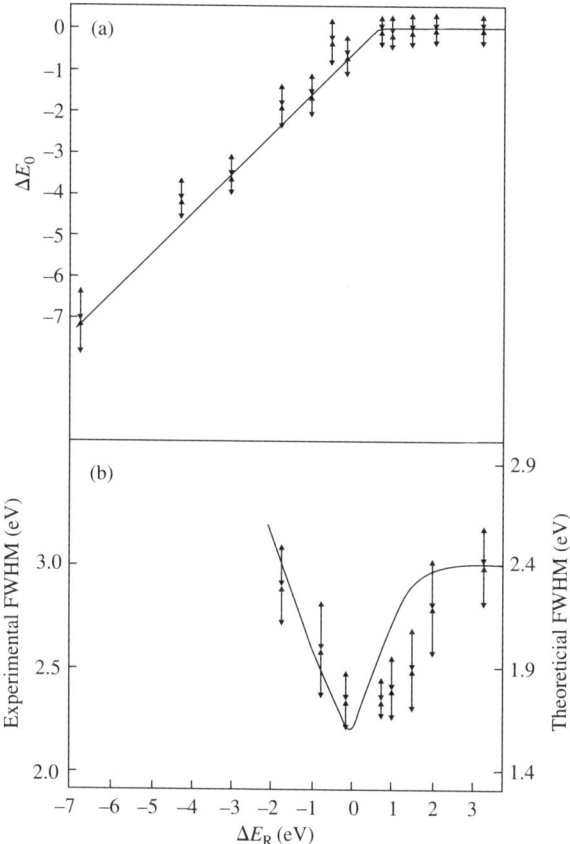

Fig. 5.6. (a) The dispersion ΔE_0, and (b) the linewidth FWHM as a function of ΔE_R. Experimental results: points; theoretical calculations: line. Note the separate scales for the experimental and theoretical linewidth. (Originally published by Eisenberger *et al.* (1976); copyright (1976) by the American Physical Society.)

These characteristics make the discrete level induced features of the RIXS spectra distinguishable from the continuum induced ones. This special behavior of the Raman shift has, for the first time, found its experimental confirmation with an investigation of Briand *et al.* (1981) performed at LURE. It was the so-called white line in the manganese K-absorption spectrum of $KMnO_4$, a large peak near the absorption discontinuity, which played the role of the discrete band in the corresponding oscillator density. This white line is interpreted (Best 1966) as being due to an excitation of a manganese 1s electron to a $3t_2$ molecular orbital of the $(MnO_4)^-$ ion with predominantly 4p character. Observing the 2p → 1s ($\alpha_{1/2}$) re-emission when varying the incident photon energy in the vicinity of

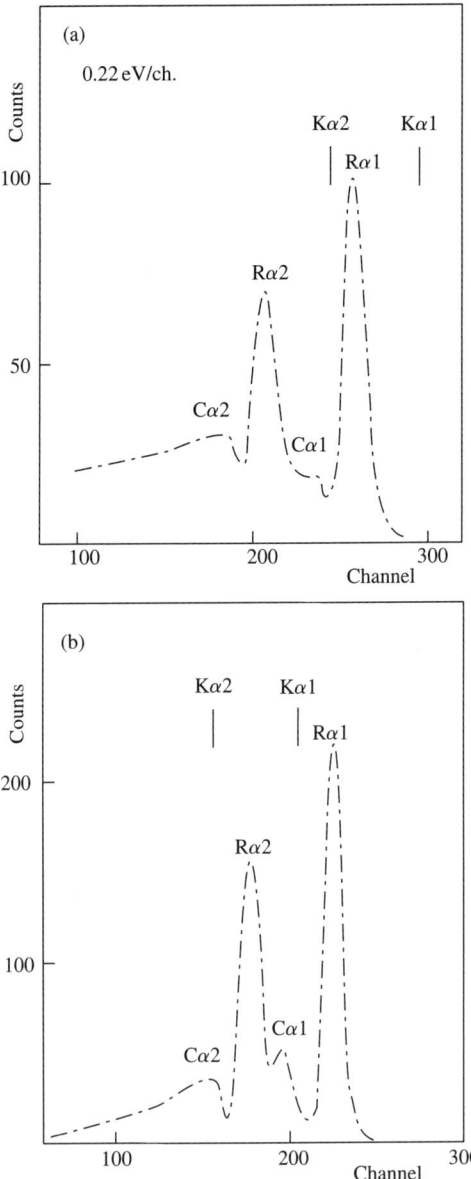

Fig. 5.7. Raman (Rα_1 and Rα_2) and Compton (Cα_1 and Cα_2) spectrum after subtraction of the X-ray Kα_1 and Kα_2 line at two different incident photon energies: (a) The incident photon energy $\hbar\omega_1$ is 8 eV smaller than the excitation energy of the white line (b) $\hbar\omega_1$ exceeds by 3.6 eV the white line excitation energy. (Originally published by Briand *et al.* (1981); copyright (1981) by the American Physical Society.)

the white line, the authors found in addition to RIXS structures attributed to the continuous part of the absorption spectrum (they called it $C\alpha_{1/2}$ (Compton) structures), peak-like structures, which they called $R\alpha_{1/2}$ (Raman) peaks. These peaks exhibited a strong resonant behavior, when the incident photon energy passed the excitation energy of the white line. Moreover, as shown in Fig. 5.7, the energetic positions of these peaks are at energies smaller than the normal fluorescence $\alpha_{1/2}$ lines, when the incident photon energy was 8 eV smaller than the excitation energy of the white line, but are found at energies larger than that of the normal fluorescence, when the incident photon energy exceeded the white-line excitation energy by 3.6 eV. This finding was a distinguishing mark of a RIXS structure due to a discrete level excitation.

MacDonald et al. (1995) have offered a comprehensive study of the evolution of X-ray RIXS into X-ray fluorescence from the excitation of xenon near the L_3 edge, performed at the X-24A beamline of the NSLS using the spectrometer device shown in Fig. 1.18. At the same beamline Miyano et al. have investigated the potassium $K\beta$ and chlorine $K\beta$ RIXS emission from KCl tuning the incident photon energy within a range of 55 eV over the potassium and the chlorine K edge respectively. Energy position, lineshape and intensity variation of the emission spectra were compared with calculations, where the absorption spectrum was modeled according to the experimental one, and an ε-independent emission was assumed. The experimental emission line peak position was found at about one eV lower energy than the calculated one for excitation energies around the first sharp peak of the absorption spectrum. This discrepancy was attributed to the breakdown of the "ε-independent-emission" approximation when core and valence excitons are involved in the excitation and the re-emission process, respectively.

As pointed out by Glatzel et al. (2002) and Glatzel et al. (2004) the complete evolution of a RIXS spectrum with varying incident photon energy is best visualized in a two-dimensional contour plot of the RIXS intensity, where the two axes are the incident energy ω_1 and the energy transfer (loss) $\omega = \omega_1 - \omega_2$. Projections along constant energy loss bring into prominence structures which are otherwise nearly or completely hidden in a normal absorption spectrum as will be shown in the next subsection. Projections along constant incident energy, on the other hand, have an energy scale which, in the case of 1s2p RIXS spectra, correspond with the excitation energy in the L-edge spectra. Also the natural linewidth in this case is the same as in L-edge spectroscopy.

Glatzel et al. (2002), for the case of Ni, and Glatzel et al. (2004), for the case of Mn, have provided interesting examples of how the 1s2p RIXS spectrum of Ni and Mn, respectively, excited at the $1s3d^{n+1}$ based pre-edge (quadrupolar) structure and plotted as a function of the energy loss $\omega = \omega_1 - \omega_2$ in various spin and oxidation states resembles the corresponding L-absorption spectrum to a certain extent. As stated above this can be easily understood, since the final state of the RIXS process is also connected with a 2p hole and an excited electron in 3d.

5.3.3 Revealing and assigning of "hidden" core excitations

The conventional way to study core excitations in order to get access to unoccupied states in condensed matter is X-ray absorption spectroscopy (XAS). However, XAS suffers from two drawbacks: (i) The information about the unoccupied density of states, more precisely about the oscillator density, is always convoluted with the lifetime broadening of the core hole involved in the final state of the XAS process. How this drawback can be overcome will be shown in the next section. (ii) Core excitations into unoccupied levels are sometimes so weak that they do not leave behind visible features in the XAS spectra, especially if they are connected with higher multipole transitions. In those cases we speak of hidden excitations. They can become detectable as peak-like structures in the corresponding RIXS spectra. According to the previous section these structures have to be assigned to transitions of electrons from a core state into a discrete unoccupied state, if (i) their intensity, as a function of the incident photon energy, goes through a resonance, and (ii) their Raman shift persists also when, with increasing incident-photon energy, the resonance has been passed over. Krisch et al. (1995) were the first to utilize this property of RIXS to identify a quadrupolar excitation channel at the L_{III}-edge of gadolinium. The experiment was performed at the wiggler beamline X21 at the NSLS using two focusing Rowland circle geometries in the horizontal scattering plane, the first one for monochromatizing, as described in Section 2.2.2.3. and a second one for analyzing, described in Section 2.2.5.1. The overall experimental resolution was 1.2 eV FWHM. The sample was a $Gd_3Ga_5O_{12}$ garnet. The Gd L_{III} edge as seen in XAS is shown in Fig. 5.9. The only detectable structure is the broad so-called white line, which is

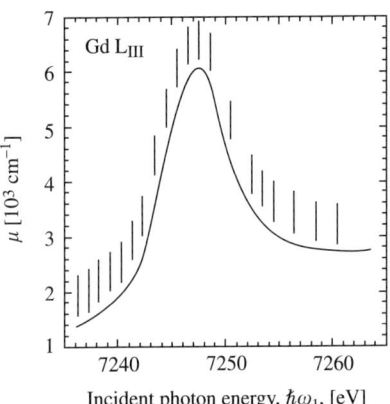

Fig. 5.8. Absorption spectrum of Gd in a $Gd_3Ga_5O_{12}$ garnet. The vertical lines correspond to the incident photon energies at which the high-resolution spectra were taken. (Reprinted with permission from Krisch et al. (1995); copyright (1995) by the American Physical Society.)

connected with the $2p^55d^1$ excitation. Other possible structures are obscured due to the about 4 eV wide 2p-core lifetime broadening. The vertical lines correspond to incident photon energies at which RIXS spectra were measured. A selection of these RIXS spectra, which all have to be attributed to intermediate virtual states with a core hole in the 2p level, are depicted in Fig. 5.9 as a function of the energy transfer $\hbar\omega_1 - \hbar\omega_2$. Three groups of structures A to C can be observed and were assigned to three different multiplet families of the final states. Group A to the $3d^94f^{n+1}5d^0$ final state of the Gd^{3+} ion, Group B to the $3d^94f^n5d^1$ final state and Group C to the $3d^94f^n5d^0\varepsilon^*$ continuum state. (It is assumed in this case, contrary to the previous sections, that the Coulomb interactions between the extra electron in the previously unoccupied levels with the 3d core hole in the final state could be different to the interactions with the 2p hole in the intermediate state.) The above assignment is based on the larger Coulomb interaction between the extra electron in the localized 4f orbital with the 3d hole, making a more tightly bound final state in comparison to an electron promoted either to the 5d orbital or to the continuum. Consequently, the excitation energy is smaller for the $3d^94f^{n+1}5d^0$ final state than that for the $3d^94f^n5d^1$ and $3d^94f^n5d^0\varepsilon^*$ final states. Group A has its resonance at 7240 eV, which is the $2p^54f^{n+1}$ intermediate excitation energy. Groups B and B* go together through a maximum at 7247.5 eV, the $2p^55d^1$ intermediate excitation energy. Therefore they must be traced back to the same multiplet family. The integrated intensity values of A and B as a function of $\hbar\omega_1$ follow the Lorentzian decay of the 2p core hole width. Both structures exhibit a Raman shift (peak positions of A and B on the $\hbar(\omega_1 - \omega_2)$ scale in Fig. 5.9 remain (nearly) fixed with increasing $\hbar\omega_1$). This qualifies both to be based on transitions into discrete levels. The group C structure, on the contrary, does not show a Raman shift when excited above threshold, so that the assignment to a transition into continuum states is compelling. The small shift of the B and B* structure on the energy transfer scale with increasing incident photon energy could be accounted for by two multiplet families separated by about 1.3 eV possibly related to the exchange splitting of the 5d band. The above assignment of the structures A and B means that the weak $2p^54f^{n+1}$ quadrupolar excitation is separated from the intense $2p^55d^1$ dipolar excitation despite the 50 times larger integrated intensity of the latter. These different excitation channels contributing to the same near-edge absorption resonance can be separated as long as their energy difference is larger than the final state core-hole lifetime broadening, in this case, the 3d and not the 2p lifetime broadening, which determines the resolution of XAS. The evidence for a quadrupolar excitation channel at the L_{III} edge was of special importance for the interpretation of the X-ray magnetic circular dichroism (XMCD) of Gd. The energy position of the excitation decaying into the group A multiplet is nearly the same as the negative feature in the L_{III} Gd XMCD spectrum, so that one has to conclude that the XMCD below the Gd L_{III} edge has a quadrupolar contribution. Bartolome *et al.* (1997) have extended the study of this quadrupolar channel at the L_{III} edge to a series of rare earth (R) compounds of the type

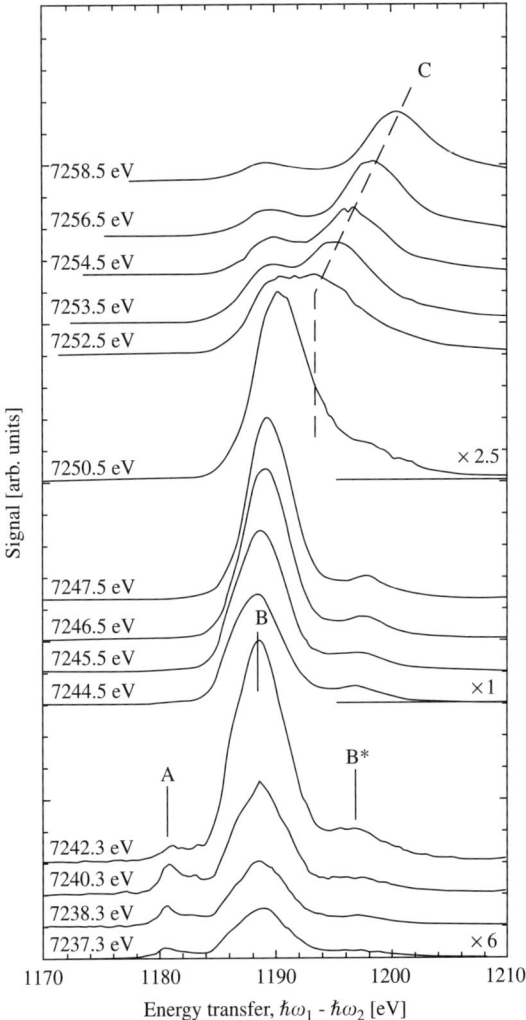

Fig. 5.9. Inelastic scattering spectra from $Gd_3Ga_5O_{12}$ garnet, plotted as a function of the energy transfer. The corresponding incident photon energy is given in the figure. The three groups of spectra are labeled by their scaling factors. The dashed line is a guide to the eye. (Reprinted with permission from Krisch et al. (1995); copyright (1995) by the American Physical Society.)

$R_2Fe_{14}B$. They found out how the energy distance between this quadrupolar $2p \rightarrow 4f$ excitation energy and the energy corresponding to the maximum of the so-called white line as derived from RIXS and XMCD varies systematically with the atomic number of the R element.

So far the "hidden" excitations, which could be made detectable, were connected with core excitations into discrete levels. Döring (2001), Hayashi et al. (2002), and Döring et al. (2004) have found in RIXS spectra of CuO indications for excitations into a narrow continuum, which are completely invisible in the corresponding XAS spectra. The experimental Cu 1s2p RIXS re-emission spectra of single-crystal CuO, measured by Döring et al. (2004) at the HARWI beamline at DESY/HASYLAB, Hamburg, using a Johann-type spectrometer with a spherically bent crystal, have already been shown in Fig. 1.17. The polarization of the incident photons relative to the crystal coordinates was determined by the angle γ between the incident beam and the crystal b-axis, where the plane of the storage ring coincided with a plane spanned by the a and b axes of the crystal. The structure A exhibits a resonance just at that incident photon energy $\hbar\omega_1$, where a small peak in the XAS spectrum of Fig. 5.10 occurs which is, according to calculations of Bocharov et al. (2001), a clear indication of quadrupolar transitions 1s $\to$ 3d. As shown in Fig. 5.11, the Raman shift of this structure persists when passing over the resonance, so that it is connected with a discrete excitation. The structure C does not reach Raman-shift saturation within the depicted range of incident photon energy and can therefore be assigned to the threshold of transitions from 1s to the 4p continuum. The structure B undergoes, as shown in Fig. 5.11, a Raman-shift saturation at 8981 eV, so that it must be ascribed to a transition into a continuum with a threshold between 8981 and

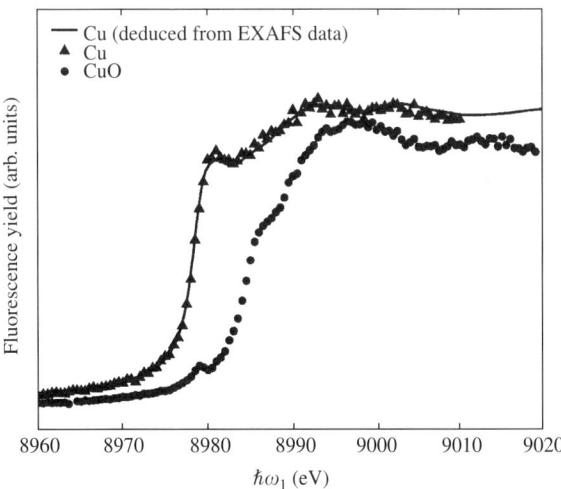

Fig. 5.10. Total fluorescence yield spectrum of Cu metal (triangles) and of CuO (dots). The solid line is the total fluorescence yield spectrum of Cu metal as deduced from an EXAFS spectrum measured in absorption. (Reprinted with permission from Döring et al. (2004); copyright (2004) by the American Physical Society.)

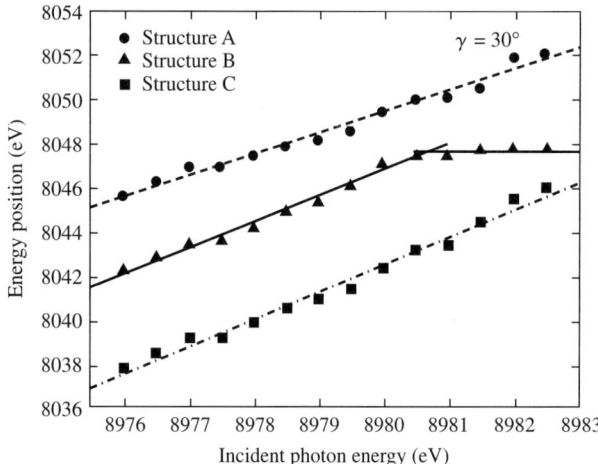

Fig. 5.11. Emission energy peak position of structure A (dots), B (triangles), and C (squares) for $\gamma = 30°$. The lines indicate linear fits to the energy position data points: A (dashed), B (solid), and C (dashed dotted). (Reprinted with permission from Döring et al. (2004); copyright (2004) by the American Physical Society.)

8982 eV. Since there is also an indication of a resonance of the structure B around 8982 eV, the continuum responsible for that structure seems to be only 2–3 eV wide. There is no feature of that kind visible in the total fluorescence XAS spectrum of Fig. 5.10. Therefore, it is justified to speak of a core excitation hidden in XAS but revealed by RIXS. Döring et al. (2004) have traced back this continuum to 4p orbitals admixed to a minority of 3d orbitals (orbitals with d_{xy}-symmetry according to calculations of Bocharov et al. 2001), which are shifted by 2–3 eV to higher energies with respect to the majority of the 3d holes, where transitions into the latter give rise to the structure A.

Journel et al. (2002) have presented another example of revealing "hidden" transitions by RIXS. Pre-edge structures not resolved in La L_{III} X-ray absorption spectra have been observed in LaS and two La-Ni intermetallics using RIXS. Figure 5.12 shows the experimental RIXS spectra of La_2Ni_7 as obtained at the beamline ID12A of the ESRF equipped with a Si(111) plane crystal monochromator, a focusing mirror and a bent quartz $(30\bar{3}0)$ analyzer. The overall resolution of this setup was 1.7 eV. Figure 5.12(a) presents the La2p3d RIXS spectra on an energy transfer scale for incident photon energies ω_1 in the pre-edge region with different distance Ω to the white line (WL) of the XAS spectrum. Although the XAS spectrum is structureless in this range of incident photon energy, the RIXS spectrum exhibits a shoulder A which develops into a maximum for a distance -4.5 eV to the white line, so as to decrease in intensity with decreasing distance to the white line. These properties of the additional structure, centered

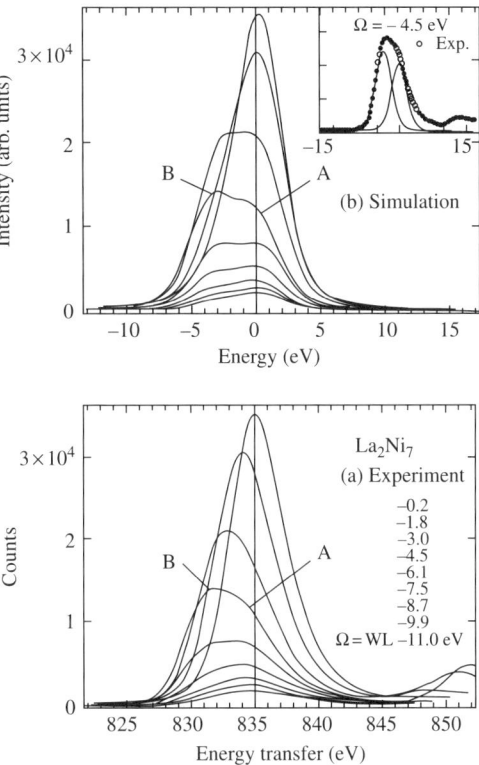

Fig. 5.12. La 2p3d RIXS spectrum for La_2Ni_7. Experimental (a) and simulated spectra (b) are on an energy transfer scale. The distance Ω of the incident photon energy from the white line is indicated. The simulation uses the experimental values of Ω and are normalized to the amplitudes of the experiment. The inset depicts the two components of the simulated spectra whose relative amplitudes depend on Ω. (Reprinted with permission from Journel *et al.* (2002); copyright (2002) by the American Physical Society.)

3 eV below the main peak B, is suggestive of a 2p core electron excitation into a discrete level The relatively high A:B intensity ratio of 0.24:0.76 for La_2Ni_7 is a point in favor of a dipole transition rather than a quadrupolar one which must be made responsible for this extra excitation. Similar additional structure in the 2p3d RIXS spectra are found in another intermetallic, LaNi, and in the covalent compound LaS. Starting from model densities of states the authors have simulated the data in an attempt to arrive at a better understanding of the origin of the RIXS data. The simulated spectra are shown in the top panel (b) of Fig. 5.12. It is assumed that the intermediate states responsible for the A structure are characterized by hybridization of the 4f states with 5d due to the strong Coulomb interactions in the presence of the core hole.

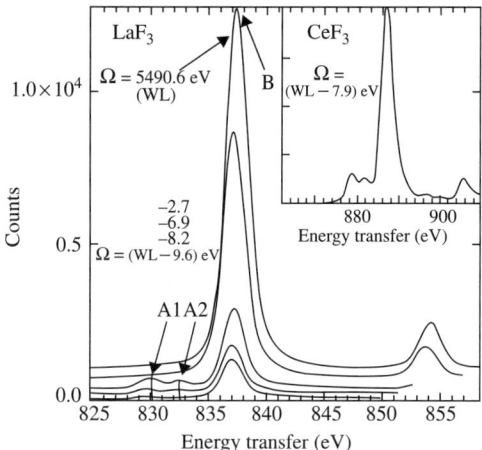

Fig. 5.13. La 2p3d RIXS spectra from LaF$_3$ showing weak resonance structures A1 and A2 attributed to quadrupole transitions. The inset shows data for CeF$_3$ taken by Gallet *et al.* (1999). (Reprinted with permission from Journel *et al.* (2002); copyright (2002) by the American Physical Society.)

The origin of the additional structure A1 and A2 in the likewise investigated RIXS spectra of the ionic compound LaF$_3$, shown in Fig. 5.13, seems to be different. They are attributed to quadrupolar transitions from the 2p^{6}4f^0(5d6s)^0 ground state to the 2p^{5}4f^1(5d6s)^0 intermediate state followed by decay to 3d^{9}4f^1(5d6s)^0, whereas the main line, as in the case of the intermetallics and the covalent compounds, is due to the normal dipole transition to 2p^{5}4f^0(5d6s)^1 with a 3d^{9}4f^0(5d6s)^1 final state. The spins of the 4f^1 and 3d^9 states in La are parallel (A1) or antiparallel (A2). This interpretation was established by means of atomic multiplet calculation, which is justified since the 5d state, as shown by the XAS measurement, becomes localized under the effect of the core-hole potential. Moreover, the very small A:B intensity ratio of = 0.04:0.96 (each measured at its resonance energy) is compatible with what one would expect as the ratio between quadrupole and dipole transition probabilities (van Veenendaal 1996). It should be noted that the RIXS spectra of CeF$_3$ measured by Gallet *et al.* (1999) exhibit very similar double-peaked pre-structures which are interpreted also as being due to quadrupolar transitions where the two possible spin orientations of the parallel spins in 4f^2 and the spin of the 3d^9 state produces the doublet (see the inset in Fig. 5.13).

5.3.4 *Suppression of the core-hole lifetime broadening*

In conventional absorption spectroscopy, the spectrum is broadened by a short lifetime of the contributing core hole, for example the 1s core hole, so that structures in the 1s absorption spectra are often smeared out. The same is true with the 2p absorption spectrum of rare earth compounds due to the short 2p core hole lifetime. As already shown in Fig. 1.16, Hämäläinen et al. (1991) attained a dramatic improvement in resolution far beyond the limit imposed by the $2p_{1/2}$ core hole lifetime in the excitation spectrum of Dy for $Dy(NO_3)_3$ by using a narrow part of the $3d_{5/2} \rightarrow 2p_{3/2}$ fluorescence line as a signal to indicate the strength of the absorption process when the incident photon energy is scanned across the $2p_{1/2}$ absorption edge. One can easily understand this improvement of resolution by rewriting (5.2) in such a way (see also Section 6.10.2) that the essential part of the RIXS intensity distribution becomes transparent:

$$F(\omega_1,\omega_2) = \sum_f \left| \sum_n \frac{\langle f|T_2^+|n\rangle\langle n|T_1|i\rangle}{E_i - E_n + \hbar\omega_1 - i\Gamma_n/2} \right|^2 \delta(E_i - E_f + \hbar\omega_1 - \hbar\omega_2), \quad (5.15)$$

where the operators T_1 and T_2 represent photon excited and radiative transitions, respectively. Γ_n denotes the spectral broadening due to the core-hole lifetime of the intermediate state $|n\rangle$ as a result of the Auger and radiative decays of the core hole.

If we replace the δ-function of (5.15) by a Lorentzian of FWHM Γ_f, in order to account for the finite lifetime of the final states, set $\hbar\omega_2$ at the peak of the fluorescence emission, and assume the interaction of the core holes corresponding to $|n\rangle$ and $|f\rangle$ with each other and with the valence states to be negligible, the matrix element $\langle f|T_2^+|n\rangle$ can be removed out of the summation over n. Furthermore, the summation over f can be skipped, since we have restricted ω_2 to a narrow part of the fluorescence line, so that $E_f + \hbar\omega_2$ can be set E_n. Then the intensity of the RIXS spectrum as a function of incident energy reads

$$I(\omega_1) = \sum_n |\langle n|T_1|i\rangle|^2 \left\{ \frac{1}{(E_i - E_n + \hbar\omega_1)^2 + \Gamma_n^2/4} \right\}$$
$$\times \left\{ \frac{\Gamma_f/\pi}{(E_n - E_i - \hbar\omega_1)^2 + \Gamma_f^2/4} \right\}. \quad (5.16)$$

If one considers this expression for the case $\Gamma_f \ll \Gamma_n$, the intensity of the RIXS emission as a function of the incident energy can be regarded as essentially the same as the absorption spectrum, which is given by

$$I(\omega_1) = \sum_n \frac{|\langle n|T_1|i\rangle|^2(\Gamma_n/2\pi)}{(E_n - E_i - \hbar\omega_1)^2 + \Gamma_n^2/4}, \quad (5.17)$$

but not broadened with the lifetime broadening Γ_n of the core hole involved in the absorption process but with a fictitious spectral broadening Γ_f, determined by the lifetime broadening of the core hole left behind in the re-emission process. We shall call, in what follows, this mode of edge spectroscopy partial fluorescence yield (PFY) XAS spectroscopy, in contrast to the total fluorescence yield (TFY) XAS, where the integrated scattered intensity as a function of ω_1 is measured.

An alternative method to overcome the core hole lifetime broadening of X-ray absorption spectroscopy by means of RIXS has been proposed by Hayashi et al. (2003). One starts with (5.14), which is written down in a slightly varied form by assuming the oscillator strength $g_{\text{I,K}}$ to be independent of ω_2:

$$\frac{\mathrm{d}^2\sigma}{\mathrm{d}\Omega^2\mathrm{d}\omega_2} \sim \left(\frac{e^2}{mc^2}\right)^2 \left(\frac{\omega_2}{\omega_1}\right) \left(\frac{\Delta\Omega_{\text{KL}}(\Omega_{\text{KL}}+\omega_1-\omega_2)}{(\Delta\Omega_{\text{KL}}-\omega_2)^2+\Gamma_{1s}^2/4\hbar^2}\right) \left(\frac{\mathrm{d}g_{1s}}{\mathrm{d}\varepsilon}\right)_{\omega_1-\omega_2-\Omega_{2p}}.$$
(5.18)

Thus it becomes transparent that one can obtain the oscillator density $(\mathrm{d}g_{1s}/\mathrm{d}\varepsilon)$ for $\varepsilon = \omega_1 - \omega_2 - \Omega_{2p}$ analytically from (5.18), of course only in relative units by using the RIXS cross-section $\mathrm{d}^2\sigma/\mathrm{d}\Omega_2\mathrm{d}\omega_2$, likewise measured in relative units as a function of ω_2 for a fixed incident photon energy $\hbar\omega_1$. All other quantities in (5.18) are well-known constants. As the RIXS cross-section, also $(\mathrm{d}g_{1s}/\mathrm{d}\varepsilon)$ is free of the lifetime broadening due to Γ_{1s}, and the width is only determined by the experimental resolution and Γ_{2p}. Since the analytical inversion of (5.18) can be performed with many different values of ω_1, the extent to what the oscillator densities $\mathrm{d}g_{1s}/\mathrm{d}\varepsilon$ obtained with different ω_1 agree can serve as an additional consistency test. This has been demonstrated by Hayashi et al. (2003) with Fig. 5.14.

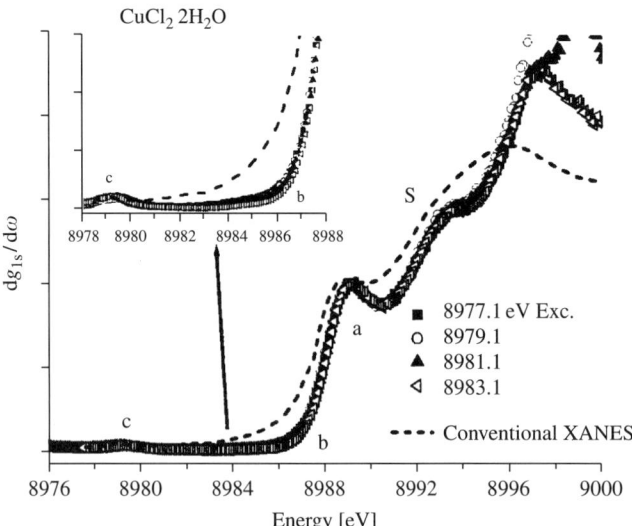

Fig. 5.14. Oscillator density of $CuCl_2$ $2H_2O$ in the XANES region of the Cu K edge as calculated from the RIXS spectra. Conventional X-ray absorption XANES spectra (dashed line) for comparison. (Reprinted with permission from Hayashi et al. (2003); copyright (2003) by the American Physical Society.)

Selected RIXS spectra of $CuCl_2$ $2H_2O$ for four different incident photon energies were used to obtain the oscillator density $dg_{1s}/d\varepsilon$ via the above-described analytical inversion. The agreement of the $dg_{1s}/d\varepsilon$ curves originating from RIXS spectra with different ω_1 is very good, at least for energies in the XANES (X-ray absorption near edge structure) range smaller than 8996 eV. For comparison, conventional XAS XANES spectra are shown. The much less broadening of the RIXS based spectra is evident. The RIXS spectra were obtained at the beamline BL47XU of SPring-8 equipped with a dispersive Si(111) two-crystal monochromator and a spherically bent Si(444) analyzer. The overall energy resolution was 1.1 eV.

5.4 RIXS applied to mixed valent systems

The relative weights of different configurations in mixed valency compounds have been extensively investigated by XAS. Nevertheless, if temperature dependence is concerned, the spectral changes are often small, and the quantitative analyses rely on the deconvolution of broad spectra and on assumptions on the underlying line shapes. Therefore, novel spectroscopies such as RIXS open up new opportunities. Since the de-excitation channels of the XAS final states are separately measured, the RIXS signal from a specific configuration in the hybrid ground state can be greatly enhanced. Utilizing these advantages of RIXS, Dallera *et al.* (2002) were able to follow with unprecedented accuracy the temperature dependence of the 4f occupation $\langle n_f \rangle$ in YbAgCu$_4$ and YbInCu$_4$. For YbInCu$_4$, measurements of the temperature dependence of physical properties such as the electronic specific heat, and the static and dynamic magnetic susceptibility, which are found indicative for a characteristic temperature dependence of $\langle n_f \rangle$, were in stark contrast with surface-sensitive photoemission (PES) data (Reinert *et al.* 1998, Susaki *et al.* 2001, Moore *et al.* 2000). Also the PES data on YbAgCu$_4$ were different according to whether they were obtained from polycrystalline (Malterre *et al.* 1996) or single crystalline material (Joyce *et al.* 1999). Thus the bulk sensitivity of RIXS along with its higher selectivity with respect to different configuration should be the right choice to solve these discrepancies. The RIXS measurements on these two materials were performed at the beamline ID16 of the ESRF, equipped with a Si(111) double-crystal monochromator, focusing mirror and a spherically bent Si(620) analyzer. The total resolution was 1.3 eV. Figure 5.15(a) shows the L_{III} XAS spectrum of YbInCu$_4$ both in partial fluorescence yield (PFY), where the intensity of the Yb $L\alpha_1$ ($2p^53d^{10} \rightarrow 2p^63d^9$) fluorescence is measured, and in total fluorescence yield (TFY), for temperatures below and above $T_V = 42$ K. At that temperature a discrete change in $\langle n_f \rangle$ should take place. Figure 5.15(b) presents the level scheme for an Yb ion in YbInCu$_4$. Thus the XAS structure 3+ has to be attributed to transitions from the mixed valency ground state to $2p^54f^{13}VB^4$ (VB=valence band), the structure 2+ to transitions to $2p^54f^{14}VB^3$. For reasons that have been discussed in Section 5.3.4, these structures are much sharper in the PFY mode, so that

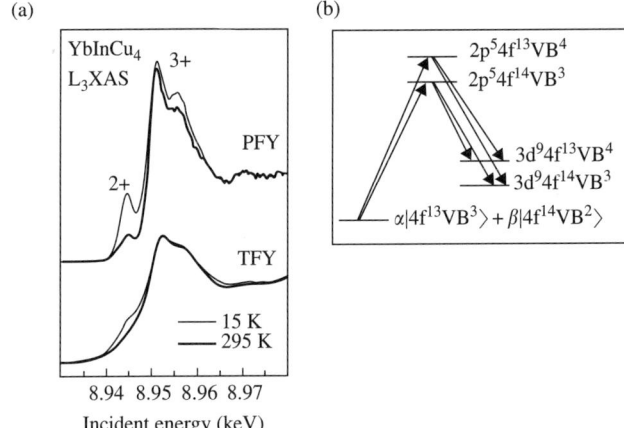

Fig. 5.15. (a) Bottom: L_3 absorption spectra of YbInCu$_4$ measured above and below $T_V = 42$ K in total fluorescence yield (TFY); and top: partial fluorescence yield (PFY) spectra according the Lα_1 fluorescence as a function of incident photon energy. (b) Energy level scheme for an Yb ion in YbInCu$_4$. The arrows show the XAS and RIXS transitions, respectively. (Reprinted with permission from Dallera et al. (2002); copyright (2002) by the American Physical Society.)

they can be used to estimate the temperature evolution of the valence. However, the corresponding deconvolution procedures would still rely on assumptions on the line shapes and would therefore be loaded with uncertainties. As depicted in Fig. 5.15(b) final states of mainly $3d^94f^{13}$ and $3d^94f^{14}$ character are reached from the XAS final states. Then the weak Yb^{2+} contribution can be selectively and resonantly enhanced by tuning the incident photon energy through the Yb^{2+} peak in the PFY XAS spectrum. This is demonstrated for the case of YbAgCu$_4$ in Fig. 5.16. One sees the dramatic enhancement of the 2+ structure, when the incident photon energy $\hbar\nu_{in} = \hbar\omega_1$ is scanned across the 2+ structure of PFY XAS spectrum as indicated in the insert. This enhanced signal is large enough to be followed continuously as a function of the temperature. Its intensity is proportional to the $4f^{14}$ weight in the initial state and therefore to $(1 - \langle n_h \rangle)$, where n_h is the number of 4f holes. Thus Dallera et al. (2002) found that YbInCu$_4$ exhibits a sharp drop of $(1 - \langle n_h \rangle)$ at $T_V = 42$ K. By contrast, PES measurements, which probe a thin and possibly perturbed layer near the surface, yield a broad transition (Reinert et al. 1998, Susaki et al. 2001). For YbAgCu$_4$, a normal Kondo system, the decrease of the RIXS 2+ structure with increasing temperature is now continuous. The plot of $(1 - \langle n_h \rangle)$ as a function of temperature exhibits a change of concavity around 70 K, which has been attributed to a Kondo temperature $T_K = 70$ K by means of a numerical calculation within the framework of the Anderson impurity model.

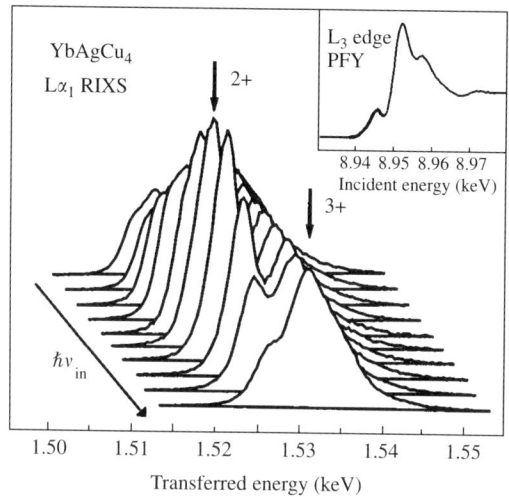

Fig. 5.16. $L\alpha_1$ RIXS spectra excited at 1 eV intervals along the pre-edge Yb^{2+} feature shown as a thick line of the PFY L_3 XANES, plotted against the transferred energy. (Reprinted with permission from Dallera et al. (2002); copyright (2002) by the American Physical Society.)

Another study of the intermediate valence in lanthanide materials by means of RIXS is presented by Dallera et al. (2004), who investigated the valence changes in YbAl$_2$ under pressure. Again these authors used a combination of PFY-XAS and RIXS in order to come to conclusive results. The measurements were performed at the ID16 beamline of the ESRF under similar conditions as the previous ones. The incident monochromatic photon entered the diamond anvil cell through the diamond and the scattered beam passed through the Be gasket. For the PFY-XAS measurements the intensity of the Yb $L\alpha_1$ fluorescence was measured while tuning the incident photon energy across the L_{III} edge. The RIXS spectra were recorded by analyzing the scattered beam at a fixed value of the incoming photon energy. The measured PFY-XAS spectra as a function of pressure between 0 and 385 kbar can be separated into a $I(2+)$ and a $I(3+)$ integrated intensity component, attributed to a $2p^54f^{14}\varepsilon d$ and a $2p^54f^{13}\varepsilon d$ XAS final state, respectively, split by the large Coulomb interaction energy between the 2p core-hole and the "spectator" 4f hole. Thus the Yb valence v can be estimated from the expression

$$v = 2 + \frac{I(3+)}{I(2+) + I(3+)}. \quad (5.19)$$

Likewise the RIXS data plotted as a function of incident photon energy can be separated into 2+ and 3+ contributions, however the fluorescence component, increasing with increasing incident photon energy, makes this separation less accurate, even if the separation procedure is done with the constraint that the fluorescence component peaks at an energy transfer that follows the incident energy. Therefore, the result of separation was tested by a completely independent analysis of the RIXS data. Pressure-dependent RIXS spectra are used, excited for an incident photon energy well below the absorption threshold so that the spectra are free of the perturbing fluorescence line. Since it can be assumed that the 0 kbar spectrum is mostly 2+, the 385 kbar spectrum mostly 3+, the former was subtracted from the latter in order to obtain a first approximation of a pure Yb^{3+} lineshape. The curve thus obtained was then subtracted from the 0 kbar spectrum to obtain a better approximation of the 2+ contribution. This procedure was repeated till stable line 2+ and 3+ lineshapes were achieved. These were used to fit the data at all pressures in order to use the integrated intensities $I(2+)$ and $I(3+)$ for calculating the Yb valence as a function of pressure according to (5.19).

By comparing XAS and RIXS data of CeO_2 and CeF_3, Hague et al. (2004) came to another assignment of structures in the CeO_2 XAS spectrum than Bianconi et al. (1987) whose interpretation was based on a multiplet calculation on the basis of the Anderson impurity approach. They attributed the structure 3 of the CeO_2 XAS spectrum shown in Fig. 5.17 to the $4f^1$-related configuration, and indicated that the shoulder to the lower energy side designated 2 was part of its multiplet structure. Contrarily, Hague et al. (2004) found out that the structures of the CeO_2 RIXS spectrum, which have to be attributed to structure 2 and 3, exhibit a distinct resonance behavior, so that they must be due to

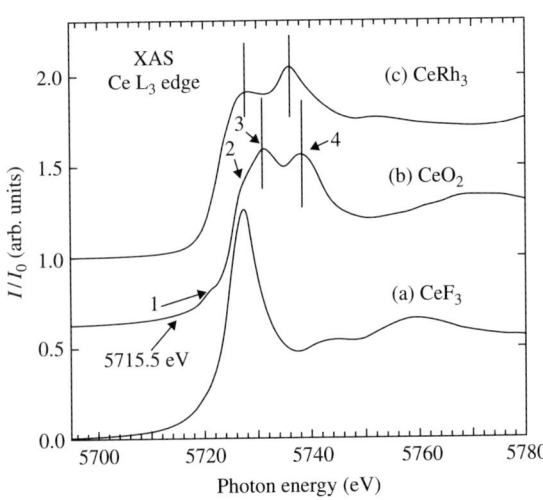

Fig. 5.17. X-ray absorption spectrum (XAS) at the Ce L_3 edge of: (a) CeF_3, (b) CeO_2, and (c) $CeRh_3$ (Reprinted with permission from Hague et al. (2004); copyright (2004) by Elsevier B.V.)

well-localized states. Additionally, the shoulder 2 of the CeO_2 XAS spectrum in Fig. 5.17 coincides well with the CeF_3 white line, which is certainly connected with a pure $4f^1$ XAS final state, according to the pure $4f^1$ ground state of the ionic CeF_3. For these two reasons, it is justified to assign structure 2 to a $4f^1L^{-1}$ related XAS final state, where L^{-1} represents a ligand hole. Peak 3 might then reflect crystal field splittings as proposed by Soldatov et al. (1994). Thus the combined application of XAS and RIXS leads to the conclusion that the core excitation spectrum of CeO_2 may not be interpreted straightforwardly in terms of either the Anderson impurity model (Bianconi et al. 1987) or a purely band structure approach (Soldatov et al. 1994).

5.5 Symmetry selective RIXS

5.5.1 *Symmetry selectivity in the excitation process*

The availability of strongly polarized synchrotron based incident radiation enables utilizing the symmetry selectivity of the excitation process as provided by the relative orientation of the incident polarization with respect to symmetry directions of the crystalline matter under investigation. The starting point for discussing the symmetry selectivity of the RIXS excitation process is the squared matrix element of (5.2)

$$|M_i, n|^2 = \left| \sum_j \langle n|\mathbf{e}_1 \cdot \mathbf{p}_j/\hbar \exp(i\mathbf{K}_1 \cdot \mathbf{r}_j)|i\rangle \right|^2. \tag{5.20}$$

Expanding the exponential function of (5.20) to the term first order in the spatial coordinate, one obtains the so-called dipole and quadrupolar contribution (now written within the limits of the one-electron approximation):

$$M_{i,n} \approx M^D_{i,n} + M^Q_{i,n} = \langle n|\mathbf{e}_1 \cdot \mathbf{p}|i\rangle + i\langle n|\mathbf{e}_1 \cdot \mathbf{p}\,\mathbf{K}_1 \cdot \mathbf{r}|i\rangle. \tag{5.21}$$

Utilizing the commutator $\mathbf{p} = (m/i\hbar)[\mathbf{r}, H_0]$ and the orthogonality of $|n\rangle$ and $|i\rangle$ one ends up with (Brouder 1990)

$$|M_{i,n}|^2 \approx |M^D_{i,n} + M^Q_{i,n}|^2 \sim |\langle n|\mathbf{e}_1 \cdot \mathbf{r}|i\rangle|^2 + (1/4)|\langle n|\mathbf{e}_1 \cdot \mathbf{r}\,\mathbf{K}_1 \cdot \mathbf{r}|i\rangle|^2, \tag{5.22}$$

since $M^D_{i,n}$ is pure imaginary and $M^Q_{i,n}$ pure real.

Choosing the real representation $Y_{l,m}(\theta, \phi)$ of the spherical harmonics

$$Y_{0,0} = s = (1/4\pi)^{1/2} \qquad Y_{2,-2} = d_{xy} = (15/4\pi)^{1/2} xy$$

$$Y_{1,1} = p_x = (3/4\pi)^{1/2} x \qquad Y_{2,-1} = d_{xz} = (15/4\pi)^{1/2} xz$$

$$Y_{1,-1} = p_y = (3/4\pi)^{1/2} y \qquad Y_{2,1} = d_{yz} = (15/4\pi)^{1/2} yz$$

$$Y_{1,0} = p_z = (3/4\pi)^{1/2} z \qquad Y_{2,2} = d_{x^2-y^2} = (15/16\pi)(x^2 - y^2)$$

$$Y_{2,0} = d_{z^2} = (5/16\pi)(3z^2 - 1) \tag{5.23}$$

(we have explicitly represented only those that matter in the following due to the dipolar and quadrupolar selection rules) to be the basis of $|i\rangle$ and $|n\rangle$, then

$$|n\rangle = \sum_{l,m} P_{l,m}\, R_l(r)\, Y_{l,m}(\theta,\phi) \quad \text{and} \quad |i\rangle = R_0(r)\, Y_{00}(\theta,\phi). \tag{5.24}$$

Then the angular-dependent part of the dipole matrix element writes

$$M^D_{i,n} \sim e_x P_{1,1} + e_y P_{1,-1} + e_z P_{1,0} \tag{5.25}$$

and that of the quadrupolar matrix element

$$M^Q_{i,n} \sim \left(\frac{4\pi}{15}\right)^{1/2} (e_x K_y + e_y K_x) P_{2,-2} + (e_x K_z + e_z K_x) P_{2,-1} + \sqrt{3}(e_z K_z) P_{2,0}$$
$$+ (e_y K_z + e_z K_y) P_{2,1} + (e_x K_x + e_y K_y) P_{2,2}. \tag{5.26}$$

Inserting (5.25) and (5.26) into (5.22), the excitation probability can be expressed as a weighted sum of $(\mathbf{e}_1, \mathbf{K}_1)$-independent partial spectral components. The weights of these components are determined by the orientation of $\mathbf{e}_1$ and $\mathbf{K}_1$ with respect to the crystal coordinates, and will be denoted partial spectral weights (Bocharov 1998). The number of individual partial spectral components depends on the full point symmetry of the crystal (Brouder 1990). For a monoclinic system, there are four independent dipole partial spectral components and nine quadrupole components. This number reduces to three dipole and five quadrupole components in the case of a tetragonal system. Then the excitation probability or, by applying these considerations to the case of X-ray absorption, the absorption coefficient m can be decomposed into a dipolar, μ_D, and a quadrupolar one, μ_Q, $\mu = \mu_D + \mu_Q$, with

$$\mu_D \sim p_x e_x^2 + p_y e_y^2 + p_z e_z^2 \tag{5.27}$$

$$\mu_Q \sim d_{xy}(e_x K_y + e_y K_x)^2 + d_{xz}(e_x K_z + e_z K_x)^2 + \sqrt{3} d_{z^2}(e_z K_z)^2$$
$$+ d_{yz}(e_y K_z + e_z K_y)^2 + d_{x^2-y^2}(e_x K_x + e_y K_y)^2, \tag{5.28}$$

where p_x, p_y, p_z, d_{xy}, d_{xz}, d_{yz}, $d_{x^2-y^2}$ and d_{z^2} denote the partial spectral components weighted by the $\mathbf{e}_1$ and $\mathbf{K}_1$ dependent partial spectral weights, so that the variation of the scattering geometry allows us to determine the relative magnitude of the partial spectral components.

Using the nearly linear polarization of hard X-ray synchrotron radiation, this polarization sensitivity of the X-ray absorption process was employed the first time by Dräger et al. (1988) in connection with XAS measurements at the Fe K edge of Fe metal, FeS_2, Fe_2O_3, and $FeCO_3$. From the angular dependence of the pre-edge K absorption the existence of predominantly ligand-field-induced dipole transitions in Fe_2O_3 and of quadrupole transitions in $FeCO_3$ was concluded. A very interesting application of the decomposition of hard X-ray absorption spectra into partial spectral components has been published by Heumann et al. (1997) in a study on $Bi_2Sr_2CaCu_2O_8$ for five selected sample orientations,

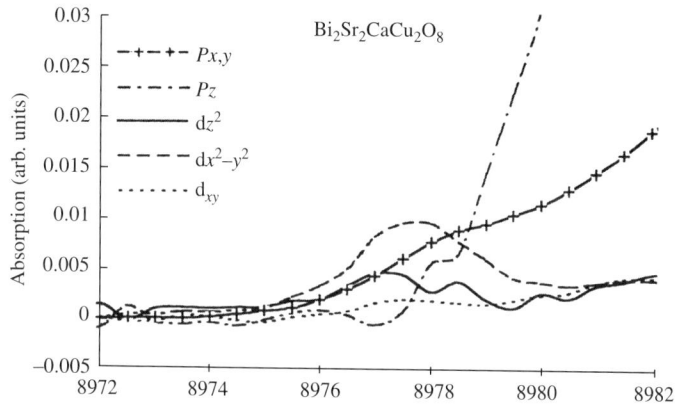

Fig. 5.18. The Cu K pre-edge of $Bi_2Sr_2CaCu_2O_8$ resolved into the contributions of different p_m- and d_m-like states as indicated. (Reprinted with permission from Heumann et al. (1997); copyright (1997) by Editions de Physique.)

which provided the partial spectral components $p_{x,y}$, p_z, d_{z^2}, $d_{x^2-y^2}$ and d_{xy} in the pre-edge region as shown in Fig. 5.18. The existence of empty $d_{x^2-y^2}$ like states is clearly shown as well as the absence of d_{xy} holes. The existence of d_{z^2} like hole states, until then controversially discussed, is proved. Later on Bocharov et al. (2001) performed a complete decomposition of the Cu K pre-edge in CuO into their partial spectral components according to (5.27) and (5.28) assuming that one can apply the tetragonal decomposition to the weekly deformed monoclinic structure of CuO. Thus, by varying e_1 and K_1 appropriately equations (5.27) and (5.28) were used as a system of linear equations to determine the partial spectral components, which were then compared with multiple scattering calculations.

The utilization of polarization dependence of the excitation process in hard X-ray RIXS studies for symmetry assignment is still rather scarce. The same mode of decomposition as performed by Bocharov et al. (2001) was also applied by Döring et al. (2004) in a RIXS study on CuO. By calculating the partial spectral weights as a function of the angle γ between the incident beam and the b-axis, shown in Figs. 5.19 and 5.20, the authors were able to assign the structure C in Fig. 1.17 to transitions into a p_z-like continuum, whereas the structure B must be attributed to transitions into a narrow continuum with both p_z and p_x symmetry. The measured γ-dependence of the structure A revealed strong admixtures of d_{yz}, $d_{x^2-y^2}$ and d_{z^2} symmetry to the d_{xy} symmetry found by Bocharov (2001) to be dominant. The orthogonal coordinate system used here is not connected with the whole unit cell but attached to the local neighborhood of the absorbing Cu atom, oriented according to the oxygen parallelogram, so that the x-axis is parallel to its longer side, the y-axis lies within the CuO_4 parallelogram, perpendicular to x, and the z-axis is perpendicular to the xy-plane. Since in the real CuO structure two interpenetrating CuO_4 parallelograms with

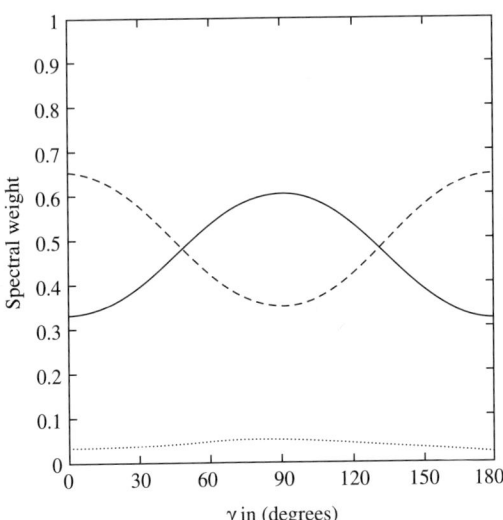

Fig. 5.19. Partial spectral weights with p_z (dashed line), p_y (dotted line) and p_z (solid line) symmetry as a function of the incident angle γ. (Reprinted with permission from Döring et al. (2004); copyright (2204) by the American Physical Society.)

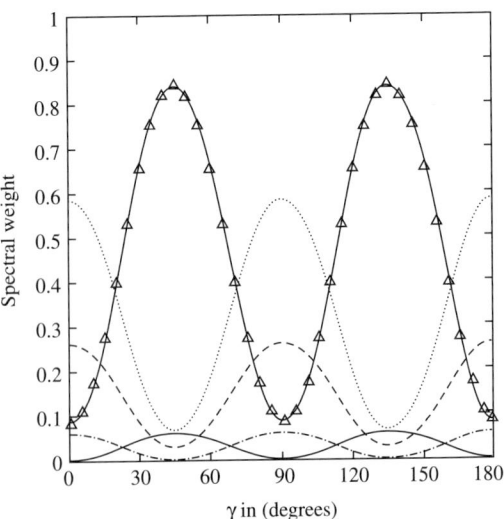

Fig. 5.20. Partial spectral weights with d_{xy} (solid line), d_{xz} (solid line with triangles), d_{yz} (dot-dashed line), $d_{x^2-y^2}$ (dashed line), and d_{3z^2-1} (dotted line) symmetry as a function of incident angle γ. (Reprinted with permission from Döring et al. (2004); copyright (2004) the American Physical Society.)

different orientations exist, the calculated partial spectral weights of Figs. 5.19 and 5.20 have to be understood as an average of these two orientations.

A strong polarization dependence of the charge transfer scattering (see Section 5.7.2) was found by Hämäläinen et al. (2000) in Nd_2CuO_4. Investigating the integral intensity of the charge transfer excitation as a function of the incident photon energy, they found a single peak at 8999.5 eV, if the incident polarization was parallel to the CuO planes ($\mathbf{e}_1 \| ab$ plane), whereas for $\mathbf{e}_1 \| c$-axis such a peak was established by Hill et al. (1998) at 8990 eV. Measurements of the polarization dependent absorption by monitoring the fluorescence yield showed that the former case had to be associated with $1s \to 4p\sigma$ transitions, the latter with $1s \to 4p\pi$ transitions, as was confirmed by calculations. Details of this experiment are discussed in Section 5.7.2.

In a cubic system the d orbitals are split into the e_g and t_{2g} symmetries according to $e_g = d_{x^2-y^2} + d_{z^2}$ and $t_{2g} = d_{xy} + d_{yz} + d_{xz}$. As shown by Dräger et al. (2001), the $1s \to 3d(e_g)$ transitions can be separated from the $1s \to 3d(t_{2g})$ by using two different $\mathbf{e}_1$ and $\mathbf{K}_1$ orientations with respect to the single-crystal sample. This symmetry selection was utilized to investigate the spin character of the two 3d states in MnO and NiO, as shown in Section 5.6.

5.5.2 Symmetry selectivity in the re-emission process

The symmetry selectivity in the re-emission process of RIXS is based on the squared matrix element of (5.2) between the intermediate and the final state

$$|M_{n,f}|^2 = \left| \sum_j \langle f | \mathbf{e}_2 \cdot \mathbf{p}_j / \hbar \exp(-i\mathbf{K}_2 \cdot \mathbf{r}_j) | n \rangle \right|^2, \quad (5.29)$$

which we will treat only in the dipole and one-electron approximation:

$$|M_{n,f}|^2 \sim |\langle f | \mathbf{e}_2 \cdot \mathbf{r} | n \rangle|^2, \quad (5.30)$$

so that it can be cast into the following form

$$|M_{n,f}|^2 \sim \mathbf{e} \cdot \mathbf{T} \cdot \mathbf{e} \quad (5.31)$$

with the tensor $\mathbf{T}$. In its principal coordinate system this tensor is reduced to the diagonal components

$$T_\eta = |\langle f | \eta | n \rangle|^2, \quad (5.32)$$

where $\eta = x, y, z$ are the coordinates of the principal axes of the tensor. These axes are determined by the crystallographic axes of the emitting crystal except the monoclinic and triclinic systems. Thus we have for an orthorhombic crystal

$$|M_{n,f}|^2 = T_x \sin^2 \vartheta \cos^2 \phi + T_y \sin^2 \vartheta \sin^2 \phi + T_z \cos^2 \vartheta. \quad (5.33)$$

Here the principal x, y, z axes of the tensor are parallel to the orthorhombic a, b, c axes, respectively, and (ϑ, ϕ) are the polar coordinates of the polarization vector $\mathbf{e}_2$ with respect to these axes. Therefore, a diagonal tensor component T_η of the principal squared matrix element is a contributory factor to the intensity of re-emitted radiation with a polarization parallel to the principal η-axis. By applying these considerations to a singular X-ray emission process, the tensor component T_η determines via

$$I_\eta(\omega) = \sum_f T_\eta \, \delta(E_f - E_n - \hbar\omega) \qquad (5.34)$$

the intensity of the X-ray emission polarized parallel to the principal η-axis. The principal intensities $I_\eta(\omega)$ can be measured separately by means of a suitable experimental arrangement, which has to include experimental information about the polarization ratio q of the analyzer, which gives the ratio of its reflecting power for the p and s component of the radiation. (p/s component means radiation with polarization vector parallel/perpendicular to the diffraction plane of the analyzer.) The symmetry selectivity of hard X-ray emission spectroscopy was employed for the first time by Brümmer et al. (1969), and found later on many applications (see Dräger and Brümmer 1984 for a review). In RIXS studies, the symmetry selectivity of the re-emission process has been utilized so far only to investigate the symmetry related properties of the emission of molecules in the gas phase, "aligned" by selection due to polarization and energy of the exciting radiation. This shall be described in the following subsection.

5.5.3 RIXS of aligned molecules

As already stressed, the full potential of RIXS becomes evident whenever the intimate connection between excitation and re-emission process is employed. This especially applies to making use of the symmetry selectivity of both the excitation and the re-emission process. This means to observe the polarization of the re-emission to infer the alignment of the molecular ions in the randomly oriented gas phase following polarized core-level excitation. In contrast to a similar procedure in connection with the polarized excitation of valence electrons using ultraviolet and visible fluorescence polarization (see for example Poliakoff et al. 1981), the core-level studies offer two distinct advantages: (i) Re-emission is observed on a time-scale ($\leq 10^{-14}$ s) that is short compared with normal molecular tumbling periods ($10^{-11} - 10^{-12}$ s). Valence-shell hole states, on the contrary, are sufficiently long lived ($\geq 10^{-10}$ s) that rotational motion can occur prior to radiative decay. (ii) Core-level excitation is highly selective to specific parts of the near-edge fine structure, i.e. to the same atomic species in different environments. Measurements of the fluorescence polarization following polarized core-level excitation can be realized by means of the device (Southworth et al. 1991a) shown in Fig. 1.18. Such an instrumentation needs (i) a source of polarized synchrotron X-rays, sufficiently monochromatized to resolve the near-edge fine structure, and diffracted by the monochromator under a Bragg angle such as to essentially suppress the component of linear polarization having its electric vector parallel to the plane of incidence; (ii) a high-throughput spectrometer to analyze the re-emmitted radiation, in Fig. 1.18 a curved crystal analyzer in connection with a position sensitive detector, in order to record the entire emission spectrum simultaneously; (iii) polarization sensitivity of the secondary spectrometer, achieved by Bragg reflection under a nearly 45° Bragg angle at the analyzer; (iv) the possibility to change the direction of the polarization accepted by the secondary spectrometer with respect to the polarization of the incident radiation from parallel to perpendicular, in Fig. 1.18 attained by rotating the secondary spectrometer around an axis in the direction of the emitted radiation; and/or (v) the ability to record the emission spectra at different emission angles ϕ with respect to the polarization vector of the incident radiation. This flexibility allows us also to change the polarization direction of the re-emitted beam with respect to that of the incident beam. Thus, in general, one can measure the emitted X-ray intensities as a function of ϕ with the emission spectrometer positioned to detect X-rays either polarized parallel, $I_\parallel(\phi)$, or perpendicular, $I_\perp(\phi)$, to the plane which contains the polarization vector of the incident beam and which is normal to the propagation direction of the incident X-rays.

In Section 6.10.4 a general relation (6.320) for the squared RIXS amplitude $|F_{\nu n}(\omega_1)|^2$ as a function of the incident frequency is derived, where one assumes transition from various localized or delocalized core orbitals enumerated by the index k to unoccupied molecular orbitals ν in the absorption process and transitions from occupied orbitals n to the core levels, so that $F_{\nu n}$ stands for the amplitude of an optical transition $n \to \nu$. This relation was derived by taking into account the polarization (unit) vectors $\mathbf{e}_1$ and $\mathbf{e}_2$ of the incident and the scattered radiation, respectively, the directional cosines with respect to the laboratory frame of the absorption and re-emission dipole matrix elements, respectively, and expressions, which contain both the moduli of these matrix elements and the resonance denominator of the Kramers–Heisenberg relation (5.1). This way we have in one factor of this general relation information about polarization, in a second factor information about the orientation of the scattering molecules, and in a third factor information about the molecules themselves. Thus for structures with fixed molecular orientation one can use the knowledge about two of these factors to obtain information about the third. But the most spectacular application of this general relation is to molecules in random orientation, this means to vapors or liquids. By averaging over the orientation of the molecules expressions for three representative polarization ratios $P_1 \equiv \lambda(\text{ln})/\lambda(\text{lp})$, $P_2 \equiv \lambda(\text{cp})/\lambda(\text{lp})$, and $P_3 \equiv \lambda(\text{cp})/\lambda(\text{ln})$ were found, see equation (6.327). Here $\lambda(\text{ln})$ and $\lambda(\text{lp})$ are the orientation averaged squared RIXS amplitudes with polarization perpendicular and parallel to the incident polarization, respectively, and $\lambda(\text{cp})$ the orientation averaged RIXS amplitude for the case that the absorbed and emitted photons have circular polarization. Using these relations it is possible to assign the symmetries of occupied or unoccupied molecular orbitals (MOs), as Luo *et al.* (1994) have elucidated, by means of the calculated Table 6.1 of Section 6.10.4, which enables the symmetry assignment of occupied (unoccupied) MOs for the

RIXS transitions, when the unoccupied (occupied) MO is of symmetry Γ_0, by means of measured polarization ratios P_1, P_2, P_3. Unfortunately, the analysis of the re-emitted radiation with respect to its degree of circular polarization was, until now, not achieved, so that a direct application of this table was not possible. But it has been shown by Luo et al. (1994) that, for the case of only one participating core orbital ($k = j$), symmetry assignments are possible without using circular polarization. Thus the determination of the polarization ratio P_1 makes it possible to find the angle $\phi_{\nu n}$ between the dipole moments $\mathbf{d}_{\nu j}$ and $\mathbf{d}_{jn}(\nu)$ (the ν in the brackets accounts for screening of the transition matrix element $\mathbf{d}_{jn}$, due to the "spectator" electron in the formerly unoccupied orbital ν) via

$$P_1 = \frac{2 - \cos^2 \phi_{\nu n}}{2 \cos^2 \phi_{\nu n} + 1}, \tag{5.35}$$

a relation, which can be deduced from (6.319), (6.324), and (6.327) as a sufficient approximation. Another equation, deduced by Gel'mukhanov and Ågren (1994) for randomly oriented molecules, by averaging the expression (6.320) over the directions of the photon propagation under fixed angle ϑ between the real polarization vectors $\mathbf{e}_1$ and $\mathbf{e}_2$, instead of averaging over all molecular orientations, writes, again specified for the case of a single core orbital,

$$\langle |F_{\nu n}|^2 \rangle \sim 2 - \cos^2 \vartheta + (3 \cos^2 \vartheta - 1) \cos^2 \phi_{\nu n}, \tag{5.36}$$

and relates another observable, namely the polarization P of the re-emission, defined as

$$P \equiv \frac{I_{0°} - I_{90°}}{I_{0°} + I_{90°}}, \tag{5.37}$$

where $I_{0°}$ and $I_{90°}$ are the ω_2-integrated re-emission intensities for $\theta = 0°$ and $\theta = 90°$, respectively, to $\phi_{\nu n}$:

$$P = \frac{3 \cos^2 \phi_{\nu n} - 1}{\cos^2 \phi_{\nu n} + 3} \tag{5.38}$$

with limiting values $P = 1/2$ and $-1/3$ for $\phi_{\nu n} = 0°$ and $90°$, respectively. As shown in what follows, (5.36) and (5.38) are widely used for the interpretation of experimental data.

The following examples of the application of polarized re-emission followed polarized excitation with the aim to get experimental information about symmetry assignment of molecular orbitals in systems of randomly oriented molecules are all taken from harder X-ray studies. Soft X-ray studies suffer from special problems connected with dynamical symmetry breaking due to vibronic coupling, as will be briefly discussed at the end of this subsection.

The first harder X-ray experiments on the polarization of molecular fluorescence have been performed by Lindle et al. (1988) on CH_3Cl molecules in the gas

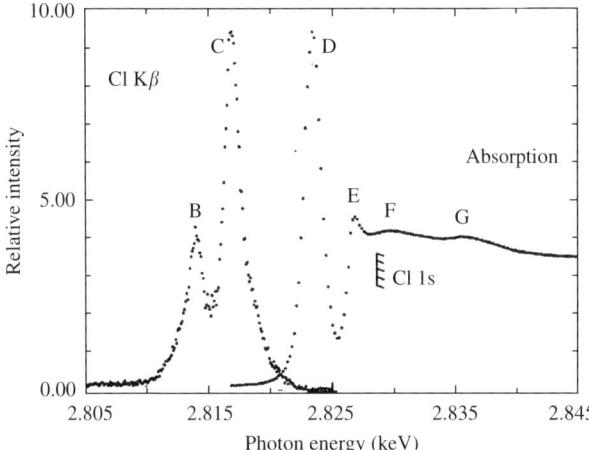

Fig. 5.21. Right side: Cl K-edge absorption spectrum of CH_3Cl; left side: Cl Kβ emission spectrum of CH_3Cl taken at an excitation energy of 2880 eV. (Originally published by Lindle *et al.* (1988); copyright (1988) by the American Physical Society.)

phase using a spectrometer as sketched in Figure 1.18 at the beamline X-24A of the NSLS. Figure 5.21 shows on the right-hand side the X-ray absorption spectrum with a strong pre-absorption structure denoted D. On the left-hand side the Cl Kβ emission spectrum is depicted as taken with a resolution of 0.9 eV at an excitation energy of 2880 eV well above the Cl 1s threshold. This emission spectrum exhibits no detectable change with the fluorescence polarization. Excitation to continuum states constituting a mixture of symmetries yields excited molecules with little alignment. If the Cl Kβ emission is excited with 2823.4 eV photon energy, centered at the feature D of the absorption spectrum, the intensity ratio between the structures B and C of the Cl Kβ emission is strongly dependent on whether the orientation of the measured fluorescence polarization is parallel or perpendicular (see the corresponding labels) to the incident E vector, as shown in Fig. 5.22. The **E**-vector orientation can be varied in the spectrometer of Fig. 1.18. This is characteristic for the emission of aligned occupied molecular orbitals. The "alignment" occurs due to the preferred excitation of electrons into unoccupied orbitals with symmetry axes oriented in a distinct way relative to the polarization vector of the incident radiation, specifically, molecules excited with incident energies on absorption feature D ($1a_1 \to 8a_1$) have their symmetry axes oriented largely in the incident synchrotron-radiation polarization direction. Consequently, the fluorescence peak B, which partly has to be ascribed to a $7a_1 \to 1a_1$ transition with a dipole matrix element also parallel to the symmetry axis of the molecule, should be stronger relative to the fluorescence peak C (ascribed to a $3e \to 1a_1$ transition) when observed in parallel rather than in

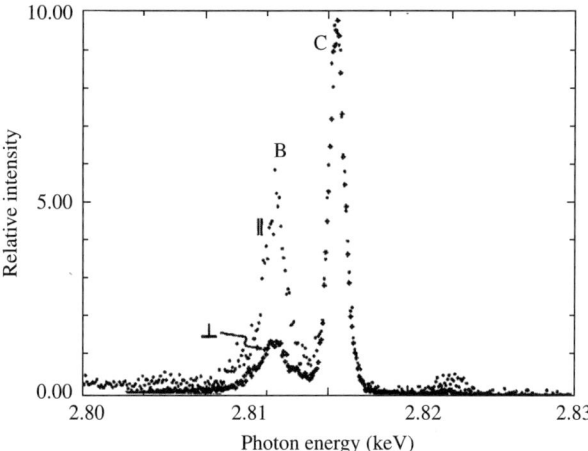

Fig. 5.22. Cl Kβ Fluorescence spectra from CH_3Cl following the Cl 1s $\rightarrow$ $8a_1$ excitation with 2823.4 eV photon energy, centered on feature D of the absorption spectrum in Fig. 5.21. The labels parallel and perpendicular refer to orthogonal orientations of the measured fluorescence polarization relative to the incident **E** vector. The two spectra are scaled to have identical areas of the peak C. (Originally published by Lindle *et al.* (1988); copyright (1988) by the American Physical Society.)

perpendicular polarization, because the 3e $\rightarrow$ $1a_1$ transition matrix element is oriented perpendicular to the symmetry axis. Partly the emission peak B contains also the 2e $\rightarrow$ $1a_1$ transition. Thus the polarization P, according to (5.38), should be, in the case of peak-D excitation, $P = -1/3$ for the transition 2e $\rightarrow$ $1a_1$ and 3e $\rightarrow$ $1a_1$ and $P = 1/2$ for the transition $7a_1 \rightarrow 1a_1$. On this basis, Lindle *et al.* (1988) have calculated a B-to-C intensity ratio of 0.9 and 0.24 for parallel and perpendicular polarizations, which are in rather good agreement with the experimental values of 0.86(5) and 0.29(2), respectively.

The results of a similar investigation on randomly oriented CF_3Cl molecules by Southworth *et al.* (1991b) are shown in Fig. 1.19. In this experiment the emission angle θ of Fig. 1.18 was changed from 0° to 90° with the secondary spectrometer positioned such that the accepted polarization vector was parallel to a plane perpendicular to the propagation direction of the incident beam (reference plane). Thus the 0° case corresponds to an orientation of the emitted polarization perpendicular to the incident beam polarization, the 90° case to a parallel orientation of the two polarization directions. Since the excitation energy was tuned to the Cl 1s to valence transition, $1a_1 \rightarrow 11a_1$, it becomes clear, for the same reasons as in our first example, that in the 0° case the fluorescence peak assigned to a $10a_1 \rightarrow 1a_1$ transition has a much smaller intensity relative to the 7e $\rightarrow$ $1a_1$ emission than in the 90° case. Of course, changing θ with an accepted polarization vector perpendicular to the reference plane has not exhibited any angular dependence of the $10a_1/7e$ intensity ratio.

More chlorofluoromethanes in the gas phase, namely CF_2Cl_2 and $CFCl_3$ in addition to CF_3Cl, were studied by Lindle et al. (1991) in regard to the polarization of the emission followed by polarized excitation. The authors found also in these cases highly polarized molecular K-valence X-ray fluorescence, when the excitation occurred in the pre-edge range of the absorption spectrum.

The near edge S K absorption spectrum of H_2S in the gas phase exhibits a broad subthreshold peak the width of which suggests that it is an unresolved two-peak structure. MO calculations of Mazalov et al. (1973) indicate that there exists a MO with a_1 symmetry about 0.5 eV above the dominant MO with b_2 symmetry which both make a contribution to this peak. On the other hand, Schwarz's calculations indicate the opposite order with an energy separation of 0.7 eV. A RIXS study of Mayer et al. (1991) helped to solve this puzzle by measuring the S K-valence emission spectrum for various excitation energies around this subthreshold peak. The relative intensity of the three emission peaks A, B, C, the assignment of which was, according to their polarization, A: $2b_1 \to 1a_1$, B: $5a_1 \to 1a_1$, and C $2b_2 \to 1a_1$, measured as a function of the excitation energy, favored clearly the results of the former calculations. A theoretical RIXS study of Gel'mukhanov and Ågren (1994), performed on H_2S has convincingly demonstrated how the recording of the S K excitation spectra at three different angles θ between the incident and the re-emitted polarization, and with the photon emission energy tuned to the A, B, and C resonance, respectively, could solve in an unambiguous way the problem of the right order of the near lying unoccupied MO's. Higher energy resolution of incident and re-emitted radiation should additionally lead to lifetime suppression as described in Section 5.3.4, so that these MO's could be resolved.

The sulfur $K\beta$ X-ray emission with resonant excitation at the main absorption peak, the 4π resonance, performed by Miyano et al. (1998) on carbonyl sulfide at the beamline X24A using the spectrometer of Fig. 1.18, exhibits a strong influence of the 4π electron on the final valence-hole state. The $[3\pi](4\pi)$ state is split into Σ^+, Σ^- and Δ states. This energy separation allows observation of differences between parallel and perpendicular polarization. The observed polarization differences are consistent with the molecular symmetries of the calculated intermediate and final states.

It might suggest itself to utilize the interconnection of the symmetry properties of absorption and re-emission also for EXAFS (extended X-ray absorption fine structure) studies. This has been proposed by Gel'mukhanov and Ågren (1994). They pointed out that the oscillatory part of the X-ray absorption cross-section in the EXAFS region is proportional to $(\mathbf{e}_1 \cdot \mathbf{R}_{\alpha\alpha'})^2$, where $\mathbf{R}_{\alpha\alpha'}$ is the internuclear axis direction, so that structural information of well-ordered systems can be obtained by utilizing the polarization dependence of the absorption process. However, if one observes the EXAFS oscillations in the polarized re-emission from a distinct molecular orbital $|n\rangle$ with a dipole moment $\mathbf{d}_n$ the corresponding cross-section is proportional to $(\mathbf{e}_2 \cdot \mathbf{d}_n)^2 (\mathbf{e}_1 \cdot \mathbf{R}_{\alpha\alpha'})^2$. Thus directional information as well as information on nuclear distances can be extracted. It has been shown by Gel'mukhanov and Ågren (1994) that this information can also be obtained from disordered systems, since the oscillatory part of the RIXS cross-section, then averaged over molecular orientations, depends on $\mathbf{e}_1$, $\mathbf{e}_2$, $\mathbf{d}_n$, and $\mathbf{R}_{\alpha\alpha'}$ through the two scalar products $(\mathbf{e}_1 \cdot \mathbf{e}_2)$ and $(\mathbf{d}_n \cdot \mathbf{R}_{\alpha\alpha'})$. Unfortunately, this additional source of information has not yet been used in a real experiment. Therefore, the reader is referred to the literature (Gel'mukhanov and Ågren 1994) for further details.

As already mentioned, one problem in interpreting RIXS spectra of molecules must be discussed, even though not that relevant to harder X-ray applications we restrict ourselves to. It is the problem of symmetry breaking upon core electron excitation. The existence of nontotally symmetric vibrational modes, as in polyatomic molecules, can introduce dynamical symmetry breaking and vibronic coupling into forbidden symmetries (Skytt et al. 1996). This would manifest itself by the appearance of forbidden spectral lines and could lead to misinterpretation in the course of symmetry assignment by means of RIXS. Let us consider CO_2 as a typical example. The addition of a carbon atom between the two oxygen atoms introduces an antisymmetric stretch mode, which can couple the allowed $|1\sigma_u^{-1} 2\pi_u\rangle {}^1\Pi_u$ and the forbidden $|1\sigma_g^{-1} 2\pi_u\rangle {}^1\Pi_g$ oxygen core excited states, contrary to the case of O_2, where the inverse symmetry is retained upon core excitation. Thus the resonant emission spectra connected with the excitation to the first unoccupied level $2\pi_u$ of CO_2, shown in Fig. 5.23, consists of two peaks: the high-energy peak derived from transitions to the forbidden $|1\pi_g^{-1} 2\pi_u\rangle {}^1\Pi_u$ final state and the low-energy peak to the allowed $|1\pi_u^{-1} 2\pi_u\rangle {}^1\Pi_g$ final state. The significant intensity in the high-energy peak under exact resonance condition is a clear indication for the violation of the parity selection rule due to symmetry breaking. By detuning the excitation energy from the exact resonance, as shown in Fig. 5.23, the intensity of the high-energy peak decreases drastically compared with the low-energy feature. This can easily be understood on the basis of the Kramers–Heisenberg formula (5.2) if we identify there the intermediate states $|n\rangle$ with the excited vibrational levels, which might have an effective width Δ, and replace the scattering operators by the dipole operators D', D, according to the dipole approximation we are restricted to. Let us define the detuning energy Ω as the difference between the excitation photon energy $\hbar\omega_1$ and the center of the vibrational envelope. When the detuning energy is much larger than the effective width of the absorption band, $\Omega \gg \Delta$, the denominator of (5.2) can be removed from the sum over the intermediate states $|n\rangle$. This means that the summation over the intermediate vibrational levels can be made complete $(\sum_n |n\rangle\langle n| = 1)$. The scattering amplitude takes the simple form $F_\mathrm{f} \sim \langle f | D'D | i \rangle$ and becomes this way independent of the symmetry broken intermediate state. It restores the parity selection rule. This is the physical reason for symmetry purifying by detuning. But it must be stressed that the same derivation can be carried out if the lifetime width Γ_n in (5.2) is large, $\Gamma_n \gg \Delta$, in which case

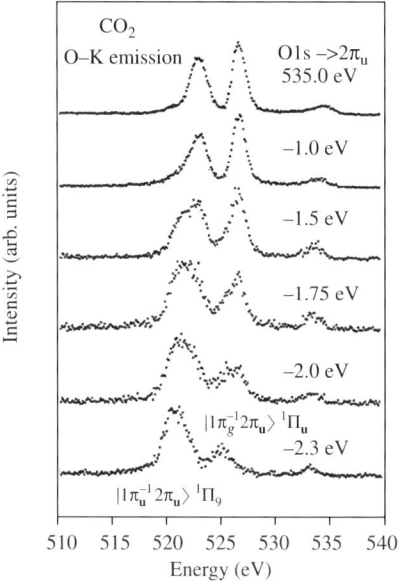

Fig. 5.23. Oxygen K $2\pi_u$ resonant X-ray emission spectra of CO_2 with different detuning energies below the $2\pi_u$ resonance. (Reprinted with permission from Skytt *et al.* (1996); copyright (1996) by the American Physical Society.)

the resonant X-ray transition also becomes symmetry purified. Of course we can also argue with a concept of time by saying that the duration of the intermediate state is short compared with the vibrational time period, so that the molecule has no time to execute the antisymmetric vibration that introduces forbidden parity in the electronic wavefunctions. This constitutes the advantage of applying harder X-rays for symmetry assignment of randomly oriented molecules by means of RIXS as shown above. Certainly, there is another phenomenon, which can lead to violation of parity rules without symmetry breaking, for example in the case of homonuclear diatomic molecules, namely channel interferences, as will be described in the next subsection.

5.6 Coherence of absorption and re-emission process in RIXS

5.6.1 *Fluorescence interferometry*

It was first shown by Ma *et al.* (1992) and later on in more detail by Ma (1994b) that the coherence of the absorption and the re-emission process of RIXS has a consequence, which is best demonstrated in terms of the classical Young double slit experiment as depicted in Fig. 5.24. Let us follow Ma and Blume (1995) and assume a diatomic molecule with interatomic distance **R** hit by the incident X-ray and excited in one of the two atoms into an intermediate state consisting of

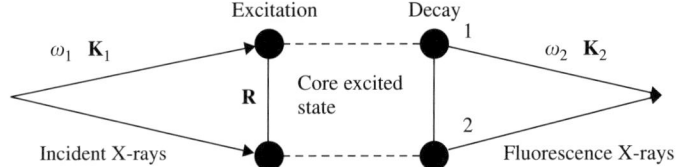

Fig. 5.24. RIXS from a diatomic molecule. $\hbar\omega_1$ and $\hbar\omega_2$, $\mathbf{K}_1$ and $\mathbf{K}_2$ are the energy and wavevector of the incident and fluorescence photon, respectively. (Reprinted with permission from Ma and Blume (1995); copyright (1995) by the American Institute of Physics.)

a core hole and an electron in a former unoccupied valence level. A fluorescence photon is emitted when the core hole is filled with a valence electron. Because of the uncertainty principle, the criterion for the observation of an interference pattern is that one cannot determine which "slit" the light passed through. As long as the photons taking the two paths in Fig. 5.24 leave the sample in the same final state, they should interfere. The sample itself is left in an excited state, which has to be delocalized over the two atoms for interference to occur. The electronic structure of the diatomic model can be thought to be composed of a core level ϕ_1, an occupied valence level ϕ_2 and an unoccupied valence level ϕ_3. It will turn out that interference will occur only if the valence levels are interacting, so that they can be approximated by symmetric and antisymmetric molecular orbitals:

$$\chi_2^\pm = (1/2)^{-1/2}[\phi_2(1) \pm \phi_2(2)] \quad \text{and} \quad \chi_3^\pm = (1/2)^{-1/2}[\phi_3(1) \pm \phi_3(2)].$$

The energy splitting between the bonding and the antibonding levels might be $2\Delta_2$, $2\Delta_3$, respectively, the energy of the unsplit levels ε_2 and ε_3, respectively. The core levels are assumed to be strongly localized, and that there is no interatomic transition, i.e. $\int d^3 r \phi_1(\mathbf{r}) \mathbf{p} \cdot \mathbf{e_2} \phi_{2/3}(\mathbf{r} - \mathbf{R}) \approx 0$. We will also neglect the effect of the core holes on the electronic orbitals of the intermediate state. The double differential scattering cross-section shall be calculated according to (5.2), where the first sum is over the four possible final states $|\bar{\chi}_2^+ \chi_3^+\rangle$, $|\bar{\chi}_2^- \chi_3^+\rangle$, $|\bar{\chi}_2^+ \chi_3^-\rangle$, $|\bar{\chi}_2^- \chi_3^-\rangle$ (the bar above the wavefunction indicates a hole state), the second sum ranges over the intermediate state, which corresponds to a core hole located on one of the two atoms and an electron in the χ_3 state. Thus we end up with

$$\frac{d^2\sigma}{d\Omega_2 d\hbar\omega_2} = \left(\frac{e^2}{mc^2}\right)^2 \left(\frac{\omega_2}{\omega_1}\right) \left(\frac{1}{m}\right) (F^-[2 + 2\cos(\mathbf{q} \cdot \mathbf{R})] \delta(\hbar\bar{\omega} - \Delta_2 - \Delta_3)$$
$$+ F^-[2 - 2\cos(\mathbf{q} \cdot \mathbf{R})] \delta(\hbar\bar{\omega} + \Delta_2 - \Delta_3)$$
$$+ F^+[2 - 2\cos(\mathbf{q} \cdot \mathbf{R})] \delta(\hbar\omega - \Delta_2 + \Delta_3)$$
$$+ F^+[2 + 2\cos(\mathbf{q} \cdot \mathbf{R})] \delta(\hbar\omega + \Delta_2 + \Delta_3)), \tag{5.39}$$

where $\mathbf{q} = \mathbf{K}_1 - \mathbf{K}_2$, $\hbar\bar{\omega} = \hbar\omega_2 - \hbar\omega_1 + \varepsilon_3 - \varepsilon_2$, $F^{\pm} = |M_{21}|^2|M_{13}|^2/[4(\varepsilon_3 - \varepsilon_1 - \hbar\omega_1 \pm \Delta_3)^2 + \Gamma^2]$, and M_{21}, M_{13} are the dipole transition matrix elements between the core and valence levels. Γ is the lifetime width of the intermediate state. If $\Gamma_{1,2}$ is the energy resolution of the monochromator and of the X-ray spectrometer, respectively, then the necessary condition for observing the interference is $\Gamma_{1,2} < \Delta_2, \Delta_3$. The physics behind this condition is easily understood: Initially the photoelectron and the valence hole are created in the region of the localized core hole. This interaction with the photons takes $\hbar/\Gamma_{1,2}$, time enough for the created hole to become delocalized over the whole molecule (time $\hbar/\Delta_{2,3}$). Thus it is not possible to determine from the final state which path the fluorescence photon has taken. Only in that case can the interference effects be observed. If, on the other hand $\Delta_{2,3}/\Gamma_{1,2} \to 0$, no interference term is present, and one recovers the normal emission spectra, where the fluorescence is regarded as emitted from individual atoms.

In this respect resonant inelastic X-ray scattering is different from elastic scattering, since in the latter case also noninteracting atoms contribute coherently to the scattering process. Contrarily, with $\Delta = 0$, the interference terms in (5.39) are absent. Noninteracting atoms do not contribute coherently to RIXS.

Ma (1994b) has called attention to the possibility to extract structural information using the interference terms of (5.39). Again we treat the simple case of two identical atoms with a distance given by $\mathbf{R}$. Let us assume that in the absorption process a monochromatic incident X-ray photon excites a 1s electron into the antibonding orbital χ_3^-. A fluorescence X-ray photon is later emitted when an electron from the bonding orbital χ_2^+ fills the core hole. Using the dipole approximation we obtain, according to (5.39), the cross-section for the excitation–decay process:

$$\frac{d\sigma}{d\Omega} = 2\left(\frac{e^2}{mc^2}\right)^2 \left(\frac{\omega_2}{m\omega_1}\right) |\langle\phi_1|\mathbf{p}\cdot\mathbf{e}_2|\chi_2^+\rangle|^2 |\langle\chi_3^-|\mathbf{p}\cdot\mathbf{e}_1|\phi_1\rangle|^2 [1 + \cos(\mathbf{q}\cdot\mathbf{R})]$$
$$\times [(\varepsilon_3 - \varepsilon_1 - \hbar\omega_1 + \Delta_3)^2 + \Gamma^2]^{-1}. \quad (5.40)$$

The $\cos(\mathbf{q}\cdot\mathbf{R})$ term describes the interference of the fluorescence X-rays emitted from the two atoms. If the atom pairs are aligned, such as molecules in a single crystal or molecular adsorbates on a stepped surface, the interatomic distance $a = |\mathbf{R}|$ can be easily obtained by analyzing the interference pattern. Ma (1994b) has shown that even for a randomly oriented samples the interference pattern will not be smeared out, when the cross-section of (5.40) is averaged over the orientation of $\mathbf{R}$. Cowan (1990) has shown that due to the polarization-dependent matrix element in (5.40) the absorption process creates an ensemble of excited molecules that are aligned to the polarization of the incident X-rays. The radiative decay of the excited molecules occurs within the lifetime of the core levels (in femtoseconds) much faster than the reorientation of these molecules by molecular rotation (in picoseconds). Thus the fluorescence pattern is determined by these partially aligned molecules.

If we assume the bonding and the antibonding orbitals to be both of σ character then the dipole matrix elements of (5.40) have only components along the axis of the molecule, so that we have to calculate the following average:

$$\langle (1+\cos(\mathbf{q}\cdot\hat{\mathbf{R}}))(\mathbf{e}_1\cdot\hat{\mathbf{R}})^2(\mathbf{e}_2\cdot\hat{\mathbf{R}})^2\rangle_\Omega = \left(\frac{1}{15}\right)[1+2(\mathbf{e}_1\cdot\mathbf{e}_2)^2]$$

$$+\left[\frac{j_2(x)}{x^2}\right][1+2(\mathbf{e}_1\cdot\mathbf{e}_2)^2]+\left[\frac{j_1(x)}{x}-\frac{5j_2(x)}{x^2}\right][4(\mathbf{e}_1\cdot\mathbf{e}_2)(\mathbf{e}_1\cdot\hat{\mathbf{q}})(\mathbf{e}_2\cdot\hat{\mathbf{q}})$$

$$+(\mathbf{e}_1\cdot\hat{\mathbf{q}})^2+(\mathbf{e}_2\cdot\hat{\mathbf{q}})^2]+\left[\frac{j_0(x)-10j_1(x)}{x}+\frac{35j_2(x)}{x^2}\right](\mathbf{e}_1\cdot\hat{\mathbf{q}})^2(\mathbf{e}_2\cdot\hat{\mathbf{q}})^2,$$
(5.41)

where $x = qa$, and the $j_l(x)$ are the spherical Bessel functions. To get an impression of how sensitive the interference patterns are to the interatomic distance we show in Fig. 5.25 the X-ray emission intensity of randomly oriented diatomic molecules as a function of the scattering angle for different values of a/λ (λ is the X-ray wavelength), and for a scattering geometry where the incident X-rays are linearly polarized in the horizontal plane and the fluorescence X-rays are detected in the horizontal plane with polarization either in the horizontal (Fig. 5.25a) or vertical (Fig. 5.25b) plane. From Figs. 5.25a,b it appears that this method can be effectively utilized for determining interatomic distances in systems with $a/\lambda \geq 0.2$ with an accuracy of a few percent of the wavelength. It is expected by Ma (1994b) that, comparable with the application of anomalous X-ray diffraction, a complete structure determination for a group of identical atoms should be possible, but compared with the anomalous diffraction there is no problem with the elastically scattered background. Additionally a symmetry analysis of the final states involved in the RIXS process is possible as has been demonstrated in Section 5.5.3. Compared to the application of extended X-ray absorption fine structure (EXAFS), where the determination of interatomic distances beyond the first coordination shell becomes difficult, the sensitivity of RIXS interferometry increases with increasing distance of the identical atoms. Thus distances in larger molecules or clusters become accessible.

Mills et al. (1997) have pointed out that, in the case of a homonuclear diatomic molecule, these channel interferences can give rise to strong nondipole two-photon features as found in the RIXS spectra of Cl_2 gas. Thus the transitions $X \to K_u/K_g \to A^1\Pi_u$ and $X \to K_u/K_g \to C^1\Sigma_u^+$, which connect *gerade* (g) initial with *ungerade* (u) final states and are therefore forbidden by the parity selection rules have integrated intensities comparable with the allowed transitions $X \to K_u \to X^1\Sigma_g^+$ and $X \to K_u \to B^1\Pi_g$. Utilizing the averaging procedure which leads, for the $\sigma \to \sigma$ transitions, to (5.41) one obtains for the integrated RIXS emission intensity in the case of perpendicular polarization and 90° scattering,

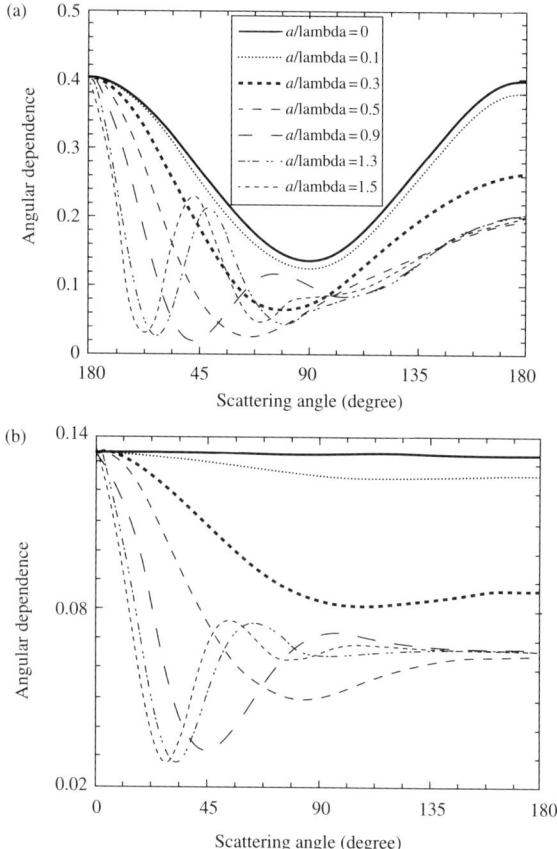

Fig. 5.25. X-ray emission intensity as a function of the scattering angle for a scattering geometry, where the incident X-rays are linearly polarized in the horizontal plane and the fluorescence X-rays are detected in the horizontal plane with polarization (a) in the horizontal or (b) in the vertical plane. The curves are for a/λ values as indicated in (a). (Originally published by Ma (1994b); copyright (1994) by Elsevier Science B.V.)

as shown in Fig. 1.18:

$$I_{\rm em}^{\perp}\begin{pmatrix} {\rm X} \to {\rm X}^1\Sigma_{\rm g}^+ \\ {\rm X} \to {\rm C}^1\Sigma_{\rm u}^+ \end{pmatrix} \sim |\langle 1{\rm s}|\mu_z|_{5\sigma_{\rm g}}^{5\sigma_{\rm u}}\rangle|^2 \left\{ \left(\frac{1}{8}\right) \pm \left(\frac{15}{16}\right)\frac{j_1(qa)}{qa} \mp \left(\frac{45}{16}\right)\frac{j_2(qa)}{(qa)^2} \right\}$$

$$I_{\rm em}^{\perp}\begin{pmatrix} {\rm X} \to {\rm A}^1\Pi_{\rm u} \\ {\rm X} \to {\rm B}^1\Pi_{\rm g} \end{pmatrix} \sim |\langle 1{\rm s}|\mu_x|_{5\pi_{\rm g}}^{5\pi_{\rm u}}\rangle|^2 \left\{ \left(\frac{1}{2}\right) \mp \left(\frac{15}{16}\right)\frac{j_1(qa)}{x} \pm \left(\frac{45}{16}\right)\frac{j_2(qa)}{(qa)x^2} \right\}$$

(5.42)

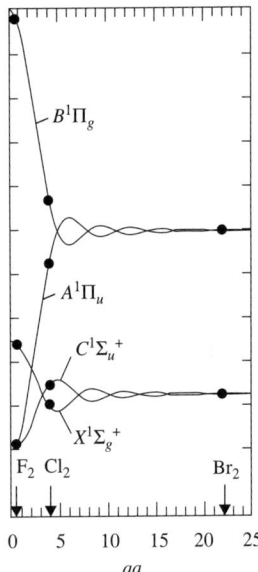

Fig. 5.26. Variation with qa of the structure factors of perpendicular polarized dipole allowed contributions $X \to K_u \to X^1\Sigma_g$ and $X \to K_u \to B^1\Pi_g$ as well as nondipole two-photon features $X \to K_u/K_g \to A^1\Pi_u$ and $X \to K_u/K_g \to C^1\Sigma_u^+$ of the RIXS spectrum for homonuclear diatomic halogens. (Originally published by Mills *et al.* (1997); copyright (1997) the American Physical Society.)

$\mu_{x/z}$ are the components of the one-electron dipole momentum operator with z along the body-frame internuclear axis, q is the momentum transfer, a the equilibrium molecular bond distance, and 1s refers to a single atomic component of the $1\sigma_g$ and $1\sigma_u$ symmetry orbitals. Figure 5.26 depicts the qa dependence of the structure factor (expression within the { }-brackets) in (5.42). Arrows indicate the situation for the three halogen molecules F_2, Cl_2 and Br_2. F_2 can be considered as an example for the dipole limit ($q \to 0$), where the nondipole two-photon features nearly vanish. Br_2 stands for the hard-X-ray limit with nondipole components with equal intensity as the dipole ones. Cl_2 is settled in between of both limiting cases. As a result of this study, the authors concluded that the appearance of nondipole features in the re-emission spectra of diatomic molecules is not an indication for symmetry breaking as asserted by Gel'mukhanov and Ågren (1994a) and by Glans *et al.* (1996) in a study on O_2.

5.6.2 *Momentum conservation; Bloch-**k** selectivity*

Let us now consider the consequences of the coherence between absorption and re-emission processes in RIXS for a periodic lattice of identical atoms. Following the more detailed derivation of Section 6.10.5 we shall assume that during the absorption of a photon one electron is excited from a tightly bound core state into a formerly unoccupied state above the Fermi energy, and that the emission of a photon is connected with refilling the previously created core hole by an electron from the valence band. Within the limits of the one-electron approximation, the following single-hole wavefunctions characterize the initial and the final state, respectively, of the RIXS process:

$$|i\rangle = \exp(i\mathbf{k}_i \cdot \mathbf{r}) u_{\nu_c \mathbf{k}_i}(\mathbf{r}) \quad \text{and} \quad |f\rangle = \exp(i\mathbf{k}_f \cdot \mathbf{r}) u_{\nu_v \mathbf{k}_f}(\mathbf{r}). \quad (5.43)$$

They are written as states in the conduction band and in the valence band, respectively, where ν is the band index, $\mathbf{k}$ is the Bloch wavevector and $u_{\nu \mathbf{k}}$ is the lattice periodic function. The intermediate state is then represented by

$$|n\rangle = |m\mathbf{R}_j\rangle = \psi_m(\mathbf{r} - \mathbf{R}_j), \quad (5.44)$$

where ψ_m is an atomic wavefunction with core-level index m centered at the atomic position $\mathbf{R}_j$. Moreover, we are utilizing the so-called "frozen core approximation" assuming that the excitation or deexcitation of an electron does not affect the other electrons of the sample via Coulomb interaction. Inserting (5.43) and (5.44) into (5.2) (where the summation over j, j' is skipped in line with the one-electron approximation), and performing the summation over initial, final and intermediate states, one ends up with

$$\frac{d^2\sigma}{d\Omega_2 d\hbar\omega_2} = 2\pi \left(\frac{\omega_2}{\omega_1}\right) r_0^2 \sum_{\nu_i \mathbf{k}_i} \sum_{\nu_f \mathbf{k}_f} |M_1(\nu_i, \mathbf{k}_i, m)|^2 \delta(E_{\nu_i \mathbf{k}_i} - E_m - \hbar\omega_1)$$
$$\times \delta_{g,(\mathbf{K}_1 - \mathbf{K}_2 - \mathbf{k}_i + \mathbf{k}_f)} |M_2(\nu_f, \mathbf{k}_f, m)|^2 \delta(E_{\nu_f \mathbf{k}_f} - E_m - \hbar\omega_2). \quad (5.45)$$

Here $\mathbf{g}$ is a reciprocal lattice vector, $E_{\nu \mathbf{k}}$ the energy eigenvalues of the Bloch state $|\nu \mathbf{k}\rangle$, and E_m the energy of the core state. $M_{1/2}$ are the matrix elements representing the transitions between the Bloch state $|\nu \mathbf{k}\rangle$ and the core state $|\psi_m\rangle$ in the absorption/re-emission process:

$$M_{1/2}(\nu, \mathbf{k}, n) = \langle u_{\nu \mathbf{k}}(\mathbf{r}) | \exp(i\mathbf{K}_{1/2} \cdot \mathbf{r}) \mathbf{e}_{1/2} \cdot \mathbf{p} | \psi_m(\mathbf{r}) \rangle. \quad (5.46)$$

These matrix elements are determined by the overlap between the strongly localized core wavefunctions $\psi_n(\mathbf{r}')$ with the conduction band and the valence band wavefunctions $u_{\nu \mathbf{k}}(\mathbf{r}')$, respectively, and make both processes element-specific.

In deriving (5.45) we have taken into account that the intermediate state encloses all equivalent atoms within the coherent X-ray field, so that the summation over the intermediate states is a summation over the atomic positions $\mathbf{R}_j$, which gives rise to the Kronneker delta in (5.45). It is this Kroneker delta which indicates the interference between the fluorescence emission from equivalent atoms due to the fact that it is not possible to determine from the final state which path the fluorescence photon has taken, just as in the previous case of the Young-type scattering experiment on a diatomic molecule. On the other hand, one can interpret this Kronneker delta also as an indication for momentum conservation in the coherent absorption/re-emission process. As depicted in Fig. 5.27 for the case of Si, the momentum $\mathbf{q} = \mathbf{K}_1 - \mathbf{K}_2$ is transferred to the system of Bloch electrons, where an electron from the valence band with crystal momentum $\mathbf{k}_f$ is excited into a Bloch state with crystal momentum $\mathbf{k}_i$ in the conduction band. (Of course, the conservation relation is only valid modulo a reciprocal lattice vector $\mathbf{g}$, since the $\mathbf{k}$'s are crystal momenta or Bloch wavevectors). In other words, if a core electron is transferred to the conduction band, and selects one or a few Bloch states $|\nu_i \mathbf{k}_i\rangle$ according to the delta function $\delta(E_{\nu_i \mathbf{k}_i} - E_m - \hbar\omega_1)$ in (5.45), then the hole state $|\nu_f \mathbf{k}_f\rangle$ in the valence band is not only determined by the delta function $\delta(E_{\nu_f \mathbf{k}_f} - E_m - \hbar\omega_2)$ and thus depending

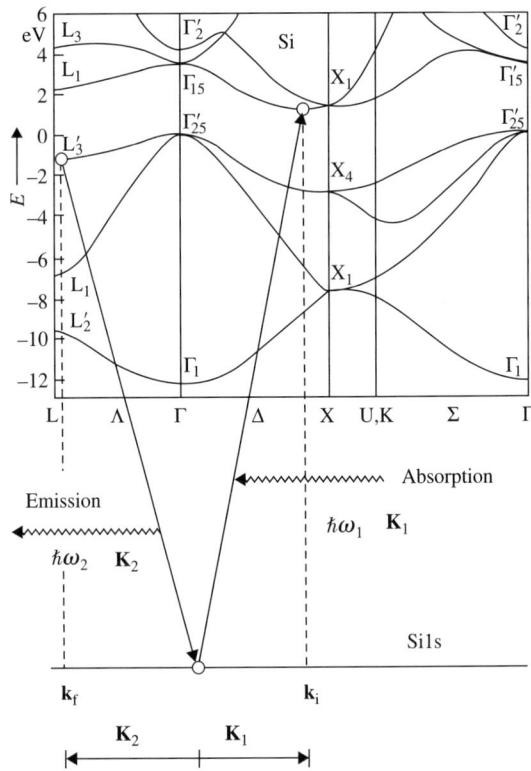

Fig. 5.27. Momentum conservation in the RIXS process, illustrated using the band structure of silicon. $\mathbf{K}_1$ and $\mathbf{K}_2$ are the wavevectors of the incident and emitted X-rays, respectively. $\mathbf{k}_i$ and $\mathbf{k}_f$ are the momenta of the photoelectron in the conduction band and of the valence electron, respectively. (Originally published by Ma et al. (1995); copyright (1995) by The American Physical Society.)

on the setting $\hbar\omega_2$ of the analyzer, but this hole state must meet the momentum conservation as expressed in the Kronecker delta of (5.45). The RIXS spectra of a crystalline solid become Bloch-**k** selective, whenever the electronic structure of this solid can be described within the limits of an energy band model. This selectivity can, under favorable conditions, be utilized for band mapping, as will be shown below.

Ma *et al.* (1992) were the first to interpret the dependence of the spectral shape of the C K emission on the incident photon energy as being due to momentum conservation. The authors measured the C K emission spectra of diamond at selected excitation energies, indicated by the arrows in Fig. 5.28(a), using a high-resolution grazing-incidence multigrating spectrometer with a two-dimensional detector at the beamline X1B of the NSLS. The spectrometer resolution was 0.9 eV. Figure 5.28(b) depicts some of the emission spectra including one which was taken with an incident energy far away from the C K threshold with 350 eV, and which resembles fluorescence emission as obtained with high-energy electron excitation. The assignment of the critical points for both the absorption and the emission spectra, using the results of Painter *et al.* (1971), Chelikowski and Louie (1984), and Himpsel *et al.* (1980), are indicated in Figs. 5.28(a) and (b). In order to emphasize the excitation energy dependence, Fig. 5.29 shows the difference spectra between 60% of the 350 eV spectrum and other spectra of Fig. 5.28(b). The labels for each spectrum are used to indicate the symmetry point in the conduction band where the core electron is mostly excited. The bars at the bottom specify the decay of the valence electron to the core hole. Now it becomes transparent that, whenever the photon excites the photoelectron near to a specific symmetry point, let's say X in Fig. 5.28, a valence electron is created preferentially at a point of the same symmetry, i.e an X point. In other words, they are the same Bloch-**k** points of the Brillouin zone which are involved in the absorption and emission process thus reflecting the Bloch **k** conservation indicated by the Kronneker delta of (5.45) for the case that $|\mathbf{K}_1 - \mathbf{K}_2|$ are very small compared with the Brillouin-zone dimensions. These conclusion were later verified by Johnson and Ma (1994) based on results of a tight binding calculation.

Very similar conclusion were drawn by Skytt *et al.* (1994) from the results of a RIXS study on highly oriented pyrolytic graphite (HOPG) likewise performed at the X1B beamline of the NSLS. This investigation has not only shown a strong incident-photon energy dependence of the RIXS spectra, which again could be traced back to Bloch-**k** conservation, but has additionally utilized the angular momentum dependence of the absorption and emission related matrix elements of (5.45). When the polarization vector of the incident X-ray beam is parallel to the basal plane of graphite, only excitations to σ states are possible (small angles θ between the incident beam and the *c*-axis). Excitations to π states become more likely the more perpendicular to the basal plane the polarization vector is (θ angles near 90°). The emission direction had an angle $90° - \theta$ with the *c*-axis. When π-electrons fill the core vacancy, the X-rays are emitted in a dipole pattern such that no π emission is along the *c*-axis. Photons resulting

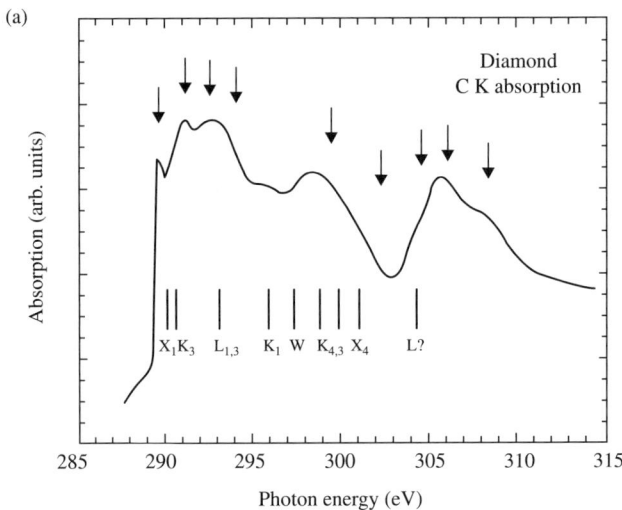

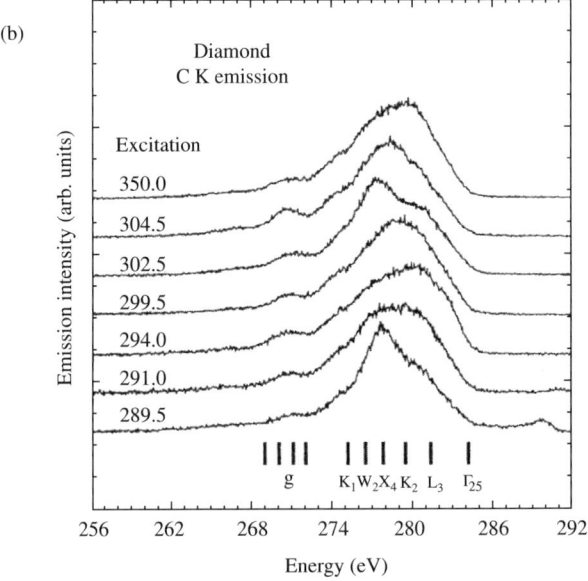

Fig. 5.28. C K absorption (a) and emission (b) spectra of diamond. The emission spectra were taken at selected excitation energies as indicated by the arrows in (a). The emission spectra are normalized by its total area. Vertical bars indicate symmetry points. g in (b) stand for a group of symmetry points, K, X, W, L. (Reprinted with permission from Ma *et al.* (1992); copyright (1992) by the American Physical Society.)

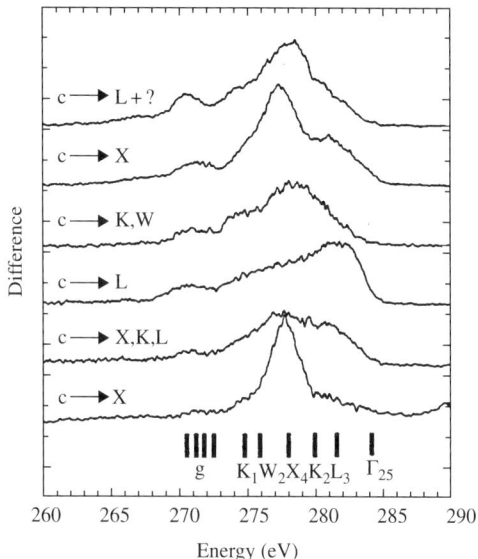

Fig. 5.29. Difference spectra. Differences between the spectra in Fig. 5.28(b) and 60% of the 350 eV spectrum. The label c → X indicated that the core electron is mostly excited to the X-point in the conduction band. The bars at the bottom specify the decay of the valence electron to the core hole. (Reprinted with permission from Ma *et al.* (1992); copyright (1992) by the American Physical Society.)

from filling the core hole from σ states are emitted in all directions but strongest in the direction of the c-axis. This way the selected emission spectra shown in Fig. 5.30 exhibit both a strong dependence on the emission direction and on the incident-photon energy. The emission direction dependence can be understood according to the above symmetry considerations. Thus the emission from π valence states is strongly suppressed with emission direction in the nearly c-direction, whereas an emission direction nearly perpendicular to the c-axis brings about both emission from π and σ states. This symmetry sensitivity of the RIXS spectra can be used to extract π and σ resolved contributions to the emission. The dependence on the incident-photon energy must again be traced back to the Bloch-**k** conservation. The critical points in the valence band are indicated in Fig. 5.30, with thicker bars for the π-symmetry with thinner for the σ symmetry. One can, for example, easily see, when going from the curve b (285.5 eV excitation energy) to curve d (286.6 eV excitation energy), that both for the π and for the σ states the emission critical points are shifting from K to M in agreement with a corresponding shift of the absorption critical points. Another important point was raised in this study. To obtain the RIXS intensity of Fig. 5.30 from the emission data, it was assumed that the fluorescence spectrum

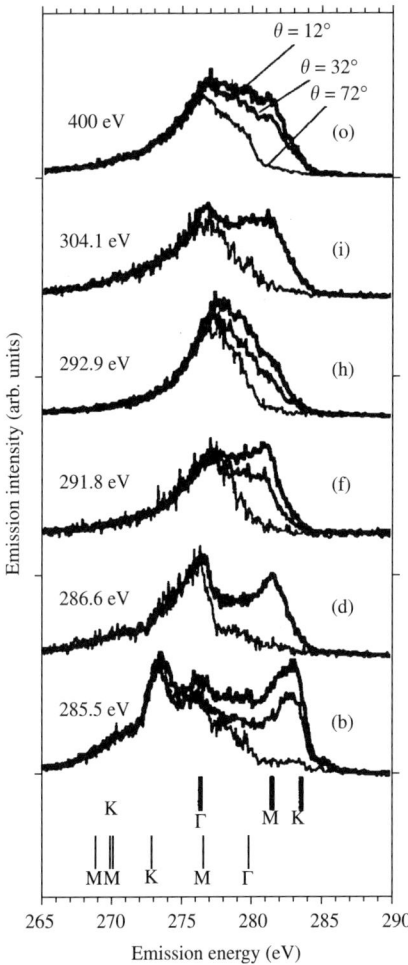

Fig. 5.30. RIXS spectra of graphite with different incidence angles and excitation energies as indicated in the figure. Thinner bars indicate the energy position of critical points in the valence band with σ symmetry, thicker bars those with π symmetry. (Reprinted with permission from Skytt et al. (1994); copyright (1994) by the American Physical Society.)

consists of a so-called coherent part, for which the momentum is conserved and an incoherent part, with a shape which is independent of the incident energy. The high-energy excited fluorescence spectrum was used as the incoherent part. Therefore, in order to obtain the coherent part, first a linear background was subtracted from each spectrum and then a certain amount of the high-energy excited spectrum, recorded at the same emission angle, was subtracted. The amount subtracted was dependent upon the excitation energy, and was chosen such that no negative features appear in the difference spectrum. This procedure was later taken over by many others working with Bloch-**k** selectivity RIXS. Its physical foundation has to be found in relaxation processes occurring in the intermediate state of the RIXS process, as first emphasized by Ma (1994). Further considerations by Eisebitt and Eberhardt (2000) placed electron–phonon interaction in the foreground, to be the most important reason for relaxation. One assumes that one electron–phonon scattering event may be sufficient to randomize **k** and make this transition show up in what Eisebitt and Eberhardt (2000) call the **k**-unselective fraction. Two time-scales are of importance in this connection: the intermediate state lifetime determined by the core-hole lifetime, or the core-hole decay rate R_c, and the electron–phonon scattering rate R_{ph}. Then the Bloch **k**-selective fraction f will be given by

$$f = \frac{R_c}{R_c + R_{ph}}. \tag{5.47}$$

Since R_{ph} is a function of the energy of the excited electron in the conduction band, also f will vary with the excitation energy in a RIXS experiment. Take, as an example, the calculations of R_{ph} on Si from Fischetti and Higman (1991), according to which the electron–phonon scattering rate in an energy range 0.1 eV above the conduction band minimum is about one order of magnitude lower than the core-hole decay rate of the Si 2p core level, so that f is nearly one. But already 1.0 eV above the conduction band minimum R_{ph} is roughly one half of the 2p core hole decay, and f reaches 0.65. This tendency carried on. Therefore, it makes sense to speak of the completely **k**-unselective valence fluorescence spectrum, when excited far above threshold. Thus f can be determined experimentally by subtracting a certain amount of this valence fluorescence spectrum from the actual RIXS spectrum so that no negative features appear, and by relating the resulting spectrum to the total RIXS spectrum (Eisebitt and Eberhardt (2000)). The experimental determined **k**-selective fraction for crystalline Si follows the excitation energy dependence of (5.47), although with smaller absolute values compared with the theory. This can be traced back to other relaxation processes like electron–electron Coulomb interactions, which are of special importance for highly correlated material as will be shown in Section 5.7.

The effect of the core–hole potential in the intermediate state, in other words, the appearance of a core exciton, can also disturb the Bloch-**k** selectivity. As Ma (1994a) has pointed out, the photoelectron in the presence of the core hole can be described by a wavepacket around the core hole, so that the Bloch state $|\mathbf{k}_i\rangle$ in (5.43) must be replaced by a superposition of Bloch waves written $\sum_\mathbf{k} A(\mathbf{k} - \mathbf{k}_i)|\mathbf{k}\rangle$ with the envelope function $A(\mathbf{k})$. If the effect of the core hole is small, $A(\mathbf{k}) \approx \delta(\mathbf{k} - \mathbf{k}_i)$, the result of (5.45) is recovered. In the other limit, when the electron is excited into a localized core-excitonic state, the whole Brillouin zone has to be sampled in (5.45), and no electronic momentum information can be extracted from this type of scattering process. Van Veenendaal and Carra (1997) have exemplified this phenomenological picture by a calculation, where the presence of a core-hole Coulomb potential was accounted for by an additional term in the Hamiltonian. Their calculated results exhibit a clear indication for additional structures appearing in the RIXS spectra of graphite at 8 eV, shown in Fig. 5.31, which can be traced back to excitations into states in the flat-band region between L and M, not governed by **k** selectivity, but emphasized by a high density of states. As depicted in the calculated X-ray absorption spectrum (inset) of Fig. 5.31, these states are pulled down by the core-hole Coulomb interaction, so that the absorption can take place closer to the threshold. The additional structure at 8 eV in the RIXS spectrum has been observed by Carlisle *et al.* (1995), however interpreted as a **k**-unselective contribution.

In the case of a shallow core exciton we can assume a Gaussian envelope function, the width of which is roughly $2\pi/D$, where D is the size of the exciton.

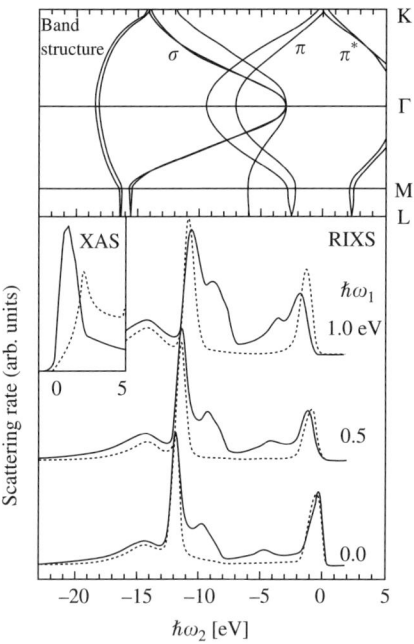

Fig. 5.31. Upper panel: band structure of graphite. Lower panel: numerical calculations of the graphite K-edge emission spectra for emission angle $\alpha = 25°$ in the definition of Carlisle *et al.* (1995). The incident and scattered photon energy $\hbar\omega_1$ and $\hbar\omega_2$, respectively are measured from the absorption threshold. The independent quasiparticle approximation is given by the dashed line. The inclusion of a local core-hole potential of -3 eV is represented by the solid line. Inset: corresponding absorption spectrum. (Reprinted with permission from van Veenendaal and Carra (1997); copyright (1997) by the American Physical Society.)

Then this width has to be added to the other uncertainty of the momentum conservation relation.

Most of the applications of the Bloch-**k** selectivity to band mapping of systems that follow the conceptions of the band structure model have been performed within the soft X-ray energy range. This is mainly due to the much longer lifetime of the contributing core hole compared with that utilized in the case of harder X-rays. Thus the influence of the core-hole width on the energy resolution and consequently on the amount of **k**-selectivity could be reduced. On the other hand, harder X-rays offer a larger freedom to play with the amount and the direction of the momentum transfer **q**, utilizing in this way the orientation dependence of the spectra as an additional source of band structure information. Moreover, harder X-ray RIXS probes really the bulk, whereas soft X-ray RIXS is strongly influenced by surface effects. Additionally, the short core-hole lifetime, and consequently the high core-hole decay rate has the advantage of being much larger than the electron–phonon scattering rate, so that the relaxation in the intermediate state is strongly suppressed. Since this book is mainly dealing with the application of harder X-rays, the following examples belong to this category of **k**-selective probing. Only at the end of this subsection will a state-of-the-art soft X-ray band mapping experiment be shown.

Ma *et al.* (1995) were the first to exploit the **q**-orientation dependence of the valence electron RIXS spectra of Si taken at the Si K edge in order to obtain direct evidence of their **k**-selective behavior. Figure 5.32 shows the spectra recorded at various points of the Si K absorption edge in each case with different crystal orientations as indicated. The spectra were measured at the beamline X24A of the NSLS using a bent InSb crystal and a position-sensitive detector to energy-analyze the radiation scattered under 90° from the Si sample. Especially, the spectra excited with an incident energy of 1840.8 eV exhibit a strong orientation dependence. Figure 5.27 illustrates how this orientation dependence can be understood. For the near-threshold excitation of 1840.8 eV, the photoelectron is excited to the conduction band minimum near the X_1 point at about $\Delta_1 = (2\pi/a)(0.8, 0, 0)$, where a is the lattice constant of Si. Momentum conservation then predicts that the emission can occur only from points in the Brillouin zone given by $\mathbf{k}_2 = \Delta_1 + (\mathbf{K}_2 - \mathbf{K}_1)$. For the (111) sample there are six points that comply with the above relation, three of them are very close to the L, and three close to the K type critical point. Indeed, the most pronounced peaks of the (111) curve are at -7, -4.5, -2 eV, which agree nicely with the calculated positions of the L_1, K_4 and L_3 points.

Another example of Bloch-**k** selective RIXS with hard X-rays has been offered by Enkisch *et al.* (1999). These authors have investigated the ordered stoichometric NiAl alloy using excitation energies near the Ni K edge. The RIXS measurements were performed at the beamline ID28 of the ESRF using a 1 m spherical Rowland spectrometer. The overall energy resolution was 1.0 eV. Figure 5.34 shows measured spectra with **q**∥[110] and an excitation energy 1.7 eV above the Ni K edge for different values of the momentum transfer |**q**| in units

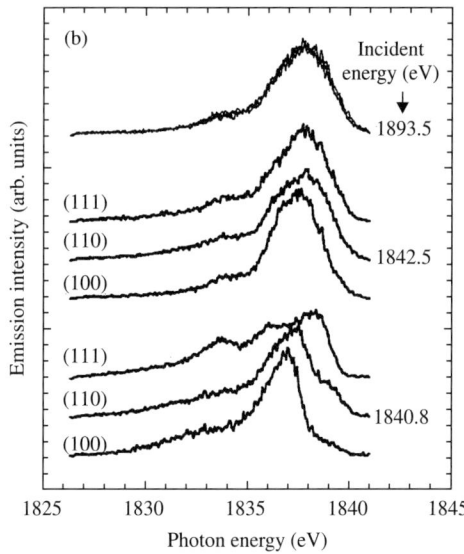

Fig. 5.32. Si K emission spectra excited at two different near K-edge excitation energies and an excitation energy far above the K-edge. Spectra for three different orientations of the momentum transfer $\hbar\mathbf{q}$ are shown. The spectra were normalized to have unit area. (Originally published by Ma et al. (1995); copyright (1995) by the American Physical Society.)

of $(2\pi/a)(1,1,0)$. The shape of the spectra depend sensitively on the magnitude of $\mathbf{q}$. The arrows illustrate the strong dispersion of a distinct spectral feature. In order to find out the band structure related origin of this dispersion, one must have in mind that this spectroscopy offers a very selective image of the NiAl band structure. The predominance of the dipole transitions according to the matrix elements in (5.45), on the one hand, and the resonance behavior at the Ni K edge, on the other hand, let only the Ni projected band structure with p-symmetry contribute to the spectra. Taking this into account corresponding band structure calculations of the spectra based on linearized augmented plane wave (LAPW) show an overall good agreement with experiment. Thus the dispersion visible in Fig. 5.33 can be explained in the following way. For the 1.7 eV spectrum, the possible $\mathbf{k}_i$ vectors are distributed around the R point. With increasing $|\mathbf{q}|$, the $\mathbf{k}_f$ are shifted parallel to the [110] direction from the R point to the point X. Thus the lowest band with p character around the X point causes the dispersing shoulder. It must be stressed that the good agreement between calculation and measurement could only be achieved by adding a so-called shakeup satellite to the valence line, sometimes described as a radiative Auger effect (Åberg 1971), being the excitation of another valence electron into the conduction band during the re-emission process resulting in an energy loss of the emitted photon. Since the minimal energy loss of such a process is zero, whereas the maximum energy

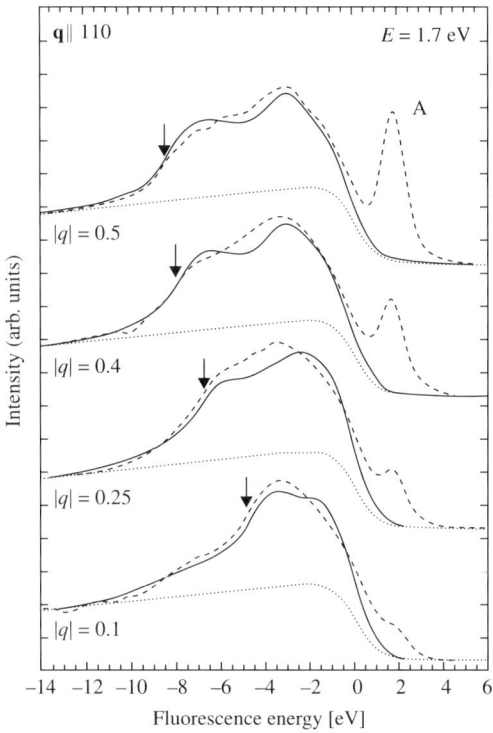

Fig. 5.33. Measured (dashed line) and calculated (solid line) RIXS spectra of NiAl with constant incident photon energy $E_{\text{edge}} + 1.7$ eV. $\mathbf{q} \| [110]$, q in units $|2\pi(1,1,0)/a|$. Peak A is the quasielastic line. Dotted line: estimated shakeup satellite. (Reprinted with permission from Enkisch et al. (1999); copyright (1999) by the American Physical Society.)

loss is not limited, the corresponding shakeup satellite shows a steep drop on its high-energy side and a slowly decreasing tail on its low-energy side as shown in Fig. 5.33. Contrary to applications of soft X-rays to band mapping, this hard X-ray study did not need the subtraction of a **k**-unselective contribution to the spectra in order to come into agreement with calculations. This is apparently due to the fact that the core–hole decay rate is large compared with the electron–phonon scattering rate, a very important advantage of the hard X-ray application of this spectroscopy.

How, in the general case, information about band dispersion (so-called band mapping) can be extracted from RIXS measurements which is exemplarily demonstrated guided by soft X-ray experiments on diamond, performed by Sokolov et al. (2003). In Fig. 5.34(a) the C Kα X-ray emission spectrum of diamond is shown as excited with an incident photon energy of 289.4 eV, just above the threshold. The structures denoted 1 to 6 have been made better visible

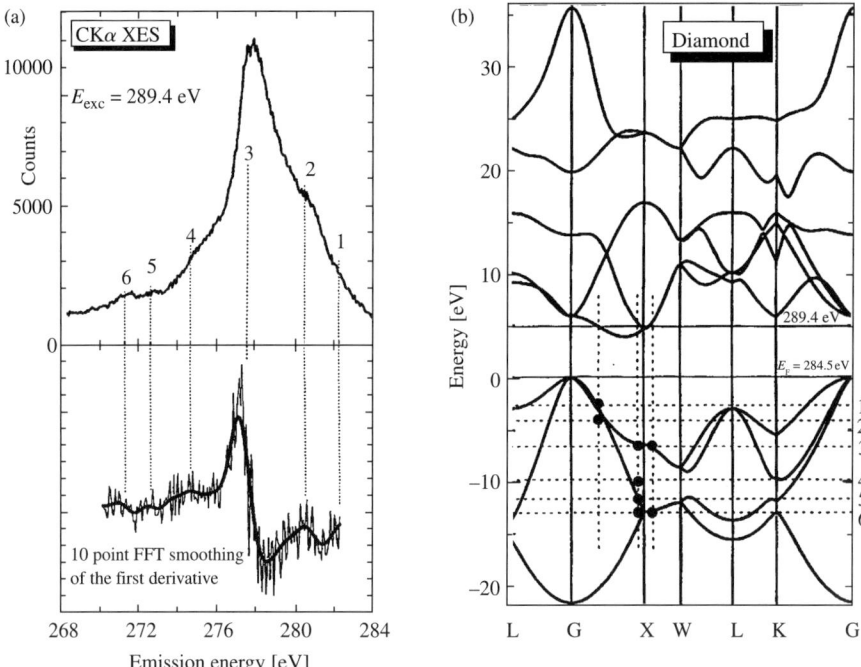

Fig. 5.34. (a) Upper panel: the C Kα X-ray emission spectrum (XES) of diamond measured at the exciting photon energy $E_{exc} = 289.4$ eV. Lower penal: FFT smoothing of the first derivative of the above XES spectrum. (b) Selection of excitation points in the unoccupied band structure for the incident energy of 289.4 eV. Experimental points on the dispersion curves for occupied states according to the features 1–6 of the XES. (Reprinted with permission from Sokolov (2003); copyright (2003) by IOP Publishing Ltd.)

by means of a 10 point fast Fourier transform (FFT) smoothing of the first derivative. The horizontal line in Fig. 5.34(b) corresponding to the incident-photon energy, and intersecting the calculated band structure of unoccupied states, defines the selected **k** points on the high-symmetry directions within the Brillouin zone. Vertical lines (direct transitions!) extending from these selected **k** points intersect horizontal lines corresponding to the marked features on the emission spectrum thus giving the experimental points on the dispersion curves for occupied states. This procedure is repeated for many different excitation energies as shown in Fig. 5.35, ending up with a densely populated band structure of occupied states with predominantly p symmetry. One can easily realize that the band mapping by means of RIXS is less direct than band structure determination using **a**ngular **r**esolved **p**hotoemission **s**pectroscopy (ARPES). In the RIXS case the crystal momentum $\mathbf{k}_i$ of the unoccupied state reached by the photoelectron must be deduced from a calculated band structure, unless this

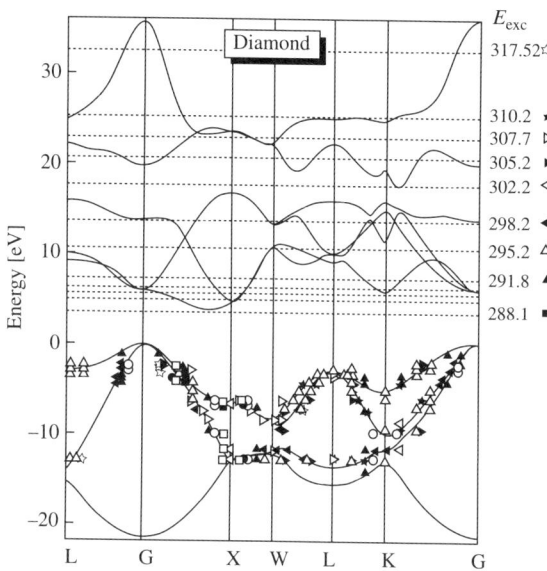

Fig. 5.35. Comparison of experimental and calculated dispersion curves of diamond based on the treatment of experimental spectra presented in Fig. 5.34. Each symbol corresponds to one excitation energy. (Reprinted with permission from Sokolov *et al.* (2003); copyright (2003) IOP Publishing Ltd.)

state can be unambiguously identified to be a high-symmetry critical point by a structure in the core-electron absorption spectrum. On the contrary, ARPES provides two components of $\mathbf{k}_i$ simply by the value of the kinetic energy of the photoelectron. Of course, the determination of the third component of $\mathbf{k}_i$ needs also the identification of high-symmetry critical points. Concerning the symmetry information, both methods are complementary, since ARPES probes the itinerant band symmetry, whereas RIXS reflects the local angular momentum character of the investigated states. Moreover, the bulk band structure information of RIXS is not disturbed by surface states as in the case of ARPES. Finally, ARPES needs conducting single crystals, whereas RIXS can be performed even on insulators in electromagnetic fields.

5.6.3 *Spin conservation; internal spin reference*

We have seen in the previous section that it is the coherence between absorption and re-emission as parts of the RIXS process, which gave rise to momentum conservation for the whole process and consequently to Bloch $\mathbf{k}$ selectivity, very useful for band mapping. Of similar importance is the validity of spin conservation for the whole RIXS process, since it enables spin-selective information about the density of unoccupied states. This was proposed for the first time by

Hämäläinen et al. (1992) with an investigation of the spin-dependent absorption of MnO and MnF_2. The principle of this application of spin conservation has already been shown schematically with Fig. 1.22 of Section 1.3.4.4 for the case of resonantly excited Mn $K\beta_{1,3}\beta'$ (1s3p) emission in MnO. Multiplet calculations (Hermsmeier et al. 1988, Peng et al. 1994) and the evaluation of the 3p photoemission (Sinkovic and Fadley 1985) have shown that the $3p^53d^5$ final state configuration, which is the same for the $K\beta$ X-ray emission and the 3p photoemission, exhibits two groups of multiplets as a consequence of the strong exchange coupling between the 3p hole and the 3d electrons. The main component, $K\beta_{1,3}$, which is responsible for the higher-energy part of the $K\beta$-line, corresponds to a final state with a missing 3p electron, which was spin down relative to the spin orientation of the 3d electrons, as depicted in Fig. 1.22. On the contrary, the "satellite" component $K\beta'$, shifted down in energy by 17 eV relative to the main line, must be ascribed to a final state with a missing spin-up 3p electron. If now an electron from the 1s level, spin down with respect to the internal spin reference, namely the oriented 3d spins, is excited into the MnO conduction band, an electron with the same spin orientation from the 3p level must refill the 1s hole. This means that excitation of an electron into a spin-down state of the MnO conduction band is always connected with a re-emission into the main component of the $K\beta$ line. An excitation into a spin-up state of the conduction band leads to re-emission into the $K\beta'$ satellite. Therefore, monitoring the absorption process by measuring the integral intensity of the main component of the $K\beta$ line, one obtains information about the spin-down partial density of unoccupied states (minority spin), if one assumes that the corresponding matrix element is independent on the spin polarization. Using the integral intensity of the $K\beta'$ satellite to monitor the absorption process, the spin-up partial density of unoccupied states (majority spin) is observed. It must be stressed that the spin orientation considered here is always the spin orientation with respect to the orientation of the 3d spins at the excited atom. The internal spin reference is strictly local, independent of the collective spin orientation of the 3d electrons, and therefore, the same in a ferromagnetic, antiferromagnetic or paramagnetic state.

Wang et al. (1997) have emphasized that not only the $K\beta$ X-ray emission of 3d transition metal ions is spin polarized, but also the $K\alpha$ emission. They offered a systematic analysis of the first-row transition-metal ions using the ligand-field multiplet calculations for $K\alpha$ and $K\beta$ emission spectra. This way it can be derived from the simulations that the high-energy side of $K\alpha_1$ is more favorable for spin-down transitions, while the low-energy side of $K\alpha_1$ represents mostly spin-up transitions.

The result of such a spin-dependent X-ray absorption study on MnO performed by Hämäläinen et al. (1992) at the double focusing 25 hybrid wiggler beamline of the NSLS is shown in Fig. 1.23. The minority (spin-down) and majority (spin-up) spectra are normalized to be the same well above the absorption edge. The clear difference in the spin-dependent density of states (DOS) extends about 10 eV above the edge, where also the absorption edges exhibit a slight energy shift between the different spin states (0.4 eV for MnO). This shift reflects the exchange splitting, where the majority spin states have a lower energy than the minority states. The pre-edge peak below the absorption edge can be attributed to quadrupolar transitions to unoccupied 3d states. This pre-peak is strongly damped in the majority spin spectrum, an indication for the fact that, according to Hund's rule, all five available spin-up 3d states are already occupied while all five remaining spin-down states are empty. As shown later for the EuO case, one must take into account that neither the main Kβ line is pure spin-up nor the satellite pure spin-down. This means that also the DOS spectra shown in Fig. 1.23 cannot be considered to be pure majority and minority spin, respectively.

Full multiple-scattering calculations of Soldatov *et al.* (1995) have confirmed the spin-dependent absorption studies of Hämäläinen *et al.* (1992). Their approach was based on a two-step procedure. In order to obtain two sets of potentials for both spin-up and spin-down electrons, the authors have performed a self-consistent spin-polarized calculation of a MnO_6 (MnF_2F_4) cluster that is a fragment of the MnO (MnF_2) crystal. The two sets of potentials, corresponding to two different spin states, have been used to calculate two sets of the partial phase shifts of the photoelectron and two dipole transition matrix elements. Then a full multiple scattering XANES calculation in real space of Soldatov *et al.* (1993) was applied, where, in the case of MnO, a cluster of nine shells around the central Mn atom, and in the case of MnF_2 a cluster of five shells, was considered. The result of such a calculation is shown for MnF_2 in Fig. 5.36 and compared with the measurements of Hämäläinen *et al.* (1992). Notice that the dipolar matrix elements shown in the lower panel of Fig. 5.36 exhibit a distinct spin dependence. Thus one has to conclude that the spin sensitivity of the edge spectra is not only due to the spin polarization of the unoccupied DOS, but also influenced by the spin-dependent matrix elements. The agreement of the calculated edge spectra with experiment is rather good, even if the spin-dependent shift of the edge seems to be overestimated by theory. But one should have in mind that, as already stated above, the experimental spectra are not pure spin-up and spin-down, respectively.

De Groot *et al.* (1995) have corrected their local-spin-selective absorption spectra of MnP by taking into account the incomplete spin polarization of the Kβ emission spectra due to the overlap of the Lorentzian tail of the spin-down main line with the spin-up Kβ satellite. The intensity of this tail is less than 1% at 17 eV from the peak to lower energies, but because the satellite intensity is rather low, the influence of this tail becomes visible as the onset of the spin-up spectrum at 17 eV lower energy than the spin-down spectrum, so that it must be corrected for. This correction is performed by subtracting an appropriate fraction of the spin-down spectrum shifted over 17 eV from the spin-up spectrum. De Groot *et al.* (1995) have drawn attention to another peculiarity of spin-selective X-ray absorption: As shown in Section 5.3.4, one can suppress the influence of

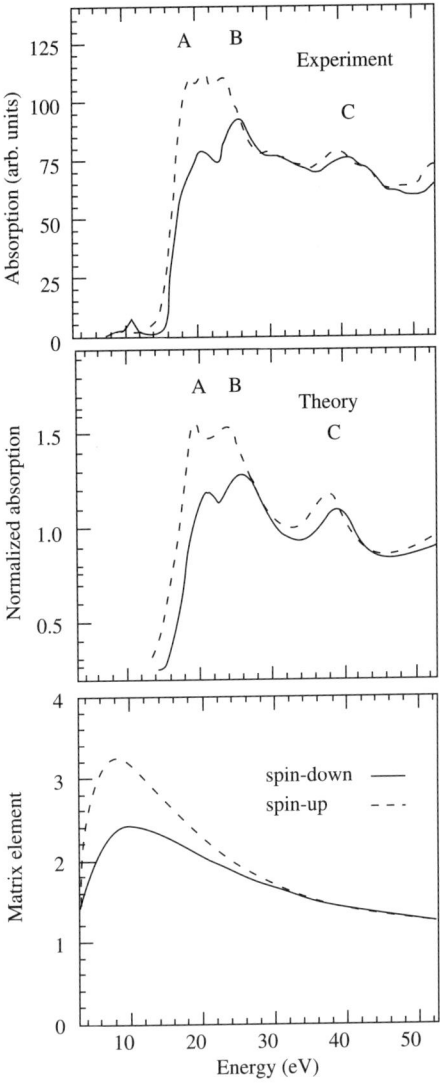

Fig. 5.36. Upper panel: experimental spin-dependent XANES spectrum at the Mn K edge in MnF_2 from Hämäläinen et al. (1992). Middle panel: calculated normalized Mn K edge XANES of MnF_2 for two spin configurations. Spectra have been normalized to the value of the absorption coefficient at high energy. Lower panel: energy-dependent normalized squared dipole transition matrix elements, corresponding to dipole Mn 1s → εp electronic transition in MnF_2 for two spin configurations. In all curves: dashed: spin-up; solid: spin-down. (Originally published by Soldatov et al. (1995); copyright (1995) by the American Physical Society.)

the core-hole lifetime broadening on the absorption spectrum, when utilizing the partial fluorescence yield (PFY) mode by monitoring the absorption cross-section looking at X-rays re-emitted due to refilling the core hole by an electron of a shallow core state. This mechanism works well, if the final state hole has a sharp energy. In cases, where the hole states are split into multiplet components distributed over an energy range comparable with the lifetime broadening of the intermediate state core hole, also the absorption spectrum monitored by this multiplet ensemble is broadened correspondingly. This applies to the Kβ satellite, while the main line is essentially single component. Therefore, in order to compare appropriately the spin-up and the spin-down edge spectra one has to convolute the latter by the additional broadening, that is the 1s core-hole lifetime broadening. The study of de Groot et al. (1995) on MnP has pioneered another interesting application of spin-selective RIXS, namely the determination of the energy dependence of the so-called Fano factor by comparing the spin-selective X-ray absorption as measured by means of RIXS with the X-ray magnetic circular dichroism (XMCD) (Schütz et al. 1987) of these spectra. The former is given by

$$\mu_\uparrow(\omega_1) - \mu_\downarrow(\omega_1) = [R_\uparrow^2(\omega_1)\rho_\uparrow(\omega_1)] - [R_\downarrow^2(\omega_1)\rho_\downarrow(\omega_1)], \quad (5.48)$$

where R^2 is the squared dipole matrix element coupling the 1s wavefunction with the p wavefunction and ρ is the density of empty states. The XMCD is defined as the difference between the absorption coefficient of left (μ_L) and right (μ_R) circularly polarized X-rays which is given by

$$\mu_L(\omega_1) - \mu_R(\omega_1) = P(\omega_1)([R_\uparrow^2(\omega_1)\rho_\uparrow(\omega_1)] - [R_\downarrow^2(\omega_1)\rho_\downarrow(\omega_1)]). \quad (5.49)$$

$P(\omega_1)$ is the so-called Fano factor (Fano 1969), which is indicative of the physical reason for XMCD: With circularly polarized X-rays one excites to p states with either $j = 3/2$ or $j = 1/2$. Only because of spin–orbit coupling in theses p states are the radial matrix elements for $p_{3/2}$ and $p_{1/2}$ not exactly equal, so that a dichroic effect is visible. For small differences between the matrix elements the Fano factor P can be approximated by

$$P \approx \frac{8}{3}\left[\frac{R_{3/2} - R_{1/2}}{R_{3/2} + R_{1/2}}\right], \quad (5.50)$$

where R_j are the radial matrix elements. By comparing (5.48) with (5.49) one can easily recognize that the Fano factor can be determined experimentally via

$$P(\omega_1) = \frac{\mu_L(\omega_1) - \mu_R(\omega_1)}{\mu_\uparrow(\omega_1) - \mu_\downarrow(\omega_1)}. \quad (5.51)$$

Figure 5.37 shows the result of such a measurement on MnP.

So far the knowledge about the spin polarization of the Kα or Kβ emission has been used to obtain the spin-selective absorption spectra at the K edge. Especially in the case of Mn(II) complexes, the spin polarization of the Kβ satellite was nearly completely spin-up, and spin polarization of the main K$\beta_{1,3}$

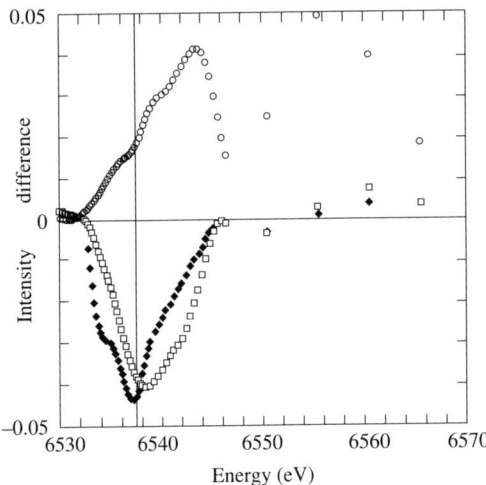

Fig. 5.37. Open circles: local spin-selective X-ray absorption difference spectrum. Open squares: X-ray magnetic circular dichroism (XMCD) spectrum multiplied with 50. Their ratio is the Fano factor (solid diamonds). Above 6546 eV the 5 eV average values are plotted. (Reprinted with permission from de Groot *et al.* (1995); copyright (1995) by the American Physical Society.)

line was strongly dominated by spin-down. Thus the influence of slightly mixed polarization could be corrected for empirically (de Groot *et al.* 1996). The situation is more complicated at the L edges of rare-earth compounds, where the spin polarization of the 2p4d X-ray emission ($L\beta_{2,15}$) is utilized. Making the crude assumption that the main line of the $2p_{3/2}4d$ emission of Eu in EuO is pure spin-down (in the literal sense used above), the satellite shifted by 29 eV to lower energies is pure spin-up, Magdans (2001) obtained from the integrated intensities of the main (HP) peak and the satellite (SAT) as a function of the incident energy a "spin selective" L-edge spectrum at two different temperatures as shown in Fig. 5.38(a). EuO is paramagnetic at ambient temperatures 293 K and ferromagnetic at 10 K. All spectra were normalized to the same peak intensity. However, multiplet calculations and XMCD measurements at the $L\beta_{2,15}$ emission line (Wittkop *et al.* 2000) have clearly shown that the two components of the emission spectrum are not pure spin-down and spin-up, respectively, but reveal the following mixture: satellite: 17.7% down, 82.3% up; main peak: 60.1% down, 39.9% up. Moreover, the integrated intensity ratio of the main line to satellite is 82.6 to 17.4. Therefore, the measured edge spectra are composed the following way of the pure spin polarized ones:

$$\begin{aligned} I^{\text{Sat}}(\omega_1) &= a \cdot I^{\uparrow}(\omega_1) + b \cdot I^{\downarrow}(\omega_1) \\ I^{\text{Main}}(\omega_1) &= c \cdot I^{\uparrow}(\omega_1) + d \cdot I^{\downarrow}(\omega_1) \end{aligned} \quad (5.52)$$

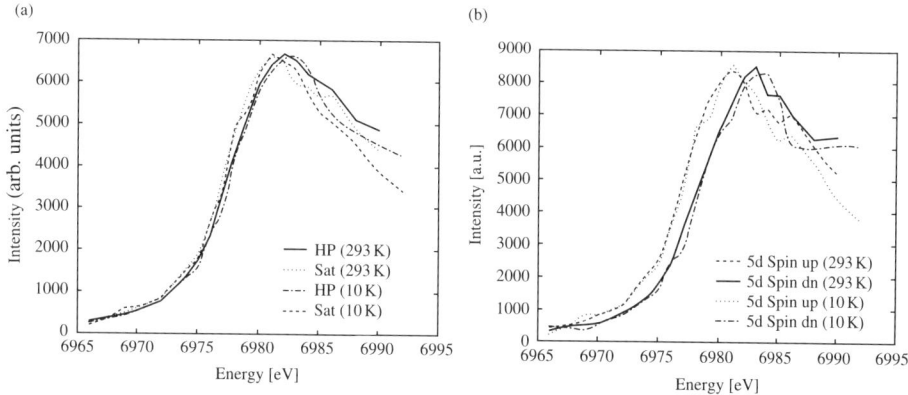

Fig. 5.38. (a) Integrated intensities of satellite (SAT) and main peak (HP) of the Eu L$\beta_{2,15}$ fluorescence spectrum of EuO after background subtraction and normalization to the same peak intensity as a function of the incident photon energy in the range of the Eu L$_3$ edge, for two different temperatures as indicated. (b) Spin polarized Eu L$_3$ edge spectrum of EuO for two different temperatures as obtained from (a) taking into account the spin polarization of the 4d multiplet. (Reprinted with permission from Magdans (2001).)

with $a = 0.143$, $b = 0.031$, $c = 0.330$, and $d = 0.496$. By applying these relations to the spectra of Fig. 5.29(a), Magdans (2001) ends up with the pure spin-selective L-edge spectra of EuO as shown in Fig. 5.38(b). On the average the spin-up spectra are shifted by 1.7 ± 0.1 eV for 293 K and by 1.6 ± 0.1 eV for 10 K to lower energies with respect to the spin-down ones. This is in rather good agreement with LMTO (linear muffin-tin orbital) calculations of Attenkofer (2001) which gave an average shift of 1.5 eV of the 5d bands. The small shift of the spectra to lower energies with increasing temperature (0.3 ± 0.1 eV for spin up and 0.2 ± 0.1 eV for spin down) can be traced back to a combined action of lattice expansion, electron–phonon interaction and the real part of the self-energy of the 5d electrons due to magnon exchange between 5d spin down and the aligned 4f in the ferromagnetic state (d-f model of Nolting *et al.* 1987a,b).

An alternative way to achieve spin-selective absorption spectroscopy without specific knowledge about the spin polarization of the re-emission used to monitor the absorption process has been proposed by Dräger *et al.* (2001). Supposing only spin-conserving transitions in the RIXS process the complementary spin parts x for spin up and $(1 - x)$ for spin down in the K absorption spectrum should be the same for the resonantly scattered K$\alpha_{1,2}$ radiation, thus writing for the emitted intensity

$$I(\omega_1, \omega_2) = x(\omega_1) \cdot u(\omega_1, \omega_2) + [1 - x(\omega_1)] \cdot d(\omega_1, \omega_2), \quad (5.53)$$

where $u(\omega_1,\omega_2)$ and $d(\omega_1,\omega_2)$ are the spin up and spin down component of the resonantly scattered radiation. $I(\omega_1,\omega_2)$ has to be normalized to the integral intensity. Then one has to solve the nonlinear equations like (5.53). For this reason one has to analyze two or more RIXS emission spectra measured at one ω_1 but angular dependently for several sample orientations or, alternatively, some spectra measured at one orientation and excited at several, but closely neighboring ω_1. In this case the comparatively small ω_1 dependence of $u(\omega_1,\omega_2)$ and $d(\omega_1,\omega_2)$ can be neglected. The spectra can be resolved into the u and d thus determining the x and $(1-x)$ by using special methods of the principal component analysis (PCA) (Jolliffe 2002). This way Dräger et al. (2001) have analyzed the spin character of the quadrupole allowed 1s → 3d transition in MnO and NiO resolving the pre-edge structure with respect to the orbital symmetry into t_{2g} and e_g contributions as described in Section 5.5.1. Both components exhibited spin-down character.

5.7 Excitations in the intermediate state

5.7.1 *Theoretical approach*

As shown in Chapter 2, nonresonant inelastic X-ray scattering is an excellent tool for investigating collective and single-particle excitation in condensed matter, where the dynamic structure factor $S(\mathbf{q},\omega)$ was the quantity one can obtain in a very direct way from experiment. Above all, it was the dependence of the measured spectra on the momentum transfer $\mathbf{q}$, which was unrestricted in its magnitude and lead to invaluable information about dispersion of the excitation, that has attracted attention. Unfortunately, these experiments suffer from the rather unfavorable ratio between photoabsorption and inelastic scattering cross-section, especially when taking into account that this ratio is increasing dramatically with the atomic number of the scatterer. Therefore, most of the experiments described in Chapter 2 were restricted to low-Z elements and compounds. Excitations in the very interesting highly correlated systems which are the parent compounds of high-T_c superconductors could not be investigated with nonresonant IXS because of the significant fraction of high-Z elements. But the first experimental studies of Kao et al. (1996) on NiO and of Hill et al. (1998) on Nd_2CuO_4 as well as a thorough theoretical study of Platzman and Isaacs (1998) have opened new spectroscopic access to electronic excitations in these materials by means of RIXS. They considered a RIXS process, where a 1s core electron of the transition metal is excited into unoccupied states with 4p symmetry. The existence of the core hole in the intermediate state can lead to an excitation of the transition metal 3d electron system, so that the final state, after refilling the core hole by the 4p electron, can be characterized by an excited state of the transition metal 3d system with energy $\hbar\omega = \hbar\omega_1 - \hbar\omega_2$. The spectrum of scattered radiation will exhibit, in addition to the elastic line, a spectrum which is shifted relative to the elastic line to lower energies by the excitation energy $\hbar\omega$ transferred to the 3d system. Moreover, the momentum $\hbar\mathbf{q} = \hbar(\mathbf{K}_1 - \mathbf{K}_2)$ is transferred to the 3d excitation, so that its dispersion can be measured. Compared with the same excitation in nonresonant inelastic scattering, the intensity of the corresponding feature in the resonant case is much larger when the incident photon energy is tuned to the 1s binding energy in order to make use of the strong resonance denominator of (5.2). Of course, in the resonant case, there does not exist such a direct relationship between the double differential scattering cross-section and the dynamic structure factor $S(\mathbf{q},\omega)$ as in the nonresonant case, especially because of the complicated matrix elements that connect the absorption and the re-emission process. Therefore, it is mainly the information about magnitude and dispersion of the excitation energy which matters.

Within the last decade, four different modes of interpreting this type of RIXS spectroscopy were developed, each related to specific applications. The first one is the so-called Anderson impurity model as originally proposed by Anderson (1961) in order to discuss the magnetic moment of an impurity atom of a 3d transition element in nonmagnetic host metals. The application of this model is based on the experience that energy-band calculations using the Kohn–Sham equation (Kohn and Sham 1965) of the density functional theory (Hohenberg and Kohn 1964) is certainly applicable to the analysis of RIXS for semiconductors and ionic insulators, but for f- and d-electron systems it has not often been used successfully in the RIXS analysis. In the Anderson impurity model, we consider only a single atom with f or d states, and at the same time the other more extended electrons such as conduction or valence electrons are described with the band model. Hybridization between the localized f or d states and extended conduction or valence electron states is taken into account. Thus the Hamiltonian of this model writes for the case of localized 3d electrons and extended oxygen 2p bands (using the notation of Kotani and Shin 2001):

$$H = \sum_{\Gamma,k,\sigma} \varepsilon_{\Gamma k} a^+_{\Gamma k \sigma} a^+_{\Gamma k \sigma} \sum_{\Gamma,\sigma} \varepsilon_{d\Gamma} a^+_{d\Gamma\sigma} a_{d\Gamma\sigma} + \sum_{\mu} \varepsilon_{p} a^+_{p\mu} a_{p\mu}$$

$$+ \sum_{\Gamma,k,\sigma} V(\Gamma^k)(a^+_{d\Gamma\sigma} a_{\Gamma k \sigma} + a^+_{\Gamma k \sigma} a_{d\Gamma\sigma})$$

$$+ U_{dd} \sum_{(\Gamma,\sigma)\neq(\Gamma',\sigma')} a^+_{d\Gamma\sigma} a_{d\Gamma\sigma} a^+_{d\Gamma'\sigma'} a_{d\Gamma'\sigma'} - U_{dc} \sum_{\Gamma,\sigma,\mu} a^+_{d\Gamma\sigma} a_{d\Gamma\sigma}(1 - a^+_{p\mu} a_{p\mu})$$

$$+ \frac{1}{2} \sum_{\nu_1,\nu_2,\nu_3,\nu_4} g_{dd}(\nu_1,\nu_2,\nu_3,\nu_4) a^+_{d\nu_1} a_{d\nu_2} a^+_{d\nu_3} a_{d\nu_4}$$

$$+ \sum_{\nu_1,\nu_2,\mu_1,\mu_2} g_{pd}(\nu_1,\nu_2,\mu_1,\mu_2) a^+_{d\nu_1} a_{d\nu_2} a^+_{d\mu_1} a_{d\mu_2} + \zeta_d \sum_{\nu_1,\nu_2} (\mathbf{1} \cdot \mathbf{s})_{\nu_1\nu_2} a^+_{d\nu_1} a_{d\nu_2}$$

$$+ \zeta_p \sum_{\mu_1,\mu_2} (\mathbf{1} \cdot \mathbf{s})_{\mu_1\mu_2} a^+_{p\mu_1} a_{p\mu_2}. \quad (5.54)$$

Here $\varepsilon_{\Gamma k}$, $\varepsilon_{d\Gamma}$, and ε_p are energies of the oxygen 2p valence band, the transition-metal 3d state and a core level (with p symmetry), respectively. a^+ and a are the creation and annihilation operators, respectively, for states indicated by the indices. ν denotes the combined indices representing the spin (σ) and orbital (Γ) states, and μ stands for the spin and orbital states of the core level. The interactions V, U_{dd}, and $-U_{dc}$, respectively, are the hybridization between 3d and valence band states, the Coulomb interaction between the 3d electrons, and the core-hole potential acting on the 3d electrons. g_{pd} represents the multipole components of the Coulomb interaction between 3d and core states, and g_{dd} that between 3d states, with both including the Slater integrals in their explicit form. ζ_d and ζ_p are the spin–orbit coupling constants. In the calculation of RIXS spectra for 3d metal oxides, according to (5.2), first the Hamiltonian is diagonalized for the nominally ground state d^n, taking into account the interatomic configuration interaction with a sufficient number of configurations d^n, $d^{n+1}\underline{L}$, $d^{n+2}\underline{L}^2$, ... for the calculated results to converge, where $\underline{L}$ represents a hole in the oxygen 2p valence band. Similar calculations are done for the intermediate states, which include a core hole, and the final states in (5.2). We have to account for the whole spectrum of excitations in the intermediate and final states in order to interpret appropriately excitations found in the RIXS spectra as stressed above. The main parameters of the Anderson impurity model are V, U_{dd}, U_{dc}, and the charge-transfer energy defined by

$$\Delta \equiv E(d^{n+1}\underline{L}) - E(d^n), \tag{5.55}$$

where $E(d^n)$ is the energy averaged over multiplet terms of the d^n configuration. It should be noted that application of the Anderson impurity model is strictly a second-order perturbation treatment and does without the third-order perturbation treatment, which is presented below. Of course, the Hamilton formalism given in (5.54) can also be applied to 4f metal ions surrounded by oxygens. Also the 2p core can be replaced by a 1s core. It should be mentioned that the Anderson impurity model, as given in (5.54), is for RIXS not as applicable as it is for the description of X-ray absorption and photoemission, since often the single-metal-ion approximation breaks down, so that one has either to take into account the periodic arrangement of the metal ions, or a larger cluster model.

The second mode of interpreting the RIXS process has been introduced by Tsutsui et al. (1999) in order to describe transitions in the 3d valence system of copper oxide type Mott insulators, and is based on an extended Hubbard model, where the 3d system is represented by the following Hamiltonian with second and third neighbor hoppings:

$$H_{3d} = -t \sum_{(i,j)_{1st},\sigma} d_{i,\sigma}^+ d_{j,s} - t' \sum_{(i,j)_{2nd},\sigma} d_{i,\sigma}^+, d_{j,\sigma}$$
$$- t'' \sum_{(i,j)_{3rd},\sigma} d_{i,\sigma}^+ d_{j,\sigma} + \text{H.C.} + U \sum_i n_{i,\uparrow}^d n_{i,\downarrow}^d, \tag{5.56}$$

where $d_{i,\sigma}^+$ is the creation operator of a 3d electron with spin σ at site $\mathbf{i}$, $n_{i,\sigma}^d = d_{i,\sigma}^+ d_{i,\sigma}$, the summations $(\mathbf{i,j})_{1st}$, $(\mathbf{i,j})_{2nd}$, and $(\mathbf{i,j})_{3rd}$ run over first, second and third nearest neighbor pairs, respectively. t, t', and t'' are the corresponding hopping parameters. (5.56) has been applied to 2D oxides. In the case of 1D oxides second and third neighbor hoppings were dropped (Tsutsui et al. 2000). The on-site Coulomb energy U corresponds to the charge transfer energy of cuprates. In the intermediate state, there are a 1s core hole and a 4p electron, with which the 3d electrons interact. This interaction is written as

$$H_{1s-3d} = -V \sum_{i,\sigma,\sigma'} n^s_{i,\sigma} n^s_{i,\sigma'}, \qquad (5.57)$$

where $n^s_{i,\sigma}$ is the number operator of the 1s core hole with spin σ at site i. Since the 4p electron is strongly delocalized, its interaction with the 3d electron is very weak so that it can be neglected. Starting with the assumption that the incident photon energy is set to the threshold of the $1s \to 4p_z$ absorption spectrum, one lets enter the photoexcited 4p electron into the bottom of the $4p_z$ band with momentum $\mathbf{k}_0$, where the z-axis is perpendicular to the CuO_2 planes. Thus the total Hamiltonial of the model is given by

$$H = H_{3d} + H_{1s-3d} + H_{1s,4p}, \qquad (5.58)$$

where $H_{1s,4p}$ being the kinetic and the on-site energy terms of a 1s hole and a 4p electron, and the RIXS spectrum can be expressed as

$$I(\mathbf{q},\omega) = \sum_f \left| \langle f | \sum_\sigma s_{\mathbf{k}_0-\mathbf{K}_2,\sigma} p_{\mathbf{k}_0,\sigma} \left[\frac{1}{H - E_i - \hbar\omega_1 - i\Gamma/2} \right] p^+_{\mathbf{k}_0,\sigma} s^+_{\mathbf{k}_0-\mathbf{K}_1,\sigma} | i \rangle \right|^2$$
$$\times \delta(\hbar\omega - E_f + E_0). \qquad (5.59)$$

Here $s^+_{\mathbf{k},\sigma}$ ($p^+_{\mathbf{k},\sigma}$) is the creation operator of the 1s-core hole (4p electron) with momentum $\mathbf{k}$ and spin σ. $|i\rangle$ is the ground state of the half-filled system with energy E_i, $|f\rangle$ is the final state of the RIXS process with energy E_f, and $\Gamma/\hbar$ is the inverse of the relaxation time in the intermediate state. For practical purposes the terms $H_{1s,4p}$ in (5.58) are replaced by $\varepsilon_{1s,4p}$, the energy difference between the 1s level and the bottom of the $4p_z$ band.

The RIXS spectrum of (5.59) can then be calculated on quadratic clusters of a given size with periodic boundary conditions in the 2D case or on a ring in the 1D case, by utilizing the conjugate-gradient method together with the Lanczös technique. The parameters t, t', t'' U, V, and Γ have to be chosen according to the experiences with other spectroscopies (e.g. ARPES, angular resolved photo-emission spectroscopy).

The third mode of interpreting the RIXS process connected with excitations in the valence electron system, namely the third-order perturbation treatment, was first proposed by Platzman and Isaacs (1998) and Abbamonte et al. (1999), and is derived in Section 6.10.6 in greater detail. The derivation describes the intermediate state $|n\rangle$ with energy E_n as being characterized by an electron–hole pair with a hole in the inner-shell level. This pair, which can be considered as a virtual exciton, takes up the momentum of the incident photon and can Coulomb interact with the valence electron system via the Hamiltonian H_C, so that an additional scattering process, a so-called shakeup process in the intermediate state, takes place, which promotes the system into a new intermediate state $|m\rangle$ with energy E_m. If the virtual exciton recombines, the total energy loss $\hbar(\omega_1 - \omega_2)$ of the RIXS process and the momentum transfer $\hbar\mathbf{q} = \hbar(\mathbf{K}_1 - \mathbf{K}_2)$ is imparted to the excited valence electron system. Contrary to the preceding procedure, where H_C was included into the total Hamiltonian, H_C is now treated as a perturbing Hamiltonian. Thus one ends up, as shown in Section 6.10.6, with the following third-order expression for the double differential scattering cross-section describing the shakeup part of the resonant scattering processes:

$$\left(\frac{d^2\sigma}{d\Omega_2 d\hbar\omega_2}\right)_{\text{shakeup}} = r_0^2 \left(\frac{\omega_2}{\omega_1}\right)$$

$$\times \sum_f \left| \sum_{m,n} \frac{\langle f|b_2|m\rangle \langle m|H_C|n\rangle \langle n|b_1|i\rangle}{(E_n - E_i - \hbar\omega_1 - i\Gamma_n/2)(E_m - E_i - \hbar\omega_1 - i\Gamma_m/2)} \right|^2$$

$$\times \delta(E_f - E_i - \hbar\omega), \tag{5.60}$$

where $b_{1/2}$ stand for the scattering operators $\sum_j \mathbf{e}_{1/2} \cdot \mathbf{p}_j/\hbar \exp(\mp i\mathbf{K}_{1/2} \cdot \mathbf{r}_j)$ in (5.2). This expression is characterized by the existence of the double-resonance denominator and the inserted matrix element of the Coulomb interaction Hamiltonian H_C. The double-resonance denominator can lead to anomalies in the dependence of the shakeup satellites' energy position on the incident photon energy. We will come to this point later. According to Döring et al. (2004), the matrix element $M_C \equiv \langle m|H_C|n\rangle$ in a system with lattice translation invariance can be calculated to

$$M_C = \sum_{\mathbf{g}} \left(\frac{4\pi e^2}{|\mathbf{q}+\mathbf{g}|^2}\right) F_e(\mathbf{g}, \mathbf{e}_1) \langle v'|\rho_v(\mathbf{q}+\mathbf{g})|v\rangle, \tag{5.61}$$

where the summation is over all reciprocal lattice vectors $\mathbf{g}$. F_e is the static $\mathbf{g}$th structure factor of the virtual exciton, $\rho_v(\mathbf{q})$ is the $\mathbf{q}$th Fourier transform of the valence electron density operator, and $|v\rangle, |v'\rangle$ are the state vectors of the valence electron system before and after the excitation induced by the Coulomb interaction in the intermediate state. Here it is assumed that the virtual exciton remains unaltered upon the Coulomb interaction in the intermediate state. The matrix element $\langle v'|\rho_v(\mathbf{q}+\mathbf{g})|v\rangle$ in (5.61) squared, multiplied by $\delta(\omega - E_{v'} + E_v)$ and summed over the final states $|v'\rangle$ is nothing else than the dynamic structure factor $S(\mathbf{q}+\mathbf{g},\omega)$ of the valence electron system.

The fourth mode of interpretation has been introduced by van den Brink and van Veenendaal (1993), who have derived, on the basis of a second-order perturbation treatment and a series expansion of the Kramers–Heisenberg formula, a similar relation between the RIXS scattering amplitude A_{fi} and the matrix element $\langle f|\rho_v(\mathbf{q})|i\rangle$ for the case that the final state of the RIXS process is characterized by the excitation of the conduction electrons described by a single band of spinless fermions (sf), which writes:

$$A_{\text{fi}}^{\text{sf}} = P_1(\hbar\omega, U)\langle f|\rho_v(\mathbf{q})|i\rangle, \quad (5.62)$$

where U is the local core-hole potential, $\Delta = \hbar\omega_1 - E_B - i\Gamma_n$, and $P_1(\hbar\omega, U) \equiv UE_B[(\Delta - U)(\Delta - \hbar\omega)]^{-1}$. E_B is the binding energy of the empty state the core electron is excited to in the absorption part of the RIXS process. This derivation was performed under the following limitations: (i) a strong core-hole potential is assumed so that $t/U < 1$, where t is the hopping parameter of the valence electrons; (ii) the core-hole potential is treated to be strongly local. Under these conditions the resonant double differential scattering cross-section can be expressed in terms of the dynamic structure factor $S_v(\mathbf{q}, \omega)$ of the valence electrons:

$$\frac{d^2\sigma}{d\Omega_2 d\hbar\omega_2} \sim |P_1(\hbar\omega, U)|^2 S_v(\mathbf{q}, \omega). \quad (5.63)$$

These calculations can be generalized to the situation where the electrons have an additional spin degree of freedom. This leads to

$$A_{\text{fi}} = [P_1(\hbar\omega, U) - P_2(\hbar\omega, U)]\langle f|\mathbf{S}_\mathbf{q}^2|i\rangle + P_2(\hbar\omega, U)\langle f|\rho_v(\mathbf{q})|i\rangle, \quad (5.64)$$

where $P_2(\hbar\omega, U) = P_1(2\hbar\omega, 2U)/2$. $\mathbf{S}_\mathbf{q}^2$ is the longitudinal spin density correlation function: $\mathbf{S}_\mathbf{q}^2 \equiv [S(S+1)]^{-1}\sum_\mathbf{k}\mathbf{S}_{\mathbf{k}+\mathbf{q}}\mathbf{S}_{-\mathbf{k}}$. The spin and charge correlation function have different resonant enhancements. For Re$\Delta = U$ the scattering amplitude is dominated by the longitudinal spin response function. At incident energies, where Re$\Delta = 2U$, the contributions to the scattering amplitude of charge and spin are approximately equal.

The application of RIXS for investigating excitations in the intermediate state as presented in the subsections 5.7.2 and 5.7.3 are restricted to harder X-rays, where the advantage of true bulk sensitivity and of a wide range of accessible q-values can be fully employed.

5.7.2 Bonding–antibonding charge transfer excitations

Charge transfer excitations in the intermediate state of the RIXS process were first observed by Kao et al. (1996) on NiO at the X21 beamline of the NSLS using a horizontally focusing Si(220) monochromator and a spherically bent backscattering analyzer with a total energy resolution of 1.5 eV. In Fig. 5.39(a) the K absorption spectrum of a powdered NiO sample is shown together with (b) the normal valence band X-ray emission spectrum of NiO excited with an incident photon energy of 8460 eV, far beyond the Ni K absorption edge. When tuning the incident photon energy at the peak of the absorption spectrum (c) a new structure ∼6 eV below the quasielastically scattered line becomes visible, which

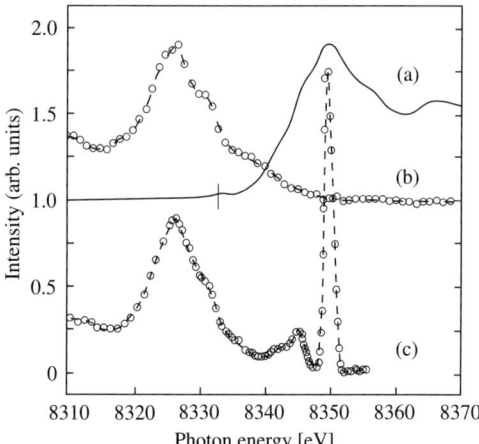

Fig. 5.39. (a) Ni K absorption spectrum of NiO. (b) Normal valence-band X-ray emission spectrum of NiO. (c) RIXS spectrum of NiO taken at the incident photon energy tuned to the main peak of the absorption spectrum. (Reprinted with permission from Kao et al. (1996); copyright (1996) by the American Physical Society.)

has to be attributed to charge transfer scattering in the intermediate state: Based on the Anderson impurity model the authors account for covalency between the metal ion and the ligand and represent the ground state by constructing bonding and antibonding linear combination essentially of the configurations $3d^8$ and $3d^9\underline{L}$, where $\underline{L}$ denotes a ligand hole. The ground state is then the bonding state with predominantly $3d^8$ configuration. The intermediate state of the RIXS process (the final state of the corresponding absorption process) can be represented by bonding and antibonding combinations of $\underline{c}\,3d^8 k$ and $\underline{c}\,3d^9\underline{L}k$, where $\underline{c}$ denotes the 1s core hole and k the excited photoelectron. The core hole potential reverses the balance between the $3d^8$ and $3d^9\underline{L}$ configurations, so that the bonding state is now predominately $\underline{c}\,3d^9\underline{L}\,k$, and is lower in energy than the antibonding state with mostly $\underline{c}\,3d^8 k$ configuration, as well known from other core-level spectroscopies. It is the more effective screening of the core hole by the additional electron at the Ni site, which lowers the energy of the $3d^9\underline{L}$ configuration ("well-screened") compared with the $3d^8$ configuration ("poorly-screened"). If one of these intermediate states decays by direct recombination of the excited electron with the core hole, the system can return to either the ground state or the excited state with mainly $3d^9\underline{L}$ configuration, so that it is justified to speak of a net scattering process where charge is transferred from the ligand to the nickel. This interpretation is supported by the fact that the energy transfer of the experimental loss features, which is between 4.9 and 7.8 eV depending on the incident photon energy, was found close to the reported charge transfer energy of NiO, 6.2 eV (Geunseop Lee and Oh 1991).

RIXS measurements on Nd_2CuO_4 by Hill *et al.* (1998), also performed at the X21 beamline of the NSLS on single crystals with **q** parallel to the *c*-axis in horizontal scattering geometry, found very similar interpretation. An excitation observed at 6 eV energy loss, when the incident photon energy was tuned to 8990 eV, was attributed to a transition from a bonding ground state, which is a linear combination of $3d^9$ and $3d^{10}\underline{L}$ configuration ($\underline{L}$ is again a hole in the O 2p ligand band) with predominant contributions of the former to an antibonding state with mainly $3d^{10}\underline{L}$ contributions. Calculations of the amplitude of the 6 eV excitation as a function of the incident photon energy done by using the Anderson impurity model, and shown in Fig. 5.40, exhibit resonances both at 8983 and 8990 eV, where these energies belong to the bonding (mainly $\underline{1s}\ 3d^{10}\underline{L}\ 4p_\pi$ configuration) and the antibonding (mainly $\underline{1s}\ 3d^9\ 4p_\pi$ configuration) intermediate state (final state of the absorption process), respectively. Figure 5.40 depicts the calculated and measured Nd_2CuO_4 absorption spectrum, where the peaks corresponding to these final states of the absorption process are clearly

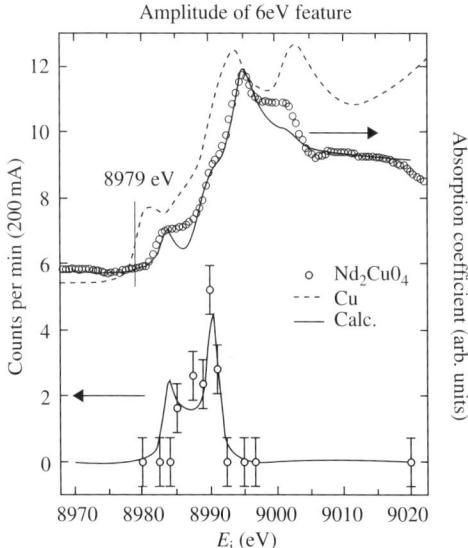

Fig. 5.40. Lower half: the amplitude of the 6 eV feature of the Nd_2CuO_4 RIXS spectrum as a function of the incident photon energy (open circles, only one peak). The solid line is the result of a calculation on the basis of the Anderson impurity model (two peaks). Upper half: open circles: measured X-ray absorption spectrum of polycrystalline Nd_2CuO_4; solid line: calculated absorption spectrum using the same parameter as for the scattered intensity. The absorption of Cu metal (dashed line) is shown for reference. (Reprinted with permission from Hill *et al.* (1998); copyright (1998) by the American Physical Society.)

visible. It was found remarkable by the authors that the experimental data in Fig. 5.40 show only one resonance at 8990 eV. The authors make the Zhang–Rice (ZR) singlet (Zhang and Rice 1988) responsible for this behavior, where the ZR singlet is formed, when, after a charge transfer process (absorption transition at 8983 eV into a 1s $3d^{10}\underline{L}$ $4p_\pi$ configuration), the hole at the oxygen which is bound to both a copper at site i and a copper at site j hybridizes with the other three oxygens appertaining to the CuO_4 plaquette at j. Thus the hole-spins at the four oxygens can form a singlet state with the hole-spin at the copper site j. Because of the small transition probability between the strong local antibonding final state of the RIXS process and the more delocalized ZR singlet the 6 eV excitation appears suppressed at the 8983 eV excitation. This interpretation was supported by the findings of Hämäläinen et al. (2000), who calculated the amplitude of the 6 eV feature in the RIXS spectrum of Nd_2CuO_4 as a function of the incident photon energy for two cases: One is a single copper site calculation on the basis of the Anderson impurity model with many oxygen sites (256), where two resonances were visible, one at the absorption transition at 8995 eV into a 1s $3d^{10}\underline{L}$ $4p_\sigma$ configuration and a second one at the absorption transition into a 1s $3d^9$ $4p_\sigma$ configuration. The other calculation is with a Cu_5O_{16} cluster model, in which five CuO_4 plaquettes are arranged in the ab plane exhibits only one resonance of the 6 eV excitation amplitude connected with the absorption transition into a 1s $3d^9$ $4p_\sigma$ configuration, in agreement with the experiment. It was the admission of more extended states in the latter calculation, which made the difference. Measurements with different values of the momentum transfer q between 3.5 and 7.9 Å^{-1} at a fixed value of the incident photon energy of Hämäläinen et al. (2000) have revealed to what extent the remaining 6 eV excitation possesses strong local character.

Neither dispersion of the energy position nor changes in the integrated intensity could be observed. But it should be stressed that these measurements have been performed in vertical scattering geometry, where it is possible to change the value of q by an appropriate rotation of the crystal, and keeping both the orientation of **q** and of the incident polarization $\mathbf{e}_1$ (assumed to be parallel to the horizontal plane) fixed with respect to the crystal coordinates. It has been emphasized by Döring et al. (2004) in a RIXS study on CuO single crystals that this is not the case in horizontal scattering geometry, where fixing the orientation of the incident polarization with respect to the crystal means variation of both value and direction of **q** when changing the scattering angle. Under these conditions, the intensity of the bonding–antibonding charge transfer excitation exhibited a strong dependence on the scattering angle, which was interpreted utilizing an expression for the resonant scattering amplitude introduced by Hannon et al. (1988) for an electric dipole resonance.

In the aforementioned RIXS investigations of NiO and Nd_2CuO_4 and their interpretation via the Anderson impurity model, it was tacitly assumed that the energy loss position of the measured feature coincides with the corresponding excitation energy. The consequent application of the third-order perturbation treatment by Abbamonte et al. (1999) in a study on La_2CuO_4, and by Döring et al. (2004) on CuO have revealed a more complicated behavior, due to the double resonance denominator in (5.60). The RIXS spectra for different incident photon energies of Abbamonte et al. (1999), shown already in Fig. 1.24, exhibit features with peaks between 2.8 and 6.5 eV energy loss. But these spectra collapse to the same function, if one follows Abbamonte et al. (1999), and rewrites (5.60) using self-evident shorthand for the matrix elements, and assuming that there is only one excitation to $|m\rangle$ in the intermediate state

$$\left(\frac{d^2\sigma}{d\Omega_2 d\hbar\omega_2}\right)_{\text{shakeup}} \sim \sum_f \left|\sum_{1s,4p} \frac{M_{em}M_{\text{Coul}}M_{\text{abs}}}{(E_{1s,4p} - \hbar\omega_1 - i\Gamma_K/2)(E_{1s,4p} - \hbar\omega_2 - i\Gamma_K/2)}\right|^2$$

$$\times \delta(E_f - E_i - \hbar\omega). \quad (5.65)$$

Here the intermediate state $|n\rangle$ is specified by all possible states of the 1s hole and the 4p electrons. $E_{1s,4p}$ is their energy and $\Gamma_K/\hbar$ their inverse lifetime, which is assumed to be independent of the state of the 3d system. If one additionally postulates that the intermediate states are approximately degenerate with energy $E_{1s,4p} = E_K$, we can factorize the energy denominator from the sum and write

$$\left(\frac{d^2\sigma}{d\Omega_2 d\hbar\omega_2}\right)_{\text{shakeup}} \sim S_K(\mathbf{q},\omega,\mathbf{e}_1,\mathbf{e}_2)[(E_K - \hbar\omega_1)^2 + \Gamma_K^2/4]^{-1}$$

$$\times [(E_K - \hbar\omega_2)^2 + \Gamma_K^2/4]^{-1}, \quad (5.66)$$

where

$$S_K = \sum_f \left|\sum_{1s,4} M_{em}M_{\text{Coul}}M_{\text{abs}}\right|^2 \delta(E_f - E_i - \hbar\omega). \quad (5.67)$$

Thus by dividing each spectrum by its respective denominator, and using E_K and Γ_K as flexible parameters in a fitting procedure which minimizes the total variation among the curves, Abbamonte et al. (1999) found, as shown in Fig. 5.41, that for $E_K = (8995.14 \pm 1.61)$ eV and $\Gamma_K/2 = (2.38 \pm 0.542)$ eV the spectra collapse onto a single curve with a single peak at 6.1 eV energy loss and a width of 3.9 eV. Along with the comments to (5.61) we see that this curve is strictly related to the dynamic structure factor of the 3d system weighted by the form factor of the core states. Cluster calculations of Simon et al. (1996) indicate that this feature is a transition from the b_{1g} ground state to an a_{1g} excited state composed of symmetric combinations of a central $Cu3d_{x^2-y^2}$ orbital and the surrounding O $2p_\sigma$ orbitals.

Döring et al. (2004) have found in a RIXS study on single crystals of CuO performed at the HARWI beamline of HASYLAB and at the X21 beamline of the NSLS spectral features for incident energies between 8981 and 8988 eV with a peak-like structure with an energy position shifting from 3 eV to 6eV, and

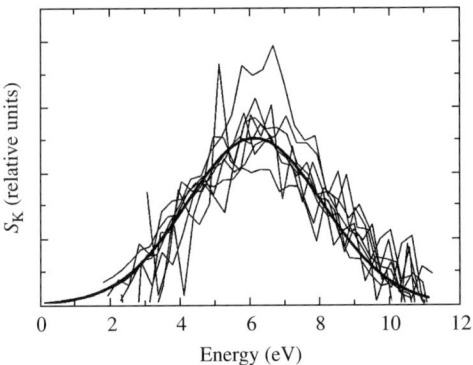

Fig. 5.41. The result of dividing the energy denominators in equation (5.66) from the experimental RIXS spectra. For $E_K = (8995.14 \pm 1.61)$ eV and $\Gamma_K = (2.38 \pm 0.542)$ eV all spectra collapse onto a single curve. The thick line is a Gaussian fit. (Reprinted with permission from Abbamonte *et al.* (1999); copyright (1999) by the American Physical Society.)

for incident energies between 8991 and 9001 eV a peak-like structure, which stays more or less fixed at 5.5 eV energy loss. Also this dependence of the peak position on the incident energy could only be explained when the double resonance denominator of (5.60) was taken into account. Assuming the matrix elements $\langle f|b_2|m\rangle$ and $\langle n|b_1|i\rangle$ to be independent of energy and writing $E_m - E_i = E_n - E_f = E_{\text{ex}}$, then the intensity of the shakeup satellites is given by

$$I_{\text{shakeup}} \sim L_1(E_{\text{ex}} - \hbar\omega_1)L_2(E_{\text{ex}} - \hbar\omega_2)G_{\text{s}}(\omega), \tag{5.68}$$

where $L(E)$ stands for the Lorentzian $L_{1/2}(E) = (E_{1/2}^2 + \Gamma_{1/2}^2/4)^{-1}$ and $G_{\text{s}} = M_C^2$ is represented by a single Gaussian, centered at the energy loss $\Delta \equiv E_m - E_n$ and with a lifetime broadening of Γ_{s}. It turned out that the dependence of the satellites' position on the incident energy could not be fitted to a single expression like (5.68), but one needs a sum of two with different E_{ex} and different $\Gamma_{1/2}$. The shakeup structure between 8981 and 8988 eV incident energy with its shifting energy position could be fitted to $E_{\text{ex}}^A = 8981.4$ eV, $\Gamma_{A2} < \Gamma_{A1} \approx 2.5$ eV, $\Delta = 5.4$ eV and $\Gamma_{\text{s}} = 1.8$ eV, whereas the stable feature, stable in energy position, between 8991 and 9001 eV incident energy, was fitted to $E_{\text{ex}}^B = 8995.8$ eV, $\Gamma_{B1} \approx \Gamma_{B2} \approx 3.5$ eV, but with the same values of Δ and Γ_{s}. Thus the shakeup satellites must be traced back to a single excitation of the 3d system, namely, according to cluster calculations of Eskes *et al.* (1990), to a transition from the b_{1g} bonding ground state to antibonding states with a_{1g}, b_{2g}, b_{1g} or e_g symmetry. It is apparently the different size and ratio of Γ_1 and Γ_2 which gives rise the very different behaviour of the structures denoted A and B. The experimental shakeup satellite structures together with the simulation using the above fit parameters

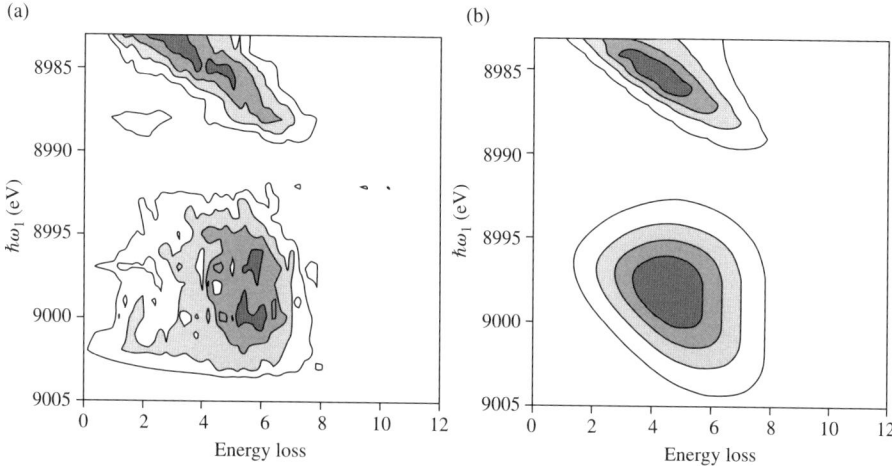

Fig. 5.42. Dependence on the incident photon energy $\hbar\omega_1$ of the intensity of the shakeup satellites (peak intensity normalized to one) in the form of a level diagram. (a) Experiment; (b) simulation. The distance of the level lines is 0.2. (Reprinted with permission from Döring et al. (2004); copyright (2004) by the American Physical Society.)

are shown in Fig. 5.42 in the form of contour maps. Thus it must be concluded that it depends sensitively on the lifetime width both of the absorption and the re-emission channel, how the double-resonance denominator in (5.60) takes effect on the shift of the shakeup satellites' position. It needs in every case a careful test before the energy position of such a shakeup satellite can be identified with an excitation energy in the 3d system. By the way, the absorption channel with $E_e^A = 8981.4$ eV seems to be identical with that which gave rise to the structure B in the RIXS Kα_1 spectra of CuO from Döring et al. (2004) shown in Fig. 1.17.

More bonding–antibonding charge transfer excitations have been reported by authors whose main interest was the investigation of transition across the Mott gap, as will be reported later in Section 5.7.3. There a structure was found at 5.8 eV seen by Hasan et al. (2000) in an investigation on $Ca_2CuO_2Cl_2$, and interpreted as a transition from the ground state to a high-energy excited state composed of symmetric contribution of a central Cu-$3d_{x^2-y^2}$ orbital and the surrounded O $2p_\sigma$ orbitals. Hasan et al. (2002) reported a 5.6 eV feature in the RIXS spectra of the 1D systems Sr_2CuO_3 and $SrCuO_2$, which again was attributed to a charge transfer excitation from the ground state to the antibonding-type excited state. Kim et al. (2002) found a structure at 7.2 eV in the RIXS spectra of La_2CuO_4, when the absorption was to the poorly screened $1s3d^94p$. Hasan et al. (2003) have detected a 6.4 eV feature in the RIXS spectra of $CuGeO_3$, and ascribed it also to excitations to the unoccupied copper orbitals with antibonding character. RIXS measurements on Li_2CuO_2 of Hasan et al. (2003) exhibited a 5.5 eV feature which they also assign to excitations to antibonding states. Also Kim et al. (2004b) have observed, in an investigation of Li_2CuO_2, a strong feature around 5.4 eV and a weaker one at 7.6 eV. The former exhibits an incident energy dependence of its energy position, which is explained with a double-resonance denominator as in (5.63). The intermediate state for these two features was the poorly screened $1s3d^94p$ excited state in the absorption spectrum. Kim et al. (2004c) found in $La_{2-x}Sr_xCuO_4$ a bonding–antibonding charge transfer scattering feature with a relative large energy loss of 8 eV. Finally Ishii et al. (2005) showed, in an investigation of $Nd_{1.85}Ce_{0.15}CuO_4$, a charge transfer excitation to antibonding states around 6 eV energy loss.

A survey of all charge transfer exitation from a bonding to an antibonding molecular orbital in a copper oxygen plaquette as investigated by means of RIXS is offered by Kim et al. (2004d), where measurements of the authors together with those of other groups are considered under a common point of view, namely the influence of the copper oxygen bond length on the excitation energy and the amount of dispersion. The consideration starts with the strong hybridization between the Cu $3d_{x^2-y^2}$ level and the O $2p_\sigma$ level, where p_σ denotes p_x or p_y orbitals pointing towards the Cu ions. (x,y) is the plane of the copper oxygen plaquettes. As a result of this hybridization, the Cu-O bond has a strong covalent character, and a large energy splitting $\Delta_{\sigma\sigma^*}$ exists between the bonding (σ) and antibonding (σ^*) molecular orbital. Thus $\Delta_{\sigma\sigma^*}$ should increase with increasing p-d hybridization and could serve as an independent route to determine the hopping matrix element t_{pd}, since it is believed that $\Delta_{\sigma\sigma^*}$ is directly related to t_{pd}. The result of this comprehensive study is shown in Fig. 5.43, where both the measured peak position for $\mathbf{q_r} = (\pi,0)$ ($\mathbf{q_r}$ denotes the reduced momentum transfer $\mathbf{q_r} = \mathbf{q} - \mathbf{g}$) and the dispersion along the Cu-O bond direction is given for a large variety of cuprates. There exists indeed a systematic dependence of the excitation energy on the local structure, and that is apparently insensitive to whether the crystal has planar, corner-sharing, or edge-sharing chain structure. The dependence on the Cu-O distance d follows a power law $d^{-\eta}$ with $\eta = 8$. This is much larger than that expected from tight binding theory, for which $\sim d^{-3.5}$ is predicted (Harrison 1989). It is remarkable that at least the cuprates with smaller Cu-O distances (< 1.93 Å) exhibit a clear dispersion, so that one has to conclude that the localized picture breaks down for small bond distances. The minimum excitation energy is occurring at the zone boundary, implying an indirect gap.

Whereas the use of other spectroscopic methods for investigating electronic excitation in highly correlated systems either rule out the variation of pressure or impose severe constraints on many of them, it was shown by Shukla et al. (2003), in a pioneering study on NiO, that RIXS experiments can be done under ultrahigh pressure up to 100 GPa. The experiment was carried out at the beamline ID 16 of the ESRF with a monochromator resolution of 1.4 eV. The incident energy was tuned to the pre-peak of the absorption spectrum, associated with quadrupolar transitions, with the advantage of lowering the absorption ensuring

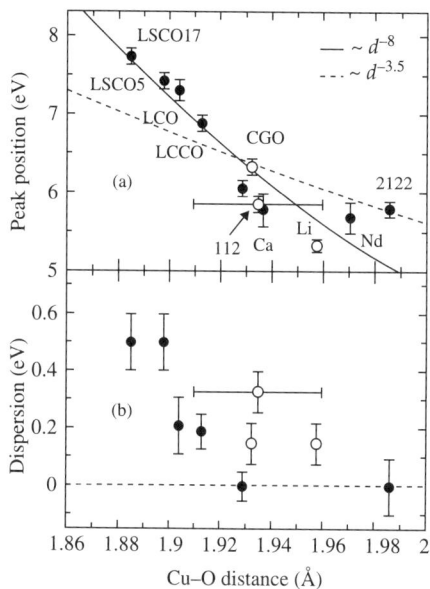

Fig. 5.43. (a) The value of the peak position $\Delta_{\sigma\sigma^*}$ at $(\pi,0)$ and (b) the amount of dispersion along the Cu–O bond direction for each sample is plotted as a function of $d_{\text{Cu-O}}$. Solid symbols: perfect square plaquettes; open symbols: distorted plaquettes. The solid line and dashed lines are fits to a power law $\Delta_{\sigma\sigma^*} \sim d_{\text{Cu-O}}^{-8}$ and $d_{\text{Cu-O}}^{-3.5}$, respectively. 2122: $Sr_2CuO_2Cl_2$; Nd: Nd_2CuO_4; Ca: $Ca_2CuO_2Cl_2$; 2342: $Sr_2Cu_3O_4Cl_2$; LCCO: $La_{1.9}Ca_{1.1}Cu_2O_6$; LCO: La_2CuO_4; LSCO5: $La_{1.95}Sr_{0.05}CuO_4$; LSCO17: $La_{1.83}Sr_{0.17}CuO_4$; Li: Li_2CuO_2; CGO: $CuGeO_3$; 112: $SrCuO_2$ (Reprinted with permission from Kim *et al.* (2004d); copyright (2004) by the American Physical Society.)

sizeable penetration. The polycrystalline sample was loaded in a symmetrical Mao–Bell cell with 300 μm culet diamonds using a 5 mm diameter high-strength Be gasket. No pressure medium was used. The scattering angle was 90°. The RIXS spectra at different pressure are shown in Fig. 5.44. At room pressure one sees two structures: The first one peaking at 5.3 eV with an edge at 4.3, which is directly related to the charge transfer gap in NiO (Sawatzki and Allen 1984), associated with the metal–ligand transition leading to a $d^{n+1}\underline{L}$ excited state, where d^n is the metal ground-state configuration. The second peak at 8.5 eV is interpreted by the authors to be a metal–metal transition leading to $d^{n+1}d^{n-1}$, related to the correlation energy $U_{\text{d-d}}$. The 5.3 eV feature does not shift appreciably as pressure is increased, but its width (determined by fitting three Gaussians to the spectra) doubles from 2 to 4 eV. This behavior is traced back to an increase of the band width, so that the sharp resonances seen at low pressure become ill defined. The 8.5 eV peak shifts to above 10 eV up to 50 GPa and then remains constant, while its width also increases from 4 to 5 eV.

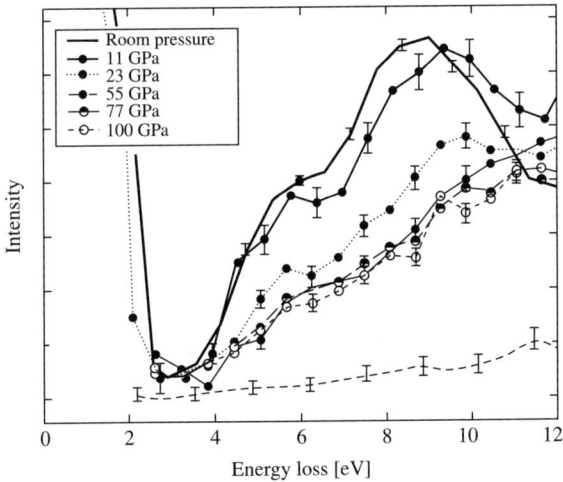

Fig. 5.44. RIXS spectra of NiO as a function of pressure. Bottom dashed line: nonresonant background. As pressure is increased the resonant intensity decreases and the 5.3 and 8.5 eV features smear out. (Reprinted with permission from Shukla et al. (2003); copyright (2003) by the American Physical Society.)

This behavior would suggest an initial increase in the d-d Coulomb interaction with pressure, although screening increases with pressure. A possible closing of the charge transfer gap is discussed. Signs of metallic luster were visible around 100 GPa, possibly a result of the pressure gradient. Further experiments using a pressure medium as He seem to be necessary, in order to settle all questions.

5.7.3 Excitations in quasi-low-dimensional Mott insulators

Certainly the most spectacular application of excitations in the intermediate state of the RIXS process is the investigation of excitations across the Mott gap in quasi-low-dimensional Mott insulators, and of their dispersion. This very successful branch of RIXS spectroscopy was pioneered by Hasan et al. (2000) with an investigation on the 2D Mott insulator $Ca_2CuO_2Cl_2$ after theoretical predictions of Tsutsui et al. (1999) and Tsutsui et al. (2000) and some preliminary experiments of Abbamonte et al. (1999) on $Sr_2CuO_2Cl_2$. The experiments of Hasan et al. (2000), the first thorough study of the dispersion of the charge transfer scattering across the Mott gap by RIXS, were performed at the X21 triple axis spectrometer at the NSLS with an overall energy resolution of 440 meV by tuning the incident photon energy to 8996 eV (peak of the Cu K absorption spectrum). The RIXS spectra were measured for a range of the reduced momentum transfer along the [110] direction from $(0,0)$ to (π,π), taken in the second Brillouin zone, and along the [100] direction from $(0,0)$ to $(\pi,0)$ taken in the third Brillouin zone. The spectra in the [110] direction exhibit a low-lying feature with a dispersion between 2.4 and 3.7 eV, the spectra in [100] direction a feature with less dispersion between 2.4 and 3.0 eV, shown, after subtraction of the quasielastic line in Fig. 5.45. These low-energy features are interpreted as **q**-resolved excitations across the effective Mott gap from the occupied band consisting of Cu $3d_{x^2-y^2}$ and O $2p_\sigma$ orbitals (Zhang-Rice band; ZRB) to the unoccupied upper Hubbard band (UHB). The shakeup of the valence electron system in the intermediate state creates a particle–hole pair across the gap, which carries energy and momentum. This pair propagates in a background of an antiferromagnetically ordered lattice. In a local picture, the created hole forms a Zhang–Rice singlet and an electron is excited onto the neigboring Cu site occupying the UHB. The calculations to this experiment, already published by Tsutsui et al. (1999), follow

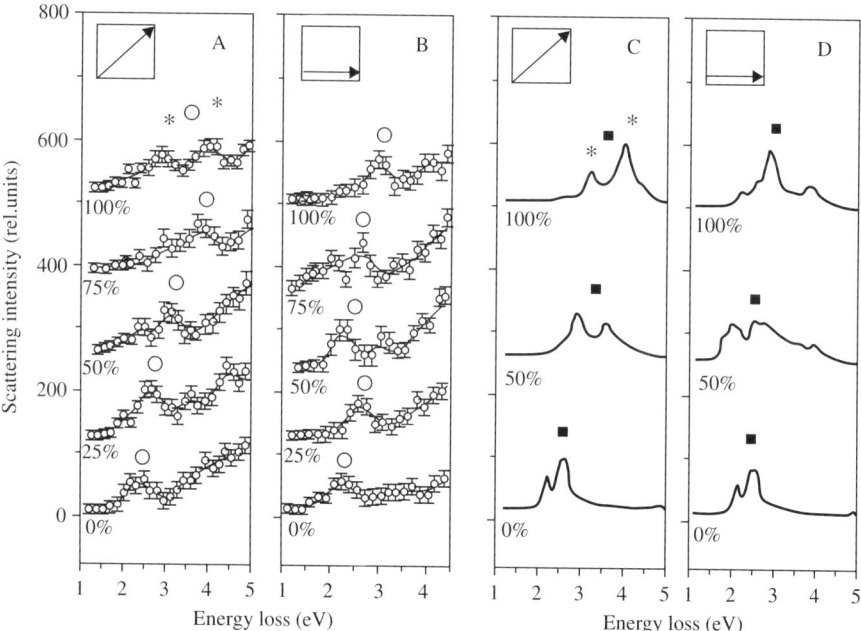

Fig. 5.45. (A) and (B) experimental RIXS spectra (with elastic line removed) of $Ca_2CuO_2Cl_2$ along the [110] and [100] direction, respectively. (C) and (D) calculated spectra (low-energy component only) along the [110] and [100] direction, respectively. The centers of gravity of spectral weights are indicated by open circles for the experimental data and by closed squares for the model calculation. The asterisks on the topmost spectra of (A) and (D) mark the center of gravity of two features. The percentages show the normalized **q** coordinates. (Reprinted with permission from Hasan et al. (2000); copyright (2000) by the American Association for the advancement of Science.)

the Hubbard model with long-range hopping (t-t'-t''-U model) as described in Section 5.7.1, where the lower Hubbard band is regarded as the ZRB. The values of the model parameter are taken to fit ARPES data of $Sr_2CuO_2Cl_2$ namely $t = 0.35$ eV, $U/t = 10$, $t'/t = -0.34$, and $t''/t = 0.23$. In Fig. 5.45, the result of these calculations for the two directions of momentum transfer are depicted. The agreement with respect to the dispersion of the center of gravity found for the calculations with the experiment are rather good. Even the doublet at (π, π) is resolved in the measurement. It has been stressed by Tsutsui et al. (1999) and already earlier by Romberg et al. (1990), that the dispersion of the particle–hole excitations in the Mott system, as shown in Fig. 5.45, cannot be simply understood as independent excitation of single-particle states being the convolution of the single-particle spectral density functions $A(\mathbf{k}, \omega)$ (see equation 2.84 in Section 2.3.3.1). One has, on the contrary, to treat the process of particle–hole excitation exactly by introducing the spectral function $B(\mathbf{k}, \mathbf{q_r}, \omega)$ of the two-body Green's function, which describes the particle–hole excitation, defined as

$$B(\mathbf{k}, \mathbf{q_r}, \omega) = \sum_{\alpha} \left| \langle \alpha | \sum_{\sigma} d^+_{\mathbf{k}+\mathbf{q_r}, \sigma} d_{\mathbf{k}, \sigma} | 0 \rangle \right|^2 \delta(\hbar\omega - E_\alpha + E_0), \qquad (5.69)$$

where the states $|\alpha\rangle$ have the same point group symmetry as that of the RIXS final state, and $d^+_{\mathbf{k}, \sigma}$ is the creation operator of a 3d electron with spin σ and momentum $\mathbf{k}$. In this sense, RIXS provides more information about highly correlated systems than the combination of ARPES with inverse ARPES.

Kim et al. (2002) have investigated the dispersion of charge transfer excitations in La_2CuO_4 in a RIXS study at the 9IDB beamline of the APS. They found in the 2–4 eV range of energy loss a double-peak feature, which shows two-dimensional characteristics. One peak with lower energy loss in the range of 2.5 eV exhibits a strong dispersion of nearly 1 eV when going from $(0,0)$ to $(\pi, 0)$ and vanishes at (π, π). The other peak has a zone center of 3.9 eV and a dispersion of 0.5 eV in the $(\pi, 0)$ and (π, π) direction. Both peaks are ascribed to a bound exciton formed by an electron–hole pair created by exciting an electron from the ZR band to the conduction band across the charge transfer gap. Since such an exciton has zero spin it can move without disturbing the antiferromagnetic order of the copper oxide planes, so that it exhibits a large dispersion. The energy difference between the two excitons could be due to different symmetries of the hole in the ZR band. Nomura and Igarashi (2005) have provided another interpretation for the above double-peak structure using the perturbative method developed by Platzman and Isaacs (1998). The antiferromagnetic ground state was described using the Hatree–Fock theory. Electron correlation in the scattering process was taken into account within the random phase approximation (RPA). The two peaks were ascribed to charge transfer transition, where the Cu 3d electrons hybridized with the O 2p band are excited to the upper Cu 3d band. The two-peak structure is attributable to the structure of the

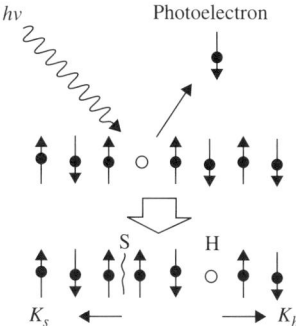

Fig. 5.46. Schematic representation of the photoemission process in a chain with short-range antiferromagnetic ordering. A photohole created in the photoemission process decays into a spin excitation (spinon, S) and a charge excitation (holon, H). Spinon and holon propagate independently with different speed. (Reprinted with permission from Kim *et al.* (1997); copyright (1997) by the American Physical Society.)

Cu 3d partial density of states mixed with the broad O 2p band, rather than some kind of a bound exciton mode.

It is commonly believed that $SrCuO_2$ and Sr_2CuO_3 are prototypes of a 1D half-filled spin-1/2 quantum system which should exhibit spin–charge separation as first predicted by Lieb and Wu (1968), so that, in these systems, charge fluctuations propagate rather freely and independently of the spin fluctuations. Figure 5.46 from Kim *et al.* (1997) depicts a simplified picture of the spin–charge separation in such half-filled 1D antiferromagnetic insulators: When an electron is removed by a photon, a hole is left behind. Hopping of this hole to a neighboring site, or equivalently, hopping of a neighboring electron into the hole site, creates a magnetic excitation (wavy line in Fig. 5.46). Additional hopping of the hole in the same direction does not create further magnetic excitations. Thus the motion of the charge vacancy is free from magnetic interactions (with the exception of the first step). The single photon hole decays into two separate "defects" in the chain. The motion of the charge H is governed by the hopping energy t, the propagation of the magnetic excitation S by the exchange interaction J. This leads to two different speeds; these defects can be considered as two separate particles, a holon and a spinon. As a consequence, the band dispersion of a 1D half-filled antiferromagnetic insulator should exhibit quite another behavior than a corresponding 2D system, where the bandwidth is mainly controlled by J and should be smaller by a factor of three. First experimental verifications of these theoretical predictions were found by Kim *et al.* (1997) by means of ARPES and by Neudert *et al.* (1998) using EELS, based on calculations of Stephan and Penc (1996). Hasan *et al.* (2002) and Hasan *et al.* (2003) have published the first RIXS measurements on $SrCuO_2$ and Sr_2CuO_3, where they were

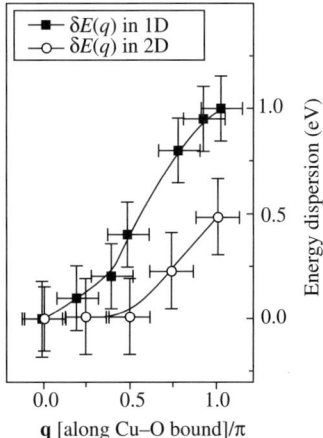

Fig. 5.47. Comparison of the **q** dependence of charge excitations along the Cu–O bond direction in 1D (Sr_2CuO_3 and $SrCuO_2$) and 2D ($Ca_2CuO_2Cl_2$) Mott insulators. (Reprinted with permission from Hasan *et al.* (2002); copyright (2002) by the American Physical Society.)

able to follow the dispersion of the low-energy excitations with $\mathbf{q} \parallel$ Cu–O chain, where the excitation was ascribed to transitions across the Mott gap. This dispersion shown in Fig. 5.47 was compared with that of a 2D cuprate $Ca_2CuO_2Cl_2$ (Hasan *et al.* 2002). The fact that the charge excitations were found to be more dispersive in 1D than in 2D along the Cu–O bond direction was considered as direct evidence of holons in a quantum antiferromagnetic spin-1/2 chain. Band structure calculation have predicted the opposite behavior. It was stressed by the authors that, contrary to EELS, RIXS experiments provide a natural way to identify excitations originating from the Cu–O planes. A somewhat different interpretation of RIXS spectra from $SrCuO_2$ was provided by Kim *et al.* (2004a), who supported their interpretation by dynamical density-matrix renormalization group calculations of the dynamical density–density correlation function $N(\mathbf{q}, \omega)$ on the basis of an extended Hubbard model, the parameters of which were fitted to optical conductivity measurements, and were compatible with the value $J \approx 0.23$ eV for the Heisenberg exchange coupling J obtained from neutron scattering. The RIXS measurements at the beamline 9IDB of the APS were performed at an incident photon energy of 8982 eV in 0.2π steps of the momentum transfer. The authors found a broad continuum of excitations which they attribute to holon–antiholon pairs, arguing that the RIXS process conserves the total charge, so that an electron is moved from one site to another, creating a hole and a doubly occupied site. The decay of the hole creates a holon and a spinon, while the doubly occupancy decays into an antiholon and a spinon. The excitation continuum exhibited a peak with sinusoidal dispersion as shown in Fig. 5.48. The same dispersion is found at the peak of the calculated dynamical

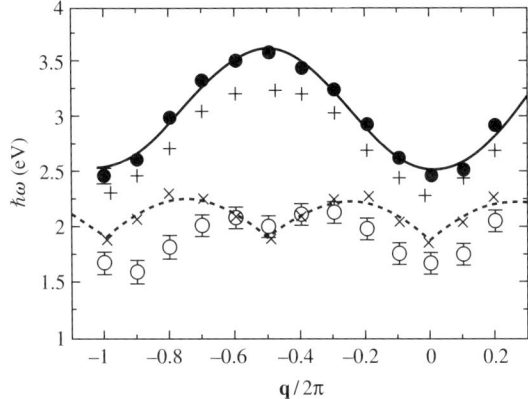

Fig. 5.48. Filled circles: dispersion relation of the peak position of the RIXS spectra of SrCuO$_2$; open symbols: dispersion of the onset energy of the RIXS spectra. The solid line is $\hbar\omega(q) = 3.07 - 0.55\cos(q)$. Calculated peaks and onset energies in $N(\mathbf{q},\omega)$ are plotted as $+$ and $\times$ symbol, respectively. (Reprinted with permission fromy Kim et al. (2004a); copyright (2004) by the American Physical Society.)

density–density correlation function $N(\mathbf{q},\omega)$. The amount of dispersion is 1.1 eV comparable with that seen by Hasan et al. (2002). New was the observation that the onset of spectral weight in this excitation as well as the onset of $N(\mathbf{q},\omega)$ follows a spinon dispersion relation (also depicted in Fig. 5.48) with a bandwidth of $\sim \pi J/2$ and a local minimum at $q = \pi$, consistent with low-energy field theory (Controzzi and Essler 2002). Contrary to the findings of Neudert et al. (1998) no indication for holon–antiholon excitons near $q = \pi$ was detected.

Nearly zero dispersion of the charge excitations across the effective Mott gap was found by Hasan et al. (2003) in a quasi-zero-dimensional model cuprate Li$_2$CuO$_2$.

In a theoretical study, Tsutsui et al. (2003) have investigated the influence of hole (electron) doping on the Mott gap excitation in RIXS. By using the extended Hubbard model with second- and third-nearest-neighbor hoppings, as presented in Section 5.7.1, they found that the center of gravity of the RIXS spectra of electron-doped cuprates ($\langle n \rangle = 18/16$) is shifted to higher energy losses by roughly $2t$, whereas the character and the amount of the dispersion remains unaltered. On the contrary, the dispersion of the center of gravity of the Mot gap induced RIXS spectra of hole-doped cuprates vanishes nearly completely, where the zero-momentum spectrum is also shifted to higher energies by $\sim 2t$ upon hole doping ($\langle n \rangle = 14/16$). These shifts can be traced back to shifting of the Fermi energy to the lower Hubbard band (upper Hubbard band) upon hole (electron) doping. The different doping dependence of the dispersion must be connected with changes in the antiferromagnetic (AF) correlation. In the electron-doped case the short-range AF correlation is kept, whereas the correlation is strongly suppressed in the hole-doped case. Thus the RIXS spectra in the electron-doped case are similar to the undoped case, while the spectra in the hole-doped system are different.

5.7.4 Orbital excitations

The orbital degree of freedom is believed to be one of the important ingredients for colossal magnetoresistance (CMR) and various complex phenomena observed in manganites. The long-range orbital order and its implications in the magnetic interaction have been discussed a long time ago. But due to limitations in the observation technique, orbital order has been recognized for a long time as a hidden degree of freedom. It was not until 1998 that Murakami et al. (1998) succeeded in the first experimental proof of orbital order in $La_{0.5}Sr_{1.5}MnO_4$ by means of resonant X-ray diffraction. The alternate orbital alignment was confirmed by observation of the superlattice reflection induced by the anisotropy of the anomalous atomic scattering factor near the Mn K edge. The type of the orbital ordering can be deduced from the polarization dependence of the anomalous X-ray scattering. Thus Ishihara and Maekawa (1998) have numerically calculated the X-ray scattering intensity in several types of orbital ordering as a function of azimuthal and analyzer angles. A typical example for orbital order is provided by the CMR progenitor $LaMnO_3$, which is a Mott insulator where the electronic configuration of the Mn^{+3} ions is $t_{2g}^3 e_g^1$ with spin quantum number $S=2$. A band gap appears between two e_g bands of the Mn ions hybridized with the O 2p orbitals. The occupied e_g orbitals of $3d_{3x^2-r^2}$ and $3d_{3y^2-r^2}$ are alternately ordered in the ab plane accompanied by the lattice distortion below 780 K (Murakami et al. 1998b). The experimental technique of resonant X-ray diffraction was regozized as a powerful probe to detect the orbital order through intensive experimental and theoretical studies. In contrast, the dynamics of orbital degree of freedom has less intensively been investigated so far. We can classify the dynamics of the orbitals into two different excitations: (i) a collective orbital excitations named an orbital wave corresponding to the spin wave in magnetically ordered materials; this orbital wave has been observed in a Raman-scattering experiment in $LaMnO_3$ at 120–160 meV at zero momentum transfer (Saitoh et al. 2001); (ii) an individual orbital excitation, the particle–hole excitation, which can be related to the Stoner excitation in ordered magnetic systems. Since the highest occupied and the lowest unoccupied electronic states of orbitally ordered insulators have different orbital character, the particle–hole excitation across the Mott gap changes the symmetry of the electronic system. The first experimental investigation of this individual orbital excitation was performed by Inami et al. (2003) at the BL11XU of SPring-8 on $LaMnO_3$ single crystals, using a diamond (111) double-crystal monochromator in connection with a horizontally bent mirror. The scattered radiation was analyzed by means of a spherically bent ($R = 2$ m) Ge (531) crystal. The total energy resolution was 0.5 eV. The RIXS spectra for two different directions of the momentum transfer $\mathbf{q}$ ($\mathbf{q} = (h,h,0)$ and $(h,0,0)$), and different values of q are shown in Fig. 5.49. Three peaks centered at 2.5, 8, and 11 eV are distinguishable, exhibiting only rather flat dispersion, as shown for the 2.5 eV peak in Fig. 5.49. It is the 2.5 eV peak which was attributed by the authors to transitions from the effective lower Hubbard band (LHB: O 2p + $e_g^1 \uparrow$) to the upper Hubbard band (UHB: $e_g^2 \uparrow$) across the Mott gap caused by the Coulomb interaction between the excited 4p electron and the 3d electron at the same site. This excitation must be regarded as an orbital excitation. As depicted in Fig. 5.50, LaMnO$_3$ shows the $3d_{3x^2-r^2}/3d_{3y^2-r^2}$-type orbital order, where the occupied $3d_{3x^2-r^2}$ and $3d_{3y^2-r^2}$

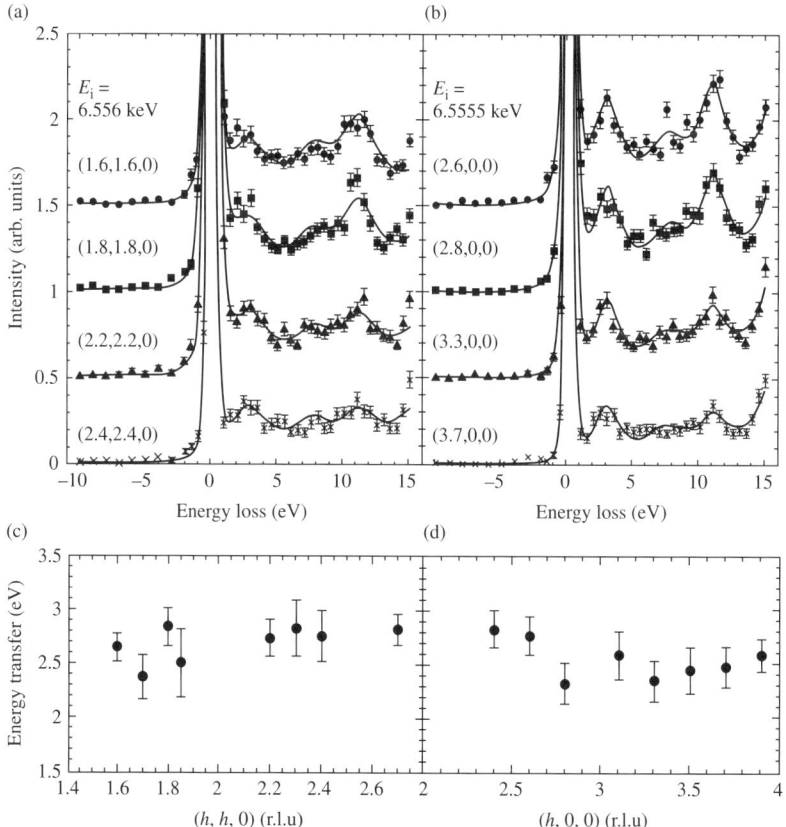

Fig. 5.49. q-dependence of the RIXS spectra of LaMnO$_3$ along $\mathbf{q} = (h,h,0)$ (a) and $\mathbf{q} = (h,0,0)$ (b) at incident energies 6.556 and 6.5555 keV, respectively. Solid lines are guides to the eye. The dispersion of the peak positions of the 2.5 eV peak is shown in (c) and (d). (Reprinted with permission from Inami et al. (2003); copyright (2003) by the American Physical Society.)

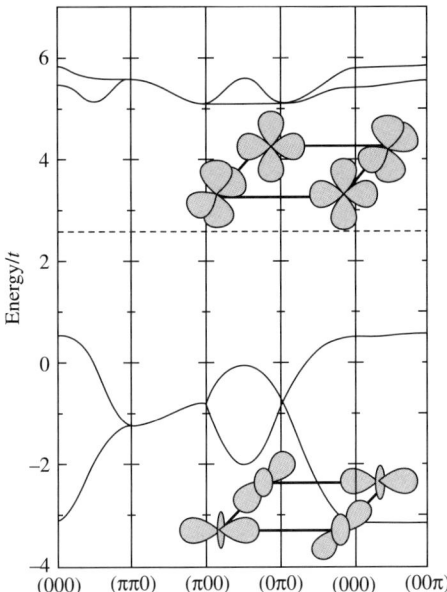

Fig. 5.50. The electronic band structure of the orbital ordered LaMnO$_3$. The broken line indicates the chemical potential. The inset depicts a schematic illustration of the orbital ordered state: the $3d_{y^2-z^2}$ and $3d_{z^2-x^2}$ orbitals (top) constituting the upper Hubbard band, and the $3d_{3x^2-r^2}/3d_{3y^2-r^2}$-type orbital order of the lower Hubbard band. (Reprinted with permission from Inami et al. (2003); copyright (2003) by the American Physical Society.)

orbitals are strongly hybridized with O 2p orbitals. In contrast, the lowest unoccupied e$_g$ band (UHB) consists of the $3d_{y^2-z^2}$ and $3d_{z^2-x^2}$ orbitals, so that it has different orbital symmetry from that of the LHB. The excitation from the LHB to the UHB changes the symmetry of the orbitals. This assignment is supported by looking at the azimuthal angle (ψ) dependence of the intensity of the 2.5 eV peak as shown in Fig. 5.51. The definition of ψ is shown in the inset of this figure, where the azimuthal angle ψ is defined as 90°, when the c-axis is perpendicular to the scattering plane. The tendency of the measured ψ-dependence with a π-periodicity and minima at $\psi = \pm\pi/2$ is well reproduced by calculations on the basis of a theory of Kondo et al. (2001). The leading equation of this theory, as deduced from the Kramers–Heisenberg formula (5.2), extended by enclosing the spin degree of freedom, and reformulated by the Liouville operator method is:

$$\frac{d^2\sigma}{d\Omega_2 d\hbar\omega_2} = r_0^2 \left(\frac{\omega_2}{\omega_1}\right) \sum_{\gamma\gamma'\sigma\sigma'} P_{\gamma'\sigma'}^{\lambda_2\lambda_1*} P_{\gamma\sigma}^{\lambda_2\lambda_1} \Pi(\omega, \mathbf{q}) \qquad (5.70)$$

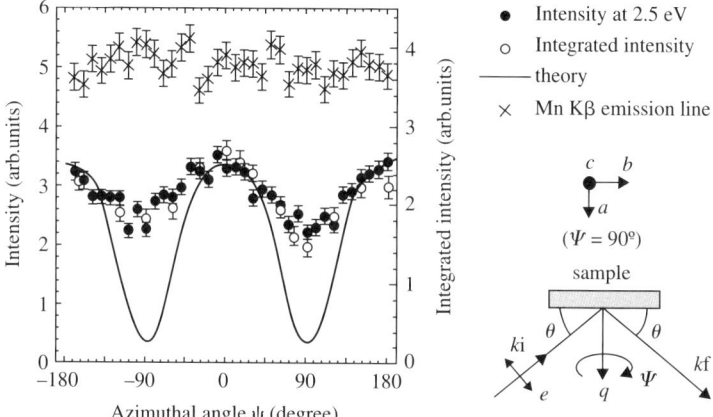

Fig. 5.51. Left-hand panel: dependence of the intensity of the 2.5 eV peak in the LaMnO$_3$ RIXS peak on the azimuthal angle ψ. Open circles: integrated experimental intensity; solid circles: experimental intensity at 2.5 eV. Solid line: integrated intensity of the theoretically calculated intensity of the RIXS spectrum. The crosses give the azimuthal dependence of the Mn Kβ_5 emission line. The intensity is independent of ψ. Right-hand panel: The experimental setup for the measurement of the azimuthal dependence. At $\psi = 90°$ and $0°$ the c-axis is perpendicular and in the scattering plane, respectively. (Reprinted with permission from Inami *et al.* (2003); copyright (2003) by the American Physical Society.)

where

$$P_{\gamma\sigma}^{\lambda_2\lambda_1} = \sum_{\alpha=x,y,z} (\mathbf{e}_{K_2\lambda_2})_\alpha D_{\gamma\sigma\alpha} (\mathbf{e}_{K_1\lambda_1})_\alpha \qquad (5.71)$$

describes the polarization part with the X-ray polarization vector $\mathbf{e}_{K\lambda}$ and the amplitude $D_{\gamma\sigma\alpha}$ of the orbital excitation from the 3d$_\gamma$ orbital by X-rays with polarization α defined by

$$D_{\gamma\sigma\alpha} = (E_{\gamma\sigma\alpha} - \hbar\omega_1 + i\Gamma)^{-1} C_{\gamma\alpha}(E_\alpha - \hbar\omega_1 + i\Gamma)^{-1}. \qquad (5.72)$$

Here E_α and $E_{\gamma\sigma\alpha}$ are the excitation energies of the states $|4p_\alpha^1 3d_{\gamma\sigma} 1s\rangle$ and $|4p_\alpha^1 3d_{-\gamma\sigma} 1s\rangle$, respectively, where $-\gamma$ indicates the counterpart of orbital γ. $C_{\gamma\alpha}$ is the amplitude of the Coulomb interaction between the excited 4p$_\alpha$ electron and the 3d$_\gamma$ electrons causing the orbital excitation. $\Pi(\omega, \mathbf{q})$ is the Fourier transform of the correlation function of the orbital pseudospin operators defined by

$$\Pi(t, \mathbf{R}_{l'} - \mathbf{R}_l) = (|B|^4/m^2) \langle T_{l'\gamma'\sigma'}^x(t) T_{l\gamma\sigma}^x(0) \rangle, \qquad (5.73)$$

where the x-component of the pseudospin operator $T_{l\gamma\sigma}^x = d_{l-\gamma\sigma}^+ d_{l\gamma\sigma}$ represents the orbital excitation. $d_{l\gamma\sigma}$ is the annihilation operator of the d electron at site

l with spin σ and orbital γ. B is the matrix element of the dipole transition: $B = (1/2)\langle 4p_\alpha | - i\nabla_\alpha | 1s \rangle$.

Inserting the parameters of the actual experimental arrangement into (5.71), taking into account in (5.73) the change of the orbital symmetry due to the transition from the LHB to the UHB as indicated above, and averaging the differential cross-section over the π and σ polarization of the scattered beam, one ends up with an azimuthal angle dependence of the differential cross-section divided by $r_0^2 |B|^4 \omega_2 / (4m^2 \omega_1)$ for the 2.5 eV peak as shown by the solid line in Fig. 5.51, in qualitative agreement with experiment. The larger amplitude of the periodic oscillations may be attributed to orbital fluctuations in the ground state that has been neglected in the calculations. The small dispersion of the 2.5 eV feature has to be ascribed to the fact that hopping to the nearest-neighbor site in the UHB is forbidden in the ab plane because of the orthogonality of the wavefunctions. Thus the small dispersion is a direct consequence of orbital ordering. A more detailed presentation of the above theory is given by Ishihara and Maekawa (2000).

The two other peaks in the RIXS spectrum at 8 and 11 eV are attributed to charge transfer excitations from the O 2p bands to the empty Mn 3d and 4s/4p bands, respectively.

An extended study on orbital excitation by RIXS was presented by Ishii et al. (2004) on hole-doped manganites $La_{1-x}Sr_x MnO_3$ ($x = 0.2, 0.4$) and compared with the results on undoped $LaMnO_3$. The undoped sample was measured in the paramagnetic insulating phase at 300 K, the $x = 0.2$ hole-doped sample both in the ferromagnetic phase (7 K) and in the insulating phase at 400 K, and the $x = 0.4$ hole-doped sample in the ferromagnetic phase at 300 K. The RIXS spectra were taken at the 11XU beamline of SPring-8 with a total energy resolution of 230 meV. The effect of hole-doping can be summarized as follows:

(i) A peak in the 2 eV energy-loss region of the RIXS spectra appears indicating the persistence of the Mott gap.

(ii) The energy gap is partly filled, the spectral weight shifts toward lower energies.

(iii) The low-energy spectral features show little change with momentum transfer, as already seen in the undoped case.

(iv) In the $x = 0.2$ sample a clear temperature dependence in intensity around 2–4 eV is found.

The RIXS intensity of the $\mathbf{q} = (2.7, 0, 0)$ spectra increases with decreasing temperature, whereas the scattering intensity of $\mathbf{q} = (2.2, 2.2, 0)$ is independent of temperature. This is in contrast with optical conductivity, whose spectral weight at 2 eV decreases with decreasing temperature. The authors attribute this anisotropic temperature dependence to an anisotropy of the ferromagnetic exchange interaction. The change in RIXS intensity along $[h00]$ may be due to the evolution of the double-exchange interaction, since the temperature dependence of the $\mathbf{q} = (2.7, 0, 0)$ scattering intensity is in qualitative accord with that of bulk magnetization: the intensity begins to increase at T_C (~ 300 K), and saturates around 150 K. On the contrary, the scattering intensity is independent of temperature along $[hh0]$, the Mn–O–Mn direction, which can be traced back to the ferromagnetic superexchange interaction which is active both in the metallic and in the insulating phase.

Grenier et al. (2005) found a similar temperature dependence upon transition into a ferromagnetic metallic phase of the $[0h0]$ intensity of RIXS spectra between 1.5 and 5 eV in a series of manganites exhibiting a variety of magnetic, orbital, charge, and structural ground states, while the transition into an antiferromagnetic insulating phase with decreasing temperature diminishes the RIXS intensity of these features. Contrary to Ishii et al. (2003), Grenier et al. (2005) found that this excitation spectrum does not depend on the presence or absence of orbital order, and that its temperature dependence reflects the magnetic order of the sample. Based on local density approximation + Hubbard U (LDA + U) calculations on $LaMnO_3$ of local density of states and of Wannier functions the authors explain the magnetic order dependent features as arising from intersite 3d–3d transitions between Mn atoms (mediated via the O 2p): For ferromagnetic neighbors, hopping between neighboring sites, and thus low-energy excitations are allowed, while for antiferromagnetic neighbors such excitations require an improbable spin flip and are therefore suppressed.

5.8 Magnetic circular and linear dichroism in RIXS

Let us classify the intensity of an X-ray absorption (XAS) or a RIXS signal, in simple terms, according to the change Δm_j of the magnetic quantum number connected with the transition between initial and final state in XAS or between initial and intermediate state in RIXS. Three different linear combinations of intensities have to be considered. (i) The so-called isotropic intensity which is the sum of intensities with $\Delta m_j = +1$, $\Delta m_j = 0$ and $\Delta m_j = -1$. (ii) The magnetic circular dichroism (MCD) is the $\Delta m_j = +1$ minus $\Delta m_j = -1$ intensity. (iii) The linear dichroism (MLD) is the sum of the $\Delta m_j = +1$ and $\Delta m_j = -1$ intensities minus twice the $\Delta m_j = 0$ intensity. The $\Delta m_j = +1$ intensity is caused by right circularly, the $\Delta m_j = 0$ intensity by Z linearly and the $\Delta m_j = -1$ intensity by left circularly polarized light.

X-ray magnetic circular dichroism (XMCD) in X-ray absorption spectroscopy (XAS) is a very often used technique to study magnetic materials exploiting the difference in the number of created core holes for parallel and antiparallel alignment of the magnetization and helicity vector of the incident light (see e.g. Ebert and Schütz 1996). It was especially the use of sum rules (Thole et al. 1992), which enables us to relate the intensity of the dichroic signal to groundstate expectation values of orbital and spin-dependent effective operators acting on the valence electrons, and made XMCD to a powerful method. On the other hand it was stressed by Thole and van der Laan (1991,1993,1994) that the charge distribution of the excited core hole is not only characterized by its integral which is proportional to the absorption but also by its spatial distribution which is in general not spherical. Thus information about the deviation from the spherical symmetry, i.e. about the core-hole polarization, can be obtained by measuring the angular distribution of the particles emitted by the decay of the core hole. These particles can either be Auger electrons from the autoionization process connected with the core hole decay or photons re-emitted in a RIXS process. Since the measured deviation from the spherical symmetry of the core hole is directly connected with ground-state properties via the transition matrix elements of the RIXS process, important information about material properties not accessible by XMCD in XAS could be expected. (A theoretical foundation on the basis of a multipolar expansion is provided in Section 6.10.3) This additional information can be found in pure form, when the incident light is perpendicular to the magnetization, since in this case the dichroism in absorption vanishes. In this connection we will speak of coplanar perpendicular geometry if the plane defined by the incident photon and the emitted products detected in the experiment contains the magnetization. The detection of the emitted products breaks the mirror symmetry of the system, thus getting information about the nonspherical part of the core-hole distribution. Until 1999 this approach has been used only in experiments based on electron detection (Thole et al. 1995, Dürr et al. 1996). Braicovich et al. (1999) were the first to measure the circular dichroism in RIXS with incidence perpendicular to the magnetization, where the absorption dichroism vanishes. The geometry of the experiment (coplanar perpendicular) is sketched in Fig. 5.52. The measurements were performed using the circularly polarized soft X-rays from the beamline ID12B of the ESRF with a monochromator and detector bandwidth of 1.2 and 1.3 eV, respectively, on Ni ferrite ($NiFe_2O_4$) and Co metal, in each case

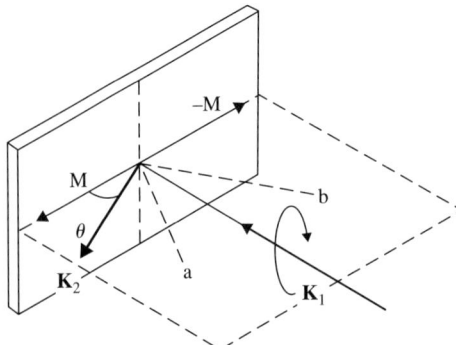

Fig. 5.52. Schematic geometry of coplanar perpendicular RIXS. $\mathbf{K}_1$ is the direction of the incident radiation, $\mathbf{K}_2$ the detection direction. The magnetization **M** is in the surface plane. (Reprinted with permission from Braicovich et al. (1999); copyright (1999) by the American Physical Society.)

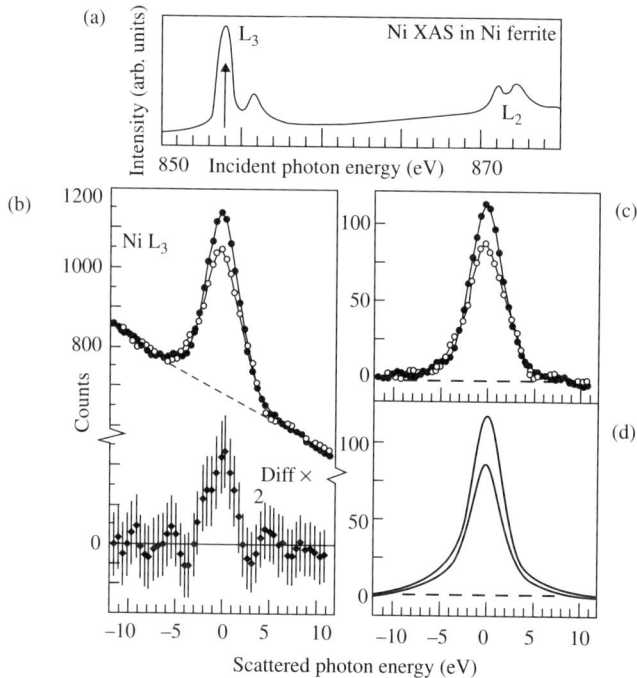

Fig. 5.53. XMCD in the perpendicular RIXS at Ni L_3 in Ni ferrite. Panel (a): $L_{2,3}$ absorption spectrum, the arrow indicates the excitation energy. Panel (b): Raw scattering data with reversed magnetization or helicity vector; difference curve after correction for incomplete circular polarization. Panel (c): Experimental data after background subtraction and polarization correction. Panel (d): Theoretical spectra. (Reprinted with permission from Braicovich *et al.* (1999); copyright (1999) by the American Physical Society.)

at the L_2 and L_3 edge of the transition metal. The excitation and de-excitation of type $2p^6 3d^n \rightarrow 2p^5 3d^{n+1} \rightarrow 3s^1 3d^{n+1}$ is utilized. The results for Ni excited at the L_3 white line of the absorption spectrum (a) is shown in Fig. 5.53 with the energy of the scattered photons measured relative to the peak maximum (final Ni-$3s^1$ state) which is on the decreasing tail of the nonresonant Fe $L_{2,3}$ signal (b). A large change in the signal is found when either the polarization or the magnetization is reversed. (c) is the spectrum after background subtraction and correction for incomplete circular polarization. The theoretical curves (d) are based on calculations according to a model of Braicovich *et al.* (1997) which takes into account the polarization and angles in the matrix element of the Kramers–Heisenberg formula (5.2). Qualitatively the appearance of dichroism can be understood looking at Fig. 5.52 considering two opposed effects: Spin–orbit interaction would align the expectation value of the core-hole orbital moment along its spin direction which has a tendency to be oriented parallel to the magnetization **M**. On the other hand, the selection rules for the excitation in the RIXS process lead to a preferred alignment of the orbital moment parallel to the helicity vector of the circularly polarized light. Thus the core-hole orbital moment is aligned along an intermediate direction depicted by the dashed lines a and b in Fig. 5.52 for the two opposite magnetization directions, so that the photon spectrum scattered along an off-normal direction must change when either the magnetization or the helicity vector is reversed. A detailed analysis of the angular and polarization dependence of the double differential RIXS cross-section for the general case was presented by Ferriani et al. (2004), whose derivations we will sketch in what follows, where also the decisive role of interference effects in the intermediate state will be stressed.

We start with the Kramers–Heisenberg formula (5.2) for the double differential scattering cross-section which we write in the dipole approximation as follows:

$$\frac{d^2\sigma}{d\Omega_2 d\hbar\omega_2} \sim \sum_f \left| \sum_n \frac{\langle f|T^{(2)}|n\rangle\langle n|T^{(1)}|i\rangle}{E_i - E_n - \hbar\omega_1 + i\Gamma_n/2} \right|^2 \delta(E_i - E_f + \hbar\omega), \qquad (5.74)$$

where $T^{(1)} = \mathbf{e}_1 \cdot \mathbf{r}$ and $T^{(2)} = \mathbf{e}_2 \cdot \mathbf{r}$ are the dipole transition operators for excitation and decay, respectively, of type $2p^63d^n \to 2p^53d^{n+1} \to 3s^13d^{n+1}$. For the general case, we introduce the angular notation of Fig. 5.54 with three Cartesian coordinate systems, one (x, y, z) connected with the incident beam, (x', y', z') with the scattered beam and (x'', y'', z'') with the sample and its magnetization.

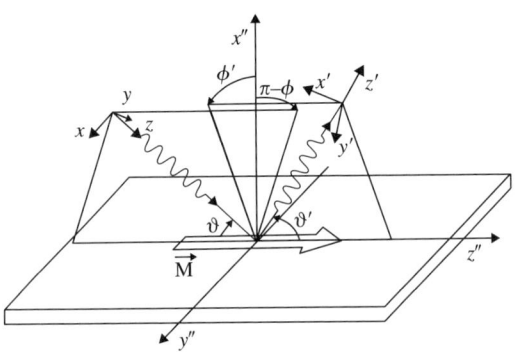

Fig. 5.54. Three coordinate systems necessary to describe the geometry of the scattering process. Two are related to the incident (xyz) and emitted $(x'y'z')$ photons. The third is connected with the sample with z'' parallel to the magnetization $\mathbf{M}$. x'' and y'' are parallel to the octahedron axes in the case of an O_h crystal field. (Originally published by Ferriani et al. (2004); copyright (2004) by the American Physical Society.)

Writing the scalar products in spherical coordinates $\mathbf{e} \cdot \mathbf{r} = r \sum_\mu (-1)^\mu e_{-\mu} C_\mu^{(1)}(\hat{\mathbf{r}})$ with the spherical tensor operator $C_m^{(l)} = [4\pi/(2l+1)]^{1/2} Y_{lm}$ and the spherical harmonics Y_{lm}, one ends up with

$$\frac{d^2\sigma^{\mathbf{e}_1 \mathbf{e}_2}}{d\Omega_2 d\hbar\omega_2} \sim \sum_f \left| \sum_{\mu\mu'} (-1)^{\mu'} e_{1,\mu}(e_{2,-\mu'})^* \sum_n \frac{\langle i|rC_\mu^{(1)}(\hat{\mathbf{r}})|n\rangle \langle n|rC_{\mu'}^{(2)}(\hat{\mathbf{r}})|f\rangle}{E_i - E_n - \hbar\omega_1 + i\Gamma_n/2} \right|^2$$
$$\times \delta(E_i - E_f + \hbar\omega), \qquad (5.75)$$

where the index 1 and 2 on $\mathbf{e}$ and ω refers in each case to the incident and scattered radiation, respectively. The polarization covariant components e_μ are then obtained in terms of the Cartesian coordinates of different frames. For circular polarization one has

$$(\mathbf{e}^+)_\mu = \begin{cases} \frac{1}{2}(1-\cos\vartheta)\exp(i\phi), & \mu = 1 \\ -\frac{1}{\sqrt{2}}\sin\vartheta, & \mu = 0 \\ \frac{1}{2}(1+\cos\vartheta)\exp(-i\phi), & \mu = -1 \end{cases}$$

and

$$(\mathbf{e}^-)_\mu = \begin{cases} -\frac{1}{2}(1+\cos\vartheta)\exp(i\phi), & \mu = 1 \\ -\frac{1}{\sqrt{2}}\sin\vartheta, & \mu = 0 \\ -\frac{1}{2}(1-\cos\vartheta)\exp(-i\phi), & \mu = -1. \end{cases}$$

The same expression holds for the scattered photon as a function of ϑ' and ϕ'.

For the case that the polarization of the scattered beam is not measured, one has to sum over the $\mathbf{e}_2$ and obtains:

$$\frac{d^2\sigma^{\mathbf{e}_1}}{d\Omega_2 d\hbar\omega_2} \sim \sum_f F^{\mathbf{e}_1}(\hbar\omega_1) \delta(E_i - E_f + \hbar\omega)$$

$$F^{\mathbf{e}_1}(\hbar\omega_1) = \sum_{\mathbf{e}_2} \sum_{\mu_1\mu_2\mu_3\mu_4} (-1)^{\mu_2+\mu_4} e_{1,\mu_1}(e_{2,-\mu_2})^* (e_{1,\mu_3})^* e_{2,-\mu_4} f_{\mu_1\mu_2} f^*_{\mu_3\mu_4}$$

with

$$f_{\mu\mu'} = \sum_n \frac{\langle i|rC_\mu^{(1)}(\hat{\mathbf{r}})|n\rangle \langle n|rC_{\mu'}^{(2)}(\hat{\mathbf{r}})|f\rangle}{E_i - E_n + \hbar\omega_1 + i\Gamma_n/2}. \qquad (5.76)$$

Due to selection rules, only transitions with $M' = M + \mu$ are allowed, so that only the $f_{\mu_1\mu_2} f^*_{\mu_3\mu_4}$ products with $\mu_1 + \mu_2 = \mu_3 + \mu_4$ will contribute to $F^{\mathbf{e}_1}(\hbar\omega_1)$. Thus the intensity for the sum, $F^{\text{sum}}(\hbar\omega_1) = F^+ + F^-$, and the difference, $F^{\text{dich}}(\hbar\omega_1) = F^+ - F^-$, over the incident photon circular polarization makes the angular dependence explicit

$$F^{\text{sum}}(\hbar\omega_1) = (1+\cos^2\vartheta)(1+\cos^2\vartheta')S_1 + (1+\cos^2\vartheta)\sin^2\vartheta' S_2$$
$$+ \sin^2\vartheta(1+\cos^2\vartheta')S_3$$
$$+ \sin^2\vartheta\sin^2\vartheta'\{S_4 + \text{Re}\{S_5\exp[2i(\phi-\phi')]\}\}$$
$$+ \sin 2\vartheta \sin 2\vartheta'\{\text{Re}\{S_6\exp[i(\phi-\phi')]\}\} \quad (5.77)$$

$$F^{\text{dichr}}(\hbar\omega_1) = \cos\vartheta(1+\cos^2\vartheta')D_1 + \cos\vartheta\sin^2\vartheta' D_2$$
$$+ \sin\vartheta\sin 2\vartheta'\{\text{Re}\{D_3\exp[i(\phi-\phi')]\}\}, \quad (5.78)$$

with

$$S_1 = \frac{1}{4}(|f_{11}|^2 + |f_{1-1}|^2 + |f_{-11}|^2 + |f_{-1-1}|^2); \quad S_2 = \frac{1}{2}(|f_{10}|^2 + |f_{-10}|^2);$$

$$S_3 = \frac{1}{2}(|f_{01}|^2 + |f_{0-1}|^2); \quad S_4 = |f_{00}|^2; \quad S_5 = \frac{1}{2}(f_{1-1}f^*_{-11});$$

$$S_6 = \frac{1}{4}(f_{10}f^*_{01} - f_{1-1}f^*_{00} - f_{00}f^*_{-11} + f_{0-1}f^*_{-10});$$

$$D_1 = \frac{1}{2}(-|f_{11}|^2 - |f_{1-1}|^2 + |f_{-11}|^2 + |f_{-1-1}|^2); \quad D_2 = -|f_{10}|^2 + |f_{-10}|^2;$$

$$D_3 = \frac{1}{2}(-f_{10}f^*_{01} + f_{1-1}f^*_{00} - f_{00}f^*_{-11} + f_{0-1}f^*_{-10}),$$

which include the atomic structure.

In perpendicular geometry (incident wave vector $\mathbf{K}_1$ perpendicular to $\mathbf{M}$) $\cos\vartheta$ is zero, so that, according to (5.78) only D_3 contributes to the dichroism. D_3 is solely composed of $f_{\mu_1\mu_2}f^*_{\mu_3\mu_4}$ products with $\mu_1 \neq \mu_3$ and $\mu_2 \neq \mu_4$ which means that the dichroism is only due to the interference between different excitation–emission paths.

In the case of X-ray absorption circular dichroism, the dichroic signal can be extracted either by reversing the magnetization vector $\mathbf{M}$ or reversing the direction of the helicity vector $\mathbf{P}$. In both cases the signal is the same. As Ferriani et al. (2004) have shown, this is, in the general, not the case with dichroism in RIXS with perpendicular geometry, namely whenever the imaginary part of $f_{\mu_1\mu_2}f^*_{\mu_3\mu_4}$ products contributing to F^{dich}

$$\text{Im}(f_{\mu_1\mu_2}f^*_{\mu_3\mu_4}) = -\frac{\Gamma}{2}\sum(E_m - E_n)\langle i|rC^{(1)}_{\mu_1}(\hat{\mathbf{r}})|m\rangle\langle m|rC^{(1)}_{\mu_2}(\hat{\mathbf{r}})|f\rangle$$
$$\times \langle i|rC^{(1)}_{\mu_3}(\hat{\mathbf{r}})|m\rangle\langle m|rC^{(1)}_{\mu_4}(\hat{\mathbf{r}})|f\rangle\{[(E_i - E_m + \hbar\omega_1)^2 + \Gamma^2/4)]$$
$$\times [(E_i - E_n + \hbar\omega_1)^2 + \Gamma^2/4)]\}^{-1} \quad (5.79)$$

cannot be neglected with respect to $\text{Re}(f_{\mu_1\mu_2}f^*_{\mu_3\mu_4})$, where the intermediate state lifetime broadening Γ is assumed to be the same for both intermediate states.

Neglect of $\text{Im}(f_{\mu_1\mu_2} f^*_{\mu_3\mu_4})$ is possible if the multiplet structure of the intermediate configuration is such that the only states which interfere, i.e. those with energy distance smaller than Γ, are nearly degenerate with respect to Γ ($|E_m - E_n| \ll \Gamma$). This means, with other words, if the prerequisites of the so-called fast collisions approximations are accomplished.

A further proof that magnetic circular dichroism can be found with RIXS in perpendicular geometry was provided by Fukui et al. (2001) in an investigation of the amorphous alloy $Gd_{33}Co_{67}$ at the Gd $L_{2/3}$ edge. The XMCD measurements were performed at the elliptical multipole wiggler beamline 28B of the Photon factory, Japan. The scattered radiation was analyzed by means of cylindrically bent Ge crystals in connection with a position-sensitive proportional counter. A strong XMCD signal was found in the $L_3 - M_{4,5}$ X-ray emission spectrum (XES), a weaker one in the $L_2 - M_4$ XES, both with resonant excitation at the $L_{2/3}$ white line (7247 eV) and the pre-edge region (7240 eV), as well as far above the white line (7277 eV), which the authors called the normal position. The XMCD signal in perpendicular geometry (incident radiation perpendicular to the magnetization) exhibited significant differences to the signal in longitudinal geometry (incident radiation parallel to the magnetization). Especially the sign of the $L_3 - M_4$ MCD is negative in perpendicular, and positive in longitudinal geometry. Moreover, the MCD in perpendicular geometry showed a pronounced dependence on the scattering angle. These findings were found in good agreement with theoretical results which were obtained on the same basis as those of Ferriani et al. (2004), certainly only for the coplanar geometry (the scattered beam is in the plane formed by the incident beam and the magnetization). Nevertheless Fukui et al. (2001) stressed that it is the interference between different excitation–emission paths which gives rise to XMCD in perpendicular geometry. The XMCD upon "normal" excitation was very similar to that detected by de Groot et al. (1997) for Gd metal.

Furthermore, we will discuss a special experimental geometry, the so-called conical scan, depicted in Fig. 5.55. The incident beam with helicity vector **P** is oriented perpendicular to the magnetization **M** (perpendicular geometry). The scattered beam is detected along a direction ε that can rotate around the incident one describing the surface of a cone with half-aperture α and axis $\mathbf{K}_1$. The position of ε on the cone is measured by β with $\beta = 0$ on the plane containing **P** and **M**. In this special experimental configuration, changing only β with fixed α, the self-absorption is constant and the experimental results need not be corrected for self-absorption. Moreover, this geometry is best suited to apply the RIXS sum rules as derived for the general case in Section 6.10.3. In this section, it was shown that the RIXS cross-section integrated over the scattered photon energy $\hbar\omega_2$ can be represented as the ground-state expectation value of a linear combination of double-tensor operators $w_\rho^{(a,b)r}$, which describe the multipole moments of the charge and magnetic distribution of the empty valence states. The index $a \in \{0, \ldots, 2l\}$ denotes the orbital, the index $b \in \{0, 1\}$ the spin parts, whereas r represents the ground-state multipole order. We use, in what

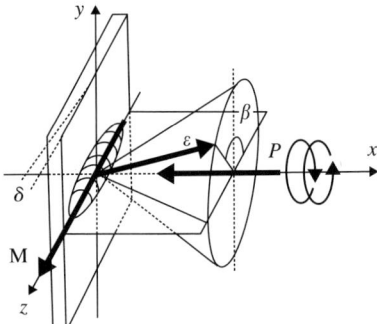

Fig. 5.55. Schematics for measuring integrated XMCD RIXS in a conical scan. **P** is the helicity vector of the incident circularly polarized photon beam, ε is the direction of the detected photon beam, and **M** is the magnetization. The detector for measuring the integrated inelastic scattered radiation moves along the cone with angular coordinate β. In pure perpendicular geometry, $\delta = 0$. (Reprinted with permission from Braicovitch et al. (2003); copyright (2003) by the American Physical Society.)

follows, the expressions in SO$_2$ symmetry (neglect of the crystal field), so that the index $\rho = 0$ can be skipped. As shown by Borgatti et al. (2004) the RIXS sum and dichroic signal, respectively, integrated over the incident photon energy, and recorded in conical geometry, for the case of pure circular polarization, is given by

$$J^{\text{RIXS}}_{\text{sum}} \sim \frac{1}{6}\left[C_0 + (C_1 - C_2)\left(\frac{3\sin^2\alpha - 2}{2}\right) + (3C_1 + C_2)\left(\frac{\sin^2\alpha\cos 2\beta}{2}\right)\right], \tag{5.80}$$

$$J^{\text{RIXS}}_{\text{dichr}} \sim -C_3 \sin 2\alpha \cos\beta, \tag{5.81}$$

where, for an electric dipole transition (E_1), the coefficients C_i are given in Table 5.2.

For excitation of a p$_{3/2}$ core hole and decay via s$_{1/2}$ → p$_{3/2}$, as used in most of the examples described below, the numerical coefficients B^0 and B^2 in Table 5.2 are equal to 1. A p$_{1/2}$ core hole and a decay via s$_{1/2}$ → p$_{1/2}$ requires $B^0 = 1$ and $B^2 = 0$, so that the circular dichroism disappears at the L$_2$ edge, assuming SO$_2$ symmetry. For other cases of excitation and decay, the reader can use the table given by Borgati et al. (2004). The definition of the factors $A_i^{E_1}$ is compiled in Table 5.3.

Table 5.2. Explicit form of the coefficients C_i in (5.80) and (5.81) for electric dipole transitions (E_1).

C_0	$8A_0^{E_1} B^0$
C_1	$2A_1^{E_1} B^2$
C_2	$A_2^{E_1} B^2$
C_3	$A_3^{E_1} B^2$

Table 5.3. Definition of the factors $A_i^{E_1}$ in Table 5.2.

$A_0^{E_1}$	$4E^{000} - E^{022}$
$A_1^{E_1}$	$4E^{202} - (E^{220} + E^{222} + E^{224})$
$A_2^{E_1}$	$6E^{220} - 6E^{222} + E^{224}$
$A_3^{E_1}$	$3E^{121} - 2E^{123}$

Table 5.4. Explicit expression for $E^{zz'r}$ in Table 5.3 as a linear combination of the ground-state moments $w^{(ab)r}$ for a core hole p_j.

	$j = 3/2$	$j = 1/2$
E^{000}	$2w^{000} + w^{110}$	$w^{000} - w^{110}$
E^{202}	$(1/5)(2w^{112} + 10w^{202} + 3w^{312})$	$(-1/5)(2w^{112} - 5w^{202} + 3w^{312})$
E^{022}	$2w^{112} + w^{202}$	
E^{220}	$(1/5)(w^{000} + 2w^{110})$	
E^{222}	$(2/35)(7w^{112} + 5w^{202} + 3w^{312})$	
E^{224}	$(18/35)(2w^{314} + w^{404})$	
E^{101}	$(1/3)(w^{011} + 2w^{211} + 6w^{101})$	
E^{121}	$(2/15)(5w^{011} + 3w^{101} + w^{211})$	
E^{123}	$(3/5)(2w^{213} + w^{303})$	

The corresponding definitions for quadrupole transitions (E_2) are made up by Borgatti *et al.* (2004). The coefficients $E^{zz'r}$ are linear combinations of the expectation values of double-tensor operators $w^{(ab)r}$ listed in Table 5.4 for the case of a $p_{3/2}$ and a $p_{1/2}$ core hole in the intermediate state. (We skip the $\langle\ \rangle$ brackets which

Table 5.5. Relation between the double-tensor operators $w^{(ab)r}$ and standard ground-state operators, $S_z = \sum_i s_{z,i}$, $L_z = \sum_i l_{z,i}$, $T_z = (1/4)\sum_i (3[l_z(\boldsymbol{\iota}\cdot\mathbf{s})]_+ - 2l^2 s_z)_i$, $Q_{zz} = \sum_i (l_z^2 - (1/3)\boldsymbol{\iota}^2)_i$, $P_{zz} = \sum_i (l_z s_z - (1/3)\boldsymbol{\iota}\cdot\mathbf{s})_i$, $R_{zz} = (1/3)\sum_i [5l_z(\boldsymbol{\iota}\cdot\mathbf{s})l_z - (l^2-2)\boldsymbol{\iota}\cdot\mathbf{s} - (2l^2+1)l_z s_z]_i$, referring to the hole properties of shells with angular quantum number l; physical meaning of the $w^{(ab)r}$.

$w^{(ab)r}$	Relation to ground state operators	Physical meaning
w^{000}	n_h	number of holes
w^{110}	$(ls)^{-1}\sum_i \boldsymbol{\iota}_i \cdot \mathbf{s}_i$	spin–orbit coupling
w^{001}	$s^{-1} S_z$	spin moment
w^{101}	$l^{-1} L_z$	orbital moment
w^{211}	$(2l+3)l^{-1} T_z$	magnetic dipole term
w^{202}	$3[l(2l-1)]^{-1} Q_{zz}$	charge quadrupole
w^{112}	$3l^{-1} P_{zz}$	anisotropic spin–orbit
w^{312}	$3[(l-1)(2l-1)]^{-1} R_{zz}$	quadrupole from orbit octupole and spin

should designate the expectation value.) For other cases the reader is referred to Borgatti et al. (2004). The list of ground-state multipoles $w^{(ab)r}$, as far as they are involved in (5.80) and (5.81), with their relation to standard ground-state operators and their physical meaning is found in Table 5.5. The equation (5.80) takes a rather simple form at the "magic" cone-angle $\sin^2 \alpha = 2/3$, since the part which is modulated with β does not include w^{000}.

According to the notation used with the double-tensor operators $w^{(ab)r}$, the physical meaning of the higher-order operators, the explicit representations of which are not given here, is the following: w^{303}: charge octupole; w^{213}: octupole from orbit quadrupole and spin; w^{404}: charge hexadecapole; w^{314}: hexadecapole from the orbital octupole and spin coupling.

According to the definition of the linear magnetic dichroism as given at the beginning of this subsection, the sum signal $J_{\text{sum}}^{\text{RIXS}}$ can be represented as being composed of the isotropic intensity and the linear dichroism signal:

$$J_{\text{sum}}^{\text{RIXS}} = \left(\frac{2}{3}\right) J_{\text{isotr}}^{\text{RIXS}} + \left(\frac{1}{3}\right) J_{\text{lin. dichr.}}^{\text{RIXS}}. \tag{5.82}$$

Thole and van der Laan (1995) have calculated the isotropic intensity of the resonant photoemission and its linear dichroism in a study on the core-hole polarization on the basis of a multipolar expansion of the corresponding cross-sections. As far as the angular dependence of the photoemission intensity and its composition of the double-tensor operators are concerned the expressions for the isotropic intensity and the linear dichroism are formally equivalent, and can be transferred to describe the RIXS process, here already written for the special geometry of a conical scan at the magic cone angle:

$$J_{\text{isotr}}^{\text{RIXS}} \sim [8E^{000}B^0 + 6E^{022}B^2 \sin^2\alpha \cos 2\beta] \quad (5.83)$$

$$J_{\text{lin. dichr}}^{\text{RIXS}} \sim -4E^{202}B^0 - [8E^{222} + (15/6)E^{224}]B^2 \sin^2\alpha \cos 2\beta. \quad (5.84)$$

The sum-rule expression (5.80) and (5.81) can be used to extract information about the ground-state multipoles $w^{(ab)r}$, from the sum-rule analysis of dipolar RIXS XMCD up to order $r=3$, from the dipolar RIXS XMLD up to order $r=4$. The advantage of RIXS magnetic dichroism over XAS dichroism thus becomes evident, since the sum-rule analysis of XAS XMCD on a dipolar level yields only information about multipoles up to order $r=1$, the XAS XMLD on a dipolar level up to order $r=2$.

Braicovich et al. (2003) were the first to utilize a RIXS conical scan for such a sum-rule analysis. The idea underlying this work was to take RIXS measurements in a suitable geometry allowing them to gather specific terms in (5.80) and (5.81) with the aim to recover the ground-state expectation values of the charge, spin and orbital moments up to order 4. The measurements were performed on beamline ID08 at the ESRF, which has complete polarization control and delivers $\sim 100\%$ circular polarization. The bandpass of the incident photons was 0.6 eV. The RIXS signal as a function of the incident photon energy from the channel $2p^63d^n \rightarrow 2p^53d^{n+1} \rightarrow 2p^63s^13d^{n+1}$ of Co ferrite, integrated over the scattered photon energy, was recorded at various fixed angled β in a conical geometry as shown in Fig. 5.55 by means of a photodiode with a filter to remove the unwanted valence scattering (channel $2p^63d^n \rightarrow 2p^53d^{n+1} \rightarrow 2p^63d^n$). With a sample of atomic number Z, a $(Z-1)$ filter is used. The valence scattering photons are absorbed since they are above the filter $L_{2,3}$ threshold. The photons from the $3s^1$ channel are below this threshold and thus detected with high efficiency. The XMCD and XMLD in absorption were also measured in situ using total electron yield. Figure 5.56 shows the measured $\hbar\omega_2$-integrated RIXS intensity as a function of the incident photon energy for right and left circular polarized incident radiation, together with the corresponding dichroism. As predicted by theory in SO_2, there is no dichroism at the L_2 edge. This supports the use of SO_2 expressions in the course of the sum-rule analysis. Without going into too much detail of this sum-rule analysis, the steps to obtain values of the multipole moments $w^{(ab)r}$ can be sketched as follows:

(i) The quadrupolar moment ($r=2$) w^{112} was obtained with the "magic" cone aperture and normalizing the measurement of J_{sum} at the L_3 edge for different values of β to $J_{\text{sum}}(\beta=45°)$, where w^{000} was set to 2.8, w^{110} to 1, and, assuming $E^{224}=0$, w^{202} was obtained by XAS XMLD along L_3 and L_2 (Carra et al. 1993). The central panel of Fig. 5.56 depicts the experimental values of $J_{\text{sum}}/J_{\text{sum}}(\beta=45°)$ (points) together with the theory

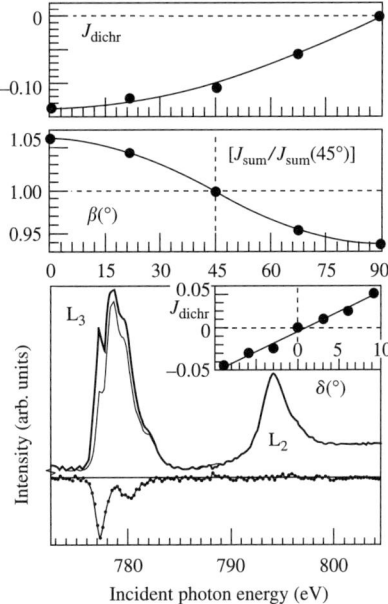

Fig. 5.56. Measured $\hbar\omega_2$-integrated RIXS intensity as a function of the incident photon energy. Thick and thin lines correspond to opposite circular polarization, the experimental points are the difference (dichroic) spectrum. Central panel: β-scan of normalized J_{sum}. Upper panel β-scan of J_{dichr}. Inset: δ-scan of J_{dichr}. (Reprinted with permission from Braicovitch et al. (2003); copyright (2003) by the American Physical Society.)

using w^{112} and w^{202} extracted from experimental results. The quadrupolar moment w^{312} was then recovered from the linear dichroic contribution to $J_{\text{sum}}(\beta = 90°)$ at the L_2 edge, which is solely determined by E^{202}.

(ii) The total hexadecapole moment $(r = 4)$, $(35/18)E^{224} = 2w^{314} + w^{404}$, was extracted from the linear dichroic contribution to $J_{\text{sum}}(\beta = 0°)$ at the magic cone angle, where all other participating moments are already known from the previous measurements.

(iii) In order to obtain the octupolar moments $(r = 3)$ one relies on the dipolar moments already known from the XAS XMCD. Then a small deviation from the perpendicular geometry by a sample rotation δ (see Fig. 5.55) using "magic" aperture and $\beta = 90°$, the relative weights of the first order and third order contributions to the L_3 RIXS MCD can be changed for the general case of the circular dichroism of resonant photoemission given by Thole and van der Laan (1995) with δ according to:

$$J_{\text{dichr}} = \left(E^{101} + \frac{2}{3}E^{123}\right)\delta, \quad (5.85)$$

where E^{101} is completely known from XAS XMCD, so that the total octupole moment $E^{123} = (3/5)(2w^{213} + w^{303})$ can be deduced. The inset of Fig. 5.56 illustrates the δ-dependence of J_{dichr}, as normalized to J_{sum}.

The remarkable result of this study is the persistence of higher order atom-like multipoles of the Co 3d hole states upon solid state formation.

5.9 References

Abbamonte, P., C.A. Burns, E.D. Isaacs, P.M. Platzman, L.L. Miller, S.W. Cheong, and M.V. Klein (1999). *Phys. Rev. Lett.* **83** 860

Åberg, T. (1971). *Phys. Rev. A* **4** 1735

Åberg, T. and J. Tulkki (1985). In *Atomic Inner-Shell Physics*, ed. by B. Crasemann (Plenum, New York) Chap. 10

Anderson, P.W. (1961). *Phys. Rev.* **124** 41

Attenkofer, K. (2001). *Die magnetische Kopplung in ausgewählten Verbindungen- Neue Möglichkeiten und Entwicklungen der Rumpfanregungsspektroskopie mit zirkular polarisierten Photonen.* Thesis, University of Hamburg

Bartolome, F., J.M. Tonnerre, L. Seve, D. Raoux, J. Chaboy, L.M. Garcia, M.H. Krisch, and C.-C. Kao (1997). *Phys. Rev. Lett.* **79** 3775

Best, P.E. (1966). *J. Chem. Phys.* **44** 3248

Bianconi, A., A. Marcelli, H. Dexpert, R. Kamatak, A. Kotani, T. Jo, and J. Petiau (1987). *Phys. Rev. B* **35** 806

Bocharov, S., G. Dräger, D. Heumann, A. Simunek, and O. Sipr (1998). *Phys. Rev. B* **58** 7668

Bocharov, S., Th. Kirchner, G. Dräger, O. Sipr, and A. Simunek (2001). *Phys. Rev. B* **63** 045104

Borgatti, F., G. Ghiringhelli, P. Ferriani, G. Ferrari, G. Van der Laan, and C.M. Bertoni (2004). *Phys. Rev B* **69** 134420

Braicovich, L., A. Tagliaferri, G. van der Laan, G. Ghiringhelli, and N.B. Brookes (2003) *Phys. Rev. Lett.* **90** 117401

Braicovich, L., C. Dallera, G. Ghiringhelli, N.B. Brookes, J.B. Goedkoop, and M.A. van Veenendaal (1997). *Phys. Rev. B* **55** R15989

Braicovich, L., G. van der Laan, G. Ghiringhelli, A. Tagliaferri, M.A. van Veenendaal, N.B. Brookes, M.M. Chervinskii, C. Dallera, B.D. Michelis, and H.A. Dürr (1999). *Phys. Rev. Lett.* **82** 1566

Briand, J.P., D. Girard, V.O. Kostroum, P. Chevalier, K. Wohrer, and J.P. Mosse (1981). *Phys. Rev. Lett.* **46** 1625

Brouder, C. (1990). *J. Phys.: Condens. Matter* **2** 701

Brümmer, O., G. Dräger, and K. Machlitt (1969). In *Proc. Internat. Symp. X-Ray Spectra and Electronic Structure of Matter, Vol. II*, ed. V.V. Nemoshkalenko, Kiev, p. 300

Carlisle, J.A., E.L. Shirley, E.A. Hudson, L.J. Terminello, T.A. Calcott, J.J. Jia, D.L. Ederer, R.C.C. Perera, and F.J. Himpsel (1996). *Phys. Rev. Lett.* **74** 4054

Carra, P., H. König, B.T. Thole, and M. Altarelli (1993). *Physica B* **192** 182
Chelikowsky, J.R. and S.G. Louie (1984). *Phys. Rev. B* **29** 3470
Controzzi, D. and F.H.L. Essler (2002). *Phys. Rev. B* **66** 165112
Dallera, C., M. Grioni, A. Shukla, G. Vanko, J.L. Sarrao, J.P. Rueff, and D.L. Cox (2002). *Phys. Rev. Lett.* **88** 196403
Dallera, C., E. Annese, J.-P. Rueff, A. Palenzona, G. Vanko, L. Braicovich, A. Shukla, and M. Grioni (2005). *J. Electr. Spectr. and Rel. Phenom.* **137–140** 651
de Groot, F.M.F., M. Nakazawa, A. Kotani, M.H. Krisch, and F. Sette (1997). *Phys. Rev. B* **56** 7285
De Groot, F.M.F., S. Pizzini, A. Fontaine, K. Hämäläinen, C.C. Kao, and J.B. Hastings (1995). *Phys. Rev. B* **51** 1045
Döring, G. (2001). *Untersuchung der elektronischen Struktur von CuO mit resonanter Röntgen-Raman-Streuung*. Thesis, University of Dortmund
Döring, G., C. Sternemann, A. Kaprolat, A. Mattila, K. Hämäläinen, and W. Schülke (2004). *Phys. Rev. B* **70** 085115
Dräger, G. and O. Brümmer (1984). *Phys. Stat. Sol.* (b) **124** 11
Dräger, G.R. Frahm, G. Materlik, and O. Brümmer (1988). *Phys. Stat. Sol. b* **146** 287
Dräger, G., Th. Kirchner, S. Bocharov, and C.-C. Kao (2001). *Appl. Phys. A* **73** 687
Dürr, H.A., G. van der Laan, and M. Surman (1996). *J. Phys. Condens. Matter* **8** L7
Ebert, H. and G. Schütz, edit. (1996). *Spin-Orbit Influenced Spectroscopies of Magnetic Solids* (Springer Verlag, Berlin)
Eisebitt, S. and W. Eberhardt (2000). *J. Electron Spectr. and Rel. Phenom.* **110–111** 335
Eisenberger, P., P.M. Platzman, and H. Winick (1976a). *Phys. Rev. Lett.* **36** 623
Eisenberger, P., P.M. Platzman, and H. Winick (1976b). *Phys. Rev. B* **13** 2377
Enkisch, H., A. Kaprolat, W. Schülke, H.M. Krisch, and M. Lorenzen (1999). *Phys. Rev. B* **60** 8624
Eskes, H., L.H. Tjeng, and G.A. Sawatzky (1990). *Phys. Rev. B* **41** 288
Fano, U. (1969). *Phys. Rev.* **178** 131
Ferriani, P., C.M. Bertoni, and G. Ferrari (2004). *Phys. Rev. B* **69** 104433
Fischetti, M.V. and J.M. Higman (1991). in: K. Hess (ed.), *Monte Carlo Device Simulation: Full Band and Beyond* (Klower, Boston)
Fukui, K., H. Ogasawara, A. Kotani, T. Iwazumi, H. Shoji, and T. Nakamura (2001). *J. Phys. Soc.*, Japan **70** 3457
Gallet, J.-J., J.-M-Mariot, L. Journel, C.F. Hague, A. Rogalev, H. Ogasawara, A. Kotani, and M. Sacchi (1999). *Phys. Rev. B* **60** 14128
Gel'mukhanov, F. and H. Ågren (1994a). *Phys. Rev. A* **49** 4378
Gel'mukhanov, F. and H. Ågren (1994b). *Phys. Lett. A* **185** 407
Geunseop Lee and S.-J. Oh (1991). *Phys. Rev. B* **43** 14674

Glans, P., K. Gunnelin, P. Skytt, J.H. Guo, N. Wassdahl, J. Nordgren, H. Agren, F. Gel'mukhanov, T. Warwick, and E. Rotenberg (1996). *Phys. Rev Lett.* **76** 2448

Glatzel, P., U. Bergmann, W. Gu, H. Wang, S. Stepanov, B.S. Mandimutsira, C.G. Riordan, C.P. Horwitz, T. Collins, and S. Cramer (2002). *J. Am. Chem. Soc.* **124** 9668

Glatzel, P., U. Bergmann, J. Yano, H. Visser, J.H. Robblee, W. Gu, F.M.F. de Groot, G. Christou, V.L. Pecoraro, St. P. Cramer, and V.K. Yachandra (2004). *J. Am. Chem. Soc.* **126** 9946

Grenier, S., J.P. Hill, V. Kiryukhin, W. Ku, Y.-Y. Kim, K.J. Thomas, S.-W. Cheong, Y. Tokura, Y. Tomioka, D. Casa, and T. Gog (2005). *Phys. Rev. Lett.* **94** 047203

Hämäläinen, K., S. Manninen, S.P. Collins, and M.J. Cooper (1990). *J. Phys. Condens. Matter* **2** 5619

Hague, C.F., J.-M. Mariot, R. Delaunay, J.-J. Gallet, L. Journel, and J.-P. Rueff (2004). *J. Electr. Spectr. and Rel. Phenom.* **136** 179

Hämäläinen, K., D.P. Siddons, J.B. Hastings, and L.E. Berman (1991). *Phys. Rev. Lett.* **67** 2850

Hämäläinen, K., S. Manninen, S.P. Collins, and M.J. Cooper (1990). *J. Phys.: Condens. Matter* **2** (1990) 5619

Hämäläinen, K., S. Manninen, P. Suortti, S.P. Collins, M.J. Cooper, and D. Laundy (1989). *J. Phys.: Condens. Matter* **1** 5955

Hämäläinen, K., C.-C. Kao, J.B. Hastings, D.P. Siddons, L.E. Berman, V. Stojanoff, and S.P. Cramer (1992). *Phys. Rev. B* **46** 14274

Hämäläinen, K., J.P. Hill, S. Huotari, C.-C. Kao, L.E. Berman, A. Kotani, T. Ide, J.L. Peng, and R.L. Greene (2000). *Phys. Rev. B* **61** 1836

Hannon, J.P., G.T. Trammell, M. Blume, and D. Gibbs (1988). *Phys. Rev. Lett.* **61** 1245

Harrison, W.A. (1989). *Electronic Structure and the Properties of Solids* (Dover, New York)

Hasan, M.Z., E.D. Isaacs, Z.-X. Shen, L.L. Miller, K. Tsutsui, T. Tohyama, and S. Maekawa (2000). *Science* **288** 1811

Hasan, M.Z., P.A. Montano, E.D. Isaacs, Z.-X. Shen, H. Eisaki, S.K. Sinha, Z. Islam, N. Motoyama, and S. Ushida (2002). *Phys. Rev. Lett.* **88** 177403

Hasan, M.Z., Y.-D. Chuang, Y. Li, P. Momtano, M. Beno, Z. Hussain, H. Eisaki, S. Uchida, T. Gog, and D.M. Casa (2003). *Int. J. Mod. Phys. B* **17** 3479

Hasan, M.Z., Y. Li, Y.-D. Chuang, P.A. Montano, Z. Hussain, H. Eisaki, N. Motoyama, and S. Uchida (2003). *Int. J. Mod. Phys.* **17** 3513

Hasan, M.Z., Y.-D. Chuang, Y. Li, P.A. Montano, Z. Hussain, G. Dhalenne, A. Revocolevschi, H. Eisaki, N. Motoyama, and S. Uchida (2003). *Int. J. Mod. Phys. B* **17** 3519

Hayashi, H., Y. Udagawa, W.A. Caliebe, and C.-C. Kao (2002). *Phys. Rev. B* **66** 033105

Hayashi, H., R. Takeda, Y. Udagawa, T. Nakamura, H. Miyagawa, H. Shoji, S. Nanao, and N. Kawamura (2003). *Phys. Rev. B* **68** 045122

Hayashi, H., M. Kawata, R. Takeda, Y. Udagawa, Y. Watanabe, T. Takano, S. Nanao, and N. Kawamura (2004). *J. Electron Spectrosc. and Relat. Phenom.* **136** 191

Hermesmeier, B., C.S. Fadley, M.O. Krause, J. Jimenez-Mier, P. Gerard, and S.T. Manson (1988). *Phys. Rev. Lett.* **61** 2592

Heumann, D., G. Dräger, and S. Bocharov (1997). *J. Phys. IV France* **7** C2-481

Hill, J.P., C.-C. Kao, W.A.L. Caliebe, M. Matsubara, A. Kotani, J.L. Peng, and R.L. Greene (1998). *Phys. Rev. Lett.* **80** 4967

Himpsel, F.J., J.F. van der Veen, and D.E. Eastman (1980). *Phys. Rev. B* **22** 1967

Hohenberg, P. and W. Kohn (1964). *Phys. Rev.* **136** B864

Inami, T., T. Fukuda, J. Mizuki, S. Ishihara, H. Kondo, H. Nakao, T. Matsumara, K. Hirota, Y. Murakami, S. Maekawa, and Y. Endoh (2003). *Phys. Rev. B* **67** 045108

Ishihara, S. and S. Maekawa (1998). *Phys. Rev. B* **58** 13442

Ishihara, S. and S. Maekawa (2000). *Phys. Rev. B* **62** 2338

Ishii, K., T. Inami, K. Ohwada, K. Kuzushita, J. Mizuki, Y. Murakami, S. Ishihara, Y. Endoh, S. Maekawa, K. Hirota, and Y. Moritomo (2004). *Phys. Rev. B* **70** 224437

Ishii K., K. Tsutsui, Y. Endoh, T. Tohyama, S. Maekawa, M. Hoesch, K. Kuzushita, M. Tsubota, T. Inami, J. Mizuki, Y. Murakami, and K. Yamada (2005). *Phys. Rev. Lett.* **94** 207003

Johnson, P.D. and Y Ma (1994). *Phys. Rev. B* **49** 5024

Joyce, J.J., A.J. Arko, J.L. Sarrao, K.S. Graham, Z. Fisk, and P.S. Riseborough (1999). *Philos. Mag. B* **79** 1

Jolliffe, I.T. (2002) *Principal Component Analysis* (Springer, New York)

Journel, L., J.-M. Mariot, J.-P. Rueff, C.F. Hague, G. Krill, M. Nakazawa, A. Kotani, A. Rogalev, F. Wilhelm, J.-P. Kappler, and G. Schmerber (2002). *Phys. Rev. B* **66** 045106

Kao, C.-C., W.A.L. Caliebe, J.B. Hastings, and J.-M. Gillet (1996). *Phys. Rev. B* **54** 16361

Kim, C., Z.-X. Shen, N. Motoyama, E. Eisaki, S. Uchida, T. Tohayama, and S. Maekawa (1997). *Phys. Rev. B* **56** 15589

Kim, Y.-J., J.P. Hill, C.A. Burns, S. Wakimoto, R.J. Birgeneau, D. Casa, T. Gog, and C.T. Venkataraman (2002). *Phys. Rev. Lett.* **89** 177003

Kim, Y.-J., J.P. Hill, H. Benthien, H.H.L. Essler, E. Jeckelman, H.S. Choi, T.W. Noh, N. Motoyama, K.M. Kojima, S. Uchida, D. Casa, and T. Gog (2004a). *Phys. Rev. Lett.* **92** 137402

Kim, Y.-J., J.P. Hill, F.C. Chou, D. Casa, T. Gog, and C.T. Venkataraman (2004b). *Phys. Rev. B* **69** 155105

Kim, Y.-J, J.P. Hill, S. Komiya, Y. Ando, D. Casa, T. Gog, and C.T. Venkataraman (2004c). *Phys. Rev. B* **70** 094524

Kim, Y.-J., J.P. Hill, G.D. Gu, F.C. Chou, S. Wakimoto, R.J. Birgeneau, S. Komya, Y. Ando, N. Motoyama, K.M. Kojima, S. Uchida, D. Casa, and T. Gog (2004d). *Phys. Rev. B* **70** 205128

Kohn, W. and L.J. Sham (1965). *Phys. Rev.* **140** A1133

Kondo, H., S. Ishihara, and S. Maekawa (2001). *Phys. Rev. B* **64** 014414

Kotani, A. and S. Shin (2001). *Rev. Mod. Phys.* **73** 203

Krisch, M., C.-C. Kao, F. Sette, W.A. Caliebe, K. Hämäläinen, and J.B. Hastings (1995). *Phys. Rev. Lett.* **74** 4931

Lieb, E.H. and F.Y. Wu (1968). *Phys. Rev. Lett.* **20** 1445

Lindle, D.W., P-L-Cowan, T. Jach, R.E. LaVilla, R.D. Deslattes, and R.C.C. Perera (1991). *Phys. Rev. A* **43** 2353

Lindle, D.W., P.L. Cowan, R.E. LaVilla, T. Jach, R.D. Deslattes, B. Karlin, J.A. Sheehy, T.J. Gil, and P.W. Langhoff (1988). *Phys. Rev. Lett.* **60** 1010

Luo. Y., H. Agren, and F. Gel'mukhanov (1994). *J. Phy. B: At. Mol. Opt. Phys.* **27** 4169

Ma, Y. (1994a). *Phys. Rev. B* **49** 5799

Ma, Y. (1994b). *Chem. Phys. Lett.* **230** 451

Ma, Y. and M. Blume (1995). *Rev. Sci. Instrum.* **66** 1543

Ma, Y., N. Wassdahl, P. Skytt, J. Guo, J. Nordgren, P.D. Johnson, J.E. Rubensson, T. Boske, W. Eberhardt, and S. Kevan (1992). *Phys. Rev. Lett.* **69** 2598

Ma, Y., K.E. Miyano, P.L. Cowan, Y. Aglitzkiy, B.A. Karlin (1995). *Phys. Rev. Lett.* **74** 478

MacDonald, M.A., S.C. Southworth, J.C. Levin, A. Henins, R.D. Deslattes, T. LeBrun, Y. Azuma, P.L. Cowan, and B.A. Karlin (1995). *Phys. Rev. A* **51** 3598

Magdans, U. (2001). *Ausnutzung der internen Spinreferenz bei der Messung der unbesetzten 5d-Zustandsdichte des EuO mit resonant inelastischer Röntgenstreuung (RIXS)*, Diploma Thesis, University of Dortmund

Malterre, D., M. Grioni, and Y. Baer (1996). *Adv. Phys.* **45** 299

S. Manninen, P. Suortti, M.J. Cooper, J. Chomilier, and G. Loupias (1986). *Phys. Rev. B* **34** 8351

Mayer, R., D.W. Lindle, S.H. Southworth, and P.L. Cowan (1991). *Phys. Rev. A* **43** 235

Mazalov, L.N., A.P. Sadovskii, P.I. Vadash, and F.G. Gel'mukhanov (1973). *J. Struct. Chem.* **14** 234

Mills, J.D., J.A. Sheehy, T.A. Ferret, s.H. Southworth, R. Mayer, D.W. Lindle, and P.W. Langhoff (1997). *Phys. Rev. Lett.* **79** 383

Miyano, K.E., Y. Ma. S.H. Southworth, P.L. Cowan, and B.A. Karlin (1996). *Phys. Rev. B* **54** 12022

Moore, D.P., J.J. Joyce, A.J. Arko, J.L. Sarrao, L. Morales, M. Hochst, and Y.D. Chuang (2000). *Phys. Rev. B* **62** 16492

Neudert, R.M. Knupfer, M.S. Golden, J. Fink, W. Stephan, K. Penc, N. Motoyama, H. Eisaki, and S. Uchida (1998). *Phys. Rev. Lett.* **81** 657

Murakami, Y., H. Kawada, H. Kawata, M. Tanaka, T. Arima, H. Morimoto, and Y. Tokura (1998a). *Phys. Rev. Lett.* **80** 1932

Murakami, Y., J.P. Hill, D. Gibbs, M. Blume, I. Koyama, M. Tanaka, H. Kawata, T. Arima, Y. Tokura, K. Hirota, and Y. Endoh (1998b). *Phys. Rev. Lett.* **81** 582

Nolting, W., G. Borstel, and W. Borgiel (1987a). *Phys. Rev. B* **35** 7015

Nolting, W., W. Borgiel, and G. Borstel (1987b). *Phys. Rev. B* **35** 7025

Nomura, T. and J. Igarashi (2005). *Phys. Rev. B* **71** 035110

Painter, G.S., D.E. Ellis, and A.R. Lubinsky (1971). *Phys. Rev. B* **4** 3610

Peng, G., F.M.F. de Groot, K. Hämäläinen, J.A. Moore, X. Wang, M.M. Grush, J.B. Hastings, D.P. Siddons, W.H. Armstrong, O.C. Mullins, and S.P. Cramer (1994). *J. Am. Chem. Soc.* **116** 2914

Platzman, P.M. and E.D. Isaacs (1998). *Phys. Rev. B* **57** 107

Poliakoff, E.D., J.L. Dehmer, D. Dill, A.C. Parr, K.H. Jackson, and R.N. Zare (1981). *Phys. Rev. Lett.* **46** 907

Reinert, F., R. Claessen, G. Nicolay, D. Ehm, S. Hüfner, W.P. Ellis, G.H. Gweo, J.W. Allen, B. Kindler, and W. Assmus (1998). *Phys. Rev. B* **58** 12808

Saitoh, E., S. Okamoto, T. Takahashi, K. Tobe, K. Yamamoto, T. Kimura, S. Ishihara, S. Maekawa, and Y. Tokura (2001). *Nature (London)* **410** 180

Sawatzky, G. and J.W. Allen (1984). *Phys. Rev. Lett.* **53** 2339

Schülke, W. (2001). *J. Phys. Condens. Matter* **13** 7557

Schütz, G, W. Wagner, W. Wilhelm, P. Kienle, R. Zeller, R. Frahm, and G. Materlik (1987). *Phys. Rev. Lett.* **58** 737

Schwarz, W.H.E. (1975). *Chem Phys.* **11** 217

Shukla, A., J.-P. Rueff, J. Badro, G. Vanko, A. Mattila, F.M.F. de Groot, and F. Sette (2003). *Phys. Rev. B* **67** 081101

Simon, M.E., A.A. Aligi, C.D. Batista, E.R. Gagliano, and F. Lema (1996). *Phys. Rev. B* **54** R3780

Sinkovic, B. and C.S. Fadley (1985). *Phys. Rev. B* **31** 4665

Skytt, P., P. Glans, D.C. Mancini, J.-H. Guo, N. Wassdahl, J. Nordgren, and Y. Ma (1994). *Phys. Rev. B* **50** 10457

Skytt, P., P. Glans, J.-H. Guo, K. Gunnelin, C. Såthe, J. Nordgren, F.Kh. Gel'mukhanov, A. Cesar, and H. Ågren (1996). *Phys. Rev. Lett.* **77** 5035

Sokolov, A.V., E.K. Kurmaev, S. Leitch, A. Moewes, J. Kortus, L.D. Finkelstein, N.A. Skorikov, C. Xiao, and A. Hirose (2003). *J. Phys.:Condens. Matter* **15** 2081

Soldatov, A.V., T.S. Ivanchenkov, S. Della Longa, A. Kotani, Y. Iwamoto, and A. Bianconi (1994). *Phys. Rev. B* **50** 5074

Soldatov, A.V., T.S. Ivanchenko, A.P. Kovtun, S. Della Longa, and A. Bianconi (1995). *Phys. Rev. B* **52** 11757

Southworth, S.H., D.W. Lindle, R. Mayer, and P.L. Cowan (1991a). *Nucl. Instrum. & Methods in Phys. Res. B* **56/57** 304

Southworth, S.H., D.W. Lindle, R. Mayer, and P.L. Cowan (1991b). *Phys. Rev. Lett.* **67** 1098

Sparks, Jr., C.J. (1974). *Phys. Rev. Lett.* **33** 262

Stephan, W. and K. Penc (1996). *Phys. Rev. B* **54** R17269

Suortti, P. (1979). *Phys. Stat. Sol. (b)* **91** 657

Susaki, T., A. Fujimori, M. Okusawa, J.L. Sarrao, and Z. Fisk (2001). *Solid State Commun.* **118** 413

Thole, B.T. and G. van der Laan (1991). *Phys. Rev. B* **44** 12424

Thole, B.T. and G. van der Laan (1993). *Phys. Rev. Lett.* **70** 2499

Thole, B.T. and G. van der Laan (1994). *Phys. Rev. B* **49** 9613

Thole, B.T., H.A. Dürr, and G. van der Laan (1995). *Phys. Rev. Lett* **74** 237

Thole, B.T., P. Carra, F. Sette, and G. van der Laan (1992). *Phys. Rev. Lett.* **68** 1943

Tsutsui, K., T. Tohyama, and S. Maekawa (1999). *Phys. Rev. Lett.* **83** 3705

Tsutsui, K., T. Tohyama, and S. Maekawa (2000). *Phys. Rev. B* **61** 7180

Tsutsui, K., T. Tohyama, and S. Maekawa (2003). *Phys. Rev. Lett.* **91** 117001

van den Brink, J. and M. van Veenendaal (2003). cond-mat/0311446 v1

van der Laan, G. (1995). *Phys. Rev. B* **51** 240

van der Laan, G. (1998). *Phys. Rev. B* **57** 112

van der Laan, G. and B.T. Thole (1995). *J. Phys.: Condens. Matter* **7** 9947

van Veenendaal, M. and P. Carra (1997). *Phys. Rev. Lett.* **78** 2839

van Veenendaal, M., P. Carra, and B.T. Thole (1994). *Phys. Rev. B* **54** 16010

Wittkop, C., W. Schülke, and F.M.F. deGroot (2000). *Phys. Rev. B* **61** 7176

Zhang, F. and T. Rice (1988). *Phys. Rev. B* **37** 3759

6

Theoretical foundation

At many points in this book, we have referred to basic relations in order to interpret certain aspects of inelastic X-ray scattering experiments, without deriving the corresponding equations by starting from basic principles. We have chosen this way of presenting the evidences drawn from inelastic scattering experiments because otherwise the reader might be too much discouraged by the necessary mathematical formalism. On the other hand, a more theoretically interested reader, of course, will have missed a unifying concept that binds together all the different branches of the field. Therefore, it is the aim of this chapter to add such a theoretical foundation as a supplement.

A firm theoretical basis for all aspects of inelastic X-ray scattering can be found in a generalized Kramers–Heisenberg formula extended to include the electron spin, derived from a perturbation treatment using a Hamiltonian which describes adequately the interaction of the photon field with the scattering electron system. The Hamiltonian chosen to reach this goal within the framework of a quantum mechanical first- and second-order perturbation theory has been proved to deliver results, which are identical with those of a more rigorous quantum electrodynamical treatment performed to the second order in (v/c). Moreover, this Hamiltonian has an appearance suited to let shine through the origin of various parts of the interaction Hamiltonian in classical electrodynamics.

In further sections the generalized Kramers–Heisenberg formula will first be specified to the case of nonresonant charge scattering from a many-particle ensemble by introducing both the correlation functions and the theory of linear response both for an homogeneous and an inhomogeneous electron system. Then the nonresonant cross-section, which includes magnetic scattering, is extracted out of the generalized Kramers–Heisenberg formula and implications are discussed stemming from the interference between charge and magnetic scattering, where the magnetic scattering not only comprise spin but also orbital scattering.

In order to treat the Compton limit of charge and magnetic scattering from many-particle systems adequately, a rigorous theoretical foundation of the impulse approximation is offered in a further section, including the adequate handling of exchange and correlation within the framework of density functional theory. The properties of momentum space quantities and their relation to position space quantities are thoroughly discussed. Peculiar features of magnetic Compton scattering are presented. Moreover, the fully relativistic treatment of the Compton limit is sketched so that the approximations undertaken to evaluate high-energy photon Compton scattering were brought on a firmer basis.

Further on the resonant part of the generalized Kramers–Heisenberg formula is discussed. In the course of this discussion the coherency of (virtual) absorption and re-emission process is emphasized, the most important aspect of resonant inelastic X-ray scattering (RIXS). On the basis of this coherency we deal with the so-called Raman shift, and the special polarization dependence of RIXS, together with the information offered by a coupled-multipolar expansion of the RIXS amplitude. Moreover, the consequences of momentum and spin conservation of the RIXS process are discussed. This chapter ends with an extension of the Kramers–Heisenberg formula to formally include the appearance of shakeup satellites via a third-order perturbation treatment.

In addition it is the aim of this theoretical chapter to provide so many details about the matter in question that the reader will become qualified to get admittance to theoretical papers, and to go on to develop the theory for the particular cases of interest.

6.1 Hamiltonian for the photon–electron interaction

As already mentioned, the double differential cross-section for inelastic X-ray scattering should be calculated, for the sake of physical transparency, in the first instance on the basis of a nonrelativistic perturbation treatment. This procedure is justified by the fact, as shown by Gell-Mann and Goldberger (1954), that by using the photon–electron interaction Hamiltonian as given below within a perturbation scheme restricted to second order in v/c, the result of a rigorous quantum field calculation can be reproduced up to terms linear in the photon frequency. Therefore, the derivation of the generalized Kramers–Heisenberg formula for the double differential scattering cross-section starts, by following the procedure of Blume (1985), with writing down the following Hamiltonian for electrons in a quantized electromagnetic field, which contains the radiation field, the electrons and interaction terms:

$$H = \sum_j \frac{1}{2m}[\mathbf{p}_j - (e/c)\mathbf{A}(\mathbf{r}_j)]^2 + \sum_{jj'} V(r_{jj'}) - \frac{e\hbar}{2mc} \sum_j \boldsymbol{\sigma}_j \cdot \nabla \times \mathbf{A}(\mathbf{r}_j)$$
$$- \frac{e\hbar}{4m^2c^2} \sum_j \boldsymbol{\sigma}_j \cdot \mathbf{E}(\mathbf{r}_j) \times [\mathbf{p}_j - (e/c)\mathbf{A}(\mathbf{r}_j)]$$
$$+ \sum_{\mathbf{K}\lambda} \hbar\omega_\mathbf{K} \left[c^+(\mathbf{K}\lambda)c(\mathbf{K}\lambda) + \frac{1}{2}\right], \tag{6.1}$$

where the summation j, j' is over all electrons of the scattering system, the summation $\mathbf{K}\lambda$ over all modes of the photon field. $\mathbf{p}$ is the momentum operator and $\mathbf{A}(\mathbf{r})$ the operator of the vector potential of the electromagnetic wave at the position $\mathbf{r}$ of the electron. This vector potential operator can be expressed in terms of photon creation and annihilation operators, $c^+(\mathbf{K}\lambda)$ and $c(\mathbf{K}\lambda)$, respectively according to

$$\mathbf{A}(\mathbf{r}) = \sum_{\mathbf{K}\lambda} \left(\frac{2\pi\hbar c^2}{V\omega_{\mathbf{K}}}\right)^{\frac{1}{2}} [\mathbf{e}(\mathbf{K}\lambda) c(\mathbf{K}\lambda) \exp(i\mathbf{K}\cdot\mathbf{r} - i\omega_{\mathbf{K}}t)$$
$$+ \mathbf{e}^*(\mathbf{K}\lambda) c^+(\mathbf{K}\lambda) \exp(-i\mathbf{K}\cdot\mathbf{r} + \omega_{\mathbf{K}}t)], \quad (6.2)$$

where λ counts the two orthogonal polarization states of the photon field corresponding to the wavevector $\mathbf{K}$ with the polarization unit vector $\mathbf{e}(\mathbf{K}\lambda)$, which can either describe two orthogonal linear polarized states or, when complex valued, left- and right-hand circular polarization. Since the photon field is transversal

$$\mathbf{K}\cdot\mathbf{e}(\mathbf{K}\lambda) = 0 \quad (6.3)$$

is always valid.

$\boldsymbol{\sigma}$ in equation (6.1) is the spin vector operator whose components are the well-known Pauli matrices. $\mathbf{E}(\mathbf{r})$ is the operator of the electric field at the electron position $\mathbf{r}$, which can be expressed in terms of the vector potential operator $\mathbf{A}(\mathbf{r})$ via the gauge equation:

$$\mathbf{E} = -\nabla\phi - \frac{1}{c}\dot{\mathbf{A}}, \quad (6.4)$$

where ϕ is the Coulomb potential.

It might promote the physical transparency to take a look at the classical counterpart of the terms of the above Hamiltonian. The first term represents the kinetic energy of the electron system in the presence of the radiation field. The second term is the potential energy of the interacting electrons. The third term is, in the classical limit, nothing else than the potential energy of the magnetic moment $(e\hbar/2mc^2)\boldsymbol{\sigma}$, connected with the spin $\boldsymbol{\sigma}$, in the magnetic field $\nabla\times\mathbf{A}$ of the radiation. The fourth term can be understood as one-half of the potential energy of the magnetic moment $(e\hbar/2mc^2)\boldsymbol{\sigma}$ in a magnetic field $(1/c)(\mathbf{v}\times\mathbf{E})$, which one finds in the rest frame, when the electron is moving with the velocity $\mathbf{v}$, now expressed in terms of the canonical momentum according to $\mathbf{v} = (1/m)[\mathbf{p} - (e/c)\mathbf{A}]$. This term is called the spin–orbit term, since part of it, as will be shown below, is responsible for the spin–orbit coupling in an atom. The factor $1/2$ is the so-called Thomas factor, which is known to be a correction factor for the classical spin–orbit interaction of atomic electrons. Finally, the fifth term represents the energy of the photon field (sum of harmonic oscillators with frequencies $\omega_{\mathbf{K}}$).

In order to use the above Hamiltonian for a perturbation treatment of the scattering cross-section, one has to separate that part which represents the interaction between the radiation field and the electrons. This means to separate the following terms containing explicitly the vector potential $\mathbf{A}$ after having calculated the square in the first term, inserted $\mathbf{E}$ from (6.4), and evaluated the spin–orbit term. Since the spin–orbit term is in contrast to the other $\mathbf{A}$ containing expressions of order $(v/c)^2$, we will, for reasons to be discussed later, omit linear terms in $\mathbf{A}$ and keep only the quadratic terms and those independent of $\mathbf{A}$:

$$H_{i1} = \frac{e^2}{2mc^2} \sum_j \mathbf{A}^2(\mathbf{r}_j) \tag{6.5}$$

$$H_{i2} = -\frac{e}{mc} \sum_j \mathbf{A}(\mathbf{r}_j) \cdot \mathbf{p}_j \tag{6.6}$$

$$H_{i3} = -\frac{e\hbar}{2mc} \sum_j \boldsymbol{\sigma}_j \cdot [\boldsymbol{\nabla} \times \mathbf{A}(\mathbf{r}_j)] \tag{6.7}$$

$$H_{i4} = -\frac{e^2\hbar}{4m^2c^4} \sum_j \boldsymbol{\sigma}_j \cdot [\dot{\mathbf{A}}(\mathbf{r}_j) \times \mathbf{A}(\mathbf{r}_j)]. \tag{6.8}$$

The remaining parts of the Hamiltonian describe on the one hand the electron system

$$H_0 = \sum_j \frac{1}{2m}\mathbf{p}_j^2 + \sum_{jj'} V(\mathbf{r}_{jj'}) + \frac{e\hbar}{4m^2c^2} \sum_j \boldsymbol{\sigma}_j \cdot (\boldsymbol{\nabla}\phi \times \mathbf{p}_j), \tag{6.9}$$

where the $\mathbf{A}$ free part of the spin–orbit term, the ordinary spin–orbit coupling term for electrons, has been included. On the other hand they stand for the radiation field

$$H_\mathbf{R} = \sum_{\mathbf{K}\lambda} \hbar\omega_\mathbf{K} \left[c^+(\mathbf{K}\lambda)c(\mathbf{K}\lambda) + \frac{1}{2} \right]. \tag{6.10}$$

6.2 The generalized Kramers–Heisenberg formula

In this section we will make use of the interaction Hamiltonian with its terms H_{i1} to H_{i4} with the aim to calculate the double differential scattering cross-section via Fermi's golden rule. Since the scattering process we are going to describe is a two-photon process, and since the vector potential $\mathbf{A}$ as represented in (6.2) is linear in the creation and annihilation operators we need a second-order perturbation treatment for all terms of the interaction Hamiltonian, that are linear in $\mathbf{A}$, i.e. H_{i2} and H_{i3}. Only the terms quadratic in $\mathbf{A}$, i.e. H_{i1} and H_{i4}, are appropriate for first-order perturbation calculation. Now it becomes clear why we have omitted there parts of the spin–orbit term which were linear in $\mathbf{A}$. Their contributions to the scattering cross-section calculated in second-order perturbation theory would be of higher order than $(v/c)^2$, thus being out of the limits set to our nonrelativistic calculation.

According to Fermi's golden rule the transition probability per unit time is given to second order by

$$w = \frac{2\pi}{\hbar} |\langle F|H_{i1} + H_{i4}|I\rangle$$

$$+ \sum_N \langle F|H_{i2} + H_{i3}|N\rangle \langle N|H_{i2} + H_{i3}|I\rangle/(E_I - E_N)|^2 \delta(E_I - E_F), \quad (6.11)$$

where the initial state $|I\rangle$ and the final state $|F\rangle$ have to be represented as products of the state vectors of the incident photon field $|\mathbf{K}_1 \lambda_1\rangle$ and the scattered photon field $|\mathbf{K}_2 \lambda_2\rangle$, respectively, with the state vectors of the electron system in its initial state $|i\rangle$ and final state $|f\rangle$, respectively:

$$|I\rangle = |i, \mathbf{K}_1\lambda_1, 0\rangle; \quad |F\rangle = |f, 0, \mathbf{K}_2\lambda_2\rangle. \quad (6.12)$$

The corresponding energies are

$$E_I = E_i + \hbar\omega_1; \quad E_F = E_f + \hbar\omega_2. \quad (6.13)$$

The intermediate state $|N\rangle$ can be, according to Fig. 1.3, either a state where the initial photon has been annihilated first, so that

$$|N\rangle = |n, 0\rangle \quad E_N = E_f \quad (6.14)$$

or the final photon has first been created, so that

$$|N\rangle = |n, \mathbf{K}_1\lambda_1, \mathbf{K}_2\lambda_2\rangle \quad E_N = E_f + \hbar\omega_1 + \hbar\omega_2, \quad (6.15)$$

By inserting (6.14) and (6.15) into (6.11) one sees that only the first diagram of Fig. 1.3 leads to a resonant term. In order to account for the final lifetime of the intermediate state, we will add an imaginary part, $i\Gamma_n/2$, to the energy denominator.

In order to calculate the matrix elements in (6.11) we make use of the orthonormality relation, which the state vectors of the photon fields have to satisfy

$$\langle \mathbf{K}_\nu \lambda_\nu | \mathbf{K}_\mu \lambda_\mu \rangle = \delta_{\nu\mu} \quad (6.16)$$

as well as the equation of motion, which reads, e.g. for the creation operator:

$$\dot{c}^+(\mathbf{K}\lambda) = \frac{i}{\hbar}[c^+(\mathbf{K}\lambda), H_R]. \quad (6.17)$$

The double differential scattering cross-section $d^2\sigma/d\Omega_2 d\hbar\omega_2$ is obtained from the transition probability by multiplying w both by the density $V/(2\pi)^3$ of states in $\mathbf{K}_2$ space, and by the volume of $\mathbf{K}_2$-space, which can be reached by the scattered photons in the solid angle element $d\Omega_2$ and the energy element $d\hbar\omega_2$, i.e. $\mathbf{K}_2^2 d\Omega_2 dK_2$, and by dividing w both by the incident flux c/V and by $d\Omega_2 d\hbar\omega_2$. Together with

$$\mathbf{K}_2^2 dK_2 = \frac{\omega_2^2}{c^3} d\omega_2 \qquad (6.18)$$

one ends up with

$$\frac{d^2\sigma}{d\Omega_2 d\hbar\omega_2} = \frac{wV^2\omega_2^2}{8\pi^3 \hbar c^4}. \qquad (6.19)$$

This way the expression of the generalized Kramers–Heisenberg formula for the double differential scattering cross-section connected with a transition of the scattering electron system from an initial state $|i\rangle$ to a final state $|f\rangle$ reads, if one introduces the definitions:

$$\omega \equiv \omega_1 - \omega_2; \quad \mathbf{q} \equiv \mathbf{K}_1 - \mathbf{K}_2, \qquad (6.20)$$

$$\left(\frac{d^2\sigma}{d\Omega_2 d\hbar\omega_2}\right)_{|i\rangle \to |f\rangle} = \left(\frac{e^2}{mc^2}\right)^2 \left(\frac{\omega_2}{\omega_1}\right) \left| \langle f | \sum_j \exp(i\mathbf{q} \cdot \mathbf{r}_j) | i \rangle (\mathbf{e}_1 \cdot \mathbf{e}_2^*) - i \left[\frac{\hbar(\omega_1 + \omega_2)}{2mc^2} \right] \right.$$

$$\times \langle f | \sum_j \exp(i\mathbf{q} \cdot \mathbf{r}_j)(\boldsymbol{\sigma}_j/2) | i \rangle (\mathbf{e}_2^* \times \mathbf{e}_1) + \frac{\hbar^2}{m} \sum_n \sum_{jj'}$$

$$\times \left\{ \frac{\langle f | [\mathbf{e}_2^* \cdot \mathbf{p}_j/\hbar - i(\mathbf{K}_2 \times \mathbf{e}_2) \cdot \boldsymbol{\sigma}_j/2] \exp(-i\mathbf{K}_2 \cdot \mathbf{r}_j) | n \rangle}{E_i - E_n + \hbar\omega_1 - i\Gamma_n/2} \right.$$

$$\times \langle n | [\mathbf{e}_1 \cdot \mathbf{p}_{j'}/\hbar + i(\mathbf{K}_1 \times \mathbf{e}_1) \cdot \boldsymbol{\sigma}_{j'}/2] \exp(i\mathbf{K}_1 \cdot \mathbf{r}_{j'}) | i \rangle$$

$$+ \frac{\langle f | [\mathbf{e}_1 \cdot \mathbf{p}_j/\hbar + i(\mathbf{K}_1 \times \mathbf{e}_1) \cdot \boldsymbol{\sigma}_j/2] \exp(i\mathbf{K}_1 \cdot \mathbf{r}_j) | n \rangle}{E_i - E_n - \hbar\omega_2}$$

$$\left. \times \langle n | [\mathbf{e}_2^* \cdot \mathbf{p}_{j'}/\hbar - i(\mathbf{K}_2 \times \mathbf{e}_2) \cdot \boldsymbol{\sigma}_{j'}/2] \exp(-i\mathbf{K}_2 \cdot \mathbf{r}_{j'}) | i \rangle \right\} \Bigg|^2$$

$$\times \delta(E_i - E_f + \hbar\omega). \qquad (6.21)$$

One can easily recognize that the spin-dependent contributions to the matrix elements in (6.21) are by $\hbar\omega_{1/2}/mc^2$ smaller than the spin-independent ones. For that reason, in inelastic X-ray scattering with incident photon energies in the 10 keV range, the spin-dependent parts of the matrix elements become experimentally accessible only in the form of interference terms with the spin-independent ones.

6.3 Correlation functions

Assuming that the incident photon energy of an inelastic X-ray scattering experiment is far from the excitation energy of an inner-shell level, one can certainly neglect the resonance part (third term) of the generalized Kramers–Heisenberg formula. Moreover, in a first step, we will also pass over the spin-dependent contributions as well as the fourth term of (6.19), which will give rise, as shown later, to orbital magnetic scattering. Thus one is left with the following expression for the total double differential scattering cross-section (Pines and Nozières 1966):

$$\frac{d^2\sigma}{d\Omega_2 d\hbar\omega_2} = r_0^2 \left(\frac{\omega_2}{\omega_1}\right) |\mathbf{e}_1 \cdot \mathbf{e}_2^*|^2 \sum_{if} \sum_{jj'} g_i \langle i| \exp(-i\mathbf{q}\cdot\mathbf{r}_j)|f\rangle$$

$$\times \langle f| \exp(i\mathbf{q}\cdot\mathbf{r}_{j'})|i\rangle \, \delta(E_i - E_f + \hbar\omega), \quad (6.22)$$

where we have summed over all possible final states and have averaged over all initial states by introducing the probability of initial state $|i\rangle$, given in the usual manner by

$$g_i = Z^{-1} \exp(-E_i/k_B T), \quad (6.23)$$

Z being the partition function, k_B the Boltzman factor and r_0 the so-called classical electron radius $r_0 \equiv e^2/mc^2$. Thus the double differential scattering cross-section is determined, on the one hand, by the so-called Thomson differential scattering cross-section

$$\left(\frac{d\sigma}{d\Omega_2}\right)_{Th} \equiv r_0^2 \left(\frac{\omega_2}{\omega_1}\right) |\mathbf{e}_1 \cdot \mathbf{e}_2^*|^2, \quad (6.24)$$

that describes the strength of the photon–electron coupling, and by the strength of the excitation the scattering system is undergoing, expressed in terms of the so-called dynamic structure factor

$$S(\mathbf{q},\omega) \equiv \sum_{if}\sum_{jj'} g_i \langle i| \exp(-i\mathbf{q}\cdot\mathbf{r}_j)|f\rangle \langle f| \exp(i\mathbf{q}\cdot\mathbf{r}_{j'})|i\rangle \, \delta(E_i - E_f + \hbar\omega), \quad (6.25)$$

where these excitations have to satisfy energy conservation.

Following van Hove (1954), one can change into another picture, when describing scattering from a many-particle electron system. In this alternative picture, scattering is understood as being caused by density fluctuations of the system, more precisely, by the correlation of the electron density in one space/time point with the density in another point of space and time. Intuitively, this description connects us much more intimately with the dynamics of the scattering system than the excitation picture, and should therefore be elaborated in some detail in what follows.

In order to switch to this correlation-mode of describing inelastic scattering we will introduce, into (6.22), time-dependent operators in the Heisenberg representation. If A is any operator in the so-called Schrödinger representation, and

H is the system Hamiltonian, then

$$A(t) = \exp(iHt/\hbar)\, A\, \exp(-iHt/\hbar) \qquad (6.26)$$

is the corresponding operator in the time-dependent Heisenberg representation.

Let us now write the energy conserving δ-function of (6.22) in the integral representation

$$\delta(E_i - E_f + \hbar\omega) = \frac{1}{2\pi\hbar} \int_{-\infty}^{\infty} dt\, \exp(-i\omega t)\, \exp[i(E_f - E_i)/\hbar]. \qquad (6.27)$$

Next we define the electron density operator

$$\rho(\mathbf{r}) \equiv \sum_j \delta(\mathbf{r} - \mathbf{r}_j) \qquad (6.28)$$

and its spatial Fourier transform, the electron density fluctuation operator

$$\rho(\mathbf{q}) = \int d^3 r \sum_j \exp(-i\mathbf{q}\cdot\mathbf{r})\, \delta(\mathbf{r} - \mathbf{r}_j)$$

$$= \sum_j \exp(-i\mathbf{q}\cdot\mathbf{r}_j) \qquad (6.29)$$

(we take the quantization volume to be unity throughout the whole subsection).

Then by introducing these definitions into the matrix elements of (6.25) and by utilizing (6.27) we can write

$$\sum_f \langle i|\rho(\mathbf{q})|f\rangle \langle f|\rho^+(\mathbf{q})|i\rangle\, \delta(E_i - E_f + \hbar\omega)$$

$$= \frac{1}{2\pi\hbar} \int_{-\infty}^{\infty} dt\, \exp(-i\omega t) \sum_f \langle i|\rho(\mathbf{q})|f\rangle \langle f|\rho^+(\mathbf{q})|i\rangle \exp[i(E_f - E_i)t/\hbar]$$

$$= \frac{1}{2\pi\hbar} \int_{-\infty}^{\infty} dt\, \exp(-i\omega t) \sum_f \langle i|\rho(\mathbf{q})|f\rangle \langle f|\exp(iE_f t/\hbar)\rho^+(\mathbf{q})\exp(-iE_i t/\hbar)|i\rangle.$$

$$(6.30)$$

Now we make use of the fact that $|i\rangle$ and $|f\rangle$ are eigenvectors of the system Hamiltonian H, so that

$$\exp(-iE_i t/\hbar)|i\rangle = \exp(-iHt/\hbar)|i\rangle; \quad \langle f|\exp(iE_f t/\hbar) = \langle f|\exp(Ht/\hbar). \qquad (6.31)$$

In this way, the time-dependent representation (see 6.26) of the density fluctuation operator $\rho^+(\mathbf{q})$ can be introduced into (6.30):

$$\sum_f \langle i|\rho(\mathbf{q})|f\rangle \langle f|\rho^+(\mathbf{q})|i\rangle \, \delta(E_i - E_f + \hbar\omega)$$

$$= \frac{1}{2\pi\hbar} \int_{-\infty}^{\infty} dt \exp(-i\omega t) \sum_f \langle i|\rho(\mathbf{q},0)|f\rangle \langle f|\rho^+(\mathbf{q},t)|i\rangle$$

$$= \frac{1}{2\pi\hbar} \int_{-\infty}^{\infty} dt \exp(-i\omega t) \, \langle i|\rho(\mathbf{q},0)\rho^+(\mathbf{q},t)|i\rangle, \tag{6.32}$$

where we have used the completeness property to sum over f. If we finally define the time-dependent correlation function of the electron density fluctuation operators according to

$$\langle \rho(\mathbf{q},0)\rho^+(\mathbf{q},t)\rangle \equiv \sum_i g_i \langle i|\rho(\mathbf{q},0)\rho^+(\mathbf{q},t)|i\rangle, \tag{6.33}$$

we can write the dynamic structure factor of (6.25) as follows:

$$S(\mathbf{q},\omega) = \frac{1}{2\pi\hbar} \int_{-\infty}^{\infty} dt \exp(-i\omega t) \, \langle \rho(\mathbf{q},0)\rho^+(\mathbf{q},t)\rangle,$$

$$= \frac{1}{2\pi\hbar} \int_{-\infty}^{\infty} dt \exp(-i\omega t) \left\langle \sum_{i,j} \exp[-i\mathbf{q}\cdot\mathbf{r}_i(0)] \exp[i\mathbf{q}\cdot\mathbf{r}_j(t)] \right\rangle \tag{6.34}$$

In the second line of (6.34) we have used (6.29). This expression for $S(\mathbf{q},\omega)$ is best suited to discuss the role of interference phenomena, when probing a multiple-particle system with a certain momentum and energy transfer. Since, in the classical limit, $\exp[i\mathbf{q}\cdot\mathbf{r}_j(t)]$ is the phase of a wave scattered from the jth electron at time t, the answer for the system will depend strongly on both how $2\pi/q$ compares with characteristic lengths, l_c, (interparticle distances) and how ω compares with characteristic frequencies, ω_c, of the system. If, on the one hand, ql_c is small compared with 2π, and ω is comparable with ω_c, interference between waves scattered from many particles at different time is of importance. Therefore, it is mainly the collective behavior of the multiple-particle system which we are probing under those conditions. We will call this domain of energy and momentum transfer the "regime of characteristic energy losses", which will be the matter of Sections 6.4–6.6. If, on the other hand, ql_c is large compared with 2π, and ω is large compared with ω_c, the waves scattered from different particles do not interfere, so that we are probing the position of **one** particle at different times. In other words, the measurement of $S(\mathbf{q},\omega)$ will yield information about the single-particle momenta, provided the condition $\omega \gg \omega_c$ guarantees that the time scale of probing is small enough to prevent the remaining system becoming rearranged. In what follows, we will call this domain the "Compton regime", which will be the matter of Section 6.8. The condition $\omega \gg \omega_c$ will turn out to be the requirement for treating the scattering process within the limits of the so-called impulse approximation.

One can also write the dynamical structure factor in terms of the so-called van Hove space–time correlation function of an electron system

$$G_c(\mathbf{r}, t) \equiv \sum_{jj'} \int d^3r' \langle \delta(\mathbf{r}' - \mathbf{r}_j(0)) \, \delta(\mathbf{r}' + \mathbf{r} - \mathbf{r}_{j'}(t)) \rangle,$$

$$= \int d^3r' n_2(\mathbf{r}', \mathbf{r}, t) \tag{6.35}$$

where $n_2(\mathbf{r}', \mathbf{r}, t)$, the two-particle density correlation function, can be interpreted, in the classical limit, to be the probability to find at the time t a particle in the vector distance $\mathbf{r}$ from the reference point $\mathbf{r}'$, if at the time $t = 0$ a particle (the same or another one) came across at this reference point. The correlation function is then the result of averaging over all reference points. Introducing $G_c(\mathbf{r}, t)$ into (6.32) we obtain

$$S(\mathbf{q}, \omega) = \frac{1}{2\pi\hbar} \int_{-\infty}^{\infty} dt \exp(-i\omega t) \int d^3r \exp(i\mathbf{q} \cdot \mathbf{r}) \, G_c(\mathbf{r}, t). \tag{6.36}$$

Note that the integration with respect to the reference coordinate $\mathbf{r}'$ in (6.35) limits the information about the two-particle correlation, which can be obtained by measuring the dynamic structure factor of inhomogeneous electron systems. We shall close this gap later by introducing the so-called coherent inelastic scattering in Section 6.11.

Integrating the final equation for the double differential cross-section

$$\frac{d^2\sigma}{d\Omega_2 d\hbar\omega_2} = r_0^2 \left(\frac{\omega_2}{\omega_1} \right) |\mathbf{e}_1 \cdot \mathbf{e}_2^*|^2 \, S(\mathbf{q}, \omega), \tag{6.37}$$

over the scattered photon energy $\hbar\omega_2$ (equivalent to integrating over $\hbar\omega$ for fixed incident energy), we obtain for the differential cross-section

$$\frac{d\sigma}{d\Omega_2} = r_0^2 |\mathbf{e}_1 \cdot \mathbf{e}_2^*|^2 \, S(\mathbf{q}) \tag{6.38}$$

where the static structure factor $S(\mathbf{q})$ is given by

$$S(\mathbf{q}) = \hbar \int_{-\infty}^{\infty} d\omega S(\mathbf{q}, \omega). \tag{6.39}$$

(Here we have assumed that the energy transfers $\hbar\omega$ are always small compared with the incident X-ray photon energy, so that ω_2/ω_1 in (6.35) can be replaced by unity, i.e. the quasielastic approximation applies.) From (6.36), we have

$$S(\mathbf{q}) = \langle \rho(\mathbf{q},0)\, \rho^+(\mathbf{q},0) \rangle \equiv \langle \rho(\mathbf{q})\, \rho(\mathbf{q}) \rangle. \tag{6.40}$$

The absence of an explicit time dependence in a correlation function should always mean an equal-time correlation function.

By introducing the pair-correlation function $g(r)$ of an homogeneous electron system to be the probability density of finding two electrons r apart, one can easily deduce from (6.36) the following relationship between the static structure factor and $g(r)$

$$S(\mathbf{q}) = 1 + \rho \int [g(r) - 1] \exp(-i\mathbf{q} \cdot \mathbf{r})\, \mathrm{d}^3 r, \tag{6.41}$$

where ρ is the spatially averaged electron density.

6.4 Response functions, dielectric functions, homogeneous electron system

Let us now introduce the frequency-dependent response function, which is based on the electron density fluctuation operators and therefore intimately connected with the dynamical structure factor. We will assume that an external weak time-dependent potential

$$\phi_{\mathrm{ext}}(\mathbf{r}, t) = \sum_{\mathbf{q}\omega} \phi_{\mathrm{ext}}(\mathbf{q}, \omega) \exp(i\mathbf{q} \cdot \mathbf{r} - i\omega t) \tag{6.42}$$

couples to the electron system represented by the electron density fluctuation operator $\rho(\mathbf{q})$, defined in (6.29), where the interaction should be switched on adiabatically, that is, very slowly. Then real transitions between the electron states are not introduced over a long period of time. Hence the interaction Hamiltonian in Fourier representation reads (see Pines 1964)

$$H_{\mathrm{int}} = -\lim_{\varepsilon \to 0_+} \sum_{\mathbf{q}\omega} e\, \rho(\mathbf{q})\, \phi_{\mathrm{ext}}(\mathbf{q}, \omega) \exp[-i(\omega + i\varepsilon)t]. \tag{6.43}$$

As a consequence of this interaction, which we shall allow, in what follows, as a weak perturbation, the expectation value of the electron density fluctuation operator to change, where we will assume that this change $\delta\langle \rho(\mathbf{q}') \rangle(t)$, or we say also the response, is only linearly dependent on the perturbation H_{int}, so that we can write

$$\delta\langle \rho(\mathbf{q}') \rangle(t) = (-e) \lim_{\varepsilon \to 0_+} \sum_{\mathbf{q}} \chi(\mathbf{q}', \mathbf{q}, \omega)\, \phi_{\mathrm{ext}}(\mathbf{q}, \omega) \exp[-i(\omega + i\varepsilon)t]. \tag{6.44}$$

To be as general as possible we have assumed that the perturbing potential with the spatial periodicity $2\pi/q$ can induce an electron density fluctuation with another spatial periodicity $2\pi/q'$. Of course, this can happen only in an inhomogeneous electron system. We will make use of this fact when we treat the linear response of electrons in a periodic lattice. For the time being we will restrict ourselves to the homogeneous case, so that we can shorten the diagonal terms of χ

$$\chi(\mathbf{q}, \omega) \equiv \chi(\mathbf{q'}, \mathbf{q}, \omega).$$

Moreover, when calculating $\langle \rho(\mathbf{q}, \omega) \rangle$ later on in first-order perturbation theory, it suffices to keep in (6.44) only those terms of H_{int} which vary with a frequency ω. The other Fourier components of the external potential do not contribute in first-order.

The mediating coefficient $\chi(\mathbf{q}, \omega)$ is called the frequency-dependent density-response function or polarization function.

If we decompose $\chi(\mathbf{q}, \omega)$ into real and imaginary parts according to

$$\chi(\mathbf{q}, \omega) = \chi'(\mathbf{q}, \omega) + i\chi''(\mathbf{q}, \omega) \tag{6.45}$$

χ' and χ'' have to satisfy, as any causal response function, the famous Kramers–Kronig dispersion relations (for a derivation of these relations the reader is referred to Brauer (1972) or to many other relevant textbooks):

$$\chi'(\mathbf{q}, \omega) = \frac{-1}{\pi} P \int_{-\infty}^{\infty} d\omega' \, \chi''\frac{(\mathbf{q}, \omega')}{(\omega - \omega')} \tag{6.46}$$

$$\chi''(\mathbf{q}, \omega) = \frac{1}{\pi} P \int_{-\infty}^{\infty} d\omega' \, \chi'\frac{(\mathbf{q}, \omega')}{(\omega - \omega')}, \tag{6.47}$$

where P stands for the principal part of the integral.

We now calculate explicitly the frequency-dependent density response function within the limits of a first-order time-dependent perturbation treatment. Let us assume that the system was in a particular eigenstate ψ_n before the perturbation. The system eigenfunction $\psi(t)$ at a time t after the perturbation can be expanded into unperturbed eigenstates as

$$\psi(t) = \sum_{n'} a_{n'}(t) \exp(-iE_{n'}t/\hbar)\psi_{n'}, \tag{6.48}$$

where the coefficients satisfy, for $t \leq 0$, the following relations:

$$a_n(t) = 1; \quad a_{n'}(t) = 0 \quad \text{for } n' \neq n. \tag{6.49}$$

Hence the expectation value of the electron density fluctuation operator $\rho(\mathbf{q'})$ after the perturbation can be written as

$$\langle \rho(\mathbf{q}) \rangle (t) = \langle \psi(t) | \rho(\mathbf{q}) | \psi(t) \rangle$$

$$= \langle n | \rho(\mathbf{q}) | n \rangle + \sum_{n' \neq n} a_{n'}(t) \exp[i(E_n - E_{n'})t/\hbar] \langle n' | \rho^+(\mathbf{q}) | n \rangle + \text{c.c.} \tag{6.50}$$

The coefficients $a_{n'}(t)$ are now evaluated according to the first-order time-dependent perturbation scheme, where the perturbation Hamiltonian is given by (6.43), thus obtaining

$$a_{n'}(t) = -e\phi_{\text{ext}}(\mathbf{q},\omega)\langle n'|\rho^+(\mathbf{q})|n\rangle \left(\frac{\exp\{-\mathrm{i}[E_n - E_{n'} + \hbar(\omega + \mathrm{i}\varepsilon)]t/\hbar\}}{E_n - E_{n'} + \hbar(\omega + \mathrm{i}\varepsilon)}\right)$$

$$+ e\phi_{\text{ext}}^+(\mathbf{q},\omega)\langle n'|\rho(\mathbf{q})|n\rangle \left(\frac{\exp\{-\mathrm{i}[E_n - E_{n'} - \hbar(\omega - \mathrm{i}\varepsilon)]t/\hbar\}}{E_n - E_{n'} - \hbar(\omega - \mathrm{i}\varepsilon)}\right). \quad (6.51)$$

Substituting $a_{n'}(t)$ of (6.51) into (6.50), and noting that, when a state $|n'\rangle$ is coupled to the ground state $|n\rangle$ by $\rho(\mathbf{q})$, then the same state cannot be coupled to the ground state by $\rho^+(\mathbf{q})$, we get for the variation of the expectation value of $\rho(\mathbf{q})$ due to the perturbation

$$\delta\langle\rho(\mathbf{q})\rangle(t) = -e\phi_{\text{ext}}(\mathbf{q},\omega)\exp[-\mathrm{i}(\omega + \mathrm{i}\varepsilon)t]\left[\sum_n g_n \sum_{n'\neq n} \frac{\langle n|\rho(\mathbf{q})|n'\rangle\langle n'|\rho^+(\mathbf{q})|n\rangle}{E_n - E_{n'} + \hbar(\omega + \mathrm{i}\varepsilon)}\right.$$

$$\left. + \sum_{n'\neq n} \frac{\langle n|\rho^+(\mathbf{q})|n'\rangle\langle n'|\rho(\mathbf{q})|n\rangle}{E_n - E_{n'} - \hbar(\omega - \mathrm{i}\varepsilon)}\right], \quad (6.52)$$

where we deal, instead of a single eigenstate before perturbation, with the thermodynamical average over initial states as in (6.25), in both equations (6.51) and (6.52) the limit as $\varepsilon \to 0_+$ being understood.

Now we can compare (6.52) with (6.44) and obtain immediately an expression for the frequency-dependent density response function

$$\chi(\mathbf{q},\omega) = \lim_{\varepsilon\to 0_+} \sum_{n,n'}{}' g_n \left\{\frac{\langle n|\rho(\mathbf{q})|n'\rangle\langle n'|\rho^+(\mathbf{q})|n\rangle}{E_n - E_{n'} + \hbar(\omega + \mathrm{i}\varepsilon)} - \frac{\langle n|\rho^+(\mathbf{q})|n'\rangle\langle n'|\rho(\mathbf{q})|n\rangle}{E_{n'} - E_n + \hbar(\omega + \mathrm{i}\varepsilon)}\right\}, \quad (6.53)$$

where the prime over the summation means the term with $n'=n$ must be excluded. Because

$$\lim_{\varepsilon\to 0_+}\frac{1}{x \pm \mathrm{i}\varepsilon} = \mathrm{P}\frac{1}{x} \mp \mathrm{i}\pi\delta(x) \quad (6.54)$$

we may write for the imaginary part of $\chi(\mathbf{q},\omega)$

$$\chi''(\mathbf{q},\omega) = -\pi\sum_{nn'}{}' g_n[\langle n|\rho(\mathbf{q})|n'\rangle\langle n'|\rho^+(\mathbf{q})|n\rangle\,\delta(E_n - E_{n'} + \hbar\omega)$$

$$- \langle n|\rho^+(\mathbf{q})|n'\rangle\langle n'|\rho(\mathbf{q})|n\rangle\,\delta(E_{n'} - E_n + \hbar\omega)]. \quad (6.55)$$

We interchange the summation indices in the second term on the right-hand side of (6.55) and obtain

$$\sum_{nn'}{}' g_n [\langle n|\rho^+(\mathbf{q})|n'\rangle\langle n'|\rho(\mathbf{q})|n\rangle\, \delta(E_{n'}-E_n+\hbar\omega)]$$

$$= \sum_{n'n}{}' g_{n'} \exp(-\hbar\omega/k_\mathrm{B}T)\langle n|\rho^+(\mathbf{q})|n'\rangle\langle n'|\rho(\mathbf{q})|n\rangle\, \delta(E_{n'}-E_n+\hbar\omega)$$

$$= \sum_{nn'}{}' g_n \exp(-\hbar\omega/k_\mathrm{B}T)\langle n|\rho(\mathbf{q})|n'\rangle\langle n'|\rho^+(\mathbf{q})|n\rangle\, \delta(E_n-E_{n'}+\hbar\omega), \quad (6.56)$$

so that we have the final result

$$\frac{1}{\pi}\left\{\frac{1}{1-\exp(-\hbar\omega/k_\mathrm{B}T)}\right\}\chi''(\mathbf{q},\omega)$$

$$= -\sum_{nn'}{}' g_n \langle n|\rho(\mathbf{q})|n'\rangle\langle n'|\rho^+(\mathbf{q})|n\rangle\, \delta(E_n-E_{n'}+\hbar\omega). \quad (6.57)$$

If we replace $|n\rangle$ and $|n'\rangle$ by the initial state $|\mathrm{i}\rangle$ and the final state $|\mathrm{f}\rangle$, respectively, of our scattering experiment, the right-hand side of (6.57) reads

$$-\sum_\mathrm{i}\sum_\mathrm{f} g_\mathrm{i}\langle\mathrm{i}|\rho(\mathbf{q})|\mathrm{f}\rangle\langle\mathrm{f}|\rho^+(\mathbf{q})|\mathrm{i}\rangle\, \delta(E_\mathrm{i}-E_\mathrm{f}+\hbar\omega), \quad (6.58)$$

One recognizes easily that (6.58) is nothing else than the negative value of the dynamic structure factor $S(\mathbf{q},\omega)$ of (6.25). Hence we have derived the famous fluctuation–dissipation theorem, which connects the dynamical structure factor, a quantity related, according to equation (6.34), to the density fluctuations of the system, with the imaginary part of the frequency-dependent density response function, which stands for the energy dissipation in the electron system upon scattering:

$$S(\mathbf{q},\omega) = -\frac{1}{\pi}\left\{\frac{1}{1-\exp(-\hbar\omega/k_\mathrm{B}T)}\right\}\chi''(\mathbf{q},\omega). \quad (6.59)$$

In what follows the frequency-dependent density response function shall be denoted by the polarization function $\chi(\mathbf{q},\omega)$.

We are now going to derive a relation between the polarization function and the macroscopic dielectric function $\varepsilon(\mathbf{q},\omega)$ by starting with the Fourier transform of the macroscopic Poisson equations which, on the one hand, connects the displacement-field $\mathbf{D}(\mathbf{q},\omega)$ with the charge density $z\rho_\mathrm{ext}(\mathbf{q},\omega)$ associated with the external potential $\phi_{\text{ext}}(\mathbf{q}, \omega)$ introduced above

$$i\mathbf{q} \cdot \mathbf{D}(\mathbf{q}, \omega) = 4\pi z \rho_{\text{ext}}(\mathbf{q}, \omega), \tag{6.60}$$

and, on the other hand, relates the electric field $\mathbf{E}(\mathbf{q}, \omega)$ to the sum of the external charge density $z\rho_{\text{ext}}(\mathbf{q}\omega)$ and the induced charge density $-e\delta\langle\rho(\mathbf{q},\omega)\rangle$:

$$i\mathbf{q} \cdot \mathbf{E}(\mathbf{q}, \omega) = 4\pi[-e\delta\langle\rho(\mathbf{q}, \omega)\rangle + z\rho_{\text{ext}}(\mathbf{q}, \omega)], \tag{6.61}$$

where $\delta\langle\rho(\mathbf{q},\omega)\rangle$ is connected with the perturbation induced variation of the expectation value of the electron density fluctuation operator $\delta\langle\rho(\mathbf{q})\rangle(t)$ of (6.52) via

$$\delta\langle\rho(\mathbf{q})\rangle(t) = \delta\langle\rho(\mathbf{q}, \omega)\rangle \exp(-i\omega t). \tag{6.62}$$

(For the sake of simplicity we have presupposed, for the time being, a homogeneous electron system reacting to the external potential.)

Since $\mathbf{D}$ an $\mathbf{E}$ are assumed to be purely macroscopic longitudinal fields one can define a macroscopic scalar dielectric function by means of the relation

$$\mathbf{D}(\mathbf{q}, \omega) = \varepsilon(\mathbf{q}, \omega)\, \mathbf{E}(\mathbf{q}, \omega), \tag{6.63}$$

so that we obtain by inserting (6.61) and (6.62) into (6.63) the following expression for the dielectric function:

$$\frac{1}{\varepsilon(\mathbf{q},\omega)} = 1 - \frac{e\delta\langle\rho(\mathbf{q},\omega)\rangle}{z\rho_{\text{ext}}(\mathbf{q},\omega)}. \tag{6.64}$$

By using now (6.62) together with (6.52) and by making use of the relation between the external charge density and the corresponding external potential

$$\phi_{\text{ext}}(\mathbf{q}, \omega) = \left(\frac{4\pi}{q^2}\right) z\rho_{\text{ext}}(\mathbf{q}, \omega) \tag{6.65}$$

we end up with an equation which connects the dielectric function with the polarization function:

$$\frac{1}{\varepsilon(\mathbf{q},\omega)} = 1 + v(\mathbf{q})\,\chi(\mathbf{q},\omega); \quad v(\mathbf{q}) = \frac{4\pi e^2}{q^2}, \tag{6.66}$$

where $v(\mathbf{q})$ is the Fourier transformed Coulomb potential.

Hence the dynamic structure factor $S(\mathbf{q}, \omega)$ is now, according to (6.59), connected with the imaginary part of the reciprocal dielectric function via

$$S(\mathbf{q}, \omega) = \left\{\frac{1}{1 - \exp(-\hbar\omega/k_B T)}\right\}\left(\frac{q^2}{4\pi^2 e^2}\right) \text{Im}\left[\frac{-1}{\varepsilon(\mathbf{q},\omega)}\right]. \tag{6.67}$$

As shown by Pines (1964), $1/\varepsilon(\mathbf{q},\omega)$ as well as $\varepsilon(\mathbf{q},\omega)$ satisfy the Kramers–Kronig dispersion relations, here written down for $1/\varepsilon(\mathbf{q},\omega)$:

$$\mathrm{Re}\left[\frac{1}{\varepsilon(\mathbf{q},\omega)}\right] - 1 = \left(\frac{1}{\pi}\right) \mathrm{P} \int d\omega' \frac{\mathrm{Im}[1/\varepsilon(\mathbf{q},\omega)]}{\omega' - \omega} \tag{6.68}$$

$$\mathrm{Im}\left[\frac{1}{\varepsilon(\mathbf{q},\omega)}\right] = -\left(\frac{1}{\pi}\right) \mathrm{P} \int d\omega' \frac{\mathrm{Re}[\{1/\varepsilon(\mathbf{q},\omega)\} - 1]}{\omega' - \omega}, \tag{6.69}$$

and, as a consequence of particle number conservation, the following sum rules:

$$\int_0^\infty d\omega\, \omega\, \mathrm{Im}\left[\frac{-1}{\varepsilon(\mathbf{q},\omega)}\right] = \frac{\pi}{2}\omega_{\mathrm{p}}^2 \tag{6.70}$$

$$\int_0^\infty d\omega\, \omega\, \mathrm{Im}[\varepsilon(\mathbf{q},\omega)] = \frac{\pi}{2}\omega_{\mathrm{p}}^2, \tag{6.71}$$

where $\omega_{\mathrm{p}} = (4\pi\rho e^2/m)^{1/2}$ is the so-called plasmon frequency (ρ = spatially averaged electron density).

6.5 Lindhard polarization function, self-consistent field approximation

In Section 6.4 the general scheme for calculating the polarization function $\chi(\mathbf{q},\omega)$ has been provided, but in order to make use of it we have to know the eigenstates $|n\rangle$ and the eigenvalues of the energy E_n of the unperturbed many-particle problem; this means solving the stationary many-particle Schrödinger equation

$$H_0|n\rangle = E_n|n\rangle. \tag{6.72}$$

Since this is not possible in general, we have to restrict ourselves to approximations and simplified models.

The simplest model for the valence electrons of (simple) metals is a gas of free electrons, where the positive ions are smeared to a structureless positively charged background, which has to neutralize the negative charge of the electrons (so-called jellium model). But even the Schrödinger equation of this rather simplified model cannot be solved exactly because of the long-range Coulomb interaction between the electrons. An approximation that has held good in many branches of solid state physics and has helped to a better understanding of the underlying physics is the so-called self-consistent field approximation, which we will provide in what follows.

The starting point for this approximation is the so-called Lindhard polarization function (Lindhard 1954), which is nothing else than the polarization function according to (6.53) calculated for a system of independent particles of

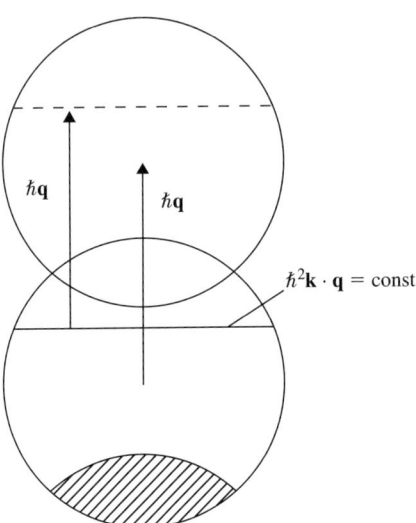

Fig. 6.1. Transitions from occupied free-electron states into unoccupied states, induced by the momentum transfer $\hbar\mathbf{q}$. All transitions from the plane $\hbar^2\mathbf{k}\cdot\mathbf{q} = \hbar m\omega - \hbar^2 q^2/2$ (solid line) to the plane $\hbar^2\mathbf{k}\cdot\mathbf{q} \equiv \hbar m\omega + \hbar^2 q^2/2$ (dashed line) are connected with the same energy transfer. Transitions from the hatched region of the Fermi sphere are forbidden by Pauli's principle.

a free electron gas. Thus one takes the states $|n\rangle$ to be plane-wave functions $\exp(i\mathbf{k}\cdot\mathbf{r})$, appropriate for a Slater determinant of one-electron wavefunctions. The energies associated with states characterized by $\mathbf{k}$ are simply the kinetic energies of particles with momentum $\hbar\mathbf{k}$. In this case, one can easily calculate the matrix elements of (6.53) and find that the transitions introduced by $\rho^+(\mathbf{q})$ are those in which an electron is taken from a state with momentum $\hbar\mathbf{k}$ within the Fermi sphere with radius k_F to a state with momentum $\hbar(\mathbf{k}+\mathbf{q})$ outside the Fermi sphere, where the latter limitation arises from the Pauli principle (see Fig. 6.1). The excitation energy associated with such a transition is the given by

$$E_{n'} - E_n = E_{\mathbf{k}+\mathbf{q}} - E_{\mathbf{k}} = \frac{\hbar^2(\mathbf{k}+\mathbf{q})^2}{2m} - \frac{\hbar^2 k^2}{2m} = -\hbar^2\left(\frac{\mathbf{k}\cdot\mathbf{q}}{m}\right) + \frac{\hbar^2 q^2}{2m} \quad (6.73)$$

so that the polarization function of (6.53) takes the form

$$\chi_L(\mathbf{q},\omega) = -\frac{2}{\hbar} \sum_{\substack{k \leq k_F \\ |\mathbf{k}+\mathbf{q}| > k_F}} \left[\frac{1}{\omega + \hbar(\mathbf{k}\cdot\mathbf{q}/m) + \hbar q^2/2m + i\varepsilon} \right.$$

$$\left. - \frac{1}{\omega - \hbar(\mathbf{k}\cdot\mathbf{q}/m) - (\hbar q^2/2m) + i\varepsilon} \right]$$

$$= -\frac{2}{\hbar} \sum_{\mathbf{k}} f(E_{\mathbf{k}})[(1 - f(E_{\mathbf{k+q}}))] \left[\frac{1}{\omega + \hbar(\mathbf{k} \cdot \mathbf{q}/m) + (\hbar q^2/2m) + i\varepsilon} \right.$$

$$\left. - \frac{1}{\omega - \hbar(\mathbf{k} \cdot \mathbf{q}/m) - (\hbar q^2/2m) + i\varepsilon} \right], \quad (6.74)$$

where $f(E_{\mathbf{k}})$ is the Fermi function and the spin degeneracy was taken into account. By changing the summing index $\mathbf{k}$ in the first term of (6.74) to $-\mathbf{k} - \mathbf{q}$, and further putting $f(E_{\mathbf{k}}) = f(E_{-\mathbf{k}})$ and adding the second term, one ends up with the somewhat more compact form of the above equation

$$\chi_{\mathrm{L}}(\mathbf{q}, \omega) = \frac{2}{\hbar} \sum_{\mathbf{k}} \frac{f(E_{\mathbf{k}}) - f(E_{\mathbf{k+q}})}{\omega - \hbar(\mathbf{k} \cdot \mathbf{q}/m) - \hbar q^2/2m + i\varepsilon} \quad (6.75)$$

an expression for the so-called Lindhard polarization function. Of course, in the last two equations the limit as $\varepsilon \to 0_+$ is understood.

Equation (6.75) can also be written in terms of "undressed" one-particle Green's functions:

$$\chi_{\mathrm{L}}(\mathbf{q}, \omega) = 2 \int \frac{\mathrm{d}E}{2\pi i} \int \left(\frac{\mathrm{d}^3 k}{8\pi^3} \right) G^0(\mathbf{k}, E) \, G^0(\mathbf{k} + \mathbf{q}, E + \hbar\omega), \quad (6.76)$$

so that $\chi_{\mathrm{L}}(\mathbf{q}, \omega)$ can be represented by the very instructive and compact diagram of Fig. 6.2. (Nolting 1991). The "undressed" Green's function $G^0(\mathbf{k}, \varepsilon)$ is

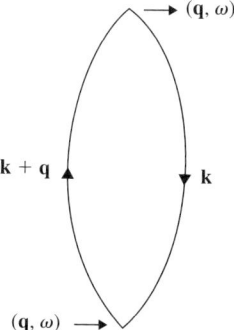

Fig. 6.2. Zero-order diagram belonging to an expansion of the polarization function $\chi(\mathbf{q}, \omega)$ into irreducible diagrams (Nolting 1991). The propagators correspond to free one-particle Green's functions.

given by

$$G^0(\mathbf{k}, \varepsilon) = \Theta_<(\mathbf{k})/(E - E(\mathbf{k}) - i\varepsilon)$$
$$+ \Theta_>(\mathbf{k})/(E - E(\mathbf{k}) + i\varepsilon); \quad \varepsilon \to 0_+ \tag{6.77}$$

$$\Theta_<(\mathbf{k}) = \begin{cases} 1 & \text{for } k \le k_F \\ 0 & \text{for } k > k_F \end{cases} \quad k_F = \text{Fermi momentum} \tag{6.78}$$

$$\Theta_>(\mathbf{k}) = 1 - \Theta_<(\mathbf{k}).$$

We will designate $G^0(\mathbf{k})$ as the hole propagator, and $G^0(\mathbf{k}+\mathbf{q})$ as the particle propagator.

Switching in (6.75) from summation to integration in momentum space one finds the following analytical expression for $\chi_L(\mathbf{q},\omega)$, which originates from Lindhard (1954):

$$\chi_L(\mathbf{q},\omega) = -\left[\frac{3q^2}{512\pi\gamma^2 Z^3 e^2}\right] \left\{ 4Z + [1 - (U-Z)^2] \ln\frac{U-Z-1}{U-Z+1} \right.$$
$$\left. - \left[1 - (U+Z)^2 \ln\frac{U+Z-1}{U+Z+1}\right] \right\} \tag{6.79}$$

$$Z = \left|\frac{q}{2k_F}\right|, \quad U = \frac{\hbar\omega}{4ZE_F}, \quad E_F = \frac{\hbar^2 k_F^2}{2m}, \quad \gamma = \frac{E_F}{\hbar\omega_p}.$$

Principle values are to be used for the logarithms in case of negative or complex arguments.

Of course, one cannot expect that $\chi_L(\mathbf{q},\omega)$ might already be a good approximation for the response of the electron system, which does not consist of independent but, on the contrary, of strongly interacting particles. To take this fact into account, at least in an approximate way, we write, instead of the equation

$$\delta\langle\rho(\mathbf{q},\omega)\rangle = (-e)\,\chi(\mathbf{q},\omega)\,\phi_{\text{ext}}(\mathbf{q},\omega), \tag{6.80}$$

which for $\mathbf{q}' = \mathbf{q}$, is equivalent to (6.44), the following relation

$$\delta\langle\rho(\mathbf{q},\omega)\rangle = (-e)\,\chi_0(\mathbf{q},\omega)[\phi_{\text{ext}}(\mathbf{q},\omega) + \phi_{\text{ind}}(\mathbf{q},\omega)]. \tag{6.81}$$

This relation means that we consider independent electrons in an effective field, which is composed of the external field and an induced field. We denote a polarization function used to couple the total potential $\phi_{\text{tot}} = \phi_{\text{ext}} + \phi_{\text{ind}}$ to the electron density fluctuations a proper polarization function. One can approach more realistic cases, when one discards the assumption of independent electrons by taking into account both the interaction of the excited particles/holes with the surrounding medium of interacting electrons (so-called self-energy effects) and the particle–hole interaction (vertex corrections).

In the self-consistent field (SCF) approximation, the induced field of equation (6.81) can, in a first step, be identified with the macroscopic field raised by the induced density fluctuations $\delta\langle\rho(\mathbf{q},\omega)\rangle$ via the Poisson equation

$$(-e)\,\phi_{\text{ind}}(\mathbf{q},\omega) = v(\mathbf{q})\,\delta\langle\rho(\mathbf{q},\omega)\rangle, \qquad (6.82)$$

so that $\delta\langle\rho(\mathbf{q},\omega)\rangle$ satisfies the following self-consistent equation

$$\delta\langle\rho(\mathbf{q},\omega)\rangle = (-e)\,\chi_0(\mathbf{q},\omega)\phi_{\text{ext}}(\mathbf{q},\omega) + v(\mathbf{q})\,\delta\langle\rho(\mathbf{q},\omega)\rangle\chi_0(\mathbf{q},\omega). \qquad (6.83)$$

and yields, being solved,

$$\delta\langle\rho(\mathbf{q},\omega)\rangle = (-e)\left(\frac{\chi_0(\mathbf{q},\omega)}{1 - v(\mathbf{q})\chi_0(\mathbf{q},\omega)}\right)\phi_{\text{ext}}(\mathbf{q},\omega). \qquad (6.84)$$

Comparing (6.84) with (6.80) we find the polarization function of the SCF approximation

$$\chi^{\text{SCF}}(\mathbf{q},\omega) = \frac{\chi_0(\mathbf{q},\omega)}{1 - v(\mathbf{q})\chi_0(\mathbf{q},\omega)}. \qquad (6.85)$$

If we replace the proper polarization function $\chi_0(\mathbf{q},\omega)$ in (6.84) by the Lindhard polarization function $\chi_L(\mathbf{q},\omega)$, the SCF polarization function is often denoted, for more historical reasons, as the random phase approximation (RPA) of the polarization function.

It was already mentioned that a better approximation of the response for a system of interacting electrons can be achieved by improving the proper polarization function. Moreover, one can approach the many-particle problem by including exchange and correlation effects into (6.82). This is done by introducing, in spirit of the time-dependent density functional theory (TDDFT) (Runge and Gross 1984), the Fourier transform $f_{\text{xc}}(\mathbf{q},\omega)$ of the dynamical exchange-correlation kernel defined by the equation (Burke and Gross 1998)

$$f_{\text{xc}}[\rho_0](\mathbf{r}t;\mathbf{r}'t') = \frac{dv_{\text{xc}}[\rho](\mathbf{r}t)}{d\rho(\mathbf{r}'t')}, \qquad (6.86a)$$

where v_{xc} is the exchange-correlation potential. The functional derivative is to be evaluated at the ground-state density, thus obtaining

$$(-e)\phi_{\text{ind}}(\mathbf{q},\omega) = v(\mathbf{q})\,\delta\langle\rho(\mathbf{q},\omega)\rangle + f_{xc}(\mathbf{q},\mathbf{q},\omega)\,\delta\langle\rho(\mathbf{q},\omega)\rangle. \qquad (6.86\text{b})$$

Within the so-called adiabatic local-density approximation (ALDA) of density functional theory (DF) (Kohn and Sham 1965) f_{xc} in (6.86b) becomes frequency and $\mathbf{q}$ independent

$$f_{xc} = \lim_{\omega\to 0} \int_{-\infty}^{\infty} dt\, \exp(i\omega t) \int d^3r \int d^3r'\, \delta(\mathbf{r}-\mathbf{r}')$$

$$\times \exp[i\mathbf{q}\cdot(\mathbf{r}-\mathbf{r}')] dv_{xc}[\rho](\mathbf{r}t)/d\rho(\mathbf{r}'0)|_{\rho=\rho_0(\mathbf{r})}$$

$$= \int d^3r\, dv_{xc}[\rho](\mathbf{r})/d\rho(\mathbf{r})|_{\rho=\rho_0(\mathbf{r})}, \qquad (6.87)$$

where $v_{xc}[\rho]$ is the electron density dependent exchange and correlation potential of the free-electron gas which can be determined by means of Monte Carlo simulations (Ceperley and Adler 1980). More generally one writes formally

$$f_{xc}(\mathbf{q},\mathbf{q},\omega) = -v(\mathbf{q})\,\tilde{G}(\mathbf{q}), \qquad (6.88)$$

where $\tilde{G}(\mathbf{q})$ is denoted as the local-field correction function. Within the ALDA $\tilde{G}(\mathbf{q})$ necessarily is proportional to q^2. This behavior is, for small q, also established in non-local-field theories. For larger q, these theories often predict a leveling out to constant values.

By inserting (6.88) into (6.86b), and by replacing the proper polarization function $\chi_0(\mathbf{q},\omega)$ by the Lindhard polarization function, $\chi_L(\mathbf{q},\omega)$, one ends up with the following expression for the SCF polarization function

$$\chi^{\text{SCF}}(\mathbf{q},\omega) = \frac{\chi_L(\mathbf{q},\omega)}{1 - v(\mathbf{q})[1-\tilde{G}(\mathbf{q})]\chi_L(q,\omega)}. \qquad (6.89)$$

By using $\chi^{\text{SCF}}(\mathbf{q},\omega)$ in the definition of the macroscopic dielectric function, as given in (6.66), one obtains the SCF approximation of the dielectric function

$$\varepsilon^{\text{SCF}}(\mathbf{q},\omega) = \frac{1 - v(\mathbf{q})(1-\tilde{G}(\mathbf{q})\chi_L(\mathbf{q},\omega))}{1 + v(\mathbf{q})\tilde{G}(\mathbf{q})\chi_L(\mathbf{q},\omega)}, \qquad (6.90)$$

or, by setting $\tilde{G}(\mathbf{q}) = 0$, the RPA counterpart:

$$\varepsilon^{\text{RPA}}(\mathbf{q},\omega) = 1 - v(\mathbf{q})\chi_L(\mathbf{q},\omega), \qquad (6.91)$$

so that, according to (6.67), the RPA dynamic structure factor reads

$$S^{\text{RPA}}(\mathbf{q},\omega) = \left\{ \frac{1}{1-\exp(-\hbar\omega/k_\text{B}T)} \right\} \left(\frac{q^2}{4\pi^2 e^2} \right) \text{Im} \left[\frac{-1}{\varepsilon^{\text{RPA}}(\mathbf{q},\omega)} \right]$$

$$= \left\{ \frac{1}{1-\exp(-\hbar\omega/k_\text{B}T)} \right\} \left(\frac{q^2}{4\pi^2 e^2} \right) \text{Im} \left[\frac{\varepsilon^{\text{RPA}}(\mathbf{q},\omega)}{|\varepsilon^{\text{RPA}}(q,\omega)|^2} \right]. \quad (6.92)$$

This expression for $S^{\text{RPA}}(\mathbf{q},\omega)$ reveals two sources for a peak structure of the dynamic structure factor: (i) Peaks of $-\text{Im}\chi_0(\mathbf{q},\omega)$ that occur, according to (6.92), whenever certain values of $(\mathbf{q},\omega)$ bring about a large number of single-particle excitations, where holes are created in the Fermi sea at $k < k_\text{F}$ and particles at $(k+q) > k_\text{F}$; we say the peaks are due to particle–hole excitation. (ii) Zero passages of $|\varepsilon^{\text{RPA}}(\mathbf{q},\omega)|^2$ that are found at $\omega = \omega_\text{p}$ for small q, since, according to (6.79), $\varepsilon^{\text{RPA}}(\mathbf{q},\omega) = 1 - v(\mathbf{q})\chi^{\text{RPA}}(\mathbf{q},\omega)$ becomes simply

$$\varepsilon^{\text{RPA}}(\mathbf{q},\omega) = 1 - \frac{\omega_\text{p}}{\omega} \quad (6.93)$$

for $U \gg Z+1$ or $\hbar\omega \gg 4E_\text{F}(q/2k_\text{F})[(q/2k_\text{F})+1]$.

Therefore, peaks of $S^{\text{RPA}}(\mathbf{q},\omega)$ can also be traced back to collective excitations of the electron system with frequencies near ω_p, the free-electron plasma frequency.

Finally, we shall extend the SCF approximation to electron systems in a periodic lattice potential, so that the single-particle states are Bloch states $|\mathbf{k},\nu\rangle$ ($\mathbf{k}$ = Bloch wavevector restricted to the first Brillouin zone, ν = band index) with energy eigenvalues $E(\mathbf{k},\nu)$. Thus the equation (6.53) must be modified accordingly obtaining in the limit $T = 0$

$$\chi_{\text{EC}}(\mathbf{q},\omega) = -\lim_{\varepsilon \to 0_+} 2 \sum_{\substack{\mathbf{k}\nu \\ \mathbf{k}'\nu'}}{}' \left\{ \frac{\langle \mathbf{k}\nu | \exp(i\mathbf{q}\cdot\mathbf{r})|\mathbf{k}'\nu'\rangle \langle \mathbf{k}'\nu'|\exp(-i\mathbf{q}\cdot\mathbf{r})|\mathbf{k}\nu\rangle}{E(\mathbf{k},\nu) - E(\mathbf{k}',n') - \hbar(\omega - i\varepsilon)} \right.$$

$$\left. + \frac{\langle \mathbf{k}\nu | \exp(-i\mathbf{q}\cdot\mathbf{r})|\mathbf{k}'\nu'\rangle \langle \mathbf{k}'\nu'|\exp(i\mathbf{q}\cdot\mathbf{r})|\mathbf{k}\nu\rangle}{E(\mathbf{k},\nu) - E(\mathbf{k}',\nu') + \hbar(\omega + i\varepsilon)} \right\}, \quad (6.94)$$

and by following the steps of the derivation to equation (6.75) and (6.91) we get the final result, known as the SCF dielectric function of Ehrenreich and Cohen (1959) and equivalent to the RPA free-electron dielectric function:

$$\varepsilon^{\text{SCF}}(\mathbf{q},\omega) = 1 - v(\mathbf{q}) \lim_{\varepsilon \to 0_+} \frac{2}{V} \sum_{\substack{\mathbf{k}\nu \\ \mathbf{k}'\nu'}}{}' |\langle \mathbf{k}'\nu' | \exp(i\mathbf{q}\cdot\mathbf{r})|\mathbf{k}\nu\rangle|^2$$

$$\times \frac{f(E(\mathbf{k},\nu)) - f(E(\mathbf{k}',\nu'))}{\hbar\omega + E(\mathbf{k},\nu) - E(\mathbf{k}',\nu') + i\varepsilon}. \quad (6.95)$$

6.6 Density response and macroscopic dielectric function of periodic solids

As already stated in Section 6.4, in general, an inhomogeneous electron system can respond with a density fluctuation $\sim \exp(i\mathbf{q}' \cdot \mathbf{r})$ to an external perturbation $\sim \exp(i\mathbf{q} \cdot \mathbf{r})$, where $\mathbf{q}' \neq \mathbf{q}$. We will now consider a special case of an inhomogeneous medium, namely an electron system under the influence of a periodic lattice potential, so that the electron density can be expanded into the following Fourier series:

$$\rho(\mathbf{r}) = \sum \rho(\mathbf{g}) \exp(i\mathbf{r} \cdot \mathbf{g}), \tag{6.96}$$

where $\mathbf{g}$ is a reciprocal lattice vector. Doing so has the consequence that a Fourier component $\phi_{\text{ext}}(\mathbf{q}_r + \mathbf{g}, \omega)$ of the external potential can also induce density fluctuations belonging to another wavevector $\mathbf{q}_r + \mathbf{g}'$, where $\mathbf{q}_r$ is restricted to lie in the first Brillouin zone (Wiser 1963, Saslow and Reiter 1973). Thus we have to rewrite (6.44)

$$\delta\langle\rho(\mathbf{q}_r + \mathbf{g}')\rangle(t) = (-e) \lim_{\varepsilon \to 0_+} \sum_{\mathbf{g}} \chi(\mathbf{q}_r + \mathbf{g}', \mathbf{q}_r + \mathbf{g}, \omega)$$

$$\times \phi_{\text{ext}}(\mathbf{q}_r + \mathbf{g}, \omega) \exp[-(i\omega + \varepsilon)t]. \tag{6.97}$$

This way the polarization function changes into a $(\mathbf{g}, \mathbf{g}')$-polarization matrix, which again can be calculated using the first-order perturbation scheme of Section 6.4 accordingly. Within the limits of the one-electron approximation and representing the one-electron states by Bloch wavefunctions, we obtain the Lindhard type polarization matrix $\mathbf{X}$ with its elements

$$\mathbf{X}^{\text{L}}_{\mathbf{gg}'} = 2 \lim_{\varepsilon \to 0_+} \sum_{\substack{\mathbf{k}\nu \\ \mathbf{k}'\nu'}}{}' \langle \mathbf{k}\nu | \exp[-i(\mathbf{q}_r + \mathbf{g}) \cdot \mathbf{r}] | \mathbf{k}'\nu' \rangle \langle \mathbf{k}'\nu' | \exp[i(\mathbf{q}_r + \mathbf{g}') \cdot \mathbf{r}] | \mathbf{k}\nu \rangle$$

$$\times \frac{f(E(\mathbf{k}, \nu)) - f(E(\mathbf{k}', \nu'))}{\hbar\omega + E(\mathbf{k}, \nu) - E(\mathbf{k}', \nu') + i\varepsilon}. \tag{6.98}$$

Proceeding now to the SCF approximation we rewrite (6.81), the fundamental equation of self-consistency, in terms of the proper polarization matrix

$$\delta\langle\rho(\mathbf{q}_r + \mathbf{g}', \omega)\rangle = (-e) \sum_{\mathbf{g}} X^0_{\mathbf{gg}'} \phi_{\text{tot}}(\mathbf{q}_r + \mathbf{g}, \omega), \tag{6.99}$$

with

$$\phi_{\text{tot}}(\mathbf{q}_r + \mathbf{g}, \omega) = \phi_{\text{ind}}(\mathbf{q}_r + \mathbf{g}, \omega) + \phi_{\text{ext}}(\mathbf{q}_r + \mathbf{g}, \omega). \tag{6.100}$$

According to (6.85), the Fourier transform of the induced potential reads, if we include the exchange-correlation contribution,

$$(-e)\phi_{\text{ind}}(\mathbf{q}_{\text{r}} + \mathbf{g}, \omega) = v(\mathbf{q}_{\text{r}} + \mathbf{g}) \, \delta\langle\rho(\mathbf{q}_{\text{r}} + \mathbf{g}, \omega)\rangle$$
$$+ \sum_{\mathbf{g}''} f_{\text{xc}}(\mathbf{q}_{\text{r}} + \mathbf{g}, \mathbf{q}_{\text{r}} + \mathbf{g}'', \omega) \, \delta\langle\rho(\mathbf{q}_{\text{r}} + \mathbf{g}, \omega)\rangle \quad (6.101)$$

so that

$$\delta\langle\rho(\mathbf{q}_{\text{r}} + \mathbf{g}', \omega)\rangle = -\sum_{\mathbf{g}} T_{\mathbf{gg}'} \, \delta\langle\rho(\mathbf{q}_{\text{r}} + \mathbf{g}, \omega)\rangle - e \sum_{\mathbf{g}} X^0_{\mathbf{gg}'} \, \phi_{\text{ext}}(\mathbf{q}_{\text{r}} + \mathbf{g}, \omega), \quad (6.102)$$

where we have introduced the so-called dielectric matrix $\mathbf{1} + \mathbf{T}$ with its elements

$$\varepsilon_{\mathbf{gg}'} \equiv \delta_{\mathbf{gg}'} - v(\mathbf{q}_{\text{r}} + \mathbf{g}) X^0_{\mathbf{gg}'} - \sum_{\mathbf{g}''} f_{\text{xc}}(\mathbf{q}_{\text{r}} + \mathbf{g}, \mathbf{q}_{\text{r}} + \mathbf{g}'', \omega) X^0_{\mathbf{gg}'}$$

$$\equiv \delta_{\mathbf{gg}'} - v(\mathbf{q}_{\text{r}} + \mathbf{g}) X^0_{\mathbf{gg}'} - X^{\text{xc}}_{\mathbf{gg}'}. \quad (6.103\text{a})$$

Within the adiabatic local-density approximation (ALDA) f_{xc} in (6.103a) becomes frequency and $\mathbf{q}_{\text{r}}$ independent:

$$f_{\text{xc}}(\mathbf{g}, \mathbf{g}'') = \int d^3 r \exp[-i(\mathbf{g} - \mathbf{g}'') \cdot \mathbf{r}] dv_{\text{xc}}[\rho](\mathbf{r})/d\rho(\mathbf{r})|_{\rho=\rho_0(\mathbf{r})}. \quad (6.103\text{b})$$

Using (6.101) we can write

$$\sum_{\mathbf{g}} (\delta_{\mathbf{gg}'} + T_{\mathbf{gg}'}) \, \delta\langle\rho(\mathbf{q}_{\text{r}} + \mathbf{g}, \omega)\rangle = -e \sum_{\mathbf{g}} X^0_{\mathbf{gg}'} \, \phi_{\text{ext}}(\mathbf{q}_{\text{r}} + \mathbf{g}, \omega) \quad (6.104)$$

which can be formally solved with respect to $\delta\langle\rho(\mathbf{q}_{\text{r}} + \mathbf{g})\rangle$

$$\delta\langle\rho(\mathbf{q}_{\text{r}} + \mathbf{g}', \omega)\rangle = (-e) \sum_{\mathbf{g}} [(\mathbf{1} + \mathbf{T})^{-1}]_{\mathbf{gg}'} X^0_{\mathbf{gg}'} \, \phi_{\text{ext}}(\mathbf{q}_{\text{r}} + \mathbf{g}, \omega), \quad (6.105)$$

where $\mathbf{1}$ is the unit matrix.

Now we can use (6.64) which defines the reciprocal macroscopic dielectric function and which we write, in order to emphasize that the $\mathbf{q}, \omega$ dependent terms are to be understood as macroscopic quantities

$$\left[\frac{1}{\varepsilon(\mathbf{q}, \omega)}\right]_{\text{M}} = 1 - \left[\frac{e\delta\langle\rho(\mathbf{q}, \omega)\rangle_{\text{M}} v(\mathbf{q})}{\phi_{\text{ext}}(\mathbf{q}, \omega)_{\text{M}}}\right]. \quad (6.106)$$

so that, with

$$\mathbf{q} = \mathbf{q}_{\text{r}} + \mathbf{g}_{\text{o}}, \quad (6.107)$$

we obtain

$$\left[\frac{1}{\varepsilon(\mathbf{q},\omega)}\right]_{\mathrm{M}} = [\mathbf{1} + \mathbf{V}\mathbf{X}^0(\mathbf{1}+\mathbf{T})^{-1}]_{\mathbf{g}_o\mathbf{g}_o}, \qquad (6.108)$$

where $\mathbf{V}$ is a diagonal matrix built by the Fourier components of the Coulomb potential.

To derive (6.108) we have made use of the fact that, as pointed out by Wiser (1963), the $\mathbf{q}$th Fourier component of any macroscopic quantity is just the $\mathbf{q}$, $\mathbf{g} = 0$ component of the corresponding microscopic quantity:

$$\delta\langle\rho(\mathbf{q},\omega)\rangle_{\mathrm{M}} = \delta\langle\rho(\mathbf{q}_{\mathrm{r}}+\mathbf{g}_o+\mathbf{0},\omega)\rangle \qquad (6.109)$$

$$\phi_{\mathrm{ext}}(\mathbf{q},\omega)_{\mathrm{M}} = \phi_{\mathrm{ext}}(\mathbf{q}_{\mathrm{r}}+\mathbf{g}_o+\mathbf{0},\omega). \qquad (6.110)$$

If we specialize (6.108) for the RPA case by setting all $f_{\mathrm{xc}}(\mathbf{q}_{\mathrm{r}}+\mathbf{g},\mathbf{q}_{\mathrm{r}}+\mathbf{g}'',\omega) = 0$, and replacing $X^0_{\mathbf{gg}'}$ by $X^{\mathrm{L}}_{\mathbf{gg}'}$ of (6.98), then

$$\mathbf{1} = (\mathbf{1}+\mathbf{T})^{-1}(\mathbf{1}+\mathbf{T}) = (\mathbf{1}+\mathbf{T})^{-1}(\mathbf{1}-\mathbf{V}\mathbf{X}^{\mathrm{L}}) \qquad (6.111)$$

$$\left[\frac{1}{\varepsilon(\mathbf{q},\omega)}\right]_{\mathrm{M,\ RPA}} = [(\mathbf{1}+\mathbf{T})^{-1}]_{\mathbf{g}_o\mathbf{g}_o}, \qquad (6.112)$$

whereas the inclusion of exchange correlation according to (6.103a) yields

$$\left[\frac{1}{\varepsilon(\mathbf{q},\omega)}\right]_{\mathrm{M,\ SCF}} = \{[\mathbf{1}-\mathbf{X}^{\mathrm{xc}}](\mathbf{1}+\mathbf{T})^{-1}\}_{\mathbf{g}_o\mathbf{g}_o}. \qquad (6.113)$$

6.7 Nonresonant spin- and orbital-magnetic scattering

Until now, when treating the nonresonant case, we have neglected the spin-dependent terms in the generalized Kramers–Heisenberg formula and also terms, which originate from the second-order terms being, also out of resonance, of the same order of magnitude as the nonresonant spin-dependent ones. We will now drop this neglect and will thus arrive at an expression for the double differential scattering cross-section which includes in addition to the nonresonant charge scattering also spin and orbital magnetic scattering (Blume 1985).

We start with the generalized Kramers–Heisenberg formula of equation (6.21). Far from resonance, this means for $\omega_1 \approx \omega_2 \gg (E_n - E_\mathrm{i})/\hbar$, we can neglect $(E_n - E_\mathrm{i})$ in the denominator of the third and fourth term of (6.19) compared with $\hbar\omega_1 \approx \hbar\omega_2$. Then we can use closure in the sum over the intermediate states

$|n\rangle$, and these third and fourth terms read

$$\left(\frac{\hbar^2}{m}\right)\sum_{jj'}\langle f|\left(\frac{1}{\hbar\omega_1}\right)\{[\mathbf{e}_2^*\cdot\mathbf{p}_j/\hbar - i(\mathbf{K}_2\times\mathbf{e}_2^*)\cdot\boldsymbol{\sigma}_j/2]\exp(-i\mathbf{K}_2\cdot\mathbf{r}_j)$$

$$\times[\mathbf{e}_1\cdot\mathbf{p}_{j'}/\hbar + i(\mathbf{K}_1\times\mathbf{e}_1)\cdot\boldsymbol{\sigma}_{j'}/2]\exp(i\mathbf{K}_1\cdot\mathbf{r}_{j'})\}$$

$$-\left(\frac{1}{\hbar\omega_2}\right)\{[\mathbf{e}_1\cdot\mathbf{p}_j/\hbar + i(\mathbf{K}_1\times\mathbf{e}_1)\cdot\boldsymbol{\sigma}_j/2]\exp(i\mathbf{K}_1\cdot\mathbf{r}_j)$$

$$\times[\mathbf{e}_2^*\cdot\mathbf{p}_{j'}/\hbar - i(\mathbf{K}_2\times\mathbf{e}_2^*)\cdot\boldsymbol{\sigma}_{j'}/2]\exp(-i\mathbf{K}_2\cdot\mathbf{r}_{j'})\}|i\rangle, \qquad (6.114)$$

and after some algebra, where we are using $\mathbf{p}_j = (\hbar/i)\frac{\partial}{\partial\mathbf{r}_j}$, $\mathbf{K}_1\cdot\mathbf{e}_1 = \mathbf{K}_2\cdot\mathbf{e}_2 = 0$, the well known vector identity $(\mathbf{a}\times\mathbf{b})\cdot(\mathbf{c}\times\mathbf{d}) = (\mathbf{a}\cdot\mathbf{c})(\mathbf{b}\cdot\mathbf{d}) - (\mathbf{b}\cdot\mathbf{c})(\mathbf{a}\cdot\mathbf{d})$ and the following generalized commutator relation for the spin vector operator:

$$\omega_2(\mathbf{a}\cdot\boldsymbol{\sigma})(\mathbf{b}\cdot\boldsymbol{\sigma}) - \omega_1(\mathbf{b}\cdot\boldsymbol{\sigma})(\mathbf{a}\cdot\boldsymbol{\sigma}) = i(\omega_1+\omega_2)(\mathbf{a}\times\mathbf{b})\cdot\boldsymbol{\sigma}, \qquad (6.115)$$

we obtain

$$\left(\frac{\hbar^2}{m}\right)\sum_j\langle f|\exp(i\mathbf{q}\cdot\mathbf{r}_j)\left\{\left(\frac{1}{\hbar c^2}\right)\{(\mathbf{e}_1\times\mathbf{e}_2^*)\cdot[\mathbf{p}_j\times(\hat{\mathbf{K}}_1-\hat{\mathbf{K}}_2)/\hbar]\}\right.$$

$$+\left(\frac{i}{\hbar c^2}\right)[\omega_1(\mathbf{e}_2^*\cdot\hat{\mathbf{K}}_1)(\hat{\mathbf{K}}_1\times\mathbf{e}_1) - \omega_2(\mathbf{e}_1\cdot\hat{\mathbf{K}}_2)(\hat{\mathbf{K}}_2\times\mathbf{e}_2^*)]\cdot\boldsymbol{\sigma}_j/2$$

$$+\left.\left(\frac{i}{\hbar c^2}\right)[(\omega_1+\omega_2)(\hat{\mathbf{K}}_2\times\mathbf{e}_2^*)\times(\hat{\mathbf{K}}_1\times\mathbf{e}_1)]\cdot\boldsymbol{\sigma}_j/4\right\}|i\rangle, \qquad (6.116)$$

where $\hat{\mathbf{K}}_{1/2}$ is a unit vector in the $\mathbf{K}_{1/2}$ direction.

By inserting this expression into the generalized Kramers–Heisenberg formula of equation (6.21) on ends up with the following off-resonance double differential scattering cross-section:

$$\left(\frac{d^2\sigma}{d\Omega_2 d\hbar\omega_2}\right)_{|i\rangle\to|f\rangle} = \left(\frac{e^2}{mc^2}\right)^2\left(\frac{\omega_2}{\omega_1}\right)\bigg|\langle f|\sum_j\exp(i\mathbf{q}\cdot\mathbf{r}_j)|i\rangle(\mathbf{e}_1\cdot\mathbf{e}_2^*)$$

$$-i\left(\frac{\hbar}{mc^2}\right)\langle f|\sum_j\exp(i\mathbf{q}\cdot\mathbf{r}_j)\left\{(ic)[\mathbf{p}_j\times(\hat{\mathbf{K}}_1-\hat{\mathbf{K}}_2)/\hbar]\cdot\mathbf{C}\right.$$

$$+\mathbf{D}\cdot\boldsymbol{\sigma}_j/2\bigg\}|i\rangle\bigg|^2 \delta(E_i - E_f + \hbar\omega), \qquad (6.117)$$

where

$$\mathbf{C} = -(\mathbf{e}_2^* \times \mathbf{e}_1)$$

$$\mathbf{D} = \frac{1}{2}(\omega_1 + \omega_2)[\mathbf{e}_2^* \times \mathbf{e}_1 - (\hat{\mathbf{K}}_2 \times \mathbf{e}_2^*) \times (\hat{\mathbf{K}}_1 \times \mathbf{e}_1)]$$
$$+ [-\omega_1(\mathbf{e}_2^* \cdot \hat{\mathbf{K}}_1)(\hat{\mathbf{K}}_1 \times \mathbf{e}_1) + \omega_2(\mathbf{e}_1 \cdot \hat{\mathbf{K}}_2)(\hat{\mathbf{K}}_2) \times \mathbf{e}_2^*]. \quad (6.118)$$

The second term of the matrix element of (6.117), the magnetic scattering amplitude, is roughly by $\hbar\omega_1/mc^2$ smaller than the first term, the charge scattering amplitude. This magnetic term consists of an orbital contribution whose polarization dependence is given by $\mathbf{C}$, and a spin-dependent contribution the polarization and photon wave vector dependence of which is much more complicated, defined by $\mathbf{D}$. This offers, in principle, a way to distinguish between both in an experiment that utilizes this different polarization behavior.

The pure magnetic constituent of the double differential scattering cross-section is at least by $(\hbar\omega_1/mc^2)^2$, this means by $4 \cdot 10^{-4}$ for 10 keV incident photon energy, smaller then the charge scattering contribution, and therefore difficult to measure separately, even if one tries to suppress charge scattering by measuring with linear polarized X-rays under a scattering angle of 90 degrees, i.e. $(\mathbf{e}_1 \cdot \mathbf{e}_2^*) = 0$. Nevertheless, if we consider only the spin part, this magnetic constituent contains very interesting information about spin correlation in solids that can be deduced, when switching from the excitation picture of (6.117) to the correlation picture, as done for the charge scattering already in Section 6.3. Then, in analogy to (6.30), the spin-dependent contribution to the pure magnetic double differential scattering cross-section writes

$$\left(\frac{d^2\sigma}{d\Omega_2 d\hbar\omega_2}\right)_m = \left(\frac{r_0^2}{2\pi}\right)\left(\frac{\hbar}{mc^2}\right)^2 \left(\frac{\omega_2}{\omega_1}\right)$$

$$\times \int_{-\infty}^{\infty} dt \exp(-i\omega t) \langle i| \sum_j \exp(i\mathbf{q} \cdot \mathbf{r}_j(0))(\boldsymbol{\sigma}_j \cdot \mathbf{D}/2)$$

$$\times \sum_j \exp(-i\mathbf{q} \cdot \mathbf{r}_j(t))(\boldsymbol{\sigma}_j \cdot \mathbf{D}/2)^* |i\rangle. \quad (6.119)$$

By introducing the spin density $\mathbf{s}(\mathbf{r},t)$ via

$$\mathbf{s}(\mathbf{r},t) \equiv \sum_j \delta(\mathbf{r} - \mathbf{r}_j(t))\, \boldsymbol{\sigma}_j/2 \quad (6.120)$$

and its Fourier transform

$$s(\mathbf{q},t) = \int d^3r \sum_j \exp(i\mathbf{q}\cdot\mathbf{r})\,\delta(\mathbf{r}-\mathbf{r}_j(t))\,\boldsymbol{\sigma}_j/2$$

$$= \sum_j \exp(i\mathbf{q}\cdot\mathbf{r}_j(t))\,\boldsymbol{\sigma}_j/2 \qquad (6.121)$$

we obtain for the spin-dependent contribution to the double differential scattering cross-section in analogy to (6.35) and (6.36)

$$\left(\frac{d^2\sigma}{d\Omega_2 d\hbar\omega_2}\right)_m = \left(\frac{r_0^2}{2\pi}\right)\left(\frac{\hbar}{mc^2}\right)^2\left(\frac{\omega_2}{\omega_1}\right)\int_{-\infty}^{\infty} dt\,\exp(-i\omega t)\int d^3r \int d^3r'\,\exp(-i\mathbf{q}\cdot\mathbf{r})$$

$$\times \sum_i g_i \langle i | \sum_{jj'} \delta(\mathbf{r}'-\mathbf{r}_j(0))(\boldsymbol{\sigma}_j\cdot\mathbf{D}/2)$$

$$\times \delta(\mathbf{r}'+\mathbf{r}-\mathbf{r}'_j(t))(\boldsymbol{\sigma}'_j\cdot\mathbf{D}/2)^*|i\rangle \qquad (6.122)$$

and by defining the density operator of spin-up (down) electrons according to

$$\rho_{\uparrow(\downarrow)}(\mathbf{r},t) \equiv \sum_{j\uparrow(\downarrow)} \delta(\mathbf{r}-\mathbf{r}_j(t)), \qquad (6.123)$$

using additionally the well-known properties of σ_z acting on spinors

$$\left(\frac{d^2\sigma}{d\Omega_2 d\hbar\omega_2}\right)_m = \left(\frac{r_0^2}{2\pi}\right)\left(\frac{\hbar}{mc^2}\right)^2\left(\frac{\omega_2}{\omega_1}\right)\int_{-\infty}^{\infty} dt\,\exp(-i\omega t)|D_z|^2$$

$$\times \int d^3r \int d^3r'\,\exp(-i\mathbf{q}\cdot\mathbf{r}) \sum_a g_a \langle a|[\rho_\uparrow(\mathbf{r}',0)-\rho_\downarrow(\mathbf{r}',0)]$$

$$\times [\rho_\uparrow(\mathbf{r}'+\mathbf{r},t)-\rho_\downarrow(\mathbf{r}'+\mathbf{r},t)]|a\rangle \qquad (6.124)$$

where $|a\rangle$ is the spin-free ground-state vector of the electron system. Therefore, we can state that the spin-dependent part of the pure magnetic contribution to the double differential scattering cross-section is the Fourier transform in space and time of the time-dependent correlation function made up of the density fluctuation operators of the electrons with unpaired spins. We can summarize this statement in analogy to (6.36) by introducing the dynamic structure factor of the electrons with unpaired spins

$$\left(\frac{d^2\sigma}{d\Omega_2 d\hbar\omega_2}\right)_m = \left(\frac{r_0^2}{2\pi}\right)\left(\frac{\hbar}{mc^2}\right)^2\left(\frac{\omega_2}{\omega_1}\right)|D_z|^2 S_{\text{unpaired}}(\mathbf{q},\omega). \qquad (6.125)$$

As already mentioned, it might be rather difficult to measure separately the pure magnetic double differential scattering cross-section with sufficient statistical accuracy. However, looking at (6.117) one realizes that the much larger interference term between charge and magnetic scattering could become measurable, provided this term gets real-valued contributions. According to (6.119), this happens, if e.g. the polarization vector $\mathbf{e}_1$ of the incident radiation becomes complex, in other words, if the incident radiation is (partly) circularly polarized,

$$\mathbf{e}_1 = \frac{1}{2}(\mathbf{e}_x \pm i\mathbf{e}_y), \tag{6.126}$$

so that the imaginary part of $\mathbf{D}(\mathbf{e}_1 \cdot \mathbf{e}_2^*)$ change its sign, when going from right-hand to left-hand circular polarization of the incident beam. The explicit expression for the spin-dependent part of the interference term of the double differential cross-section reads

$$\left(\frac{d^2\sigma}{d\Omega_2 d\hbar\omega_2}\right)_i = \left(\frac{-r_0^2}{2\pi}\right)\left(\frac{\hbar}{mc^2}\right)\left(\frac{\omega_2}{\omega_1}\right)\left\{i(\mathbf{e}_1 \cdot \mathbf{e}_2^*)\langle i|\sum_j \exp(-i\mathbf{q}\cdot\mathbf{r}_j)|f\rangle\right.$$

$$\left. \times \langle f|\sum_j \exp(i\mathbf{q}\cdot\mathbf{r}_j)(\boldsymbol{\sigma}_j \cdot \mathbf{D}/2)|i\rangle + \text{c.c.}\right\}\delta(E_i - E_f + \hbar\omega),$$

$$\tag{6.127}$$

or by switching again from the excitation picture to the correlation picture

$$\left(\frac{d^2\sigma}{d\Omega_2 d\hbar\omega_2}\right)_i = \left(\frac{-r_0^2}{2\pi}\right)\left(\frac{\hbar}{mc^2}\right)\left(\frac{\omega_2}{\omega_1}\right)\int_{-\infty}^{\infty} dt\,\exp(-i\omega t)\int d^3r \int d^3r'\,\exp(-i\mathbf{q}\cdot\mathbf{r})$$

$$\times \left\{\langle i|\sum_{jj'}\delta(\mathbf{r}' - \mathbf{r}_j(0))\delta(\mathbf{r}' + \mathbf{r} - \mathbf{r}_{j'}(t))\right.$$

$$\left. \times [i(\mathbf{e}_1 \cdot \mathbf{e}_2^*)(\boldsymbol{\sigma}_j \cdot \mathbf{D})]|i\rangle + \text{c.c.}\right\}, \tag{6.128}$$

and by writing

$$\sum \delta(\mathbf{r} - \mathbf{r}_j(t)) = \rho_\uparrow(\mathbf{r},t) + \rho_\downarrow(\mathbf{r},t) \tag{6.129}$$

$$\left(\frac{d^2\sigma}{d\Omega_2 d\hbar\omega_2}\right)_i = \left(\frac{r_0^2}{2\pi}\right)\left(\frac{\hbar\omega_1}{mc^2}\right)\left(\frac{\omega_2}{\omega_1}\right) \mathrm{Im}[(\mathbf{e}_1\cdot\mathbf{e}_2^*)D_z]\int_{-\infty}^{\infty} dt\, \exp(-i\omega t)$$

$$\times \int d^3r \int d^3r'\, \mathrm{ep}(-i\mathbf{q}\cdot\mathbf{r})\big\{\langle a|\rho_\uparrow(\mathbf{r}',0)\,\rho_\uparrow(\mathbf{r}'+\mathbf{r},t)$$

$$-\rho_\downarrow(\mathbf{r}',0)\,\rho_\downarrow(\mathbf{r}'+\mathbf{r},t)|a\rangle\big\}$$

$$= \left(\frac{r_0^2}{2\pi}\right)\left(\frac{\hbar\omega_1}{mc^2}\right)\left(\frac{\omega_2}{\omega_1}\right) \mathrm{Im}[(\mathbf{e}_1\cdot\mathbf{e}_2^*)D_z][S_\uparrow(\mathbf{q},\omega) - S_\downarrow(\mathbf{q},\omega)]. \tag{6.130}$$

Therefore, we can state that the interference term built up by the charge scattering amplitude and the spin-dependent part of the magnetic scattering amplitude is the Fourier transform in space and time of the difference of the time-dependent correlation function made up of the density fluctuation operators of electrons with opposite spins. In other words, the interference term is proportional to the difference of the dynamic structure factors of electrons with opposite spins.

A better insight into the propagation of the photon polarization states upon scattering, especially for the case of magnetic scattering, enables the representation of the scattering cross-sections in terms of the scattering amplitude operator $\mathbf{G}$, a 2×2 matrix in the space spanned by the polarization states, and the density matrix $\boldsymbol{\mu}$, a 2×2 matrix describing the incident polarization state. It was shown by Lipps and Tolhoek (1954) that the double differential scattering cross-section can be written in terms of the charge, $\mathbf{G}_c$, and magnetic, $\mathbf{G}_m$, contribution to the scattering amplitude operator

$$\left(\frac{d^2\sigma}{d\Omega_2 d\hbar\omega_2}\right) = r_0^2\left(\frac{\omega_2}{\omega_1}\right) \mathrm{Tr}\langle f|\mathbf{G}_c + \mathbf{G}_m|i\rangle\boldsymbol{\mu}\langle f|\mathbf{G}_c + \mathbf{G}_m|i\rangle^*, \tag{6.131}$$

where the density matrix $\boldsymbol{\mu}$ is defined so that $\mathbf{e}^*\boldsymbol{\mu}\mathbf{e}$ with

$$\boldsymbol{\mu} = \begin{pmatrix} \mu_{11} & \mu_{12} \\ \mu_{21} & \mu_{22} \end{pmatrix}; \quad \mu_{21} = \mu_{12}^*; \quad \mathrm{Tr}\boldsymbol{\mu} = 1 \tag{6.132}$$

gives the probability to find a photon with the polarization $\mathbf{e}$. In terms of the components P_1, P_2, P_3 of the Stoke's vector $\mathbf{P}$ (Berestetzki et al. 1986), the density matrix $\boldsymbol{\mu}$ reads

$$\boldsymbol{\mu} = \frac{1}{2}\begin{pmatrix} 1+P_3 & P_1 - iP_2 \\ P_1 + iP_2 & 1-P_3 \end{pmatrix} = \frac{(1+\mathbf{P}\cdot\boldsymbol{\sigma})}{2}, \tag{6.133}$$

where $P_1 = 1$ describes complete linear polarization of a photon in a plane inclined by $45°$ against the reference plane (here the scattering plane spanned by $\mathbf{K}_1$ and $\mathbf{K}_2$), $P_2 = +1(-1)$ stands for a photon completely circularly polarized with positive (negative) helicity, and $P_3 = +1(-1)$ characterizes a photon completely linear polarized perpendicular to (in) the reference plane. Thus P_2 is the mean helicity of the incident radiation. Complete linear polarization is characterized by $P_2 = 0$; $P_1^2 + P_3^2 = 1$, an unpolarized photon state by $P_1 = P_2 = P_3 = 0$. The components of the vector $\boldsymbol{\sigma}$ are the well-known Pauli matrices. The diagonal elements $G_{11}(G_{22})$ of the 2×2 scattering amplitude operator $\mathbf{G}$ are amplitudes of a scattering process, where the polarization $\mathbf{e}_1$ of the incident photon and the polarization $\mathbf{e}_2$ of the scattered photon both are perpendicular to (in) the scattering plane. The off-diagonal element G_{21} of $\mathbf{G}$ stands for an event, where $\mathbf{e}_1$ is perpendicular to and $\mathbf{e}_2$ in the scattering plane, and vice versa for G_{12}. By means of (6.117) and following the above definitions one finds easily that the total scattering amplitude operator writes

$$\mathbf{G} = \sum_j \exp(i\mathbf{q} \cdot \mathbf{r}_j) \begin{pmatrix} 1 & 0 \\ 0 & \hat{\mathbf{K}}_2 \cdot \hat{\mathbf{K}}_1 \end{pmatrix} - \left(\frac{i\hbar}{mc^2}\right)$$

$$\times \sum_j \exp(i\mathbf{q} \cdot \mathbf{r}_j) \left\{ ic\, \mathbf{p}_j \times (\hat{\mathbf{K}}_1 - \hat{\mathbf{K}}_2) \begin{pmatrix} 0 & \hat{\mathbf{K}}_1 \\ -\hat{\mathbf{K}}_2 & \hat{\mathbf{K}}_2 \times \hat{\mathbf{K}}_1 \end{pmatrix} \right.$$

$$\left. + \frac{\boldsymbol{\sigma}_j}{2} \cdot \begin{pmatrix} \frac{1}{2}(\omega_1 + \omega_2)(\hat{\mathbf{K}}_1 \times \hat{\mathbf{K}}_2) & \frac{1}{2}(\omega_1 - \omega_2)\hat{\mathbf{K}}_1 + [\omega_2 \hat{\mathbf{K}}_1 \cdot \hat{\mathbf{K}}_2 \\ & -(\omega_1 + \omega_2)/2]\hat{\mathbf{K}}_2 \\ \frac{1}{2}(\omega_1 - \omega_2)\hat{\mathbf{K}}_2 - [\omega_1 \hat{\mathbf{K}}_1 \cdot \hat{\mathbf{K}}_2 & \frac{1}{2}(\omega_1 + \omega_2)(\hat{\mathbf{K}}_1 \times \hat{\mathbf{K}}_2) \\ -(\omega_1 + \omega_2)/2]\hat{\mathbf{K}}_1 & \end{pmatrix} \right\}.$$

(6.134)

6.8 The Compton scattering regime

In what follows we will define "Compton scattering" to be that regime of inelastic X-ray scattering, where we are looking at scattering events with high momentum transfer q and high energy transfer $\hbar\omega$. In this respect, high momentum transfer means that $2\pi/q \gg l_c$, where l_c is the interparticle distance of the scattering electron system. We have already discussed in Section 6.3 that with such a large momentum transfer we can neglect all interference effects between waves scattered from different particles at different times, when calculating the dynamic structure factor according to (6.34). Therefore, one is probing positions of the same particle at different times, so that it is mainly single-particle properties we are looking at, and we will see in the next section that it is the momentum density distribution of the electron system we get information from. But in order to obtain this information we have to ensure that the probing of the single-particle positions occurs in such short time distances that the system will not become rearranged during the scattering process, which requires an energy transfer, $\hbar\omega$, much greater than characteristic energies of the scattering system (i.e. binding energies for core electrons, Fermi energy for quasi-free electrons in metals). Thus the definition of the Compton scattering regime depends to a large extent on the scattering system we are looking at. As long we are interested in obtaining information about the momentum density of valence electrons separately, the energy transfer $\hbar\omega$ need not be that large as for cases, where also the knowledge of the momentum distribution of tightly bound core electrons is required simultaneously. In the following section we will investigate more accurately in what way the DDSCS yields information about the electron momentum density, provided the experimental conditions for working within the Compton scattering regime are met.

6.8.1 *Impulse approximation, Compton profiles*

Following Eisenberger and Platzman (1970), we will now derive an approximate expression for the nonresonant and nonrelativistic DDSCS according to (6.25) by assuming an energy and momentum transfer valid for the Compton regime as defined in Section 6.3. We will call this treatment of the DDSCS its impulse approximation.

Inserting into (6.25) again the integral representation of the δ-function as given in (6.27) and making use of

$$H\,|x\rangle = E_x\,|x\rangle, \tag{6.135}$$

where H is the Hamiltonian and $|x\rangle$ is any state vector of the system, we get the following expression for $S(\mathbf{q},\omega)$:

$$S(\mathbf{q},\omega) = \frac{1}{2\pi\hbar}\int_{-\infty}^{\infty} dt\,\exp(-i\omega t) \sum_f \langle i|\sum_j \exp(-i\mathbf{q}\cdot\mathbf{r}_j)|f\rangle$$

$$\times \langle f|\exp(iHt/\hbar)\sum_j \exp(i\mathbf{q}\cdot\mathbf{r}_j)\exp(-iHt/\hbar)|i\rangle, \tag{6.136}$$

where we have dropped the averaging over initial states.

Now we write in (6.136) the Hamilton operator H of the scattering system as composed of a kinetic energy term H_0 and a potential energy part V

$$H = H_0 + V. \tag{6.137}$$

Furthermore, we expand $\exp(iHt/\hbar)$ in (6.135) in the following manner:

$$\exp(iHt/\hbar) = \exp(iH_0 t/\hbar)\exp(iVt/\hbar)\exp(-[H_0,V]t^2/2\hbar^2)\ldots, \tag{6.138}$$

where the higher order terms in (6.138) contain multiple commutators and are of higher order in powers of the time.

The essence of the impulse approximation consists of the assumption that, whenever

$$\hbar\omega \gg (\langle [H_0, V] \rangle)^{\frac{1}{2}} \tag{6.139}$$

one can set

$$\exp(-[H_0, V]t^2/2\hbar^2) \cong 1, \tag{6.140}$$

in (6.138), since appreciable contributions to the time integral in (6.136) occur only for $t \leq 1/\omega$. The relation (6.139) is valid if the transferred energy $\hbar\omega$ is large compared with characteristic energies of the system described by the Hamiltonian H, so that this relation just represents one of the requirements we have defined as essential for the Compton regime. We shall show that, at least for a system of free electrons, the prerequisites claimed for the momentum transfer in Section 6.3 are then met automatically.

Let us now see what changes are occurring with the expression (6.136) for the DDSCS, when we implement (6.138) together with (6.140). For the sake of physical transparency we shall treat the following approximation in the first instance for a one-electron atom. This allows us to drop the index j in (6.136). Since V commutes with $\mathbf{r}$, one obtains by inserting (6.138) and (6.140) into (6.136), and by utilizing closure,

$$\frac{d^2\sigma}{d\Omega_2 d\hbar\omega_2} = \left(\frac{d\sigma}{d\Omega_2}\right)_{Th} \frac{1}{2\pi\hbar} \int dt \exp(-i\omega t) \langle i | \exp(-i\mathbf{q} \cdot \mathbf{r}) \exp(iH_0 t/\hbar)$$

$$\times \exp(i\mathbf{q} \cdot \mathbf{r}) \exp(-iH_0 t/\hbar) | i \rangle. \tag{6.141}$$

The insertion of a complete set of eigenfunctions $|\mathbf{p}_f\rangle$ of the kinetic energy part H_0 of the Hamiltonian into (6.141) together with

$$\exp(iH_0 t/\hbar)|\mathbf{p}_f\rangle = \exp[i\varepsilon(\mathbf{p}_f)t/\hbar]|\mathbf{p}_f\rangle, \tag{6.142}$$

where

$$\varepsilon(\mathbf{p}_f) \equiv \frac{\mathbf{p}_f^2}{2m}, \tag{6.143}$$

transforms (6.141) into

$$\frac{d^2\sigma}{d\Omega_2 d\hbar\omega_2} = \left(\frac{d\sigma}{d\Omega_2}\right)_{Th} \sum_{\mathbf{p}_f} |\langle i | \exp(-i\mathbf{q}\cdot\mathbf{r})|\mathbf{p}_f\rangle|^2 \delta[\varepsilon(\mathbf{p}_f) - \varepsilon(\mathbf{p}_f - \hbar\mathbf{q}) - \hbar\omega], \tag{6.144}$$

where the integral representation of the δ-function has been revoked. Finally we introduce the following change in definition

$$\mathbf{p} \equiv \mathbf{p}_f - \hbar\mathbf{q}, \tag{6.145}$$

and switch over from a sum over the final states $|\mathbf{p}_f\rangle$ to a $\mathbf{p}$-integration, ending up with

$$\frac{d^2\sigma}{d\Omega_2 d\hbar\omega_2} = \left(\frac{d\sigma}{d\Omega_2}\right)_{\text{Th}} \left(\frac{1}{2\pi\hbar}\right)^3 \int |\langle i|\mathbf{p}\rangle|^2 \delta\left(\frac{\hbar^2 q^2}{2m} + \hbar\mathbf{p}\cdot\mathbf{q}/m - \hbar\omega\right) d\mathbf{p}, \tag{6.146}$$

since, according to (6.143) and (6.145)

$$\varepsilon(\mathbf{p}_f) - \varepsilon(\mathbf{p}_f - \hbar\mathbf{q}) = \varepsilon(\mathbf{p} + \hbar\mathbf{q}) - \varepsilon(\mathbf{p}) = \frac{\hbar^2 q^2}{2m} + \frac{\hbar\mathbf{p}\cdot\mathbf{q}}{m}. \tag{6.147}$$

Let $\psi(\mathbf{r})$ be the single-particle position space wavefunction representing the initial state $|i\rangle$, and let $\chi(\mathbf{p})$ be its Fourier transform, the momentum-space wave function. Then the quantity

$$\rho(\mathbf{p}) \equiv \left(\frac{1}{2\pi\hbar}\right)^3 |\langle i|\mathbf{p}\rangle|^2 = |\chi(\mathbf{p})|^2 = \left(\frac{1}{2\pi\hbar}\right)^3 \left|\int \psi(\mathbf{r})\exp(-i\mathbf{p}\cdot\mathbf{r}/\hbar)d\mathbf{r}\right|^2 \tag{6.148}$$

under the integral of (6.146) represents the momentum space density of the scattering system in the ground state $|i\rangle$, and gives the probability of finding the initial electron with a given momentum $\mathbf{p}$. Due to the δ-function, the $\mathbf{p}$-space integral in (6.146) is over a plane in momentum space perpendicular to $\mathbf{q}$, where the distance p_q of this plane from the origin of momentum space is determined by

$$p_q = \frac{\omega m}{q} - \frac{\hbar q}{2}$$

$$\approx mc\left(\frac{[\hbar\omega_1 - \hbar\omega_2 - (\hbar^2\omega_1\omega_2/mc^2)(1-\cos\theta)]}{[\hbar^2\omega_1^2 + \hbar^2\omega_2^2 - 2\hbar^2\omega_1\omega_2\cos\theta]^{1/2}}\right). \tag{6.149}$$

Here the definition of $\mathbf{q} \equiv \mathbf{K}_1 - \mathbf{K}_2$ has been used, and $\hbar^2(\omega_1-\omega_2)^2/2mc^2$ has been omitted.

Choosing $\mathbf{q}$ to lie in the z-direction, so that $p_q = p_z$, we can write (6.146) in the convenient form:

$$\frac{d^2\sigma}{d\Omega_2 d\hbar\omega_2} = \left(\frac{d\sigma}{d\Omega_2}\right)_{\text{Th}} \left(\frac{m}{\hbar q}\right) \iint \rho(p_x, p_y, p_z = p_q) dp_x\, dp_y$$

$$= \left(\frac{d\sigma}{d\Omega_2}\right)_{\text{Th}} \left(\frac{m}{\hbar q}\right) J(p_q), \tag{6.150}$$

where we have introduced the so-called (directional) Compton profile

$$J(p_q) \equiv \iint \rho(p_x, p_y, p_z = p_q) dp_x\, dp_y, \tag{6.151}$$

which we can interpret as the projection of the momentum density $\rho(\mathbf{p})$ along the direction of the scattering vector $\mathbf{q}$. Figure 1.2b presents schematically a Compton profile both on the energy loss and the p_z-scale, together with the quasi elastically scattered line. If the system under investigation is isotropic, one can define a radial momentum distribution $I(p) = 4\pi p^2 \rho(p)$ and rewrite (6.150) as

$$J(p_q) = \frac{1}{2}\int_{|p_q|}^{\infty} \left(\frac{1}{p}\right) I(p)\mathrm{d}p. \tag{6.152}$$

The essence of the impulse approximation can be made physically somewhat more transparent by looking at the changes which happen in the course of this approximation, when going from (6.25) to (6.144). The only things we have changed, apart of dropping averaging of the initial states, were: In the first place, to replace both the total initial and final state energy in the argument of the δ-function by the corresponding kinetic energies, thus dropping the potential energy in the energy difference; in the second place, to assume the electron to be free with momentum $\mathbf{p}_f$ in its final state, where one must be aware that this stipulation is a direct consequence of the potential energy cancellation and works so well because of the completeness of the eigenfunctions $|\mathbf{p}_f\rangle$; in the third place, one can interpret the momentum space transformation (6.145) as having utilized momentum conservation in order to obtain $\mathbf{p}_f - \hbar\mathbf{q}$ for the momentum of the initial state. But also this procedure is nothing else than a consequence of the potential energy cancellation in the energy-conserving δ-function of (6.25). In other words, we have assumed the scattering process to be so fast that the scattering atom has no time to rearrange (to change the potential) with the consequence that the transferred energy (momentum) is so large that the recoil electron can be considered as free. Thus we have reduced the scattering off a bound electron to a scattering process, which is sketched in Fig. 6.3 and where the electron hit by the photon is assumed to be free, moving with a certain initial momentum $\mathbf{p}$, and where the momentum $\hbar\mathbf{q}$ is transferred to this electron, so that its energy becomes $(\mathbf{p} + \hbar\mathbf{q})^2/2m$. Then energy conservation makes the

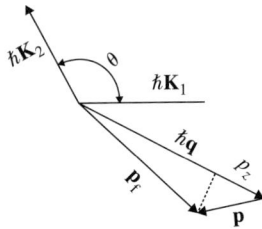

Fig. 6.3. Vector diagram of an inelastic scattering process off a free electron with an initial momentum $\mathbf{p}$. The final electron momentum $\mathbf{p}_f = \mathbf{p} + \hbar\mathbf{q}$; $\mathbf{q} = \mathbf{K}_1 - \mathbf{K}_2$.

energy

$$\hbar\omega = \frac{(\mathbf{p}+\hbar\mathbf{q})^2}{2m} - \frac{\mathbf{p}^2}{2m} = \frac{\hbar^2\mathbf{q}^2}{2m} + \frac{\hbar\mathbf{p}\cdot\mathbf{q}}{m} \qquad (6.153)$$

be transferred from the photon to the scattering atom. Of course, the bonding (or the potential V) of the electron is not neglected completely by this approximate treatment. Bonding, on the contrary, comes into play via the probability of finding the electron with a certain momentum $\mathbf{p}$, where this probability is determined by the Fourier transform of the single-particle wavefunction, which represents a bound state (a solution of the Schrödinger equation including the potential V). Equation (6.153) describes the energy loss of the scattered photon and can be understood as composed of the classical Compton shift $\hbar^2\mathbf{q}^2/2m$, which is valid for scattering from a free electron at rest, and an additional Doppler shift $\hbar\mathbf{p}\cdot\mathbf{q}/m$, due to the movement with momentum p of the scattering electron.

Let us now proceed to the scattering from a many-electron system. If we can understand the system in the first instance as composed of independent particles described by a single Slater determinant

$$\psi(\mathbf{r}_1,\mathbf{r}_2,\ldots,\mathbf{r}_N) = (N!)^{-1/2} \begin{vmatrix} \psi_a(\mathbf{r}_1) & \cdots & \psi_a(\mathbf{r}_j) & \cdots & \psi_a(\mathbf{r}_N) \\ \vdots & & \vdots & & \vdots \\ \psi_k(\mathbf{r}_1) & \cdots & \psi_k(\mathbf{r}_j) & \cdots & \psi_k(\mathbf{r}_N) \\ \vdots & & \vdots & & \vdots \\ \psi_t(\mathbf{r}_1) & \cdots & \psi_t(\mathbf{r}_j) & \cdots & \psi_t(\mathbf{r}_N) \end{vmatrix}, \qquad (6.154)$$

the Compton profile can be simply composed of the contributions of occupied individual orbitals $\psi_k(\mathbf{r}_j)$ to the momentum density

$$\rho(\mathbf{p}) \equiv \left(\frac{1}{2\pi\hbar}\right)^3 |\langle i|\mathbf{p}\rangle|^2 = \sum_{k(\text{occ})} |\chi_k(\mathbf{p})|^2$$

$$= \left(\frac{1}{2\pi\hbar}\right)^3 \sum_{k(\text{occ})} \left|\int \psi_k(\mathbf{r})\exp(-i\mathbf{p}\cdot\mathbf{r}/\hbar)d\mathbf{r}\right|^2. \qquad (6.155)$$

Therefore, a system of independent bound electrons, investigated by inelastic X-ray scattering under the conditions of the impulse approximation, is considered to consist of free particles with a certain momentum distribution, where both energy and momentum are conserved during the collision. The momentum density for any particular momentum in this noninteracting electron system is obtained from the square of the Fourier transform of its single-particle wavefunction.

The simplest system in this respect would be a system of really free electrons at zero temperature being subject only to the Pauli principle, so that

$$\rho(\mathbf{p}) = \begin{cases} \dfrac{3}{4\pi p_F^3} & \text{for } p \leq p_F \quad (p_F = \text{Fermi momentum}) \\ 0 & \text{for } p > p_F. \end{cases} \qquad (6.156)$$

Then the momentum space integration in (6.151) simply yields the area of slices through the Fermi sphere with radius p_F, multiplied by $3/(4\pi p_\mathrm{F}^3)$, in order to normalize the resulting Compton profile

$$J(p_q) = \frac{3}{4p_\mathrm{F}^3}(p_\mathrm{F}^2 - p_q^2) \quad \text{for } |p_q| \leq p_\mathrm{F} \tag{6.157}$$

to unity.

The spectral distribution of the inelastically scattered radiation is determined by a Compton-shifted and Doppler-broadened band, where the intensity distribution of this Doppler-broadened spectral distribution contains the required information about the momentum distribution of the scattering electron system.

Finally, we will apply the formalism of the impulse approximation also to an arbitrary many-electron system. By following Benesch and Smith (1973), let us again introduce the conditions of the impulse approximation, as formulated in (6.140), into (6.136), but let us now keep the j-summation over the particles of the system. Thus we obtain instead of (6.141):

$$\frac{d^2\sigma}{d\Omega_2 d\hbar\omega_2} = \left(\frac{d\sigma}{d\Omega_2}\right)_\mathrm{Th} \left(\frac{1}{2\pi\hbar}\right) \int dt \exp(-i\omega t) \langle i| \sum_{jj'} \exp(-i\mathbf{q} \cdot \mathbf{r}_j) \exp\left(\frac{iH_0 t}{\hbar}\right)$$

$$\times \exp(i\mathbf{q} \cdot \mathbf{r}_{j'}) \exp\left(-\frac{iH_0 t}{\hbar}\right) |i\rangle. \tag{6.158}$$

We now introduce the (spin-free) one-particle density matrix $\Gamma_1(\mathbf{r}_1|\mathbf{r}_1')$ and the two-particle density matrix $\Gamma_2(\mathbf{r}_1, \mathbf{r}_2|\mathbf{r}_1', \mathbf{r}_2')$ (Löwdin 1956), defined as follows:

$$\Gamma_1(\mathbf{r}_1|\mathbf{r}_1') \equiv N \int \Gamma_N(\mathbf{r}_1, \mathbf{r}_2, \ldots, \mathbf{r}_N|\mathbf{r}_1', \mathbf{r}_2', \ldots, \mathbf{r}_N') d\mathbf{r}_2 \ldots d\mathbf{r}_N \tag{6.159}$$

$$\Gamma_2(\mathbf{r}_1, \mathbf{r}_2|\mathbf{r}_1', \mathbf{r}_2') \equiv \binom{N}{2} \int \Gamma_N(\mathbf{r}_1, \mathbf{r}_2, \ldots, \mathbf{r}_N|\mathbf{r}_1', \mathbf{r}_2', \ldots, \mathbf{r}_N') d\mathbf{r}_3 \ldots d\mathbf{r}_N, \tag{6.160}$$

where Γ_N is the N-particle density matrix derived from the N-particle wave functions Ψ_N via

$$\Gamma_N(\mathbf{r}_1, \mathbf{r}_2, \ldots, \mathbf{r}_N|\mathbf{r}_1', \mathbf{r}_2', \ldots, \mathbf{r}_N')$$
$$\equiv \Psi_N(\mathbf{r}_1, \mathbf{r}_2, \ldots, \mathbf{r}_N) \Psi_N^*(\mathbf{r}_1', \mathbf{r}_2', \ldots, \mathbf{r}_N'), \tag{6.161}$$

and where, according to Löwdin's (1956) convention, the integration $\int d\mathbf{r}_j$ in an expression of the form

$$\int \mathbf{F} \Psi_N(\mathbf{r}_1, \mathbf{r}_2, \ldots, \mathbf{r}_N) \Psi_N^*(\mathbf{r}_1', \mathbf{r}_2', \ldots, \mathbf{r}_N') d\mathbf{r}_1, \ldots, d\mathbf{r}_N \tag{6.162}$$

with an arbitrary Hermitian operator $\mathbf{F}$ means: let $\mathbf{F}$ operate on $\Psi_N(\mathbf{r}_1, \mathbf{r}_2, \ldots, \mathbf{r}_N)$, set $\mathbf{r}_j' = \mathbf{r}_j$ and integrate with respect to $\mathbf{r}_j$.

We can now implement the one- and two-particle density matrices into (6.158) by assuming the ground state $|i\rangle$ to be represented by an N-particle wavefunction Ψ_N, and by decomposing the operator within the $\langle\rangle$ brackets of (6.158) into a part diagonal in the electron coordinates and a corresponding nondiagonal term

$$\frac{d^2\sigma}{d\Omega_2 d\hbar\omega_2} = \left(\frac{d\sigma}{d\Omega_2}\right)_{Th}\left(\frac{1}{2\pi\hbar}\right)\int dt\,\exp(-i\omega t)(N\int \exp(-i\mathbf{q}\cdot\mathbf{r}_1)\exp\left(\frac{iH_0 t}{\hbar}\right)$$

$$\times \exp(i\mathbf{q}\cdot\mathbf{r}_1)\exp(-iH_0 t/\hbar)\,\Gamma_1(\mathbf{r}|\mathbf{r}'_1)d\mathbf{r}_1 + 2\binom{N}{2}$$

$$\times \int \exp(-i\mathbf{q}\cdot\mathbf{r}_1)\exp(iH_0 t/\hbar)\exp(i\mathbf{q}\cdot\mathbf{r}_2)\,\Gamma_2(\mathbf{r}_1,\mathbf{r}_2|\mathbf{r}'_1,\mathbf{r}'_2)d\mathbf{r}_1 d\mathbf{r}_2),$$

(6.163)

where we have used that Ψ_N is antisysmmetric and that the operator within the $\langle\rangle$ brackets is symmetric in the electron coordinates.

It has been pointed out by Benesch and Smith (1973) that the second term in (1.163), which contains the two-particle density matrix, can be neglected, when $2\pi/q$ is very small compared with the interparticle distance $\mathbf{r}_2 - \mathbf{r}_1$; this means, under the conditions we have formulated for the validity of the impulse approximation.

Insertion of a complete set of eigenfunctions of H_0 into (6.163) yields

$$\frac{d^2\sigma}{d\Omega_2 d\hbar\omega_2} = \left(\frac{d\sigma}{d\Omega_2}\right)_{Th}\left(\frac{1}{2\pi\hbar}\right)\int dt\,\exp[-i\omega t + i\varepsilon(\mathbf{p}_{f1})t/\hbar - i\varepsilon(\mathbf{p}_{f1} - \hbar\mathbf{q})t/\hbar]$$

$$\times \sum_{\mathbf{p}_f}\exp\left[i\left(\frac{\mathbf{p}_{f1}}{\hbar} - \mathbf{q}\right)\cdot\mathbf{r}_1\right]\exp\left[-i\left(\frac{\mathbf{p}_{f1}}{\hbar} - \mathbf{q}\right)\cdot\mathbf{r}'_1\right]\Gamma_1(\mathbf{r}_1|\mathbf{r}'_1)d\mathbf{r}_1\,d\mathbf{r}'_1.$$

(6.164)

Following Benesch and Smith (1971), we define the one-particle (spin-free) density matrix in momentum space, $\Gamma_1(\mathbf{p}_1|\mathbf{p}'_1)$, to be the six-dimensional Fourier-transform of the one-particle (spin-free) density matrix $\Gamma_1(\mathbf{r}_1|\mathbf{r}'_1)$ in position space:

$$\Gamma_1(\mathbf{p}_1|\mathbf{p}'_1) \equiv \left(\frac{1}{2\pi\hbar}\right)^3 \int \Gamma_1(\mathbf{r}_1|\mathbf{r}'_1)\exp[-i(\mathbf{p}_1\cdot\mathbf{r}_1 - \mathbf{p}'_1\cdot\mathbf{r}'_1)/\hbar]d\mathbf{r}_1\,d\mathbf{r}'_1. \quad (6.165)$$

We insert this definition into (6.164), define a new momentum variable

$$\mathbf{p}_1 = \mathbf{p}_{f1} - \hbar\mathbf{q} \qquad (6.166)$$

pass over to $\mathbf{p}_1$ integration, and invoke the integral definition of the δ-function, so that we end up with the following expression for the DDSCS

$$\frac{d^2\sigma}{d\Omega_2 d\hbar\omega_2} = \left(\frac{d\sigma}{d\Omega_2}\right)_{\text{Th}} \int \Gamma_1(\mathbf{p}|\mathbf{p})\,\delta(\hbar\omega - \hbar^2 q^2/2m - \hbar\mathbf{p}\cdot\mathbf{q}/m)d\mathbf{p}, \quad (6.167)$$

where we have used (6.147). The index 1 on the momentum space variable has been skipped. Equation (6.167) is the many-electron equivalent to (6.146), since the diagonal element of the one-particle density matrix in momentum space just describes the momentum space density:

$$\rho(\mathbf{p}) = \Gamma_1(\mathbf{p}|\mathbf{p}). \quad (6.168)$$

Eisenberger and Platzman (1970) have estimated the corrections to the impulse approximation by comparing the third frequency moment of $S(\mathbf{q},\omega)$ as computed for the exact Hamiltonian with the corresponding third frequency moment of the impulse approximation. (The zeroth through second moments do not differ.) They found the corrections to be of the order $(E_B/E_R)^2$, where E_B is the binding energy of the initial electron state and $E_R = \hbar^2 q^2/2m$ is the recoil energy.

6.8.2 Densities in position and momentum space, kinetic energy

It was the most important result of the preceding Chapter that, within certain experimental limits, inelastic X-ray scattering can yield information about the momentum space density $\rho(\mathbf{p})$ in the form of the Compton profile

$$J(p_z) = \int\int \rho(p_x, p_y, p_z) dp_x\, dp_y, \quad (6.169)$$

where we have represented $\rho(\mathbf{p})$ as the modulus squared of the one-electron momentum space wavefunction $\chi(\mathbf{p})$ and as the diagonal term of the one-particle density matrix in momentum space $\Gamma_1(\mathbf{p}|\mathbf{p})$, respectively.

$$\rho(\mathbf{p}) = \begin{cases} \chi(\mathbf{p})\,\chi^*(\mathbf{p}) \\ \Gamma_1(\mathbf{p}|\mathbf{p}). \end{cases} \quad (6.170)$$

We will now investigate, in somewhat more detail, how this information about wavefunction and density matrix, respectively, in momentum space is related to the corresponding position space quantities and to the experimental information one can get from those. We will learn that the measurement of the momentum space density is complementary to the investigation of the position space density by means of X-ray diffraction, and will lead to a better understanding of ground-state properties of condensed matter. Eventually it will be shown what is still missing, when the knowledge about the full one-particle density matrix, this means its diagonal and its nondiagonal terms, is aspired to.

The measurement of one directional Compton profile of an anisotropic system does not provide the full three-dimensional momentum space density but only its projection along one direction. Somewhat similar is true with the investigation of position space electron densities by means of X-ray diffraction: The measured quantity is, in the case of crystalline matter, the intensity of one Bragg reflection, which is proportional to the modulus squared form factor

$$F(\mathbf{q}) \equiv \int \rho(\mathbf{r}) \exp(i\mathbf{q} \cdot \mathbf{r}) \, d\mathbf{r}, \tag{6.171}$$

the Fourier transform of the position space electron density, represented in terms of the one-electron wavefunction $\psi(\mathbf{r})$ and the one-particle density matrix $\Gamma_1(\mathbf{r}|\mathbf{r}')$, respectively

$$\rho(\mathbf{r}) = \begin{cases} \psi(\mathbf{r}) \, \psi^*(\mathbf{r}) \\ \Gamma_1(\mathbf{r}|\mathbf{r}), \end{cases} \tag{6.172}$$

where $\mathbf{q}$ is restricted to the discrete reciprocal lattice vectors $\mathbf{g}$ because of the lattice translation symmetry of the crystal lattice. Let us assume that the so-called phase problem of X-ray diffraction can be solved, so that one obtains a set of experimental form factors $F(\mathbf{g})$ for a large number of $\mathbf{g}$'s. Then the position space density $\rho(\mathbf{r})$ can be obtained, at least approximately because of the limited number of $\mathbf{g}$'s, by means of Fourier synthesis, according to

$$\rho(\mathbf{r}) = \frac{1}{V} \sum_{\mathbf{g}} F(\mathbf{g}) \exp(-i\mathbf{g} \cdot \mathbf{r}). \tag{6.173}$$

This procedure of getting the position space density from experiment can serve as a hint for a way to reconstruct the full momentum space density from a set of directional Compton profiles.

To this end, let us define the momentum space analog of the form factor, the so-called reciprocal form factor (Pattison and Williams 1976, Pattison et al. 1977)

$$B(\mathbf{r}) \equiv \int \rho(\mathbf{p}) \exp(-i\mathbf{r} \cdot \mathbf{p}/\hbar) \, d\mathbf{p}, \tag{6.174}$$

the Fourier transform of the momentum space density $\rho(\mathbf{p})$ in one of the representations given in (6.167). One can easily realize that, by choosing the z-axis in the q-direction, the one-dimensional Fourier transform of the Compton profile just gives the reciprocal form factor along that direction:

$$B(0,0,z) = \int J(p_z) \exp(-izp_z/\hbar) \, dp_z. \tag{6.175}$$

Therefore, by measuring Compton profiles for a large number of directions, one finds values of the reciprocal form factor on a fine mesh in position space, so that it becomes feasible to reconstruct the full three-dimensional momentum density by Fourier synthesis (again only approximately because of the finite number of discrete values of $B(\mathbf{r})$):

$$\rho(\mathbf{p}) = \left(\frac{1}{2\pi\hbar}\right)^3 \int B(\mathbf{r}) \exp(i\mathbf{p} \cdot \mathbf{r}/\hbar) \, d\mathbf{r}. \tag{6.176}$$

It is of importance to realize how the reconstruction of the position space density could be achieved by using a momentum space quantity, namely the form factor, and likewise the reconstruction of the momentum space density was attained by utilizing a position space quantity, the reciprocal form factor. Therefore, it is justified to ask whether there exist also relations between both form factors on the one hand and density related quantities defined in their own space on the other hand.

One can easily show by utilizing the well-known convolution theorem of the Fourier transform, that the form factor is given by the autocorrelation of the momentum space wavefunction in the one-electron representation or as the momentum space average of a nondiagonal element of the momentum space one-electron density matrix in the many-electron representation:

$$F(\mathbf{p}) = \begin{cases} \int \chi(\mathbf{p}') \chi^*(\mathbf{p}' + \mathbf{p}) \, d\mathbf{p}' \\ \int \Gamma_1(\mathbf{p}'|\mathbf{p}' + \mathbf{p}) \, d\mathbf{p}'. \end{cases} \qquad (6.177)$$

Likewise one finds that the reciprocal form factor is the autocorrelation of the position space wavefunction and the position space average of a nondiagonal element of the one-electron density matrix in position space, respectively:

$$B(\mathbf{r}) = \begin{cases} \int \psi(\mathbf{r}') \psi^*(\mathbf{r}' + \mathbf{r}) \, d\mathbf{r}' \\ \int \Gamma_1(\mathbf{r}'|\mathbf{r}' + \mathbf{r}) \, d\mathbf{r}'. \end{cases} \qquad (6.178)$$

In Fig. 6.4, which originates from Weyrich *et al.* (1979), all the relationships between various position and momentum space functions mentioned so far are illustrated for the case of a one-electron wavefunction, where FT means Fourier transform, MS denotes modulus squared and AC stands for autocorrelation. Notice that an arrow headed at both ends indicates the reversibility of the procedure which connects the two quantities, whereas a single headed arrow refers to irreversibility. For sake of completeness we have also included the Patterson function $P(\mathbf{r})$, defined in the following equation to be the autocorrelation of the electron density in position space, and its momentum space analog, the reciprocal Patterson function $P(\mathbf{p})$:

$$P(\mathbf{r}) \equiv \int d\mathbf{r}' \, \rho(\mathbf{r}') \rho(\mathbf{r}' + \mathbf{r}), \qquad (6.179)$$

$$P(\mathbf{p}) \equiv \int d\mathbf{p}' \, \rho(\mathbf{p}') \rho(\mathbf{p}' + \mathbf{p}). \qquad (6.180)$$

In Fig. 6.5 basically the same relationships are shown, but now based on the one-particle density matrix in momentum and position space. Here DE (NDE) means diagonal (nondiagonal) element and AV stands for average within the corresponding space.

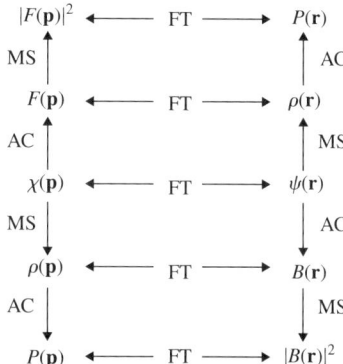

Fig. 6.4. Relationship between various position and momentum space quantities as deduced from wavefunctions. FT means Fourier transform, AC stands for autocorrelation and MS for modulus squared. Double (single) headed arrows indicate reversibility (nonreversibility) of the procedure connecting two quantities.

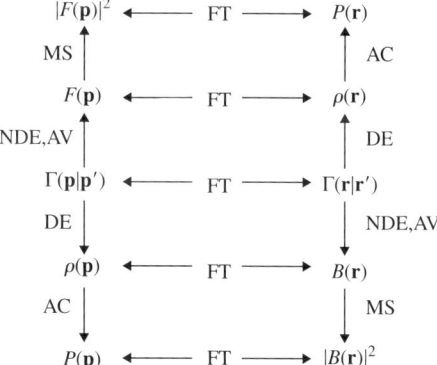

Fig. 6.5. Relationship between various position and momentum space quantities as deduced from one-particle density matrices. FT means Fourier transform, DE stands for diagonal element, and NDE AV for the average within the corresponding space of nondiagonal elements. Double (single) headed arrows indicate reversibility (nonreversibility) of the procedure connecting two quantities.

Summarizing we can state that the measurement of Compton profiles not only yields information about the momentum space, namely about the diagonal elements of its one-particle density matrix, but also limited information about a position space quantity, namely the spatial average of the nondiagonal elements of its density matrix. The information drawn from diffraction experiments is just complementary, knowledge about the diagonal terms of the density matrix in position space is connected with limited information about the density matrix in momentum space in the form of a spatial average of its nondiagonal elements. Although both sources of experimental information about the ground state are complementary, we do not get full knowledge about all one-particle properties of the system under investigation. In order to get the latter, we would need the full one-particle density matrix in any space this means its diagonal and its nondiagonal elements, either in momentum or in position space (Löwdin 1956). As will be shown in Section 6.11, one can fill this gap in the information about the one-particle ground state properties of condensed matter, at least in principle, by an appropriate combination of X-ray diffraction and Compton scattering.

So far we have introduced the momentum space wavefunction $\chi(\mathbf{p})$ in (6.146) and (6.155) to be the Fourier transform of the corresponding position space wavefunction $\psi(\mathbf{r})$. But it is also instructive to understand $\chi(\mathbf{p})$ as the solution of the Schrödinger equation in the momentum space representation

$$\left(\frac{p^2}{2m} - E\right)\chi(\mathbf{p}) + V(i\hbar\boldsymbol{\nabla})\chi(\mathbf{p}) = 0, \tag{6.181}$$

where the kinetic energy operator is a simple multiplicative factor. The potential energy operator $V(i\hbar\boldsymbol{\nabla})$, on the other hand, stands for the Fourier transform of the position space potential energy operator, where the position space coordinate has to be replaced by $i\hbar\boldsymbol{\nabla}$, which makes (6.181) a very complicated integro-differential equation, which has been solved only in a few cases (Fock 1935, McWeeny and Coulson 1949). However, the fact that the kinetic energy operator in (6.181) is simply multiplicative leads directly to the following relation between the momentum space density $\rho(\mathbf{p})$ and the expectation value $\langle T \rangle = \langle p^2/2m \rangle$ of the kinetic energy:

$$\langle T \rangle = \frac{1}{2m}\int \chi(\mathbf{p})p^2\chi^*(\mathbf{p})\,\mathrm{d}\mathbf{p} = \frac{1}{2m}\int \rho(\mathbf{p})p^2\,\mathrm{d}\mathbf{p}. \tag{6.182}$$

It was first pointed out by Hicks (1940) and later on by Epstein (1973) and Weiss (1978) that the above relation between $\langle T \rangle$ and $\rho(\mathbf{p})$ can be transformed into a relation between $\langle T \rangle$ and the second moment of the Compton profile. For an isotropic system this can easily be deduced by inverting (6.152), the relationship between the Compton profile $J(p_q)$ and the radial momentum density $I(p) = 4\pi p^2 \rho(p)$, where $\rho(p)$ is the spherical average of $\rho(\mathbf{p})$, so that we obtain

$$I(p) = -2p\frac{\mathrm{d}J(p)}{\mathrm{d}p}. \tag{6.183}$$

By inserting (6.183) into (6.182) we end up with

$$\langle T \rangle = -\frac{1}{m}\int_0^\infty p^3 \left(\frac{\mathrm{d}J(p)}{\mathrm{d}p}\right)\mathrm{d}p. \tag{6.184}$$

Partial integration together with the assumption $\lim_{p\to\infty} p^3 J(p) = 0$ yields

$$\langle T \rangle = \frac{3}{m} \int_0^\infty p^2 J(p) \mathrm{d}p \qquad (6.185)$$

for isotropic systems. This result can easily be extended to anisotropic systems, as has been pointed out by Weyrich (1978), by inserting

$$p^2 = p_x^2 + p_y^2 + p_z^2; \quad \mathrm{d}\mathbf{p} = \mathrm{d}p_x \mathrm{d}p_y \mathrm{d}p_z \qquad (6.186)$$

into (6.182). Then we obtain

$$\langle T \rangle = \frac{1}{m} \sum_{i=x,y,z} \int_0^\infty p_i^2 J(p_i) \, \mathrm{d}p_i. \qquad (6.187)$$

The physical significance of the above relationship between Compton profiles and expectation values of the kinetic energy follows from the application of the virial theorem, which connects the expectation values of kinetic and potential energy of a system subject only to Coulomb interaction with the total energy E by

$$\langle T \rangle = \frac{-\langle V \rangle}{2} = -E. \qquad (6.188)$$

It was first shown by Benesch and Smith (1971) that the application of the virial theorem together with (6.185) and (6.187), respectively, will offer the possibility, at least in principle, of either determining differences between the total energy of two stable states of a system from differences of the corresponding Compton profiles, or of finding values for the cohesive energy from differences of calculated free atomic and measured condensed matter state Compton profiles. The practical application of this potentiality is limited by the large difficulties with handling the high weight of the high-momentum portion of the Compton profile when calculating the second moment from experimental results, a problem discussed thoroughly by Epstein (1973). But there is still another problem, which must not be forgotten. $\langle T \rangle$ in (6.188) is not only determined by the electrons, which are the subject of (6.185) and (6.187), but must also include the kinetic energy of the nuclei, connected with their thermal motion. Even if this contribution is small when looking at the total kinetic energy, it can be important when energy differences are concerned, since thermal energies and e.g. energy differences between two phases of a solid state system are of the same order of magnitude. Nevertheless, the relationship between Compton profiles and total energy can be rather helpful, when deviations from this relation are used as a measure of the quality of a model calculation of momentum space quantities, as has been proposed by Roux and Epstein (1973). Weiss (1978) used the virial relationship to calculate cohesive energies from theoretical band structure calculated Compton profiles. He found in some cases inadequacies of the band model by the fact that the cohesive energy not only had the wrong magnitude but also the wrong sign.

6.8.3 Momentum densities, Compton profiles and reciprocal form factors in condensed matter physics

After having defined all physical quantities which are related to Compton profile measurements together with their mutual relationship in Section 6.8.2, the following section is devoted to a general understanding of the physics behind the investigation of momentum space densities in atoms, molecules and solids without referring to special systems. We will start with atoms in order to give a feeling for the momentum space representation of wavefunctions and densities for various atomic states, since we are well aware that most readers are much less familiar with momentum space representation than with position space. We will then proceed to molecules, where emphasis is laid on the role which chemical bonds play in momentum space properties and how they will influence the reciprocal form factor. Finally the peculiarities of the solid state are stressed, mainly their band structure, as long as they are relevant for momentum space densities and the reciprocal form factor, respectively. In all cases the influence of electron correlation on the interpretation of momentum space properties is stressed. The more general considerations of this section will help to understand better the experimental and theoretical results for very different systems as presented in other chapters of the book.

6.8.3.1 Atoms

In order to get insight into the physical nature of the momentum density of an isolated atom, let us assume that its electronic state can be represented by an independent-particle model wavefunction, i.e. by a single Slater determinant.

As already found, in this case, the Compton profile can be simply composed of the contributions of the individual orbitals to the momentum density. These individual orbitals might be written in the form of hydrogenic Slater type wavefunctions, separable into radial and angular parts

$$\psi(\mathbf{r}) = f_{nl}(\mathrm{r}) Y_{lm}(\theta, \phi). \tag{6.189}$$

As pointed out by Podolsky and Pauling (1929), the corresponding momentum space wavefunctions $\chi(\mathbf{p})$ retain the same angular dependency

$$\chi(\mathbf{p}) = u_{nl}(p) Y_{lm}(\theta_p, \phi_p), \tag{6.190}$$

so that only the radial part of (6.189) needs to be transformed. This transformation is found from the general relation of (6.148) by utilizing the well-known expansion of a plane wave into spherical harmonics (in the following equations (6.191)–(6.193) atomic units $e = m = \hbar = 1$ are used):

$$u_{nl}(p) = \left(\frac{2}{\pi}\right)^{\frac{1}{2}} (-\mathrm{i})^l \int_0^\infty j_l(pr) f_{nl}(r) r^2 \, \mathrm{d}r, \tag{6.191}$$

where $j_l(pr)$ are spherical Bessel functions.

For further discussion we will list the transforms of the radial parts of a 1s and 2p orbital, respectively:

$$f_{1s}(r) = 2\gamma_{1s}^{3/2} \exp(-\gamma_{1s} r) \qquad u_{1s}(p) = \left(\frac{32\gamma_{1s}^5}{\pi}\right)^{\frac{1}{2}} (\gamma_{1s}^2 + p^2)^{-2} \qquad (6.192)$$

$$f_{2p}(r) = \left(\frac{4\gamma_{2p}^5}{3}\right)^{\frac{1}{2}} r \exp(-\gamma_{2p} r) \qquad u_{2p}(p) = \left(\frac{512\gamma_{2p}^7}{3\pi}\right)^{\frac{1}{2}} i^{-1} p(\gamma_{2p}^2 + p^2)^{-3}.$$

The exponents γ of the basis functions together with the coefficients to be used in analytic wavefunctions expanded in the Roothan–Hartree–Fock method were presented for $Z \leq 54$ by Clementi and Roetti (1974).

The corresponding Compton profiles read as follows

$$J_{1s}(p_q) = \left(\frac{8\gamma_{1s}^5}{3\pi}\right)(\gamma_{1s}^2 + p_q^2)^{-3}$$

$$J_{2p}(p_q) = \left(\frac{512\gamma_{2p}^7}{15\pi}\right)(\gamma_{2p}^2 + 5p_q^2)(\gamma_{2p}^2 + p_q^2)^{-5}.$$

(6.193)

The above examples have the advantage to demonstrate in a manner which can be read directly from the analytical result, the following very general statement: Electron states, characterized by a large value of γ, so that they are well localized in position space, are widely spread in momentum space, and vice versa. Therefore, the order of electron shells with increasing radial coordinate r in position space is K, L, M, ... , whereas the order in momentum space with increasing coordinate p is just the inverse. The fact that the delocalized electron states are found concentrated at small momenta makes Compton measurements, in which momentum space densities are the subject of experiment, very sensitive to all properties, which are determined by the more or less delocalized valence electrons, such as for instance all kinds of chemical bonding. Measurements of the position space electron density as performed with X-ray diffraction, on the other hand, is much less sensitive to valence electron properties, and a very high accuracy is necessary, if, for instance, effects of chemical bonding on X-ray diffraction data should be detected. Additionally, another important feature of the momentum space density can be read from the analytical result of (6.193). The "radial" part of the momentum density of s sates, $|u_s(p)|^2$, always has its maximum at $p = 0$, whereas the momentum density of p states, $|u_p(p)|^2$, starts with the value 0 at $p = 0$ and reaches a maximum value at a certain momentum p_{max}. Due to the two-dimensional integration in momentum space, when calculating Compton profiles from momentum space densities according to (6.151), this possibility to distinguish between s and p states is somewhat washed out and has to be restored by means of a three-dimensional reconstruction procedure.

Atomic Compton profiles numerically calculated using Hartree–Fock wavefunctions for atomic numbers $1 \leq Z \leq 36$ and relativistic Dirac–Hartree–Fock wavefunctions for atomic numbers $36 \leq Z \leq 102$ are presented by Biggs *et al.* (1975).

6.8.3.2 Molecules, bonds

The next step in studying the momentum density and the corresponding observables such as Compton profiles or reciprocal form factors, is to investigate molecular bonds in their most fundamental form, i.e. diatomic bonds. In order to make the physics transparent, we are dealing with, we again choose the simplest representation of such a bond, written as a linear combination of two identical atomic orbitals (molecular orbital (MO) model)

$$\psi(\mathbf{r}) = (2 \pm 2S)^{-1/2} \{\psi_a(\mathbf{r}) \pm \psi_a(\mathbf{r} - \mathbf{R})\}, \tag{6.194}$$

where $\mathbf{R}$ is the internuclear separation and S is the overlap integral defined by

$$S \equiv \int \psi_a(\mathbf{r}) \psi_a^*(\mathbf{r} - \mathbf{R}) \, d\mathbf{r}. \tag{6.195}$$

Provided $S > 0$, the positive sign in (6.194) refers to a bonding, the negative to an antibonding molecular orbital. For $S < 0$ the opposite signs are valid. The charge density of this model molecular orbital is given by

$$e\rho(\mathbf{r}) = (2 \pm 2S)^{-1} e\{|\psi_a(\mathbf{r})|^2 + |\psi_a(\mathbf{r} - \mathbf{R})|^2 \pm \text{Re}[\psi_a(\mathbf{r}) \psi_a^*(\mathbf{r} - \mathbf{R})]\}. \tag{6.196}$$

The interference term in (6.196), which represents the so-called bonding charge when looking to a bonding orbital, will peak at $\mathbf{r} = \mathbf{R}/2$ midway between the nuclear positions of the bond. But be aware that this bonding charge is small compared with the charge $e|\psi_a(0)|^2$ concentrated at the position of each nucleus. Therefore, one needs high accuracy when one has to measure bonding charges in an X-ray diffraction experiment.

As first discussed by Epstein and Tanner (1977), the momentum density corresponding to the model bond of (6.194) is

$$|\chi(\mathbf{p})|^2 = (1 \pm S)^{-1} |\chi_a(\mathbf{p})|^2 [1 \pm \cos(\mathbf{p} \cdot \mathbf{R}/\hbar)], \tag{6.197}$$

where $\chi_a(\mathbf{p})$ is the Fourier transform of the atomic orbital $\psi_a(\mathbf{r})$. First of all, one sees that the bond in momentum space is single-centered, which is a general property of momentum space densities. Furthermore, the build-up in momentum space density is in the direction perpendicular to the bond, since there $\mathbf{p} \cdot \mathbf{R} = 0$. Finally, the momentum space density is modulated with period $2\pi/\mathbf{R}$. The increase in kinetic energy, which has to accompany total energy minimization due to bond formation, should manifest itself in an increased width of the bond Compton profile compared with the sum of the contributing isolated atom profile.

It is also very instructive to consider the reciprocal form factor $B(\mathbf{r})$ of our diatomic model bond. By using (6.174) we obtain

$$B(\mathbf{r}) = (1 \pm S)^{-1}(B_\mathrm{a}(\mathbf{r}) \pm \frac{1}{2}[B_\mathrm{a}(\mathbf{r}+\mathbf{R}) + B_\mathrm{a}(\mathbf{r}-\mathbf{R})]), \qquad (6.198)$$

where $B_\mathrm{a}(\mathbf{r})$ are the atomic reciprocal form factors. As pointed out by Weyrich et al. (1979), $B(\mathbf{r})$ possesses, contrary to the charge density of (6.196), a secondary peak at $\mathbf{r} = \mathbf{R}$, which is of the same order of magnitude as the first peak at $r = 0$. The sign of the extremum will depend on the nature of the wavefunctions. For 1s functions, for example, a positive maximum corresponds to bonding, a negative maximum to antibonding. Since $B(\mathbf{r})$ is an observable that can be obtained by Compton profile measurements, this obviously very elementary example may elucidate the special sensitivity of Compton scattering to chemical bonding.

An alternative model for a covalent bond is the so-called valence bond (VB) model, in which products of atomic orbitals are superimposed:

$$\psi(\mathbf{r}_1,\mathbf{r}_2) = [2(1 \pm S^*S)]^{-1/2}(\psi_\mathrm{a}(\mathbf{r}_1)\psi_\mathrm{b}(\mathbf{r}_2-\mathbf{R}) \pm \psi_\mathrm{a}(\mathbf{r}_2)\psi_\mathrm{b}(\mathbf{r}_1-\mathbf{R})) \quad (6.199)$$

In order to calculate the momentum density and the reciprocal form factor of this model bond, it is useful to perform the integration with respect to $\mathbf{r}_2$ and to convert only the coordinates of electron 1 to momentum space (Coulson 1941). This way one obtains for real ψ_a and ψ_b and for the special case $\psi_\mathrm{a} = \psi_\mathrm{b}$ the following reciprocal form factor of the VB model

$$B(\mathbf{r}) = (1 \pm S^2)^{-1}(B_\mathrm{a}(\mathbf{r}) \pm \frac{S}{2}[B_\mathrm{a}(\mathbf{r}+\mathbf{R}) + B_\mathrm{a}(\mathbf{r}-\mathbf{R})]). \qquad (6.200)$$

Apart from the normalization factor, the VB reciprocal form factor differs from the MO result only by the factor S of the interference term, so that the conclusions reached for the MO $B(\mathbf{r})$ are equally valid for the $B(\mathbf{r})$ from VB wavefunctions.

In practice, it is difficult to observe these effects of bonding on $B(\mathbf{r})$ in the purest form, because experiments cannot be performed on isolated and strictly oriented molecules. Measurements on gases or liquids yield only the spherically averaged part of the momentum space density and of the reciprocal form factor of the bond, respectively. Even in solids there exists rarely a unique orientation of a bond, so that also in this case the anisotropy is diminished by averaging over bond-orientation according to the actual crystal structure.

6.8.3.3 *Solids*
As already shown in Section 6.8.1, a system of free independent electrons at zero temperature is rather simple to treat when its Compton profile is concerned. The Compton profile is built up by slices through the Fermi sphere

of radius p_F in momentum space, which result in an inverted parabola

$$J(p_q) = \frac{3}{4p_F^3}(p_F^2 - p_q^2) \quad \text{for } |p_q| \leq p_F. \tag{6.201}$$

The corresponding reciprocal form factor $B(0,0,z)$ reads

$$B(0,0,z) = \frac{3}{k_F^2 z^2}([\sin(k_F z)/k_F z] - \cos(k_F z)) \tag{6.202}$$

so that its zero passages at $k_F z = 4.493, 7.725, 10.904, \ldots$ are directly related to the Fermi momentum $p_F = \hbar k_F$, and it is noteworthy, as indicated by Pattison and Williams (1976) that even the convolution of the Compton profile with the experimental resolution will not shift the zero passages of $B(0,0,z)$. They can be used as very accurate measures of the Fermi momentum of the free-electron gas.

It is well known (Luttinger 1960) that electron correlation in a Fermi liquid does not destroy the step-like behavior of the Fermi distribution at the Fermi momentum, but reduces the step height to a value smaller then unity, so that the momentum density within the Fermi sphere is diminished compared to that of the independent-particle system, and momentum density tails appear as shown schematically in Fig. 6.6. Correspondingly also the Compton profile is changed, acquiring a tail, but retaining a discontinuity of its slope at $p_z = \pm p_F$.

Let us now switch on the interaction of the valence electrons of a solid with the ion cores of a crystal lattice. Then the momentum density $\rho(\mathbf{p})$ of the inhomogeneous electron system can be expressed in terms of electron field operators $\Psi(\mathbf{r},t)$

$$\rho(\mathbf{p}) = (2\pi\hbar)^{-3} \int d\mathbf{r} \int d\mathbf{r}' \exp[i\mathbf{p}\cdot(\mathbf{r}-\mathbf{r}')/\hbar] \langle \Psi^+(\mathbf{r},0)\Psi(\mathbf{r}',0) \rangle, \tag{6.203}$$

where $\langle \ldots \rangle$ means the thermal average for the system of N electrons in the volume V of the crystal. We expand the field operators into Bloch waves,

$$\Psi(\mathbf{r},t) = \sum_{\mathbf{k}} \sum_{\nu} a_{\mathbf{k},\nu}(t)\,\phi_{\mathbf{k},\nu}(\mathbf{r}). \tag{6.204}$$

$a_{\mathbf{k},\nu}$ annihilates an electron with wavevector $\mathbf{k}$ in the band ν. The Bloch wavefunction, $\phi_{\mathbf{k},\nu}(\mathbf{r})$, can be expanded into plane waves according to

$$\phi_{\mathbf{k},\nu}(\mathbf{r}) = \left(\frac{1}{2\pi\hbar}\right)^{3/2} \sum_{\mathbf{g}} \alpha_\nu(\mathbf{k}+\mathbf{g})\exp[i(\mathbf{k}+\mathbf{g})\cdot\mathbf{r}], \tag{6.205}$$

where $\mathbf{g}$ is a reciprocal lattice vector. The electron momentum density can then be written as follows:

$$\rho(\mathbf{p}) = \sum_{\nu,\nu'} \sum_{\mathbf{k}} \sum_{\mathbf{g}} n_{\nu\nu'}(\mathbf{k})\,\alpha_\nu^*(\mathbf{k}+\mathbf{g})\,\alpha_{\nu'}(\mathbf{k}+\mathbf{g})\,\delta(\mathbf{k}+\mathbf{g}-\mathbf{p}/\hbar). \tag{6.206}$$

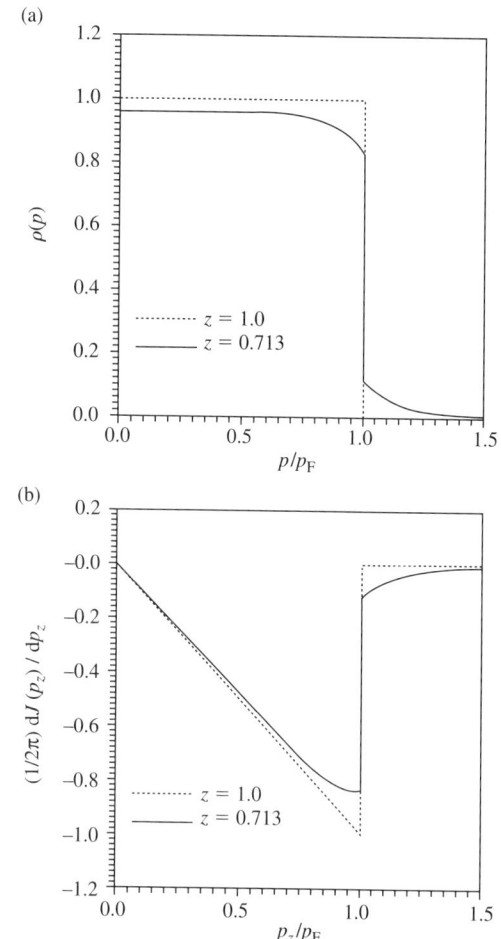

Fig. 6.6. (a) Momentum space electron density $\rho(p)$. Solid line: interacting electron gas with a renormalization constant $z = 0.713$, as calculated according to an interpolation scheme of Schülke et al. (1996); Dashed line: noninteracting electron gas ($z = 1$); p_F = Fermi momentum. (b) First derivative of the free electron Compton profiles corresponding to the momentum space densities of (a).

The function

$$n_{\nu\nu'}(\mathbf{k}) \equiv \langle a^+{}_{\mathbf{k},\nu}(0)\, a_{\mathbf{k},\nu'}(0) \rangle \tag{6.207}$$

can be interpreted as the mean occupation number density of Bloch states, where the nondiagonal elements of the occupation number density are due to mixing between different bands caused by electron–electron interaction. In what follows

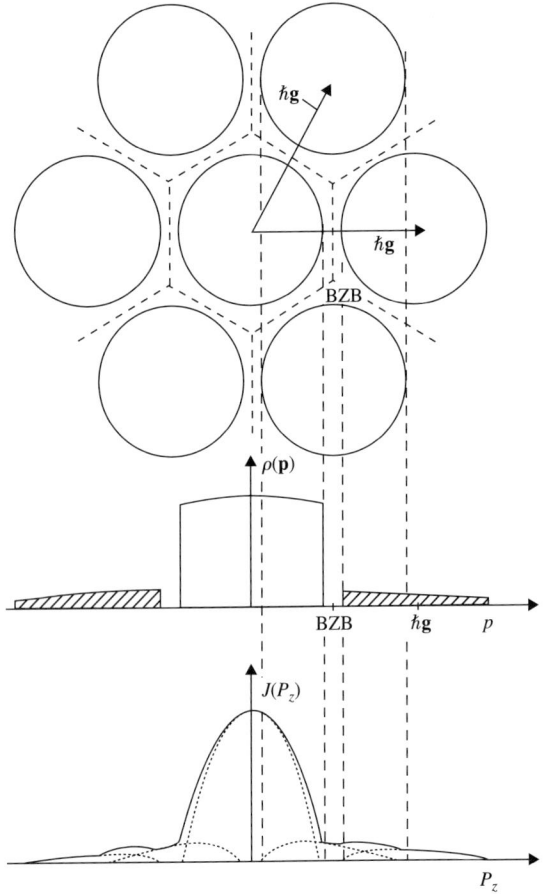

Fig. 6.7. *Upper part:* Momentum space density of a monovalent metal, taking into account electron–ion interaction but neglecting Coulomb correlation of the electrons. Momentum density is accumulated within Fermi spheres centered at each reciprocal lattice point **g**. (BZB = Brillouin-zone boundary). *Middle part:* Valence electron density $\rho(\mathbf{p})$ for one direction of **p**, taking into account electron–ion interaction; the higher momentum components (HMC's) are hatched. *Lower part:* Partial Compton profiles (dashed lines), corresponding to the momentum components represented in the upper part, together with the total Compton profile (solid line).

we shall consider the Bloch electrons to be independent scattering particles so that the nondiagonal elements of the mean occupation number density vanish and the diagonal elements represent the Brillouin zone **k**-space occupation for the different bands ν. They describe the shape of the Fermi surfaces in the case of metals.

Equation (6.206) inserted into (6.151) tells us that the Compton profile of solid state electrons is determined, on the one hand, by the occupation number density, and whenever one can neglect its nondiagonal elements, by the shape of the Fermi surface or that of the Brillouin zone. On the other hand, the Bloch wavefunctions, represented by their plane wave expansion coefficients $a_\nu(\mathbf{k}+\mathbf{g})$, define additionally the shape of the Compton profile. According to (6.206), the contribution of the **g**th plane wave expansion coefficient to $\rho(\mathbf{p})$ is centered at **g** in momentum space, where the $\mathbf{g} \neq \mathbf{0}$ contributions are called higher momentum components (HMC). We demonstrate in Fig. 6.7, for a monovalent metal, that, similar to the Fermi surface around $\mathbf{g} = \mathbf{0}$, the Fermi surfaces around the **g**'s (so-called secondary Fermi surfaces), also produce discontinuities of $\rho(\mathbf{p})$ in the case of metals, causing discontinuities in the first derivative of the Compton profiles according to (6.151). These higher momentum components are the direct consequence of the fact that one measures, in a Bloch state $|\mathbf{k}\rangle$, not only the electron momentum $\hbar\mathbf{k}$ but, with a certain probability given by $|a_\nu(\mathbf{k}+\mathbf{g})|^2$, also momenta $\hbar(\mathbf{k}+\mathbf{g})$.

Let us now consider the reciprocal form factor $B(\mathbf{r})$ of a solid in Bloch representation. By inserting (6.206) into (6.174) we obtain, neglecting the nondiagonal elements of the mean occupation number density,

$$B(\mathbf{r}) = \sum_\nu \sum_\mathbf{g} \sum_\mathbf{k} n_{\nu\nu}(\mathbf{k})\, a_\nu^*(\mathbf{k}+\mathbf{g})\, a_\nu(\mathbf{k}+\mathbf{g}) \exp[i(\mathbf{k}+\mathbf{g})\cdot\mathbf{r}]. \qquad (6.208)$$

If we now look at the values of $B(\mathbf{r})$ at lattice translation vectors **R**, we find, as a consequence of the normalization of the Bloch waves,

$$B(\mathbf{R}) = \sum_\nu \sum_\mathbf{k} n_{\nu\nu}(\mathbf{k}) \exp(i\mathbf{k}\cdot\mathbf{R}). \qquad (6.209)$$

This relation reveals two important properties of $B(\mathbf{R})$ for solids (Schülke 1977): First, for insulators $n_{\nu\nu}(\mathbf{k})$ is either 1 or 0 throughout the whole Brillouin zone, so that $B(\mathbf{R})$ vanishes for all **R**, a very good test for the reliability of theoretical calculations of Compton profiles of insulators. Secondly, for metals the function

$$N(\mathbf{k}) \equiv \sum_\nu n_{\nu\nu}(\mathbf{k}) \qquad (6.210)$$

characterizing the shape of the Fermi surface of monovalent metals can be reconstructed by the following Fourier series:

$$N(\mathbf{k}) = \frac{1}{N_\mathbf{R}} \sum_\mathbf{R} B(\mathbf{R}) \exp(-i\mathbf{k}\cdot\mathbf{R}), \qquad (6.211)$$

where $N_\mathbf{R}$ is the number of **R**-values taken into account in the Fourier series of (6.211). To be successful with this series one needs sufficient $N_\mathbf{R}$, which means measured Compton profiles for many directions of **q**.

6.8.4 *Exchange and correlation corrected momentum density of solids*

Until now we have derived the fundamental relations for momentum density, Compton profiles and reciprocal form factors of solids, assuming scattering by independent particles. This means that the wavefunctions used are solutions of a one-electron Schrödinger equation, of course including some approximate treatment of exchange and correlation in the form of a local potential. It was the Bloch-**k** dependent occupation of these one-electron states and the corresponding plane wave expansion coefficients which determined the above quantities. We shall now sketch an approximate many-electron treatment, which is directly related to the so-called local density approximation (LDA) of solving band structures on the basis of the Kohn–Sham equation (Kohn and Sham 1965)

$$(-\nabla^2 + V_{\text{ext}}(\mathbf{r}) + V_{\text{H}}[\rho](\mathbf{r}) + \delta E_{\text{xc}}[\rho]/\delta\rho(\mathbf{r}))\Phi_i(\mathbf{r}) = \varepsilon_i\Phi_i(\mathbf{r}), \qquad (6.212)$$

where $V_{\text{ext}}(\mathbf{r})$ is the Coulomb potential of the nuclei, $V_{\text{H}}[\rho](\mathbf{r})$ the local Hartree potential, written as a functional of the electron density ρ, and $E_{\text{xc}}[\rho]$ is the nonlocal exchange-correlation energy functional of the electron density. It is well known that the solutions $\Phi_i(\mathbf{r})$ are not one-particle wavefunctions as in the Hartree–Fock method, so that they cannot be used simply to calculate the momentum density according to (6.155). They only constitute the ground-state position space electron density. The local density approximation (LDA) of the exchange-correlation energy functional is based on the corresponding exchange-correlation energy per electron $\varepsilon_{\text{xc}}[\rho(\mathbf{r})]$ of the homogeneous interacting electron gas defined by

$$\varepsilon_{\text{xc}}[\rho(\mathbf{r})] \equiv \varepsilon_0^{\text{h}}[\rho(\mathbf{r})] - \varepsilon_0^{\text{f}}[\rho(\mathbf{r})], \qquad (6.213)$$

where $\varepsilon_0^{\text{h}}[\rho(\mathbf{r})]$ is the ground-state energy per electron of the interacting electron system of density $\rho(\mathbf{r})$, and $\varepsilon_0^{\text{f}}[\rho(\mathbf{r})]$ the ground-state energy of the electron gas of density $\rho(\mathbf{r})$ constituted of noninteracting particles subjected only to the Pauli principle. Using these quantities, which are available in analytical form (see e.g. Hedin and Lundquist 1971), the LDA of the exchange-correlation energy reads

$$E_{\text{xc}}^{\text{LDA}}[\rho] = \int \rho(\mathbf{r})\,\varepsilon_{\text{xc}}[\rho(\mathbf{r})]\,d\mathbf{r}. \qquad (6.214)$$

According to Feynman's theorem (Feynman 1939), as applied to the LDA, the ground-state expectation value of an operator O is composed of the expectation value $O_0[\rho]$ as calculated using a Slater determinant built by solutions $\Phi_i(\mathbf{r})$ of the LDA Kohn–Sham equation, and a correction term $\Delta O[\rho]$, which is given in LDA by

$$\Delta O[\rho] = \int \rho(\mathbf{r})(O_0^{\text{h}}[\rho(\mathbf{r})] - O_0^{\text{f}}[\rho(\mathbf{r})])\,d\mathbf{r}, \qquad (6.215)$$

where O_0^{h} is the expectation values per electron of the operator O for the homogeneous interacting electron gas and O_0^{f} the corresponding quantity for the

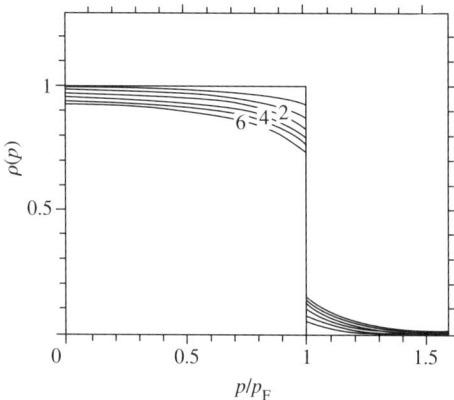

Fig. 6.8. Electron momentum densities of an interacting homogeneous electron system for different density parameters r_s as indicated. (Originally published by Lundqvist (1967b); copyright (1967) by Springer–Verlag, Heidelberg.)

non-interacting electron system. This general relation can now be applied to the momentum density (Lam and Platzman 1974), so that we get finally the desired LDA approximated correction to the momentum density of an interacting solid state electron system, when its ground state "wavefunctions" are solutions of the LDA Kohn–Sham equation:

$$\Delta_\rho^{\mathrm{LDA}}(p) = \int \rho(\mathbf{r})(n_0^{\mathrm{h}}(p)[\rho(\mathbf{r})] - n_0^{\mathrm{f}}(p)[\rho(\mathbf{r})])\,\mathrm{d}\mathbf{r}. \tag{6.216}$$

Here $n_0^{\mathrm{h}}(p)$ and $n_0^{\mathrm{f}}(p)$ are the momentum density of an interacting and a non-interacting homogeneous electron system, respectively, which are available in the literature (Lundqvist 1967a,b, Farid et al. 1993, Barbiellini and Bansil 2001). In Fig. 6.8 we have plotted $\rho(p) \equiv n_0^{\mathrm{h}}(p)$ for different electron density parameters $r_s \equiv (3/4\pi\rho)^{1/3}$, where ρ is the average electron density in atomic units ($\hbar = e = m = 1$).

6.8.5 Magnetic Compton profiles, spin density in momentum space

In Section 6.7 we have derived an expression for the DDSCS (6.117 and 6.118), which contained a magnetic contribution due to both orbital and spin magnetic moments. We found, restricting ourselves to the spin contribution, that the interference term between the charge and the magnetic scattering offers interesting information about the difference of the dynamic structure factors of electrons with opposite spins (6.124 and 6.125), which is experimentally accessible, when using circularly polarized incident radiation. In Section 6.8.1 we have seen how the application of the impulse approximation converts the dynamic structure factor into the Compton profile with its information about the electron momentum density. Thus it suggests itself to extend the impulse approximation to (6.124) and (6.125), leading immediately to the following relation for the DDSCS, as first derived by Platzman and Tzoar (1970):

$$\frac{d^2\sigma}{d\Omega_2 d\hbar\omega_2} = r_0^2 \left(\frac{\omega_2}{\omega_1}\right) \int d\mathbf{p}\{(\mathbf{e}_1 \cdot \mathbf{e}_2^*)^2 (\rho_\uparrow(\mathbf{p}) + \rho_\downarrow(\mathbf{p}))$$

$$+ 2\left(\frac{\hbar\omega_1}{mc^2}\right)(\text{Im}[(\mathbf{e}_1 \cdot \mathbf{e}_2^*)^* D_z])(\rho_\uparrow(\mathbf{p}) - \rho_\downarrow(\mathbf{p}))\}$$

$$\times \delta\left(\hbar\omega - \frac{\hbar^2 q^2}{2m} - \hbar\mathbf{p} \cdot \frac{\mathbf{q}}{m}\right) \quad (6.217)$$

where $\rho_{\downarrow(\uparrow)}(\mathbf{p})$ means the spin-dependent electron momentum density. If the polarization of the scattered photon is not observed, so that one has to average with respect to $\mathbf{e}_2$, then the prefactor of the second term of (6.217) reduces to $(\hbar\omega_1/mc^2)P_2(\cos\theta - 1)(\hat{\mathbf{K}}_1 \cos\theta + \hat{\mathbf{K}}_2)_z$, whereas the prefactor of the first term can be written as $(1/2)(1 + \cos^2\theta + P_3 \sin^2\theta)$. The parameter

$$P_2 \equiv 2\mathbf{e}_x \mathbf{e}_y \sin\delta \quad (6.218)$$

is the second Stokes parameter (Berestetzki et al. 1986), where δ is the phase difference between the oscillations in the x- and y-directions of the incident radiation. The x-direction is perpendicular to the scattering plane, and the y-direction is in this plane. Thus P_2 describes the mean helicity of the incident photon. The parameter

$$P_3 \equiv |\mathbf{e}_x|^2 - |\mathbf{e}_y|^2 \quad (6.219)$$

is the third Stokes parameter and represents the degree of linear polarization. $P_3 = +1(-1)$ means radiation with complete linear polarization perpendicular to (in) the scattering plane. By the way, the first Stokes parameter P_1 characterizes the linear polarization with respect to a direction at an angle of 45° to the x-axis.

In order to obtain the desired information about the momentum distribution of unpaired-spin electrons $[\rho_\uparrow(\mathbf{p}) - \rho_\downarrow(\mathbf{p})]$ or, in other words, information about the spin density in momentum space, one has either to reverse the magnetic field, which magnetizes a ferromagnetic sample, or to reverse the handedness of the circularly polarized incident photon, so that one obtains by recording the difference spectrum

$$\Delta\left[\frac{d^2\sigma}{d\Omega_2 d\hbar\omega_2}\right] = 2r_0^2 \left(\frac{\omega_2}{\omega_1}\right)\left(\frac{\hbar\omega_1}{mc^2}\right) P_2 \cos\theta(\cos\theta - 1) \int d\mathbf{p}[\rho_\uparrow(\mathbf{p}) - \rho_\downarrow(\mathbf{p})]$$

$$\times \delta\left(\hbar\omega - \frac{\hbar^2 q^2}{2m} - \hbar\mathbf{p} \cdot \frac{\mathbf{q}}{m}\right). \quad (6.220)$$

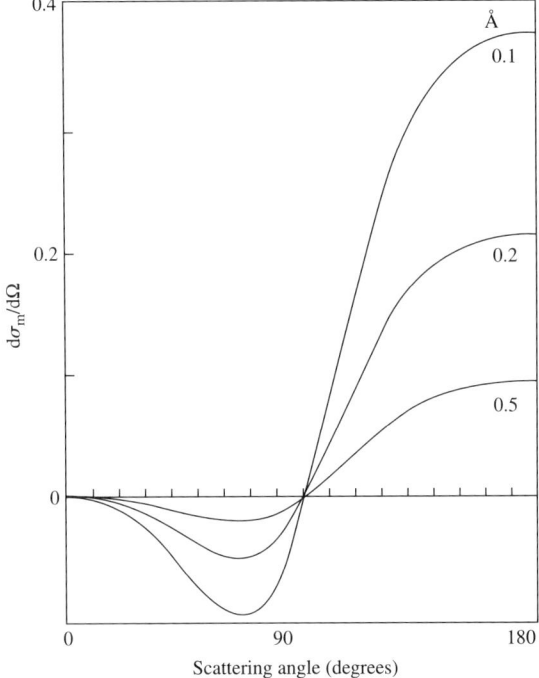

Fig. 6.9. Magnetic scattering cross-section $d\sigma_m/d\Omega$ as a fraction of the spinless cross-section $d\sigma/d\Omega$, plotted as a function of the scattering angle for different incident photon energies (wavelength in Å). (Reprinted with permission from Cooper (1985); copyright (1985) by IOP Publishing.)

The **p** integral in this equation is called the magnetic Compton profile. The equation (6.220) is already specialized to the case in which $\mathbf{K}_1$ is parallel to the quantization axis.

In order to get an impression about the relative magnitude of magnetic Compton scattering, when compared with charge Compton scattering, we have plotted in Fig. 6.9 the magnetic scattering cross-section $d\sigma_m/d\Omega$ as a fraction of the spinless cross-section $d\sigma/d\Omega$ for different incident photon energies (wavelengths) as a function of scattering angle. Having in mind that the density of unpaired electrons is only a small fraction of the total electron density, one sees that the difference spectrum is, even under the optimal condition $P_2 = \pm 1$, only a few percent of the total spectrum.

Remembering that we have so far confined our considerations about magnetic Compton scattering to the spin contribution in (6.117) and (6.118), one may wonder to what extent the orbital term will contribute to magnetic Compton scattering and could also give information about the density in momentum space of the orbital magnetic moments.

In order to estimate the orbital contribution we shall restrict ourselves again to the interference term between the charge and the orbital magnetic scattering in (6.117) and (6.118). For the time being we only consider the one-electron case. By making use of the ingredients of the impulse approximation, namely using plane waves $\exp(i\mathbf{p}_f\cdot\mathbf{r})$ for the final states and cancellation of the potential energy in the energy conserving δ-function, one ends up with the following contribution of orbital magnetic Compton scattering:

$$\left(\frac{d^2\sigma}{d\Omega_2 d\hbar\omega_2}\right)_{\text{orb mag}} = r_0^2 \left(\frac{\omega_2}{\omega_1}\right)\left(\frac{\hbar\omega_1}{mc^2}\right)\sum_{\mathbf{p}}|\langle i|\mathbf{p}\rangle|^2(\mathbf{e}_1\cdot\mathbf{e}_2^*)\left(\mathbf{q}\times\frac{\mathbf{p}}{\hbar\mathbf{K}_1^2}\right)$$

$$\times (\mathbf{e}_2^*\times\mathbf{e}_1)\,\delta[\varepsilon(\mathbf{p}_f) - \varepsilon(\mathbf{p}_f - \hbar\mathbf{q}) - \hbar\omega]. \quad (6.221)$$

Following the discussion of Sakai (1994), we will assume that the atomic orbitals of electrons in the state $|i\rangle$ possess inversion symmetry in space, so that $|\langle i|\mathbf{p}\rangle| = |\langle i|-\mathbf{p}\rangle|$. Therefore, when transforming the $\mathbf{p}$ summation into an integral and taking into account the δ-function, after the integration over p_x and p_y (see the definitions leading to (6.150)), the contribution of (6.221) to the magnetic Compton scattering reduces to zero. Thus it is justified to neglect the orbital contributions to the magnetic scattering in the Compton limit. This was also confirmed experimentally by Timms et al. (1993).

In Sections 6.7 and 6.8.5 we have based our discussion of magnetic (spin dependent) inelastic scattering on a Hamiltonian which contained quasirelativistic correction terms to an extent that spin-dependent terms of the scattering amplitude, which are of higher order than $\hbar\omega_1/mc^2$ or $(v/c)^2$, did not appear or were omitted. It should be mentioned that Grotch et al. (1983) have extended the quasirelativistic Hamiltonian, so that the above limitations are dropped certainly at the price of no longer conserving the direct relationship between cross-section and Compton profile scattering. These quasirelativistic calculations found their subsequent justification by a more rigorous relativistic treatment of spin-dependent Compton scattering by Bhatt et al. (1983). Nevertheless, we will renounce these higher order treatments of magnetic Compton scattering, since they seem to be without practical importance in view of the low intensity of magnetic scattering.

6.9 Relativistic treatment of Compton scattering

The first applications of Compton scattering for studies of electron momentum density were using X-rays with energies in the 10 keV range, so that the recoil electrons were far from being relativistic. But in the early 1970s γ-rays with energies of some 100 keV came into play, in order to obtain a sufficient good momentum space resolution and to extend the range of application to elements of higher atomic numbers. This way a relativistic treatment of the basic relationship between Compton scattering cross-section and electron momentum distribution became indispensable, particularly because we have already used relativistic corrections of the interaction Hamiltonian when dealing with magnetic scattering. Such a relativistic treatment was performed the first time by Eisenberger and Reed (1974). Their approach was a heuristic one, based on an analytic expression of Jauch and Rohrlich (1976), and can be characterized shortly as follows. Taking the relativistic cross-section for colliding beams of electrons and photons, neglecting in the spirit of the impulse approximation binding effects, they considered the case of scattering against stationary wavepackets composed of plane wave states, characterized by their momentum distribution function $\rho(\mathbf{p})$. By revising the flux factor, the double differential scattering cross-section was derived in the form of an integral over free particle scattering events weighted by $\rho(\mathbf{p})$. With some justified approximations it was possible to preserve the direct relationship between double differentials scattering cross-section and the Compton profile, at least for scattering angles very near to 180°. We shall base our presentation on studies of Ribberfors (1975a,b), who has extended the above heuristic approach to cases of arbitrary scattering angles and polarized radiation. This heuristic procedure and the approximations used found their final justification in a rigorous relativistic treatment by Holm (1988), which we will shall present with due conciseness.

6.9.1 *Heuristic approach*

We will use units such that $\hbar = 1$, $c = 1$ in the following sections in order to simplify the equations. The starting point of our derivation of the relativistic double differential Compton scattering cross-section is an expression for the total relativistic scattering cross-section σ for colliding beams, given by Jauch and Rohrlich (1976), where the initial photons of the beam are characterized by the four-vector $\kappa_1 = (\mathbf{K}_1, i\omega_1)$ and the scattering free electron beam consists of electrons described by the four-vector $\pi_1 = (\mathbf{p}_1, iE_1)$. After scattering the electron is in the state $\pi_2 = (\mathbf{p}_2, iE_2)$ and the photon in $\kappa_2 = (\mathbf{K}_2, i\omega_2)$. Then σ reads

$$\sigma = m^2 r_0^2 \int d\mathbf{K}_2 \, d\mathbf{p}_2 (\tfrac{1}{2} k_1 E_2 \omega_2) \, X(k_1, k_2) \, \delta(\pi_1 + \kappa_1 - \pi_2 - \kappa_2), \qquad (6.222)$$

where

$$k_1 = E_1 \omega_1 - \mathbf{p}_1 \cdot \mathbf{K}_1 \qquad (6.223)$$

$$k_2 = E_1 \omega_2 - \mathbf{p}_1 \cdot \mathbf{K}_2 = k_1 + \kappa_1 \kappa_2 = k_1 - \omega_1 \omega_2 (1 - \cos\theta), \qquad (6.224)$$

and

$$X(k_1, k_2) = \left(\frac{k_1}{k_2}\right) + \left(\frac{k_2}{k_1}\right) + 2m^2 \left(\frac{1}{k_1} - \frac{1}{k_2}\right)$$
$$+ m^4 \left(\frac{1}{k_1} - \frac{1}{k_2}\right)^2. \qquad (6.225)$$

The delta function

$$\delta(\pi_1 + \kappa_1 - \pi_2 - \kappa_2) = \delta(\mathbf{p}_1 + \mathbf{K}_1 - \mathbf{p}_2 - \mathbf{K}_2)\,\delta(E_1 + \omega_1 - E_2 - \omega_2) \quad (6.226)$$

represents momentum and energy conservation in the scattering process, so that the projection of $\mathbf{p}_1$ on the scattering vector $\mathbf{q} \equiv \mathbf{K}_1 - \mathbf{K}_2$ can easily be calculated to be

$$p_1 \cos\beta \equiv \frac{\mathbf{p}_1 \cdot \mathbf{q}}{q} = \frac{E_1(\omega_1 - \omega_2) - \omega_1\omega_2(1 - \cos\theta)}{[\omega_1^2 + \omega_2^2 - 2\omega_1\omega_2\cos\theta]^{1/2}}. \quad (6.227)$$

For the time being, we have assumed that neither the polarization of the incident photon nor that of the scattered photon is observed, a case which applies to γ-Compton experiments.

In the spirit of the heuristic approach we are dealing with, and in accordance with the essential points of the impulse approximation, we will now deduce the double differential cross-section for the scattering of photons against electrons in bound states, characterized by the momentum distribution $\rho(\mathbf{p}_1)$, where the scattering system is at rest, which means $\langle \rho(\mathbf{p}) \rangle = 0$. We use $d\mathbf{K}_2 = \omega_2^2 d\omega_2 d\Omega_2$ in (6.222), where $d\Omega_2$ is the solid angle element in the direction of $\mathbf{K}_2$. Moreover, the flux factor $k_1/\mathbf{E}_1\omega_1$ in (6.222), which applies to colliding beams, has to be replaced by one ($c = 1$). In this way we obtain

$$\frac{d^2\sigma}{d\omega_2 d\Omega_2} = \frac{m^2 r_0^2}{2\omega_1} \int d\mathbf{p}_1 d\mathbf{p}_2 \rho(\mathbf{p}_1) \left[\frac{\omega_2 X(k_1, k_2)}{E_1 E_2}\right] \delta(\pi_1 + \kappa_1 - \pi_2 - \kappa_2), \quad (6.228)$$

and by integrating over $\mathbf{p}_2$

$$\frac{d^2\sigma}{d\omega_2 d\Omega_2} = \frac{m^2 r_0^2 \omega_2}{2\omega_1} \int d\mathbf{p}_1 \rho(\mathbf{p}_1) \left[\frac{X(k_1, k_2)}{E_1 E_2}\right] \delta(E_1 + \omega_1 - E_2 - \omega_2). \quad (6.229)$$

If we restrict ourselves for the time being to an isotropic electron momentum distribution, some algebra modifies (6.229), so that we finally obtain:

$$\frac{d^2\sigma}{d\omega_2 d\Omega_2} = \frac{2\pi m^2 r_0^2 \omega_2}{2\omega_1 q} \int_{p_{\min}}^{\infty} dp\, \frac{p\rho(p) X_{\text{int}}}{E(p)}, \quad (6.230)$$

where

$$p_{\min} = \frac{|E(p_{\min})(\omega_1 - \omega_2) - \omega_1\omega_2(1 - \cos\theta)|}{q};$$

$$E(p_{\min}) = (p_{\min}^2 + m^2)^{\frac{1}{2}}, \quad (6.231)$$

and

$$X_{\text{int}} \equiv 2 + F([(E_1 - W - D)^2 - H^2]^{-\frac{1}{2}}$$
$$- [(E_1 - D)^2 - H^2]^{-\frac{1}{2}}) + \frac{m^4}{\omega_1^2}((E_1 - D)[(E_1 - D)^2 - H^2]^{-\frac{3}{2}}$$
$$+ (E_1 - D - W)[(E_1 - D - W)^2 - H^2]^{-\frac{3}{2}}) \quad (6.232)$$

with
$$D \equiv \frac{(\omega_1 - \omega_2 \cos\theta)(p_1 \cos\beta)}{q} \tag{6.233}$$

$$H \equiv \frac{(\omega_2 \sin\theta \sin\beta)}{q} \tag{6.234}$$

$$W \equiv \omega_2(1 - \cos\theta) \tag{6.235}$$

$$F \equiv W - \frac{2m^2}{\omega_1} - \frac{2m^4}{\omega_1^2 W}. \tag{6.236}$$

We shall use these expressions for a comparison with the results of a more rigorous relativistic derivation of the differential Compton scattering cross-section in Section 6.9.2.

Allowing also for anisotropic electron momentum distributions one finds after a lot of algebra (for details the reader is referred to Eisenberger and Reed (1974), and to Ribberfors (1975a,b))

$$\frac{d^2\sigma}{d\omega_2 d\Omega_2} = \frac{mr_0^2 \omega_2}{2\omega_1[q - (\omega_1 - \omega_2)p_{1z}/m]} \int dp_{1x} dp_{1y} \rho(\mathbf{p}_1) X(k_1, k_2), \tag{6.237}$$

where the Cartesian coordinate system has its z-axis parallel to $\mathbf{q}$. In deriving (6.233) we have approximated E_1 by m, so that

$$p_{1z} = \frac{m(\omega_1 - \omega_2) - \omega_1 \omega_2(1 - \cos\theta)}{[\omega_1^2 + \omega_2^2 - 2\omega_1 \omega_2 \cos\theta]^{1/2}} \tag{6.238}$$

and

$$k_1 = m\omega_1 - \mathbf{p}_1 \cdot \mathbf{K}_1 \tag{6.239}$$

$$k_2 = k_1 - \omega_1 \omega_2(1 - \cos\theta). \tag{6.240}$$

With help of a considerable amount of algebra (for details see Ribberfors 1975b) we can bring (6.237) into a form, where, at least in a good approximation, a direct relation between scattering cross-section and Compton profile exists so that an anisotropic Compton profile can be extracted from the measurements:

$$\frac{d^2\sigma}{d\omega_2 d\Omega_2} = \frac{mr_0^2 \omega_2}{(2\omega_1[q - (\omega_1 - \omega_2)p_{1z}/m])} \left(\tilde{X} J(p_{1z}) + C \int_{|p_{1z}|}^{\infty} p\langle J(p)\rangle dp \right), \tag{6.241}$$

where

$$\tilde{X} = \frac{R_1}{R_2} + \frac{R_2}{R_1} + 2m^2 \left(\frac{1}{R_1} - \frac{1}{R_2} \right) + m^4 \left(\frac{1}{R_1} - \frac{1}{R_2} \right)^2 \tag{6.242}$$

with

$$R_1 = \omega_1[m - (\omega_1 - \omega_2\cos\theta)p_{1z}/q] \tag{6.243}$$

$$R_2 = R_1 - \omega_1\omega_2(1-\cos\theta) \tag{6.244}$$

$$C = \left(\frac{\omega_2\sin\vartheta}{q}\right)^2\left[F\left(\frac{1}{(m-D-W)^3} - \frac{1}{(m-D)^3}\right)\right.$$
$$\left. + \left(\frac{3m^4}{\omega_1^2}\right)\left(\frac{1}{(m-D-W)^4} + \frac{1}{(m-D)^4}\right)\right]. \tag{6.245}$$

$\langle J(p_1)\rangle$ is the isotropic Compton profile defined by

$$\langle J(p)\rangle \equiv 2\pi\int_p^\infty p_1\langle\rho(p_1)\rangle\mathrm{d}p_1. \tag{6.246}$$

It has been shown by Ribberfors (1975a) that the second term of (6.241) is very small (of the order p_{1z}/m) compared to the first one, and that it can be neglected in most cases. (For $\theta = 180°$ the second term vanishes completely.) In this sense, (6.241) is the desired result. We are able to calculate anisotropic Compton profiles from experimental double differential scattering cross-section data. But even if we want to calculate the Compton profile $J(p_{1z})$ with higher accuracy, we can use the complete equation (6.241) to get $J(p_{1z})$ by a rapidly converging iteration with

$$J_0(p_{1z}) = \left(\frac{2\omega_1[q-(\omega_1-\omega_2)p_{1z}/m]}{(mr_0^2\omega_2)\widetilde{X}}\right)\frac{\mathrm{d}^2\sigma}{\mathrm{d}\omega_2\mathrm{d}\Omega_2} \tag{6.247}$$

as a first approximation.

For practical purposes we write the prefactor included in [ ]-brackets in (6.247) and p_{1z} in a manner independent of any unit system:

$$[\cdots] = \frac{2\omega_1\hbar[q-(\omega_1-\omega_2)p_{1z}/mc^2]}{(mr_0^2\omega_2\widetilde{X})} \tag{6.248}$$

$$\widetilde{X} = \frac{R_1}{R_2} + \frac{R_2}{R_1} + 2m^2c^4\left(\frac{1}{R_1} - \frac{1}{R_2}\right) + m^4c^8\left(\frac{1}{R_1} - \frac{1}{R_2}\right)^2$$

$$R_1 = \hbar\omega_1[mc^2 - (\omega_1-\omega_2\cos\theta)p_{1z}/q]$$

$$R_2 = R_1 - \hbar^2\omega_1\omega_2(1 - \cos\theta)$$

$$p_{1z} = \frac{mc[(\omega_1 - \omega_2) - \hbar\omega_1\omega_2(1 - \cos\theta)/mc^2]}{[\omega_1^2 + \omega_2^2 - 2\omega_1\omega_2\cos\theta]^{1/2}}, \tag{6.249}$$

which is equal to the nonrelativistic (approximated) expression of equation (6.149).

Let us now investigate the relativistic differential cross-section for Compton scattering of polarized photons. Those studies are of importance, when synchrotron radiation with its special polarization properties is concerned, or multiple scattering processes are to be examined, in which the polarization of the scattered photon becomes the polarization of the incident photon of the secondary process. We have now to start with the Jauch–Rohrlich (1976) expression for polarized photons, where the X-factor in (6.218) has to be written:

$$X(k_1, k_2) = \frac{1}{2}\left[\frac{k_1}{k_2} + \frac{k_2}{k_1}\right] - 1$$

$$+ 2\left(\frac{\mathbf{e}_1 \cdot \mathbf{e}_2 + (\mathbf{e}_1 \cdot \mathbf{p}_1)(\mathbf{e}_2 \cdot \mathbf{p}_2)}{k_1} - \frac{(\mathbf{e}_2 \cdot \mathbf{p}_1)(\mathbf{e}_1 \cdot \mathbf{p}_2)}{k_2}\right)^2. \tag{6.250}$$

If one limits oneself to isotropic momentum distributions one ends up, after a considerable amount of algebra and by neglecting terms of an order of magnitude comparable with the second term of (6.241), with

$$\frac{d^2\sigma}{d\omega_2 d\Omega_2} = \left(\frac{mr_0^2\omega_2}{2\omega_1 q}\right) J(|p_{1z}|)X, \tag{6.251}$$

where

$$X = \frac{1}{2}\left(\frac{R_1}{R_2} + \frac{R_2}{R_1}\right) - 1 + 2(\mathbf{e}_1 \cdot \mathbf{e}_2)^2. \tag{6.252}$$

It is of importance to know (Ribberfors 1975b) that the X-factor of (6.250) applies also to anisotropic systems, if the very small $1/k_1$ and $1/k_2$ terms of (6.250) are neglected.

6.9.2 *Relativistic differential Compton cross-section for central-field HF wavefunctions*

In order to gain confidence in the heuristic approach of Section 6.9.1 we will follow Holm (1988) and sketch the derivation of the differential Compton scattering cross-section in a rigorous relativistic treatment within the limits of the impulse approximation. We will restrict ourselves to electron states in closed shells described by relativistic central-field Hartree–Fock wavefunctions, so that we can compare the results of this calculation with the differential scattering cross-section for isotropic momentum distributions as presented above in (6.230).

The relativistic perturbation Hamiltonian, which, in the Dirac theory, describes the interaction of the electron system with the transverse electromagnetic field $\mathbf{A}(t)$, can be written as (Sakurai 1982):

$$H_I(t) = -e\boldsymbol{\alpha} \cdot \mathbf{A}(t), \tag{6.253}$$

where

$$\boldsymbol{\alpha} = i\gamma_4 \boldsymbol{\gamma} = i\gamma_4(\gamma_1, \gamma_2, \gamma_3) \tag{6.254}$$

and $\gamma_\mu (\mu = 1, 2, 3, 4)$ stands for a gamma matrix. The vector potential $\mathbf{A}(t)$ can be represented as

$$\mathbf{A}(t) = \sum_{\mathbf{K},\alpha} \left(\frac{1}{2V\omega_1}\right)^{1/2} (a_{\mathbf{K}\alpha} \mathbf{e}^{(\alpha)} \exp[i(\mathbf{K} \cdot \mathbf{r} - \omega t)]$$

$$+ a_{\mathbf{K}\alpha}^+ \mathbf{e}^{(\alpha)} \exp[-i(\mathbf{K} \cdot \mathbf{r} - \omega t)]), \tag{6.255}$$

where V is the normalization volume, and $\mathbf{e}^{(\alpha)}$ is the polarization vector of the transverse photon. $a_{\mathbf{K}\alpha}$ and $a_{\mathbf{K}\alpha}^+$ are, respectively, the annihilation and creation operators for a transverse photon.

In the Dirac theory, Compton scattering is described in second-order perturbation theory, so that the second-order transition amplitude $S_{\text{fi}}^{(2)}(\tau)$ is given by

$$S_{\text{fi}}^{(2)}(\tau) = -\sum_n \int_0^\tau dt \int_0^t dt' \langle f|H_I(t)|n\rangle \exp[i(E_{\text{f}} - E_n)t]$$

$$\times \langle n|H_I(t')|i\rangle \exp[i(E_n - E_{\text{i}})t'], \tag{6.256}$$

where $|f\rangle$, $|n\rangle$ and $|i\rangle$ represent the final, intermediate and initial states, respectively, τ is the time duration of the perturbation, and E_{f}, E_n and E_{i} are the total relativistic energies of the electron system in the final, intermediate and initial state, respectively.

Assuming that the spectator electrons have the same states in the initial and final states, we only have to consider the electron which is involved in the scattering process. Moreover, the electron and the photon are acting in different subspaces, so that we can ignore the intermediate state so far as the photon is concerned. Having these points in mind, and utilizing eigenvalue equations of the form $\exp(iHt)|i\rangle = \exp(iE_{\text{i}}t)|i\rangle$, where H is the relativistic Hamiltonian of the electron, we obtain the following expression for the second-order transition amplitude, utilizing the completeness of the intermediate states

$$S_{\text{fi}}^{(2)}(\tau) = -\left(\frac{e^2}{2V}\right)(\omega_1\omega_2)^{-1/2} \int_0^\tau dt \int_0^t dt' [C_0(H, \mathbf{K}_1, \mathbf{K}_2, \omega_1, \omega_2, \mathbf{e}_1, \mathbf{e}_2)$$

$$+ C_0(H, -\mathbf{K}_1, -\mathbf{K}_2, -\omega_1, -\omega_2, \mathbf{e}_1, \mathbf{e}_2)], \tag{6.257}$$

where

$$C_0(H, \mathbf{K}_1, \mathbf{K}_2, \omega_1, \omega_2, \mathbf{e}_1, \mathbf{e}_2)$$
$$= \langle \psi_f | \exp(iHt)(\boldsymbol{\alpha} \cdot \mathbf{e}_1) \exp([i(\mathbf{K}_1 \cdot \mathbf{r} - \omega_1 t)] \exp(-iHt) \exp(iHt')$$
$$\times (\boldsymbol{\alpha} \cdot \mathbf{e}_2) \exp([-i(\mathbf{K}_2 \cdot \mathbf{r} - \omega_2 t')] \exp(-iHt') | \psi_i \rangle. \tag{6.258}$$

ψ_f and ψ_i is the wavefunction of the electron in the final and in the initial state, respectively.

We now follow the way how the impulse approximation has been introduced in Section 6.8.1, namely by approximating $\exp(iHt)$ and $\exp(-iHt)$ in (6.254) by $\exp(iH_0 t)\exp(iV(\mathbf{r})t)$ and $\exp(-iV(\mathbf{r})t)\exp(-iH_0 t)$, respectively. H_0 is the free-electron Hamiltonian and $V(\mathbf{r})$ the potential, $H = H_0 + V(\mathbf{r})$. In this way the potential vanishes and we obtain

$$S_{fi}^{(2)}(\tau) = -\left(\frac{e^2}{2V}\right)(\omega_1\omega_2)^{-\frac{1}{2}} \int_0^\tau dt \int_0^t dt' [C_0(H_0, \mathbf{K}_1, \mathbf{K}_2, \omega_1, \omega_2, \varepsilon_1, \varepsilon_2)$$
$$+ C_0(H_0, -\mathbf{K}_1, -\mathbf{K}_2, -\omega_1, -\omega_2, \varepsilon_1, \varepsilon_2)]. \tag{6.259}$$

In order to be able to move the exponential energy operators outside the matrix element, we have to expand the initial state into plane waves, to insert a complete set of plane waves as intermediate states and to assume that the final state of the electron is a plane wave. Whereas the first two procedures are not approximations the third step is approximate, and can only be justified when the energy transfer to the electron is large compared to the binding energy of the initial state. To have an impression of what the expansion of a relativistic electron state into plane waves looks like, we show the expansion of the initial state:

$$\psi_i(\mathbf{r}) = \sum_{\mathbf{p}_1, s_1} \left(\frac{m}{VE}\right)^{\frac{1}{2}} [A_{\mathbf{p}_1 s_1} u^{(s)}(\mathbf{p}_1) \exp(i\mathbf{p}_1 \cdot \mathbf{r}) + B_{\mathbf{p}_1 s_1} v^{(s)}(\mathbf{p}_1) \exp(-i\mathbf{p}_1 \cdot \mathbf{r})], \tag{6.260}$$

where

$$A_{\mathbf{p}_1 s_1} = \left(\frac{m}{VE}\right)^{\frac{1}{2}} \int d\mathbf{r} u^{(s)+}(\mathbf{p}_1) \exp(-i\mathbf{p}_1 \cdot \mathbf{r}) \psi_i(\mathbf{r}) \tag{6.261}$$

$$B_{\mathbf{p}_1 s_1} = \left(\frac{m}{VE}\right)^{\frac{1}{2}} \int d\mathbf{r} v^{(s)+}(\mathbf{p}_1) \exp(i\mathbf{p}_1 \cdot \mathbf{r}) \psi_i(\mathbf{r}). \tag{6.262}$$

$u^{(s)}(\mathbf{p})$, with $s = 1, 2$ are the four-component electron spinors, $v^{(s)}(\mathbf{p})$, with $s = 1, 2$ are the four-component "positron spinors", and E is the relativistic energy of a free particle, $E = (m^2 + p^2)^{1/2}$. The derivation then follows the lines indicated above, using the relation for the total cross-section σ

$$\sigma = \sum_{\mathbf{p}_2,s_2} \sum_{\mathbf{K}_2} \frac{V}{\tau} |S_{fi}^{(2)}(\tau)|^2, \qquad (6.263)$$

and utilizing $d\mathbf{K}_2 = \omega_2^2 d\omega_2 d\Omega_2$ in order to pass over from the total to the double differential scattering cross-section. One then takes the limit $\tau \to \infty$, going through some lengthy algebra, using trace techniques employed earlier by Sakurai (1982), and introducing the Fourier transforms $\psi_{nljm_j}(\mathbf{p})$ of the general relativistic wavefunctions in the central-field approximation. The final result is

$$\frac{d^2\sigma}{d\omega_2 d\Omega_2} = \left(\frac{m^2 r_0^2 \omega_2}{2\omega_1 q}\right) \int_{p_{\min}}^{\infty} \left(\frac{X_{\text{int}}(p)}{E(p)}\right)\left(\frac{1}{4}\right)(2j+1)\left(\left[1+\left(\frac{m}{E_1}\right)\right]^{\frac{1}{2}} \chi_{nlj}^{G}(p)\right.$$

$$\left. \pm(-1)\left[1-\left(\frac{m}{E_1}\right)\right]^{\frac{1}{2}} \chi_{nlj}^{F}(p)\right)^2 p\,dp. \qquad (6.264)$$

The positive sign is valid for $j = l + \frac{1}{2}$, the negative for $j = l - \frac{1}{2}$. $\chi_{nlj}^{G}(p)$ and $\chi_{nlj}^{F}(p)$ are constituents of $\psi_{nljm_j}(\mathbf{p})$ and defined as follows

$$\chi_{nlj}^{G}(p) = 4\pi \left(\frac{1}{2\pi}\right)^{\frac{3}{2}} \int_0^{\infty} g_{nlj}(r) j_l(pr) r^2 \, dr \qquad (6.265)$$

$$\chi_{nlj}^{F}(p) = 4\pi \left(\frac{1}{2\pi}\right)^{\frac{3}{2}} \int_0^{\infty} f_{nlj}(r) j_{l'}(pr) r^2 \, dr, \qquad (6.266)$$

where j_l and $j_{l'}$ stand for spherical Bessel functions. The $g_{nlj}(r)$ and $f_{nlj}(r)$ are the radial parts of $\psi_{nljm_j}(\mathbf{r})$, the general relativistic wavefunctions in the central-field approximation, according to

$$\psi_{nljm_j}(\mathbf{r}) = \begin{bmatrix} g_{nlj}(r) & 0 \\ 0 & i f_{nlj}(r) \end{bmatrix} \begin{bmatrix} y_{jl}^{m_j}(\theta,\phi) \\ y_{jl'}^{m_j}(\theta,\phi) \end{bmatrix}, \qquad (6.267)$$

where the quantum number l' can be expressed in terms of the other quantum numbers as follows:

$$l' = \begin{cases} l+1 & \text{when } j = l + \frac{1}{2} \ (l \geq 0) \\ l-1 & \text{when } j = l - \frac{1}{2} \ (l > 0). \end{cases} \qquad (6.268)$$

The function $y_{jl}^{m_j}(\theta,\phi)$ is connected with the spherical harmonics $Y(\theta,\phi)$ by:

$$y_{jl}^{m_j}(\theta,\phi) = (2l+1)^{-\frac{1}{2}} \begin{bmatrix} (l + m_j + \frac{1}{2})^{\frac{1}{2}} Y_{l,m_j-\frac{1}{2}}(\theta,\phi) \\ (l - m_j + \frac{1}{2})^{\frac{1}{2}} Y_{l,m_j+\frac{1}{2}}(\theta,\phi) \end{bmatrix} \qquad (6.269)$$

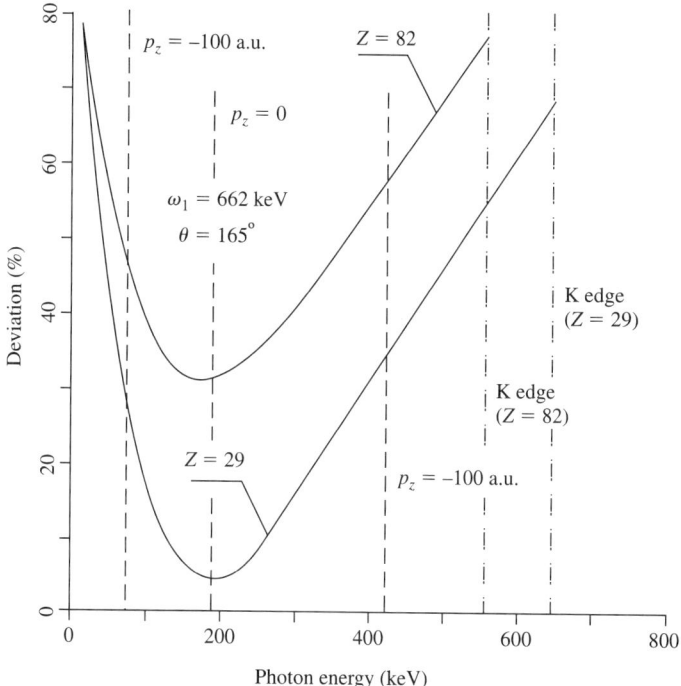

Fig. 6.10. Relative deviation (in %) of the expression (6.241) (heuristic approach) from the expression (6.264) for hydrogen-like systems in the ground state with different atomic numbers Z as a function of the scattered photon energy. Corresponding p_z values together with the position of the K edges are indicated. (Originally published by Holm (1988); copyright (1988) by the American Physical Society.)

when $j = l + \frac{1}{2}$, and

$$y_{jl}^{m_j}(\theta, \phi) = (2l+1)^{-\frac{1}{2}} \begin{bmatrix} -\left(l - m_j + \frac{1}{2}\right)^{\frac{1}{2}} Y_{l,m_j-\frac{1}{2}}(\theta, \phi) \\ \left(l + m_j + \frac{1}{2}\right)^{\frac{1}{2}} Y_{l,m_j+\frac{1}{2}}(\theta, \phi) \end{bmatrix} \quad (6.270)$$

when $j = l - \frac{1}{2}$.

Equation (6.264) refers to the case where the polarization of the photons is not observed. Therefore it can directly be compared with (6.230).

To give a typical example for such a comparison, the relative difference (in percent) between the expression (6.264) and (6.241) is plotted in Figs. 6.10 and 6.11 as a function of the photon energy of the scattered radiation, where the incident photon energy was 662 keV and the scattering angle 165°. The momentum transform used in both equations corresponds to the radial part of a relativistic wavefunction for a hydrogen-like 1s state (Mukoyama 1982). The calculation are for Cu ($Z = 29$) and Pb ($Z = 82$). The scattered photon energies corresponding to different values of p_z, as well as the position of the K edges, are indicated. One can conclude that, at least for scattering angles larger than 100°, for elements with lower Z, and within a small p_z range around $p_z = 0$ the deviations of the heuristic approach from the more rigorous relativistic treatment of the differential Compton scattering cross-section are not too large, so that the application of the heuristic scheme, often described in the literature, seems to be justified. Furthermore, the deviation becomes still much smaller for electron states with lower binding energy. Moreover, the rigorous relativistic treatment of Holm (1988) provides a good reference, when another problem is concerned, namely, to what extent the concept of a Compton profile, directly related to measurements of the double differential scattering cross-section, will survive the

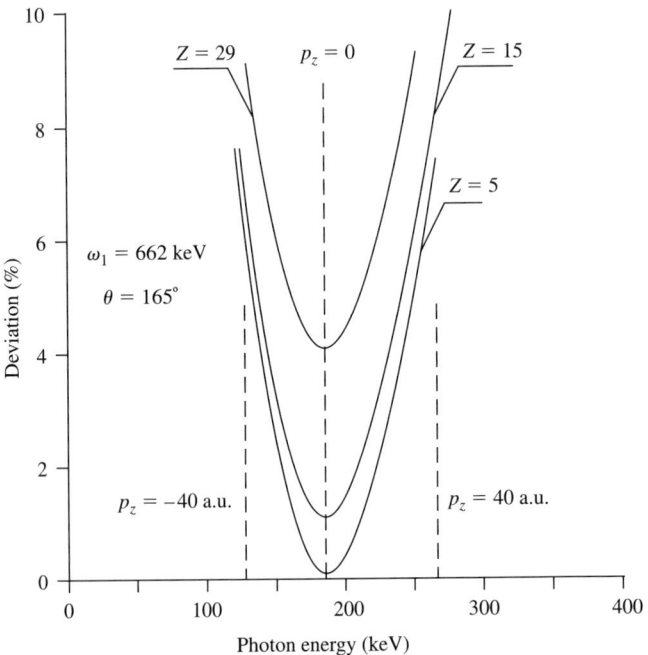

Fig. 6.11. Relative deviation of the expression, (6.241) with the second term neglected, from equation (6.264) plotted as a function of the scattered photon energy. The same wavefunctions were used as for Fig. 6.10. Corresponding p_z values are indicated. (Originally published by Holm (1988); copyright (1988) by the American Physical Society.)

relativistic treatment. One realizes that the approximations which enable us to retain the concept of a Compton profile are valid, at least for such high incident photon energies, only for electrons with lower binding energies and mainly near $p_z = 0$. Of course, what is important is not the rather big difference at $p_z = 0$ for high Z elements, since Compton profiles are not derived from differential scattering cross-sections measured in absolute units, but from measurements in arbitrary units (counts), which were renormalized according to the theoretical norm of the contributing Compton profiles. Therefore, what counts are the variations of the differences with p_z, which can be read from Figs. 6.10 and 6.11. These are not that dramatic in a narrow range around $p_z = 0$, which is used for the renormalization in most experiments.

Nevertheless, it seems worthwhile to test in every case the validity of those approximations, which enables us to express the results of inelastic scattering experiments in simple terms of Compton profiles. The equations presented above should give a sufficient formal background to do this.

6.10 Resonant inelastic scattering

Until now we have discussed the generalized Kramers–Heisenberg formula (see 6.21) only under the assumption that the incident photon energy is far from the excitation energy of an inner-shell level. If we now, contrary to this assumption, let the incident photon energy fall into the direct vicinity of an inner-shell excitation energy, the third, the resonant part of (6.21), becomes dominant, so that we can neglect the other parts and write for the spin-independent part, which shall solely be analyzed

$$\frac{d^2\sigma}{d\Omega_2 d\hbar\omega_2} = \left(\frac{e^2}{mc^2}\right)^2 \left(\frac{\omega_2}{\omega_1}\right) \Bigg| \sum_f \frac{\hbar^2}{m} \sum_n \sum_{jj'} \langle f| \mathbf{e}_2 \cdot \mathbf{p}_j/\hbar \, \exp(-i\mathbf{K}_2 \cdot \mathbf{r}_j)|n\rangle$$

$$\times \langle n| \mathbf{e}_1 \cdot \mathbf{p}_{j'}/\hbar \, \exp(i\mathbf{K}_1 \cdot \mathbf{r}_j)|i\rangle / (E_i - E_n + \hbar\omega_1 - i\Gamma_n/2) \Bigg|^2$$

$$\times \delta(E_i - E_f + \hbar\omega). \tag{6.271}$$

This equation determines the resonant inelastic X-ray scattering (RIXS) spectrum, especially its dependence on the incident and re-emitted photon energy and polarization, on the scattering geometry (momentum conservation), and on the type of excited intermediate and final state (selection rules due to angular momentum conservation). We must always have in mind that RIXS has to be considered as a coherent coupling of the (virtual) absorption with the re-emission process as indicated by the two matrix elements in (6.271). The consequences of this most important aspect of RIXS shall be discussed in the following subsections.

6.10.1 Dependence of the energy position of the RIXS spectra on the incident photon energy (Raman shift)

Removing unimportant factors in (6.271) the essential part of the RIXS spectrum can be described in the following form

$$F(\omega_1,\omega_2) = \sum_f \left| \sum_n \frac{\langle f|T_2^+|n\rangle\langle n|T_1|i\rangle}{E_i - E_n + \hbar\omega_1 - (i\Gamma_n/2)} \right|^2 \delta(E_i - E_f + \hbar\omega_1 - \hbar\omega_2), \quad (6.272)$$

where the operators T_1 and T_2 represent photon excited and radiative transitions, respectively. Γ_n denotes the spectral broadening due to the core-hole lifetime of the intermediate state $|n\rangle$ as a result of the Auger and radiative decays of the core hole, which can be taken approximately to be constant, independent of the index n. In order to see how the energy position $\hbar\omega_2$ of the emitted photon generally depends on the incident photon energy, we assume exemplarily that in the intermediate state $|n\rangle$ a core electron, let say a 1s electron, is excited above the Fermi level E_F into the conduction band of a metal, and that in the final state a shallow core electron, let say a 2p electron, makes a radiative transition to the 1s state. If we change the incident energy $\hbar\omega_1$ continuously around $E_F - E_K$, the binding energy of the 1s electron, and observe the emitted photon around $E_F - E_L$, where $E_F - E_L$ is the binding energy of the 2p electron, then $F(\omega_1,\omega_2)$ is written, apart from factors unimportant in the context of this consideration, disregarding electron–electron interaction, as

$$F(\omega_1,\omega_2) \sim \sum_{\mathbf{k}(k>k_F)} \left| \frac{1}{E_K - E_\mathbf{k} + \hbar\omega_1 - (i\Gamma_K/2)} \right|^2 \delta(E_L - E_\mathbf{k} + \hbar\omega_1 - \hbar\omega_2),$$

$$= \begin{cases} \dfrac{\rho}{(\Delta E_{\text{emit}})^2 + (\Gamma_K^2/4)} & \text{for } \Delta E_{\text{emit}} \leq \Delta E_{\text{inc}} \\ 0 & \text{for } \Delta E_{\text{emit}} > \Delta E_{\text{inc}} \end{cases}, \quad (6.273)$$

where

$$\Delta E_{\text{inc}} = \hbar\omega_1 - (E_F - E_K) \quad (6.274)$$

$$\Delta E_{\text{emit}} = \hbar\omega_2 - (E_L - E_K) \quad (6.275)$$

Here Γ_K is the lifetime broadening of the 1s core level, and ρ the density of states of the conduction band. One sees that for $\Delta E_{\text{inc}} < 0$ the peak of $F(\omega_1,\omega_2)$ occurs at $\Delta E_{\text{emit}} = \Delta E_{\text{inc}}$, so that $\hbar\omega_2$ increases by the same amount as $\hbar\omega_1$ increases. We will call this shift of the emitted photon energy the Raman shift. But since for $\Delta E_{\text{inc}} > 0$ the peak of $F(\omega_1,\omega_2)$ occurs at $\Delta E_{\text{emit}} = 0$, the Raman shift disappears, the reemitted photon energy levels out to its value known for excitation energies far beyond the 1s core binding energy. Furthermore, the full width at half maximum (FWHM) of the re-emitted spectral line has its minimum value for $\Delta E_{\text{inc}} = 0$. Its shape is only half of the Lorentzian with FWHM = $2\Gamma_K$, which determines the shape of the emitted line for excitation energies far beyond the 1s binding energy (line narrowing). The experimental results of Eisenberger *et al.* (1976a, b) confirmed these theoretical predictions. Until now we have assumed that in the intermediate state a core electron is excited into a conduction band with continues energy distribution. Let us now, on the contrary, assume the intermediate state consists of a core electron excited into a discrete unoccupied level with energy E_d, so that E_d replaces E_F in (6.274). Then the Raman shift persists also for $\Delta E_{\text{inc}} > 0$, where, of course, the intensity of the corresponding emission line is fading out with increasing incident photon energy.

6.10.2 *High-resolution edge spectroscopy by means of RIXS*

Normal X-ray absorption spectroscopy (**X**-ray **a**bsorption **n**ear **e**dge **s**pectroscopy XANES) with the aim to obtain symmetry selected information about the density of unoccupied states is limited in resolution by the core-hole lifetime width Γ_n. The absorption probability as a function of the incident photon energy, as measured for instance by means of the total fluorescence yield, is given by

$$I(\omega_1) = \sum \frac{|\langle n|T_1|i\rangle|^2(\Gamma_n/2\pi)}{(E_n - E_i - \hbar\omega_1)^2 + \Gamma_n^2/4}, \tag{6.276}$$

so that the resolution of the spectra are limited by the lifetime width Γ_n of the core hole. On the other hand, the essential part of the RIXS spectrum as a function of $\hbar\omega_1$, according to (6.272), taken at a certain value $\hbar\omega_2$ of the re-emitted spectrum, can be written as

$$I(\omega_1, \omega_2) = \sum_f \left| \sum_n \frac{\langle f|T_2^+|n\rangle\langle n|T_1|i\rangle}{E_i - E_n + \hbar\omega_1 - (i\Gamma_n/2)} \right|^2$$

$$\times \frac{\Gamma_f/\pi}{(E_f + \hbar\omega_2 - E_i - \hbar\omega_1)^2 + \Gamma_f^2/4}, \tag{6.277}$$

where the energy-conservation δ-function of (6.272) was broadened into a Lorentzian of FWHM Γ_f, in order to account for the finite lifetime of the final states. Following Tanaka *et al.* (1994) and Carra *et al.* (1995) we set $\hbar\omega_2$ at the peak of the fluorescence emission. Moreover, we assume the interaction of the core holes corresponding to $|n\rangle$ and $|f\rangle$ with each other and with the valence states to be negligible, so that the matrix element $\langle f|T_2^+|n\rangle$ is independent of the energy position in the conduction band the core-electron is excited into, and can be removed out of the summation over n. Furthermore, we have to consider only one final state, so that the summation over f can be skipped, and $E_f + \hbar\omega_2$ can be set E_n. Then the essential part of (6.277) reads

$$I(\omega_1) = \sum_n |\langle n|T_1|i\rangle|^2 \left\{ \frac{1}{(E_i - E_n + \hbar\omega_1)^2 + \Gamma_n^2/4} \right\}$$

$$\times \left\{ \frac{\Gamma_f/\pi}{(E_n - E_i - \hbar\omega_1)^2 + \Gamma_f^2/4} \right\}. \tag{6.278}$$

If one considers this expression for the case $\Gamma_f \ll \Gamma_n$, the intensity of the RIXS emission as a function of the incident energy can be regarded as essentially the same as the absorption spectrum of (6.276) with a fictitious spectral broadening Γ_f. This possibility to overcome in this way the resolution limit of conventional absorption spectroscopy given by the core-hole lifetime broadening Γ_n, provided the spectral distribution of the incident radiation together with the spectrometer resolution of the analyzer are small enough, has already been proposed by Hämäläinen et al. (1991), but without indicating the restrictions formulated above.

6.10.3 Multipolar expansion of RIXS cross-sections, sum rules

The huge amount of information, as offered by (6.271), about the full electron–electron interaction in systems under investigation, for instance in rare-earth systems with the nearly localized 4f electrons and the broad 5d bands, can be fully analyzed by means of a multipolar expansion of the DDSCS as proposed by van Veenendaal et al. (1996), whose notation is used in what follows. According to (6.271), the RIXS amplitude $f_{LL'}$ is calculated by means of a coupled multipolar expansion in the limit $Kr \ll 1$ near the resonance (for details of the derivation see van Veenendaal et al. 1996) as a linear combination of pairs of tensors of increasing rank z:

$$f_{LL'} = 4\pi K_1^{-1} \sum_z \sum_{\zeta=-z}^{z} T_\zeta^{(z)*}(\mathbf{e}_1, \mathbf{K}_1, \mathbf{e}_2^*, \mathbf{K}_2) \langle f|F_\zeta^{(z)}(L, K_1^{-1}, L', K_2^{-1})|i\rangle, \tag{6.279}$$

where L and L' is the change of the total angular momentum connected with the transition $|i\rangle \to |n\rangle$ and $|n\rangle \to |f\rangle$, respectively. The angular factor on the right-hand side of (6.279) is given by

$$T_\zeta^{(z)*}(L, L') = \sum_{M,M'} C_{L'M'LM}^{z\zeta} [\mathbf{e}_1 \cdot \mathbf{Y}_{LM}^*(\hat{\mathbf{K}}_1)][\mathbf{e}_2^* \cdot \mathbf{Y}_{L'M'}(\hat{\mathbf{K}}_2)], \tag{6.280}$$

with the coupling defined via the well-known Clebsch–Gordan coefficients and $\mathbf{Y}_{LM}(\hat{\mathbf{K}})$ is a vector spherical harmonic of electric type (see Akhiezer and

Berestetsky 1957). The frequency-dependent transition operator in (6.279) takes the form

$$F_\zeta^{(z)}(L, K_1^{-1}, L', K_2^{-1}) = R_{LK_1^{-1}}^{L', K_2^{-1}}(c_1, l; c_2, c_1)$$

$$\times \sum_{l_z, \sigma, \text{all } m} S_\zeta^{(z)}(L, L') c_{j_1 m_1}^+, c_{j_2 m_2} G(\omega_1) l_{l_z \sigma}^+ c_{j_1 m_1}$$

(6.281)

with

$$G(\omega_1) = \sum_n |n\rangle (E_i + \hbar\omega_1 - E_n + (i\Gamma_n/2))^{-1} \langle n| \qquad (6.282)$$

and

$$R_{LK_1^{-1}}^{L', K_2^{-1}}(c_1, l; c_2, c_1) = K(c_1, L, l, K_1^{-1}) K(c_2, L', c_1, K_2^{-1}) \langle R_{n_1 c_1 j_1}(r)|$$

$$\times r^{L'} |R_{n_2 c_2 j_2}(r)\rangle \langle R_{n_l}(r)| r^L |R_{n_1 c_1 j_1}(r)\rangle \qquad (6.283)$$

with

$$K(c, L, l, K_1^{-1}) = -(eK_1^{L+\frac{1}{2}}) i^L \left[\frac{C_{c0; L0}^{l0}}{(2L+1)!!}\right] \left\{\frac{(2c+1)(2L+1)(L+1)}{L(2l+1)}\right\}^{\frac{1}{2}}$$

(6.284)

and $\langle R_{ncj}(r)|r^L|R_{n'c'j'}(r)\rangle$ the radial matrix element.

The angular dependence of the transition operator $S_\zeta^{(z)}(L, L')$ takes the form

$$S_\zeta^{(z)}(L, L') = \sum_{\substack{M, M' \\ \gamma_1 \gamma_2 \gamma_1' \sigma'}} C_{L'M'; LM}^{z\zeta} (-1)^{L'-M'} C_{c_1 \gamma_1'; \sigma'/2}^{j_1 m_1'}$$

$$\times C_{c_2 \gamma_2; \sigma'/2}^{j_2 m_2} C_{c_1 \gamma_1; \sigma/2}^{j_1 m_1} C_{c_2 \gamma_2; L'-M'}^{c_1 \gamma_1'} C_{c_1 \gamma_1; LM'}^{ll_z}. \qquad (6.285)$$

The operators $c_{j_1 m_1}$ and $c_{j_2 m_2}$ in (6.281) create a core hole in the intermediate and final states, respectively. The core electrons are specified by spin–orbit coupled quantum numbers ($j_i = c_i \pm \frac{1}{2}$, $i = 1, 2$). The operator $l_{l_z \sigma}^+$ creates an electron in the conduction band, where these are labeled by uncoupled orbitals l_z and spin σ quantum numbers. The intermediate states $|n\rangle$ with energy E_n are assumed to be of the form

$$|n\rangle = \sum_{l_z \sigma m_1} a_{l_z \sigma m_1} l_{l_z \sigma}^+ c_{j_1 m_1} |i\rangle. \qquad (6.286)$$

In order to account for core-hole decay processes, which might occur before the core-hole (j_1) is filled by a core electron (j_2), a finite width Γ_n has been introduced. The final state $|f\rangle$ is characterized by the presence of a particle–hole excitation.

Rewriting (6.285) by use of standard techniques of angular momentum theory yields:

$$S_\zeta^{(z)}(L, L') = (-1)^{j_1+j_2+1}[j_1][j_2 c_1 z\, l]^{\frac{1}{2}} \begin{Bmatrix} j_1 & j_2 & L' \\ c_2 & c_1 & \frac{1}{2} \end{Bmatrix}$$

$$\times \sum_{jm}[j] \begin{Bmatrix} j_1 & j & L \\ l & c_1 & \frac{1}{2} \end{Bmatrix} \begin{Bmatrix} L & L' & z \\ j_2 & j & j_1 \end{Bmatrix}$$

$$\times \begin{pmatrix} \frac{1}{2} & l & j \\ \sigma & l_z & -m \end{pmatrix} \begin{pmatrix} j & z & j_2 \\ m & -\zeta & -m_2 \end{pmatrix}, \tag{6.287}$$

where $\{..\}$ and $(...)$ are the well-known $6j$ and $3j$ symbols, respectively. Moreover the notation $[a \ldots b] = (2a+1)\ldots(2b+1)$ has been utilized.

When applying the multipolar expansion of the RIXS amplitude, it is often justified to make the fast collision approximation, which amounts to neglecting the dispersion of the intermediate states, so that the expansion of the resonant denominator

$$(E_n - E_i - \hbar\omega_1 - (i\Gamma_n/2))^{-1} = (\overline{E}_n - E_i - \hbar\omega_1 - (i\overline{\Gamma}_n/2))^{-1}$$

$$\times \sum_{k}^{\infty} \left(\frac{\overline{E}_n - (i\overline{\Gamma}_n/2) - E_n + i\Gamma_n/2}{\overline{E}_n - E_i - \hbar\omega_1 - (i\overline{\Gamma}_n/2)} \right)^k \tag{6.288}$$

can be truncated at $k = 0$. Collisions are fast when

$$\text{Max}(|\hbar\omega_1 - E_n + E_i|, \Gamma_n) \ll [\overline{(E_n - \overline{E}_n)^2}]^{\frac{1}{2}}. \tag{6.289}$$

Then by utilizing closure with respect to the intermediate states $|n\rangle$ in (6.279)–(6.283) the j_1 core hole drops out of the amplitude

$$\langle f|c^+_{m_1 m'_1} c_{j_2 m_2} l^+_{l_z \sigma} c_{j_1 m_1}|i\rangle = \delta_{m_1 m'_1} \langle f|c_{j_2 m_2} l^+_{l_z \sigma}|i\rangle, \tag{6.290}$$

and $G(\omega_1)$ in (6.281) has to be replaced by

$$\overline{G}(\omega_1) = (E_i + \hbar\omega_1 - \overline{E}_n + i\overline{\Gamma}_n)^{-1}. \tag{6.291}$$

Then we can consider the RIXS process, according to the last $3j$ symbol of equation (6.287), as a 2^z pole absorption of an effective photon of energy $\hbar(\omega_1 - \omega_2)$ connected with an excitation from the core-level $j_2 m_2$ to the conduction band state $(l\,\frac{1}{2})jm$. But even in the absence of spin–orbit coupling in the ground and final state, the transition operator of (6.287) is not purely orbital in this case; it is also spin-dependent via spin–orbit interaction in the intermediate state.

Within the limits of the fast collision approximation we can apply the same formalism of time-dependent scattering operators (van Hove 1954) to (6.271), which led us from (6.21) (excitation picture) to (6.34) (correlation picture).

Inserting the multipolar expansion into (6.271) we end up with the following expression for the double differential scattering cross-section (van Veenendaal et al. 1996):

$$\frac{d^2\sigma}{d\Omega d\hbar\omega_2} = \left(\frac{8\pi\omega_2}{\hbar\omega_1}\right) K_1^{-2} \int_{-\infty}^{\infty} dt\, \exp[i(\omega_1 - \omega_2)t] \sum_{r\rho} \sum_{zz'} T_\rho^{(zz')r*} \langle i|O_\rho^{(zz')r}(t)|i\rangle, \quad (6.292)$$

where the geometrical factor, $T_\rho^{(zz')r}$, is given by

$$T_\rho^{(zz')r} = \sum_{\zeta\zeta'} C_{z\zeta;z'\zeta'}^{r\rho} T_\zeta^{(z)} T_{\zeta'}^{(z')}. \quad (6.293)$$

The scattering operator, $O_\rho^{(zz')r}(t)$, has to be written as

$$O_\rho^{(zz')r}(t) = \left| G(\overline{\omega}_1) R_{L,K_1^{-1}}^{L',K_2^{-1}} \right|^2 \sum_{\zeta\zeta'} C_{z\zeta;z'\zeta'}^{r\rho}$$

$$\times \sum_{\substack{l_z l_z' \sigma \sigma' \\ m_1 m_1' m_2 m_2'}} S_\zeta^{(z)+} S_{\zeta'}^{(z')} c_{j_2 m_2'}^+(t) l_{l_z' \sigma'}(t) l_{l_z \sigma}^+ c_{j_2 m_2}. \quad (6.294)$$

Since we shall discuss, in what follows, experimental arrangements with different ingoing and outgoing photon states, it is useful to redefine the geometry factors by working with each photon coupled to itself rather than coupling ingoing and outgoing photons together as done in the definition of (6.280). Therefore, we introduce the following transformation for the geometry factor:

$$T_\rho^{(zz')r} \to \tilde{T}_\rho^{(zz')r} = \sum_{\zeta\zeta'} C_{z-\zeta;z'-\zeta'}^{r\rho} \tilde{T}_\zeta^z(L) \tilde{T}_{\zeta'}^{z'}(L') \quad (6.295)$$

with

$$\tilde{T}_\zeta^z(L) = \sum_{M,M'} C_{LM';LM}^{z\zeta} [\mathbf{e}_1 \cdot \mathbf{Y}_{LM}^*(\widehat{\mathbf{K}}_1)][\mathbf{e}_1^* \cdot \mathbf{Y}_{LM'}(\widehat{\mathbf{K}}_1)]. \quad (6.296)$$

The scattering operator is transformed accordingly $O_\rho^{(zz')r} \to \tilde{O}_\rho^{(zz')r}$.

It is the goal of the above multipole expansion that the RIXS cross-section, integrated over ω_2 in the quasielastic approximation, namely

$$\frac{d\sigma}{d\Omega} \cong 8\pi K_1^{-2} \sum_{zz'r\rho} T_\rho^{(zz')r*}(L,L') \langle i|\tilde{O}_\rho^{(zz')r}(0)|i\rangle \quad (6.297)$$

with

$$\tilde{O}_\rho^{(zz')r}(t) = \left|\overline{G}(\omega_1) R_{L,K_1^{-1}}^{L',K_2^{-1}}\right|^2 \sum_{\zeta\zeta'} C_{z\zeta;z'\zeta'}^{r0}$$
$$\times \sum_{\substack{l_z l'_z \sigma \sigma' \\ m_1 m'_1 m_2}} S_\zeta^{(z)+} S_{\zeta'}^{(z')} l_{l_{z'}\sigma'}(t) l_{l_z\sigma}^+ \qquad (6.298)$$

can be represented as the ground-state expectation value of a linear combination of double-tensor operators $\langle w_\rho^{(ab)r}\rangle$ (see van Veenendaal et al. (1996) for a detailed derivation of the relation given below by using a considerable amount of angular momentum algebra) by rewriting the scattering operator in the following form

$$O_\rho^{(zz')r}(0) = \left|\overline{G}(\omega_1) R_{L,K_1^{-1}}^{L',K_2^{-1}}\right|^2 \sum_{ab} C_{j_1}^{abrzz'}(c_1, L, l) w_\rho^{(ab)r}$$
$$\times \frac{B_{j_1 j_2}^{z'}(c_1, L', c_2) n_{Lz} n_{L'z'} n_{zz'r}^{-1}[zz']^{\frac{1}{2}}}{[lc_1^2 c_2 r]^{\frac{1}{2}}} \qquad (6.299)$$

with

$$C_{j_1}^{abrzz'}(c_1, L, l) = \sum_x [j_1 c_1 labrx] \begin{Bmatrix} a & b & r \\ z & z' & x \end{Bmatrix} \begin{Bmatrix} c_1 & \frac{1}{2} & j_1 \\ c_1 & \frac{1}{2} & j_1 \\ x & b & z' \end{Bmatrix}$$
$$\times \begin{Bmatrix} L & l & c_1 \\ L & l & c_1 \\ z & a & x \end{Bmatrix} n_{la} n_{sb} n_{abr} n_{Lz}^{-1} n_{j_1 z'}^{-1} n_{zz'r}, \qquad (6.300)$$

$$B_{j_1 j_2}^{z'}(c_1, L', c_2) = (-1)^{j_1+j_2+c_1+c_2}[j_1 j_2 c_1 c_2]$$
$$\times \begin{Bmatrix} j_1 & j_2 & L' \\ c_2 & c_1 & \frac{1}{2} \end{Bmatrix}^2 \begin{Bmatrix} L' & L' & z' \\ j_1 & j_1 & j_2 \end{Bmatrix} n_{j_1 z'} n_{L'z'}^{-1}, \qquad (6.301)$$

where the normalization factors are given by

$$n_{st} = \begin{pmatrix} s & t & s \\ -s & 0 & s \end{pmatrix}; \quad n_{stu} = \begin{pmatrix} s & t & u \\ 0 & 0 & 0 \end{pmatrix}. \qquad (6.302)$$

The essential parts of (6.299) are the double tensor operators $w_\rho^{(ab)r}$, which are written in terms of the second-quantization operators (Judd 1967):

$$w_\rho^{(ab)r} = -(-1)^{a-b+r}[abr]^{-\frac{1}{2}} n_{la}^{-1} n_{sb}^{-1} n_{abr}^{-1}$$

$$\times \sum_{\substack{l_z l'_z \sigma \sigma' \\ \alpha \beta}} C_{a-\alpha;b-\beta}^{r\rho} C_{l'_z;l_z}^{a\alpha} C_{\sigma'/2;\sigma/2}^{b\beta} l_{l'_z \sigma'}^{+} l_{l_z \sigma}. \quad (6.303)$$

These operators describe the multipole moments of the charge and magnetic distribution of the valence l electrons. The indices a and b denote the orbital and spin parts, respectively. The values of a and b are limited to $a = 0, \ldots, 2l$ and $b = 0, 1$. For $b = 0$ the operators are purely orbital, for $b = 1$, they depend on spin. r is the ground-state multipole order. If these relations are derived for SO_2 symmetry, that of a magnet with negligible crystal fields, all scattering operators O_ρ^r branch to the totally symmetric representation $\rho = 0$ (only totally symmetric transition operators yield a non-zero ground-state expectation value). The double tensor operators $w_\rho^{(ab)r}$ with $\rho = 0$ can be related to standard ground-state operators (see for example van der Laan 1998) of the valence empty states, as there are the number (of holes) operator n via

$$w^{000} = n \quad (6.304)$$

the spin–orbit coupling operator via

$$w^{110} = (ls)^{-1} \sum_i l_i \cdot s_i \quad (6.305)$$

the spin moment via

$$w_0^{011} = -s^{-1} S_z = -s^{-1} \sum_i s_{z,i} \quad (6.306)$$

the orbital moment via

$$w_0^{101} = -l^{-1} L_z = -l^{-1} \sum_i l_{z,i} \quad (6.307)$$

and the magnetic dipole term via

$$w_0^{211} = \left(\frac{1}{4}\right)(2l+3)l^{-1} \sum_i (3[l_z(l \cdot s)]_+ - 2l^2 s_z)_i. \quad (6.308)$$

Similar relations can be obtained for the higher-rank tensors.

This means that (6.297) together with (6.299) play a similar role in RIXS as the famous sum rules of the X-ray magnetic circular and linear dichroism in XAS (Thole et al. 1992). Utilizing (6.297), which gives the integrated intensity of RIXS spectra for different angular factors, this means for different combinations of incident and/or re-emitted photon polarization with incident and/or re-emitted photon directions, enables one to extract expectation values of different ground-state operators of the valence empty states. Indeed, by reducing (6.299) to the case of an isotropic photon (setting $z' = 0$), then absorption and re-emission are decoupled, so that the standard sum rules of X-ray absorption and dichroism are recovered. However, when the angular dependence of the outgoing photon is detected ($z' \neq 0$), other linear combinations of $w^{(ab)r}$ are selected, and information about higher multipole moments is offered. This fact has been stressed by Braicovich et al. (1999).

Note that the cross-section of (6.297) is a linear combination of pairs of tensors of increasing rank r, which transform according to the irreducible representation of the spherical group (SO_3). One has to bear in mind that the operator $\tilde{O}_\rho^{(zz')r}$ in (6.297) must be totally symmetric, in order to yield a nonzero result for the cross-section. In spherical symmetry only $\rho = 0$ has this property. In lower symmetry, also $r > 0$ tensors which branch to the totally symmetric representation contribute to the cross-section.

6.10.4 Symmetry assignment of molecular orbitals by polarized RIXS

The fact that RIXS must be considered as a process where the virtual absorption and the fluorescence re-emission are coherently connected, leads to the possibility to perform symmetry assignment of unoccupied molecular orbitals when the symmetry of the occupied orbitals is known, or vice versa, and this is feasible even in systems of randomly oriented molecules, this means in molecular gases or liquids. The absorption process with a distinct polarization of the incident photon is selective to certain symmetries of the unoccupied orbitals. In this way the absorption process creates excited molecules that are aligned to the polarization of the incident X-rays. Since the radiative decay of these excited molecules occurs much faster than the reorientation of these molecules, the pattern of the re-emitted radiation reflects that of these partially aligned molecules. In other words, we find information about the symmetry of the excited molecules in the polarization of the re-emitted photon. Of course, this polarization must be measured. In what follows, we shall establish a formal frame for this idea, which gives an unequivocal identification of the symmetry species of molecular orbitals.

Following Luo et al. (1994), we start by writing down the RIXS amplitude $F_{\nu n}(\omega_1)$, according to (6.271), within the limits of the dipole approximation (setting $\exp(i\mathbf{K}_{1/2} \cdot \mathbf{r}) = 1$) by using

$$\langle \eta | \mathbf{e} \cdot \mathbf{p} | \mu \rangle = i[m(E_\eta - E_\mu)/\hbar](\mathbf{e} \cdot \mathbf{d}_{\eta\mu}), \qquad (6.309)$$

where $\mathbf{d}_{\eta\mu}$ is shorthand for the dipole matrix element

$$\mathbf{d}_{\eta\mu} = \langle \eta | \mathbf{r} | \mu \rangle, \qquad (6.310)$$

so that we obtain within the one-electron approximation

$$F_{\nu n}(\omega_1) = \sum_k f_{\nu n}^k(\omega_1) \tag{6.311}$$

$$f_{\nu n}^k(\omega_1) = \frac{r_0 \hbar^2 \omega_{\nu k} \omega_{nk}(\nu)(\mathbf{e}_1^* \cdot \mathbf{d}_{\nu k})(\mathbf{e}_2 \cdot \mathbf{d}_{kn}(\nu))}{\hbar\omega_1 - \hbar\omega_{\nu k} + (i\Gamma_{\nu k}/2)}, \tag{6.312}$$

where we use the following notation: $\hbar\omega_{\nu k} = E(k^{-1}\nu) - E_0$; $\hbar\omega_{nk}(\nu) = E(k^{-1}\nu) - E(n^{-1}\nu)$; $\mathbf{d}_{\nu k} = \langle 0|\mathbf{r}|k^{-1}\nu\rangle$; $\mathbf{d}_{kn}(\nu) = \langle k^{-1}\nu|\mathbf{r}|n^{-1}\nu\rangle$. $|k^{-1}\nu\rangle$ is the intermediate state with a hole in the core orbital, enumerated by the index k, and an electron in the formerly unoccupied molecular orbital ν. $|n^{-1}\nu\rangle$ is an optical excited state with a hole in the formerly occupied molecular orbital n and an electron in the formerly unoccupied molecular orbital ν, so that $\omega_{\nu n} = (E(n^{-1}\nu) - E_0)/\hbar$ is the frequency of the optical excitation $n \to \nu$. $|0\rangle$ is the ground state. The polarization vectors $\mathbf{e}_1$ and $\mathbf{e}_2$ should be defined in laboratory coordinates X, Y, Z.

Then the double differential scattering cross-section writes

$$\frac{d^2\sigma}{d\Omega_2 d\hbar\omega_2} = \sum_\nu \sum_n \left(\frac{\omega_2}{\omega_1}\right) |F_{\nu n}(\omega_1)|^2 \Delta(\omega_1 - \omega_2 - \omega_{\nu n}, \Gamma_{\nu n}), \tag{6.313}$$

where we have already taken into account the final state lifetime broadening $\Gamma_{\nu n}$ by replacing the δ-function of (6.267) by a Lorentzian

$$\Delta(\omega, \Gamma) \equiv \frac{\Gamma}{[2\pi(\hbar^2\omega^2 + \Gamma^2/4)]}. \tag{6.314}$$

In a realistic experiment we have to convolute the double differential scattering cross-section with the spectral distribution function $\Phi(\omega_1 - \omega_0)$ of the incident X-rays:

$$\frac{d\sigma(\omega_2, \omega_0)}{d\Omega_2} = \int d\omega_1 \left(\frac{d^2\sigma}{d\Omega_2 d\hbar\omega_2}\right) \Phi(\omega_1 - \omega_0). \tag{6.315}$$

Since the lifetime broadening, $\Gamma_{\nu n}$, of the optical transition $n \to \nu$ is much smaller than the width of the X-ray transition $\Gamma_{\nu k}$, we can replace the Lorentzian in (6.313) by the δ-function $\delta(\omega_1 - \omega_2 - \omega_{\nu n})$. Therefore, the width of the convoluted emission line

$$\frac{d\sigma(\omega_2, \omega_0)}{d\Omega_2} = \sum_\nu \sum_n \left(\frac{\omega_2}{\omega_1}\right) |F_{\nu n}(\omega)|^2 \Phi(\omega_2 + \omega_{\nu n} - \omega_0) \tag{6.316}$$

is restricted only by the width of the spectral distribution function of the incident X-rays, and the instrumental resolution.

For samples in gaseous or liquid phases we have to average the cross-section of (6.316) over all molecular orientations. In order to do this, let us express the dipole moments $\mathbf{d}_{\nu k}$ and $\mathbf{d}_{kn}$, originally defined in the molecular coordinate system ξ, η, ζ, in laboratory coordinate axes by means of the direction cosine transformation

$$d_k^X = d_k^\xi c_{\xi X} + d_k^\eta c_{\eta X} + d_k^\zeta c_{\zeta X} = d_k^\alpha c_{\alpha X},$$

$$\vdots$$
$$\vdots \qquad (6.317)$$

where the Greek index α means summation over the values ξ, η and ζ. $c_{\xi X}$ is the cosine of the angle between the ξ-axis of the molecular coordinate system and the X-axis of the laboratory frame. The other direction cosines are defined correspondingly, so that a Greek letter stands for the molecular axis and a capital italic letter for the laboratory axis. By inserting this transformation into (6.311) one obtains

$$F_{\nu n} = (c_{\beta A} c_{\gamma B})(e_{1A}^* e_{2B}) r_0 \sum_k \frac{\hbar^2 \omega_{\nu k} \omega_{nk}(\nu) d_{\nu k}^\beta d_{kn}^\gamma(\nu)}{(\hbar\omega_1 - \hbar\omega_{\nu k} + i\Gamma_{\nu k}/2)} \qquad (6.318)$$

Introducing the following second-rank tensor

$$F_{\nu n}^{\beta\gamma} = r_0 \sum_k \frac{\hbar^2 \omega_{\nu k} \omega_{nk}(\nu) d_{\nu k}^\beta d_{kn}^\gamma(\nu)}{(\hbar\omega_1 - \hbar\omega_{\nu k} + (i\Gamma_{\nu k}/2))}. \qquad (6.319)$$

we can write $|F_{\nu n}|^2$ as

$$|F_{\nu n}|^2 = (e_{1A}^* e_{2B} e_{1R} e_{2S}^*)(c_{\beta A} c_{\gamma B} c_{\rho R} c_{\sigma S})(F_{\nu n}^{\beta\gamma} F_{\nu n}^{\beta\sigma *}). \qquad (6.320)$$

This way we have collected all information on polarization in the first factor, a Cartesian fourth-rank tensor, the information about orientation of the molecules in the middle factor and the information about the molecules themselves in the last factor (also a Cartesian fourth-rank tensor). As long we are dealing with molecules which exhibit fixed orientation (for example surface adsorbates) then we can use two of these factors to obtain information about the third. However, in the case of gas phase samples, the orientation factor must be averaged. By using the averaging procedure of Monson and McClain (1970) we obtain

$$\langle c_{\beta A} c_{\gamma B} c_{\rho R} c_{\sigma S}\rangle = \frac{1}{30}[\delta_{AB}\delta_{RS}(4\delta_{\beta\gamma}\delta_{\rho\sigma} - \delta_{\beta\rho}\delta_{\gamma\sigma} - \delta_{\beta\sigma}\delta_{\gamma\rho})$$
$$+ \delta_{AR}\delta_{BS}(-\delta_{\beta\gamma}\delta_{\rho\sigma} + 4\delta_{\beta\rho}\delta_{\gamma\sigma} - \delta_{\beta\sigma}\delta_{\gamma\rho})$$
$$+ \delta_{AS}\delta_{BR}(-\delta_{\beta\gamma}\delta_{\rho\sigma} - \delta_{\beta\rho}\delta_{\gamma\sigma} + 4\delta_{\beta\sigma}\delta_{\gamma\rho})]. \qquad (6.321)$$

Inserting (6.321) into (6.320) and (6.316), and putting the summation back, we can write the orientation average of $|F_{\nu n}|^2$

$$\langle |F_{\nu n}|^2\rangle \equiv \lambda_{\nu n} = F\lambda_{\nu n}^F + G\lambda_{\nu n}^G + H\lambda_{\nu n}^H \qquad (6.322)$$

and thus the orientation average of the cross-section:

$$\left\langle \frac{d\sigma(\omega_2,\omega_0)}{d\Omega_2} \right\rangle = \sum_{\nu n}\left(\frac{\omega_2}{\omega_1}\right)(F\lambda^F_{\nu n} + G\lambda^G_{\nu n} + H\lambda^H_{\nu n})\Phi(\omega_2 + \omega_{\nu n} - \omega_0) \quad (6.323)$$

with

$$\lambda^F_{\nu n} = \sum_\beta F^{\beta\beta}_{\nu n} \sum_\gamma F^{\gamma\gamma*}_{\nu n}$$

$$\lambda^G_{\nu n} = \sum_{\beta\gamma} F^{\beta\gamma}_{\nu n} F^{\beta\gamma*}_{\nu n}$$

$$\lambda^H_{\nu n} = \sum_{\beta\gamma} F^{\beta\gamma}_{\nu n} F^{\gamma\beta*}_{\nu n} \quad (6.324)$$

and

$$F = -|\mathbf{e}_1\cdot\mathbf{e}_2^*|^2 + 4|\mathbf{e}_1\cdot\mathbf{e}_2|^2 - 1$$

$$G = -|\mathbf{e}_1\cdot\mathbf{e}_2^*|^2 - |\mathbf{e}_1\cdot\mathbf{e}_2|^2 + 4$$

$$H = 4|\mathbf{e}_1\cdot\mathbf{e}_2^*|^2 - |\mathbf{e}_1\cdot\mathbf{e}_2|^2 - 1. \quad (6.325)$$

One easily realizes that the molecular parameters $\lambda^F_{\nu n}$, $\lambda^G_{\nu n}$ and $\lambda^H_{\nu n}$ depend on the symmetries of the unoccupied, ν, and occupied, n, molecular orbitals. Transitions between orbitals with certain symmetries will give rise to a distinct polarization dependence of the cross-section, that can be used for symmetry assignment. To clarify this point, let us denote the angle between the linear polarization vectors of absorbed and emitted photons by ϑ. Then $\lambda_{\nu n}$ can be expressed by

$$\lambda_{\nu n} = (-\lambda^F_{\nu n} + 4\lambda^G_{\nu n} - \lambda^H_{\nu n}) + (3\lambda^F_{\nu n} - 2\lambda^G_{\nu n} + 3\lambda^H_{\nu n})\cos^2\vartheta. \quad (6.326)$$

By measuring the scattering cross-section for different values of ϑ, the molecular parameters can be determined.

Luo et al. (1994) exemplify the symmetry assignment by proposing three different combinations of incoming and outgoing photon polarization, namely, $\lambda_{\nu n}(\text{lp})$, $\lambda_{\nu n}(\text{ln})$ and $\lambda_{\nu n}(\text{cp})$ for absorbed and emitted photons having parallel linear, perpendicular linear and circular polarization, respectively. The latter can be arbitrary clockwise or counterclockwise. Thus they obtain the following polarization ratios (which must not be mixed up with the Stokes parameter):

$$P_1 = \frac{\lambda_{\nu n}(\text{ln})}{\lambda_{\nu n}(\text{lp})} = \frac{-\lambda^F_{\nu n} + 4\lambda^G_{\nu n} - \lambda^H_{\nu n}}{2\lambda^F_{\nu n} + 2\lambda^G_{\nu n} + 2\lambda^H_{\nu n}}$$

$$P_2 = \frac{\lambda_{\nu n}(\text{cp})}{\lambda_{\nu n}(\text{lp})} = \frac{-2\lambda^F_{\nu n} + 3\lambda^G_{\nu n} + 3\lambda^H_{\nu n}}{2\lambda^F_{\nu n} + 2\lambda^G_{\nu n} + 2\lambda^H_{\nu n}}$$

$$P_3 = \frac{\lambda_{\nu n}(\text{cp})}{\lambda_{\nu n}(\text{ln})} = \frac{-2\lambda^F_{\nu n} + 3\lambda^G_{\nu n} + 3\lambda^H_{\nu n}}{-\lambda^F_{\nu n} + 4\lambda^G_{\nu n} - \lambda^H_{\nu n}}. \quad (6.327)$$

Using the results of a group theoretical analysis of McClain (1971), who has performed for the case of two-photon absorption, a case which is formally identical

Table 6.1. Symmetry assignment of occupied (unoccupied) MOS for the RIXS transitions, when the unoccupied (occupied) MO is symmetry Γ^0. For the inversion symmetry Γ^0 can be gerade (g) or ungerade (u). All other symmetry elements should be labelled gerade (g)

	Polarization ratios						Groups				
Case	P_1	P_2	P_3	C_1 (C_i)	C_2 (C_{2h}, C_s)	C_{2v}	D_2 (D_{2h})	C_4 (S_4, C_{4h})	C_{4v} (D_{2d}, D_4, D_{4h})	C_3 (S_6)
1	×	×	×	Γ^0	Γ^0	Γ^0	Γ^0	Γ^0	Γ^0	Γ^0
2	×	3/2	×		$b \times \Gamma^0$	$a_2 \times \Gamma^0$ $b_1 \times \Gamma^0$ $b_2 \times \Gamma^0$	$b_1 \times \Gamma^0$ $b_2 \times \Gamma^0$ $b_3 \times \Gamma^0$	$e \times \Gamma^0$	$e \times \Gamma^0$	$e \times \Gamma^0$
3	3/4	3/2	2					$b \times \Gamma^0$	$b_1 \times \Gamma^0$ $b_2 \times \Gamma^0$	
4	∞	3/2	0						$a_2 \times \Gamma^0$	

	Polarization ratios						Groups		
Case	P_1	P_2	P_3	C_{3v} (D_3, D_{3d})	C_6 (C_{3h}, C_{6h})	C_{6v} (D_{3h}, D_{6h}, D_6)	$C_{\infty v}$ ($D_{\infty h}$)	T^a (T_h)	O^a (O_h, T_d)
1	×	×	×	Γ^0	Γ^0	Γ^0	Γ^0	Γ^0	Γ^0
2	×	3/2	×	$e \times \Gamma^0$	$e_1 \times \Gamma^0$ $e_2 \times \Gamma^0$	$e_1 \times \Gamma^0$ $e_2 \times \Gamma^0$	$\pi \times \Gamma^0$ $\delta \times \Gamma^0$	$t \times \Gamma^0$ $e \times \Gamma^0$	$e \times \Gamma^0$ $t_2 \times \Gamma^0$ $t_1 \times \Gamma^0$
3	3/4	3/2	2						
4	∞	3/2	0	$a_2 \times \Gamma^0$		$a_2 \times \Gamma^0$			

^{a} $\omega = \exp(2\pi/3)$.

with RIXS spectroscopy, Luo et al. (1994) have set up the Table 6.1, very helpful for symmetry assignment. This Table gives for four different combinations of P_1, P_2 and P_3 the allowed symmetry combination of occupied and unoccupied molecular orbitals as calculated for each of the 32 point groups of the involved molecule and for the two groups of linear molecules, under the assumption that the original state is a totally symmetric one. Thus an unequivocal symmetry assignment of the occupied (unoccupied) molecular orbital can be achieved when the symmetry of the unoccupied (occupied) molecular orbital is known. Of course, this table reflects the general symmetry selection rule for the RIXS process: the product of the irreducible representations $\Gamma_\nu \times \Gamma_\alpha \times \Gamma_\beta \times \Gamma_n$, where ν and n denote the molecular orbital, α and β the components of the dipole moments, must contain the totally symmetric representation.

It has been pointed out by Luo et al. (1994) that the symmetry assignment is also possible by measuring only P_1, when there is only a single core orbital involved in the absorption, since in this case $\lambda_{\nu n}^F = \lambda_{\nu n}^H$. For details, the reader is referred to Gel'mukhanov and Ågren (1994).

6.10.5 Momentum conservation and magnetic quantum number selection rules in RIXS, Bloch-**k** selectivity, spin selectivity

We will demonstrate in what follows that the coherence between absorption and re-emission process in RIXS has a further consequence, namely the momentum conservation, valid for the combination of both processes. It will turn out that the momentum conservation, which brings about the Bloch-**k** selectivity of RIXS, comes from the summation of the coherent absorption and re-emission event among all equivalent atoms within the coherent X-ray field.

We start again with (6.271) and assume that only one electron interacts with a photon at a time, i.e. during the absorption of a photon one electron is excited from a bound state into a formerly unoccupied state above the Fermi energy, and during the emission of a photon, one electron from the valence band refills the previously created core hole. Moreover, one assumes that the excitation and the de-excitation of an electron does not affect the other electrons of the sample via Coulomb interaction but that they remain "frozen" in their states. We will refer to this procedure as the "frozen core approximation". It should be noted that this approximation is only valid as long as the creation of holes in the valence states is concerned. If a hole in the valence band is created the resulting change of the potential is uniformly distributed over the whole sample, due to the delocalization of the valency states. Therefore, the states of the other valence electrons are only marginally affected. If, on the other hand, a strongly bound core electron is removed from its shell, all electrons within the outer shells screen the hole thus changing their states. By applying the frozen core approximation one can replace the many-particle wavefunctions in (6.271) by single-particle wavefunctions. Thus the sum over j, j' vanishes and the states were characterized by a single hole wavefunction, so that

$$|i\rangle = |\nu_c \mathbf{k}_i\rangle = \exp(i\mathbf{k}_i \cdot \mathbf{r}) u_{\nu_c \mathbf{k}_i}(\mathbf{r}) \text{ and } |f\rangle = |\nu_v \mathbf{k}_f\rangle = \exp(i\mathbf{k}_f \cdot \mathbf{r}) u_{\nu_v \mathbf{k}_f}(\mathbf{r}) \quad (6.328)$$

are states in the conduction band and in the valence band, respectively, where ν is the band index, **k** is the Bloch vectors and $u_{\nu\mathbf{k}}$ is the lattice periodic function. The intermediate state may then be represented by

$$|n\rangle = |n\mathbf{R}_j\rangle = \psi_n(\mathbf{r} - \mathbf{R}_j), \quad (6.329)$$

where ψ_n is an atomic wavefunction with core level index n centered at the atomic position $\mathbf{R}_j$. Having in mind that the energy of a hole state is the negative of the corresponding electron state, and that we have to introduce additionally a sum over the initial states $|i\rangle$, since the hole state in the conduction band is not known a priori, (6.271) reads

$$\frac{d^2\sigma}{d\Omega_2 d\hbar\omega_2} = \left(\frac{\omega_2}{\omega_1}\right) r_0^2 \sum_{\nu_i \mathbf{k}_i} \sum_{\nu_f \mathbf{k}_f} \left|\sum_{n\mathbf{R}_j} M_2(\nu_f, \mathbf{k}_f, n, \mathbf{R}_j)\right.$$

$$\left. \times \frac{M_1(\nu_i, \mathbf{k}_i, n, \mathbf{R}_j)}{E_{\nu_i \mathbf{k}_i} - E_{n\mathbf{R}_j} - \hbar\omega_1 - (i\Gamma_{n\mathbf{R}_j}/2)}\right|^2 \delta(E_{\nu_i \mathbf{k}_i} - E_{\nu_f \mathbf{k}_f} - \hbar\omega_1 + \hbar\omega_2),$$

$$(6.330)$$

where

$$M_1(\nu_i, \mathbf{k}_i, n, \mathbf{R}_j) = \langle \exp(i\mathbf{k}_i \cdot \mathbf{r}) u_{\nu_i \mathbf{k}_i}(\mathbf{r}) | (\mathbf{e}_1 \cdot \mathbf{p}/\hbar) \exp(i\mathbf{K}_1 \cdot \mathbf{r}) | \psi_n(\mathbf{r} - \mathbf{R}_j) \rangle \quad (6.331)$$

$$M_2(\nu_f, \mathbf{k}_f, n, \mathbf{R}_j) = \langle \psi_n(\mathbf{r} - \mathbf{R}_j) | (\mathbf{e}_2^* \cdot \mathbf{p}/\hbar) \exp(-i\mathbf{K}_2 \cdot \mathbf{r}) | \exp(i\mathbf{k}_f \cdot \mathbf{r}) u_{\nu_f \mathbf{k}_f}(\mathbf{r}) \rangle \quad (6.332)$$

are the transition matrix elements. These can be evaluated by (i) replacing $\mathbf{r} - \mathbf{R}_j$ by $\mathbf{r}'$, (ii) making use of the lattice periodicity of $u(\mathbf{r})$, and (iii) assuming that the spatial extent of the core wavefunction is small compared with the lattice spacing, so that $\exp(-i\mathbf{k}_i \cdot \mathbf{r}') \simeq 1$ and $\exp(i\mathbf{k}_f \cdot \mathbf{r}') \simeq 1$ within the range of the core:

$$\begin{aligned} M_1(\nu_i, \mathbf{k}_i, n, \mathbf{R}_j) &= \exp(-i\mathbf{k}_i \cdot \mathbf{R}_j) \exp(i\mathbf{K}_1 \cdot \mathbf{r}') \langle u_{\nu_i \mathbf{k}_i}(\mathbf{r}') | \\ &\quad \times \exp(i\mathbf{K}_1 \cdot \mathbf{r}') \mathbf{e}_1 \cdot \mathbf{p} | \psi_{n_j}(\mathbf{r}') \rangle \\ &= \exp(-i\mathbf{k}_i \cdot \mathbf{R}_j) \exp(i\mathbf{K}_1 \cdot \mathbf{r}') M_1'(\nu_i, \mathbf{k}_i, n) \end{aligned} \quad (6.333)$$

$$\begin{aligned} M_2(\nu_f, \mathbf{k}_f, n, \mathbf{R}_j) &= \exp(i\mathbf{k}_f \cdot \mathbf{R}_j) \exp(-i\mathbf{K}_2 \cdot \mathbf{r}') \langle \psi_{n_j}(\mathbf{r}') | \\ &\quad \times \exp(-i\mathbf{K}_2 \cdot \mathbf{r}') \mathbf{e}_2 \cdot \mathbf{p} | u_{\nu_f \mathbf{k}_f}(\mathbf{r}') \rangle \\ &= \exp(-i\mathbf{k}_f \cdot \mathbf{R}_j) \exp(i\mathbf{K}_2 \cdot \mathbf{r}') M_2'(\nu_f, \mathbf{k}_f, n). \end{aligned} \quad (6.334)$$

Since the energy and the lifetime broadening of a core hole is independent of the position of the atom, one finds that $E_{n\mathbf{R}_j} = E_n$ and $\Gamma_{n\mathbf{R}_j} = \Gamma_n$ and the sum over $\mathbf{R}_j$ can be carried out:

$$\sum_{\mathbf{R}_j} \exp[i(\mathbf{K}_1 - \mathbf{K}_2 - \mathbf{k}_i + \mathbf{k}_f) \cdot \mathbf{R}_j] = \delta_{\mathbf{g},(\mathbf{K}_1 - \mathbf{K}_2 - \mathbf{k}_i + \mathbf{k}_f)}, \quad (6.335)$$

where $\mathbf{g}$ is a reciprocal lattice vector. If we additionally convert the denominator of (2.271), which constitutes a Lorentzian, into a δ-function assuming the lifetime broadening of the intermediate state to vanish, we end up with

$$\begin{aligned} \frac{d^2\sigma}{d\Omega_2 d\hbar\omega_2} &= 2\pi \left(\frac{\omega_2}{\omega_1}\right) r_0^2 \sum_{\nu_i \mathbf{k}_i} \sum_{\nu_f \mathbf{k}_f} |M_1'(\nu_i, \mathbf{k}_i, n)|^2 \delta(E_{n_i \mathbf{k}_i} - E_n - \hbar\omega_1) \\ &\quad \times \delta_{\mathbf{g},(\mathbf{K}_1 - \mathbf{K}_2 - \mathbf{k}_i + \mathbf{k}_f)} |M_2'(\nu_f, \mathbf{k}_f, n)|^2 \delta(E_{\nu_f \mathbf{k}_f} - E_n - \hbar\omega_2), \end{aligned} \quad (6.336)$$

where we have skipped the summation over n, assuming that only a single core level is involved. The first matrix element and the first δ-function together with the sum over the initial states $|n_i \mathbf{k}_i\rangle$ stand for the absorption part of the scattering process and open access to the density of unoccupied conduction electron states. The second matrix element, the second δ-function and the sum over the final states $|n_f \mathbf{k}_f\rangle$ represent the emission part of the scattering process. Both processes are determined by matrix elements whose values reflect the overlap between the strongly localized core wavefunctions $\psi_n(\mathbf{r}')$ with the conduction band and the valence band wavefunctions $u_{\nu \mathbf{k}}(\mathbf{r}')$, respectively, and make both processes element specific. What is unusual in (6.336) is the additional Kroneker delta, that reflects crystal momentum conservation for the whole inelastic scattering process, and makes this process Bloch-$\mathbf{k}$ selective. This means that, whenever the absorption process has filled a certain unoccupied state in the conduction band, characterized by $|\nu_i \mathbf{k}_i\rangle$, the re-emission process cannot occur from an arbitrary valence state $|\nu_f \mathbf{k}_f\rangle$, but this state is fixed by crystal momentum conservation: $\mathbf{k}_i - \mathbf{k}_f$ must be equal to the momentum transfer $\mathbf{K}_1 - \mathbf{K}_2$ modulo the reciprocal lattice vector g. It should again be stressed that, though the absorption and emission process occur on individual atoms, momentum conservation as indicated by equation (6.336) comes from the summation of the coherent absorption and emission events among all the equivalent atoms within the coherent X-ray field, contributing $\exp[i\mathbf{R}_j \cdot (\mathbf{K}_1 - \mathbf{k}_i)]$ and $\exp[-i\mathbf{R}_j \cdot (\mathbf{K}_2 - \mathbf{k}_f)]$ to the overall scattering process. It has been stressed by Ma (1994) that this feature of RIXS can also be used to derive structural information, exemplified on an oriented molecule with two identical atoms in a distance $\mathbf{R}$, whose initial state in the RIXS process may be represented by the wavefunction ψ_i of a hole in an unoccupied antibonding orbital, whereas the final state is characterized by the wave function ψ_f of a hole in a binding orbital, both delocalized over the two atoms. Again the intermediate state is a core hole described by (6.329) with $\mathbf{R}_j = 0$ and $\mathbf{R}$, respectively. Then a derivation, equivalent to that which led from (6.330) to (6.336) results in

$$\frac{d\sigma}{d\Omega_2 d\hbar\omega_2} = 2r_0^2 |\langle \psi_i | \exp(i\mathbf{K}_1 \cdot \mathbf{r}')\mathbf{e}_1 \cdot \mathbf{p}|\psi_n\rangle|^2 |\langle \psi_n | \exp(-i\mathbf{K}_2 \cdot \mathbf{r}')\mathbf{e}_2 \cdot \mathbf{p}|\psi_i\rangle|^2$$

$$\times \frac{(1 + \cos[(\mathbf{K}_1 - \mathbf{K}_2) \cdot \mathbf{R}])\delta(E_i - E_f - \hbar\omega_1 + \hbar\omega_2)}{(E_i - E_n - \hbar\omega_1)^2 + \Gamma_n^2/4}. \quad (6.337)$$

This expression shows that the intensity scattered from an oriented diatomic molecule depends on the scattering angle in a way comparable with that of the classical Young's experiment. This way, structural information similar to that obtained by EXAFS can be received from oriented samples. However, it was shown by Ma (1994) that even for a system of randomly oriented molecules a distinct dependence of the scattered intensity on the scattering angle remains, provided one observes both the polarization of the incident and of the scattered beam, likewise as in Section 6.10.4. For details of the averaging procedure the reader is referred to Ma (1994).

Another consequence of the coherence between the absorption and re-emission part of the RIXS process should be mentioned. The magnetic quantum number selection rules for the absorption and emission process are $m_n - m_i = m_{\mathbf{K}_1}$ and $m_f - m_n = m_{\mathbf{K}_2}$, respectively, where the indices have the same meaning as above. $m_{\mathbf{K}_1}$ and $m_{\mathbf{K}_2}$ are the quantum numbers of the photons, e.g., $m_{\mathbf{K}} = 1$, 0, −1 for left circularly, linear, and right circular polarized light, respectively.

Therefore, for the overall scattering process, the following relation holds:

$$m_i - m_f = m_{K_1} - m_{K_2}, \qquad (6.338)$$

so that knowledge of the incident and analysis of the re-emitted polarization can result in an interesting magnetic quantum number selectivity, which can be used for the investigation of magnetic materials, above all, since this information can be made both element and symmetry specific and Bloch-**k** vector selective as shown above.

The spin selection rules play a similar role. The absorption part of the RIXS process creates a spin-polarized core hole, whenever the excitation occurs into a spin-aligned unoccupied state, e.g. a 5d state in a rare-earth spin-aligned with respect to the 4f spins. Consequently, due to the spin selection rules, this deeper lying core hole can only be refilled by an electron of, e.g. a shallow 4d core level, which leaves behind a hole with the same spin polarization as the deeper core hole. When the hole states in the 4d level are spin split in energy (due to exchange interaction with the 4f), one has the choice to set the analyzer to one of the spin-split levels, so that the absorption process is selective with respect to the spin orientation of the unoccupied levels.

6.10.6 *Shakeup processes in the intermediate state of RIXS*

We will now consider shakeup processes in the intermediate state of RIXS. They are a valuable source of information about excitations of the valence electron system, since these excitations appear amplified by resonance when compared with excitations seen by nonresonant inelastic scattering. We are looking for an appropriate formal frame for the following process: Let an electron–hole pair with a hole in an inner-shell level be created in the absorption part of the RIXS process. This pair, defining the intermediate state $|n\rangle$ can be considered as an virtual exciton, which takes up the momentum of the incident photon. Thus the Coulomb interaction of this virtual exciton with the surrounding valence electron system, described by the Hamiltonian H_C, can lead to a scattering process in the intermediate state, as a result of which the valence electron system undergoes an excitation, so that a new intermediate state $|m\rangle$ appears. If the virtual exciton recombines, the photon energy loss $\hbar(\omega_1 - \omega_2)$ of the overall resonant scattering process reflects the excitation energy of the valence electron system, and its momentum transfer the momentum imparted the valence electron system in the course of this excitations. In this way also the dispersion of the excitation becomes accessible.

We will take H_C into consideration in a more general form by following Platzman and Isaac (1998) and Abbamonte *et al.* (1999), respectively, on the basis of a third-order perturbation treatment. Therefore we consider H_C as a perturbing Hamiltonian by writing

$$H_e = H_0 + H_C, \qquad (6.339)$$

where H_0 is the unperturbed system Hamiltonian. Using the relation

$$\frac{1}{H_e - z} = \sum_n \frac{|n\rangle\langle n|}{(E_n - i\Gamma_n/2 - z)} \tag{6.340}$$

we can write (6.271) in the following way:

$$\frac{d\sigma}{d\Omega_2 d\hbar\omega_2} = r_0^2 \left(\frac{\omega_2}{\omega_1}\right) \sum_f \left| \langle f|b_2 \left\{ \frac{1}{[H_e - (E_i + \hbar\omega_1)]} \right\} b_1 |i\rangle \right|^2 \delta(E_f - E_i - \hbar\omega), \tag{6.341}$$

where

$$b_1 = \mathbf{e}_1 \cdot \sum_j \mathbf{p}_j \exp(-i\mathbf{K}_1 \cdot \mathbf{r}_j)$$

$$b_2 = \mathbf{e}_2^* \cdot \sum_j \mathbf{p}_j \exp(i\mathbf{K}_2 \cdot \mathbf{r}_j)\rangle. \tag{6.342}$$

With the definitions

$$G_0(z) \equiv \frac{1}{H_0 - z} \quad \text{and} \quad G_e(z) \equiv \frac{1}{H_e - z} \tag{6.343}$$

(6.340) can easily be transformed into the Dyson-like equation

$$G_e(z) = G_0(z) - G_0(z) H_C G_e(z) \tag{6.344}$$

whose first-order iteration can be written

$$G_e(z) = G_0(z) - G_0(z) H_C G_0(z), \tag{6.345}$$

so that (6.341) becomes

$$\frac{d\sigma}{d\Omega_2 d\hbar\omega_2} = r_0^2 \left(\frac{\omega_2}{\omega_1}\right) \sum_f \left| \langle f|b_2 \left\{ \frac{1}{[H_0 - (E_i + \hbar\omega_1)]} \right\} b_1 |i\rangle \right.$$

$$+ \langle f|b_2 \left\{ \frac{1}{[H_0 - (E_i + \hbar\omega_1)]} \right\} H_C \left\{ \frac{1}{[H_0 - (E_i + \hbar\omega_1)]} \right\}$$

$$\left. \times b_1 |i\rangle \right|^2 \delta(E_f - E_i - \hbar\omega), \tag{6.346}$$

and after inserting complete sets of eigenfunctions of H_0

$$\frac{d\sigma}{d\Omega_2 \hbar d\omega_2} = r_0^2 \left(\frac{\omega_2}{\omega_1}\right) \sum_f \left| \sum_n \frac{\langle f|b_2|n\rangle\langle n|b_1|i\rangle}{E_n - E_i - \hbar\omega_1 - i\Gamma_n/2} \right.$$

$$+ \sum_{m,n} \frac{\langle f|b_2|m\rangle\langle m|H_C|n\rangle\langle n|b_1|i\rangle}{(E_n - E_i - \hbar\omega_1 - i\Gamma_n/2)(E_m - E_i - \hbar\omega_1 - i\Gamma_m/2)} \right|^2$$

$$\times \delta(E_f - E_i - \hbar\omega). \tag{6.347}$$

The first term of (6.347) is equivalent to (6.271). The second term describes just the scattering process we have outlined above: First a photon of energy $\hbar\omega_1$ is absorbed to bring the electron system from the ground state into the intermediate state $|n\rangle$, which might be considered to be a virtual exciton. This virtual exciton takes up the full momentum of the incident photon and Coulomb interacts with the valence electrons, so that a so-called shakeup process can take place, namely the excitation into a new intermediate state $|m\rangle$. This intermediate state decays into the final state $|f\rangle$, and a photon of energy $\hbar\omega_2$ is emitted. The double resonance denominator of the second term of (6.347) deserves special attention. In order to discuss the consequences of this double resonance, we shall further reduce the second term of (6.347). We assume that the valence electron system involved in the shakeup excitation is decoupled from the conduction electron bound by the core hole to build the virtual exciton. Then we can separate the excitation energy Δ of the valence electron system from the excitation energy E_{ex} of the remaining system by writing

$$E_n - E_i = E_{\text{ex},1}; \quad E_n - E_m = \Delta; \quad E_m - E_f = E_{\text{ex},2}, \quad (6.348)$$

where we additionally will assume that the excitation of the remaining system and also the deexcitation can be represented by a single energy, each with a single lifetime broadening.

Along with the energy conserving δ-function, the second term of equation (6.347), the shakeup amplitude A_{shake} takes the form:

$$A_{\text{shake}} = [(E_{\text{ex},2} - \hbar\omega_2 - i\Gamma_m/2)(E_{\text{ex},1} - \hbar\omega_1 - i\Gamma_n/2)]^{-1}$$
$$\times \sum_{m,n} \langle f|b_2|m\rangle \langle m|H_C|n\rangle \langle n|b_1|i\rangle. \quad (6.349)$$

It should be mentioned that the above scattering amplitude has the same form as the one which describes the phonon contribution to the optical resonant Raman scattering (Martin and Falicov 1975).

Of course, one has to take notice of the fact that A_{shake} provides only an additional contribution to the scattering intensity and that the first term of (6.347) produces the main contribution, whose amplitude, A_{fluor}, can now be written

$$A_{\text{fluor}} = (E_{\text{ex},1} - \hbar\omega_1 - i\Gamma_n/2)^{-1} \sum_n \langle f|b_2|n\rangle \langle n|b_1|i\rangle. \quad (6.350)$$

We shall designate contributions of A_{shake} to the scattering spectra by shakeup satellites, whereas A_{fluor} supplies the main fluorescence line.

In order to establish discussions of the relevant shakeup excitations, the matrix element of the Coulomb interaction Hamiltonian H_C should be specified by considering a resonant scattering process, where, in the absorption part, a virtual exciton is created in the state $|e\rangle$. The remaining valence electron system, formerly in the state $|v\rangle$, is excited into the state $|v'\rangle$ by means of the Coulomb interaction with the virtual exciton, so that the momentum $\mathbf{q} = \mathbf{K}_1 - \mathbf{K}_2$ is transferred to the valence electron system. During the re-emission of the photon, the virtual exciton recombines from the state $|e'\rangle$, and the valence system remains excited in the state $|v'\rangle$, so that the Coulomb interaction matrix element reads:

$$M_C = \langle m|H_C|n\rangle = \langle e', v'| \int d\mathbf{x}'d\mathbf{x} \left[\frac{e^2\rho(\mathbf{x}')\rho(\mathbf{x})}{|\mathbf{x}-\mathbf{x}'|}\right]|e, v\rangle. \quad (6.351)$$

We again assume that the valence electron system is decoupled from the conduction electrons bound by the core hole to build the exciton, so that this virtual exciton remains unaltered by the Coulomb interaction, even its energy, with exception of the delivery of the momentum $\mathbf{q}$, so that the wave vector $\mathbf{k}$ of this exciton must be substituted by $\mathbf{k}+\mathbf{q}$. Thus we have to write

$$M_C = \int d\mathbf{x}'d\mathbf{x}\langle e|\rho(\mathbf{x}')\exp(-i\mathbf{q}\mathbf{x}')|e\rangle \left[\frac{e^2}{|\mathbf{x}-\mathbf{x}'|}\right]\langle v'|\rho(\mathbf{x})|v\rangle, \quad (6.352)$$

where we have omitted corresponding exchange terms, which means that, e.g. the primed coordinate is associated solely with the exciton, the unprimed with the 3d valence system. By making use of the lattice translation symmetry

$$\rho(\mathbf{x}) = \sum_i \sum_j \delta(\mathbf{x}-\mathbf{R}_i-\mathbf{r}_j), \quad (6.353)$$

where $\mathbf{R}_i$ are the lattice translation vectors and the sum j is over the corresponding electron (hole) sites of the elementary cell, we end up, after some Fourier transform algebra, with

$$M_C = \sum_\mathbf{g} \left(\frac{4\pi e^2}{|\mathbf{q}+\mathbf{g}|^2}\right) F_e(\mathbf{g}, \mathbf{e}_1)\langle v'|\rho_v(\mathbf{q}+\mathbf{g})|v\rangle. \quad (6.354)$$

The sum is over all reciprocal lattice vectors $\mathbf{g}$, $F_e(\mathbf{g}, \mathbf{e}_1)$ is the $\mathbf{g}$th structure factor of the exciton, of course explicitly dependent on the incident polarization $\mathbf{e}_1$, and $\rho_v(\mathbf{q})$ is the $\mathbf{q}$th Fourier transform of the valence electron density operator

$$\rho_v(\mathbf{q}) = \sum_j \exp(i\mathbf{q}\cdot\mathbf{r}_{vj}). \quad (6.355)$$

6.11 Coherent inelastic scattering

6.11.1 Non-diagonal response

It has been shown in Section 6.3 that conventional inelastic X-ray scattering cannot give complete information about the two-particle density correlation function, $n_2(\mathbf{r}', \mathbf{r}, t)$, as soon as the full translational symmetry of the scattering system is broken. The physical reason for that shortcoming can be understood very easily: Since, according to (6.2), the incident wave excites all points of the scattering sample with equal probability, we cannot expect that, by sampling the spatial and temporal correlation within an inhomogeneous electron system, we will get any structural information about the reference coordinate ($\mathbf{r}'$ in 6.35). The dynamic structure factor includes only an average of the two-particle density correlation function with respect to this reference coordinate $\mathbf{r}'$ (see 6.35).

It has been pointed out by Schülke (1981) and has been confirmed experimentally by Golovchenko et al. (1981) and by Schülke et al. (1981) that, in order to overcome this shortcoming of conventional inelastic scattering, we have to prepare the initial photon state of the scattering experiment in such a way that there exists a spatial modulation of excitation, which is commensurable with the intrinsic spatial modulation of the inhomogeneous system under investigation. Such a modulation can be achieved by making the initial photon state to be a Bloch state, which in its simplest form consists in the superposition of two coherent plane waves, whose wavevectors $\mathbf{K}_{10}$ and $\mathbf{K}_{1h}$ are connected by a reciprocal lattice vector $\mathbf{g}$

$$\mathbf{K}_{1h} = \mathbf{K}_{10} + \mathbf{g}. \tag{6.356}$$

According to the dynamical theory of X-ray diffraction (von Laue 1960) such a photon Bloch state can be realized, when an incident wave with wavevector $\mathbf{K}_{10}$ meets the condition for Bragg diffraction at a set of net planes, which correspond to a reciprocal lattice vector $\mathbf{g}$. The coherent superposition of the incident wave with the diffracted wave, whose wavevector follows $\mathbf{K}_{1h} = \mathbf{K}_{10} + \mathbf{g}$, results in a photon Bloch state. The intensity distribution $I(\mathbf{r})$ of this state within the crystal is given by

$$I(\mathbf{r}) = A_0^2 + A_h^2 + 2A_0 A_h \cos(\mathbf{g} \cdot \mathbf{r} + \Delta\phi), \tag{6.357}$$

provided the condition both for the so-called Bragg-case of diffraction with σ-polarization ($\mathbf{E} \perp (\mathbf{K}_{10}, \mathbf{K}_{1h})$-plane), and the two-beam case has been met. The amplitudes A_0 and A_h of the two plane wave components of the Bloch wave field as well as their mutual phase shift $\Delta\phi$ can be fixed by the experimental conditions of the diffraction experiment. Especially the phase shift $\Delta\phi$ can be changed between 0 and π by sweeping the whole Bragg reflection range.

The DDSC for inelastic scattering of photons from a Bloch-type wave field has been derived by Schülke and Mourikis (1986) using the vector potential operator

$$A(\mathbf{r}_j) = \left(\frac{4\pi c^2}{V}\right)^{\frac{1}{2}} \{\mathbf{e}_{10}[c_0^+ \exp(i\mathbf{K}_{10} \cdot \mathbf{r}_j) + c_0 \exp(-i\mathbf{K}_{10} \cdot \mathbf{r}_j)]$$
$$+ \mathbf{e}_{1h}[c_h^+ \exp(i\mathbf{K}_{1h} \cdot \mathbf{r}_j) + c_h \exp(-i\mathbf{K}_{1h} \cdot \mathbf{r}_j)]\} \tag{6.358}$$

in place of the simple plane wave expression of (6.2) with c_0^+ and c_0 the photon creation and annihilation operator of the A_0 component of the Blochwave, respectively, and c_h^+, c_h of the A_h component correspondingly. $\mathbf{e}_{10}$ and $\mathbf{e}_{1h}$ are the polarization unit vectors of the A_o and A_h component of the Bloch wave, respectively. The corresponding photon state vector (Marcuse 1970) reads

$$|\Phi\rangle = \sum_{n_{10}} |n_{10}\rangle \left(\frac{A_0^{2n_{10}} e^{-A_0^2}}{n_{10}!}\right)^{\frac{1}{2}} \sum_{n_{1h}} |n_{1h}\rangle \left(\frac{A_h^{2n_{1h}} e^{-A_h^2}}{n_{1h}!}\right)^{\frac{1}{2}} \exp[-i(n_{1h} + \tfrac{1}{2})\Delta\phi]$$
(6.359)

with n_{10} and n_{1h} photons in the eigenstates $|n_{10}\rangle$ and $|n_{1h}\rangle$, respectively, and represents the initial photon state. Then the expectation value of the vector potential is given by

$$\langle\Phi|A(\mathbf{r}_j)|\Phi\rangle = \left(\frac{\hbar c}{2\omega_1}\right)^{\frac{1}{2}} \mathrm{Re}[A_0 \mathbf{e}_{10} \exp(i\mathbf{K}_{10}\cdot\mathbf{r}_j) + A_h \mathbf{e}_{1h} \exp(i\mathbf{K}_{1h}\cdot\mathbf{r}_j + i\Delta\phi)],$$
(6.360)

so that the desired spatial intensity distribution of (6.357) is reproduced. The scattered photon again is represented by a plane wave (see 6.2). Thus, by calculating the double differential scattering cross-section in lowest order perturbation theory, one ends up with the following expression, written for the case of real valued polarization vectors (linear polarization)

$$\frac{d^2\sigma}{d\Omega_2 d\hbar\omega_2} = \left[\frac{r_0^2}{A_0^2 + A_h^2}\right]\left(\frac{\omega_2}{\omega_1}\right) \sum_f \Bigg(A_0^2(\mathbf{e}_{10}\cdot\mathbf{e}_2)^2 |\langle f|\sum_j \exp(i\mathbf{q}_o\cdot\mathbf{r}_j)|i\rangle|^2$$

$$+ A_h^2(\mathbf{e}_{1h}\cdot\mathbf{e}_2)^2 |\langle f|\sum_j \exp(i\mathbf{q}_h\cdot\mathbf{r}_j)|i\rangle|^2 + A_0 A_h (\mathbf{e}_{10}\cdot\mathbf{e}_2)(\mathbf{e}_{1h}\cdot\mathbf{e}_2)$$

$$\left\{\langle i|\sum_j \exp(-i\mathbf{q}_o\cdot\mathbf{r}_j)|f\rangle\langle f|\sum_j \exp(i\mathbf{q}_h\cdot\mathbf{r}_j)|i\rangle \exp(-i\Delta\phi)\right.$$

$$\left.+ \langle i|\sum_j \exp(-i\mathbf{q}_h\cdot\mathbf{r}_j)|f\rangle\langle f|\sum_j \exp(i\mathbf{q}_o\cdot\mathbf{r}_j)|i\rangle \exp(i\Delta\phi)\right\}\Bigg)$$

$$\times \delta(E_f - E_i - \hbar\omega)$$
(6.361)

where

$$\mathbf{q}_o \equiv \mathbf{K}_{10} - \mathbf{K}_2 \quad \text{and} \quad \mathbf{q}_h \equiv \mathbf{K}_{1h} - \mathbf{K}_2; \; \mathbf{q}_h - \mathbf{q}_o = \mathbf{g}.$$
(6.362)

In order to elucidate the surplus of information hidden in (6.361) we shall introduce the nondiagonal dynamic structure factor $S(\mathbf{q},\mathbf{g},\omega)$ defined by

$$S(\mathbf{q},\mathbf{g},\omega) \equiv \sum_f \langle i|\sum_j \exp(-i\mathbf{q}\cdot\mathbf{r}_j)|f\rangle\langle f|\sum_j \exp[i(\mathbf{q}+\mathbf{g})\cdot\mathbf{r}_j]|i\rangle$$

$$\times \delta(E_f - E_i - \hbar\omega),$$
(6.363)

so that (6.361) writes

$$\frac{d^2\sigma}{d\Omega_2 d\hbar\omega_2} = \left[\frac{r_0^2}{A_0^2 + A_h^2}\right]\left(\frac{\omega_2}{\omega_1}\right)(A_0^2(\mathbf{e}_{10}\cdot\mathbf{e}_2)^2 S(\mathbf{q}_0,\omega)$$
$$+ A_h^2(\mathbf{e}_{1h}\cdot\mathbf{e}_2)^2 S(\mathbf{q}_h,\omega) + 2A_0 A_h(\mathbf{e}_{10}\cdot\mathbf{e}_2)(\mathbf{e}_{1h}\cdot\mathbf{e}_2)$$
$$\times \{\text{Re}[S(\mathbf{q}_0,\mathbf{g},\omega)]\cos(\Delta\phi)] + \text{Im}[S(\mathbf{q}_0,\mathbf{g},\omega)]\sin(\Delta\phi)\}). \quad (6.364)$$

Therefore, the real part of the nondiagonal dynamic structure factor can be obtained experimentally by subtracting one measurements with $\Delta\phi = 0$ from another one with $\Delta\phi = \pi$, provided the amplitudes A_0 and A_h of the two plane wave components are left unchanged. In any case, information about the nondiagonal dynamic structure factor can be obtained by measuring the DDSC for different positions within the Bragg-reflection range, i.e. different values of A_0, A_h and $\Delta\phi$.

The nondiagonal dynamic structure factor, representing the nondiagonal response of an inhomogeneous electron system, can now be written in terms of the two-particle density correlation function $n_2(\mathbf{r}',\mathbf{r},t)$ by following the derivation as outlined from (6.25) to (6.36). Thus we obtain

$$S(\mathbf{q},\mathbf{g},\omega) = \frac{1}{2\pi}\iiint dt\,\exp(-i\omega t)n_2(\mathbf{r}',\mathbf{r},t)\exp(-i\mathbf{q}\cdot\mathbf{r})\exp(i\mathbf{g}\cdot\mathbf{r}')d\mathbf{r}\,d\mathbf{r}',$$
$$(6.365)$$

Let us compare the diagonal structure factors $S(\mathbf{q},\omega)$ (see 6.36) with the nondiagonal ones (6.365). As already pointed out, the diagonal dynamic structure factor integrates with respect to the reference coordinate $\mathbf{r}'$ of the two-particle density correlation function, thus losing all information about the inhomogeneity of correlation properties in nontranslation invariant systems. Equation (6.365) clearly demonstrates that this shortcoming is canceled by measuring the nondiagonal terms of the dynamic structure factor, since the latter contains the **g**th Fourier transform of $n_2(\mathbf{r}',\mathbf{r},t)$ with respect to the reference coordinate, so that it becomes possible, at least in principle, to get complete information about the $\mathbf{r}'$-dependence of the density correlation function, this means full information about the inhomogeneity of the two-particle correlation in a nontranslation invariant system.

Finally we can derive a relation between the nondiagonal structure factor $S(\mathbf{q},\mathbf{g},\omega)$ and the dielectric matrix $\mathbf{T}$ defined in Section 6.6 for an electron system under the influence of a lattice periodic potential. Starting with the definition of $S(\mathbf{q},\mathbf{g},\omega)$ given in (6.363), we can introduce into this definition the polarization matrix $\mathbf{X}$ of (6.98), the dielectric matrix $\mathbf{T}$ of (6.103a) and the diagonal matrix $\mathbf{V}$ built by the Fourier components of the Coulomb potential, and we have to use the same algebra, ending up with

$$[S(\mathbf{q}_r + \mathbf{g}_0, \mathbf{g}, \omega)]_{\text{RPA}} = -\left(\frac{|\mathbf{q}_r + \mathbf{g}||\mathbf{q}_r + \mathbf{g}_0|}{4\pi e^2}\right)\text{Im}[(\mathbf{1} + \mathbf{T})^{-1}]_{\mathbf{g}_0\mathbf{g}} \quad (6.366)$$

for the RPA approximation of the nondiagonal dynamic structure factor, and with

$$[S(\mathbf{q}_r + \mathbf{g}_0, \mathbf{g}, \omega)]_{\text{SCF}} = -\left(\frac{|\mathbf{q}_r + \mathbf{g}||\mathbf{q}_r + \mathbf{g}_0|}{4\pi e^2}\right) \text{Im}\{[\mathbf{1} - \mathbf{X}^{xc}](\mathbf{1} + \mathbf{T})^{-1}\}_{\mathbf{g}_0\mathbf{g}} \quad (6.367)$$

for the SCF approximation of the nondiagonal dynamic structure factor, where $\mathbf{q}_r$ is restricted to the first Brillouin zone and

$$\mathbf{q} = \mathbf{q}_r + \mathbf{g}_0. \quad (6.368)$$

In this way, the experimental access to the nondiagonal dynamic structure factors, as proposed, in principle, by (6.364), opens the admittance to the full dielectric matrix.

6.11.2 Coherent Compton scattering, full one-particle momentum space density matrix

The extension of information about electron correlation in solids, which we have obtained by means of coherent inelastic scattering, suggests we expand the underlying idea to the case of large energy and momentum transfer. In other words, let us apply the formalism of the impulse approximation, as outlined in Section 6.8.1, to the double differential cross-section of a scattering experiment which utilizes the coherent superposition of two incident plane waves, as given in (6.361). Doing this we obtain (Schlke and Mourikis 1986):

$$\frac{d^2\sigma}{d\Omega_2 d\hbar\omega_2} = \left[\frac{r_0^2}{A_0^2 + A_h^2}\right]\left(\frac{\omega_2}{\omega_1}\right)(A_0^2(\mathbf{e}_{10}\cdot\mathbf{e}_2)^2 \int \Gamma_1(\mathbf{p}|\mathbf{p})\delta(\hbar\omega - \hbar^2 q_0^2/2m$$

$$- \hbar\mathbf{p}\cdot\mathbf{q}_0/m)d\mathbf{p} + A_h^2(\mathbf{e}_{1h}\cdot\mathbf{e}_2)^2 \int \Gamma_1(\mathbf{p}|\mathbf{p})\delta(\hbar\omega - \hbar^2 q_h^2/2m$$

$$- \hbar\mathbf{p}\cdot\mathbf{q}_h/m)d\mathbf{p} + \frac{1}{2}A_0 A_h(\mathbf{e}_{10}\cdot\mathbf{e}_2)(\mathbf{e}_{1h}\cdot\mathbf{e}_2)$$

$$\times \left\{\exp(i\Delta\phi)\left[\int \Gamma_1(\mathbf{p}+\mathbf{g}|\mathbf{p})\delta(\hbar\omega - \hbar^2 q_0^2/2m - \hbar\mathbf{p}\cdot\mathbf{q}_0/m)d\mathbf{p}\right.\right.$$

$$+ \int \Gamma_1(\mathbf{p}|\mathbf{p}-\mathbf{g})\,\delta(\hbar\omega - \hbar^2 q_h^2/2m - \hbar\mathbf{p}\cdot\mathbf{q}_h/m)d\mathbf{p}\right]$$

$$+ \exp(-i\Delta\phi)\left[\int \Gamma_1(\mathbf{p}-\mathbf{g}|\mathbf{p})\delta(\hbar\omega - \hbar^2 q_h^2/2m - \hbar\mathbf{p}\cdot\mathbf{q}_h/m)d\mathbf{p}\right.$$

$$\left.\left.+ \int \Gamma_1(\mathbf{p}|\mathbf{p}-\mathbf{g})\delta(\hbar\omega - \hbar^2 q_0^2/2m - \hbar\mathbf{p}\cdot\mathbf{q}_0/m)d\mathbf{p}\right]\right\}) \quad (6.369)$$

where we have written

$$\sum_f \langle i | \sum_j \exp(-i\mathbf{q}_0 \cdot \mathbf{r}_j) | f \rangle \langle f | \sum_j \exp(i\mathbf{q}_h \cdot \mathbf{r}_j) | i \rangle \delta(E_f - E_i - \hbar\omega)$$

$$= \frac{1}{2} \left\{ \sum_f \left[\int_{-\infty}^{+\infty} dt \exp(-i\omega t) \langle i | \sum_j \exp(-i\mathbf{q}_0 \cdot \mathbf{r}_j) | f \rangle \langle f | \right. \right.$$

$$\times \sum_j \exp(iE_f t/\hbar) \exp(i\mathbf{q}_h \cdot \mathbf{r}_j) \exp -(iE_i t/\hbar) | i \rangle$$

$$+ \int_{-\infty}^{+\infty} dt \exp(i\omega t) \langle i | \sum_j \exp(iE_i t/\hbar) \exp(-i\mathbf{q}_0 \cdot \mathbf{r}_j) \exp(-iE_f t/\hbar) | f \rangle \langle f |$$

$$\left. \left. \times \sum_j \exp(i\mathbf{q}_h \cdot \mathbf{r}_j) | i \rangle \right] \right\}, \quad (6.370)$$

in order to keep the interference terms in (6.369) real valued.

For practical application we shall assume that the experiment can be arranged in such a way that $\mathbf{q}_0$ and $\mathbf{q}_h$ are symmetry equivalent and that $q_0 = q_h = q$. In this case the following equation holds

$$\int \Gamma_1(\mathbf{p}+\mathbf{g}|\mathbf{p}) \, \delta(\hbar\omega - \hbar^2 q_0^2/2m - \hbar\mathbf{p} \cdot \mathbf{q}_0/m) d\mathbf{p}$$

$$= \int \Gamma_1(\mathbf{p}|\mathbf{p}-\mathbf{g}) \, \delta(\hbar\omega - \hbar^2 q_h^2/2m - \hbar\mathbf{p} \cdot \mathbf{q}_h/m) \, d\mathbf{p}. \quad (6.371)$$

If, additionally, the expressions on the right-hand and on the left-hand sides of (6.366) are real valued, and the crystal under investigation is centrosymmetric, we obtain finally:

$$\frac{d^2\sigma}{d\Omega_2 d\hbar\omega_2} = \left[\frac{r_0^2}{A_0^2 + A_h^2} \right] \left(\frac{\omega_2}{\omega_1} \right) \left([A_0^2 (\mathbf{e}_{10} \cdot \mathbf{e}_2)^2 + A_h^2 (\mathbf{e}_{1h} \cdot \mathbf{e}_2)^2] \right.$$

$$\times \int \Gamma_1(\mathbf{p}|\mathbf{p}) \, \delta(\hbar\omega - \hbar^2 q^2/2m - \hbar\mathbf{p} \cdot \mathbf{q}/m) d\mathbf{p}$$

$$+ 2 A_0 A_h (\mathbf{e}_{10} \cdot \mathbf{e}_2)(\mathbf{e}_{1h} \cdot \mathbf{e}_2) \cos(\Delta\phi)$$

$$\left. \times \int \Gamma_1(\mathbf{p}+\mathbf{g}|\mathbf{p}) \delta(\hbar\omega - \hbar^2 q^2/2m - \hbar\mathbf{p} \cdot \mathbf{q}/m) d\mathbf{p} \right). \quad (6.372)$$

In this way we obtain more direct access to the nondiagonal elements of the momentum space one-particle density matrix $\Gamma(\mathbf{p}+\mathbf{g}|\mathbf{p})$ than with (6.177). In (6.372) it is the projection of $\Gamma(\mathbf{p}+\mathbf{g}|\mathbf{p})$ on the scattering vector $\mathbf{q}$ one can get experimentally, and thus, by means of corresponding reconstruction procedures, an appropriate estimate of the full nondiagonal element. In (6.177) only the momentum space average of $\Gamma(\mathbf{p}+\mathbf{g}|\mathbf{p})$ is offered by experiment.

6.12 References

Abbamonte, P., C.A. Burns, E.D. Isaacs, P.M. Platzman, L.L. Miller, S.W. Cheong, and M.V. Klein (1999). *Phys. Rev. Lett.* **83** 860

Akhiezer, A.I. and V.B. Berestetsky (1957). *Quantum Electrodynamics* (Consultants Bureau, New York)

Barbiellini, B., and A. Bansil (2001). *J. Phys. and Chem. Solid* **62** 2181

Benesch, R. and V.H. Smith Jr. (1971). *Int. J. Quantum Chem.* **45** 131

Benesch, R. and V.H. Smith Jr. (1972). *Phys. Rev. A* **5** 114

Benesch, R. and V.H. Smith Jr. (1973). *Wave Mechanics-The First Fifty Years* ed. W.C. Price, S.S. Chissick, and T. Ravensdale (Butterworth, London)

Berestetzki, W.B., E.M. Lifshitz, and L.P. Pitajewski (1986). *Quantenelektrodynamik* (Akademie-Verlag, Berlin)

Bhatt, G., H. Grotch, E. Kazes, and D.A. Owen (1983). *Phys. Rev. A* **28** 2195

Biggs, F., L.B. Mendelsohn, and J.B. Mann (1975). *Atom. Data Nucl-Data Tables* **16** 201

Blume, M. (1985) *J. Appl. Phys.* **57** 3615

Braicowich, L., G. van der Laan, G. Ghiriaghelli, A. Tagliaferri, M.A. Veenendaal, N.B. Brookes, M.M. Chervinskii, C. Dallera. B. De Michelis, and H.A. Dürr (1999). *Phys. Rev. Lett.* **82** 1566b

Brauer, W. (1972). *Einführung in die Elektronentheorie der Metalle* (Akadem. Verl. Ges. Geest & Portig, Leipzig, p. 211

Burke, K and E.K.U. Gross (1998). *Density functionals: Theory and Applications*, ed. D. Joubert (Springer, Berlin) p. 116

Carra, P., M. Fabricius, and B.T. Thole (1995). *Phys. Rev. Lett.* **74** 3700

Ceperley D.M. and B.J. Alder (1980). *Phys. Rev. Lett.* **45** 566

Clementi, E. and C. Roetti (1974). *Atomic Data and Nuclear Data Tables* **14** 177

Cooper, M.J. (1985). *Rep. Prog. Phys.* **48** 415

Coulson, C.A. (1941). *Proc. Camb. Phil. Soc.* **37** 55

Ehrenreich, H. and M.H. Cohen (1959). *Phys. Rev.* **115** 786

Eisenberger, P., and P.M. Platzman (1970). *Phys. Rev.* **A 2** 415

Eisenberger, P. and W.A. Reed (1974). *Phys. Rev. B* **9** 3237

Eisenberger, P., P.M. Platzman, and H. Winick (1976a). *Phys. Rev. Lett* **36** 623

Eisenberger, P., P.M. Platzman, and H. Winick (1976b). *Phys. Rev. B* **13** 2377

Epstein, I.R. (1973). *Acc. Chem. Res.* **6** 145

Epstein, I.R., and A.C. Tanner (1977). *Compton Scattering* ed. B.C. Williams (New York: McGraw-Hill) p. 209

Farid, B., V. Heine, G.E. Engel, and I.J. Robertson (1993). *Phys. Rev.* **B 48** 11602

Feynmann, R.P. (1939). *Phys. Rev.* **56** 340

Fock, V. (1935). *Z. Phys.* **98** 145

Gell-Mann M. and M.L. Goldberger (1954). *Phys. Rev.* **96** 1433

Gel'mukhanov, F. and H. Agren (1994). *Phys. Rev. A* **49** 4378

Golovchenko, J., D.R. Kaplan, B.M. Kincaid, R.A. Levesque, A.E. Meixner, M.P. Robbins, and J. Felsteiner (1981). *Phys. Rev. Lett.* **46** 1454

Grotch, H., E. Kazes, G. Bhatt, and D.A. Owen (1983). *Phys. Rev. A* **27** 243

Hämäläinen, K., D.P. Siddons, J.B. Hastings, and L.E. Berman (1991). *Phys. Rev. Lett.* **67** 2850

Hedin and Lundqvist (1971) *J. Phys. C* **4** 2064

Hicks, B.L. (1940). *Phys. Rev.* **57** 665

Holm, P. (1988). *Phys. Rev. A* **37** 3706

Jauch, J.M., F. Rohrlich (1976). *The Theory of Photons and Electrons* (Springer Verlag, Berlin)

Judd, B.R. (1967). *Second Quantization in Atomic Spectroscopy* (Johns Hopkins University Press, Baltimore)

Kohn, W. and L.J. Sham (1995). *Phys. Rev.* **140** A1133

Lam, L. and P.M. Platzman (1974). *Phys. Rev. B* **9** 5122

v. Laue, M. (1960). *Röntgenstrahl-Interferenzen* (Akademische Verlagsgesellschaft, Frankfurt/Main)

Lindhard, J. (1954). *Matematisk-Fysiske Meddelel* **28** 8

Lipps, F.W. and H.A. Tolhoek (1954). *Physica* **20** 395

Lüwdin, P.O. (1956). *Adv. Phys.* **5** 1

Lundqvist, B.I. (1967a). *Phys. kondens. Materie* **6** 193

Lundqvist, B.I. (1967b). *Phys. kondens. Materie* **6** 206

Luo. Y., H. Agren, and F. Gelmukhanov (1994). *J. Phy. B: At. Mol. Opt. Phys.* **27** 4169

Luttinger, J.M. (1960). *Phys. Rev.* **119** 1153

Ma, Y. (1994). *Chem Phys. Lett.* **230** 451

Marcuse, D. (1970). *Engineering Quantum Electrodynamics* (Harcourt, Brace & World, Inc, New York)

Martin, R.M. and L.M. Falicov (1975). Resonant Raman Scattering in M Cardona (ed.) *Light Scattering in Solids* (Springer, Heidelberg)

McClain, W.M. (1971). *J. Chem. Phys.* **55** 2789

McWeeny, R. and C.A. Coulson (1949). *Proc. Phys. Soc. A* **62** 509

Monson, P.R. and W.M. McClain (1970). *J. Chem. Phys.* **53** 29

Mukoyama, T. (1982). *J. Phys. B* **15** L785

Nolting, W. (1991). *Grundkurs Theoretische Physik 7. Vielteilchentheorie* (Zimmermann-Neufang, Ulmen) p. 362

Pattison, P., B.G. Williams (1976). *Solid State Commun.* **20** 585

Pattison, P., W. Weyrich, and B.G. Williams (1977). *Solid St. Commun.* **21** 967

Pines, D. (1964). *Elementary Excitations in Solids* (Benjamin, New York)

Pines, D. and P. Nozières (1966). *The Theory of Quantum Liquids* Vol. 1 (Benjamin, New York) p. 204

Platzman P.M. and N. Tzoar (1970). *Phys. Rev. B* **2** 3556

Platzman, P.M. and E.D. Isaacs (1998). *Phys. Rev. B* **57** 107

Podolsky, B. and L. Pauling (1929). *Phys. Rev.* **34** 109

Ribberfors, R. (1975a). *Phys. Rev. B* **12** 2067

Ribberfors, R. (1975b). *Phys. Rev. B* **12** 3136
Roux, M. and R. Epstein (1973). *Chem. Phys. Lett.* **18** 18
Runge, E. and E.K.U. Gross (1984). *Phys. Rev. Lett.* **52** 997
Sakai, N. (1996). *J. Appl. Crystallogr.* **29** 81
Sakurai, J.J. (1982). *Advanced Quantum Mechanics* (Addison-Wesley, Reading, MA)
Saslow, W.M. and G.F. Reiter (1973). *Phys. Rev.* **B 7** 2995
Schülke, W. (1977). *Phys. Stat. Sol. (b)* 82 229
Schülke, W. (1981). *Phys. Lett. A* **83** 451
Schülke, W. and S. Mourikis (1986). *Acta Cryst. A* **42** 86
Schülke, W., U. Bonse, and S. Mourikis (1981). *Phys. Rev. Lett.* **47** 2065
Schülke, W., G. Stutz, F. Wohlert, and A. Kaprolat (1996). *Phys. Rev. B* **54** 14381
Tanaka, S., K. Okada, and A. Kotani (1994). *J. Phys. Soc. Jpn.* **63** 2780
Thole, B.T., P. Carra, F. Sette, and G. van der Laan (1992). *Phys. Rev. Lett.* **68** 1943
Timms, D.N., E. Zukowski, M.J. Cooper, D. Laundy, S.P. Collins, F. Itoh, H. Sakurai, T. Iwazumi, H. Kawata, M. Ito, N. Sakai, and Y. Tanaka (1993). *J. Phys. Soc. Jpn.* **62** 1716
van Hove, L. (1954). *Phys. Rev.* **95** 249
van der Laan, G. (1998). *Phys. Rev. B* **57** 112
van Veenendaal, M. and P. Carra (1997). *Phys. Rev. Lett* **78** 2839
van Veenendaal, M., P. Carra, and B.T. Thole (1996). *Phys. Rev. B* **54** 16010
Weiss, R.J. (1978). *Phil. Mag.* **37** 659
Weyrich, W. (1978). Habilitationsschrift, Darmstadt
Weyrich, W., P. Pattison, and B.G. Williams (1979). *Chem. Phys.* **41** 271
Wiser, N. (1963). *Phys. Rev.* **129** 62

Index

Bold numbers denote references to figures, italic numbers denote references to tables

3D electron momentum distribution 272, 349, **4.62**
3D reconstruction methods *1.3*, 272
    Fourier–Bessel 272, 298
    Fourier–Hankel 273
    Cormack 273
  momentum space density *1.3*, *1.4*, 272, 298
  occupation number density *1.3*, 274, **4.31**, 295, 298
  reciprocal form factor *1.3*
absorption coefficient
  dipolar 406
  quadrupolar 406
adiabatic local-density approximation (ALDA) 506
Ag *1.2*, 321, **4.48**
Al **1.6**, **1.12**, *1.1*, *1.4*, *1.5*, **2.13**, **2.15**, **2.17**, **2.28**, **2.29**, **2.46**, 142, **2.48a/b**, 175, 246, 293, 301, 304
Al (liquid metal) *1.1*, 106, **2.41**
aligned molecules RIXS 15, 410
AlLi alloy *1.4*, 301, **4.36**
AlLiCu 331
$Al_{72}Ni_{12}Co_{16}$ *1.4*, 331, **4.36**
analyzer 90–9
  dispersion compensating 91, **2.8**
  double-crystal 98, **2.12**
  energy resolution 91–9
  focusing crystal/Rowland geometry 90
  focussing/position-sensitive detector 94, **2.9**, **2.10**
  transmission 91–9
  ultrahigh-resolution 96, **2.11**
angular correlation of annihilation radiation (ACAR) *1.4*, 248, 301
  long-slit geometry 248
  point-slit geometry 248, 351
antibonding covalent interaction 291
antisymmetrized geminal product (AGP) 234, 304
Ar *1.3*, 269

a-Si:H 328
autocorrelation 526

band gap 145
  indirect *1.1*, 147, **2.49**
band mapping with RIXS **1.20**, *1.7*, 433, **5.34**
B_4C (icosahedral) *1.2*, 209
Be *1.1*, *1.2*, *1.3*, *1.4*, 104, **2.16**, 112, **2.24**, **2.25**, **2.26**, **2.27**, **2.43**, 172, **2.59**, **2.60**, 175, **3.5**, **3.6**, 197, 200, 204, **3.11**, **3.13**, 293, 298, **4.34**, **4.35**, **4.37**, 341
BeO *1.2*, 213, **3.18**, 288
binding energy 9, **1.2**
$Bi_{1.6}Pb_{0.4}Sr_2Ca_2Cu_3O_{10-\delta}$ 329
$Bi_2Sr_2CaCu_2O_8$ *1.4*, 406, **5.18**
Bloch-**k**-vector selective RIXS 16, 48–50, **1.21**, *1.7*, 422–35, 566–9
Bloch state 424, 507
Bloch vector 424
BN *1.1*, *1.2*, *1.7*, 282, **4.23**
  high pressure 230
Born approximation 336
Bragg case of diffraction 574
Bragg position 343
Bragg-reflection range 576
bremsstrahlung by photoelectrons 234
Brillouin zone 425, 507

C_{60} *1.1*, *1.2*, *1.4*, *1.5*, 151
$Ca_2CuO_2Cl_2$ *1.8*, 453, 455, **5.43**, 456, **5.45**, **5.47**, 460
CdS 282
CdTe 280
$Cd_{84}Y_{16}$ *1.4*, 332
CeF_3 **1.6**, **5.13**, 404, **5.17**
CeO_2 404, **5.17**
$CeRh_3$ 404, **5.17**
$CFCl_3$ 415
CF_2Cl_2 415
CF_3Cl 414

584 Index

CH$_3$Cl 412, **5.21**, **5.22**
C$_6$H$_6$ 173, **2.63**
C$_6$H$_{12}$ 173, **2.63**
characteristic energy losses 494
characteristic valence electron excitations 71–179
charge moment 477
charge transfer scattering 56, *1.1*
charge transfer upon alloying *1.3*, *1.4*, 319
CH$_3$Cl 173, **2.63**
CH$_3$CN 173, **2.63**
chemical bonds *1.4*, 271
  bonding/antibonding orbitals 274, **4.17**, **4.18**
  covalent bonding *1.3*, 274
  molecular orbital model (MO) 271, 532
  valence bond model (VB) 271, 533
CH$_3$OH 173, **2.63**
circularly polarized radiation 10, 514
  helicity 355
  oriented nuclei 353
  quarter-wave plates 38, 354
  synchrotron/elliptical multipole wiggler 353, 354
Cl$_2$ 420, **5.36**
classical electron radius 492
Clebsch–Gordan coefficients 556
Co *1.6*, 469
CO$_2$ 416, **5.23**
Co ferrite 477, **5.56**
cohesive energy 172, 529
coincidence technique 23, *1.2*, 231–4
collective excitation 102, **2.14**
colossal magnetoresistance (CMR) 462
combined density of states 148, **2.50**
Compton data processing 262–7
  background subtraction 264
  energy dependent corrections
    analyzer transmission 265
    relativistic scattering cross-section 265
    sample absorption 264
    scale correction 265
    windows and air absorption 264
  Monte Carlo simulation of multiple scattering 264, 265
Compton limit 7
Compton profile (CP) calculations
  APW (augmented plane wave) 310
  FLAPW (full-potential linearized augmented plane wave) 282, 316, 357
  free atomic 345, 530–2

GGW (generalized gradient approximation) 292
GW approximation 302
Hartree–Fock *1.3*, 282
KKR-CPA (Koringa–Kohn–Rostoker coherent potential approximation) 298, 301
LCAO (linear combination of atomic orbitals) 282
LDA (local density approximation) **1.9**, 282, 301
LMTO (linear muffin-tin orbital) 331
  normalization 345
  pseudopotential 345
  self-interaction correction *1.4*, 281
Compton profile (CP) experimental 8, **1.2**, **1.3**, **1.4**, 253
  core profile asymmetry 338, **4.57**, **4.58**
  coincident 351
  directional **1.10**, *1.3*, *1.4*, *1.6*, 238, 519
  directional differences 279, **4.21**, 280, **4.22**
  magnetic **1.14**, 539–42
  statistical accuracy 253
Compton scattering 9
  beyond impulse approximation 334–42
  bonds 271–93, 532–3
  bound electrons 521
  coherent 28, *1.3*, 343–6, **4.60**, 577–8
  final state self-energy effects *1.4*, 340, **4.59**
  high pressure 29, *1.4*, 333
  magnetic 34–43, *1.6*, 352–65
    normalization 355
    orbital contribution 542
    population of d-orbitals 364
  many-electron system 522
  molecules 532–3
  nonrelativistic limit 237
  particle–hole interaction 340
  regime 237–365, 516–42
  relativistic treatment 239–40, 486, 542–53
    central-field Hartree–Fock wavefunctions 547–53
    heuristic approach 239, 543–7
  solids 533–7
  total cross-section 250
Compton scattering experiments
  analyzer energy resolution 258
  analyzer transmission 258
  conventional X-ray sources 17, 254
  dispersion compensation 259, **4.11**
  γ-ray sources, solid state detectors 26, 254–5, **4.8**, 344

high-pass/low-pass filter combination 262, **4.12**
  synchrotron radiation sources 256, **4.9**, **4.10**
conical scan 475, **5.55**
coplanar perpendicular geometry 468, **5.52**
core-electron excitations 186–234
  "hidden" 392
core exciton *1.2*, 214, 429
core-hole decay rate 429
core-hole–electron interaction 190–4, 211–15
core-hole lifetime 14, 384
  broadening 14, 398–401
core-level resonant Raman scattering 43
correlation effects 27, *1.4*, *1.5*, 168
  anisotropic *1.3*, 313, **4.43**
correlation energy 164–6, 171, **2.61**
correlation function 486, 492
correlation picture 71, 514, 558
$CoSi_2$ *1.4*, 319
Coulomb interaction Hamiltonian 572
Cr *1.1*, *1.3*, *1.4*, **2.53**, 152, 310, **4.41**
critical energy 79
critical point 425, 427
$CsRb_2C_{60}$ *1.4*, 326
Cu **1.9**, *1.2*, *1.3*, *1.4*, *1.5*, *1.7*, **3.32**, **3.33**, 313, **4.43**, **4.44**, 385, 388, **5.6**, **5.10**, *5.1*
$CuCl_2$ $2H_2O$ 401, **5.14**
CuAl alloy *1.4*, *1.5*, 318
$CuGeO_3$ 453
cumulant expansion correction 128, **2.36**, **2.37**
CuNi alloy 316
CuO **1.17**, *1.8*, 395, **5.10**, **5.11**, 406, **5.19**, **5.20**, 451, **5.42**
CuO_4 parallelogram/plaquette 407, 450
CuPd alloy *1.4*, 316, **4.45**

Darwin width 343
d-d transitions *1.1*, 152, **253**
deadtime losses 354
Debye–Waller factor 304
deexcitation 401
density fluctuation 492
density functional cluster calculations 217, 221
density functional theory 506
  adiabatic local density approximation (ALDA) 506, 509
  local density approximation (LDA) 243
  time-dependent (TDLDA) 505

density matrix 28, 515
density of states (DOS) 19, *1.2*
density response function 498
diamond *1.1*, *1.3*, *1.7*, **2.49**, 202, **3.9**, **3.10**, **4.19**, 276, **4.21**, 279, 425, **5.28**, **5.29**, **5.34**, 433
dielectric function **2.14**
  imaginary part **2.14**, 496–501
  macroscopic 76, 124
  real part **2.14**, 496–501
dielectric matrix *1.1*, 76, 576
dipolar transition (E_1) 211, 474
dipole approximation 19, 562
dipole matrix element 522
disordered materials 15
Doppler broadening 33, 248
Doppler shift 521
double-crystal setting 79
  nondispersive 79, 344
double differential scattering cross-section (DDSCS) 5, 237,489
  charge contribution 352
  magnetic contribution 352
    orbital magnetic moment 352
    spin magnetic moment 352
double tensor operator 473, 560
dynamical exchange-correlation kernel 505
dynamical symmetry breaking 418
dynamical theory of X-ray diffraction 574
dynamic structure factor 6, **1.6**, *1.1*, 99–145, 340, 492
  diagonal 156
  electrons with unpaired spins 176, 513
  nondiagonal 156, 343, 575
$Dy(NO_3)_3$ **1.16**, 399
Dyson-like equation 571

edge singularity 190
effective Hamiltonian
  attractive electron–hole interaction 140, **2.45**
effective two-body interaction 126
electron density fluctuation operator 73, 493
electron density operator 72, 493
electron energy loss spectroscopy (EELS) 104
electron field operators 534
electron momentum density 7, **1.12**, **4.1**, **4.62**
  state selectively 351
electron–phonon scattering rate 429

electron–positron ($e^- - e^+$) interaction 249, **4.3**
elliptically polarized radiation 36
energy-dependent occupation function 307
energy transfer 5
envelope function 529
equation of motion 490
EuO 440, **5.38**
exchange-correlation energy functional 538
exchange-correlation potential (energy) 164–6, 243, 506
exchange coupling 436
excitation gap 108, **2.20**, **2.21**
excitation in the intermediate state (shakeup satellites) 442–67
  Anderson impurity model 443
  bonding–antibonding charge transfer excitation 447–56
  extended Hubbard model 444
  high pressure 454, **5.44**
  orbital excitations 462–7
  quasi-low-dimensional Mott insulators 456–62
    dispersion 456
    Mott-gap excitation 461
  series expansion of the Kramers–Heisenberg formula 446
  third-order perturbation treatment 446
excitation picture 71, 492, 514, 558
exciton-like prepeak 190
extended X-ray absorption fine structure (EXAFS) 21, *1.2*, 199, **3.7**

Fano factor of XMCD 439, **5.37**
fast collision approximation 558
Fe *1.2*, *1.6*, 310, 406
FeCO$_3$ 406
FeAl alloy **1.10**, *1.3*, *1.4*, 311, **4.42**
Fe$_2$O$_3$ 406
Fe 3wt% Si **1.15**, *1.6*
Fermi frequency (energy) 74
Fermi function 503
Fermi momentum 251
Fermi's golden rule 490
Fermi sphere 502, **6.1**
Fermi surface 26, *1.4*, 242
  break 295, **4.30**
  neck diameter 298, **4.33**
  nesting *1.4*, 318, **4.46**
  radii anisotropy *1.4*, 294, 296, **4.32**
  secondary 313, **6.7**, 5.37

FeS$_2$ 406
f-sum rule 75
  generalized 157
FeTi hydride 327
Feynman's theorem 243, 538
fluctuation–dissipation theorem 6, 499
fluorescence interferometry 417–22, **5.25**
fluorescence spectrum **1.17**, 428
  coherent part 428
  incoherent part 428
fluorescence yield 385
form factor 525
Frenkel-type exciton *1.2*, 229, **3.29**
  dispersion 230, **3.30**
frozen core approximation 567
fullerenes 221
  polymerized 223, **3.25**, **3.26**
fused-ring hydrocarbons *1.2*, 221, **3.27**

GaAs 280
Gd *1.6*, 359, **4.67**, 398
Gd$_{33}$Co$_{67}$ amorphous alloy 473
Gd$_3$Ga$_5$O$_{12}$ 392, **5.8**, **5.9**
Ge *1.3*, *1.7*, **4.21**, 279
γ-eγ experiments 32–4, **1.11**, *1.5*, 346, **4.61**
  momentum resolution 349
graphite *1.1*, *1.2*, *1.5*, *1.7*, **2.50**, **3.7**
  highly oriented pyrolytic (HOPG) 151, **2.52**, 200, **3.8**, 425, **5.30**, **5.31**
  intercalation compounds (GIC's) 150, 322
  π-bands 148, **2.50**
  σ-bands 148, **2.50**
ground-state operators 476, 561
group theoretical analysis 566
GW approximation 244, 341

H$_2$ *1.3*, 267
He *1.3*, 267
Heisenberg representation 492
^{4}He solid *1.2*, 329–31, **3.29**, **3.30**
higher momentum components (HMC) 27, *1.4*, 537, **6.7**
highest occupied molecular orbital (HOMO) 292
highly correlated superconductors 329–31
high-resolution edge spectroscopy by RIXS 555
H$_2$O *1.1*, *1.2*, *1.4*, 173, **3.20**, **3.21**, **3.22**, **3.23**, 217–21, **4.29**, 292
H$_2$S 415

Ho *5.1*, 385
HoFe$_2$ *1.6*, 362, **4.69**, **4.70**
Hume–Rothery mechanism *1.4*, 331
hydrides 326
  anionic model 326
  band structure model 327
  neutral atomic model 326
  protonic model 326
hydrogen bonds *1.2*, *1.4*, 215–21, **3.19**, 288–93
  broken 218
  covalency *1.4*, 290, **4.27**
  hydrogen related 215
  oxygen related 215
hydrogenic continuous waves (HCW) 206, 336

ice Ih 219, 288, **4.27**, **4.28**, 291
imaging electron dynamics 162–4, **2.58**
impulse approximation 7, 238, 486, 517
  first-order correction 337
inelastic X-ray scattering (IXS) 7, 17–43
  coherent 155–62, 573–8
inner-shell excitation 8, *1.2*
InSb 280
interaction Hamiltonian 5, 378
  linear in **A** (vector potential) 5, 378
  quadratic in **A** (vector potential) 5, 378
interband transitions 19, *1.1*
interference between charge and magnetic scattering 10, 514
interlayer state *1.1*, *1.2*, 226
internal spin reference 435–42
inverse dielectric matrix 303
inverse geometry 98, 204
inversion into time and real space *1.1*, 162–4, **2.58**
irreducible representation 566

jellium 104
joint refinement from Compton/Bragg scattering 274
Jones zone 280

K$_3$C$_{60}$ *1.4*, 326
K$_4$C$_{60}$ *1.4*, 326
K$_6$C$_{60}$ 325, **4.52**
KC$_8$ *1.1*, *1.4*, 325
KCl 288, 391
kinetic energy 528
  expectation value 528

Klein–Nishina scattering cross-section 12, 250
KMnO$_4$ 389, **5.7**
Kohn–Sham equation 243, 538
Kondo system 402
Kr *1.3*, 268
Kramers–Heisenberg formula 14, 378
  generalized 486, 489
Kramers–Kronig dispersion relation 19, 75, 497
Kronecker delta 423
k-unselective fraction/contribution 429

La$_{1.9}$Ca$_{1.1}$Cu$_2$O$_6$ 455, **5.43**
La$_2$CuO$_4$ **1.24**, *1.8*, 451, **5.41**, 455, **5.43**, 458
LaMnO$_3$ *1.8*, 462, **5.49**, **5.50**, **5.51**, 466
LaNi 396, **5.12**
Lam–Platzman correction *2.8*, *1.3*, 243, 539
LaS 396
La$_{2-x}$Sr$_x$CuO$_4$ 454, 455, **5.43**
La$_{1-x}$Sr$_x$MnO$_3$ *1.8*, 466
La$_{2-2x}$Sr$_{1+2x}$Mn$_2$O$_7$ *1.6*, 364, **4.71**
lattice translation symmetry 573
Li **1.7**, *1.1*, *1.2*, *1.3*, *1.4*, 102, 110, **2.22**, **2.23**, **2.39**, 175, **3.1**, **3.4**, **3.7**, 200, **3.14**, **3.15**, **3.16**, **3.17**, 211, **3.28**, 246, 293, **4.30**, 294, **4.31**, **4.32**, 304, 333, **4.56**, 341, **4.59**
LiC$_6$ *1.1*, *1.2*, *1.4*, 149, **2.51**, **3.28**, 322, **4.49**, **4.50**, **4.51**
Li$_2$CuO$_2$ *1.8*, 453, 455, **5.43**
LiF *1.1*, *1.2*, **2.45**, 172, **2.62**, 209, 213, 288
LiH *1.3*, *1.4*, **4.24**, 283
Li (liquid metal) *1.1*, 106, **2.18**
LiMg alloy *1.4*, 293, 297, **4.33**
Li$_3$N **1.8**, *1.3*, 286
Lindhard polarization function 100, 501–8
linear muffin-tin orbital (LMTO) calculations 441
linear response 486
Li-NH$_3$ *1.1*, 108, **2.19**
LiNiO$_2$ 326
local density approximation (LDA) *1.3*, 538
local-field correction 117, 135–40, **2.42**, **2.43**, 506
  dynamic 135
  semiempirical *1.1*, 138, **2.44**
  static approximation 135
Lock–Chrisp–West (LCW) folding theorem **4.36**, 301
Löwdin's symmetrical orthogonalization 285

588 Index

long-range orbital order 462
lower Hubbard band (LHB) 463
lowest unoccupied molecular orbital
   (LUMO) 292

magnetic quantum number selection rule 569
magnetic scattering amplitude
   orbital contribution 10
   spin contribution 10
matrix element
   dipolar 405
   quadrupolar 405
maximum entropy method (MEM) 274
metal–insulator transitions 153, **2.54**
Mg 293
MgB$_2$ *1.1*, 118, **2.30**, **2.31**
MgF 288
MgO *1.3*, 287, **4.25**, **4.26**
mirrors 86–8, **2.5**
mixed valency compounds 401–5
Mn 391
MnF$_2$ 436, **5.36**
MnO **1.23**, 436, 442
MnP 437
molecular orbitals (MO's) *1.3*
momentum conservation 15, 422–35, **5.27**, 569
momentum space density *1.3*, *1.5*, 242, 519
   electrons with unpaired spin 353
   exchange-correlation corrected 243, 538
momentum space resolution 251, **4.5**, **4.6**
momentum space wavefunction 524
momentum transfer dependence of XRS 203–11
monochromator 79–88
   cooling 81
   cylindrically bent crystal 83–6, **2.4**
   double crystal 79–82
   energy resolution 79
   four-bounce 82, **2.3**
   plane crystal 86
   transmission 80
monopolar transitions 211
Monte Carlo simulation 34, 506
multiple scattering *1.5*, 175
   XANES calculation 437
multiplet calculations 436
   ligand field 436
multipolar expansion 476, 556–62
multipole moment 473, 561
   charge distribution 473, 561
   magnetic distribution 473, 561

N$_2$ *1.3*, 268, 298
Na *1.1*, *1.4*, 293
NaCl 288
Na (liquid metal) *1.1*, 106
NbD$_{0.77}$ 327
NbH$_{0.3}$, NbH$_{0.76}$, NbH$_{1,2}$ 327–8
Nd$_{1.85}$Ce$_{0.15}$CuO$_4$ 454
Nd$_2$CuO$_4$ *1.8*, 409, **5.40**, 455, **5.43**
Ne *1.3*, 268
Ni *1.6*, 391
Ni$_2$MnSn 360, **4.68**
NiAl alloy *1.3*, *1.7*, 431, **5.33**
NiFe$_2$O$_4$ (nickel ferrite) 469, **5.53**
NiO *1.8*, 442, 447, **5.39**
nodal structure 304
nondiagonal response 573–7
N-particle density matrix 522
N-particle wavefunction 522
number density 165

O$_2$ *1.3*, 268
occupation number density 242, 535
   diagonal elements 242, 535
   non-diagonal elements 242, 535
off-diagonal response 155
one-electron approximation 423, 563
one-particle density matrix 522
   diagonal elements 525
   momentum space 343, 577
   nondiagonal elements 526
   position space 522
one-particle Green's function 125, 503, **6.2**
   "fully dressed" (self-energy corrected) 125
   "undressed" (non-self-energy corrected) 128
on-site approximation (OSA) 206
O-O-correlation 215
orbital degree of freedom 462
orbital magnetic scattering 510–6
orbital moment 477
orbital ordering in RIXS 57, 462–7
orthogonalized plane waves (OPW) 191, 206
orthonormality relation 490
oscillator density 380

pair-correlation function 164, **2.60**, 496
pair momentum density 247

Index

parity selection rule 416
partial fluorescence yield (PFY) **1.16**, 399, 401
partial spectral weights 406
particle–hole continuum 100, 121
particle–hole excitation 507
particle number conservation 501
partition function 492
Patterson function 526
Pauli matrices 10
Pauli principle 502, **6.1**
$Pd_{1-x}Co_x$ alloys 359
$PdH(D)_{0.71}$ *1.3*, 327
perturbation theory 5, 486
  first-order 5, 486
  second-order 6, 486
  third-order 56, 446
photoemission spectroscopy (PES) 401
photoabsorption cross-section 250, **4.4**
photon Bloch state 574
photon creation/annihilation operator 574
photon–electron interaction Hamiltonian 487
photon state vector 575
plasmaron peak 341
plasmon 100
  bands 19, *1.1*, 121, **2.32**, 159, **2.56**, **2.57**
    shape 161
    gap 161
  cutoff vector (critical momentum transfer) 102
  dispersion *1.1*, 104
  double-plasmon excitation *1.1*, 144, **2.47**, **2.48a/b**
  Fano (anti) resonances *1.1*, 121, **2.33**, **2.34**, **2.35**
  frequency (energy) 74
  lifetime 104
  pole model 303
  poles 101
  two-plasmon-band model 121
Poisson equation 505
polarization analysis 15
polarization function 74, 126, 341, 497, 499
polarization matrix 76
positional selective investigation of band structure 226–8, **3.28**
position-sensitive detector 348
positron annihilation 246–9, **4.2**
$PrBa_2Cu_3O_{7-\delta}$ *1.4*, 329, **4.54**
principal component analysis (PCA) 442
proper polarization function 75, 504

pseudogap 331
pseudopotential calculation 280, **4.22**

quadrupolar excitations 475
quadrupole transition (E_2) 475
quantum Monte Carlo (QMC) calculations 281, 304
quasicrystals 331–3
  decagonal 331
  icosahedral 331
quasiparticle
  lifetime model 129
  spectral density function 129, 302, 340

radial momentum distribution 520
Raman shift 388, 554
  excitation into continuous states 554
  excitation into discrete levels 555
random phase approximation (RPA) 17
Rb_4C_{60} *1.4*, 326
real-space multiple-scattering approach 210
reciprocal form factor 26, **1.8**, *1.3*, 240–2, 525
reciprocal macroscopic dielectric function 509
reciprocal Patterson function 526
recoiling electron 33, 346
  energy 351, **4.63**
  incoherent elastic scattering 348
  momentum distribution 346
relaxation processes in RIXS 429
renormalization constant 295, 303
resonant inelastic X-ray scattering (RIXS) **1.4**, 14, 43–57, 553–73
response function 496–501
$R_2Fe_{14}B$ 394
RIXS yield 385
rotational motion 410

sample scattering power 88
Sc *1.1*, **2.53**, 151
scattering (amplitude) operator 515, 559
scattering cross-section
  relativistic 543
  Thomson 72
scattering vector 237
Schrödinger equation 528
  momentum space representation 528
Schrödinger representation 492

selection rules 19
  dipolar 406
  quadrupolar 406
self-consistent field (SCF) approximation 75, 501–8
self-energy effects *1.1*, 126, 505
self-energy insertions 303
  off-shell 134
  on-shell 134
  self-consistent 141
shakeup processes in RIXS **1.5**, 54–8, 570–3
shakeup satellite 432
Si **1.21**, *1.1*, *1.3*, *1.4*, *1.7*, **2.34**, **2.35**, 172, 277, **4.21**, 279, **4.20**, **5.27**, 431, **5.32**
SiC *1.1*, *1.7*, 282
single-particle self-energy 126
singlet pair wavefunction (geminal) 244
Slater determinant 502, 521
Sn 280
soft X-ray absorption spectroscopy (SXAS) 189, **3.23**, 219
solvation effect on hydrogen bonding 221
SO_2 symmetry 474, 561
"spectator" electron 412
spherical symmetry of the core hole
  deviations from 468
spin band 39
  majority 39
  minority 39
spin-charge separation 459, **5.46**
  holon 459
  spinon 459
spin conservation 15, 435–42
  majority 436
  minority 436
spin correlation 512
spin degeneracy 503
spin density in momentum space 353, 356, **4.66**, 539–42
  3D reconstruction 356, **4.65**
spin magnetic moment *1.6*, 355, 359
  element specific *1.6*, 359, **4.69**
  orbital specific *1.6*, 360, **4.67**, **4.68**, **4.70**
spin-magnetic scattering 510–6
spin moment 477
spin–orbit coupled quantum numbers 557
spin–orbit coupling 488
spin polarization of energy bands 16, 357–9
spin selective RIXS 51–4, **1.22**, **1.25**, 566–70
spin vector operator 488
Sr_2CuO_3 *1.8*, 453, 459, **5.47**

$SrCuO_2$ *1.8*, 453, 455, **5.43**, 459, **5.47**
$Sr_2CuO_2Cl_2$ *1.8*, 455, **5.43**, 456, 458
$Sr_2Cu_3O_4Cl_2$ 455, **5.43**
static structure factor 164–79, **2.59**, 495
  spin dependent 175, **2.65**
  charge fluctuation induced 176
Stoke's parameter 540
Stoke's vector 515
stretch mode 416
sum rule 75, 467, 556–63
symmetry assignment of MO's 411, 562–6
symmetry purifying by detuning 416
symmetry selective RIXS 15, 45, 405–17
symmetry selectivity 405
  excitation process 405
  re-emission process 409
synchrotron radiation sources 17, 78
  multipole wiggler 78
  undulator 79

Ta 385, **5.4**, **5.5**, *5.1*
thermal disorder 304
Thomson charge scattering 10, 238
threshold singularity 211
Ti *1.1*, *1.7*, 118, 327
TiC *1.1*, 118
TiH_2 327, **4.53**
TiNi alloy 318, **4.46**
time-dependent density-functional theory (TDDFT) 116, 135, 505
time-of-flight analysis 34, **1.13**
total energy 529
total transmission 99
transferability of bonds *1.3*, 282–4
transition amplitude 548
transition matrix element 380
transition operator 557
transition probability 490
triple differential scattering cross-section 12, 347
two-body Green's function 458
two-particle density correlation function 6, 73, 343, 495, 573
two-particle density matrix 522

Umklapp processes 121, 345
upper Hubbard band (UHB) 457
urea *1.4*, 292

Index

V *1.3*, *1.4*, 306, **4.38**, **4.39**, **4.40**
van Hove space-time correlation function 73, 495
$V_{2-x}Cr_xO_3$ 154, **2.54**
$VD_{0.77}$ 327
vector potential operator 5, 574
vector spherical harmonic 556
vertex correction 129, **2.38**, 130–5, **2.41**, 341, **4.59**, 505
vertex function
  self-consistent 141
vertical scattering geometry 382
$VH_{0.71}$ *1.3*, 327
vibronic coupling 416
virial theorem 529
virtual exciton 570, 572
V_2O_3 *1.1*

Wannier functions
  orthogonality 277

Xe 391
X-ray absorption near edge structure (XANES) 401
X-ray absorption spectroscopy (XAS) *1.2*, 399, 467

X-ray magnetic circular dichroism (XMCD) 398, 439
  RIXS 467–78
  XAS 398, 439, 467
X-ray magnetic linear dichroism (XMLD)
  RIXS 467–78
  XAS 477–9
X-ray Raman scattering (XRS) 8, **1.7**, *1.2*, 186–234
  high pressure *1.2*, 228–31

Y *1.4*, 319, **4.47**
Yb *5.1*, 385, **5.5**
$YBa_2Cu_3O_{7-\delta}$ *1.4*, 329, **4.54**
$YbAgCu_4$ 401, **5.16**
$YbAl_2$ 403
$YbInCu_4$ 401, **5.15**
Young double slit experiment 417, **5.24**

Zhang–Rice band 450, 457
Zhang–Rice (ZR) singlett 450
Zn 385, *5.1*
ZnSe 280
zone boundary collective state (ZBCS) *1.1*, 108, **2.20**, **2.21**